MEMOIRS
of the
American Mathematical Society

Number 888

Basic Global Relative Invariants for Nonlinear Differential Equations

Roger Chalkley

November 2007 • Volume 190 • Number 888 (first of 3 numbers) • ISSN 0065-9266

American Mathematical Society
Providence, Rhode Island

2000 *Mathematics Subject Classification.* Primary 34A34; Secondary 34M15.

Library of Congress Cataloging-in-Publication Data

Chalkley, Roger, 1931–
    Basic global relative invariants for nonlinear differential equations / Roger Chalkley.
      p. cm. — (Memoirs of the American Mathematical Society, ISSN 0065-9266 ; no. 888)
    "November 2007, volume 190, number 888 (first of three numbers)."
    Includes bibliographical references.
    ISBN 978-0-8218-3991-1 (alk. paper)
    1. Differential equations, Nonlinear. 2. Invariants. I. Title.

QA371.C435   2007
515'.355—dc22   2007060779

Memoirs of the American Mathematical Society

This journal is devoted entirely to research in pure and applied mathematics.

Subscription information. The 2007 subscription begins with volume 185 and consists of six mailings, each containing one or more numbers. Subscription prices for 2007 are US$649 list, US$519 institutional member. A late charge of 10% of the subscription price will be imposed on orders received from nonmembers after January 1 of the subscription year. Subscribers outside the United States and India must pay a postage surcharge of US$38; subscribers in India must pay a postage surcharge of US$43. Expedited delivery to destinations in North America US$53; elsewhere US$130. Each number may be ordered separately; *please specify number* when ordering an individual number. For prices and titles of recently released numbers, see the New Publications sections of the *Notices of the American Mathematical Society*.

Back number information. For back issues see the *AMS Catalog of Publications*.

Subscriptions and orders should be addressed to the American Mathematical Society, P. O. Box 845904, Boston, MA 02284-5904, USA. *All orders must be accompanied by payment.* Other correspondence should be addressed to 201 Charles Street, Providence, RI 02904-2294, USA.

Copying and reprinting. Individual readers of this publication, and nonprofit libraries acting for them, are permitted to make fair use of the material, such as to copy a chapter for use in teaching or research. Permission is granted to quote brief passages from this publication in reviews, provided the customary acknowledgment of the source is given.

Republication, systematic copying, or multiple reproduction of any material in this publication is permitted only under license from the American Mathematical Society. Requests for such permission should be addressed to the Acquisitions Department, American Mathematical Society, 201 Charles Street, Providence, Rhode Island 02904-2294, USA. Requests can also be made by e-mail to reprint-permission@ams.org.

Memoirs of the American Mathematical Society is published bimonthly (each volume consisting usually of more than one number) by the American Mathematical Society at 201 Charles Street, Providence, RI 02904-2294, USA. Periodicals postage paid at Providence, RI. Postmaster: Send address changes to Memoirs, American Mathematical Society, 201 Charles Street, Providence, RI 02904-2294, USA.

© 2007 by the American Mathematical Society. All rights reserved.
This publication is indexed in *Science Citation Index*®, *SciSearch*®, *Research Alert*®, *CompuMath Citation Index*®, *Current Contents*®/*Physical, Chemical & Earth Sciences*.
Printed in the United States of America.

∞ The paper used in this book is acid-free and falls within the guidelines established to ensure permanence and durability.
Visit the AMS home page at http://www.ams.org/

10 9 8 7 6 5 4 3 2 1   12 11 10 09 08 07

Contents

Preface ix

Part 1. Foundations for a General Theory 1

Chapter 1. Introduction 3
 1.1. Historical motivation 3
 1.2. Context and definitions used throughout Chapters 1–20 4
 1.3. Main Theorem 8
 1.4. Notational abbreviations employed in Chapters 7–10 10
 1.5. Illustrations for the use of formulas (1.18)–(1.28) when $m \geq 2$ 10
 1.6. Completion of Paul Appell's research in [7] about $Q_2 = 0$ 13
 1.7. Inclusion of homogeneous linear differential equations 17
 1.8. Order of presentation 20

Chapter 2. The Coefficients $c_{i,j}^*(z)$ of (1.3) 23

Chapter 3. The Coefficients $c_{i,j}^{**}(\zeta)$ of (1.5) 29

Chapter 4. Isolated Results Needed for Completeness 35
 4.1. Nonsolutions of nontrivial equations 35
 4.2. Semi-invariants of the second kind are isobaric 37
 4.3. Substitutions in regard to the derivation ' for $\mathcal{S}_m$ 38
 4.4. All of the relative invariants for (1.1) when $m = 1$ 39
 4.5. Isobaric semi-invariants of weight 2 when $m \geq 2$ 43
 4.6. Further semi-invariants of the second kind when $m \geq 2$ 45

Chapter 5. Composite Transformations and Reductions 47
 5.1. The composite of substitutions (1.2) and (1.4) 47
 5.2. The condition $d_{0,1}(\zeta) \equiv d_{0,2}(\zeta) \equiv 0$ for (5.1) when $m \geq 2$ 49
 5.3. Laguerre-Forsyth canonical forms for linear equations 50
 5.4. A Laguerre-Forsyth canonical form for (1.1) when $m \geq 2$ 51
 5.5. There are no relative invariants in $\mathbb{Q}\{w_{0,1}, w_{0,2}\}$ 52

Chapter 6. Related Laguerre-Forsyth Canonical Forms 53
 6.1. Two Laguerre-Forsyth forms related by a transformation 54
 6.2. Identities for the coefficients of related canonical forms 58

Part 2. The Basic Relative Invariants for $Q_m = 0$ when $m \geq 2$ 67

Chapter 7. Formulas That Involve $L_{i,j}(z)$ 69
 7.1. The coefficients of (5.1) when $d_{0,1}(\zeta) \equiv d_{0,2}(\zeta) \equiv 0$ 69

7.2.	Derivatives of the coefficients in (5.1) when $d_{0,1}(\zeta) \equiv d_{0,2}(\zeta) \equiv 0$	73
7.3.	Special combinations of the coefficients for (5.1)	75

Chapter 8. Basic Semi-Invariants of the First Kind for $m \geq 2$ 87

Chapter 9. Formulas That Involve $V_{i,j}(z)$ 93
    9.1. The coefficients of (5.1) when $d_{0,1}(\zeta) \equiv d_{0,2}(\zeta) \equiv 0$ 93
    9.2. Derivatives of the coefficients in (5.1) when $d_{0,1}(\zeta) \equiv d_{0,2}(\zeta) \equiv 0$ 97
    9.3. Special combinations of the coefficients for (5.1)   99

Chapter 10. Basic Semi-Invariants of the Second Kind for $m \geq 2$ 111

Chapter 11. The Existence of Basic Relative Invariants   119

Chapter 12. The Uniqueness of Basic Relative Invariants   121
    12.1. Some polynomials that are not relative invariants   121
    12.2. The uniqueness of basic relative invariants   130
    12.3. Algebraic independence   131

Chapter 13. Real-Valued Functions of a Real Variable   135
    13.1. A suitable context for the evaluations when $m \geq 2$ 135
    13.2. Appropriate hypotheses when $m = 1$ 140

Part 3. Supplementary Results   141

Chapter 14. Relative Invariants via Basic Ones for $m \geq 2$ 143
    14.1. Relative invariants in terms of basic ones and $\boldsymbol{a}_{m,2}$ 143
    14.2. Combinations of invariants that yield other invariants   146
    14.3. The relative invariants of weight ≤ 9 for the equations $Q_2 = 0$ 153

Chapter 15. Results about Q_m as a Quadratic Form   157
    15.1. For Q_m to have a nontrivial factorization   157
    15.2. Relative invariants defined by determinants   162

Chapter 16. Machine Computations   167
    16.1. Expansion of $\boldsymbol{D}_2$ in terms of $\boldsymbol{a}_{m,2}, \boldsymbol{\mathcal{I}}_{m,1,1}, \boldsymbol{\mathcal{I}}_{m,1,2}$, and $\boldsymbol{\mathcal{I}}_{m,2,2}$ 167
    16.2. The expansions for $\boldsymbol{E}_6$ and $\boldsymbol{E}_7$ in (1.81) and (1.82)   168
    16.3. A comprehensive check on the consistency of (1.14)–(1.38)   169
    16.4. The relative invariants of weight ≤ 9 for the equations $Q_2 = 0$ 172
    16.5. Entry of keyboard instructions   177

Chapter 17. The Simplest of the Fano-Type Problems for (1.1)   179
    17.1. Historical motivation   179
    17.2. Results for (1.1) analogous to those in Example 17.1 for (17.1)   180
    17.3. An equation (1.1) that has special polynomial solutions   183

Chapter 18. Paul Appell's Condition of Solvability for $Q_m = 0$ 185
    18.1. Context and historical motivation   185
    18.2. The equivalent condition of Theorem 18.1   186
    18.3. Solutions of $Q_m = 0$ when Appell's condition is satisfied   189

Chapter 19. Appell's Condition for $Q_2 = 0$ and Related Topics   193

19.1.	Consequences of Chapter 18 for the equations $Q_2 = 0$	193
19.2.	An improvement for Proposition 19.6	197
19.3.	An example to illustrate Theorem 19.7	198
19.4.	Other forms for the nonsingular solutions in Theorem 19.7	199
19.5.	Conditions of the type $(u_1(z))^2 - 4u_0(z)u_2(z) \not\equiv 0$	201
19.6.	Equations constructed to have given nonsingular solutions	203
19.7.	Absence of movable branch points	206
19.8.	Two results for third-order linear equations	208
19.9.	Extensions to linear equations of higher order	210
19.10.	Linear substitutions in binary forms	213
19.11.	Properties of the polynomial $\boldsymbol{\Gamma}_n$ in (19.120)	216

Chapter 20. Rational Semi-Invariants and Relative Invariants 219
   20.1. Terminology for an extended context 219
   20.2. The integer s in Definition 20.2 220
   20.3. A context for the remainder of this chapter 223
   20.4. A technical construction needed for Section 20.5 225
   20.5. Rational semi-invariants of the first kind 230
   20.6. A technical construction needed for Section 20.7 234
   20.7. Rational semi-invariants of the second kind 240
   20.8. The structure of rational relative invariants 243
   20.9. Substitutions into rational functions of $\mathcal{Q}_m$ 244

Part 4. Generalizations for $H_{m,n} = 0$ 247

Chapter 21. Introduction to the Equations $H_{m,n} = 0$ 249
   21.1. Transformations produced by changing the variables in $H_{m,n} = 0$ 249
   21.2. Context and definitions 251
   21.3. A summary of results and a derivation ′ for $\mathcal{S}_{m,n}$ 253
   21.4. Inclusion of relative invariants for $H_{m,q} = 0$ when $1 \leq q \leq n$ 253
   21.5. Nonsolutions of nontrivial equations 255

Chapter 22. Basic Relative Invariants for $H_{1,n} = 0$ when $n \geq 2$ 257
   22.1. Existence 257
   22.2. Some polynomials that are not relative invariants 260
   22.3. Uniqueness of $\boldsymbol{B}_2, \ldots, \boldsymbol{B}_n$ as basic relative invariants 264

Chapter 23. Laguerre-Forsyth Forms for $H_{m,n} = 0$ when $m \geq 2$ 265
   23.1. The composite of substitutions (21.2) and (21.4) 265
   23.2. Laguerre-Forsyth reductions when $m \geq 2$ 267
   23.3. Related Laguerre-Forsyth canonical forms 268
   23.4. Identities for the coefficients of related canonical forms 271
   23.5. Use of computer algebra to check Theorem 23.8 272
   23.6. Properties of $\boldsymbol{a}_{m,2}$ in (23.14) and $\boldsymbol{b}_{m,2}$ in (23.16) 273

Chapter 24. Formulas for Basic Relative Invariants when $m \geq 2$ 275
   24.1. Formulas for (21.1) that are analogous to (1.18)–(1.38) 275
   24.2. Notational abbreviations employed in Chapters 25–26 277
   24.3. The special situations when $n = 2$ and when $n = 1$ 277
   24.4. Plan to evaluate Laguerre-Forsyth sums 278

Chapter 25. Extensions of Chapter 7 to $H_{m,n} = 0$, when $m \geq 2$ — 279
  25.1. The coefficients of (23.1) when $d_{0,\ldots,0,1}(\zeta) \equiv d_{0,\ldots,0,2}(\zeta) \equiv 0$ — 279
  25.2. Evaluation of Laguerre-Forsyth sums — 280

Chapter 26. Extensions of Chapter 9 to $H_{m,n} = 0$, when $m \geq 2$ — 293
  26.1. The coefficients of (23.1) when $d_{0,\ldots,0,1}(\zeta) \equiv d_{0,\ldots,0,2}(\zeta) \equiv 0$ — 293
  26.2. Evaluation of Laguerre-Forsyth sums — 294

Chapter 27. Basic Relative Invariants for $H_{m,n} = 0$ when $m \geq 2$ — 307
  27.1. Preliminary results — 307
  27.2. Principal results — 314
  27.3. Some polynomials that are not relative invariants — 316
  27.4. Uniqueness of basic relative invariants — 316
  27.5. The basic relative invariant of index $(l_1, \ldots, l_n)$ when $l_n = 1$ — 317
  27.6. The number of basic relative invariants — 318
  27.7. Relative invariants via basic ones for $m \geq 2$ — 320
  27.8. Rational semi-invariants for various classes of equations — 321

Part 5. Additional Classes of Equations — 323

Chapter 28. The Class of Equations Specified by $y''(z) y'(z)$ — 325
  28.1. Notation and terminology — 325
  28.2. Principal formulas — 326
  28.3. The relative invariants of weight ≤ 9 for the equations (28.1) — 328
  28.4. Computational procedure for Section 28.3 — 330
  28.5. Laguerre-Forsyth reductions for the equations (28.1) — 331
  28.6. All of the relative invariants for the equations (28.1) — 332

Chapter 29. Formulations of Greater Generality — 335
  29.1. Equations characterized by a single monic term — 335
  29.2. Relative invariants for some nonhomogeneous equations — 337

Chapter 30. Invariants for Simple Equations unlike (29.1) — 347
  30.1. Equations without a dominant term — 347
  30.2. Another class of homogeneous quadratic equations — 353
  30.3. The character of x_0 as a polynomial absolute invariant — 356

Bibliography — 357

Index — 359

Abstract

The problem of deducing the basic relative invariants possessed by monic homogeneous linear differential equations of order m was initiated in 1879 with Edmund Laguerre's success for the special case $m = 3$. It was solved in Number 744 of the AMS Memoirs, March 2002, by a procedure that explicitly constructs, for any $m \geq 3$, each of the $m - 2$ basic relative invariants. During that 123-year time span, only a few results were published about the basic relative invariants for other classes of ordinary differential equations.

The research presented here establishes that, *for any fixed integer $m \geq 2$, there are distinct classes $\mathcal{C}_{m,1}, \mathcal{C}_{m,2}, \mathcal{C}_{m,3}, \ldots$ of ordinary algebraic differential equations of order m such that:* (1) $\mathcal{C}_{m,1}$ *is the class of monic homogeneous linear differential equations of order m;* (2) *the class $\mathcal{C}_{m,n}$ has $\binom{m+n}{n} - 3$ basic relative invariants;* (3) *each of the basic relative invariants for these classes is known through simple explicit formulas; and,* (4) *when $m \geq 3$ and $n \geq 2$, a minor change of notation in some $m - 2$ of the basic relative invariants for $\mathcal{C}_{m,n}$ yields the $m - 2$ basic relative invariants for the class $\mathcal{C}_{m,1}$.*

With respect to any fixed integer $m \geq 1$, we begin by explicitly specifying the basic relative invariants for the class $\mathcal{C}_{m,2}$ that contains equations like $Q_m = 0$ in which Q_m is a quadratic form in $y(z)$, ..., $y^{(m)}(z)$ having meromorphic coefficients written symmetrically and the coefficient of $\left(y^{(m)}(z)\right)^2$ is 1. Then, in terms of any fixed positive integers m and n, we explicitly specify the basic relative invariants for the class $\mathcal{C}_{m,n}$ that contains equations like $H_{m,n} = 0$ in which $H_{m,n}$ is an nth-degree form in $y(z)$, ..., $y^{(m)}(z)$ having meromorphic coefficients written symmetrically and the coefficient of $\left(y^{(m)}(z)\right)^n$ is 1. These results enable us to obtain the basic relative invariants for additional classes of ordinary differential equations.

Received by the editor February 26, 2003; and in revised form June 15, 2005.
2000 *Mathematics Subject Classification.* Primary 34A34; Secondary 34M15.
Key words and phrases. relative invariant, semi-invariant.

Preface

Monic homogeneous linear differential equations of order m in $y(z)$ are given by

$$L_m = 0: \qquad y^{(m)}(z) + \sum_{i=1}^{m} c_i(z)\, y^{(m-i)}(z) = 0,$$

for various coefficients $c_1(z)$, ..., $c_m(z)$. After E. Laguerre discovered in 1879 the basic relative invariant for the monic third-order homogeneous linear equations, the main problem was to obtain, for each fixed integer $m \geq 3$, all of the basic relative invariants for the equations $L_m = 0$. While there was considerable activity that led to the discoveries of a few additional basic relative invariants, this problem remained unsolved until the publication of [**24**] in 2002. There, for each fixed integer $m \geq 3$, simple explicit formulas were presented for all $m-2$ of the basic relative invariants possessed by the equations $L_m = 0$.

The absence of general results about basic relative invariants for the equations $L_m = 0$ prior to [**24**] in 2002 is easily explained. Namely, when a change $z = f(\zeta)$ of the independent variable from z to ζ is made to transform an equation $L_m = 0$ involving z into a corresponding equation $L_m^{**} = 0$ involving ζ, explicit formulas are needed for the coefficients of $L_m^{**} = 0$; but, only unsatisfactory formulas for these coefficients existed in publications before [**21**] of 1992. Lacking adequate formulas in the 1880's, those researchers who sought general results about relative invariants for homogeneous linear equations found the computations overwhelmingly difficult after the simplest situations for $m = 3$ and $m = 4$. However, by using infinitesimal transformations in [**31**], A. R. Forsyth derived the necessary structure of each basic relative invariant of weight s for equations of order m when $3 \leq s \leq 7$ and $s \leq m$. He hoped that such computations would enable general formulas to be discovered. When such failed to materialize, researchers unwisely identified the global subject of relative invariants with the local technique of applying infinitesimal transformations. This attitude is clearly expressed in [**12**] and largely prevailed until the publication of [**19**] in 1989. It was the research of F. Neuman in [**46**] and numerous other publications that focused our attention on the global character of our subject and the undesirable local techniques associated with infinitesimal transformations.

From 1879 until recent years, almost all of the research about basic relative invariants for ordinary differential equations was restricted to homogeneous linear ones. As exceptions, we note [**3, 4, 7, 29**]. The prominent use of infinitesimal transformations made it seem unlikely that satisfactory results about basic relative invariants could be developed for general classes of nonlinear differential equations. It was not suspected that various classes of nonlinear equations provide a richer and more interesting area for research. In fact, all of the basic relative invariants for homogeneous linear differential equations can be easily deduced from the more extensive results in Parts 1–4.

We shall show in this memoir that the techniques of [24] can be extended to obtain the basic relative invariants for general classes of nonlinear differential equations. With respect to any fixed integer $m \geq 1$, we shall first obtain all of the basic relative invariants for the class specified by the equations having the form

$$Q_m = 0: \qquad \left(y^{(m)}(z)\right)^2 + \sum_{\substack{0 \leq i,j \leq m \\ (i,j) \neq (0,0)}} c_{i,j}(z)\, y^{(m-i)}(z)\, y^{(m-j)}(z) = 0,$$

where the coefficients $c_{i,j}(z)$ are meromorphic functions that satisfy $c_{j,i}(z) \equiv c_{i,j}(z)$. These results will prepare us for the greater challenge of presenting explicitly the basic relative invariants for the class specified by the equations $H_{m,n} = 0$, where m, n are fixed integers ≥ 1 and $H_{m,n}$ is an nth degree homogeneous polynomial combination of $y(z)$, $y^{(1)}(z)$, ..., $y^{(m)}(z)$ having meromorphic coefficients written symmetrically such that the coefficient of $\left(y^{(m)}(z)\right)^n$ is 1.

Through an adjustment of the subscript notation explained in Section 1.7, each relative invariant for the equations $L_m = 0$ specifies a corresponding relative invariant for the equations $Q_m = 0$. Thus, the relative invariants for the equations $L_m = 0$ are given by particular ones of the more numerous relative invariants for the equations $Q_m = 0$.

Similarly, when $m \geq 1$ and $n \geq 2$, the notation for Proposition 21.6 on page 253 with $q = 2$ shows how each relative invariant for the equations $Q_m = 0$ specifies a corresponding relative invariant for the equations $H_{m,n} = 0$. Moreover, when $n \geq 3$, the equations $H_{m,n} = 0$ possess numerous additional relative invariants.

In Part 1, we present results about transformations of $Q_m = 0$. Also, all of the relative invariants for the equations $Q_1 = 0$ are provided by Theorem 4.10 on page 41. Part 2 develops simple explicit formulas for all of the basic relative invariants of the equations $Q_m = 0$ when $m \geq 2$. The principal results are summarized in our Main Theorem on page 8. Part 3 gives constructive details to show that each relative invariant can be expressed in terms of the basic relative invariants and a particular semi-invariant of the first kind. It explains how the basic relative invariants can be generated and represented in a system of computer algebra. Also, it extends to the equations $Q_m = 0$ the Appell condition of solvability for the equations $Q_2 = 0$. Moreover, the polynomial semi-invariants and relative invariants are used to characterize the rational functions that possess analogous properties.

The equations $L_m = 0$ and $Q_m = 0$ are the special cases having $n = 1$ and $n = 2$ of the equations $H_{m,n} = 0$. Our results about the equations $H_{m,n} = 0$ are presented in Chapters 21–27 and constitute Part 4. The transition to Part 4 is eased by having the special case $n = 2$ of Parts 1–3 as a guide. All of the basic relative invariants for the equations $H_{m,n} = 0$ are provided by means of simple explicit formulas. Theorem 27.7 on page 314 and Theorem 27.13 on page 316 apply when $m \geq 2$; Theorem 22.1 on page 257 and Theorem 22.6 on page 264 are applicable for the remaining case where $n > m = 1$. For $m = n = 1$, the equations $H_{1,1} = 0$ are first-order homogeneous linear ones and have no relative invariants.

Part 5 considers other classes of ordinary algebraic differential equations that possess relative invariants. Chapters 28 and 30 complete our investigation of the various classes of homogeneous quadratic differential equations of order ≤ 2 that have relative invariants. Section 29.1 introduces a unifying context for the various classes of equations considered throughout Chapters 1–28. In Section 29.2, we exhibit more general classes for which the basic relative invariants are known.

In the next paragraph, we shall give a precise definition for *the class $\mathcal{C}_{3,1}$ of monic third-order homogeneous linear differential equations having meromorphic coefficients on some region of the complex plane*. The details can be easily modified to also obtain satisfactory definitions for each of the classes of differential equations examined in this memoir. However, by avoiding such natural but space-consuming considerations later and by employing a simplified terminology, we can more easily focus on the essentials.

To each subregion Ω of the complex plane there is a corresponding field $\mathfrak{F}_\Omega$ of meromorphic functions on Ω. We define $\mathcal{C}_{3,1}$ by means of

$$\mathcal{C}_{3,1} = \{(c_1, c_2, c_3) \mid \text{for some region } \Omega \text{ and some } c_1,\, c_2,\, c_3 \text{ in } \mathfrak{F}_\Omega\}.$$

Each (c_1, c_2, c_3) in $\mathcal{C}_{3,1}$ with c_1, c_2, c_3 in $\mathfrak{F}_\Omega$ specifies a unique monic third-order homogeneous linear differential equation on Ω that has representations such as

(A) $\qquad y'''(z) + c_1(z)\, y''(z) + c_2(z)\, y'(z) + c_3(z)\, y(z) = 0, \quad \text{for } z \text{ in } \Omega,$

as well as

(B) $\qquad p'''(t) + c_1(t)\, p''(t) + c_2(t)\, p'(t) + c_3(t)\, p(t) = 0, \quad \text{for } t \text{ in } \Omega,$

and many others that correspond to the names selected for the dependent and independent variables. When ρ is a not-identically-zero meromorphic function on Ω, there are unique meromorphic functions c_1^*, c_2^*, c_3^* on Ω such that: the substitution $y(z) = \rho(z)\, v(z)$ transforms (A) into

$$v'''(z) + c_1^*(z)\, v''(z) + c_2^*(z)\, v'(z) + c_3^*(z)\, v(z) = 0, \quad \text{for } z \text{ in } \Omega;$$

the substitution $p(t) = \rho(t)\, q(t)$ transforms (B) into

$$q'''(t) + c_1^*(t)\, q''(t) + c_2^*(t)\, q'(t) + c_3^*(t)\, q(t) = 0, \quad \text{for } t \text{ in } \Omega;$$

and etc. In this manner ρ transforms the object (c_1, c_2, c_3) of $\mathcal{C}_{3,1}$ with c_1, c_2, c_3 in $\mathfrak{F}_\Omega$ into the object (c_1^*, c_2^*, c_3^*) of $\mathcal{C}_{3,1}$ with c_1^*, c_2^*, c_3^* in $\mathfrak{F}_\Omega$. When g is a univalent analytic function on Ω with inverse designated by f on $\Omega^{**} = g(\Omega)$, there are unique meromorphic functions $c_1^{**}, c_2^{**}, c_3^{**}$ in $\mathfrak{F}_{\Omega^{**}}$ such that: with $u(\zeta) = (y \circ f)(\zeta)$, the substitution $z = f(\zeta)$ transforms (A) into

$$u'''(\zeta) + c_1^{**}(\zeta)\, u''(\zeta) + c_2^{**}(\zeta)\, u'(\zeta) + c_3^{**}(\zeta)\, u(\zeta) = 0, \quad \text{for } \zeta \text{ in } \Omega^{**};$$

with $r(\tau) = (p \circ f)(\tau)$, the substitution $t = f(\tau)$ transforms (B) into

$$r'''(\tau) + c_1^{**}(\tau)\, r''(\tau) + c_2^{**}(\tau)\, r'(\tau) + c_3^{**}(\tau)\, r(\tau) = 0, \quad \text{for } \tau \text{ in } \Omega^{**};$$

and etc. In this manner, f transforms the object (c_1, c_2, c_3) of $\mathcal{C}_{3,1}$ with c_1, c_2, c_3 in $\mathfrak{F}_\Omega$ into the object $(c_1^{**}, c_2^{**}, c_3^{**})$ of $\mathcal{C}_{3,1}$ with $c_1^{**}, c_2^{**}, c_3^{**}$ in $\mathfrak{F}_{\Omega^{**}}$. Since any member of this class $\mathcal{C}_{3,1}$ of third-order homogeneous linear differential equations can be represented in the form $L_3 = 0$, we shall simply refer to $\mathcal{C}_{3,1}$ as *the class of the equations $L_3 = 0$* and say that the equations of $\mathcal{C}_{3,1}$ are *the equations $L_3 = 0$*. This will not cause confusion in contexts where a particular equation $L_3 = 0$ is involved.

Precise interpretations can be similarly given for: *the class $\mathcal{C}_{m,1}$ of the equations $L_m = 0$* and *the equations $L_m = 0$*, *the class $\mathcal{C}_{m,2}$ of the equations $Q_m = 0$* and *the equations $Q_m = 0$*, *the class $\mathcal{C}_{m,n}$ of the equations $H_{m,n} = 0$* and *the equations $H_{m,n} = 0$*, *the class of the equations (28.1)* and *the equations (28.1)*, etc.

For the preceding context about $\mathcal{C}_{3,1}$, pencil-and-paper computations can be easily done to express each of c_1^*, c_2^*, c_3^* in terms of c_1, c_2, c_3, ρ, $\rho^{(1)}$, $\rho^{(2)}$, and $\rho^{(3)}$ as well as to express each of c_1^{**}, c_2^{**}, c_3^{**} in terms of $c_1 \circ f$, $c_2 \circ f$, $c_3 \circ f$, $f^{(1)}$, $f^{(2)}$, and $f^{(3)}$. Then, little additional effort is needed to verify that

$$\begin{bmatrix} c_3^*(z) - \frac{1}{3} c_1^*(z)\, c_2^*(z) \\ + \frac{2}{27}\left(c_1^*(z)\right)^3 - \frac{1}{2} c_2^{*(1)}(z) \\ + \frac{1}{3} c_1^*(z)\, c_1^{*(1)}(z) \\ + \frac{1}{6} c_1^{*(2)}(z) \end{bmatrix} \equiv \begin{bmatrix} c_3(z) - \frac{1}{3} c_1(z)\, c_2(z) \\ + \frac{2}{27}\left(c_1(z)\right)^3 - \frac{1}{2} c_2^{(1)}(z) \\ + \frac{1}{3} c_1(z)\, c_1^{(1)}(z) \\ + \frac{1}{6} c_1^{(2)}(z) \end{bmatrix},$$

for each z in Ω,

and

$$\begin{bmatrix} c_3^{**}(\zeta) - \frac{1}{3} c_1^{**}(\zeta)\, c_2^{**}(\zeta) \\ + \frac{2}{27}\left(c_1^{**}(\zeta)\right)^3 - \frac{1}{2} c_2^{**(1)}(\zeta) \\ + \frac{1}{3} c_1^{**}(\zeta)\, c_1^{**(1)}(\zeta) \\ + \frac{1}{6} c_1^{**(2)}(\zeta) \end{bmatrix} \equiv \left(f'(\zeta)\right)^3 \begin{bmatrix} c_3\big(f(\zeta)\big) - \frac{1}{3} c_1\big(f(\zeta)\big)\, c_2\big(f(\zeta)\big) \\ + \frac{2}{27}\left(c_1\big(f(\zeta)\big)\right)^3 - \frac{1}{2} c_2^{(1)}\big(f(\zeta)\big) \\ + \frac{1}{3} c_1\big(f(\zeta)\big)\, c_1^{(1)}\big(f(\zeta)\big) \\ + \frac{1}{6} c_1^{(2)}\big(f(\zeta)\big) \end{bmatrix},$$

for each ζ in Ω^{**}.

The two preceding identities were stated with different notation by E. Laguerre in [42] of 1879. They are the characteristic properties of the basic relative invariant that he discovered. This invariant is given in [24, page 7, Equation (1.33)] as a polynomial $\mathcal{I}$ into which substitutions can be made. We represented it as

$$\mathcal{I} \equiv w_3 - \frac{1}{3} w_1\, w_2 + \frac{2}{27}(w_1)^3 - \frac{1}{2} w_2^{(1)} + \frac{1}{3} w_1\, w_1^{(1)} + \frac{1}{6} w_1^{(2)}.$$

For more information about the history of relative invariants for homogeneous linear differential equations, see [24].

We view relative invariants for particular classes of differential equations as objects that are more remarkable than the classical invariants of algebra. Namely, relative invariants for differential equations enjoy their exalted status with respect to two completely different general types of transformations. In contrast, the familiar invariants of algebra are associated with merely a single type of transformation.

For nonlinear differential equations, the previous history of relative invariants is essentially a record of discoveries that Paul Appell made about the equations $Q_2 = 0$ along with his observations for the equations (28.1) of page 325. Our initial interest in the equations $Q_m = 0$ developed in 1957 when we sought to understand Paul Appell's research in [7] about the equations $Q_2 = 0$. The main problems this presented are explained in Section 1.1 on pages 3–4. Our completion in Section 1.6 of the most challenging part of Paul Appell's research in [7] about $Q_2 = 0$ led us to the present investigation of basic relative invariants for the equations $Q_m = 0$ and basic relative invariants for the equations $H_{m,n} = 0$. Progress depended on having the satisfactory results about basic relative invariants for the equations $L_m = 0$ developed in [24]. Our present investigation is a natural continuation and extension of that research.

Roger Chalkley

Part 1

Foundations for a General Theory

CHAPTER 1

Introduction

1.1. Historical motivation

The differential equations $Q_m = 0$ are given by (1.1) on page 4. Our main goal in Parts 1 and 2 is to obtain all of the basic relative invariants for them. Previous studies about relative invariants for such equations were limited to the special case $m = 2$. Those differential equations $Q_2 = 0$ have the form

$$(1.0\text{--A}) \quad \bigl(y''(z)\bigr)^2 + 2\,c_{0,1}(z)\,y''(z)\,y'(z) + 2\,c_{0,2}(z)\,y''(z)\,y(z) + c_{1,1}(z)\,\bigl(y'(z)\bigr)^2$$
$$+ 2\,c_{1,2}(z)\,y'(z)\,y(z) + c_{2,2}(z)\,\bigl(y(z)\bigr)^2 = 0.$$

Interesting results about (1.0–A) were presented in [**3, 5, 6, 7, 50, 29, 13, 14, 16, 27, 28, 49, 18, 20, 23**] and [**24**, pp. 151–153]. However, only [**3, 7, 29, 24**] include some information about relative invariants for the equations (1.0–A). Our use of the terminology *semi-invariant* and *relative invariant* for these equations is explained in Section 1.2 by regarding (1.0–A) as the special case $m = 2$ of (1.1).

Two problems about relative invariants for (1.0–A) remained unsolved since their introduction in [**7**] of 1889. The first of them seeks a minimal set of relative invariants and semi-invariants for the equations $Q_2 = 0$ in terms of which all of the relative invariants for such equations can be expressed. Section 1.6 presents three basic relative invariants $\mathcal{I}_{2,1,1}$, $\mathcal{I}_{2,1,2}$, $\mathcal{I}_{2,2,2}$ and a semi-invariant $a_{2,2}$ that solve this problem. Computer algebra has enabled us to explicitly express each of the relative invariants of weight ≤ 9 for the equations $Q_2 = 0$ as differential-polynomial combinations over $\mathbb{Q}$ of $\mathcal{I}_{2,1,1}$, $\mathcal{I}_{2,1,2}$, $\mathcal{I}_{2,2,2}$, and $a_{2,2}$; for this, see pages 153–155.

Our phrasing of the second problem at the end of the next paragraph requires several results from [**7**]. For this, we suppose that the coefficients of (1.0–A) belong to the field $\mathfrak{F}_\Omega$ of meromorphic functions defined on a region Ω of the complex plane. Then, the left member Q_2 of (1.0-A) is the monic quadratic form in $y''(z)$, $y'(z)$, and $y(z)$ specified over $\mathfrak{F}_\Omega$ by the symmetric matrix

$$M(z) \equiv \begin{bmatrix} 1 & c_{0,1}(z) & c_{0,2}(z) \\ c_{0,1}(z) & c_{1,1}(z) & c_{1,2}(z) \\ c_{0,2}(z) & c_{1,2}(z) & c_{2,2}(z) \end{bmatrix}.$$

We have $det\bigl(M(z)\bigr) \equiv 0$ if and only if each subregion $\mathcal{U}$ of Ω contains a subregion $\mathcal{V}$ on which there are meromorphic functions $\alpha(z)$, $\beta(z)$, $\gamma(z)$, $\delta(z)$ such that

$$(1.0\text{-B}) \qquad Q_2 \equiv \bigl(y'' + \alpha(z)\,y' + \beta(z)\,y\bigr)\bigl(y'' + \gamma(z)\,y' + \delta(z)\,y\bigr), \quad \text{on } \mathcal{V}.$$

This is the special case $m = 2$ of Theorem 15.3 considered in Example 15.6 on page 161. Since the equations $Q_2 = 0$ admit the same transformations as do homogeneous linear differential equations, Paul Appell recognized in [**7**] that the left member for each transform of an equation $Q_2 = 0$ possesses a factorization like (1.0-B) if and only if Q_2 has such a factorization. This made it plausible that

$det(M(z))$ is the value of a relative invariant at the coefficients of (1.0-A). In fact, we have $det(M(z)) \equiv D_2(z)$, where $D_2(z)$ is the value at (1.0-A) of the relative invariant $\boldsymbol{D}_2$ in (1.64) on page 14.

For any $Q_2 = 0$ that satisfies $D_2(z) \not\equiv 0$, Paul Appell indicated in [**7**] the following result. *The condition that $(Q_2)' + \lambda Q_2$ has a nontrivial factorization over $\mathfrak{F}_\Omega$ for some λ in $\mathfrak{F}_\Omega$ is equivalent to the condition that each subregion $\mathcal{U}$ of Ω contains a subregion $\mathcal{V}$ on which there are meromorphic functions $v_0(z)$, $v_1(z)$, $v_2(z)$ such that: $(v_1(z))^2 - 4 v_0(z) v_2(z) \not\equiv 0$; the functions $v_0(z)$, $v_1(z)$, $v_2(z)$ are linearly independent over the field $\mathbb{C}$ of complex numbers; and, the nonsingular solutions of $Q_2 = 0$ on $\mathcal{V}$ are $y(z) \equiv C^2 v_0(z) + CK\, v_1(z) + K^2 v_2(z)$, for C, K in $\mathbb{C}$ not both zero.* Since $Q_2 = 0$ has local solutions of this type if and only if each transform of $Q_2 = 0$ has similar ones, he correctly suspected in [**7**] of 1889 the existence of relative invariants into which the coefficients of Q_2 and their derivatives can be substituted to provide a necessary and sufficient condition for the existence of λ in $\mathfrak{F}_\Omega$ such that $(Q_2)' + \lambda Q_2$ has a nontrivial factorization over $\mathfrak{F}_\Omega$. The task of explicitly specifying such relative invariants is the second unsolved problem in [**7**].

Instead of solving the second problem, Paul Appell restricted his attention in [**7**, pages 412–413] to the equations (1.0-A) having $c_{0,1}(z) \equiv c_{1,2}(z) \equiv 0$ and conditions on their coefficients. His results were repeated in [**29**, page 382] of 1903. Our research in [**18**] has enabled us to obtain a satisfactory solution for Appell's second problem by means of two new relative invariants $\boldsymbol{E}_6$ and $\boldsymbol{E}_7$ that we present in Theorem 1.10 on page 15.

For a proof of the italicized assertion above in lines 6–12, see page 22.

1.2. Context and definitions used throughout Chapters 1–20

In Parts 1 through 3, the differential equations to which we shall apply various transformations are ones expressible in the form

$$\left(y^{(m)}(z)\right)^2 + \sum_{i=1}^{m} c_{i,i}(z) \left(y^{(m-i)}(z)\right)^2 + \sum_{0 \leq i < j \leq m} 2\, c_{i,j}(z)\, y^{(m-i)}(z)\, y^{(m-j)}(z) = 0,$$

where m is a fixed positive integer. To simplify various formulas, we introduce $c_{j,i}(z) \equiv c_{i,j}(z)$, for $0 \leq i < j \leq m$. Thus, in terms of meromorphic coefficients on a region Ω of the complex plane, we begin with equations having the form

(1.1) $\quad \left(y^{(m)}(z)\right)^2 + \sum_{\substack{0 \leq i,j \leq m \\ (i,j) \neq (0,0)}} c_{i,j}(z)\, y^{(m-i)}(z)\, y^{(m-j)}(z) = 0, \quad$ on Ω,

where $c_{j,i}(z) \equiv c_{i,j}(z)$, for $0 \leq i < j \leq m$.

The left member Q_m of (1.1) can be viewed as a particular kind of quadratic form in $y^{(m)}(z)$, $y^{(m-1)}(z)$, ..., $y'(z)$, $y(z)$. Moreover, due to the monic requirement with respect to $\left(y^{(m)}(z)\right)^2$, the coefficients for each such (1.1) are unique. Throughout, except for Chapter 13, the coefficients of the differential equations are meromorphic functions on a region of the complex plane. Chapter 13 presents another context for (1.1) in which our construction of basic relative invariants is applicable. There, with minimal conditions about derivatives, we show that the coefficients $c_{i,j}(z)$ may be real-valued functions of a real variable on an open interval of the real line.

For each meromorphic function $\rho(z) \not\equiv 0$ on Ω, there are unique meromorphic functions $c_{i,j}^*(z)$ on Ω such that the change

(1.2) $$y(z) = \rho(z)\,v(z)$$

of the dependent variable from y to v transforms (1.1) on Ω into

(1.3) $$\left(v^{(m)}(z)\right)^2 + \sum_{\substack{0 \leq i,j \leq m \\ (i,j) \neq (0,0)}} c_{i,j}^*(z)\, v^{(m-i)}(z)\, v^{(m-j)}(z) = 0, \quad \text{on } \Omega,$$

where $c_{j,i}^*(z) \equiv c_{i,j}^*(z)$, for $0 \leq i < j \leq m$.

The coefficients $c_{i,j}^*(z)$ for (1.3) are given by (2.5) in Proposition 2.1 on page 23.

For each univalent analytic function $\zeta = g(z)$ on Ω with inverse designated by $z = f(\zeta)$ on $\Omega^{**} = g(\Omega)$, there are unique meromorphic functions $c_{i,j}^{**}(\zeta)$ on Ω^{**} such that the change

(1.4) $$z = f(\zeta)$$

of the independent variable from z to ζ transforms (1.1) into

(1.5) $$\left(u^{(m)}(\zeta)\right)^2 + \sum_{\substack{0 \leq i,j \leq m \\ (i,j) \neq (0,0)}} c_{i,j}^{**}(\zeta)\, u^{(m-i)}(\zeta)\, u^{(m-j)}(\zeta) = 0, \quad \text{on } \Omega^{**},$$

where $u(\zeta) = (y \circ f)(\zeta)$ and $c_{j,i}^{**}(\zeta) \equiv c_{i,j}^{**}(\zeta)$, for $0 \leq i < j \leq m$.

The coefficients $c_{i,j}^{**}(\zeta)$ for (1.5) are given by (3.9) in Theorem 3.2 on page 30.

To define relative invariants, we assume that the symbols

(1.6) $$\boldsymbol{w}_{i,j}^{(k)}, \quad \text{for } 1 \leq j \leq m,\, 0 \leq i \leq j,\, \text{and } k \geq 0,$$

are algebraically independent variables over the field $\mathbb{Q}$ of rational numbers and we let $\mathcal{S}_m$ denote the ring of polynomials in these variables over $\mathbb{Q}$. We set

(1.7) $$\boldsymbol{w}_{j,i}^{(k)} \equiv \boldsymbol{w}_{i,j}^{(k)}, \quad \text{for } 0 \leq i < j \leq m \text{ and } k \geq 0,$$

and

(1.8) $$\boldsymbol{w}_{0,0}^{(0)} \equiv 1, \quad \text{in } \mathcal{S}_m,$$

so that various later formulas, such as (1.21) and (1.33), can be written simply. We also introduce $\boldsymbol{w}_{i,j} \equiv \boldsymbol{w}_{i,j}^{(0)}$, when $0 \leq i,\, j \leq m$.

For any polynomial $\boldsymbol{P}$ in $\mathcal{S}_m$, we let $P(z)$ denote the unique function on Ω obtained by replacing $\boldsymbol{w}_{i,j}^{(k)}$ in $\boldsymbol{P}$ with $c_{i,j}^{(k)}(z)$ from (1.1); we let $P^*(z)$ denote the unique function on Ω obtained by replacing $\boldsymbol{w}_{i,j}^{(k)}$ in $\boldsymbol{P}$ with $c_{i,j}^{*(k)}(z)$ from (1.3); and we let $P^{**}(\zeta)$ denote the unique function on Ω^{**} obtained by replacing $\boldsymbol{w}_{i,j}^{(k)}$ in $\boldsymbol{P}$ with $c_{i,j}^{**(k)}(\zeta)$ from (1.5).

When $\boldsymbol{P}$ is initially expressed as a polynomial combination over $\mathbb{Q}$ of the symbols in both (1.6) and (1.7), we obtain $P(z)$ according to the preceding method by first using (1.7) to express $\boldsymbol{P}$ as a polynomial in the variables (1.6) over $\mathbb{Q}$ and then substituting $c_{i,j}^{(k)}(z)$ from (1.1) for $\boldsymbol{w}_{i,j}^{(k)}$. However, due to $c_{j,i}^{(k)}(z) \equiv c_{i,j}^{(k)}(z)$ for the coefficients of (1.1), we obtain the same function $P(z)$ by substituting $c_{i,j}^{(k)}(z)$ from (1.1) directly for $\boldsymbol{w}_{i,j}^{(k)}$ in the initial expression for $\boldsymbol{P}$ as a polynomial combination over $\mathbb{Q}$ of the symbols in both (1.6) and (1.7). Similar observations apply to substitutions from (1.3) and (1.5).

DEFINITION 1.1. A polynomial $\boldsymbol{P}$ of $\mathcal{S}_m$ not in $\mathbb{Q}$ is a *semi-invariant of the first kind for the differential equations* (1.1) *of order* m when
$$P^*(z) \equiv P(z), \tag{1.9}$$
for each (1.1) of order m and each change (1.2) of the dependent variable.

DEFINITION 1.2. A polynomial $\boldsymbol{P}$ of $\mathcal{S}_m$ not in $\mathbb{Q}$ is a *semi-invariant of the second kind for the differential equations* (1.1) *of order* m when there is an integer s such that
$$P^{**}(\zeta) \equiv \bigl(f'(\zeta)\bigr)^s P\bigl(f(\zeta)\bigr), \tag{1.10}$$
for each (1.1) of order m and each change (1.4) of the independent variable.

DEFINITION 1.3. A polynomial $\boldsymbol{P}$ of $\mathcal{S}_m$ not in $\mathbb{Q}$ is a *relative invariant for the differential equations* (1.1) *of order* m when it is both a semi-invariant of the first kind and a semi-invariant of the second kind for such equations.

We have closely followed the context in [24] for homogeneous linear differential equations. The terminology *semi-invariant* was first introduced in [43, page 225] or [35, page 424]. The use of *relative invariant* can be traced to [34, page 121] and [33, page 327] or [39, pages 107, 465]. The earliest example of a relative invariant was published by Edmond Laguerre in [42, page 117] of 1879. It was misprinted in [35, page 421]. For details, see [24, Example 1.6, pages 7–8] or (1.102) of page 20.

DEFINITION 1.4. The *weight* of $\boldsymbol{w}_{i,j}^{(k)}$ in (1.6) is the positive integer $i + j + k$; the *weight* of a nonzero rational number is 0; and the *weight* of a nonzero monomial in $\mathcal{S}_m$ is the sum of the weights of its factors. A nonzero polynomial in $\mathcal{S}_m$ is said to be *isobaric of weight* s when each of its nonzero monomials has weight s.

MODIFICATION OF AN OLD RESULT. *If $\boldsymbol{P}$ is a semi-invariant of the second kind for the equations* (1.1) *of a given order* m, *then $\boldsymbol{P}$ is an isobaric polynomial, there is a unique positive integer s for $\boldsymbol{P}$ that satisfies* (1.10), *and s for $\boldsymbol{P}$ in* (1.10) *is equal to the weight of $\boldsymbol{P}$.*

This restatement of Proposition 4.4 on page 37 is analogous to the result for homogeneous linear differential equations in [24, page 140, Theorem A.10] that can be traced to the research of G.-H. Halphen in [34, page 120] or [39, pages 106–107].

DEFINITION 1.5. A polynomial $\boldsymbol{P}$ in $\mathcal{S}_m$ is *basic* when:
(1) there are integers i_0, j_0 such that the coefficient of $\boldsymbol{w}_{i_0,j_0}$ in $\boldsymbol{P}$ is 1 and either $1 \leq i_0 \leq j_0 \leq m$ or $i_0 = 0 < 3 \leq j_0 \leq m$;
(2) $\boldsymbol{P}$ is isobaric of weight $i_0 + j_0$;
(3) if $\boldsymbol{w}_{i,j}^{(k)}$ is any variable from (1.6) that $\boldsymbol{P}$ effectively involves, then $i \leq i_0$ and $j \leq j_0$ (in addiction to $1 \leq j \leq m$, $0 \leq i \leq j$, $k \geq 0$); and
(4) for any variables $\boldsymbol{w}_{i,j}^{(k)}$ and $\boldsymbol{w}_{p,q}^{(r)}$ from (1.6), if the coefficient of $\boldsymbol{w}_{i,j}^{(k)} \boldsymbol{w}_{p,q}^{(r)}$ in $\boldsymbol{P}$ is nonzero, then either $i = 0$ and $1 \leq j \leq 2$ or $p = 0$ and $1 \leq q \leq 2$.

A *basic polynomial* $\boldsymbol{P}$ in $\mathcal{S}_m$ has *index* (i_0, j_0) when i_0 and j_0 are the unique integers that satisfy the first and third of the previous four conditions.

In particular, a *basic relative invariant* is a relative invariant that is also a basic polynomial. The terminology *basic semi-invariant* will be similarly employed.

DEFINITION 1.6. Each relative invariant is necessarily a *global relative invariant* in the sense that the functions $P(z)$ on Ω, $P^*(z)$ on Ω, $P^{**}(\zeta)$ on Ω^{**} involved in Definition 1.3 have the same domains as the corresponding equations (1.1), (1.3), (1.5). However, the modifier *global* is useful to distinguish our relative invariants from the local constructions employed in research prior to [**19**] of 1989.

We define a polynomial $\boldsymbol{D}_1$ in $\mathcal{S}_m$ by means of

(1.11) $$\boldsymbol{D}_1 \equiv \boldsymbol{w}_{1,1} - (\boldsymbol{w}_{0,1})^2, \quad \text{for each } m \geq 1.$$

EXAMPLE 1.7. Proposition 3.5 on page 34 shows that, for each integer $m \geq 1$, $\boldsymbol{D}_1$ is a basic relative invariant of index $(1, 1)$ for the equations (1.1) of order m.

Theorem 4.10 on page 41 establishes that a polynomial $\boldsymbol{P}$ in $\mathcal{S}_1$ is a relative invariant for the equations (1.1) having order $m = 1$ if and only if it is given by

(1.12) $$\boldsymbol{P} \equiv \gamma [\boldsymbol{w}_{1,1} - (\boldsymbol{w}_{0,1})^2]^\kappa, \quad \text{for some nonzero } \gamma \text{ in } \mathbb{Q} \text{ and some } \kappa \geq 1.$$

Thus, our main task for the equations (1.1) is to construct their basic relative invariants when $m \geq 2$. To do that, a particular derivation for $\mathcal{S}_m$ will be useful.

We recall that, in analogy to a derivative, a derivation $'$ for $\mathcal{S}_m$ is a mapping $\boldsymbol{P} \mapsto \boldsymbol{P}'$ of $\mathcal{S}_m$ into $\mathcal{S}_m$ such that

$$(\boldsymbol{F} + \boldsymbol{G})' \equiv \boldsymbol{F}' + \boldsymbol{G}' \quad \text{and} \quad (\boldsymbol{F}\boldsymbol{G})' \equiv \boldsymbol{F}'\boldsymbol{G} + \boldsymbol{F}\boldsymbol{G}', \quad \text{for any } \boldsymbol{F}, \boldsymbol{G} \text{ in } \mathcal{S}_m.$$

Henceforth, in terms of the algebraically independent variables over $\mathbb{Q}$ introduced in (1.6), we let $'$ denote the unique derivation for $\mathcal{S}_m$ that satisfies

(1.13) $$\left(\boldsymbol{w}_{i,j}^{(k)}\right)' = \boldsymbol{w}_{i,j}^{(k+1)}, \quad \text{for } 1 \leq j \leq m, \; 0 \leq i \leq j, \text{ and } k \geq 0.$$

By specializing [**10**, page 139, Proposition 4] or [**11**, page A.V.130, Theorem 1], we deduce the existence and uniqueness of $'$. For any $\boldsymbol{P}$ in $\mathcal{S}_m$, we agree to write $\boldsymbol{P}^{(0)} \equiv \boldsymbol{P}$ and $\boldsymbol{P}^{(r+1)} \equiv \left(\boldsymbol{P}^{(r)}\right)'$, for $r \geq 0$. A polynomial $\boldsymbol{P}$ in $\mathcal{S}_m$ is a *constant* when $\boldsymbol{P}' \equiv 0$. Thus, the constants of $\mathcal{S}_m$ are the elements of $\mathbb{Q}$. Hence, for Definitions 1.1–1.3, the polynomials of $\mathcal{S}_m$ not in $\mathbb{Q}$ are the nonconstants of $\mathcal{S}_m$. With the derivation $'$, $\mathcal{S}_m$ is an *ordinary differential ring*; e.g., see [**41**, page 58].

Corollary 4.15 on page 45 shows that the polynomial

(1.14) $$\boldsymbol{a}_{m,2} \equiv \frac{1}{\binom{m+1}{3}}\left[\boldsymbol{w}_{0,2} - \frac{m-1}{2}\boldsymbol{w}_{0,1}^{(1)} - \frac{m-1}{2m}(\boldsymbol{w}_{0,1})^2\right]$$

is a semi-invariant of the first kind for the equations (1.1) of order $m \geq 2$. It is closely related to research of J. Cockle in [**25**] of 1862 and corresponds to notation we introduced in [**24**, page 4]. Corollary 4.15 shows that the polynomial

(1.15) $$\boldsymbol{b}_{m,2} \equiv \frac{1}{\binom{m+1}{3}}\left[\boldsymbol{w}_{0,2} - \frac{m-2}{3}\boldsymbol{w}_{0,1}^{(1)} - \frac{(3m-1)(m-2)}{6m(m-1)}(\boldsymbol{w}_{0,1})^2\right]$$

is a semi-invariant of the second kind for the equations (1.1) of order $m \geq 2$. It is closely related to research of J. Cockle in [**26**] of 1875 and corresponds to notation we introduced in [**24**, page 5]. Because $\boldsymbol{a}_{m,2}$, $\boldsymbol{b}_{m,2}$, and the polynomial

(1.16) $$\boldsymbol{b}_{m,1} \equiv \frac{1}{\binom{m}{2}}\boldsymbol{w}_{0,1}$$

are important building blocks for our constructions, we note the identity

(1.17) $$\boldsymbol{a}_{m,2} \equiv \boldsymbol{b}_{m,2} - \frac{1}{2}\boldsymbol{b}_{m,1}^{(1)} - \frac{1}{4}(\boldsymbol{b}_{m,1})^2.$$

1.3. Main Theorem

We define $\mathcal{I}_{1,1,1}$ in $\mathcal{S}_m$ through $\mathcal{I}_{1,1,1} \equiv \boldsymbol{w}_{1,1} - (\boldsymbol{w}_{0,1})^2$.

For $m \geq 2$ and integers l, n that satisfy $1 \leq l \leq n \leq m$ or $l = 0 < 3 \leq n \leq m$, we define a polynomial $\mathcal{I}_{m,l,n}$ in $\mathcal{S}_m$ by means of (1.6)–(1.8), (1.16), (1.14),

(1.18) $\quad \boldsymbol{K}_{m,i,j} \equiv 0, \quad$ for $i \leq -1$ and any j,

(1.19) $\quad \boldsymbol{K}_{m,0,j} \equiv 1, \quad$ for any j,

(1.20) $\quad \boldsymbol{K}_{m,i+1,j} \equiv \sum_{k=j+1}^{m} \left[\boldsymbol{K}_{m,i,k}^{(1)} - \frac{m-1}{2} \boldsymbol{b}_{m,1} \boldsymbol{K}_{m,i,k} + (m-i-k+1)(1-i-k) \boldsymbol{a}_{m,2} \boldsymbol{K}_{m,i-1,k} \right],$
$$\text{for } i \geq 0 \text{ and any } j,$$

(1.21) $\quad \boldsymbol{L}_{m,i,j} \equiv \sum_{p=0}^{i} \sum_{q=0}^{j} \boldsymbol{K}_{m,i-p,p} \boldsymbol{K}_{m,j-q,q} \boldsymbol{w}_{p,q}, \quad$ for $0 \leq i, j \leq m$,

(1.22) $\quad \boldsymbol{L}_{m,i,j} \equiv 0, \quad$ for $i < 0$ or $j < 0$ or $i > m$ or $j > m$,

(1.23) $\quad \boldsymbol{M}_{m,l,n,h,i} \equiv \binom{m-h}{l-h} \left[\prod_{r=1}^{l-h} (l-r) \right] \binom{m-i+h}{n-i+h} \left[\prod_{s=1}^{n-i+h} (n-s) \right] \boldsymbol{L}_{m,h,i-h},$
$$\text{for } 0 \leq h \leq l \text{ and any } i,$$

(1.24) $\quad A_{l,n,i} \equiv \dfrac{-1}{l+n+i-1}, \quad$ for $i \geq 1$,

(1.25) $\quad B_{l,n,i} \equiv \dfrac{l+n-i}{l+n+i-2}, \quad$ for $i \geq 1$,

(1.26) $\quad \boldsymbol{I}_{m,l,n,h,0} \equiv \boldsymbol{I}_{m,l,n,h,1} \equiv 0, \quad$ for $0 \leq h \leq l$,

(1.27) $\quad \boldsymbol{I}_{m,l,n,h,i+1} \equiv \boldsymbol{M}_{m,l,n,h,i+1} + A_{l,n,i} \boldsymbol{I}_{m,l,n,h,i}^{(1)} + B_{l,n,i} \boldsymbol{a}_{m,2} \boldsymbol{I}_{m,l,n,h,i-1},$
$$\text{for } 0 \leq h \leq l \text{ and } 1 \leq i \leq l+n-1,$$

and

(1.28) $\quad \mathcal{I}_{m,l,n} \equiv \sum_{h=0}^{l} \boldsymbol{I}_{m,l,n,h,l+n}.$

Theorem 8.3 on page 91 shows that $\mathcal{I}_{m,l,n}$ is a basic semi-invariant of the first kind having index (l, n). We use Theorem 11.2 on page 120, Theorem 12.2 on page 130, Corollary 12.5 on page 133, Definition 14.1 on page 143, Corollary 14.4 on page 146, and Theorem 4.10 on page 41 to establish the following result.

MAIN THEOREM. *The set $\mathcal{B}_{m,0}$ of basic relative invariants for the equations (1.1) of order $m \geq 1$ is given by*

(1.29) $\quad \mathcal{B}_{m,0} = \left\{ \mathcal{I}_{m,l,n} \middle| \text{ for } 1 \leq l \leq n \leq m \right\} \bigcup \left\{ \mathcal{I}_{m,0,n} \middle| \text{ for } 3 \leq n \leq m \right\}.$

Moreover, $\mathcal{B}_{m,0}$ is algebraically independent over $\mathbb{Q}$; each relative invariant for the equations (1.1) having order $m \geq 2$ is expressible as a differential-polynomial combination over $\mathbb{Q}$ of $\boldsymbol{a}_{m,2}$ and the polynomials in $\mathcal{B}_{m,0}$; and, when $m = 1$, the relative invariants are given by $\gamma \left[\mathcal{I}_{1,1,1} \right]^\kappa$, for $\kappa = 1, 2, \ldots$ and each $\gamma \neq 0$ in $\mathbb{Q}$.

Proposition 3.5 on page 34 shows that $\boldsymbol{\mathcal{I}}_{1,1,1}$ is a basic relative invariant of index $(1, 1)$ for the equations (1.1) having $m = 1$. We can use either Proposition 3.5 or Theorem 4.10 on page 41 to see that $\boldsymbol{\mathcal{I}}_{1,1,1}$ is the only basic relative invariant for the equations (1.1) having $m = 1$.

For $m \geq 2$ and integers l, n that satisfy $1 \leq l \leq n \leq m$ or $l = 0 < 3 \leq n \leq m$, our method of proving that $\boldsymbol{\mathcal{I}}_{m,l,n}$ is also a semi-invariant of the second kind is based on the definition of $\boldsymbol{\mathcal{J}}_{m,l,n}$ in $\mathcal{S}_m$ by means of (1.16), (1.15),

$$\tag{1.30} \boldsymbol{U}_{m,i,j} \equiv 0, \quad \text{for } i \leq -1 \text{ and any } j,$$

$$\tag{1.31} \boldsymbol{U}_{m,0,j} \equiv 1, \quad \text{for any } j,$$

$$\tag{1.32} \boldsymbol{U}_{m,i+1,j} \equiv \sum_{k=j+1}^{m} \left[\begin{array}{l} \boldsymbol{U}_{m,i,k}^{(1)} + (i+k-m)\, \boldsymbol{b}_{m,1}\, \boldsymbol{U}_{m,i,k} \\ + (m-i-k+1)(1-i-k)\, \boldsymbol{b}_{m,2}\, \boldsymbol{U}_{m,i-1,k} \end{array} \right],$$
$$\text{for } i \geq 0 \text{ and any } j,$$

$$\tag{1.33} \boldsymbol{V}_{m,i,j} \equiv \sum_{p=0}^{i} \sum_{q=0}^{j} \boldsymbol{U}_{m,i-p,p}\, \boldsymbol{U}_{m,j-q,q}\, \boldsymbol{w}_{p,q}, \quad \text{for } 0 \leq i,\, j \leq m,$$

$$\tag{1.34} \boldsymbol{V}_{m,i,j} \equiv 0, \quad \text{for } i < 0 \text{ or } j < 0 \text{ or } i > m \text{ or } j > m,$$

$$\tag{1.35} \boldsymbol{W}_{m,l,n,h,i} \equiv \binom{m-h}{l-h} \left[\prod_{r=1}^{l-h} (l-r) \right] \binom{m-i+h}{n-i+h} \left[\prod_{s=1}^{n-i+h} (n-s) \right] \boldsymbol{V}_{m,h,i-h},$$
$$\text{for } 0 \leq h \leq l \text{ and any } i,$$

$$\tag{1.36} \boldsymbol{J}_{m,l,n,h,0} \equiv \boldsymbol{J}_{m,l,n,h,1} \equiv 0, \quad \text{for } 0 \leq h \leq l,$$

$A_{l,n,i}$ in (1.24), $B_{l,n,i}$ in (1.25),

$$\tag{1.37} \boldsymbol{J}_{m,l,n,h,i+1} \equiv \boldsymbol{W}_{m,l,n,h,i+1} + A_{l,n,i}\, (\boldsymbol{J}_{m,l,n,h,i}^{(1)} + i\, \boldsymbol{b}_{m,1}\, \boldsymbol{J}_{m,l,n,h,i})$$
$$+ B_{l,n,i}\, \boldsymbol{b}_{m,2}\, \boldsymbol{J}_{m,l,n,h,i-1},$$
$$\text{for } 0 \leq h \leq l \text{ and } 1 \leq i \leq l+n-1,$$

and

$$\tag{1.38} \boldsymbol{\mathcal{J}}_{m,l,n} \equiv \sum_{h=0}^{l} \binom{m-h}{l-h} \boldsymbol{J}_{m,l,n,h,l+n}.$$

Theorem 10.3 on page 116 shows that: $\boldsymbol{\mathcal{J}}_{m,l,n}$ is a basic polynomial whose index is (l, n); and $\boldsymbol{\mathcal{J}}_{m,l,n}$ is a semi-invariant of the second kind for the equations (1.1). The identity $\boldsymbol{\mathcal{I}}_{m,l,n} \equiv \boldsymbol{\mathcal{J}}_{m,l,n}$ is proved in Theorem 11.1 on page 119. Therefore, $\boldsymbol{\mathcal{I}}_{m,l,n}$ is a basic polynomial that is both a semi-invariant of the first kind and a semi-invariant of the second kind. In this manner, the proof of Theorem 11.2 establishes that $\boldsymbol{\mathcal{I}}_{m,l,n}$ is a basic relative invariant for the equations (1.1). This technique was also the principal argument used in [**24**].

1.4. Notational abbreviations employed in Chapters 7–10

The subscripts l, n, and m of

(1.39) $\qquad \boldsymbol{K}_{m,i,j}, \boldsymbol{L}_{m,i,j}, \boldsymbol{M}_{m,l,n,h,i}, A_{l,n,i}, B_{l,n,i}, \boldsymbol{I}_{m,l,n,h,i},$

and

(1.40) $\qquad \boldsymbol{U}_{m,i,j}, \boldsymbol{V}_{m,i,j}, \boldsymbol{W}_{m,l,n,h,i}, \boldsymbol{J}_{m,l,n,h,i}$

will be fixed throughout Chapters 7-10 of Part 2. There, to save space and focus on the essentials, we shall write the abbreviations

(1.41) $\qquad \boldsymbol{K}_{i,j}, \boldsymbol{L}_{i,j}, \boldsymbol{M}_{h,i}, A_i, B_i, \boldsymbol{I}_{h,i}, \boldsymbol{U}_{i,j}, \boldsymbol{V}_{i,j}, \boldsymbol{W}_{h,i}, \boldsymbol{J}_{h,i}$

for the corresponding elements of (1.39)–(1.40). However, the notation of (1.39) and (1.40) is preferable to (1.41) in situations where specific values are given to l, n, and m as in the next two sections.

1.5. Illustrations for the use of formulas (1.18)–(1.28) when $m \geq 2$

Our formulas (1.23)–(1.28) and (1.35)–(1.38) accommodate both the situation $1 \leq l \leq n \leq m$ and the situation $l = 0 < 3 \leq n \leq m$. When $l = 0$, we must have $h = 0$ in (1.23)–(1.28). This simplification is considered further in Section 1.7 of pages 17–20 with regard to the research in [24]. We note that our formulas yield

(1.42) $\qquad \boldsymbol{M}_{m,l,n,0,i} \equiv \boldsymbol{I}_{m,l,n,0,i} \equiv \boldsymbol{W}_{m,l,n,0,i} \equiv \boldsymbol{J}_{m,l,n,0,i} \equiv 0,$

when $1 \leq l \leq n \leq m$ and $0 \leq i \leq l + n$.

Thus, when $l \geq 1$, we can rewrite (1.23)–(1.28) and (1.35)–(1.38) with $1 \leq h \leq l$.

Our Main Theorem shows that: for equations of the type (1.1) having $m = 2$, the basic relative invariants are $\boldsymbol{I}_{2,1,1}$, $\boldsymbol{I}_{2,1,2}$, and $\boldsymbol{I}_{2,2,2}$. In particular, for the research of Paul Appell described in Section 1.1, we shall obtain these basic relative invariants by specializing formulas that yield $\boldsymbol{I}_{m,1,1}$, $\boldsymbol{I}_{m,1,2}$, and $\boldsymbol{I}_{m,2,2}$, for $m \geq 2$. Computations for the latter better illustrate the use of (1.18)–(1.28).

We begin by observing that (1.18)–(1.20) and (1.16) yield

(1.43) $\qquad \boldsymbol{K}_{m,1,j} \equiv -\dfrac{m-j}{m} \boldsymbol{w}_{0,1}, \quad \text{for } m \geq 2 \text{ and } j \leq m.$

In view of (1.20), (1.43), (1.16), (1.14), and (1.19), we deduce that

(1.44) $\quad \boldsymbol{K}_{m,2,j} \equiv \dfrac{\binom{m-j}{2}}{\binom{m+1}{2}} \left[-\dfrac{m+2j+1}{m-1} \boldsymbol{w}_{0,2} + j\, \boldsymbol{w}_{0,1}^{(1)} + \dfrac{m+j+1}{m}(\boldsymbol{w}_{0,1})^2 \right],$

$\qquad\qquad\qquad\qquad\qquad\qquad\qquad\qquad\qquad\qquad$ for $m \geq 2$ and $j \leq m$.

1.5.1. $\boldsymbol{I}_{m,1,1}$, for $m \geq 2$.
We use (1.28) and (1.42) to obtain

(1.45) $\qquad \boldsymbol{I}_{m,1,1} \equiv \displaystyle\sum_{h=0}^{1} \boldsymbol{I}_{m,1,1,h,2} \equiv \boldsymbol{I}_{m,1,1,1,2}.$

For $h = 1$, we apply (1.27) and (1.26) to verify that

(1.46) $\qquad \boldsymbol{I}_{m,1,1,1,2} \equiv \boldsymbol{M}_{m,1,1,1,2} + A_{1,1,1}\, \boldsymbol{I}^{(1)}_{m,1,1,1,1} + B_{1,1,1}\, \boldsymbol{a}_{m,2}\, \boldsymbol{I}_{m,1,1,1,0}$

$\qquad\qquad\qquad \equiv \boldsymbol{M}_{m,1,1,1,2}.$

For $l = 1$, $n = 1$, $h = 1$, and $i = 2$, (1.23) gives

$$(1.47) \quad \boldsymbol{M}_{m,1,1,1,2} \equiv \binom{m-1}{0}\left[\prod_{r=1}^{0}(1-r)\right]\binom{m-1}{0}\left[\prod_{s=1}^{0}(1-s)\right]\boldsymbol{L}_{m,1,1} \equiv \boldsymbol{L}_{m,1,1}.$$

Since (1.19) and (1.43) yield $\boldsymbol{K}_{m,0,1} \equiv 1$ and $\boldsymbol{K}_{m,1,0} \equiv -\boldsymbol{w}_{0,1}$, we employ (1.21) and (1.7)–(1.8) to verify that

$$(1.48) \quad \boldsymbol{L}_{m,1,1} \equiv \boldsymbol{w}_{1,1} + \boldsymbol{K}_{m,0,1}\,\boldsymbol{K}_{m,1,0}\,\boldsymbol{w}_{1,0} + \boldsymbol{K}_{m,1,0}\,\boldsymbol{K}_{m,0,1}\,\boldsymbol{w}_{0,1} + (\boldsymbol{K}_{m,1,0})^2$$
$$\equiv \boldsymbol{w}_{1,1} - (\boldsymbol{w}_{0,1})^2 - (\boldsymbol{w}_{0,1})^2 + (-\boldsymbol{w}_{0,1})^2 \equiv \boldsymbol{w}_{1,1} - (\boldsymbol{w}_{0,1})^2.$$

By combining (1.45)–(1.48), we obtain

$$(1.49) \quad \boldsymbol{\mathcal{I}}_{m,1,1} \equiv \boldsymbol{w}_{1,1} - (\boldsymbol{w}_{0,1})^2, \quad \text{for } m \geq 2.$$

Thus, for $m \geq 2$, $\boldsymbol{\mathcal{I}}_{m,1,1}$ is the polynomial $\boldsymbol{D}_1$ of (1.11) that is independently examined in Proposition 3.5 on page 34.

1.5.2. $\boldsymbol{\mathcal{I}}_{m,1,2}$, for $m \geq 2$. We observe that (1.28) and (1.42) yield

$$(1.50) \quad \boldsymbol{\mathcal{I}}_{m,1,2} \equiv \sum_{h=0}^{1} \boldsymbol{\mathcal{I}}_{m,1,2,h,3} \equiv \boldsymbol{\mathcal{I}}_{m,1,2,1,3}.$$

Using (1.27), (1.26), and (1.23), we find that

$$\boldsymbol{\mathcal{I}}_{m,1,2,1,2} \equiv \boldsymbol{M}_{m,1,2,1,2} \equiv \binom{m-1}{0}\left[\prod_{r=1}^{0}(1-r)\right]\binom{m-1}{1}\left[\prod_{s=1}^{1}(2-s)\right]\boldsymbol{L}_{m,1,1}$$
$$\equiv (m-1)\boldsymbol{L}_{m,1,1}$$

and

$$(1.51) \quad \boldsymbol{\mathcal{I}}_{m,1,2,1,3} \equiv \boldsymbol{M}_{m,1,2,1,3} + A_{1,2,2}\,\boldsymbol{\mathcal{I}}^{(1)}_{m,1,2,1,2} + 0 \equiv \boldsymbol{L}_{m,1,2} - \frac{(m-1)}{4}\boldsymbol{L}^{(1)}_{m,1,1}.$$

We note that (1.21), (1.19), (1.43), and (1.44) give

$$(1.52) \quad \boldsymbol{L}_{m,1,2} \equiv \boldsymbol{w}_{1,2} + \boldsymbol{K}_{m,1,1}\,\boldsymbol{w}_{1,1} + \boldsymbol{K}_{m,1,0}\,\boldsymbol{w}_{0,2}$$
$$+ (\boldsymbol{K}_{m,1,0}\,\boldsymbol{K}_{m,1,1} + \boldsymbol{K}_{m,2,0})\,\boldsymbol{w}_{0,1} + (\boldsymbol{K}_{m,1,0}\,\boldsymbol{K}_{m,2,0})$$
$$\equiv \boldsymbol{w}_{1,2} - \frac{m-1}{m}\boldsymbol{w}_{0,1}\,\boldsymbol{w}_{1,1} - \boldsymbol{w}_{0,1}\,\boldsymbol{w}_{0,2} + \frac{m-1}{m}(\boldsymbol{w}_{0,1})^3.$$

Combining (1.50)–(1.52) and (1.48), we deduce that

$$(1.53) \quad \boldsymbol{\mathcal{I}}_{m,1,2} \equiv \boldsymbol{w}_{1,2} - \frac{m-1}{m}\boldsymbol{w}_{0,1}\,\boldsymbol{w}_{1,1} - \boldsymbol{w}_{0,1}\,\boldsymbol{w}_{0,2} + \frac{m-1}{m}(\boldsymbol{w}_{0,1})^3$$
$$- \frac{m-1}{4}\boldsymbol{w}^{(1)}_{1,1} + \frac{m-1}{2}\boldsymbol{w}_{0,1}\,\boldsymbol{w}^{(1)}_{0,1}, \quad \text{for } m \geq 2.$$

1.5.3. $\mathcal{I}_{m,2,2}$, for $m \geq 2$.

We apply (1.28) and (1.42) to obtain

$$\mathcal{I}_{m,2,2} \equiv \sum_{h=0}^{2} \boldsymbol{I}_{m,2,2,h,4} \equiv \boldsymbol{I}_{m,2,2,1,4} + \boldsymbol{I}_{m,2,2,2,4}. \tag{1.54}$$

Using (1.27), (1.26), and (1.23)–(1.25), we verify that

$$\boldsymbol{I}_{m,2,2,1,2} \equiv \boldsymbol{M}_{m,2,2,1,2} \equiv \binom{m-1}{1}\left[\prod_{r=1}^{1}(2-r)\right]\binom{m-1}{1}\left[\prod_{k=1}^{1}(2-k)\right]\boldsymbol{L}_{m,1,1}$$

$$\equiv (m-1)^2 \, \boldsymbol{L}_{m,1,1},$$

$$\boldsymbol{I}_{m,2,2,1,3} \equiv \boldsymbol{M}_{m,2,2,1,3} + A_{2,2,2}\,(m-1)^2\,\boldsymbol{L}_{m,1,1}^{(1)}$$

$$\equiv (m-1)\,\boldsymbol{L}_{m,1,2} - \frac{(m-1)^2}{5}\,\boldsymbol{L}_{m,1,1}^{(1)},$$

$$\boldsymbol{I}_{m,2,2,1,4} \equiv \boldsymbol{M}_{m,2,2,1,4} + A_{2,2,3}\,\boldsymbol{I}_{m,2,2,1,3}^{(1)} + B_{2,2,3}\,\boldsymbol{a}_{m,2}\,\boldsymbol{I}_{m,2,2,1,2} \tag{1.55}$$

$$\equiv 0 - \frac{1}{6}\left[(m-1)\boldsymbol{L}_{m,1,2}^{(1)} - \frac{(m-1)^2}{5}\boldsymbol{L}_{m,1,1}^{(2)}\right] + \frac{(m-1)^2}{5}\boldsymbol{a}_{m,2}\,\boldsymbol{L}_{m,1,1}$$

$$\equiv -\frac{(m-1)}{6}\boldsymbol{L}_{m,1,2}^{(1)} + \frac{(m-1)^2}{30}\boldsymbol{L}_{m,1,1}^{(2)} + \frac{(m-1)^2}{5}\boldsymbol{a}_{m,2}\,\boldsymbol{L}_{m,1,1},$$

$$\boldsymbol{I}_{m,2,2,2,2} \equiv \boldsymbol{M}_{m,2,2,2,2} \equiv 0,$$

$$\boldsymbol{I}_{m,2,2,2,3} \equiv \boldsymbol{M}_{m,2,2,2,3} \equiv (m-1)\,\boldsymbol{L}_{m,2,1} \equiv (m-1)\,\boldsymbol{L}_{m,1,2},$$

and

$$\boldsymbol{I}_{m,2,2,2,4} \equiv \boldsymbol{M}_{m,2,2,2,4} + A_{2,2,3}\,\boldsymbol{I}_{m,2,2,2,3}^{(1)} + 0 \equiv \boldsymbol{L}_{m,2,2} - \frac{(m-1)}{6}\boldsymbol{L}_{m,1,2}^{(1)}. \tag{1.56}$$

Since (1.19), (1.43), and (1.44) yield

$$\boldsymbol{K}_{m,0,2} \equiv 1, \quad \boldsymbol{K}_{m,1,1} \equiv -\frac{(m-1)}{m}\,\boldsymbol{w}_{0,1}, \quad \text{and} \quad \boldsymbol{K}_{m,2,0} \equiv -\boldsymbol{w}_{0,2} + \frac{m-1}{m}(\boldsymbol{w}_{0,1})^2,$$

we find that (1.21) and (1.7) give

$$\boldsymbol{L}_{m,2,2} \equiv \boldsymbol{w}_{2,2} + 2\boldsymbol{K}_{m,1,1}\,\boldsymbol{w}_{1,2} + (\boldsymbol{K}_{m,1,1})^2\,\boldsymbol{w}_{1,1} + 2\boldsymbol{K}_{m,2,0}\,\boldsymbol{w}_{0,2}$$

$$+ 2\boldsymbol{K}_{m,1,1}\,\boldsymbol{K}_{m,2,0}\,\boldsymbol{w}_{0,1} + (\boldsymbol{K}_{2,0})^2$$

and therefore

$$\boldsymbol{L}_{m,2,2} \equiv \boldsymbol{w}_{2,2} - \frac{2(m-1)}{m}\,\boldsymbol{w}_{0,1}\,\boldsymbol{w}_{1,2} + \frac{(m-1)^2}{m^2}\,(\boldsymbol{w}_{0,1})^2\,\boldsymbol{w}_{1,1} \tag{1.57}$$

$$- (\boldsymbol{w}_{0,2})^2 + \frac{2(m-1)}{m}\,(\boldsymbol{w}_{0,1})^2\,\boldsymbol{w}_{0,2} - \frac{(m-1)^2}{m^2}\,(\boldsymbol{w}_{0,1})^4.$$

Using (1.54)–(1.56), we obtain

$$\mathcal{I}_{m,2,2} \equiv \boldsymbol{L}_{m,2,2} - \frac{m-1}{3}\boldsymbol{L}_{m,1,2}^{(1)} + \frac{(m-1)^2}{30}\boldsymbol{L}_{m,1,1}^{(2)} + \frac{(m-1)^2}{5}\boldsymbol{a}_{m,2}\boldsymbol{L}_{m,1,1}. \tag{1.58}$$

By combining (1.58) with (1.57), (1.52), and (1.48), we conclude that

$$(1.59) \quad \mathcal{I}_{m,2,2} \equiv w_{2,2} - \frac{2(m-1)}{m} w_{0,1} w_{1,2} + \frac{6(m-1)}{5(m+1)m} w_{0,2} w_{1,1} - (w_{0,2})^2$$
$$+ \frac{(5m+2)(m-1)^2}{5(m+1)m^2} (w_{0,1})^2 w_{1,1} + \frac{2(5m+2)(m-1)}{5(m+1)m} (w_{0,1})^2 w_{0,2}$$
$$- \frac{(5m+2)(m-1)^2}{5(m+1)m^2} (w_{0,1})^4 - \frac{m-1}{3} w_{1,2}^{(1)} + \frac{(m-1)^2}{30} w_{1,1}^{(2)}$$
$$+ \frac{(m-1)^2}{3m} w_{0,1} w_{1,1}^{(1)} + \frac{(5m-4)(m-1)^2}{15(m+1)m} w_{0,1}^{(1)} w_{1,1}$$
$$+ \frac{m-1}{3} w_{0,1} w_{0,2}^{(1)} + \frac{m-1}{3} w_{0,1}^{(1)} w_{0,2} - \frac{(m-1)^2}{15} w_{0,1} w_{0,1}^{(2)}$$
$$- \frac{(m-1)^2}{15} (w_{0,1}^{(1)})^2 - \frac{(5m+2)(m-1)^2}{5(m+1)m} (w_{0,1})^2 w_{0,1}^{(1)}, \quad \text{for } m \geq 2.$$

1.6. Completion of Paul Appell's research in [7] about $Q_2 = 0$

For $m = 2$, (1.1) has the form of (1.0–A) on page 3. Paul Appell's research about (1.0–A) in [7] produced two interesting unsolved problems. Their historical motivation was presented in Section 1.1. Here, we shall give complete solutions for both of them.

1.6.1. One of the two problems and our solution. Previously, the only basic relative invariant that had been found for the equations (1.0–A) was $\mathcal{I}_{2,1,1}$ in (1.60). It was presented in [7, page 403].

THEOREM 1.8. *The basic relative invariants for the equations (1.0–A) are*

$$(1.60) \quad \mathcal{I}_{2,1,1} \equiv w_{1,1} - (w_{0,1})^2,$$
$$(1.61) \quad \mathcal{I}_{2,1,2} \equiv w_{1,2} - \frac{1}{2} w_{0,1} w_{1,1} - w_{0,1} w_{0,2} + \frac{1}{2} (w_{0,1})^3 - \frac{1}{4} w_{1,1}^{(1)} + \frac{1}{2} w_{0,1} w_{0,1}^{(1)},$$

and

$$(1.62) \quad \mathcal{I}_{2,2,2} \equiv w_{2,2} - w_{0,1} w_{1,2} + \frac{1}{5} w_{0,2} w_{1,1} - (w_{0,2})^2 + \frac{1}{5} (w_{0,1})^2 w_{1,1}$$
$$+ \frac{4}{5} (w_{0,1})^2 w_{0,2} - \frac{1}{5} (w_{0,1})^4 - \frac{1}{3} w_{1,2}^{(1)} + \frac{1}{30} w_{1,1}^{(2)} + \frac{1}{6} w_{0,1} w_{1,1}^{(1)}$$
$$+ \frac{1}{15} w_{0,1}^{(1)} w_{1,1} + \frac{1}{3} w_{0,1} w_{0,2}^{(1)} + \frac{1}{3} w_{0,1}^{(1)} w_{0,2} - \frac{1}{15} w_{0,1} w_{0,1}^{(2)}$$
$$- \frac{1}{15} (w_{0,1}^{(1)})^2 - \frac{2}{5} (w_{0,1})^2 w_{0,1}^{(1)}.$$

PROOF. By applying our Main Theorem on page 8, we substitute $m = 2$ in (1.49), (1.53), (1.59) to obtain (1.60), (1.61), (1.62) and complete the proof. □

Also, we see that (1.14) specializes for $m = 2$ to give

$$(1.63) \quad a_{2,2} \equiv w_{0,2} - \frac{1}{2} w_{0,1}^{(1)} - \frac{1}{4} (w_{0,1})^2.$$

Our Main Theorem shows that *each relative invariant for the equations (1.0–A) is expressible as a differential-polynomial combination over $\mathbb{Q}$ of $\mathcal{I}_{2,1,1}$, $\mathcal{I}_{2,1,2}$, $\mathcal{I}_{2,2,2}$, and $a_{2,2}$*. For an explanation of this terminology, see Definition 14.1 on page 143.

EXAMPLE 1.9. We illustrate our Main Theorem for $m = 2$ and the polynomial

$$(1.64) \quad D_2 \equiv \begin{vmatrix} 1 & w_{0,1} & w_{0,2} \\ w_{0,1} & w_{1,1} & w_{1,2} \\ w_{0,2} & w_{1,2} & w_{2,2} \end{vmatrix} \equiv w_{1,1} w_{2,2} - (w_{0,1})^2 w_{2,2} - (w_{1,2})^2 \\ + 2 w_{0,1} w_{0,2} w_{1,2} - (w_{0,2})^2 w_{1,1}$$

that specifies the determinant for (1.0–A) as the function $D_2(z)$ on Ω obtained when $c_{i,j}(z)$ from (1.0–A) is substituted for $w_{i,j}$ in D_2. An intuitive reason for suspecting D_2 to be a relative invariant is given on page 3. For a direct proof that D_2 is a relative invariant for the equations (1.0–A), see the special case $m = k = 2$ of Theorem 15.10 on page 164. The substitution $m = 2$ in (15.47) of Example 15.11 on page 166 yields

$$(1.65) \quad D_2 \equiv \left[\mathcal{I}_{2,2,2} \, \mathcal{I}_{2,1,1} - (\mathcal{I}_{2,1,2})^2 \right] + \frac{1}{3} \left[\mathcal{I}_{2,1,1} \, \mathcal{I}^{(1)}_{2,1,2} - \frac{3}{2} \mathcal{I}^{(1)}_{2,1,1} \, \mathcal{I}_{2,1,2} \right] \\ + \frac{1}{20} \left[\mathcal{I}_{2,1,1} \, \mathcal{I}^{(2)}_{2,1,1} - \frac{5}{4} (\mathcal{I}^{(1)}_{2,1,1})^2 - 4 a_{2,2} \, (\mathcal{I}_{2,1,1})^2 \right].$$

This expresses D_2 as a differential-polynomial combination over $\mathbb{Q}$ of $a_{2,2}$ and the basic relative invariants $\mathcal{I}_{2,1,1}, \mathcal{I}_{2,1,2}, \mathcal{I}_{2,2,2}$. Due to Proposition 14.6 on page 146 and Corollary 14.9 on page 148, we see that each of the three terms in the right member of (1.65) is a relative invariant of weight 6 for the equations (1.0–A). This provides another verification that D_2 is a relative invariant of weight 6.

The algorithmic proof for Theorem 14.3 on page 143 can be applied directly to obtain (1.65) for D_2. Namely, after solving (1.62) for $w_{2,2}$ in terms of $\mathcal{I}_{2,2,2}$ and the other $w^{(k)}_{i,j}$, we obtain an expression for D_2 from which $w_{2,2}$ is eliminated. Next, after solving (1.61) for $w_{1,2}$ and applying $'$ to obtain $w^{(1)}_{1,2}$, we successively eliminate $w^{(1)}_{1,2}$ and $w_{1,2}$. Then, solving (1.60) for $w_{1,1}$, we eliminate $w^{(2)}_{1,1}$, $w^{(1)}_{1,1}$, and $w_{1,1}$ in that order. Finally, solving (1.63) for $w_{0,2}$, we eliminate $w_{0,2}$ and find that each $w^{(k)}_{0,1}$, for $k \geq 0$, is also eliminated. The resulting expression gives (1.65). Useful machine computations for this purpose are presented in Chapter 16.

We note that (1.0–A) can be rewritten in the form

$$(1.66) \quad \bigl(y''(z) + c_{0,1}(z) \, y'(z) + c_{0,2}(z) \, y(z) \bigr)^2 \\ + B_{1,1}(z) \, \bigl(y(z) \bigr)^2 + B_{1,2}(z) \, y'(z) \, y(z) + B_{2,2}(z) \, \bigl(y(z) \bigr)^2 = 0,$$

where $B_{1,1}(z)$, $B_{1,2}(z)$, and $B_{2,2}(z)$ are the meromorphic functions on Ω obtained by substituting $c_{i,j}(z)$ from (1.0–A) for $w_{i,j}$ in the corresponding polynomials

$$(1.67) \quad B_{1,1} \equiv w_{1,1} - (w_{0,1})^2 \equiv \mathcal{I}_{2,1,1},$$
$$(1.68) \quad B_{1,2} \equiv w_{1,2} - w_{0,1} \, w_{0,2},$$

and

$$(1.69) \quad B_{2,2} \equiv w_{2,2} - (w_{0,2})^2.$$

While $B_{1,1}$ is the basic relative invariant of weight 2, the basic polynomials $B_{1,2}$ and $B_{2,2}$ are merely semi-invariants of the second kind of respective weights 3 and 4. In particular, (1.64) and (1.67)–(1.69) yield

$$(1.70) \quad D_2 \equiv B_{1,1} \, B_{2,2} - (B_{1,2})^2.$$

Thus, when (1.0–A) satisfies $D_2(z) \equiv 0$ and a subregion $\mathcal{U}$ of Ω is given, there are meromorphic functions $\xi(z)$ and $\eta(z)$ on a subregion $\mathcal{V}$ of $\mathcal{U}$ such that

$$B_{1,1}(z)\left(y'(z)\right)^2 + B_{1,2}(z)\,y'(z)\,y(z) + B_{2,2}(z)\left(y(z)\right)^2$$
$$\equiv -\left[\xi(z)\,y'(z) + \eta(z)\,y(z)\right]^2, \quad \text{on } \mathcal{V}.$$

Then, as the left member of both (1.0–A) and (1.66), Q_2 has a factorization (1.0-B) given on page 3. Conversely, when Q_2 has a factorization (1.0-B), we expand it to express $B_{1,1}(z)$, $B_{1,2}(z)$, $B_{2,2}(z)$ in terms of $\alpha(z)$, $\beta(z)$, $\gamma(z)$, $\delta(z)$ and find that (1.70) yields $D_2(z) \equiv 0$ on $\mathcal{V}$ and therefore also on Ω. Thus, *there is no nontrivial factorization for Q_2 if and only if Q_2 satisfies $D_2(z) \not\equiv 0$.* A generalization that applies to the left member Q_m of (1.1) is given in Theorem 15.3 on page 159.

1.6.2. The other problem and our solution. In terms of (1.0–A) expressed as $Q_2 = 0$, Paul Appell sought relative invariants with which to characterize those equations $Q_2 = 0$ for which $D_2(z) \not\equiv 0$ and there is a function $\lambda(z)$ such that $(Q_2)' + \lambda\,Q_2$ has a nontrivial factorization.

THEOREM 1.10. *With respect to (1.67)–(1.69) and the isobaric polynomial*

$$(1.71) \qquad \boldsymbol{H}_7 \equiv \boldsymbol{B}_{1,1}\,\boldsymbol{B}_{2,2}^{(1)} - \boldsymbol{B}_{1,1}^{(1)}\,\boldsymbol{B}_{2,2} - 2\,\boldsymbol{B}_{1,2}\,\boldsymbol{B}_{2,2}$$
$$+ 2\,\boldsymbol{w}_{0,1}\,\boldsymbol{B}_{1,1}\,\boldsymbol{B}_{2,2} - 2\,\boldsymbol{w}_{0,2}\,\boldsymbol{B}_{1,1}\,\boldsymbol{B}_{1,2}$$

in $\mathcal{S}_m$ of weight 7, the isobaric polynomial $\boldsymbol{E}_6$ of weight 6 defined in $\mathcal{S}_m$ by

$$(1.72) \qquad \boldsymbol{E}_6 \equiv \boldsymbol{B}_{1,1}\,\boldsymbol{B}_{2,2} - 2\left(\boldsymbol{B}_{1,2}\right)^2 + \boldsymbol{w}_{0,1}\,\boldsymbol{B}_{1,1}\,\boldsymbol{B}_{1,2}$$
$$- \boldsymbol{w}_{0,2}\left(\boldsymbol{B}_{1,1}\right)^2 + \boldsymbol{B}_{1,1}\,\boldsymbol{B}_{1,2}^{(1)} - \boldsymbol{B}_{1,1}^{(1)}\,\boldsymbol{B}_{1,2}$$

and the isobaric polynomial $\boldsymbol{E}_7$ of weight 7 defined in $\mathcal{S}_m$ by

$$(1.73) \qquad \boldsymbol{E}_7 \equiv \boldsymbol{H}_7 - \frac{1}{6}\left(\boldsymbol{E}_6^{(1)} + 6\,\boldsymbol{w}_{0,1}\,\boldsymbol{E}_6\right)$$

are relative invariants for the equations (1.0–A) on page 3. Moreover, for any given (1.0–A) whose left member Q_2 satisfies $D_2(z) \not\equiv 0$, the condition

$$(1.74) \qquad E_6(z) \equiv E_7(z) \equiv 0$$

is necessary and sufficient for the existence of a function λ such that $(Q_2)' + \lambda\,Q_2$ has a nontrivial factorization. Furthermore, if Q_2 satisfies both $D_2(z) \not\equiv 0$ and (1.74), then: $B_{1,1}(z) \not\equiv 0$; the meromorphic function

$$(1.75) \qquad \lambda(z) \equiv 2\,c_{0,1}(z) - \frac{B_{1,1}^{(1)}(z) + 2\,B_{1,2}(z)}{B_{1,1}(z)}, \quad \text{on } \Omega,$$

is the only function λ for which $(Q_2)' + \lambda\,Q_2$ has a nontrivial factorization; and

$$(1.76) \qquad (Q_2)' + \lambda(z)\,Q_2 \equiv 2\left[y''(z) + c_{0,1}(z)\,y'(z) + c_{0,2}(z)\,y(z)\right]\mathcal{L}_3, \quad \text{on } \Omega,$$

where

$$(1.77) \qquad \mathcal{L}_3 \equiv y'''(z) + \left[c_{0,1}(z) + \left(\lambda(z)/2\right)\right]y''(z)$$
$$+ \left[c_{0,1}^{(1)}(z) + c_{0,2}(z) + c_{0,1}(z)\left(\lambda(z)/2\right) + B_{1,1}(z)\right]y'(z)$$
$$+ \left[c_{0,2}^{(1)}(z) + c_{0,2}(z)\left(\lambda(z)/2\right) + B_{1,2}(z)\right]y(z).$$

PROOF. Because $\boldsymbol{E}_6$ and $\boldsymbol{E}_7$ have the expansions (1.81) and (1.82), Lemma 1.12 establishes that $\boldsymbol{E}_6$ and $\boldsymbol{E}_7$ are relative invariants of respective weights 6 and 7 for the equations $Q_2 = 0$. We use (1.71)–(1.73) to verify that: for any given $Q_2 = 0$, the condition $E_6(z) \equiv E_7(z) \equiv 0$ is satisfied if and only if $E_6(z) \equiv H_7(z) \equiv 0$.

To deduce the remaining assertions from [**18**, Theorem 2.2, on pages 86–87], we introduce $a(z) \equiv 1$, $b(z) \equiv 2c_{0,1}(z)$, $c(z) \equiv 2c_{0,2}(z)$, $d(z) \equiv c_{1,1}(z)$, $e(z) \equiv 2c_{1,2}(z)$, $f(z) \equiv c_{2,2}(z)$,

$$A_2(z) \equiv 4B_{1,1}(z), \quad A_3(z) \equiv 8B_{1,2}(z), \quad \text{and} \quad A_4(z) \equiv 4B_{2,2}(z).$$

Then, we have

$$(1.78) \qquad 4A_2(z)A_4(z) - (A_3(z))^2 \equiv 2^6\Big[B_{1,1}(z)B_{2,2}(z) - (B_{1,2}(z))^2\Big]$$
$$\equiv 2^6 D_2(z) \not\equiv 0,$$

$$(1.79) \qquad 2c(z)(A_2(z))^2 - b(z)A_2(z)A_3(z) - 4a(z)A_2(z)A_4(z)$$
$$+ 2a(z)(A_3(z))^2 + 2a(z)A_2^{(1)}(z)A_3(z) - 2A_2(z)A_3^{(1)} \equiv -2^6 E_6(z),$$

and

$$(1.80) \qquad c(z)A_2(z)A_3(z) - 2b(z)A_2(z)A_4(z) + 2a(z)A_3(z)A_4(z)$$
$$+ 2a(z)A_2^{(1)}A_4(z) - 2a(z)A_2(z)A_4^{(1)} \equiv -2^5 H_7(z).$$

In view of (1.78), at least one of $A_2(z)$, $A_3(z)$, $A_4(z)$ is not identically zero. Hence, we can apply [**18**, Theorem 2.2 on pages 86–87] as well as (1.79)–(1.80) to deduce that: for any given Q_2, there is function λ such that $(Q_2)' + \lambda Q_2$ has a nontrivial factorization if and only if $A_2(z) \not\equiv 0$ and $E_6(z) \equiv H_7(z) \equiv 0$. However, (1.78) and (1.79) show that we must necessarily have $A_2(z) \not\equiv 0$ whenever $E_6(z) \equiv 0$. Consequently, the condition $E_6(z) \equiv E_7(z) \equiv 0$ is satisfied if and only if $A_2(z) \not\equiv 0$ and there is a λ for which $(Q_2)' + \lambda Q_2$ has a nontrivial factorization. Moreover, when this is the situation for $Q_2 = 0$, we use [**18**, Theorem 2.2] to see that λ is unique and given by (1.75). Finally, when $E_6(z) \equiv E_7(z) \equiv 0$ and λ is given by (1.75), we apply [**18**, Theorem 3.5 on page 93] to obtain the factorization for $(Q_2)' + \lambda Q_2$ specified by (1.76) and (1.77). This completes the proof. □

REMARK 1.11. Proposition 19.1 of page 194 and Corollary 19.2 enable us to give a proof for Theorem 1.10 that is independent of the results in [**18**]. For details about this, see Remark 19.3 on page 195. In particular, the more general results of Chapter 18 specialize to yield (19.12) of page 195 as a natural motivation for the polynomials $\boldsymbol{E}_6$ in (1.72) and $\boldsymbol{H}_7$ in (1.71).

LEMMA 1.12. *In terms of $\boldsymbol{a}_{2,2}$ from (1.63) and the basic relative invariants in (1.60)–(1.62), the polynomials $\boldsymbol{E}_6$ and $\boldsymbol{E}_7$ have the representations*

$$(1.81) \qquad \boldsymbol{E}_6 \equiv \Big[\boldsymbol{\mathcal{I}}_{2,1,1}\boldsymbol{\mathcal{I}}_{2,2,2} - 2(\boldsymbol{\mathcal{I}}_{2,1,2})^2\Big] + \frac{4}{3}\Big[\boldsymbol{\mathcal{I}}_{2,1,1}\boldsymbol{\mathcal{I}}_{2,1,2}^{(1)} - \frac{3}{2}\boldsymbol{\mathcal{I}}_{2,1,1}^{(1)}\boldsymbol{\mathcal{I}}_{2,1,2}\Big]$$
$$+ \frac{3}{10}\Big[\boldsymbol{\mathcal{I}}_{2,1,1}\boldsymbol{\mathcal{I}}_{2,1,1}^{(2)} - \frac{5}{4}\big(\boldsymbol{\mathcal{I}}_{2,1,1}^{(1)}\big)^2 - 4\boldsymbol{a}_{2,2}(\boldsymbol{\mathcal{I}}_{2,1,1})^2\Big]$$

and

$$(1.82) \quad \boldsymbol{E}_7 \equiv [-2\boldsymbol{\mathcal{I}}_{2,1,2}\boldsymbol{\mathcal{I}}_{2,2,2}] + \frac{5}{6}\left[\boldsymbol{\mathcal{I}}_{2,1,1}\boldsymbol{\mathcal{I}}^{(1)}_{2,2,2} - 2\boldsymbol{\mathcal{I}}^{(1)}_{2,1,1}\boldsymbol{\mathcal{I}}_{2,2,2}\right]$$
$$+ \frac{1}{9}\left[\begin{array}{c}\boldsymbol{\mathcal{I}}_{2,1,1}\boldsymbol{\mathcal{I}}^{(2)}_{2,1,2} - \dfrac{7}{2}\boldsymbol{\mathcal{I}}^{(1)}_{2,1,1}\boldsymbol{\mathcal{I}}^{(1)}_{2,1,2} \\ + \dfrac{21}{10}\boldsymbol{\mathcal{I}}^{(2)}_{2,1,1}\boldsymbol{\mathcal{I}}_{2,1,2} - \dfrac{72}{5}\boldsymbol{a}_{2,2}\boldsymbol{\mathcal{I}}_{2,1,1}\boldsymbol{\mathcal{I}}_{2,1,2}\end{array}\right].$$

Also, $\boldsymbol{E}_6$, $\boldsymbol{E}_7$, and the bracketed terms of (1.81) and (1.82) are relative invariants.

PROOF. To deduce the expressions (1.81) and (1.82), we first expand the right members of (1.72) and (1.73) as polynomials over $\mathbb{Q}$ in the variables $\boldsymbol{w}^{(k)}_{i,j}$; then, in a manner similar to that for (1.65), we apply (1.62), (1.61), (1.60), and (1.63) by specializing the algorithmic proof given on pages 144–145 for Theorem 14.3.

The first terms in (1.81) and (1.82) are clearly relative invariants. We use Proposition 14.6 on page 146 to see that the second terms in (1.81) and (1.82) are relative invariants. For the third term of (1.81), we apply Corollary 14.9 on page 148 with $\boldsymbol{P}_p \equiv \boldsymbol{\mathcal{I}}_{2,1,1}$ and $p = 2$. For the third term of (1.82), we use Theorem 14.8 on page 147 with $\boldsymbol{P}_p \equiv \boldsymbol{\mathcal{I}}_{2,1,1}$, $p = 2$, $\boldsymbol{Q}_q \equiv \boldsymbol{\mathcal{I}}_{2,1,2}$, $q = 3$, $\alpha = -7/2$, $\beta = 21/10$, and $\gamma = -72/5$. Since each of the three bracketed expressions in the right member of (1.81) is a relative invariant, we find that $\boldsymbol{E}_6$ is a relative invariant of weight 6. Also, since each of the three bracketed expressions in the right member of (1.82) is a relative invariant, we conclude that $\boldsymbol{E}_7$ is a relative invariant of weight 7. This completes the proof. $\square$

Machine computations can be employed to easily obtain (1.81) and (1.82). For specific details, see Section 16.2 on pages 168–169. Also, after defining $\lambda(z)$ by means of (1.75), we can use computer algebra to check the identity (1.76) in terms of (1.77).

1.7. Inclusion of homogeneous linear differential equations

To explain how our study of (1.1) includes as a special case the subject of relative invariants for homogeneous linear differential equations, we introduce

$$(1.83) \quad \boldsymbol{w}^{(k)}_j = \boldsymbol{w}^{(k)}_{0,j}, \quad \text{for } 1 \leq j \leq m \text{ and } k \geq 0,$$

and we let $\mathcal{R}_m$ denote the subring of $\mathcal{S}_m$ consisting of the polynomials over $\mathbb{Q}$ in the variables (1.83). Also, for the coefficients $c_{0,j}(z)$ of (1.1) on Ω, $c^*_{0,j}(z)$ of (1.3) on Ω, and $c^{**}_{0,j}(\zeta)$ of (1.5) on Ω^{**}, we set

$$c_j(z) \equiv c_{0,j}(z), \quad c^*_j(z) \equiv c^*_{0,j}(z), \quad \text{and} \quad c^{**}_j(\zeta) \equiv c^{**}_{0,j}(\zeta), \quad \text{for } 1 \leq j \leq m.$$

Proposition 2.1 on page 23 shows that: as the substitution $y(z) = \rho(z)\,v(z)$ transforms (1.1) on Ω into (1.3) on Ω, it transforms

$$(1.84) \quad y^{(m)}(z) + \sum_{j=1}^m c_j(z)\,y^{(m-j)}(z) = 0, \quad \text{on } \Omega,$$

into

$$(1.85) \quad v^{(m)}(z) + \sum_{j=1}^m c^*_j(z)\,v^{(m-j)}(z) = 0, \quad \text{on } \Omega.$$

Thus, for the context of [**24**, page 4], a polynomial $\boldsymbol{P}$ in $\mathcal{R}_m$ is a semi-invariant of the first kind for homogeneous linear differential equations of order m if and only if it is a semi-invariant of the first kind for the equations (1.1). Theorem 3.2 of page 30 shows that: as the substitution $z = f(\zeta)$ transforms (1.1) on Ω into (1.5) on Ω^{**}, it transforms (1.84) on Ω into

$$(1.86) \quad u^{(m)}(\zeta) + \sum_{j=1}^{m} c_j^{**}(\zeta) u^{(m-j)}(\zeta) = 0, \quad \text{on } \Omega^{**} \text{ with } u(\zeta) = (y \circ f)(\zeta).$$

Thus, for the context of [**24**, page 4], a polynomial $\boldsymbol{P}$ in $\mathcal{R}_m$ is a semi-invariant of the second kind for homogeneous linear differential equations of order m if and only if it is a semi-invariant of the second kind for the equations (1.1). By combining the preceding results, we see that *a polynomial $\boldsymbol{P}$ in $\mathcal{R}_m$ is a relative invariant for homogeneous linear differential equations of order m if and only if it is a relative invariant for the equations* (1.1) *having the same order* m. In particular, each invariant presented in [**24**] for the homogeneous linear differential equations (1.84) is a corresponding invariant for the equations (1.1).

We recall from [**24**, Proposition A.15 of page 145] that the equations (1.84) possess relative invariants only when $m \geq 3$. Moreover, when $m \geq 3$, we combine [**24**, Definition 1.4 on page 5], Definition 1.5 on page 6, and Main Theorem on page 8 to conclude that: *the basic relative invariants for homogeneous linear differential equations of order* $m \geq 3$ *are* $\boldsymbol{\mathcal{I}}_{m,0,n}$, *for* $3 \leq n \leq m$. (In [**24**], they were denoted by $\boldsymbol{\mathcal{I}}_{m,3}, \ldots, \boldsymbol{\mathcal{I}}_{m,m}$.)

THEOREM 1.13. *The formulas presented in* [**24**] *for the basic relative invariants of homogeneous linear differential equations are deducible from those presented here by a minor change of notation.*

PROOF. We introduce

$$(1.87) \quad P_{m,n,i} \equiv \binom{m-i}{n-i} \prod_{r=1}^{n-i}(n-r), \quad \text{for } 3 \leq n \leq m \text{ and any } i,$$

in order to define $\widehat{\boldsymbol{I}}_{m,n,i}$ in $\mathcal{R}_m$ with respect to (1.14), (1.16), and (1.18)–(1.27) by

$$(1.88) \quad \widehat{\boldsymbol{I}}_{m,n,i} \equiv \frac{1}{P_{m,n,i}} \boldsymbol{I}_{m,0,n,0,i}, \quad \text{for } 3 \leq n \leq m \text{ and } 1 \leq i \leq n.$$

We note that (1.21) yields

$$(1.89) \quad \boldsymbol{L}_{m,0,i} \equiv \sum_{q=0}^{i} \boldsymbol{K}_{m,i-q,q} \boldsymbol{w}_{0,q} \equiv \sum_{j=0}^{i} \boldsymbol{K}_{m,i-j,j} \boldsymbol{w}_j, \quad \text{for } 0 \leq i \leq m.$$

This was designated $\boldsymbol{L}_i$ in [**24**, Equation (1.15)]. We use (1.89), (1.44), (1.43), and (1.19) to see that

$$(1.90) \quad \boldsymbol{L}_{m,0,2} \equiv 0.$$

Moreover, (1.23) and (1.87) give

$$(1.91) \quad \boldsymbol{M}_{m,0,n,0,i} \equiv P_{m,n,i} \boldsymbol{L}_{m,0,i}, \quad \text{for } 3 \leq n \leq m \text{ and } 0 \leq i \leq n.$$

By combining (1.88) with (1.26), (1.27), (1.91), and (1.90), we verify that

$$(1.92) \quad \widehat{\boldsymbol{I}}_{m,n,1} \equiv \widehat{\boldsymbol{I}}_{m,n,2} \equiv 0, \quad \text{for } 3 \leq n \leq m.$$

1.7. INCLUSION OF HOMOGENEOUS LINEAR DIFFERENTIAL EQUATIONS

In terms of (1.24), (1.25), and (1.87), we introduce

$$(1.93) \qquad \kappa_{m,n,i} \equiv A_{0,n,i} \frac{P_{m,n,i}}{P_{m,n,i+1}} \equiv \frac{-i(m-i)}{(n-i)(n+i-1)}, \quad \text{for } 1 \leq i \leq n-1,$$

and

$$(1.94) \qquad \lambda_{m,n,i} \equiv B_{0,n,i} \frac{P_{m,n,i-1}}{P_{m,n,i+1}} \equiv \frac{i(i-1)(m-i)(m-i+1)}{(n-i+1)(n+i-2)},$$
$$\text{for } 1 \leq i \leq n-1.$$

By applying the formula $\boldsymbol{I}_{m,0,n,0,i} \equiv P_{m,n,i} \widehat{\boldsymbol{I}}_{m,n,i}$ obtained from (1.88) to the restriction of (1.27) having $l = 0 < 3 \leq n \leq m$ and $2 \leq i \leq n-1$, we rewrite the resulting expression by means of (1.91) and (1.93)–(1.94) to deduce that

$$(1.95) \qquad \widehat{\boldsymbol{I}}_{m,n,i+1} \equiv \boldsymbol{L}_{m,0,i+1} + \kappa_{m,n,i} \widehat{\boldsymbol{I}}_{m,n,i}^{(1)} + \lambda_{m,n,i} \boldsymbol{a}_{m,2} \widehat{\boldsymbol{I}}_{m,n,i-1},$$
$$\text{for } 3 \leq n \leq m \text{ and } 2 \leq i \leq n-1.$$

In view of (1.92) and (1.95), we can identify $\widehat{\boldsymbol{I}}_{m,n,i}$ with [**24**, $\boldsymbol{I}_{n,i}$ in (1.18)–(1.19)]. Since (1.87) yields $P_{m,n,n} \equiv 1$, we find that (1.28) and (1.88) give

$$\boldsymbol{\mathcal{I}}_{m,0,n} \equiv \boldsymbol{I}_{m,0,n,0,n} \equiv \widehat{\boldsymbol{I}}_{m,n,n}, \quad \text{for } 3 \leq n \leq m.$$

This relates $\boldsymbol{\mathcal{I}}_{m,0,n}$ and [**24**, $\boldsymbol{I}_{n,n}$ in (1.20)]. Thus, with $m \geq 3$ and the notation of (1.83), we have checked that our Main Theorem of page 8 yields the $m-2$ basic relative invariants in [**24**] for homogeneous linear differential equations of order m.

We use (1.87), (1.15), (1.16), and (1.30)–(1.37) to define $\widehat{\boldsymbol{J}}_{m,n,i}$ in $\mathcal{R}_m$ by

$$(1.96) \qquad \widehat{\boldsymbol{J}}_{m,n,i} \equiv \frac{1}{P_{m,n,i}} \boldsymbol{J}_{m,0,n,0,i}, \quad \text{for } 3 \leq n \leq m \text{ and } 1 \leq i \leq n.$$

We note that (1.33) yields

$$(1.97) \qquad \boldsymbol{V}_{m,0,i} \equiv \sum_{q=0}^{i} \boldsymbol{U}_{m,i-q,q} \boldsymbol{w}_{0,q} \equiv \sum_{j=0}^{i} \boldsymbol{U}_{m,i-j,j} \boldsymbol{w}_j, \quad \text{for } 0 \leq i \leq m.$$

This was designated $\boldsymbol{V}_i$ in [**24**, Equation (1.24)]. We derive the formulas

$$\boldsymbol{U}_{0,2} \equiv 1, \quad \boldsymbol{U}_{1,1} \equiv -\frac{m-2}{m} \boldsymbol{w}_{0,1} \quad \text{and} \quad \boldsymbol{U}_{2,0} \equiv -\boldsymbol{w}_{0,2} + \frac{m-2}{m} (\boldsymbol{w}_{0,1})^2.$$

from (1.30)–(1.32) and combine them with (1.97) to see that

$$(1.98) \qquad \boldsymbol{V}_{m,0,2} \equiv 0.$$

Moreover, (1.35) gives

$$(1.99) \qquad \boldsymbol{W}_{m,0,n,0,i} \equiv P_{m,n,i} \boldsymbol{V}_{m,0,i}, \quad \text{for } 3 \leq n \leq m \text{ and } 0 \leq i \leq n.$$

We employ (1.96) with (1.36), (1.37), (1.99), and (1.98) to verify that

$$(1.100) \qquad \widehat{\boldsymbol{J}}_{m,n,1} \equiv \widehat{\boldsymbol{J}}_{m,n,2} \equiv 0, \quad \text{for } 3 \leq n \leq m.$$

Combining the restriction of (1.37) having $l = 0 < 3 \leq n \leq m$ and $2 \leq i \leq n-1$ with the rewritten form $\boldsymbol{J}_{m,0,n,0,i} \equiv P_{m,n,i} \widehat{\boldsymbol{J}}_{m,n,i}$ of (1.96), we then apply (1.99) and (1.93)–(1.94) to obtain

$$
(1.101) \quad \widehat{\boldsymbol{J}}_{m,n,i+1} \equiv \boldsymbol{V}_{m,0,i+1} + \kappa_{m,n,i}\left[\widehat{\boldsymbol{J}}_{m,n,i}^{(1)} + i\,\boldsymbol{b}_{m,1}\,\widehat{\boldsymbol{J}}_{m,n,i}\right]
$$
$$
+ \lambda_{m,n,i}\,\boldsymbol{b}_{m,2}\,\widehat{\boldsymbol{J}}_{m,n,i-1}, \quad \text{for } 3 \le n \le m \text{ and } 2 \le i \le n-1.
$$

Due to (1.100) and (1.101), we can identify $\widehat{\boldsymbol{J}}_{m,n,i}$ with [**24**, $\boldsymbol{J}_{n,i}$ in (1.18)–(1.19)]. Since (1.87) yields $P_{m,n,n} \equiv 1$, we find that (1.38) and (1.96) give

$$
\boldsymbol{\mathcal{J}}_{m,0,n} \equiv \boldsymbol{J}_{m,0,n,0,n} \equiv \widehat{\boldsymbol{J}}_{m,n,n}, \quad \text{for } 3 \le n \le m.
$$

This relates $\boldsymbol{\mathcal{J}}_{m,0,n}$ and [**24**, $\boldsymbol{J}_{n,n}$, in (1.27)]. Thus, for $l = 0 < 3 \le n \le m$, our Main Theorem yields the formulas of [**24**] for the basic relative invariants presented in [**24**]. This completes the proof. □

Historical information about the basic relative invariant for third-order homogeneous linear differential equations is given in [**24**, page 8]. By combining the formula we presented for it in [**24**, page 7, Equation (1.32)] with (1.83), we conclude that the right member of (1.102) is a basic relative invariant of weight 3 for the equations $Q_3 = 0$. The proof of Proposition 1.14 illustrates how our present technique yields the same result.

PROPOSITION 1.14. *For $m = 3$, the basic relative invariant $\boldsymbol{\mathcal{I}}_{3,0,3}$ is given by*

$$
(1.102) \quad \boldsymbol{\mathcal{I}}_{3,0,3} \equiv w_{0,3} - \frac{1}{3} w_{0,1}\, w_{0,2} + \frac{2}{27}(w_{0,1})^3 - \frac{1}{2} w_{0,2}^{(1)} + \frac{1}{3} w_{0,1}\, w_{0,1}^{(1)} + \frac{1}{6} w_{0,1}^{(2)}.
$$

PROOF. Formulas (1.26), (1.27), (1.23), (1.21), (1.19), (1.43), and (1.44) yield

$$
(1.103) \quad \boldsymbol{I}_{3,0,3,0,2} \equiv \boldsymbol{M}_{3,0,3,0,2} \equiv \boldsymbol{L}_{3,0,2} \equiv \sum_{q=0}^{2} \boldsymbol{K}_{3,2-q,q}\, w_{0,q} \equiv 0.
$$

Since (1.20), (1.44), and (1.43) give

$$
(1.104) \quad \boldsymbol{K}_{3,3,0} \equiv \frac{1}{2} w_{0,1}\, w_{0,2} - \frac{11}{54}(w_{0,1})^3 - \frac{1}{2} w_{0,2}^{(1)} + \frac{1}{6} w_{0,1}\, w_{0,1}^{(1)} + \frac{1}{6} w_{0,1}^{(2)},
$$

we employ (1.28), (1.103), (1.27), (1.26), (1.23), (1.21), (1.19), (1.43), (1.44), and (1.104) to obtain

$$
\boldsymbol{\mathcal{I}}_{3,0,3} \equiv \boldsymbol{I}_{3,0,3,0,3} \equiv \boldsymbol{M}_{3,0,3,0,3} \equiv \boldsymbol{L}_{3,0,3} \equiv \sum_{q=0}^{3} \boldsymbol{K}_{3,3-q,q}\, w_{0,q}
$$

and (1.102). This completes the proof. □

1.8. Order of presentation

Chapters 2 and 3 provide general information about transformations of (1.1). In particular, they enable us to easily establish in Proposition 3.5 on page 34 that: $\boldsymbol{D}_1$ in (1.11) is a basic relative invariant of index $(1, 1)$ for the equations (1.1) of order $m \ge 1$; and it is the only basic relative invariant of index $(1, 1)$. For the equations (1.1) having $m = 1$, Theorem 4.10 on page 41 shows that the relative invariants are given by $\gamma\,[\boldsymbol{D}_1]^\kappa$ as γ ranges through the nonzero rational numbers and κ ranges through the positive integers. This establishes our Main Theorem for the special situation $m = 1$ and enables us to focus on the general situation $m \ge 2$ throughout the remainder of the first twelve chapters.

1.8. ORDER OF PRESENTATION

A Laguerre-Forsyth canonical form for an equation (1.1) having $m \geq 2$ is introduced in Definition 5.6 on page 51. It is a natural extension of the familiar concept of a Laguerre-Forsyth canonical form for a homogeneous linear differential equation of order $m \geq 2$. When a Laguerre-Forsyth canonical form for an equation (1.1) is transformed into another Laguerre-Forsyth canonical form, the coefficients of the forms satisfy various identities that are presented in (6.23) of Theorem 6.3 on page 58. While (6.23) is not explicitly used in any proofs, it is our key discovery and led to the formulation of Theorem 7.10 on page 85 in terms of (1.18)–(1.28) as well as the formulation of Theorem 9.10 on page 109 in terms of (1.30)–(1.38). Theorems 7.10 and 9.10 yield Theorem 11.1 on page 119 and the deduction in Theorem 11.2 on page 120 that: for $m \geq 2$ and each pair (l, n) of integers l, n that satisfy either $1 \leq l \leq n \leq m$ or $l = 0 < 3 \leq n \leq m$, there is a basic relative invariant of index (l, n).

For $m \geq 2$, the uniqueness of a basic relative invariant having a given index is demonstrated in Theorem 12.2 on page 130. That the basic relative invariants are algebraically independent over $\mathbb{Q}$ is the content of Corollary 12.5 on page 133. Our Main Theorem also asserts that: for $m \geq 2$, each relative invariant for the equations (1.1) is expressible as a differential-polynomial combination over $\mathbb{Q}$ of $\boldsymbol{a}_{m,2}$ and the basic relative invariants for (1.1). That is established in Corollary 14.4 on page 146.

The number $\mathcal{N}_{m,2}$ of basic relative invariants in $\mathcal{S}_m$ for the equations (1.1) having order $m \geq 1$ is equal to the number of pairs (l, n) of integers that satisfy either $1 \leq l \leq n \leq m$ or $l = 0 < 3 \leq n \leq m$. In view of this, we find that $\mathcal{N}_{m,2}$ is given by $\mathcal{N}_{m,2} = \binom{m+1}{2} + (m - 2) = \binom{m+2}{2} - 3$, when $m \geq 2$, and $\mathcal{N}_{1,2} = 1$.

Throughout, with the exception of Chapter 13, the functions are meromorphic on some region of the complex plane. Chapter 13 provides a natural real-valued context in which the defining characteristics of the basic relative invariants can be directly reinterpreted.

Techniques presented in Chapter 14 combine given relative invariants for (1.1) to obtain others. They show that the sequence of operations listed by A. R. Forsyth in [31, pages 407–418] for deducing new relative invariants of homogeneous linear differential equations from given ones is considerably incomplete. In fact, the series of new results that are presented in Theorem 14.8, Propositions 14.13–14.14, and Formulations 14.15–14.16 introduce a challenging subject for further research. We apply those results in Section 14.3 to obtain the relative invariants of weight ≤ 9 for the equations $Q_2 = 0$.

When $1 \leq n \leq m$, a relative invariant $\boldsymbol{D}_n$ in $\mathcal{S}_n$ of weight $n(n + 1)$ for the equations (1.1) of order m is provided by Theorem 15.10 on page 164. The $\boldsymbol{D}_2$ mentioned on page 4 for $Q_2 = 0$ is the special case having $m = n = 2$.

Machine computations are presented in Chapter 16. Those of Sections 16.1 and 16.2 indicate how various identities were discovered. The technique of Section 16.3 provides an excellent check for our Main Theorem. Also, Section 16.3 shows how the basic relative invariants can be represented and evaluated in a system of computer algebra. For the equations (1.1) having $m \geq 3$, we use Section 1.7 to see that the machine computations in Section 16.3 for the basic relative invariants of index (l, n) subject to $l = 0$ and $3 \leq n \leq m$ are consistent with those in [24, pages 28–30].

There are problems for the equations $Q_m = 0$ that are completely analogous to the Fano-type problems studied for homogeneous linear differential equations. The simplest one of them is stated and solved in Chapter 17.

For $Q_2 = 0$ in (1.0-A) of page 3, *Paul Appell's condition* is the requirement that $(Q_2)' + \lambda Q_2$ has a nontrivial factorization for some meromorphic $\lambda(z)$ on Ω.

Suppose that a given $Q_2 = 0$ satisfies $D_2(z) \not\equiv 0$ and Appell's condition. Let $\mathcal{U}$ be a subregion of Ω. Then, in view of Proposition 19.1 on page 194, Theorem 19.7 on page 197 specifies linearly independent meromorphic functions $v_0(z)$, $v_1(z)$, $v_2(z)$ on a subregion $\mathcal{V}$ of $\mathcal{U}$ such that $(v_1(z))^2 - 4v_0(z)\,v_2(z) \not\equiv 0$ and the nonsingular solutions of $Q_2 = 0$ on $\mathcal{V}$ are given by

(1.105) $\quad y(z) \equiv C^2 v_0(z) + CK\, v_1(z) + K^2 v_2(z), \quad$ for C, K in $\mathbb{C}$ not both zero.

Conversely, for a given $Q_2 = 0$ that satisfies $D_2(z) \not\equiv 0$, suppose that its nonsingular solutions on some subregion $\mathcal{V}$ of Ω are specified by (1.105). Since the solutions of $Q_2 = 0$ on $\mathcal{V}$ are therefore free of movable branch points in the precise sense of Definition 19.14 on page 206, we apply Theorem 19.16 on page 207 as well as Corollary 19.2 and Proposition 19.1 on page 194 to conclude that $Q_2 = 0$ satisfies Appell's condition on $\mathcal{V}$ and therefore also on Ω. *This establishes rigorously the italicized assertion in lines 6–12 of page 4.*

For the left member Q_m of (1.1), Theorem 18.1 on page 186 provides a simple way to check whether there is a meromorphic function $\lambda(z)$ in terms of which $(Q_m)' + \lambda Q_m$ has a nontrivial factorization. A solution procedure for this situation is given by Proposition 18.4 on page 189 and Proposition 18.7 on page 191.

The structure of rational semi-invariants and rational relative invariants for homogeneous linear differential equations was developed in [**24**, pages 165–191]. Chapter 20 extends that theory to the equations (1.1). The main alterations appear in the proofs for Propositions 20.10 and 20.13. Theorems 20.11 and 20.14 use the modification of [**24**, Proposition A.8] presented as Proposition 4.2 on page 36.

In Chapters 21–27, we consider the differential equations $H_{m,n} = 0$ whose left members $H_{m,n}$ are forms of degree n in $y(z)$, $y'(z)$, $\ldots$, $y^{(m)}(z)$ with meromorphic coefficients written symmetrically such that the coefficient of $\left(y^{(m)}(z)\right)^n$ is 1. The equations $H_{m,1} = 0$ are the monic homogeneous linear ones of order m; and, the equations $H_{m,2} = 0$ are the ones $Q_m = 0$ studied in Parts 1–3. Proposition 21.6 on page 253 shows that: for integers q, n satisfying $1 \leq q \leq n$, the relative invariants for the equations $H_{m,n} = 0$ include in a rewritten form all of the relative invariants for the equations $H_{m,q} = 0$. We use Theorems 22.6 and 27.13 to see that the basic relative invariants for the equations $H_{m,n} = 0$ are provided by Theorem 22.1 on page 257 when $n > m = 1$ and by Theorem 27.7 on page 314 when $m \geq 2$.

Two of the five classes of homogeneous quadratic differential equations having order $m \leq 2$ that are suitable for a study of their relative invariants are specified by the equations $Q_1 = 0$ of (4.16) on page 39 and the equations $Q_2 = 0$ of (1.0–A) on page 3. The remaining three classes are provided by (28.1) on page 325, by (30.1) on page 347, and by (30.28) on page 353. Results about them are presented in Chapter 28, in Section 30.1, and in Section 30.2 respectively.

The various classes of differential equations given by (29.1) on page 335 are also suitable for the study of relative invariants. They include the class of equations $H_{m,n} = 0$ in Part 4 as well as many others. In particular, they include the class of equations (28.1) on page 325 that has all of its relative invariants conveniently specified by Theorem 28.7 on page 332.

Section 29.2 extends the subject of relative invariants to various classes of nonhomogeneous differential equations.

CHAPTER 2

The Coefficients $c_{i,j}^*(z)$ of (1.3)

PROPOSITION 2.1. *For $m \geq 1$ and $0 \leq i, j \leq m$, suppose that $c_{i,j}(z)$ and $\rho(z) \not\equiv 0$ are meromorphic functions on a region Ω of the complex plane such that $c_{j,i}(z) \equiv c_{i,j}(z)$ and $c_{0,0}(z) \equiv 1$. Then, for $0 \leq i, j \leq m$, there are unique meromorphic functions $c_{i,j}^*(z)$ on Ω subject to $c_{j,i}^*(z) \equiv c_{i,j}^*(z)$ and $c_{0,0}^*(z) \equiv 1$ such that: for any meromorphic functions $y(z)$ and $v(z)$ that are defined and related by $y(z) \equiv \rho(z)\, v(z)$ on a subregion of Ω, $y(z)$ is a local solution of*

$$(2.1) \qquad \left(Y^{(m)}\right)^2 + \sum_{\substack{0 \leq i,j \leq m \\ (i,j) \neq (0,0)}} c_{i,j}(z)\, Y^{(m-i)}\, Y^{(m-j)} = 0$$

if and only if $v(z)$ is a local solution of

$$(2.2) \qquad \left(V^{(m)}\right)^2 + \sum_{\substack{0 \leq i,j \leq m \\ (i,j) \neq (0,0)}} c_{i,j}^*(z)\, V^{(m-i)}\, V^{(m-j)} = 0;$$

and $y(z)$ is a local solution of

$$(2.3) \qquad Y^{(m)} + \sum_{j=1}^{m} c_{0,j}(z)\, Y^{(m-j)} = 0$$

if and only if $v(z)$ is a local solution of

$$(2.4) \qquad V^{(m)} + \sum_{j=1}^{m} c_{0,j}^*(z)\, V^{(m-j)} = 0.$$

Moreover, the coefficients $c_{i,j}^*(z)$ for (2.2) and (2.4) are given by

$$(2.5) \quad c_{i,j}^*(z) \equiv \sum_{\mu=0}^{i} \sum_{\nu=0}^{j} \binom{m-\mu}{i-\mu}\binom{m-\nu}{j-\nu} \frac{\rho^{(i-\mu)}(z)\, \rho^{(j-\nu)}(z)}{\left(\rho(z)\right)^2}\, c_{\mu,\nu}(z),$$

on Ω for $0 \leq i, j \leq m$.

PROOF. Let $Q\bigl[z; Y^{(0)}, \ldots, Y^{(m)}\bigr]$ denote the left member of (2.1) and set

$$Q^*\bigl[z; V^{(0)}, \ldots, V^{(m)}\bigr] \equiv \left(V^{(m)}\right)^2 + \sum_{\substack{0 \leq i,j \leq m \\ (i,j) \neq (0,0)}} c_{i,j}^*(z)\, V^{(m-i)}\, V^{(m-j)},$$

where each $c_{i,j}^*(z)$ is defined on Ω by means of (2.5). In particular, (2.5) yields

$$c_{j,i}^*(z) \equiv c_{i,j}^*(z), \quad \text{on } \Omega \text{ for } 0 \leq i, j \leq m, \quad \text{and} \quad c_{0,0}^*(z) \equiv 1.$$

For any meromorphic functions $y(z)$ and $v(z)$ that are defined on a subregion of Ω and related by $y(z) = \rho(z)\, v(z)$, we use (2.5) to obtain

$$Q[z;\, y^{(0)}(z),\, \ldots,\, y^{(m)}(z)] \equiv \sum_{0 \leq \mu,\nu \leq m} c_{\mu,\nu}(z)\, y^{(m-\mu)}(z)\, y^{(m-\nu)}(z)$$

$$\equiv \sum_{\mu=0}^{m} \sum_{\nu=0}^{m} c_{\mu,\nu}(z) \sum_{r=0}^{m-\mu} \binom{m-\mu}{r} \rho^{(r)}(z)\, v^{(m-\mu-r)}(z) \sum_{s=0}^{m-\nu} \binom{m-\nu}{s} \rho^{(s)}(z)\, v^{(m-\nu-s)}(z)$$

$$\equiv \sum_{\mu=0}^{m} \sum_{\nu=0}^{m} \sum_{i=\mu}^{m} \sum_{j=\nu}^{m} \binom{m-\mu}{i-\mu}\binom{m-\nu}{j-\nu} \rho^{(i-\mu)}(z)\, \rho^{(j-\nu)}(z)\, c_{\mu,\nu}(z)\, v^{(m-i)}(z)\, v^{(m-j)}(z)$$

$$\equiv \sum_{i=0}^{m} \sum_{j=0}^{m} \left[\sum_{\mu=0}^{i} \sum_{\nu=0}^{j} \binom{m-\mu}{i-\mu}\binom{m-\nu}{j-\nu} \rho^{(i-\mu)}(z)\, \rho^{(j-\nu)}(z)\, c_{\mu,\nu}(z) \right] v^{(m-i)}(z)\, v^{(m-j)}(z)$$

$$\equiv (\rho(z))^2 \sum_{0 \leq i,j \leq m} c^*_{i,j}(z)\, v^{(m-i)}(z)\, v^{(m-j)}(z)$$

$$\equiv (\rho(z))^2\, Q^*[z;\, v^{(0)}(z),\, \ldots,\, v^{(m)}(z)].$$

Selecting (2.2) as the equation $Q^* = 0$ based on (2.5), we observe that: $y(z)$ is a local solution of (2.1) if and only if $v(z)$ is a local solution of (2.2).

Let $L[z; Y^{(0)}, \ldots, Y^{(m)}]$ denote the left member of (2.3) and set

$$L^*[z; V^{(0)}, \ldots, V^{(m)}] \equiv V^{(m)} + \sum_{j=1}^{m} c^*_{0,j}(z)\, V^{(m-j)},$$

where each $c^*_{0,j}(z)$ is defined on Ω by means of (2.5). Then, for any meromorphic functions $y(z)$ and $v(z)$ related by $y(z) = \rho(z)\, v(z)$ on a subregion of Ω, we have

$$L[z;\, y^{(0)}(z),\, \ldots,\, y^{(m)}(z)] \equiv \sum_{\nu=0}^{m} c_{0,\nu}(z)\, y^{(m-\nu)}(z)$$

$$\equiv \sum_{\nu=0}^{m} c_{0,\nu}(z) \sum_{s=0}^{m-\nu} \binom{m-\nu}{s} \rho^{(s)}(z)\, v^{(m-\nu-s)}(z)$$

$$\equiv \sum_{\nu=0}^{m} \sum_{j=\nu}^{m} \binom{m-\nu}{j-\nu} \rho^{(j-\nu)}(z)\, c_{0,\nu}(z)\, v^{(m-j)}(z)$$

$$\equiv \sum_{j=0}^{m} \left[\sum_{\nu=0}^{j} \binom{m-\nu}{j-\nu} \rho^{(j-\nu)}(z)\, c_{0,\nu}(z) \right] v^{(m-j)}(z)$$

$$\equiv \rho(z) \sum_{j=0}^{m} c^*_{0,j}(z)\, v^{(m-j)}(z)$$

$$\equiv \rho(z)\, L^*[z;\, v^{(0)}(z),\, \ldots,\, v^{(m)}(z)].$$

Selecting (2.4) as the equation $L^* = 0$ based on (2.5), we see that: $y(z)$ is a local solution of (2.3) if and only if $v(z)$ is a local solution of (2.4).

To prove the uniqueness of the coefficients in (2.5) for (2.2) and (2.4), suppose that meromorphic function $p_{i,j}(z)$ are given on Ω for $0 \leq i, j \leq m$ such that: $p_{j,i}(z) \equiv p_{i,j}(z)$; $p_{0,0}(z) \equiv 1$; and for any meromorphic functions $y(z)$, $v(z)$ defined and related by $y(z) = \rho(z)\,v(z)$ on a subregion of Ω, $y(z)$ is a local solution of (2.1) if and only if $v(z)$ is a local solution of

$$(2.6) \qquad \left(V^{(m)}\right)^2 + \sum_{\substack{0 \leq i,j \leq m \\ (i,j) \neq (0,0)}} p_{i,j}(z)\,V^{(m-i)}\,V^{(m-j)} = 0$$

and $y(z)$ is a local solution of (2.3) if and only if $v(z)$ is a local solution of

$$(2.7) \qquad V^{(m)} + \sum_{j=1}^{m} p_{0,j}(z)\,V^{(m-j)} = 0.$$

Consequently, when the coefficients of (2.2) and (2.4) are given by (2.5), we combine our previous results about (2.1)–(2.4) with the conditions for the coefficients $p_{i,j}(z)$ of (2.6)–(2.7) to see that: equations (2.6) and (2.2) have the same local solutions; while equations (2.7) and (2.4) have the same local solutions. In particular, since (2.7) and (2.4) are monic homogeneous linear differential equations of the same order, they must coincide so that

$$(2.8) \qquad p_{0,j}(z) \equiv c_{0,j}^*(z), \quad \text{on } \Omega \text{ for } 1 \leq j \leq m.$$

For $1 \leq j \leq m$, we have

$$c_{0,j}^*(z)\,V^{(m)}\,V^{(m-j)} + c_{j,0}^*(z)\,V^{(m-j)}\,V^{(m)} \equiv 2c_{0,j}^*(z)\,V^{(m)}\,V^{(m-j)}.$$

This enables us to rewrite (2.2) as

$$(2.9) \qquad \left[V^{(m)} + \sum_{j=1}^{m} c_{0,j}^*(z)\,V^{(m-j)}\right]^2 + R_1\!\left(z;\,V^{(0)},\ldots,V^{(m-1)}\right) = 0,$$

where either $R_1\!\left(z;\,V^{(0)},\ldots,V^{(m-1)}\right)$ is a quadratic form in $V^{(0)},\ldots,V^{(m-1)}$ that has meromorphic coefficients on Ω or $R_1 \equiv 0$. Similarly, we rewrite (2.6) as

$$(2.10) \qquad \left[V^{(m)} + \sum_{j=1}^{m} p_{0,j}(z)\,V^{(m-j)}\right]^2 + R_2\!\left(z;\,V^{(0)},\ldots,V^{(m-1)}\right) = 0,$$

where either $R_2\!\left(z;\,V^{(0)},\ldots,V^{(m-1)}\right)$ is a quadratic form in $V^{(0)},\ldots,V^{(m-1)}$ that has meromorphic coefficients on Ω or $R_2 \equiv 0$. In view of (2.8)–(2.10), each local solution of (2.2) is a solution of

$$(2.11) \qquad R_2\!\left(z;\,V^{(0)},\ldots,V^{(m-1)}\right) - R_1\!\left(z;\,V^{(0)},\ldots,V^{(m-1)}\right) = 0.$$

If $m = 1$, then we use $p_{0,1}(z) \equiv c_{0,1}^*(z)$ to see that (2.11) reduces to the equation $\left(p_{1,1}(z) - c_{1,1}^*(z)\right)\left(V^{(0)}\right)^2 = 0$; it requires $p_{1,1}(z) \equiv c_{1,1}^*(z)$. Thus, to establish the uniqueness of the coefficients for (2.2), it remains to be shown that the left member of (2.11) is identically zero for $m \geq 2$.

Suppose that $m \geq 2$ and the left member of (2.11) is not identically zero. The greatest integer r such that $V^{(r)}$ is effectively involved in (2.11) must be positive; otherwise, for some meromorphic $q_{0,0}(z) \not\equiv 0$, (2.11) would reduce to $q_{0,0}(z)\left(V^{(0)}\right)^2 = 0$ and could not have as solutions all the local solutions of (2.2).

Thus, (2.11) is a differential equation of positive order r that is expressible as

$$(2.12) \quad q_{0,0}(z)\left(V^{(r)}\right)^2 + \left[\sum_{k=1}^{r} q_{0,k}(z)\, V^{(r-k)}\right] V^{(r)} \\ + \sum_{1 \leq i \leq j \leq r} q_{i,j}(z)\, V^{(r-i)}\, V^{(r-j)} = 0,$$

for some meromorphic coefficients $q_{i,j}(z)$ on Ω that satisfy either $q_{0,0}(z) \not\equiv 0$ or $\sum_{k=1}^{r} q_{0,k}(z)\, V^{(r-k)} \not\equiv 0$. The order r of (2.12) is restricted by $1 \leq r < m$; and each solution of (2.2) is a solution of (2.12). Let $\mathcal{U}_0$ be a subregion of Ω on which the coefficients of (2.2) and (2.12) are analytic. There is a real number $\delta > 0$ and there are complex numbers $z_0, a_0, a_1, \ldots a_{r-1}$ such that the neighborhood $|z - z_0| < \delta$ of z_0 is contained in $\mathcal{U}_0$ and, for any complex numbers $z_1, b_0, b_1, \ldots, b_{r-1}$ satisfying

$$(2.13) \quad |z_1 - z_0| < \delta \quad \text{and} \quad |b_i - a_i| < \delta, \quad \text{for } 0 \leq i \leq r - 1,$$

either $q_{0,0}(z_1) \neq 0$ or $\sum_{k=1}^{r} q_{0,k}(z_1)\, b_{r-k} \neq 0$. Thus, for any complex numbers z_1, $b_0, \ldots, b_{r-1}$ satisfying (2.13), there do not exist three distinct local solutions $v(z)$ about z_1 of (2.12) and

$$(2.14) \quad v^{(i)}(z_1) = b_i, \quad \text{for } 0 \leq i \leq r - 1.$$

We rewrite (2.2) in the form

$$(2.15) \quad \left(V^{(m)}\right)^2 + 2G\!\left(z; V^{(0)}, \ldots, V^{(m-1)}\right) V^{(m)} + H\!\left(z; V^{(0)}, \ldots, V^{(m-1)}\right) = 0,$$

where $\quad G\!\left(z; V^{(0)}, \ldots, V^{(m-1)}\right) \equiv \sum_{j=1}^{m} c_{0,j}^{*}(z)\, V^{(m-j)} \quad$ and

$$H\!\left(z; V^{(0)}, \ldots, V^{(m-1)}\right) \equiv \sum_{1 \leq i,j \leq m} c_{i,j}^{*}(z)\, V^{(m-i)}\, V^{(m-j)}.$$

There are two cases.

Case 1. Suppose that $G^2 - H \not\equiv 0$. Let $\widehat{z}$ and $\widehat{b}_i$, for $0 \leq i \leq m-1$, be complex numbers subject to

$$|\widehat{z} - z_0| < \frac{\delta}{2} \quad \text{and} \quad |\widehat{b}_i - a_i| < \frac{\delta}{2}, \quad \text{for } 0 \leq i \leq r - 1,$$

such that $[G(\widehat{z}; \widehat{b}_0, \ldots, \widehat{b}_{m-2}, \widehat{b}_{m-1})]^2 - H(\widehat{z}; \widehat{b}_0, \ldots, \widehat{b}_{m-2}, \widehat{b}_{m-1}) \neq 0$. Then, there is a real number ϵ satisfying $0 < \epsilon < \delta/2$ and there is an analytic function $F(z; w_0, \ldots, w_{m-1})$ of the complex variables $z, w_0, \ldots, w_{m-1}$ that yields

$$(2.16) \quad [F(z; w_0, \ldots, w_{m-1})]^2 \equiv [G(z; w_0, \ldots, w_{m-1})]^2 - H(z; w_0, \ldots, w_{m-1}),$$

when $|z - \widehat{z}| < \epsilon$ and $|w_i - \widehat{b}_i| < \epsilon$, for $0 \leq i \leq m-1$. Let β be any complex number such that $|\beta - \widehat{b}_{m-1}| < \epsilon$. For the local solutions $v(z)$ of (2.15) and

$$(2.17) \quad v^{(i)}(\widehat{z}) = \begin{cases} \widehat{b}_i, & \text{for } 0 \leq i \leq m-2, \\ \beta, & \text{for } i = m-1, \end{cases}$$

on neighborhoods of $\widehat{z}$ contained in $|z - \widehat{z}| < \epsilon$, we use (2.16) to rewrite (2.15) in a factored form. Thus, each solution $v(z)$ of

$$(2.18) \quad v^{(m)}(z) + G\big(z; v(z), \ldots, v^{(m-1)}(z)\big) - F\big(z; v(z), \ldots, v^{(m-1)}(z)\big) = 0$$

and (2.17) is a solution of (2.15), (2.2) and (2.12). By rewriting (2.18) as a system

$$(2.19) \quad \begin{cases} \dfrac{dv_0(z)}{dz} = v_1(z), \quad \dfrac{dv_1(z)}{dz} = v_2(z), \quad \ldots, \quad \dfrac{dv_{m-2}(z)}{dz} = v_{m-1}(z) \\ \dfrac{dv_{m-1}(z)}{dz} = -G\big(z; v_0(z), \ldots, v_{m-1}(z)\big) + F\big(z; v_0(z), \ldots, v_{m-1}(z)\big) \end{cases}$$

of first-order equations, the *method of limits* in [**38**, page 284] shows that: for each complex number β that satisfies $|\beta - \widehat{b}_{m-1}| < \epsilon$, there is a corresponding local solution $v(z) = v_0(z)$ about $\widehat{z}$ of (2.19) and (2.17). This yields the contradiction that (2.12) and (2.14) (with $z_1 = \widehat{z}$ and $b_i = \widehat{b}_i$, for $0 \le i \le r-1$) have an infinity of distinct local solutions about $\widehat{z}$.

Case 2. Suppose that $G^2 - H \equiv 0$. Then, the solutions of (2.2) are the solutions of the mth order homogeneous linear differential equation

$$(2.20) \quad V^{(m)} + G\big(z; V^{(0)}, \ldots, V^{(m-1)}\big) = 0.$$

In terms of the complex numbers $z_0, a_0, a_1, \ldots, a_{r-1}$ introduced for (2.13) and any complex number β, we note that there is a unique solution $v(z)$ of (2.20) and

$$v^{(i)}(z_0) = \begin{cases} a_i, & \text{for } 0 \le i \le r-1, \\ \beta, & \text{for } r \le i \le m-1. \end{cases}$$

Due to the arbitrariness of β, this yields the contradiction that (2.12) and (2.14) (with $z_1 = z_0$ and $b_i = a_i$, for $0 \le i \le r-1$) have an infinity of local solutions.

We have just shown that (2.6) must coincide with (2.2). This establishes the uniqueness of the coefficients $c_{i,j}^*(z)$ and completes the proof of Proposition 2.1. $\square$

COROLLARY 2.2. *For any integer $m \ge 1$, the polynomial $\boldsymbol{D}_1$ in (1.11) is a semi-invariant of the first kind for the differential equations (1.1) of order m.*

PROOF. We use (2.5) to deduce that

$$(2.21) \quad c_{0,1}^*(z) \equiv c_{0,1}(z) + m \frac{\rho'(z)}{\rho(z)}.$$

Also, (2.5) and a replacement of $c_{1,0}(z)$ with $c_{0,1}(z)$ yield

$$(2.22) \quad c_{1,1}^*(z) \equiv c_{1,1}(z) + 2m \frac{\rho'(z)}{\rho(z)} c_{0,1}(z) + m^2 \left[\frac{\rho'(z)}{\rho(z)}\right]^2.$$

By combining (2.22) and (2.21), we obtain

$$(2.23) \quad c_{1,1}^*(z) - \big(c_{0,1}^*(z)\big)^2 \equiv c_{1,1}(z) - \big(c_{0,1}(z)\big)^2.$$

Thus, in view of (1.11) and (2.23), $\boldsymbol{D}_1$ is a semi-invariant of the first kind. This completes the proof. $\square$

For reference in Section 4.5, we note that (2.5) yields

$$(2.24) \quad c_{0,2}^*(z) \equiv c_{0,2}(z) + (m-1)\frac{\rho^{(1)}(z)}{\rho(z)} c_{0,1}(z) + \frac{m(m-1)}{2} \frac{\rho^{(2)}(z)}{\rho(z)}.$$

CHAPTER 3

The Coefficients $c_{i,j}^{**}(\zeta)$ of (1.5)

Our deductions for $c_{i,j}^{**}(\zeta)$ in (3.9) and (3.12) of Theorems 3.2 and 3.3 require the restatement of [24, Proposition A.2, page 135] presented as Proposition 3.1. We assume that $z = f(\zeta)$ is a univalent analytic function on a region Ω^{**} of the complex plane and $\Omega = f(\Omega^{**})$. In particular, it follows that $f'(\zeta) \neq 0$, for each ζ in Ω^{**}. We define corresponding analytic functions $\alpha_{i,j}(\zeta)$ on Ω^{**} through

(3.1) $$\alpha_{0,j}(\zeta) \equiv 1, \quad \text{for any } j,$$

and

(3.2) $$\alpha_{i+1,j}(\zeta) \equiv \sum_{k=1}^{j} \left[\alpha_{i,k}^{(1)}(\zeta) - (i+k) \frac{f''(\zeta)}{f'(\zeta)} \alpha_{i,k}(\zeta) \right], \quad \text{for } i \geq 0 \text{ and any } j.$$

PROPOSITION 3.1. *Let $y(z)$ be a meromorphic function on a subregion $\mathcal{U}$ of Ω. Then, the function on $f^{-1}(\mathcal{U})$ obtained by substituting $z = f(\zeta)$ in the ith derivative $y^{(i)}(z)$ of $y(z)$ with respect to z is given by*

(3.3) $$y^{(i)}(f(\zeta)) \equiv (f'(\zeta))^{-i} \sum_{j=0}^{i} \alpha_{i-j,j}(\zeta) \, (y \circ f)^{(j)}(\zeta), \quad \text{for } i \geq 0.$$

PROOF. We shall establish (3.3) by proving the equivalent formula

(3.4) $$(f'(\zeta))^{i} y^{(i)}(f(\zeta)) \equiv \sum_{j=0}^{i} \alpha_{i-j,j}(\zeta) \, (y \circ f)^{(j)}(\zeta), \quad \text{for } i \geq 0.$$

For $i = 0$, (3.1) shows that (3.4) is valid. Let $i \geq 0$ be an integer for which (3.4) is true. We differentiate (3.4) with respect to ζ and use the consequence

$$\alpha_{i-j,j}^{(1)}(\zeta) \equiv \alpha_{i+1-j,j}(\zeta) - \alpha_{i+1-j,j-1}(\zeta) + i\frac{f''(\zeta)}{f'(\zeta)} \alpha_{i-j,j}(\zeta), \quad \text{for } i \geq j \geq 0,$$

of (3.2) to obtain

$$(f'(\zeta))^{i+1} y^{(i+1)}(f(\zeta)) + i\frac{f''(\zeta)}{f'(\zeta)} (f'(\zeta))^{i} y^{(i)}(f(\zeta))$$

$$\equiv \sum_{j=0}^{i} \alpha_{i-j,j}^{(1)}(\zeta) \, (y \circ f)^{(j)}(\zeta) + \sum_{j=0}^{i} \alpha_{i-j,j}(\zeta) \, (y \circ f)^{(j+1)}(\zeta)$$

$$\equiv \sum_{j=0}^{i} \alpha_{i+1-j,j}(\zeta) \, (y \circ f)^{(j)}(\zeta) - \sum_{j=0}^{i} \alpha_{i+1-j,j-1}(\zeta) \, (y \circ f)^{(j)}(\zeta)$$

$$+ i\frac{f''(\zeta)}{f'(\zeta)} \sum_{j=0}^{i} \alpha_{i-j,j}(\zeta) \, (y \circ f)^{(j)}(\zeta) + \sum_{j=1}^{i+1} \alpha_{i+1-j,j-1}(\zeta) \, (y \circ f)^{(j)}(\zeta).$$

By combining the preceding formulas with (3.4) and employing $\alpha_{i+1,-1}(\zeta) \equiv 0$ from (3.2) as well as $\alpha_{0,i}(\zeta) \equiv 1 \equiv \alpha_{0,i+1}(\zeta)$ from (3.1), we verify that

$$\left(f'(\zeta)\right)^{i+1} y^{(i+1)}\bigl(f(\zeta)\bigr) \equiv \left[\sum_{j=0}^{i} \alpha_{i+1-j,j}(\zeta)\, (y \circ f)^{(j)}(\zeta)\right] + \alpha_{0,i}(\zeta)\, (y \circ f)^{(i+1)}(\zeta)$$

$$\equiv \sum_{j=0}^{i+1} \alpha_{i+1-j,j}(\zeta)\, (y \circ f)^{(j)}(\zeta).$$

Thus, (3.4) is valid for each $i \geq 0$. This completes the proof. $\square$

THEOREM 3.2. *Let $z = f(\zeta)$ be a univalent analytic function on a region Ω^{**} of the complex plane. For $m \geq 1$ and $0 \leq i, j \leq m$, suppose that meromorphic functions $c_{i,j}(z)$ on $\Omega = f(\Omega^{**})$ satisfy $c_{j,i}(z) \equiv c_{i,j}(z)$ and $c_{0,0}(z) \equiv 1$. Then, for $0 \leq i, j \leq m$, there are unique meromorphic functions $c_{i,j}^{**}(\zeta)$ on Ω^{**} subject to $c_{j,i}^{**}(\zeta) \equiv c_{i,j}^{**}(\zeta)$ and $c_{0,0}^{**}(\zeta) \equiv 1$ such that: whenever meromorphic functions $y(z)$ on a subregion $\mathcal{U}$ of Ω and $u(\zeta)$ on $f^{-1}(\mathcal{U})$ are related by $u(\zeta) \equiv (y \circ f)(\zeta)$, $y(z)$ is a local solution of*

$$(3.5) \qquad \left(Y^{(m)}\right)^2 + \sum_{\substack{0 \leq i,j \leq m \\ (i,j) \neq (0,0)}} c_{i,j}(z)\, Y^{(m-i)}\, Y^{(m-j)} = 0$$

if and only if $u(\zeta)$ is a local solution of

$$(3.6) \qquad \left(U^{(m)}\right)^2 + \sum_{\substack{0 \leq i,j \leq m \\ (i,j) \neq (0,0)}} c_{i,j}^{**}(\zeta)\, U^{(m-i)}\, U^{(m-j)} = 0;$$

and $y(z)$ is a local solution of

$$(3.7) \qquad Y^{(m)} + \sum_{j=1}^{m} c_{0,j}(z)\, Y^{(m-j)} = 0$$

if and only if $u(\zeta)$ is a local solution of

$$(3.8) \qquad U^{(m)} + \sum_{j=1}^{m} c_{0,j}^{**}(\zeta)\, U^{(m-j)} = 0.$$

Moreover, the coefficients $c_{i,j}^{**}(\zeta)$ for (3.6) and (3.8) are given in terms of (3.1), (3.2), and (3.5) by

$$(3.9) \qquad c_{i,j}^{**}(\zeta) \equiv \sum_{\mu=0}^{i} \sum_{\nu=0}^{j} \bigl(f'(\zeta)\bigr)^{\mu+\nu} \alpha_{i-\mu, m-i}(\zeta)\, \alpha_{j-\nu, m-j}(\zeta)\, c_{\mu,\nu}\bigl(f(\zeta)\bigr),$$

on Ω^{**} for $0 \leq i, j \leq m$.

Also, (3.9) shows that the coefficients $c_{i,j}^{**}(\zeta)$ of (3.6) are analytic on Ω^{**} when the coefficients $c_{i,j}(z)$ of (3.5) are analytic on Ω.

3. THE COEFFICIENTS $c^{**}_{i,j}(\zeta)$ OF (1.5)

PROOF. Let $Q\bigl[z;\, Y^{(0)},\, \ldots,\, Y^{(m)}\bigr]$ denote the left member of (3.5) and set

$$Q^{**}\bigl[\zeta;\, U^{(0)},\, \ldots,\, U^{(m)}\bigr] \equiv \bigl(U^{(m)}\bigr)^2 + \sum_{\substack{0 \le i,j \le m \\ (i,j) \neq (0,0)}} c^{**}_{i,j}(\zeta)\, U^{(m-i)}\, U^{(m-j)},$$

where each $c^{**}_{i,j}(\zeta)$ is defined on Ω^{**} by (3.9). In particular, (3.9) yields

$$c^{**}_{j,i}(\zeta) \equiv c^{**}_{i,j}(\zeta), \quad \text{on } \Omega^{**} \text{ for } 0 \le i, j \le m, \quad \text{and} \quad c^{**}_{0,0}(\zeta) \equiv 1.$$

Then, whenever meromorphic functions $y(z)$ on a subregion $\mathcal{U}$ of Ω and $u(\zeta)$ on $f^{-1}(\mathcal{U})$ are related by $u(\zeta) = (y \circ f)(\zeta)$, we use the preceding notation as well as Proposition 3.1 to verify that

$$\bigl(f'(\zeta)\bigr)^{2m} \left\{ Q\bigl[z;\, y^{(0)}(z),\, \ldots,\, y^{(m)}(z)\bigr] \right\}_{z=f(\zeta)}$$

$$\equiv \bigl(f'(\zeta)\bigr)^{2m} Q\bigl[f(\zeta);\, y^{(0)}(f(\zeta)),\, \ldots,\, y^{(m)}(f(\zeta))\bigr]$$

$$\equiv \bigl(f'(\zeta)\bigr)^{2m} \sum_{0 \le \mu,\nu \le m} c_{\mu,\nu}\bigl(f(\zeta)\bigr)\, y^{(m-\mu)}\bigl(f(\zeta)\bigr)\, y^{(m-\nu)}\bigl(f(\zeta)\bigr)$$

$$\equiv \sum_{\mu=0}^{m} \sum_{\nu=0}^{m} \bigl(f'(\zeta)\bigr)^{\mu+\nu} c_{\mu,\nu}\bigl(f(\zeta)\bigr) \sum_{r=0}^{m-\mu} \alpha_{m-\mu-r,r}(\zeta)\, u^{(r)}(\zeta) \sum_{s=0}^{m-\nu} \alpha_{m-\nu-s,s}(\zeta)\, u^{(s)}(\zeta)$$

$$\equiv \sum_{\mu=0}^{m} \sum_{\nu=0}^{m} \sum_{i=\mu}^{m} \sum_{j=\nu}^{m} \frac{\alpha_{i-\mu,m-i}(\zeta)\, \alpha_{j-\nu,m-j}(\zeta)}{\bigl(f'(\zeta)\bigr)^{-\mu-\nu}}\, c_{\mu,\nu}\bigl(f(\zeta)\bigr)\, u^{(m-i)}(\zeta)\, u^{(m-j)}(\zeta)$$

$$\equiv \sum_{i=0}^{m} \sum_{j=0}^{m} \sum_{\mu=0}^{i} \sum_{\nu=0}^{j} \frac{\alpha_{i-\mu,m-i}(\zeta)\, \alpha_{j-\nu,m-j}(\zeta)}{\bigl(f'(\zeta)\bigr)^{-\mu-\nu}}\, c_{\mu,\nu}\bigl(f(\zeta)\bigr)\, u^{(m-i)}(\zeta)\, u^{(m-j)}(\zeta)$$

$$\equiv \sum_{0 \le i,j \le m} c^{**}_{i,j}(\zeta)\, u^{(m-i)}(\zeta)\, u^{(m-j)}(\zeta) \equiv Q^{**}\bigl[\zeta;\, u^{(0)}(\zeta),\, \ldots,\, u^{(m)}(\zeta)\bigr], \quad \text{on } \mathcal{U}^{**}.$$

Selecting (3.6) as the equation $Q^{**} = 0$ specified by (3.9), we see that: $y(z)$ is a local solution of (3.5) if and only if $u(\zeta)$ is a local solution of (3.6).

Let $L\bigl[z;\, Y^{(0)},\, \ldots,\, Y^{(m)}\bigr]$ denote the left member of (3.7) and set

$$L^{**}\bigl[\zeta;\, U^{(0)},\, \ldots,\, U^{(m)}\bigr] \equiv U^{(m)} + \sum_{j=1}^{m} c^{**}_{0,j}(\zeta)\, U^{(m-j)}(\zeta),$$

where each $c^{**}_{0,j}(\zeta)$ is defined on Ω^{**} by (3.9). Then, for any meromorphic functions $y(z)$ on a subregion $\mathcal{U}$ of Ω and $u(\zeta)$ on $f^{-1}(\mathcal{U})$ that are related by $u(\zeta) = (y \circ f)(\zeta)$, we use the preceding notation as well as (3.3) and (3.9) to obtain

$$\bigl(f'(\zeta)\bigr)^{m} \left\{ L\bigl[z;\, y^{(0)}(z),\, \ldots,\, y^{(m)}(z)\bigr] \right\}_{z=f(\zeta)}$$

$$\equiv \bigl(f'(\zeta)\bigr)^{m} L\bigl[f(\zeta);\, y^{(0)}(f(\zeta)),\, \ldots,\, y^{(m)}(f(\zeta))\bigr]$$

$$\equiv \bigl(f'(\zeta)\bigr)^{m} \sum_{\nu=0}^{m} c_{0,\nu}\bigl(f(\zeta)\bigr)\, y^{(m-\nu)}\bigl(f(\zeta)\bigr)$$

$$\equiv \sum_{\nu=0}^{m} c_{0,\nu}\bigl(f(\zeta)\bigr) \bigl(f'(\zeta)\bigr)^{\nu} \sum_{s=0}^{m-\nu} \alpha_{m-\nu-s,s}(\zeta)\, u^{(s)}(\zeta)$$

and

$$(f'(\zeta))^m \left\{ L\left[z;\, y^{(0)}(z),\, \ldots,\, y^{(m)}(z)\right] \right\}_{z=f(\zeta)}$$

$$\equiv \sum_{\nu=0}^{m} \sum_{j=\nu}^{m} (f'(\zeta))^\nu \alpha_{j-\nu, m-j}(\zeta)\, c_{0,\nu}(f(\zeta))\, u^{(m-j)}(\zeta)$$

$$\equiv \sum_{j=0}^{m} \left[\sum_{\nu=0}^{j} (f'(\zeta))^\nu \alpha_{j-\nu, m-j}(\zeta)\, c_{0,\nu}(f(\zeta)) \right] u^{(m-j)}(\zeta)$$

$$\equiv \sum_{j=0}^{m} c_{0,j}^{**}(\zeta)\, u^{(m-j)}(\zeta) \equiv L^{**}\left[\zeta;\, u^{(0)}(\zeta),\, \ldots,\, u^{(m)}(\zeta)\right].$$

Selecting (3.8) as the equation $L^{**} = 0$ specified by (3.9), we see that: $y(z)$ is a local solution of (3.7) if and only if $u(\zeta)$ is a local solution of (3.8).

To establish the uniqueness of (3.6) and (3.8), suppose that the coefficients of

(3.10) $$\left(U^{(m)}\right)^2 + \sum_{\substack{0 \le i,j \le m \\ (i,j) \ne (0,0)}} h_{i,j}(\zeta)\, U^{(m-i)}\, U^{(m-j)} = 0$$

and

(3.11) $$U^{(m)} + \sum_{j=1}^{m} h_{0,j}(\zeta)\, U^{(m-j)} = 0.$$

are meromorphic functions on Ω^{**} satisfying $h_{j,i}(\zeta) \equiv h_{i,j}(\zeta)$, for $0 \le i, j \le m$, and $h_{0,0}(\zeta) \equiv 1$ such that: for any meromorphic functions $y(z)$ on a subregion $\mathcal{U}$ of Ω and $u(\zeta)$ on $f^{-1}(\mathcal{U})$ that are related by $u(\zeta) = (y \circ f)(\zeta)$, $y(z)$ is a solution of (3.5) if and only if $u(\zeta)$ is a solution of (3.10) and $y(z)$ is a solution of (3.7) if and only if $u(\zeta)$ is a solution of (3.11). Consequently, the monic mth-order homogeneous linear equations (3.8) and (3.11) have the same solutions and therefore coincide. This yields, $h_{0,j}(\zeta) \equiv c_{0,j}^{**}(\zeta)$, for $1 \le j \le m$. Also, the monic mth-order differential equations (3.6) and (3.10) have same solutions. By subtracting (3.10) from (3.6), the terms of order m disappear and the resulting equation has all of the solutions that (3.6) possesses. By a slight modification of the notation for the corresponding argument in the proof of Proposition 2.1, we conclude that the difference is an equation in which all of the coefficients are identically zero. This establishes the uniqueness and completes the proof. $\square$

THEOREM 3.3. *The coefficients $c_{i,j}^{**}(\zeta)$ for (3.6), including $c_{0,0}^{**}(\zeta) \equiv 1$, are given in terms of (1.1) and $c_{0,0}(z) \equiv 1$ through*

(3.12) $$c_{i,j}^{**}(\zeta) \equiv (f'(\zeta))^{i+j} \sum_{\mu=0}^{i} \sum_{\nu=0}^{j} \beta_{i-\mu,\mu}(f(\zeta))\, \beta_{j-\nu,\nu}(f(\zeta))\, c_{\mu,\nu}(f(\zeta)),$$

on Ω^{**} for $0 \le i, j \le m$,

where $\beta_{i,j}(z)$ is defined on Ω in terms of the inverse $\zeta = g(z)$ of $z = f(\zeta)$ by

(3.13) $$\beta_{0,j}(z) \equiv 1, \quad \text{for any } j,$$

and

(3.14) $$\beta_{i+1,j}(z) \equiv \sum_{k=j+1}^{m} \left[\beta_{i,k}^{(1)}(z) + (m-i-k) \frac{g''(z)}{g'(z)} \beta_{i,k}(z) \right],$$

for $i \ge 0$ and any j.

PROOF. We shall first establish that the functions $\alpha_{i,j}(\zeta)$ defined on Ω^{**} by (3.1)–(3.2) and the functions $\beta_{i,j}(z)$ defined on Ω by means of (3.13)–(3.14) are related through

$$(3.15) \qquad \alpha_{p,m-p-q}(\zeta) \equiv \left(f'(\zeta)\right)^p \beta_{p,q}\left(f(\zeta)\right), \quad \text{for } p \geq 0 \text{ and any } q.$$

For $p = 0$ and any q, (3.1) and (3.13) show that (3.15) is valid and reduces to $1 \equiv 1$. We shall use induction on p. Suppose that (3.15) is valid for a fixed nonnegative integer p and any integer q. Then, for that p and any q, (3.2) yields

$$(3.16) \quad \alpha_{p+1,m-(p+1)-q}(\zeta) \equiv \sum_{k=1}^{m-p-1-q} \left[\alpha_{p,k}^{(1)}(\zeta) - (p+k) \frac{f''(\zeta)}{f'(\zeta)} \alpha_{p,k}(\zeta) \right]$$

$$\equiv \sum_{\nu=q+1}^{m-p-1} \left[\alpha_{p,m-p-\nu}^{(1)}(\zeta) - (m-\nu) \frac{f''(\zeta)}{f'(\zeta)} \alpha_{p,m-p-\nu}(\zeta) \right].$$

This can be written as

$$(3.17) \quad \alpha_{p+1,m-(p+1)-q}(\zeta) \equiv \sum_{\nu=q+1}^{m} \left[\alpha_{p,m-p-\nu}^{(1)}(\zeta) - (m-\nu) \frac{f''(\zeta)}{f'(\zeta)} \alpha_{p,m-p-\nu}(\zeta) \right].$$

Namely, when $p = 0$, a single term equal to 0 is added to the last sum in (3.16) to obtain the sum in (3.17). For $p \geq 1$ and $\nu \geq m - p$, we have $m - p - \nu \leq 0$ so that (3.2) yields $\alpha_{p,m-p-\nu}(\zeta) \equiv 0$ and shows that only zero terms are added to the last sum in (3.16) to obtain the sum in (3.17). We use (3.17), the hypothesis about (3.15), its consequence

$$\alpha_{p,m-p-\nu}^{(1)}(\zeta) \equiv \left(f'(\zeta)\right)^{p+1} \beta_{p,\nu}^{(1)}\left(f(\zeta)\right) + p \left(f'(\zeta)\right)^{p-1} f''(\zeta) \beta_{p,\nu}\left(f(\zeta)\right),$$

and (3.14) to deduce that

$$\alpha_{p+1,m-(p+1)-q}(\zeta) \equiv \left(f'(\zeta)\right)^{p+1} \sum_{\nu=q+1}^{m} \left[\begin{array}{l} \beta_{p,\nu}^{(1)}\left(f(\zeta)\right) \\ + (p+\nu-m) \dfrac{f''(\zeta)}{\left(f'(\zeta)\right)^2} \beta_{p,\nu}\left(f(\zeta)\right) \end{array} \right].$$

In view of $f'(\zeta) \equiv \dfrac{1}{g'(f(\zeta))}$, we have $\dfrac{f''(\zeta)}{(f'(\zeta))^2} \equiv -\dfrac{g''(f(\zeta))}{g'(f(\zeta))}$. By combining this with the preceding formula and using (3.14), we obtain

$$\alpha_{p+1,m-(p+1)-q}(\zeta) \equiv \left(f'(\zeta)\right)^{p+1} \sum_{\nu=q+1}^{m} \left[\begin{array}{l} \beta_{p,\nu}^{(1)}\left(f(\zeta)\right) \\ + (m-p-\nu) \dfrac{g''(f(\zeta))}{g'(f(\zeta))} \beta_{p,\nu}\left(f(\zeta)\right) \end{array} \right]$$

$$\equiv \left(f'(\zeta)\right)^{p+1} \beta_{p+1,q}\left(f(\zeta)\right).$$

Thus, (3.15) is valid as stated.

We apply (3.15) to verify that: for $0 \leq \mu \leq i$ and $0 \leq \nu \leq j$,

$$\left(f'(\zeta)\right)^{\mu+\nu} \alpha_{i-\mu,m-i}(\zeta)\, \alpha_{j-\nu,m-j}(\zeta) \equiv \left(f'(\zeta)\right)^{i+j} \beta_{i-\mu,\mu}\left(f(\zeta)\right) \beta_{j-\nu,\nu}\left(f(\zeta)\right).$$

By combining this with (3.9), we obtain (3.12). This completes the proof. □

PROPOSITION 3.4. *For any integer $m \geq 1$, the polynomial $\boldsymbol{D}_1$ in (1.11) is a semi-invariant of the second kind of weight 2 for the equations (1.1) of order m.*

PROOF. We use (3.1) and (3.2) to deduce

$$(3.18) \quad \alpha_{1,j}(\zeta) \equiv -\binom{j+1}{2}\frac{f''(\zeta)}{f'(\zeta)}, \quad \text{on } \Omega^{**} \text{ for } j \geq -1.$$

By combining (3.18) with (3.9) and using $c_{1,0}(z) \equiv c_{0,1}(z)$, we obtain

$$(3.19) \quad c_{0,1}^{**}(\zeta) \equiv \left(f'(\zeta)\right) c_{0,1}(f(\zeta)) - \binom{m}{2}\frac{f''(\zeta)}{f'(\zeta)}$$

and

$$(3.20) \quad c_{1,1}^{**}(\zeta) \equiv \left(f'(\zeta)\right)^2 c_{1,1}(f(\zeta)) - 2\binom{m}{2}f''(\zeta)c_{0,1}(f(\zeta)) + \left[\binom{m}{2}\frac{f''(\zeta)}{f'(\zeta)}\right]^2.$$

We observe that (3.20) and (3.19) give

$$(3.21) \quad c_{1,1}^{**}(\zeta) - \left(c_{0,1}^{**}(\zeta)\right)^2 \equiv \left(f'(\zeta)\right)^2 \left[c_{1,1}(f(\zeta)) - [c_{0,1}(f(\zeta))]^2\right].$$

Thus, in view of (1.11) and (3.21), $\boldsymbol{D}_1$ is a semi-invariant of the second kind of weight 2. This completes the proof. □

PROPOSITION 3.5. *For any fixed positive integer m, the polynomial $\boldsymbol{D}_1$ in (1.11) is a basic relative invariant of index $(1,1)$ for the equations (1.1) of order m and it is the only basic relative invariant of index $(1,1)$.*

PROOF. We use Corollary 2.2, Corollary 3.4, and Definition 1.5 to verify that $\boldsymbol{D}_1$ is a basic relative invariant of index $(1,1)$ for the equations (1.1).

Supposes that $\boldsymbol{P}$ is a basic relative invariant of index $(1,1)$ for the equations (1.1). Then, Definition 1.5 and (1.11) yield

$$\boldsymbol{P} \equiv \boldsymbol{w}_{1,1} + \tau(\boldsymbol{w}_{0,1})^2 \quad \text{and} \quad \boldsymbol{P} - \boldsymbol{D}_1 \equiv (\tau+1)(\boldsymbol{w}_{0,1})^2,$$

for some rational number τ. In terms of any (1.1) and (1.2), we have $P^*(z) \equiv P(z)$, $D_1^*(z) \equiv D_1(z)$, and

$$(3.22) \quad (\tau+1)\left(c_{0,1}^*(z)\right)^2 \equiv P^*(z) - D_1^*(z) \equiv P(z) - D_1(z) \equiv (\tau+1)\left(c_{0,1}(z)\right)^2.$$

We apply (2.21) to verify that the validity of (3.22) for each (1.1) and (1.2) requires $\tau+1 = 0$, $\tau = -1$, and $\boldsymbol{P} \equiv \boldsymbol{D}_1$. Thus, $\boldsymbol{D}_1$ is unique. This completes the proof. □

REMARK 3.6. An isobaric polynomial of weight 1 must have the form $\gamma \boldsymbol{w}_{0,1}$, for some nonzero γ in $\mathbb{Q}$. Due to (2.21) and (3.19), *an isobaric polynomial of weight 1 can not be a semi-invariant of the first kind and it is a semi-invariant of the second kind only when $m = 1$.* For the isobaric semi-invariants of weight 2, see Proposition 4.14 on page 43 when $m \geq 2$. Remarks 4.9 and 4.12 apply when $m = 1$.

We note that (3.2) and (3.18) yield

$$(3.23) \quad \alpha_{2,j}(\zeta) \equiv -\binom{j+2}{3}\frac{f'''(\zeta)}{f'(\zeta)} + 3\binom{j+3}{4}\left[\frac{f''(\zeta)}{f'(\zeta)}\right]^2, \quad \text{on } \Omega^{**} \text{ for } j \geq -2.$$

Thus, for reference in Section 4.5, we find that (3.9), (3.1), (3.18), and (3.23) give

$$(3.24) \quad c_{0,2}^{**}(\zeta) \equiv \left(f'(\zeta)\right)^2 c_{0,2}(f(\zeta)) - \binom{m-1}{2}\left(f''(\zeta)\right) c_{0,1}(f(\zeta))$$

$$- \binom{m}{3}\frac{f'''(\zeta)}{f'(\zeta)} + 3\binom{m+1}{4}\left[\frac{f''(\zeta)}{f'(\zeta)}\right]^2.$$

CHAPTER 4

Isolated Results Needed for Completeness

4.1. Nonsolutions of nontrivial equations

Corollary 4.3 will be employed in the proofs of Propositions 4.7, 4.17, 5.7, and Theorem 11.1. For those arguments, we shall use the following extension of (1.6).

CONTEXT 4.1. Suppose that the symbols $w_{i,j}^{(k)}$, for $j \geq 1$, $0 \leq i \leq j$, and $k \geq 0$, are algebraically independent variables over the field $\mathbb{Q}$ of rational numbers; and, for $\mu = 1, 2, \ldots$, let $\mathcal{S}_\mu$ denote the ring of polynomials over $\mathbb{Q}$ in those variables $w_{i,j}^{(k)}$ having $1 \leq j \leq \mu$, $0 \leq i \leq j$, and $k \geq 0$. For $\mu = m$, this gives $\mathcal{S}_m$ of page 5.

We totally order the variables $w_{i,j}^{(0)}$ having $j \geq 1$ and $0 \leq i \leq j$ according to

$$(4.1) \quad w_{0,1}^{(0)}, w_{1,1}^{(0)}, w_{0,2}^{(0)}, w_{1,2}^{(0)}, w_{2,2}^{(0)}, w_{0,3}^{(0)}, w_{1,3}^{(0)}, w_{2,3}^{(0)}, w_{3,3}^{(0)}, w_{0,4}^{(0)}, w_{1,4}^{(0)}, w_{2,4}^{(0)}, \ldots.$$

Thus, $w_{i,j}^{(0)}$ precedes $w_{p,q}^{(0)}$ if and only if either $j < q$ or $j = q$ and $i < p$. We introduce the notation

$$(4.2) \quad \mathfrak{s}_1^{(0)}, \mathfrak{s}_2^{(0)}, \mathfrak{s}_3^{(0)}, \mathfrak{s}_4^{(0)}, \mathfrak{s}_5^{(0)}, \mathfrak{s}_6^{(0)}, \mathfrak{s}_7^{(0)}, \mathfrak{s}_8^{(0)}, \mathfrak{s}_9^{(0)}, \mathfrak{s}_{10}^{(0)}, \mathfrak{s}_{11}^{(0)}, \mathfrak{s}_{12}^{(0)}, \ldots$$

for the corresponding elements of (4.1). There are functions $\phi(h)$, $\psi(h)$ such that

$$(4.3) \quad \mathfrak{s}_h^{(0)} \equiv w_{\phi(h),\psi(h)}^{(0)}, \quad \text{for } h \geq 1,$$

and there is a function $\chi(i,j)$ such that

$$(4.4) \quad w_{i,j}^{(0)} \equiv \mathfrak{s}_{\chi(i,j)}^{(0)}, \quad \text{for } j \geq 1 \text{ and } 0 \leq i \leq j.$$

It will not be necessary for us to know explicitly that:

$$\chi(i,j) \equiv i + j + \binom{j}{2} \equiv i + \binom{j+1}{2}, \quad \text{for } j \geq 1 \text{ and } 0 \leq i \leq j;$$

$\psi(h)$ is the unique integer satisfying

$$\binom{\psi(h)+1}{2} \leq h < \binom{\psi(h)+2}{2}, \quad \text{for } h \geq 1;$$

and $\phi(h) \equiv h - \binom{\psi(h)+1}{2}$, for $h \geq 1$. As a supplement to (4.2), we use (4.3) to introduce $\mathfrak{s}_h^{(k)} \equiv w_{\phi(h),\psi(h)}^{(k)}$, for $h \geq 1$ and $k \geq 1$. Now, we have

$$(4.5) \quad \mathfrak{s}_h^{(k)} \equiv w_{\phi(h),\psi(h)}^{(k)}, \quad \text{for } h \geq 1 \text{ and } k \geq 0.$$

Since the sets

$$\left\{ \mathfrak{s}_h^{(k)} \mid \text{for } h \geq 1 \text{ and } k \geq 0 \right\} \quad \text{and} \quad \left\{ w_{i,j}^{(k)} \mid \text{for } j \geq 1, 0 \leq i \leq j \text{ and } k \geq 0 \right\}$$

are equal, they are both algebraically independent over $\mathbb{Q}$.

For $\nu \geq 1$, let $\mathfrak{S}_\nu$ denote the ring of polynomials over $\mathbb{Q}$ in those variables $\mathbf{s}_h^{(k)}$ of (4.5) having $1 \leq h \leq \nu$ and $k \geq 0$. In particular, for Context 4.1, we note that $\mathcal{S}_\mu \equiv \mathfrak{S}_{\chi(\mu,\mu)}$, for $\mu \geq 1$.

PROPOSITION 4.2. *For $\nu \geq 1$, let $\boldsymbol{Q}$ be a nonzero polynomial belonging to $\mathfrak{S}_\nu$ and let z_0 be a complex number. Then, there are analytic functions $d_1(z)$, ..., $d_\nu(z)$ on a neighborhood Ω of z_0 such that the analytic function $Q(z)$ on Ω obtained by replacing $\mathbf{s}_h^{(k)}$ in $\boldsymbol{Q}$ with $d_h^{(k)}(z)$ satisfies $Q(z) \neq 0$, for each z in Ω.*

PROOF. Suppose that $\nu = 1$. The Euler Γ-function $\Gamma(z)$ is analytic on the complex plane except for poles at $z = 0, -1, -2, \ldots$. That $\Gamma(z)$ does not satisfy any algebraic differential equation over $\mathbb{Q}$ is a special case of results known from [37] or [15]. Thus, we can find a complex number z_1 and a region Ω containing z_0 such that: the function $d_1(z) \equiv \Gamma(z+z_1)$ is analytic on Ω and the analytic function $Q(z)$ on Ω obtained by replacing each $\mathbf{s}_1^{(k)}$ in $\boldsymbol{Q}$ with $d_1^{(k)}(z)$ satisfies $Q(z) \neq 0$, for each z in Ω. This shows that the assertion is true for $\nu = 1$.

Induction Hypothesis. *Suppose ν is a positive integer having the property that: for any nonzero polynomial $\boldsymbol{Q}$ in $\mathfrak{S}_\nu$ and any z_0 in $\mathbb{C}$, there are analytic functions $d_1(z)$, ..., $d_\nu(z)$ on a region Ω containing z_0 such that the analytic function $Q(z)$ on Ω obtained by substituting $d_h^{(k)}(z)$ for $\mathbf{s}_h^{(k)}$ in $\boldsymbol{Q}$ satisfies $Q(z) \neq 0$, for z in Ω.* Now, let $\widehat{\boldsymbol{Q}}$ be a nonzero polynomial in $\mathfrak{S}_{\nu+1}$ and let z_0 be a complex number. Then, we can represent $\widehat{\boldsymbol{Q}}$ by

$$\widehat{\boldsymbol{Q}} \equiv \sum_{i=1}^{r} \boldsymbol{Q}_i \boldsymbol{R}_i \quad \text{and} \quad \boldsymbol{R}_i \equiv \prod_{j=1}^{n_i} \mathbf{s}_{\nu+1}^{(\omega_{i,j})}, \quad \text{for } 1 \leq i \leq r,$$

where: r is a positive integer; each of $\boldsymbol{Q}_1, \ldots, \boldsymbol{Q}_r$ is a nonzero polynomial in $\mathfrak{S}_\nu$; $n_1, \ldots, n_r$ are nonnegative integers; for $1 \leq i \leq r$ and $1 \leq j \leq n_i$, $\omega_{i,j}$ is a nonnegative integer; and the polynomials $\boldsymbol{R}_1, \ldots, \boldsymbol{R}_r$ are linearly independent over $\mathbb{Q}$. If $r = 1$ and $n_1 = 0$, then $\boldsymbol{R}_1 \equiv 1$, $\widehat{\boldsymbol{Q}} \equiv \boldsymbol{Q}_1$, $\widehat{\boldsymbol{Q}}$ belongs to $\mathfrak{S}_\nu$, and the induction hypothesis yields suitable analytic functions $d_1(z)$, ..., $d_\nu(z)$ on a region Ω containing z_0. Then, we can arbitrarily select $d_{\nu+1}(z) \equiv 1$ on Ω. Suppose next that either $r = 1$ and $n_1 \geq 1$ or $r > 1$. When $r > 1$, n_1 and n_2 are not both zero. Thus, if necessary, we can adjust the notation so that $n_1 \geq 1$. The induction hypothesis applied to $\boldsymbol{Q}_1$ yields analytic functions $d_1(z)$, ..., $d_\nu(z)$ on a region $\mathcal{U}_1$ containing z_0 such that the analytic function $Q_1(z)$ on $\mathcal{U}_1$ obtained by replacing $\mathbf{s}_h^{(k)}$ in $\boldsymbol{Q}_1$ with $d_h^{(k)}(z)$ satisfies $Q_1(z) \neq 0$, for each z in $\mathcal{U}_1$. And, when $r > 1$, there are corresponding analytic functions $Q_2(z)$, ..., $Q_r(z)$ on $\mathcal{U}_1$ obtained by substituting $d_h^{(k)}(z)$ for $\mathbf{s}_h^{(k)}$ in $\boldsymbol{Q}_2, \ldots, \boldsymbol{Q}_r$. Introducing the complex numbers $\tau_i = Q_i(z_0)$, for $1 \leq i \leq r$, we set

$$\widehat{\boldsymbol{R}} \equiv \sum_{i=1}^{r} \tau_i \boldsymbol{R}_i.$$

Due to $\tau_1 = Q_1(z_0) \neq 0$ and the linear independence of $\boldsymbol{R}_1, \ldots, \boldsymbol{R}_r$ over $\mathbb{Q}$, $\widehat{\boldsymbol{R}}$ is a nonzero polynomial in the variables $\mathbf{s}_{\nu+1}^{(0)}$, $\mathbf{s}_{\nu+1}^{(1)}$, ... over $\mathbb{Q}$. Thus, the argument for $\nu = 1$ gives an analytic function $d_{\nu+1}(z)$ on a subregion $\mathcal{U}_2$ of $\mathcal{U}_1$ containing z_0 such that the function $\widehat{R}(z)$ on $\mathcal{U}_2$ obtained by substituting $d_{\nu+1}^{(k)}(z)$ for $\mathbf{s}_{\nu+1}^{(k)}$ in $\widehat{\boldsymbol{R}}$ satisfies $\widehat{R}(z) \neq 0$, for each z in $\mathcal{U}_2$. For $1 \leq h \leq \nu + 1$ and $k \geq 0$, the result of replacing each $\mathbf{s}_h^{(k)}$ in $\widehat{\boldsymbol{Q}}$ with the corresponding $d_h^{(k)}(z)$ yields an analytic function $\widehat{Q}(z)$ on $\mathcal{U}_2$ having

$$\widehat{Q}(z_0) = \sum_{i=1}^{r} Q_i(z_0) R_i(z_0) = \sum_{i=1}^{r} \tau_i R_i(z_0) = \widehat{R}(z_0) \neq 0.$$

Thus, by continuity, there is a subregion Ω of $\mathcal{U}_2$ containing z_0 such that $\widehat{Q}(z) \neq 0$, for each z in Ω. This completes the induction step and the proof. □

COROLLARY 4.3. *Let $\boldsymbol{P}$ be a polynomial in $\mathcal{S}_m$ subject to $\boldsymbol{P} \not\equiv 0$ and let z_0 be a complex number. Then, there is a differential equation (1.1) having analytic coefficients $c_{i,j}(z)$ on a region Ω containing z_0 such that the analytic function $P(z)$ on Ω obtained by replacing $\boldsymbol{w}_{i,j}^{(k)}$ in $\boldsymbol{P}$ with $c_{i,j}^{(k)}(z)$ satisfies $P(z) \neq 0$, for z in Ω.*

PROOF. With respect to $\nu = \chi(m, m)$, $\boldsymbol{P}$ is a polynomial in $\mathfrak{S}_\nu$. In view of Proposition 4.2, there are analytic functions $d_1(z), \ldots, d_\nu(z)$ on a region Ω containing z_0 such that the substitution of $d_h^{(k)}(z)$ for $\mathbf{s}_h^{(k)}$ in $\boldsymbol{P}$ yields an analytic function $P(z)$ on Ω that satisfies $P(z) \neq 0$, for each z in Ω. We define $c_{i,j}(z)$ on Ω through $c_{i,j}(z) \equiv d_{\chi(i,j)}(z)$, for $1 \leq j \leq m$ and $0 \leq i \leq j$. By introducing $c_{j,i}(z) \equiv c_{i,j}(z)$, when $0 \leq i < j \leq m$, we obtain a differential equation (1.1) on Ω such that the substitution of $c_{i,j}^{(k)}(z)$ for $\boldsymbol{w}_{i,j}^{(k)}$ in $\boldsymbol{P}$ yields the same function $P(z)$. This completes the proof. □

4.2. Semi-invariants of the second kind are isobaric

PROPOSITION 4.4. *If $\boldsymbol{P}$ is a semi-invariant of the second kind for the equations (1.1), then $\boldsymbol{P}$ is an isobaric polynomial, there is a unique positive integer s for $\boldsymbol{P}$ that satisfies (1.10), and s for $\boldsymbol{P}$ in (1.10) is equal to the weight of $\boldsymbol{P}$.*

PROOF. Since $\boldsymbol{P}$ in $\mathcal{S}_m$ is not a constant, there is a positive integer p such that

$$(4.6) \qquad \boldsymbol{P} \equiv \sum_{k=0}^{p} \boldsymbol{E}_k,$$

where $\boldsymbol{E}_p \not\equiv 0$ and, for $0 \leq k \leq p$, either $\boldsymbol{E}_k \equiv 0$ or $\boldsymbol{E}_k$ is an isobaric polynomial in $\mathcal{S}_m$ of weight k. For any (1.1), we restrict $z = f(\zeta)$ for (1.10) to the univalent analytic functions of the type $z = f_t(\zeta) = t\zeta$ in which t is a fixed nonzero rational number. With this, we see that (3.1), (3.2), and (3.9) yield $\alpha_{i,j}(\zeta) \equiv 0$, for $i \geq 1$ and any j, as well as $c_{i,j}^{**}(\zeta) \equiv t^{i+j} c_{i,j}(t\zeta)$ and

$$(4.7) \qquad c_{i,j}^{**(k)}(\zeta) \equiv t^{i+j+k} c_{i,j}^{(k)}(t\zeta), \quad \text{for } 0 \leq i, j \leq m \text{ and } k \geq 0.$$

If $\boldsymbol{T}$ is a nonzero term of $\boldsymbol{E}_k$, then $\boldsymbol{T}$ is expressible as

$$(4.8) \qquad \boldsymbol{T} \equiv \beta \prod_{j=1}^{q} \boldsymbol{w}_{\mu_j,\nu_j}^{(\kappa_j)}, \quad \text{subject to} \quad \sum_{j=1}^{q}(\mu_j + \nu_j + \kappa_j) = k,$$

where: β is a nonzero rational number; q is a nonnegative integer; and μ_j, ν_j, κ_j satisfy $0 \leq \mu_j, \nu_j \leq m$ and $0 \leq \kappa_j$, for $1 \leq j \leq q$. We see that (4.8) and (4.7) give

$$(4.9) \quad T^{**}(\zeta) \equiv \beta \prod_{j=1}^{q} c_{\mu_j,\nu_j}^{**(\kappa_j)}(\zeta) \equiv \beta \left[\prod_{j=1}^{q} t^{\mu_j+\nu_j+\kappa_j}\right]\left[\prod_{j=1}^{q} c_{\mu_j,\nu_j}^{(\kappa_j)}(t\zeta)\right] \equiv t^k T(t\zeta).$$

Let s be an integer for (1.10) on page 6. Since each polynomial $\boldsymbol{E}_k$ in (4.6) is either zero or a sum of expression like (4.8) for which (4.9) is valid, we have

$$(4.10) \quad t^s P(t\zeta) \equiv \bigl(f'_t(\zeta)\bigr)^s P\bigl(f_t(\zeta)\bigr) \equiv P^{**}(\zeta) \equiv \sum_{k=0}^{p} \boldsymbol{E}^{**}_k(\zeta) \equiv \sum_{k=0}^{p} t^k \boldsymbol{E}_k(t\zeta),$$

for any nonzero rational number t and any (1.1). We replace $t\zeta$ in (4.10) with z and use (4.6) to obtain

$$(4.11) \quad t^s \sum_{k=0}^{p} \boldsymbol{E}_k(z) \equiv \sum_{k=0}^{p} t^k \boldsymbol{E}_k(z), \quad \text{for any rational number } t \text{ and any (1.1).}$$

In order to establish that

$$(4.12) \quad t^s \sum_{k=0}^{p} \boldsymbol{E}_k \equiv \sum_{k=0}^{p} t^k \boldsymbol{E}_k, \quad \text{for each rational number } t,$$

suppose that (4.12) is not true for $t = t_0$ in $\mathbb{Q}$. Then, the polynomial

$$\boldsymbol{D} \equiv (t_0)^s \sum_{k=0}^{p} \boldsymbol{E}_k - \sum_{k=0}^{p} (t_0)^k \boldsymbol{E}_k$$

in $\mathcal{S}_m$ satisfies $\boldsymbol{D} \not\equiv 0$. By applying Corollary 4.3, we see that there is a differential equation (1.1) on some region Ω such that the substitution of $c_{i,j}^{(k)}(z)$ from that (1.1) for $\boldsymbol{w}_{i,j}^{(k)}$ in $\boldsymbol{D}$ yields $D(z) \not\equiv 0$. But, this contradicts (4.11). Hence, (4.12) is valid. For each $t \neq 0$ in $\mathbb{Q}$, the terms of weight p in the left and right members of (4.12) give $t^s \boldsymbol{E}_p \equiv t^p \boldsymbol{E}_p$. In view of $\boldsymbol{E}_p \not\equiv 0$, we obtain $s = p$. Regarding (4.12) as an equality of two polynomial functions in t over $\mathcal{S}_m$, we note that the coefficients of like powers of t must be equal. Thus, we deduce that $\boldsymbol{E}_k \equiv 0$, for $0 \leq k < p$, and $\boldsymbol{P} \equiv \boldsymbol{E}_p$. Consequently, $\boldsymbol{P}$ is an isobaric polynomial of weight p and $s = p$. This completes the proof. □

4.3. Substitutions in regard to the derivation $'$ for $\mathcal{S}_m$

PROPOSITION 4.5. *For $\boldsymbol{P}$ and $\boldsymbol{Q} \equiv \boldsymbol{P}'$ in $\mathcal{S}_m$, let meromorphic functions $c_{i,j}(z)$ on Ω be given for $1 \leq j \leq m$ and $0 \leq i \leq j$. Then, the meromorphic functions $P(z)$ and $Q(z)$ on Ω obtained by replacing $\boldsymbol{w}_{i,j}^{(k)}$ in $\boldsymbol{P}$ and $\boldsymbol{Q}$ with $c_{i,j}^{(k)}(z)$ are related by*

$$(4.13) \quad \frac{d}{dz} P(z) \equiv Q(z), \quad \text{on } \Omega.$$

PROOF. Clearly, (4.13) is valid when $\boldsymbol{P}$ is a constant. For γ in $\mathbb{Q}$, $1 \leq j \leq m$, $0 \leq i \leq j$, $k \geq 0$, and the situation where $\boldsymbol{P} \equiv \gamma \boldsymbol{w}_{i,j}^{(k)}$, we have $\boldsymbol{Q} \equiv \gamma \boldsymbol{w}_{i,j}^{(k+1)}$ and

$$\frac{d}{dz} P(z) \equiv \frac{d}{dz} \bigl(\gamma c_{i,j}^{(k)}(z)\bigr) \equiv \gamma c_{i,j}^{(k+1)}(z) \equiv Q(z).$$

Thus, this $\boldsymbol{P}$ satisfies (4.13). Suppose that $\boldsymbol{P}$ is a monomial in $\mathcal{S}_m$ that satisfies (4.13) with $\boldsymbol{Q} \equiv \boldsymbol{P}'$. Setting $\boldsymbol{M} \equiv \boldsymbol{w}_{i,j}^{(k)} \boldsymbol{P}$ and $\boldsymbol{N} \equiv \boldsymbol{M}'$, for some i, j, k satisfying $1 \leq j \leq m$, $0 \leq i \leq j$, and $k \geq 0$, we use $\boldsymbol{N} \equiv \boldsymbol{w}_{i,j}^{(k+1)} \boldsymbol{P} + \boldsymbol{w}_{i,j}^{(k)} \boldsymbol{Q}$ and the property (4.13) for $\boldsymbol{P}$ to deduce that

$$\frac{d}{dz} M(z) \equiv \frac{d}{dz} \bigl(c_{i,j}^{(k)}(z) P(z)\bigr) \equiv c_{i,j}^{(k+1)}(z) P(z) + c_{i,j}^{(k)}(z) Q(z) \equiv N(z).$$

Hence, any monomial in $\mathcal{S}_m$ satisfies (4.13). Since any polynomial in $\mathcal{S}_m$ is a sum of monomials in $\mathcal{S}_m$, we use the linearity of both $\dfrac{d}{dz}$ and $'$ to see that any polynomial in $\mathcal{S}_m$ also satisfies (4.13). This completes the proof. □

The notation $\boldsymbol{P}^{(0)} \equiv \boldsymbol{P}$ and $\boldsymbol{P}^{(r+1)} \equiv (\boldsymbol{P}^{(r)})'$, for $r \geq 0$, was introduced after (1.13) on page 7 for any $\boldsymbol{P}$ in $\mathcal{S}_m$.

REMARK 4.6. For each $\boldsymbol{P}$ in $\mathcal{S}_m$ and each nonnegative integer r, there is a unique function $P^{(r)}(z)$ specified by a given (1.1). When $r \geq 1$, Proposition 4.5 enables us to interpret $P^{(r)}(z)$ in several ways. We can substitute $c_{i,j}^{(k)}(z)$ from (1.1) for $w_{i,j}^{(k)}$ in $\boldsymbol{P}$ to obtain $P(z)$ and then differentiate the meromorphic function $P(z)$ r-times with respect to z to obtain $P^{(r)}(z)$. We can also start with $\boldsymbol{P}$ and successively apply the derivation $'$ for $\mathcal{S}_m$ a total of r-times; in the resulting $\boldsymbol{P}^{(r)}$, we can then replace $w_{i,j}^{(k)}$ with $c_{i,j}^{(k)}(z)$ from (1.1) to obtain $P^{(r)}(z)$. There are analogous interpretations for $P^{*(r)}(z)$ and $P^{**(r)}(\zeta)$ based on (1.3) and (1.5).

PROPOSITION 4.7. *If $\boldsymbol{P}$ in $\mathcal{S}_m$ is a semi-invariant of the first kind for the equations (1.1), then $\boldsymbol{P}'$ is a semi-invariant of the first kind for such equations. However, if $\boldsymbol{P}$ in $\mathcal{S}_m$ is a semi-invariant of the second kind, then $\boldsymbol{P}'$ is not a semi-invariant of the second kind.*

PROOF. Suppose that $\boldsymbol{P}$ in $\mathcal{S}_m$ is a semi-invariant of the first kind. Then, we have $P^*(z) \equiv P(z)$, for any transformation (1.2) of any given (1.1) into a corresponding (1.3). Setting $\boldsymbol{Q} \equiv \boldsymbol{P}'$, we use Proposition 4.5 to obtain

$$Q^*(z) = \frac{d}{dz} P^*(z) \equiv \frac{d}{dz} P(z) \equiv Q(z),$$

for any transformation (1.2) of any given (1.1) into a corresponding (1.3). Thus, $\boldsymbol{Q}$ is a semi-invariant of the first kind.

Suppose that $\boldsymbol{P}$ in $\mathcal{S}_m$ is a semi-invariant of the second kind of weight s. Then, for any transformation (1.4) of any given (1.1) into a corresponding (1.5), we have

(4.14) $$P^{**}(\zeta) \equiv (f'(\zeta))^s P(f(\zeta)).$$

For $\boldsymbol{Q} \equiv \boldsymbol{P}'$, the weight of $\boldsymbol{Q}$ is $s+1$. For any transformation (1.4) of any given (1.1) into a corresponding (1.5), (4.14) yields

(4.15) $$Q^{**}(\zeta) - (f'(\zeta))^{s+1} Q(f(\zeta)) \equiv P^{**(1)}(\zeta) - (f'(\zeta))^{s+1} P^{(1)}(f(\zeta))$$
$$\equiv s (f'(\zeta))^{s-1} f''(\zeta) P(f(\zeta)).$$

We use Corollary 4.3 to see that one can specify a particular (1.1) and a particular transformation (1.4) of (1.1) such that the corresponding right member of (4.15) is not identically zero. Thus, $\boldsymbol{P}'$ is not a semi-invariant of the second kind. This completes the proof. □

4.4. All of the relative invariants for (1.1) when $m = 1$

For $m = 1$, the differential equations (1.1) are given by

(4.16) $$(y'(z))^2 + 2 c_{0,1}(z) \, y'(z) \, y(z) + c_{1,1}(z) \, (y(z))^2 = 0, \quad \text{on } \Omega.$$

Proposition 3.5 shows that $\boldsymbol{D}_1 \equiv \boldsymbol{w}_{1,1} - (\boldsymbol{w}_{0,1})^2$ in $\mathcal{S}_1$ is the only basic relative invariant for the equations (4.16). To derive the structure of all relative invariants in $\mathcal{S}_1$ for (4.16), we first obtain the structure of all semi-invariants of the first kind.

PROPOSITION 4.8. *Suppose that P is a nonconstant polynomial in S_1. Then, P is a semi-invariant of the first kind for the equations (4.16) if and only if there is a polynomial $\Phi(X_0, X_1, \ldots, X_r)$ over $\mathbb{Q}$ in $X_0, X_1, \ldots, X_r$ such that*

$$(4.17) \qquad P \equiv \Phi\left(D_1^{(0)}, D_1^{(1)}, \ldots, D_1^{(r)}\right).$$

PROOF. We use Propositions 3.5 and 4.7 to see that $D_1^{(0)}, D_1^{(1)}, \ldots, D_1^{(r)}$ are semi-invariants of the first kind for the equations (4.16). Consequently, if P can be represented in the form (4.17), then P is a semi-invariant of the first kind.

Suppose that P is a semi-invariant of the first kind. In particular, P is a polynomial combination over $\mathbb{Q}$ of $w_{1,1}^{(0)}, w_{1,1}^{(1)}, w_{1,1}^{(2)}, \ldots$ and $w_{0,1}^{(0)}, w_{0,1}^{(1)}, w_{0,1}^{(2)}, \ldots$. For $m = 1$, we rewrite (1.11) as the identity

$$(4.18) \qquad w_{1,1}^{(0)} \equiv D_1 + \left[w_{0,1}^{(0)}\right]^2$$

and then successively apply the derivation $'$ of S_1 to (4.18). This yields a succession of identities that express each of $w_{1,1}^{(0)}, w_{1,1}^{(1)}, w_{1,1}^{(2)}, \ldots$ as a polynomial combination over $\mathbb{Q}$ of $D_1^{(0)}, D_1^{(1)}, D_1^{(2)}, \ldots$ and $w_{0,1}^{(0)}, w_{0,1}^{(1)}, w_{0,1}^{(2)}, \ldots$. By using these identities to eliminate each of $w_{1,1}^{(0)}, w_{1,1}^{(1)}, w_{1,1}^{(2)}, \ldots$ from P, we express P as

$$(4.19) \qquad P \equiv \Phi\left(D_1^{(0)}, D_1^{(1)}, \ldots, D_1^{(r)}\right) + R,$$

where $\Phi(X_0, X_1, \ldots, X_r)$ is a polynomial over $\mathbb{Q}$ in the variables $X_0, X_1, \ldots, X_r$ and R is a polynomial combination over $\mathbb{Q}$ of $D_1^{(0)}, D_1^{(1)}, D_1^{(2)}, \ldots$ as well as $w_{0,1}^{(0)}, w_{0,1}^{(1)}, w_{0,1}^{(2)}, \ldots$ such that: if T is any term of R, then at least one of the variables $w_{0,1}^{(0)}, w_{0,1}^{(1)}, w_{0,1}^{(2)}, \ldots$ is a factor of T.

Let $c_{0,1}(z)$ and $c_{1,1}(z)$ be any two meromorphic functions on a region Ω of the complex plane. Then, there is a meromorphic function $\rho_0(z) \not\equiv 0$ such that

$$(4.20) \qquad \rho_0'(z) + c_{0,1}(z)\,\rho_0(z) \equiv 0, \quad \text{on a subregion } \mathcal{U} \text{ of } \Omega.$$

In view of (4.20) and $m = 1$, we use the substitution $y(z) = \rho_0(z)\,v(z)$ to transform the restriction of (4.16) to $\mathcal{U}$ into the differential equation

$$(4.21) \qquad \left(v'(z)\right)^2 + c_{1,1}^*(z)\left(v(z)\right)^2 = 0, \quad \text{on } \mathcal{U},$$

where $c_{0,1}^*(z) \equiv 0$ and $c_{1,1}^*(z)$ are specified on $\mathcal{U}$ by (2.21) and (2.22) of page 27. The substitution of $c_{i,j}^{*(k)}(z)$ for $w_{i,j}^{(k)}$ in R yields $R^*(z) \equiv 0$, on $\mathcal{U}$. Proposition 4.7 shows that each $D_1^{(k)}$ is a semi-invariant of the first kind. Therefore, by replacing $w_{i,j}^{(k)}$ with $c_{i,j}^{*(k)}(z)$ and applying (1.9) to $D_1^{(k)}$, we deduce that, for z in $\mathcal{U}$,

$$(4.22) \qquad P(z) \equiv P^*(z) \equiv \Phi\left(D_1^{*(0)}(z), D_1^{*(1)}(z), \ldots, D_1^{*(r)}(z)\right) + 0$$

$$\equiv \Phi\left(D_1^{(0)}(z), D_1^{(1)}(z), \ldots, D_1^{(r)}(z)\right).$$

Since the functions $P(z)$ and $\Phi\left(D_1^{(0)}(z), D_1^{(1)}(z), \ldots, D_1^{(r)}(z)\right)$ are meromorphic on Ω, (4.22) is valid on Ω. Because $c_{0,1}(z)$ and $c_{1,1}(z)$ can be selected arbitrarily, we apply Corollary 4.3 to conclude that $P - \Phi\left(D_1^{(0)}, D_1^{(1)}, \ldots, D_1^{(r)}\right) \equiv 0$. This yields the representation (4.17) for P and completes the proof. $\square$

REMARK 4.9. In view of (2.21), there are no isobaric semi-invariants of the first kind of weight 1. For the equations (4.16), Proposition 4.8 shows that an isobaric polynomial in $\mathcal{S}_1$ of weight 2 is a semi-invariant of the first kind if and only if it is expressible as $\boldsymbol{P} \equiv \gamma \boldsymbol{D}_1$, for some nonzero rational number γ.

THEOREM 4.10. *A polynomial $\boldsymbol{P}$ in $\mathcal{S}_1$ is a relative invariant of the equations (4.16) if and only if it is expressible in terms of $\boldsymbol{D}_1 \equiv \boldsymbol{w}_{1,1} - (\boldsymbol{w}_{0,1})^2$ as*

$$\boldsymbol{P} \equiv \gamma \left[\boldsymbol{D}_1 \right]^\kappa, \tag{4.23}$$

for some nonzero γ in $\mathbb{Q}$ and some positive integer κ.

PROOF. For any nonzero γ in $\mathbb{Q}$ and any positive integer κ, the polynomial $\boldsymbol{P}$ defined in $\mathcal{S}_1$ by (4.23) is a relative invariant.

Conversely, suppose that $\boldsymbol{P}$ in $\mathcal{S}_1$ is a relative invariant of weight s for the equations (4.16). In particular, $\boldsymbol{P}$ is a semi-invariant of the first kind for such equations. Proposition 4.8 shows that $\boldsymbol{P}$ has a representation

$$\boldsymbol{P} \equiv \Phi\left(\boldsymbol{D}_1^{(0)}, \boldsymbol{D}_1^{(1)}, \ldots, \boldsymbol{D}_1^{(r)} \right). \tag{4.24}$$

Next, we shall give an indirect argument to establish that: for any $k \geq 1$, $\boldsymbol{P}$ does not involve $\boldsymbol{D}_1^{(k)}$.

Suppose that $\boldsymbol{P}$ involves some $\boldsymbol{D}_1^{(k)}$ having $k \geq 1$. Let k_0 denote the greatest positive integer such that $\boldsymbol{P}$ effectively involves $\boldsymbol{D}_1^{(k_0)}$. Then, for each nonzero term $\boldsymbol{T}$ of $\boldsymbol{P}$, there is a nonzero rational number β and a sequence $(e_0, e_1, \ldots, e_{k_0})$ of $k_0 + 1$ nonnegative integers such that

$$\boldsymbol{T} \equiv \beta \prod_{k=0}^{k_0} \left(\boldsymbol{D}_1^{(k)} \right)^{e_k} \quad \text{and} \quad s = weight(\boldsymbol{P}) = weight(\boldsymbol{T}) = \sum_{k=0}^{k_0} (k+2) e_k.$$

For any two sequences

$$\sigma = (\mu_0, \mu_1, \ldots, \mu_{k_0}) \quad \text{and} \quad \tau = (\nu_0, \nu_1, \ldots, \nu_{k_0}) \tag{4.25}$$

of $k_0 + 1$ nonnegative integers, we say that σ *precedes* τ, and that τ *follows* σ, when there is an integer p (necessarily unique) such that $0 \leq p \leq k_0$, $\mu_p < \nu_p$, and, for any k, if k satisfies $p < k \leq k_0$, then $\mu_k = \nu_k$. This total ordering for the sequences of $k_0 + 1$ nonnegative integers specifies a total ordering of the nonzero terms in $\boldsymbol{P}$. Thus, we can represent $\boldsymbol{P}$ by

$$\boldsymbol{P} \equiv \sum_{i=1}^{r} \boldsymbol{T}_i, \quad \text{where:} \quad \boldsymbol{T}_i \equiv \beta_i \prod_{k=0}^{k_0} \left(\boldsymbol{D}_1^{(k)} \right)^{\omega_{i,k}}, \quad \text{for } 1 \leq i \leq r; \tag{4.26}$$

r is a positive integer; each of $\beta_1, \ldots, \beta_r$ is a nonzero rational number; each $\omega_{i,k}$ is a nonnegative integer; if $1 \leq i < j \leq r$, then the sequence $(\omega_{i,0}, \omega_{i,1}, \ldots, \omega_{i,k_0})$ for $\boldsymbol{T}_i$ precedes the sequence $(\omega_{j,0}, \omega_{j,1}, \ldots, \omega_{j,k_0})$ for $\boldsymbol{T}_j$; and $\omega_{r,k_0} \geq 1$.

Let Ω^{**} be a region on which Euler's Γ-function $\Gamma(\zeta)$ is a univalent analytic function; set $f(\zeta) \equiv \Gamma(\zeta)$ on Ω^{**}; set $\Omega = f(\Omega^{**})$; and let (4.16) be the equation in which $c_{0,1}(z) \equiv 0$ and $c_{1,1}(z) \equiv 1$. This yields $D_1(z) \equiv 1$, on Ω. Since $\boldsymbol{D}_1$ is a relative invariant of weight 2, we also have $D_1^{**}(\zeta) \equiv \left(f'(\zeta) \right)^2 \equiv \left(\Gamma'(\zeta) \right)^2$, on Ω^{**}. For the function $P(z)$ on Ω obtained by substituting $c_{\kappa,1}^{(\lambda)}(z)$ for $\boldsymbol{w}_{\kappa,1}^{(\lambda)}$ in $\boldsymbol{P}$, we use

(4.26) to see that there is a rational number β_0 such that $P(z) \equiv \beta_0$. We employ $D_1^{**}(\zeta) \equiv \left(\Gamma'(\zeta)\right)^2$ to verify that

$$(4.27) \qquad D_1^{**(k)}(\zeta) \equiv 2\,\Gamma^{(1)}(\zeta)\,\Gamma^{(k+1)}(\zeta) + R_k(\zeta), \quad \text{for } k \geq 1,$$

where $R_k(\zeta)$ is a polynomial combination over $\mathbb{Q}$ of $\Gamma^{(1)}(\zeta), \ldots, \Gamma^{(k)}(\zeta)$. Since $\boldsymbol{P}$ is a semi-invariant of the second kind of weight s, we have

$$(4.28) \qquad 0 \equiv P^{**}(\zeta) - \left(f'(\zeta)\right)^s P\bigl(f(\zeta)\bigr) \equiv \sum_{i=1}^r T_i^{**}(\zeta) - \beta_0 \left(\Gamma^{(1)}(\zeta)\right)^s, \quad \text{on } \Omega^{**}.$$

We apply the properties of the last sequence $(\omega_{r,0}, \omega_{r,1}, \ldots, \omega_{r,k_0})$ for (4.26) along with (4.27) and (4.26) to see that $T_r^{**}(\zeta)$ contributes the nonzero term

$$\beta_r \left(\left(\Gamma^{(1)}(\zeta)\right)^2\right)^{\omega_{r,0}} \prod_{k=1}^{k_0} \left(2\,\Gamma^{(1)}(\zeta)\,\Gamma^{(k+1)}(\zeta)\right)^{\omega_{r,k}}$$

to the right member of (4.28) as the only term of this form. This contradicts the theorem that Euler's Γ-function $\Gamma(\zeta)$ on Ω^{**} does not satisfy any nonzero algebraic differential equation over $\mathbb{Q}$; in this regard, see [**37**] or [**15**].

Consequently, $\boldsymbol{P}$ in (4.24) is a polynomial combination over $\mathbb{Q}$ of $\boldsymbol{D}_1$. Since $\boldsymbol{P}$ is nonconstant and isobaric, $\boldsymbol{P}$ must have the form $\boldsymbol{P} \equiv \gamma \left[\boldsymbol{D}_1\right]^{\kappa}$, for some nonzero rational number γ and some positive integer κ. This yields (4.23) and completes the proof. $\qquad\square$

REMARK 4.11. Definitions 1.1–1.5 on page 6 and Theorem 4.10 show that: for $m = 1$ and (1.1) as (4.16), *there is one and only one basic relative invariant.* It is $\mathcal{I}_{1,1,1} \equiv \boldsymbol{D}_1$. The condition $\mathcal{I}_{1,1,1}(z) \equiv 0$ is equivalent to $c_{1,1}(z) \equiv \bigl(c_{0,1}(z)\bigr)^2$. Thus, we have $\mathcal{I}_{1,1,1}(z) \equiv 0$ if and only if (4.16) can be rewritten as

$$(4.29) \qquad \left[y'(z) + c_{0,1}(z)\,y(z)\right]^2 = 0, \quad \text{on } \Omega.$$

Then, the solutions of (4.16) are the solutions of a first-order homogeneous linear differential equation. An analogous condition for the equations (1.1) having $m \geq 2$ is given by condition (17.12) of page 181. The corresponding Theorem 17.2 is our solution for the simplest of the Fano-type problems about the equations (1.1).

REMARK 4.12. For $m = 1$, we use (3.19) and (3.20) to deduce

$$c_{0,1}^{**}(\zeta) \equiv f'(\zeta)\,c_{0,1}^{**}\bigl(f(\zeta)\bigr), \quad \text{and} \quad c_{1,1}^{**}(\zeta) \equiv \left(f'(\zeta)\right)^2 c_{1,1}^{**}\bigl(f(\zeta)\bigr), \quad \text{on } \Omega.$$

Thus, the polynomials $\boldsymbol{w}_{0,1}$ and $\boldsymbol{w}_{1,1}$ are semi-invariants of the second kind of respective weights 1 and 2 for the equations (4.16). Consequently, we see that *any nonconstant isobaric polynomial combination of the variables $\boldsymbol{w}_{0,1}$ and $\boldsymbol{w}_{1,1}$ over $\mathbb{Q}$ is a semi-invariant of the second kind for the equations* (4.16).

However, there are many semi-invariants of the second kind that are not of the preceding type. For instance, the proof of Proposition 14.6 on page 146 shows that: if $\boldsymbol{P}_p$ and $\boldsymbol{Q}_q$ are semi-invariants of the second kind of respective weights p and q such that the polynomial $\boldsymbol{S} \equiv p\,\boldsymbol{P}_p\,\boldsymbol{Q}_q^{(1)} - q\,\boldsymbol{Q}_q\,\boldsymbol{P}_p^{(1)}$ is nonzero, then $\boldsymbol{S}$ is a semi-invariant of the second kind of weight $p + q + 1$. As an example, the selection $\boldsymbol{P} \equiv \boldsymbol{w}_{0,1}$ with $p = 1$ and $\boldsymbol{Q} \equiv \boldsymbol{w}_{1,1}$ with $q = 2$ yields $\boldsymbol{w}_{0,1}\boldsymbol{w}_{1,1}^{(1)} - 2\boldsymbol{w}_{1,1}\boldsymbol{w}_{0,1}^{(1)}$ as a semi-invariant of the second kind for the equations (4.16). Many other examples of this type can be given. Thus, in contrast to (4.23), one should not expect a simple explicit formula that yields all of the semi-invariants of the second kind for (4.16).

4.5. Isobaric semi-invariants of weight 2 when $m \geq 2$

On pages 17–18 of Section 1.7, we showed that: *each semi-invariant of the first or second kind for homogeneous linear differential equations of order m specifies a corresponding semi-invariant of the same kind for the equations (1.1) of order m.*

When the semi-invariant of the first kind for homogeneous linear differential equations of order $m \geq 2$ given by [**24**, page 4, Equation (1.8)] is adjusted for the notation of Section 1.7 we obtain $\boldsymbol{a}_{m,2}$ in (1.14) on page 7. Thus, $\boldsymbol{a}_{m,2}$ is a semi-invariant of the first kind for equations (1.1) of order $m \geq 2$. It is also isobaric of weight 2.

The proof of Theorem 14.8 on page 147 will use the following result.

LEMMA 4.13. *For any (1.1) and any substitution $z = f(\zeta)$ from (1.4) that transforms (1.1) on Ω into (1.5) on Ω^{**}, the coefficients of (1.1) and (1.5) yield*

$$(4.30) \qquad a^{**}_{m,2}(\zeta) - \bigl(f'(\zeta)\bigr)^2 a_{m,2}\bigl(f(\zeta)\bigr) \equiv \frac{f'''(\zeta)}{2f'(\zeta)} - \frac{3}{4}\left[\frac{f''(\zeta)}{f'(\zeta)}\right]^2, \quad \text{on } \Omega^{**}.$$

PROOF. Theorem 3.2 on page 30 shows that: as the substitution $z = f(\zeta)$ from (1.4) transforms (1.1) on Ω into (1.5) on Ω^{**}, it transforms the homogeneous linear differential equation (1.84) of page 17 (with $c_j(z) \equiv c_{0,j}(z)$ on Ω) into the homogeneous linear differential equation (1.86) (with $c_j^{**}(\zeta) \equiv c_{0,j}^{**}(\zeta)$ on Ω^{**}). Hence, (4.30) is a restatement of [**24**, (A.47) in Proposition A.18 on page 147]. This completes the proof. $\square$

When the semi-invariant of the second kind for homogeneous linear differential equations of order $m \geq 2$ given by [**24**, page 5, Equation (1.9)] is adjusted for the notation of Section 1.7 we obtain $\boldsymbol{b}_{m,2}$ in (1.15) on page 7. Thus, $\boldsymbol{b}_{m,2}$ is a semi-invariant of the second kind of weight 2 for the equations (1.1) of order $m \geq 2$.

We recall from [**24**, pages 142–143, Propositions A.11] that: all of the isobaric semi-invariants of the first kind of weight 2 for homogeneous linear differential equations of order $m \geq 2$ are given by $\gamma \boldsymbol{a}_{m,2}$, as γ ranges over the nonzero rational numbers. Also, [**24**, pages 143–145, Proposition A.13] shows that: all of the semi-invariants of the second kind of weight 2 for homogeneous linear differential equations of order $m \geq 2$ are given by $\gamma \boldsymbol{b}_{m,2}$, as γ ranges over the nonzero rational numbers. For homogeneous linear differential equations, isobaric polynomials of weight 2 are the nonzero linear combinations over $\mathbb{Q}$ of $\boldsymbol{w}_{0,2}$, $\boldsymbol{w}_{0,1}^{(1)}$, and $(\boldsymbol{w}_{0,1})^2$. However, for equations of the form (1.1), isobaric polynomials of weight 2 may also involve $\boldsymbol{w}_{1,1}$ and this leads to the following result.

PROPOSITION 4.14. *The isobaric semi-invariants of the first kind of weight 2 for the equations (1.1) of order $m \geq 2$ are given by the nonzero linear combinations over $\mathbb{Q}$ of $\boldsymbol{a}_{m,2}$ from (1.14) and $\boldsymbol{D}_1$ from (1.11). Also, the semi-invariants of the second kind of weight 2 for the equations (1.1) of order $m \geq 2$ are given by the nonzero linear combinations over $\mathbb{Q}$ of $\boldsymbol{b}_{m,2}$ from (1.15) and $\boldsymbol{D}_1$ from (1.11).*

PROOF. For $m \geq 2$, each isobaric polynomial $\boldsymbol{P}$ in $\mathcal{S}_m$ of weight 2 has the form

$$(4.31) \qquad \boldsymbol{P} \equiv p\,\boldsymbol{w}_{0,2} + q\,\boldsymbol{w}_{1,1} + r\,\boldsymbol{w}_{0,1}^{(1)} + s\,(\boldsymbol{w}_{0,1})^2,$$

where p, q, r, s are rational numbers not all zero.

For any (1.1) and any transformation (1.2) of (1.1) into a corresponding (1.3), we see that (4.31), (2.24), (2.22), and (2.21) yield

$$(4.32) \qquad P^*(z) - P(z) \equiv A_1 \frac{\rho^{(2)}(z)}{\rho(z)} + A_2 \left[\frac{\rho^{(1)}(z)}{\rho(z)}\right]^2 + A_3 \frac{\rho^{(1)}(z)}{\rho(z)} c_{0,1}(z),$$

where

$$A_1 \equiv p\binom{m}{2} + rm, \quad A_2 \equiv qm^2 - rm + sm^2, \quad A_3 \equiv p(m-1) + 2qm + 2sm.$$

In view of (4.32), P in (4.31) is a semi-invariant of the first kind for the equations (1.1) of order $m \geq 2$ if and only if $A_1 \equiv A_2 \equiv A_3 \equiv 0$. The condition $A_1 \equiv 0$ is equivalent to $r \equiv -(m-1)p/2$. Thus, we have $A_1 \equiv A_2 \equiv 0$ if and only if

$$(4.33) \qquad r \equiv -\frac{m-1}{2}p \quad \text{and} \quad s \equiv \frac{r}{m} - q \equiv -\frac{m-1}{2m}p - q.$$

Moreover, when (4.33) is satisfied, we also have $A_3 \equiv 0$. Thus, P in (4.31) is a semi-invariant of the first kind for equations (1.1) of order $m \geq 2$ if and only if P is given by

$$(4.34) \qquad P \equiv p\left[w_{0,2} - \frac{m-1}{2}w_{0,1}^{(1)} - \frac{m-1}{2m}(w_{0,1})^2\right] + q\left[w_{1,1} - (w_{0,1})^2\right],$$

for some rational numbers p and q not both zero. Due to (1.14) and (1.11) on page 7, this is equivalent to the requirement that P be a nonzero linear combination over $\mathbb{Q}$ of $a_{m,2}$ and D_1.

For any (1.1) and any transformation (1.4) of (1.1) into a corresponding (1.5), we see that (4.31), (3.24), (3.20), and (3.19) yield

$$(4.35) \quad P^{**}(\zeta) - [f'(\zeta)]^2 P(f(\zeta)) \equiv B_1 \frac{f'''(\zeta)}{f'(\zeta)} + B_2 \left[\frac{f''(\zeta)}{f'(\zeta)}\right]^2 + B_3 f''(\zeta) c_{0,1}(f(\zeta)),$$

where

$$B_1 \equiv -p\binom{m}{3} - r\binom{m}{2}, \quad B_2 \equiv 3p\binom{m+1}{4} + q\left[\binom{m}{2}\right]^2 + r\binom{m}{2} + s\left[\binom{m}{2}\right]^2,$$

and

$$B_3 \equiv -p\binom{m-1}{2} - 2q\binom{m}{2} + r - 2s\binom{m}{2}.$$

In view of (4.35), P in (4.31) is a semi-invariant of the second kind for the equations (1.1) of order $m \geq 2$ if and only if $B_1 \equiv B_2 \equiv B_3 \equiv 0$. The condition $B_1 \equiv 0$ is equivalent to $r \equiv -(m-2)p/3$. Thus, we have $B_1 \equiv B_2 \equiv 0$ if and only if

$$(4.36) \qquad r \equiv -\frac{m-2}{3}p \quad \text{and} \quad s \equiv -\frac{(3m-1)(m-2)}{6m(m-1)}p - q.$$

Moreover, when (4.36) is satisfied, we also have $B_3 \equiv 0$. Thus, P in (4.31) is a semi-invariant of the second kind for the equations (1.1) of order $m \geq 2$ if and only if P is given by

$$(4.37) \quad P \equiv p\left[w_{0,2} - \frac{m-2}{3}w_{0,1}^{(1)} - \frac{(3m-1)(m-2)}{6m(m-1)}(w_{0,1})^2\right] + q\left[w_{1,1} - (w_{0,1})^2\right],$$

for some rational numbers p and q not both zero. In view of (1.15) and (1.11) on page 7, this is equivalent to the requirement that P be a nonzero linear combination over $\mathbb{Q}$ of $b_{m,2}$ and D_1. This completes the proof. □

The proof of Corollary 4.15 does not depend on the observations of pages 17–18 for Section 1.7; and it is independent of [**24**].

COROLLARY 4.15. *The polynomial $\boldsymbol{a}_{m,2}$ in (1.14) on page 7 is an isobaric semi-invariant of the first kind of weight 2 for the equations (1.1) having $m \geq 2$; and the polynomial $\boldsymbol{b}_{m,2}$ in (1.15) is a semi-invariant of the second kind of weight 2 for such equations.*

PROOF. The assertion about $\boldsymbol{a}_{m,2}$ follows by selecting $p = 1/\binom{m+1}{3}$ and $q = 0$ in (4.34). To obtain the conclusion about $\boldsymbol{b}_{m,2}$, we set $p = 1/\binom{m+1}{3}$ and $q = 0$ in (4.37). This completes the proof. □

Since the formula $c_{0,1}^*(z) \equiv c_{0,1}(z) + m \dfrac{\rho'(z)}{\rho(z)}$ from (2.21) shows that $w_{0,1}$ is not a semi-invariant of the first kind, *there are no relative invariants of weight 1 for the equations (1.1)*.

PROPOSITION 4.16. *The relative invariants of weight 2 for the equations (1.1) are given in terms of the basic relative invariant $\boldsymbol{D}_1$ from (1.11) by $\gamma \boldsymbol{D}_1$, as γ ranges over the nonzero rational numbers.*

PROOF. For $m = 1$, Theorem 4.10 on page 41 yields the desired conclusion. Suppose that $m \geq 2$. In order for the semi-invariant of the first kind in (4.34) and the semi-invariant of the second kind in (4.37) to be the same polynomial, it is necessary and sufficient that $p = 0$ and $q \neq 0$. Thus, the relative invariants of weight 2 for the equations (1.1) having $m \geq 2$ are given by $\gamma \boldsymbol{D}_1$, for each nonzero γ in $\mathbb{Q}$. This completes the proof. □

4.6. Further semi-invariants of the second kind when $m \geq 2$

Proposition 4.7 of page 39 shows that: for any $m \geq 1$ and any $\boldsymbol{P}$ in $\mathcal{S}_m$, if $\boldsymbol{P}$ is a semi-invariant of the first kind, then $\boldsymbol{P}'$ is a semi-invariant of the first kind. The following result is analogous for semi-invariants of the second kind. It requires the restriction $m \geq 2$ in order for $\boldsymbol{b}_{m,1}$ to be defined by (1.16) on page 7.

PROPOSITION 4.17. *For $m \geq 2$, suppose that $\boldsymbol{P}$ in $\mathcal{S}_m$ is a nonconstant isobaric polynomial of weight s and set $\boldsymbol{R} \equiv \boldsymbol{P}^{(1)} + s\,\boldsymbol{b}_{m,1}\boldsymbol{P}$. If $\boldsymbol{P}$ is a semi-invariant of the second kind, then $\boldsymbol{R}$ is a semi-invariant of the second kind of weight $s + 1$. If $\boldsymbol{P}$ is a semi-invariant of the first kind, then $\boldsymbol{R}$ is not a semi-invariant of the first kind.*

PROOF. Suppose that $\boldsymbol{P}$ is a semi-invariant of the second kind. Then, for any transformation (1.4) of any given (1.1) into a corresponding (1.5), (1.10) of Definition 1.2 yields

$$(4.38) \qquad P^{**}(\zeta) \equiv \big(f'(\zeta)\big)^s P\big(f(\zeta)\big), \quad \text{on } \Omega^{**}.$$

We use (1.16) and (3.19) to verify that

$$(4.39) \qquad b_{m,1}^{**}(\zeta) \equiv f'(\zeta)\,b_{m,1}\big(f(\zeta)\big) - \frac{f''(\zeta)}{f'(\zeta)}.$$

After substituting $c_{i,j}^{**(k)}(\zeta)$ from (1.5) for $w_{i,j}^{(k)}$ in $\boldsymbol{R}$, we differentiate (4.38) with respect to ζ and combine the resulting expressions with (4.39) and (4.38) to deduce

that
$$(4.40) \quad R^{**}(\zeta) \equiv P^{**(1)}(\zeta) + s\, b_{m,1}^{**}(\zeta)\, P^{**}(\zeta)$$
$$\equiv \bigl(f'(\zeta)\bigr)^{s+1} P^{(1)}\bigl(f(\zeta)\bigr) + s\bigl(f'(\zeta)\bigr)^{s-1} f''(\zeta)\, P\bigl(f(\zeta)\bigr)$$
$$+ s\bigl(f'(\zeta)\bigr)^{s+1} b_{m,1}\bigl(f(\zeta)\bigr) P\bigl(f(\zeta)\bigr) - s\bigl(f'(\zeta)\bigr)^{s-1} f''(\zeta)\, P\bigl(f(\zeta)\bigr)$$
$$\equiv \bigl(f'(\zeta)\bigr)^{s+1} \Bigl[P^{(1)}\bigl(f(\zeta)\bigr) + s\, b_{m,1}\bigl(f(\zeta)\bigr) P\bigl(f(\zeta)\bigr) \Bigr]$$
$$\equiv \bigl(f'(\zeta)\bigr)^{s+1} R\bigl(f(\zeta)\bigr), \quad \text{on } \Omega^{**}.$$

Since $\boldsymbol{P}$ is not a constant of $\mathcal{S}_m$, $\boldsymbol{R}$ is not a constant. Thus, (4.40) shows that $\boldsymbol{R}$ is a semi-invariant of the second kind of weight $s+1$.

Suppose that $\boldsymbol{P}$ is a semi-invariant of the first kind. For an indirect proof that $\boldsymbol{R}$ is not a semi-invariant of the first kind, suppose that it is. Since Proposition 4.7 shows that $\boldsymbol{P'}$ is a semi-invariant of the first kind, we find that $\boldsymbol{b}_{m,1}\boldsymbol{P}$ must be a semi-invariant of the first kind. Due to Corollary 4.3, there is an equation (1.1) such that the substitution of $c_{i,j}^{(k)}(z)$ from (1.1) for $w_{i,j}^{(k)}$ in $\boldsymbol{b}_{m,1}\boldsymbol{P}$ gives $b_{m,1}(z)\, P(z) \not\equiv 0$ on Ω. Let $\rho(z)$ be a nonzero solution of $m\,\rho'(z) + c_{0,1}(z)\,\rho(z) = 0$ on a subregion $\mathcal{U}$ of Ω. Then, for the transformation $y(z) = \rho(z)\, v(z)$ of the restriction to $\mathcal{U}$ of this (1.1) into a corresponding (1.3) on $\mathcal{U}$, we find that (2.21) and (1.16) give $b_{m,1}^{*}(z) \equiv 0$ and the contradiction
$$0 \equiv b_{m,1}^{*}(z) P^{*}(z) \equiv b_{m,1}(z)\, P(z) \not\equiv 0.$$
Thus, $\boldsymbol{R}$ is not a semi-invariant of the first kind. This completes the proof. □

Proposition 4.17 will be used on page 117 in the proof of Theorem 10.3.

CHAPTER 5

Composite Transformations and Reductions

5.1. The composite of substitutions (1.2) and (1.4)

Rather than transform the dependent variable alone or the independent variable alone, we now employ a composite substitution for (1.1) based on $y(z) = \rho(z)\,v(z)$ in (1.2) and $z = f(\zeta)$ in (1.4). Thus, (1.1), (1.2), and (1.4) specify unique meromorphic functions $d_{i,j}(\zeta)$ on Ω^{**} such that (1.2) transforms (1.1) on Ω into (1.3) on Ω and, with $w(\zeta) = (v \circ f)(\zeta)$, (1.4) transforms (1.3) on Ω into

$$(5.1) \quad \left(w^{(m)}(\zeta)\right)^2 + \sum_{\substack{0 \le i,j \le m \\ (i,j) \ne (0,0)}} d_{i,j}(\zeta)\, w^{(m-i)}(\zeta)\, w^{(m-j)}(\zeta) = 0, \quad \text{on } \Omega^{**},$$

where $d_{j,i}(\zeta) \equiv d_{i,j}(\zeta)$, for $0 \le i < j \le m$.

We recall the notation

$$(5.2) \quad c_{0,0}(\zeta) \equiv 1 \quad \text{and} \quad c_{j,i}(z) \equiv c_{i,j}(z), \quad \text{when } 0 \le i < j \le m,$$

for (1.1); and, we introduce $d_{0,0}(\zeta) \equiv 1$ as the coefficient of $\left(w^{(m)}(\zeta)\right)^2$ in (5.1).

THEOREM 5.1. *For $m \ge 1$, the coefficients of the differential equation (5.1) into which the composite of (1.2) and (1.4) transforms (1.1) via $w(\zeta) = (v \circ f)(\zeta)$ are given on Ω^{**} in terms of (1.1), (5.2), $\rho(z)$, $\zeta = g(z)$ on Ω, and $z = f(\zeta)$ on Ω^{**} by*

$$(5.3) \quad d_{i,j}(\zeta) \equiv \left(f'(\zeta)\right)^{i+j} \sum_{p=0}^{i} \sum_{q=0}^{j} \phi_{i-p,p}(f(\zeta))\, \phi_{j-q,q}(f(\zeta))\, c_{p,q}(f(\zeta)),$$

for $0 \le i, j \le m$,

where $\phi_{i,j}(z)$ is defined on Ω through

$$(5.4) \quad \phi_{0,j}(z) \equiv 1, \quad \text{for any } j,$$

and

$$(5.5) \quad \phi_{i+1,j}(z) \equiv \sum_{k=j+1}^{m} \left[\phi_{i,k}^{(1)}(z) + \left[\frac{\rho'(z)}{\rho(z)} + (m-i-k) \frac{g''(z)}{g'(z)} \right] \phi_{i,k}(z) \right],$$

for $i \ge 0$ and any j.

Moreover, the coefficients $d_{i,j}(\zeta)$ of (5.1) are analytic on Ω^{**} when the coefficients $c_{i,j}(z)$ of (1.1) are analytic on Ω and $\rho(z)$ is an analytic function on Ω that is free of zeros on Ω.

PROOF. With $c_{0,0}^*(z) \equiv 1$, we apply (3.12) of Theorem 3.3 on page 32 to (1.3) in place of (1.1) and find that

$$(5.6) \quad d_{i,j}(\zeta) \equiv \left(f'(\zeta)\right)^{i+j} \sum_{\mu=0}^{i} \sum_{\nu=0}^{j} \beta_{i-\mu,\mu}\left(f(\zeta)\right) \beta_{j-\nu,\nu}\left(f(\zeta)\right) c_{\mu,\nu}^*\left(f(\zeta)\right),$$

for $0 \leq i, j \leq m$.

With $c_{0,0}(z) \equiv 1$, we use (2.5) of Proposition 2.1 on page 23 to obtain

$$(5.7) \quad c_{\mu,\nu}^*(z) \equiv \sum_{p=0}^{\mu} \sum_{q=0}^{\nu} \binom{m-p}{\mu-p} \binom{m-q}{\nu-q} \frac{\rho^{(\mu-p)}(z) \rho^{(\nu-q)}(z)}{(\rho(z))^2} c_{p,q}(z),$$

for $0 \leq \mu, \nu \leq m$.

It is convenient to set

$$(5.8) \quad \gamma_{i,j,k}(z) \equiv \beta_{i-k,k}(z) \binom{m-j}{k-j} \frac{\rho^{(k-j)}(z)}{\rho(z)}, \quad \text{on } \Omega \text{ for } j \leq k \leq i,$$

and

$$(5.9) \quad r_{i,j}(z) \equiv \sum_{k=j}^{i} \beta_{i-k,k}(z) \binom{m-j}{k-j} \frac{\rho^{(k-j)}(z)}{\rho(z)}, \quad \text{on } \Omega \text{ for any } i \text{ and } j.$$

We combine (5.6) with (5.7) and use (5.8)–(5.9) to obtain

$$(5.10) \quad d_{i,j}(\zeta) \equiv \left(f'(\zeta)\right)^{i+j} \sum_{\mu=0}^{i} \sum_{\nu=0}^{j} \sum_{p=0}^{\mu} \sum_{q=0}^{\nu} \gamma_{i,p,\mu}\left(f(\zeta)\right) \gamma_{j,q,\nu}\left(f(\zeta)\right) c_{p,q}\left(f(\zeta)\right)$$

$$\equiv \left(f'(\zeta)\right)^{i+j} \sum_{p=0}^{i} \sum_{q=0}^{j} \left[\sum_{\mu=p}^{i} \gamma_{i,p,\mu}\left(f(\zeta)\right)\right] \left[\sum_{\nu=q}^{j} \gamma_{j,q,\nu}\left(f(\zeta)\right)\right] c_{p,q}\left(f(\zeta)\right)$$

$$\equiv \left(f'(\zeta)\right)^{i+j} \sum_{p=0}^{i} \sum_{q=0}^{j} r_{i,p}\left(f(\zeta)\right) r_{j,q}\left(f(\zeta)\right) c_{p,q}\left(f(\zeta)\right).$$

We first introduced $r_{i,j}(z)$ on Ω in [**24**, page 41, Formula (6.15)]. The argument given in [**24**, pages 41–44] established in [**24**, page 44, Formula (6.26)] that

$$(5.11) \quad r_{i,j}\left(f(\zeta)\right) \equiv \phi_{i-j,j}\left(f(\zeta)\right), \quad \text{on } \Omega^{**} \text{ for } 0 \leq j \leq i \leq m.$$

We combine (5.10) and (5.11) to obtain (5.3). Since $g(z)$ is a univalent analytic function on Ω, $g'(z)$ is analytic and free of zeros on Ω. Thus, (5.3)–(5.5) yield the last assertion of Theorem 5.1. This completes the proof. $\square$

COROLLARY 5.2. *For $m \geq 1$, (1.2) and (1.4) with $w(\zeta) = (v \circ f)(\zeta)$ transform*

$$(5.12) \quad y^{(m)}(z) + \sum_{j=1}^{m} c_{0,j}(z) y^{(m-j)}(z) = 0, \quad \text{on } \Omega,$$

into

$$(5.13) \quad w^{(m)}(\zeta) + \sum_{j=1}^{m} d_{0,j}(\zeta) w^{(m-j)}(\zeta) = 0, \quad \text{on } \Omega^{**},$$

where $d_{0,j}(\zeta)$ is given by (5.3) for $i = 0$ and $1 \leq j \leq m$.

PROOF. Proposition 2.1 on page 23 shows that the substitution $y(z) = \rho(z)\, v(z)$ from (1.2) transforms (5.12) into

$$\tag{5.14} v^{(m)}(z) + \sum_{\nu=1}^{m} c^*_{0,\nu}(z)\, v^{(m-\nu)}(z), \quad \text{on } \Omega,$$

where $c^*_{0,0}(z) \equiv 1$ and the coefficients $c^*_{0,\nu}(z)$, for $1 \leq \nu \leq m$, are provided by (5.7). Using Theorem 3.3 on page 32 with the context of Theorem 3.2, we find that the substitution $z = f(\zeta)$ from (1.4) with $w(\zeta) = (v \circ f)(\zeta)$ transforms (5.14) into the equation (5.13) whose coefficients $d_{0,j}(\zeta)$, for $1 \leq j \leq m$, are specified by (5.6). The deductions involving (5.10) and (5.11) show that the coefficients $d_{0,j}(\zeta)$ in (5.13) are given by (5.3), for $i = 0$ and $1 \leq j \leq m$. This completes the proof. □

5.2. The condition $d_{0,1}(\zeta) \equiv d_{0,2}(\zeta) \equiv 0$ for (5.1) when $m \geq 2$

Next, we impose restrictions on $y(z) = \rho(z)\, v(z)$ in (1.2) and $z = f(\zeta)$ in (1.4) so that the transformed equation (5.1) satisfies the condition $d_{0,1}(\zeta) \equiv d_{0,2}(\zeta) \equiv 0$. For this purpose, we recall from page 5 that $b_{m,1}(z)$ and $b_{m,2}(z)$ are the functions on Ω obtained by substituting $c^{(k)}_{0,j}(z)$ from (1.1) for $\boldsymbol{w}^{(k)}_{0,j}$ in $\boldsymbol{b}_{m,1}$ and $\boldsymbol{b}_{m,2}$ of (1.16) and (1.15).

PROPOSITION 5.3. *For $m \geq 2$, suppose that the composite of $y(z) = \rho(z)\, v(z)$ and $z = f(\zeta)$ with $w(\zeta) = (v \circ f)(\zeta))$ transforms (1.1) into (5.1). Then, (5.1) satisfies*

$$\tag{5.15} d_{0,1}(\zeta) \equiv 0 \quad \text{and} \quad d_{0,2}(\zeta) \equiv 0$$

if and only if $\rho(z)$ and the inverse $\zeta = g(z)$ of $z = f(\zeta)$ satisfy

$$\tag{5.16} \frac{\rho''(z)}{\rho(z)} - \left[\frac{m-2}{m-1}\right]\left[\frac{\rho'(z)}{\rho(z)}\right]^2 + b_{m,1}(z)\frac{\rho'(z)}{\rho(z)} + (m-1)\, b_{m,2}(z) \equiv 0$$

and

$$\tag{5.17} -\frac{g''(z)}{g'(z)} \equiv b_{m,1}(z) + \frac{2\rho'(z)}{(m-1)\rho(z)}.$$

Moreover, if $\rho(z)$ and $g(z)$ are related by (5.17), then (5.16) is equivalent to

$$\tag{5.18} \frac{g'''(z)}{2g'(z)} - \frac{3}{4}\left[\frac{g''(z)}{g'(z)}\right]^2 \equiv a_{m,2}(z).$$

PROOF. Corollary 5.2 shows that: as a composite substitution transforms the monic (1.1) into a monic (5.1), it also transforms (5.12) into (5.13). In order to apply [**24**, page 46, Proposition 6.7], we make the identifications $c_j(z) \equiv c_{0,j}(z)$, $c^*_j(z) \equiv c^*_{0,j}(z)$, $d_j(\zeta) \equiv d_{0,j}(\zeta)$, and $\boldsymbol{w}^{(k)}_j = \boldsymbol{w}^{(k)}_{0,j}$, for $1 \leq j \leq m$ and $k \geq 0$. By letting $\mathcal{R}_m$ denote the subring of $\mathcal{S}_m$ consisting of the polynomials over $\mathbb{Q}$ in the variables $\boldsymbol{w}^{(k)}_j$, we apply [**24**, page 46, Proposition 6.7] to complete the proof. □

PROPOSITION 5.4. *For any (1.1) on Ω and any subregion $\mathcal{W}$ of Ω, there are analytic function $\rho_0(z)$ and $g_0(z)$ on a subregion $\mathcal{U}$ of $\mathcal{W}$ such that: $\rho_0(z) \neq 0$, for each z in $\mathcal{U}$; $g_0(z)$ is univalent on $\mathcal{U}$; and $\rho_0(z)$, $g_0(z)$ satisfy (5.16)–(5.17) on $\mathcal{U}$.*

PROOF. Let $\sigma_0(z)$ be an analytic function on a subregion $\mathcal{U}_1$ of $\mathcal{W}$ such that: $\sigma_0(z) \not\equiv 0$, for each z in $\mathcal{U}_1$, and

(5.19) $\quad \sigma_0''(z) + b_{m,1}(z)\,\sigma_0'(z) + b_{m,2}(z)\,\sigma_0(z) = 0, \quad$ for each z in $\mathcal{U}_1$.

Let $h_0(z)$ be an analytic function on a subregion $\mathcal{U}_2$ of $\mathcal{U}_1$ such that $h_0(z) \not\equiv 0$ and

(5.20) $\quad h_0'(z) + \left[b_1(z) + \dfrac{2\sigma_0'(z)}{\sigma_0(z)}\right] h_0(z) = 0, \quad$ for each z in $\mathcal{U}_2$.

Let $g_0(z)$ be a univalent analytic function on a subregion $\mathcal{U}$ of $\mathcal{U}_2$ such that, for each z in $\mathcal{U}$, $g_0'(z) = h_0(z)$. We set $\rho_0(z) \equiv (\sigma_0(z))^{m-1}$ on $\mathcal{U}$ and use (5.19) to verify that $\rho_0(z)$ satisfies (5.16) on $\mathcal{U}$. Due to $(m-1)\sigma'(z)/\sigma_0(z) \equiv \rho_0'(z)/\rho(z)$ and (5.20), we see that $\rho_0(z)$ and $g_0(z)$ satisfy (5.17). This completes the proof. $\square$

5.3. Laguerre-Forsyth canonical forms for linear equations

The discovery of the explicit formulas presented in [**24**] for the basic relative invariants of homogeneous linear differential equations was made possible by the research of A. R. Forsyth in [**31**, pages 403–407]. Next, we recall those contributions of Forsyth so that they can be applied again to discover analogous explicit formulas for the basic relative invariants of the equations (1.1).

A monic homogeneous linear equation of order $m \geq 2$ in $y(z)$ is given by

(5.21) $\quad \displaystyle y^{(m)}(z) + \sum_{j=1}^{m} c_j(z)\,y^{(m-j)}(z) = 0$

when the coefficients $c_i(z)$ are meromorphic functions on a region Ω. With respect to a meromorphic function $\rho(z) \not\equiv 0$ on a subregion $\mathcal{U}$ of Ω and a univalent analytic function $\zeta = g(z)$ on $\mathcal{U}$ with inverse designated by $z = f(\zeta)$ on $\mathcal{V} = g(\mathcal{U})$, there are unique meromorphic functions $d_i(\zeta)$ on $\mathcal{V}$ such that the composite substitution

(5.22) $\quad y(z) = \rho(z)\,v(z), \quad z = f(\zeta), \quad w(\zeta) = (v \circ f)(\zeta)$

transforms the restriction of (5.21) to $\mathcal{U}$ into the equation

(5.23) $\quad \displaystyle w^{(m)}(\zeta) + \sum_{i=1}^{m} d_i(\zeta)\,w^{(m-i)}(\zeta) = 0, \quad$ on $\mathcal{V}$.

We showed in [**24**, page 42, Theorem 6.2] that the coefficients of (5.23) are

(5.24) $\quad \displaystyle d_i(\zeta) \equiv (f(\zeta)) \sum_{j=0}^{i} \phi_{i-j,j}(f(\zeta))\,c_j(f(\zeta)), \quad$ on $\mathcal{V}$ for $1 \leq i \leq m$,

where the functions $\phi_{i,j}(z)$ are the meromorphic ones on $\mathcal{U}$ defined by (5.4)–(5.5) of page 47. For the preceding context, the results in [**24**, Propositions 6.7 and 6.8] are completely analogous to those in Propositions 5.3 and 5.4. They establish that: *for any given (5.21) on Ω, there are numerous ways of selecting $\rho(z)$ and $g(z)$ for (5.22) such that (5.22) transforms the restriction of (5.21) to some subregion of Ω into an equation (5.23) that satisfies*

(5.25) $\quad d_1(\zeta) \equiv 0 \quad \text{and} \quad d_2(\zeta) \equiv 0.$

Using a different argument, A. R. Forsyth arrived at the preceding conclusion in [**31**, page 403]. Thus, each (5.21) of order $m \geq 2$ has numerous Laguerre-Forsyth canonical forms according to the following definition.

DEFINITION 5.5. *A Laguerre-Forsyth canonical form for a given homogeneous linear differential equation* (5.21) *of order* $m \geq 2$ *on* Ω *is an equation of the form*

$$(5.26) \qquad w^{(m)}(\zeta) + \sum_{i=3}^{m} d_i(\zeta)\, w^{(m-i)}(\zeta) = 0, \quad \text{having } d_1(\zeta) \equiv d_2(\zeta) \equiv 0,$$

on some region $\mathcal{V}$ *such that the restriction of* (5.21) *to some subregion* $\mathcal{U}$ *of* Ω *can be transformed into* (5.26) *on* $\mathcal{V}$.

When a Laguerre-Forsyth canonical form of order $m \geq 3$ can be transformed into another Laguerre-Forsyth canonical form, relations among the coefficients of the two equations were presented by A. R. Forsyth in [**31**, pages 403–407]. To describe these results of his with simple notation, suppose that an equation (5.21) on Ω satisfying $c_1(z) \equiv 0$ and $c_2(z) \equiv 0$ is transformed by (5.22) with $\mathcal{U} = \Omega$ and $\mathcal{V} = \Omega^{**}$ into an equation (5.23) that satisfies (5.25). In terms of the formula

$$(5.27) \qquad C_{j,q} \equiv (-1)^q \frac{\binom{j-1}{q}}{\binom{2j-2}{q}} \binom{m-j+q}{q}, \quad \text{for } 0 \leq q \leq j \text{ and } 2 \leq j \leq m,$$

that is similar to [**24**, page 49, (6.48))], Forsyth's relations are given by

$$(5.28) \qquad \sum_{\nu=0}^{j} C_{j,\nu}\, d_{j-\nu}^{(\nu)}(\zeta) \equiv \bigl(f'(\zeta)\bigr)^{j} \sum_{\nu=0}^{j} C_{j,\nu}\, c_{j-\nu}^{(\nu)}\bigl(f(\zeta)\bigr), \quad \text{for } 2 \leq j \leq m.$$

We have included the trivial $0 \equiv 0$ when $j = 2$ so that these relations can be more easily generalized. For the notation used by Forsyth and its connection to that employed above, see [**24**, pages 77–78 and page 49]. The constants $C_{j,q}$ in (5.27) and (5.28) motivated our development in [**24**] of simple formulas for the basic relative invariants of homogeneous linear differential equations. This is reflected in [**24**, Theorems 7.3 and 8.3], where the left members for (5.28) were employed with respect to a Laguerre-Forsyth canonical form (5.26) of an equation (5.21) not assumed to satisfy $c_1(z) \equiv c_2(z) \equiv 0$. While not explicitly used in [**24**], *the identities of* (5.28) *can be easily deduced from* [**24**, Theorems 7.3 and 9.1] *in the same way that* Corollary 27.8 *on page* 315 *is established by means of* Theorem 25.10 *on page* 290 *and* Theorem 27.7.

5.4. A Laguerre-Forsyth canonical form for (1.1) when $m \geq 2$

Propositions 5.3 and 5.4 show that each equation (1.1) has numerous Laguerre-Forsyth canonical forms in the sense of the following definition that we introduce through analogy with Definition 5.5.

DEFINITION 5.6. *A Laguerre-Forsyth canonical form for an equation* (1.1) *of order* $m \geq 2$ *on* Ω *is an equation of the type* (5.1) *satisfying* $d_{0,1}(\zeta) \equiv d_{0,2}(\zeta) \equiv 0$ *on a region* $\mathcal{V}$ *such that the restriction of* (1.1) *to some subregion* $\mathcal{U}$ *of* Ω *can be transformed into this* (5.1) *on* $\mathcal{V}$.

For the situation where a Laguerre-Forsyth canonical for a given equation (1.1) of order $m \geq 2$ can be transformed into another Laguerre-Forsyth canonical form, we shall show in Chapter 6 that the coefficients provided by (6.3) yield relations in (6.23) that are analogous to those of (5.28).

5.5. There are no relative invariants in $\mathbb{Q}\{w_{0,1}, w_{0,2}\}$

For $m \geq 2$, we introduce $\mathbb{Q}\{w_{0,1}, w_{0,2}\}$ as the subring of $\mathcal{S}_m$ that consists of the polynomials over $\mathbb{Q}$ in the variables $w_{0,1}^{(\kappa)}$, $w_{0,2}^{(\lambda)}$ having $\kappa, \lambda \geq 0$. For each $\boldsymbol{P}$ in $\mathbb{Q}\{w_{0,1}, w_{0,2}\}$, $\boldsymbol{P}'$ belongs to $\mathbb{Q}\{w_{0,1}, w_{0,2}\}$. Thus, $\mathbb{Q}\{w_{0,1}, w_{0,2}\}$ is the differential subring of $\mathcal{S}_m$ generated by $w_{0,1}$ and $w_{0,2}$; e.g., see [**41**]. The polynomials of $\mathcal{S}_m$ in $\mathbb{Q}\{w_{0,1}, w_{0,2}\}$ are the differential polynomials over $\mathbb{Q}$ in the variables $w_{0,1}$, $w_{0,2}$.

PROPOSITION 5.7. *For any $\boldsymbol{P}$ in $\mathbb{Q}\{w_{0,1}, w_{0,2}\}$, $\boldsymbol{P}$ is not a relative invariant for the equations* (1.1).

PROOF. Let $\boldsymbol{P}$ be a nonconstant polynomial in $\mathbb{Q}\{w_{0,1}, w_{0,2}\}$. For an indirect argument, suppose that $\boldsymbol{P}$ is a relative invariant for the equations (1.1) having order $m \geq 2$. For any given (1.1) on Ω, we apply Propositions 5.4 and 5.3 to obtain a composite substitution that transforms the restriction of (1.1) to some subregion $\mathcal{U}$ of Ω into a corresponding equation (5.1) on a region $\mathcal{V}$ such that the coefficients of (5.1) satisfy $d_{0,1}(\zeta) \equiv d_{0,2}(\zeta) \equiv 0$ on $\mathcal{V}$. Since $\boldsymbol{P}$ in $\mathbb{Q}\{w_{0,1}, w_{0,2}\}$ is an isobaric polynomial of positive weight, we use the condition $d_{0,1}(\zeta) \equiv d_{0,2}(\zeta) \equiv 0$ to deduce that the function on $\mathcal{V}$ obtained by replacing $w_{i,j}^{(k)}$ in $\boldsymbol{P}$ with $d_{i,j}^{(k)}(\zeta)$ from (5.1) is identically 0 on $\mathcal{V}$. Properties (1.10) and (1.9) of $\boldsymbol{P}$ as a relative invariant yield $P(z) \equiv 0$, on $\mathcal{U}$, for the function $P(z)$ obtained by replacing $w_{i,j}^{(k)}$ in $\boldsymbol{P}$ with $c_{i,j}^{(k)}(z)$ from (1.1). Meromorphic continuation gives $P(z) \equiv 0$ on Ω. Since the relation $P(z) \equiv 0$ is valid for any selection of (1.1), Corollary 4.3 on page 37 provides the contradiction $\boldsymbol{P} \equiv 0$. Thus, $\boldsymbol{P}$ is not a relative invariant for the equations (1.1). This completes the proof. □

COROLLARY 5.8. *Suppose that $\boldsymbol{R}$ in $\mathcal{S}_m$ is a relative invariant for the equations* (1.1). *Then, $\boldsymbol{R}$ must effectively involve at least one of the variables $w_{i,j}^{(k)}$ having $k \geq 0$ and either $1 \leq i \leq j \leq m$ or $i = 0 < 3 \leq j \leq m$.*

PROOF. If $\boldsymbol{R}$ were to not involve any such variable, then $\boldsymbol{R}$ would be a polynomial in $\mathbb{Q}\{w_{0,1}, w_{0,2}\}$ and, due to Proposition 5.7, $\boldsymbol{R}$ would not be a relative invariant for the equations (1.1). □

CHAPTER 6

Related Laguerre-Forsyth Canonical Forms

The concept of a Laguerre-Forsyth canonical form for (1.1) was introduced in Definition 5.6 on page 51. In order to develop relations among the coefficients of two Laguerre-Forsyth canonical forms for (1.1) when one of the forms can be transformed into the other, we assume throughout this chapter that one of the forms is (1.1) and the other is (5.1) on page 47. Thus, we suppose that (1.1) on Ω satisfies $c_{0,1}(z) \equiv c_{0,2}(z) \equiv 0$ and is transformed by the composite substitution $y(z) = \rho(z) v(z)$ and $z = f(\zeta)$ with $w(\zeta) = (v \circ f)(\zeta)$ into an equation (5.1) on Ω^{**} that satisfies $d_{0,1}(z) \equiv d_{0,2}(z) \equiv 0$. For this situation, Corollary 5.2 on page 48 shows that the corresponding linear equation (5.12) satisfies $c_{0,1}(z) \equiv c_{0,2}(z) \equiv 0$ and is transformed by the same composite substitution into the linear equation (5.13) that satisfies $d_{0,1}(z) \equiv d_{0,2}(z) \equiv 0$. By applying (5.28) and (5.27) to the context of (5.12) and (5.13) for (1.1) and (5.1), we obtain the relations

$$(6.1) \quad \sum_{\nu=0}^{j} C_{j,\nu}\, d_{0,j-\nu}^{(\nu)}(\zeta) \equiv \bigl(f'(\zeta)\bigr)^{j} \sum_{\nu=0}^{j} C_{j,\nu}\, c_{0,j-\nu}^{(\nu)}\bigl(f(\zeta)\bigr), \quad \text{for } 2 \leq j \leq m,$$

where the coefficients $C_{j,\nu}$ are given by (5.27). We repeat (5.27) here as

$$(6.2) \quad C_{j,q} \equiv (-1)^{q}\, \frac{\binom{j-1}{q}}{\binom{2j-2}{q}} \binom{m-j+q}{q}, \quad \text{for } 0 \leq q \leq j \text{ and } 2 \leq j \leq m.$$

While the terms corresponding to $\nu = j$, $\nu = j-1$, and $\nu = j-2$ in the left and right members of (6.1) equal 0, the writing of (6.1) in its current unsimplified form more readily leads to its generalization in (6.23).

Our main challenge was that of finding rational numbers $A_{p,q}^{i,j}$ with which the formulas in (6.23) are valid identities that include the relations (6.1). We use

$$(6.3) \quad A_{p,q}^{i,j} \equiv (-1)^{p+q}\, \frac{\binom{i-1}{p}\binom{j-1}{q}}{\binom{p+q}{q}\binom{2i+2j-2}{p+q}} \binom{m-i+p}{p}\binom{m-j+q}{q},$$

$$\text{for } 0 \leq i,\, j \leq m,\ 2 \leq i+j,\ 0 \leq p \leq i, \text{ and } 0 \leq q \leq j,$$

and (6.2) to see that $A_{0,q}^{0,j} \equiv C_{j,q}$, when $i = 0$ and $2 \leq j \leq m$. Our discovery that (6.3) is suitable for (6.23) occurred after Propositions 6.1 and 6.2 were available to simplify the expression obtained by subtracting one side of (6.23) from the other. Then, guided by the form of (6.2) for $A_{0,q}^{0,j}$, we substituted various guessed formulas for $A_{p,q}^{i,j}$ in those expressions and used machine computations to arrive at (6.3).

Arguments in Sections 7.3 and 9.3 require a less-restricted formula than (6.3). For this reason, we define $A_{p,q}^{i,j}$ by

$$(6.4) \quad A_{p,q}^{i,j} \equiv (-1)^{p+q} \frac{\left[\prod_{k=1}^{p}(i-k)\right]\left[\prod_{k=1}^{q}(j-k)\right]}{\left[\prod_{k=1}^{p+q}(2i+2j-1-k)\right]} \binom{m-i+p}{p}\binom{m-j+q}{q},$$

for any integers i, j, p, q such that $p+q \leq 2i+2j-2$.

If i, j, p, q satisfy the conditions of (6.3), then they satisfy the conditions of (6.4) and we see that the corresponding $A_{p,q}^{i,j}$ of (6.4) is also given by (6.3).

Throughout this chapter, we shall use the integers defined by

$$(6.5) \quad B_{p,q}^{i,j} \equiv p!\, q! \binom{i-1}{p}\binom{j-1}{q}\binom{m-i+p}{p}\binom{m-j+q}{q},$$

for $0 \leq p \leq i$ and $0 \leq q \leq j$.

6.1. Two Laguerre-Forsyth forms related by a transformation

Propositions 6.1 and 6.2 provide details for the situation where a substitution transforms a given Laguerre-Forsyth canonical form into a second Laguerre-Forsyth canonical form.

PROPOSITION 6.1. *For $m \geq 2$, suppose that (1.1) has $c_{0,1}(z) \equiv c_{0,2}(z) \equiv 0$ and suppose that the substitutions $y(z) = \rho(z)\, v(z)$, $z = f(\zeta)$, and $w(\zeta) = (v \circ f)(\zeta)$ transform (1.1) on Ω into (5.1) on Ω^{**} subject to $d_{0,1}(\zeta) \equiv d_{0,2}(\zeta) \equiv 0$. Then, either there are complex numbers $a \neq 0$, $b \neq 0$, z_0 such that*

$$(6.6) \quad \rho(z) \equiv a, \quad \text{on } \Omega, \quad f(\zeta) \equiv b\zeta + z_0, \quad \text{on } \Omega^{**},$$

and

$$(6.7) \quad d_{i,j}(\zeta) \equiv \bigl(f'(\zeta)\bigr)^{i+j} c_{i,j}\bigl(f(\zeta)\bigr), \quad \text{for } 0 \leq i, j \leq m,$$

or else there are complex numbers $a \neq 0$, $b \neq 0$, z_0, ζ_0 such that

$$(6.8) \quad \rho(z) \equiv a\, (z-z_0)^{m-1}, \quad \text{on } \Omega, \quad f(\zeta) \equiv \frac{b}{\zeta-\zeta_0} + z_0, \quad \text{on } \Omega^{**},$$

and

$$(6.9) \quad d_{i,j}(\zeta) \equiv \bigl(f'(\zeta)\bigr)^{i+j} \sum_{p=0}^{i}\sum_{q=0}^{j} B_{p,q}^{i,j} \left[\frac{\zeta-\zeta_0}{b}\right]^{p+q} c_{i-p,j-q}\bigl(f(\zeta)\bigr),$$

for $0 \leq i, j \leq m$,

where $B_{p,q}^{i,j}$ is defined by (6.5).

PROOF. We use $c_{0,1}(z) \equiv c_{0,2}(z) \equiv 0$, (1.16), (1.15), and (5.16) to obtain $b_{m,1}(z) \equiv b_{m,2}(z) \equiv 0$ and see that the function $\rho(z) \not\equiv 0$ must satisfy

$$(6.10) \quad \left(\frac{\rho'(z)}{\rho(z)}\right)' + \frac{1}{m-1}\left(\frac{\rho'(z)}{\rho(z)}\right)^2 \equiv 0.$$

6.1. TWO LAGUERRE-FORSYTH FORMS RELATED BY A TRANSFORMATION

Case 1. Suppose $\rho'(z) \equiv 0$. Then, (5.17) requires $g''(z) \equiv 0$ for the univalent function $g(z)$. Thus, there are complex numbers $a \neq 0$, $b \neq 0$, and z_0 such that $\rho(z) \equiv a$ on Ω, $\zeta = g(z) = (1/b)(z - z_0)$ on Ω, and $z = f(\zeta) = b\zeta + z_0$ on Ω^{**}. For this, (5.4)-(5.5) yield $\phi_{0,j}(z) \equiv 1$, for any j, as well as $\phi_{i,j}(z) \equiv 0$, for $i \geq 1$ and any j. Hence, (5.3) gives

$$d_{i,j}(\zeta) \equiv (f'(\zeta))^{i+j} \phi_{0,i}(f(\zeta)) \phi_{0,j}(f(\zeta)) c_{i,j}(f(\zeta)) \equiv (f'(\zeta))^{i+j} c_{i,j}(f(\zeta)),$$

on Ω^{**} for $0 \leq i, j \leq m$. Consequently, for Case 1, (6.6) and (6.7) are satisfied.

Case 2. Suppose that $\rho'(z) \not\equiv 0$. Then, (6.10) shows that the function $\tau(z)$ defined on Ω by $\tau(z) = \rho'(z)/\rho(z)$ satisfies

$$\tau(z) \neq 0 \quad \text{and} \quad -\frac{\tau'(z)}{(\tau(z))^2} = \frac{1}{m-1}, \quad \text{on } \Omega.$$

Hence, there are complex numbers $a \neq 0$ and z_0 such that

$$(6.11) \qquad \tau(z) = \frac{m-1}{z - z_0} \quad \text{and} \quad \rho(z) = a(z - z_0)^{m-1}, \quad \text{on } \Omega.$$

We use this and (5.17) to deduce that $g''(z)/g'(z) \equiv -2/(z - z_0)$, on Ω. Thus, there are complex numbers $b \neq 0$ and ζ_0 such that $g'(z) \equiv -b/(z - z_0)^2$,

$$\zeta = g(z) = \frac{b}{z - z_0} + \zeta_0, \quad \text{on } \Omega, \text{ and} \quad z = f(\zeta) = \frac{b}{\zeta - \zeta_0} + z_0, \quad \text{on } \Omega^{**}.$$

This and (6.11) give (6.8). For the use of (5.5) here, we have

$$(6.12) \qquad \frac{\rho'(z)}{\rho(z)} + (m - i - k)\frac{g''(z)}{g'(z)} \equiv \frac{2i + 2k - m - 1}{z - z_0}, \quad \text{on } \Omega.$$

To apply Theorem 5.1 on page 47, we first use (5.4) to see that the formula

$$(6.13) \qquad \phi_{i,j}(z) \equiv i! \binom{i+j-1}{i}\binom{m-j}{i}\frac{1}{(z-z_0)^i}, \quad \text{on } \Omega \text{ for } j \leq m,$$

is correct for $i = 0$. Suppose that i is a nonnegative integer for which (6.13) is valid. Then, we find that (5.5), (6.13), and (6.12) yield

$$(6.14) \qquad \phi_{i+1,j}(z) \equiv \sum_{k=j+1}^{m} \left[\phi_{i,k}^{(1)}(z) + \left[\frac{\rho'(z)}{\rho(z)} + (m-i-k)\frac{g''(z)}{g'(z)}\right] \phi_{i,k}(z)\right]$$

$$\equiv \frac{i!}{(z-z_0)^{i+1}} \sum_{k=j+1}^{m} \left[\binom{i+k-1}{i}\binom{m-k}{i}(i + 2k - m - 1)\right]$$

For the fixed integer $i \geq 0$, we set

$$(6.15) \qquad G_k \equiv (i+1)\binom{i+k-1}{i+1}\binom{m-k+1}{i+1}, \quad \text{for any } k,$$

and observe that

$$G_k - G_{k+1} \equiv (i+1)\left[\binom{i+k-1}{i+1}\binom{m-k+1}{i+1} - \binom{i+k}{i+1}\binom{m-k}{i+1}\right]$$

$$\equiv \binom{i+k-1}{i}\binom{m-k}{i}\left[\frac{(k-1)(m-k+1) - (i+k)(m-k-i)}{i+1}\right]$$

$$\equiv \binom{i+k-1}{i}\binom{m-k}{i}[i + 2k - m - 1], \quad \text{for any } k.$$

We combine the preceding formula with (6.14) and (6.15) to verify that

$$\phi_{i+1,j}(z) \equiv \frac{i!}{(z-z_0)^{i+1}} \sum_{k=j+1}^{m} [G_k - G_{k+1}] \equiv \frac{i!}{(z-z_0)^{i+1}} [G_{j+1} - G_{m+1}]$$
$$\equiv (i+1)! \binom{i+j}{i+1} \binom{m-j}{i+1} \frac{1}{(z-z_0)^{i+1}}.$$

Thus, (6.13) is valid for each $i \geq 0$. We use (5.3), (6.13), (6.5), and (6.8) to obtain

$$d_{i,j}(\zeta) \equiv (f'(\zeta))^{i+j} \sum_{p=0}^{i} \sum_{q=0}^{j} \phi_{i-p,p}(f(\zeta)) \phi_{j-q,q}(f(\zeta)) c_{p,q}(f(\zeta))$$
$$\equiv (f'(\zeta))^{i+j} \sum_{p=0}^{i} \sum_{q=0}^{j} \phi_{p,i-p}(f(\zeta)) \phi_{q,j-q}(f(\zeta)) c_{i-p,j-q}(f(\zeta))$$
$$\equiv (f'(\zeta))^{i+j} \sum_{p=0}^{i} \sum_{q=0}^{j} B_{p,q}^{i,j} \frac{1}{(f(\zeta)-z_0)^{p+q}} c_{i-p,j-q}(f(\zeta))$$
$$\equiv (f'(\zeta))^{i+j} \sum_{p=0}^{i} \sum_{q=0}^{j} B_{p,q}^{i,j} \left[\frac{\zeta-\zeta_0}{b}\right]^{p+q} c_{i-p,j-q}(f(\zeta)).$$

This yields (6.9) and completes the proof of Proposition 6.1. □

We define integers $D_{p,q,r}^{i,j,k}$ in terms of (6.5) through

(6.16) $\quad D_{p,q,r}^{i,j,k} \equiv \binom{k}{r} \left[\prod_{t=1}^{r}(2i+2j-p-q+k-t)\right] B_{p,q}^{i,j},$

for $0 \leq p \leq i$, $0 \leq q \leq j$, and any r, k.

PROPOSITION 6.2. *The kth derivative of $d_{i,j}(\zeta)$ in (6.9) is given on Ω^{**} in terms of (6.8) and (6.16) by*

(6.17) $\quad d_{i,j}^{(k)}(\zeta) \equiv (f'(\zeta))^{i+j+k} \sum_{p=0}^{i} \sum_{q=0}^{j} \sum_{r=0}^{k} D_{p,q,r}^{i,j,k} \left[\frac{\zeta-\zeta_0}{b}\right]^{p+q+r} c_{i-p,j-q}^{(k-r)}(f(\zeta)),$

for $0 \leq i, j \leq m$ and $k \geq 0$.

PROOF. Formula (6.9) of page 54 shows that (6.17) is true for $k = 0$. Suppose that k is a positive integer such that (6.17) is valid when k is replaced by $k-1$. In terms of $h = i+j+k$, $T(\zeta) \equiv (\zeta-\zeta_0)/b$, and $d = p+q$, we have

$$d_{i,j}^{(k-1)}(\zeta) \equiv (f'(\zeta))^{h-1} \sum_{p=0}^{i} \sum_{q=0}^{j} \sum_{r=0}^{k-1} D_{p,q,r}^{i,j,k-1} [T(\zeta)]^{d+r} c_{i-p,j-q}^{(k-1-r)}(f(\zeta)).$$

6.1. TWO LAGUERRE-FORSYTH FORMS RELATED BY A TRANSFORMATION

After differentiating this expression with respect to ζ, we use $T'(\zeta) \equiv 1/b$ as well as $1/f'(\zeta) \equiv (-b)\left[T(\zeta)\right]^2$ and $f''(\zeta)/\left(f'(\zeta)\right)^2 \equiv 2T(\zeta)$ to deduce that

$$d_{i,j}^{(k)}(\zeta) \equiv \left(f'(\zeta)\right)^h \sum_{p=0}^{i} \sum_{q=0}^{j} \sum_{r=0}^{k-1} D_{p,q,r}^{i,j,k-1} \left[T(\zeta)\right]^{d+r} c_{i-p,j-q}^{(k-r)}\!\left(f(\zeta)\right)$$

$$+ \left(f'(\zeta)\right)^{h-1} \sum_{p=0}^{i} \sum_{q=0}^{j} \sum_{r=0}^{k-1} D_{p,q,r}^{i,j,k-1} (d+r) \left[T(\zeta)\right]^{d+r-1} \frac{1}{b} c_{i-p,j-q}^{(k-1-r)}\!\left(f(\zeta)\right)$$

$$+ (h-1)\left(f'(\zeta)\right)^h \frac{f''(\zeta)}{\left(f'(\zeta)\right)^2} \sum_{p=0}^{i} \sum_{q=0}^{j} \sum_{r=0}^{k-1} D_{p,q,r}^{i,j,k-1} \left[T(\zeta)\right]^{d+r} c_{i-p,j-q}^{(k-1-r)}\!\left(f(\zeta)\right)$$

$$\equiv \left(f'(\zeta)\right)^h \sum_{p=0}^{i} \sum_{q=0}^{j} \sum_{r=0}^{k-1} D_{p,q,r}^{i,j,k-1} \left[T(\zeta)\right]^{d+r} c_{i-p,j-q}^{(k-r)}\!\left(f(\zeta)\right)$$

$$+ \left(f'(\zeta)\right)^h \sum_{p=0}^{i} \sum_{q=0}^{j} \sum_{r=0}^{k-1} (-d-r) D_{p,q,r}^{i,j,k-1} \left[T(\zeta)\right]^{d+r+1} c_{i-p,j-q}^{(k-1-r)}\!\left(f(\zeta)\right)$$

$$+ \left(f'(\zeta)\right)^h \sum_{p=0}^{i} \sum_{q=0}^{j} \sum_{r=0}^{k-1} (2h-2) D_{p,q,r}^{i,j,k-1} \left[T(\zeta)\right]^{d+r+1} c_{i-p,j-q}^{(k-1-r)}\!\left(f(\zeta)\right)$$

$$\equiv \left(f'(\zeta)\right)^h \sum_{p=0}^{i} \sum_{q=0}^{j} \sum_{r=0}^{k-1} D_{p,q,r}^{i,j,k-1} \left[T(\zeta)\right]^{d+r} c_{i-p,j-q}^{(k-r)}\!\left(f(\zeta)\right)$$

$$+ \left(f'(\zeta)\right)^h \sum_{p=0}^{i} \sum_{q=0}^{j} \sum_{r=1}^{k} (2h-d-r-1) D_{p,q,r-1}^{i,j,k-1} \left[T(\zeta)\right]^{d+r} c_{i-p,j-q}^{(k-r)}\!\left(f(\zeta)\right).$$

Since (6.16) yields $D_{p,q,r}^{i,j,k-1} = 0$, for $r = k$, and $D_{p,q,r-1}^{i,j,k-1} = 0$, for $r = 0$, we see that

$$(6.18) \qquad d_{i,j}^{(k)}(\zeta) \equiv \left(f'(\zeta)\right)^{i+j+k} \sum_{p=0}^{i} \sum_{q=0}^{j} \sum_{r=0}^{k} E_{p,q,r}^{i,j,k} \left[T(\zeta)\right]^{p+q+r} c_{i-p,j-q}^{(k-r)}\!\left(f(\zeta)\right),$$

where

$$(6.19) \qquad E_{p,q,r}^{i,j,k} \equiv D_{p,q,r}^{i,j,k-1} + (2i+2j-p-q+2k-1-r) D_{p,q,r-1}^{i,j,k-1},$$

when $0 \leq p \leq i$, $0 \leq q \leq j$, and $0 \leq r \leq k$. We set $v = 2i+2j-p-q$ and then combine (6.19) with (6.16) to establish that: for each r satisfying $1 \leq r \leq k$,

$$(6.20) \qquad E_{p,q,r}^{i,j,k} \equiv \binom{k-1}{r} \left[\prod_{t=1}^{r} (v+(k-1)-t)\right] B_{p,q}^{i,j}$$

$$+ (v+2k-1-r) \binom{k-1}{r-1} \left[\prod_{t=1}^{r-1} (v+(k-1)-t)\right] B_{p,q}^{i,j}$$

$$\equiv R \left[\prod_{t=1}^{r-1} (2i+2j-p-q+k-1-t)\right] B_{p,q}^{i,j},$$

where
$$(6.21) \quad R \equiv \binom{k-1}{r}(v+k-1-r) + \binom{k-1}{r-1}(v+2k-1-r)$$
$$\equiv \binom{k}{r}(v+k-1-r) + \binom{k-1}{r-1}k \equiv \binom{k}{r}(v+k-1-r) + \binom{k}{r}r$$
$$\equiv \binom{k}{r}(2i+2j-p-q+k-1).$$

Thus, for $1 \leq r \leq k$, we find that (6.20), (6.21), and (6.16) yield
$$(6.22) \quad E_{p,q,r}^{i,j,k} \equiv \binom{k}{r}\left[\prod_{t=0}^{r-1}(2i+2j-p-q+k-1-t)\right]B_{p,q}^{i,j}$$
$$\equiv \binom{k}{r}\left[\prod_{t=1}^{r}(2i+2j-p-q+k-t)\right]B_{p,q}^{i,j} \equiv D_{p,q,r}^{i,j,k}.$$

For $r=0$, we apply (6.19) and (6.16) to verify that
$$E_{p,q,0}^{i,j,k} \equiv D_{p,q,0}^{i,j,k-1} \equiv B_{p,q}^{i,j} \equiv D_{p,q,0}^{i,j,k}.$$

Consequently, (6.22) is true for $0 \leq r \leq k$. By replacing $E_{p,q,r}^{i,j,k}$ in (6.18) with $D_{p,q,r}^{i,j,k}$, we conclude that (6.17) is valid for any $k \geq 0$. This completes the proof. □

6.2. Identities for the coefficients of related canonical forms

We assume that i and j are fixed integers that satisfy $0 \leq i, j \leq m$ as well as $2 \leq i+j$. Machine computations and the form of (6.2) for (6.1) led us to the formulas (6.3) and (6.4) for the rational numbers $A_{p,q}^{i,j}$.

THEOREM 6.3. *For $m \geq 2$, suppose that (1.1) satisfies $c_{0,1}(z) \equiv c_{0,2}(z) \equiv 0$ and suppose that the substitutions $y(z) = \rho(z)\,v(z)$, $z = f(\zeta)$, and $w(\zeta) = (v \circ f)(\zeta)$ transform (1.1) on Ω into (5.1) on Ω^{**} subject to $d_{0,1}(\zeta) \equiv d_{0,2}(\zeta) \equiv 0$. Then, along with $d_{\nu,\mu}(\zeta) \equiv d_{\mu,\nu}(\zeta)$ and $c_{\nu,\mu}(z) \equiv c_{\mu,\nu}(z)$, for $0 \leq \mu, \nu \leq m$, the coefficients of (5.1) and (1.1) satisfy*

$$(6.23) \quad \sum_{\mu=0}^{i}\sum_{\nu=0}^{j} A_{\mu,\nu}^{i,j}\, d_{i-\mu,j-\nu}^{(\mu+\nu)}(\zeta) \equiv \bigl(f'(\zeta)\bigr)^{i+j}\sum_{\mu=0}^{i}\sum_{\nu=0}^{j} A_{\mu,\nu}^{i,j}\, c_{i-\mu,j-\nu}^{(\mu+\nu)}\bigl(f(\zeta)\bigr),$$

on Ω^{**} for $0 \leq i, j \leq m$ and $2 \leq i+j$.

OBSERVATION. For $i=0$ and $2 \leq j \leq m$, we note that (6.3) and (6.2) yield $A_{0,q}^{0,j} \equiv C_{j,q}$. Thus, (6.23) includes (6.1).

PROOF OF THEOREM 6.3. We use Proposition 6.1 on page 54 to see that the transformation of (1.1) into (5.1) must be given by either (6.6) or (6.8). If it is given by (6.6), then (6.6) and (6.7) yield

$$(6.24) \quad d_{i,j}^{(k)}(\zeta) \equiv \bigl(f'(\zeta)\bigr)^{i+j+k} c_{i,j}^{(k)}\bigl(f(\zeta)\bigr), \quad \text{for } 0 \leq i, j \leq m \text{ and } k \geq 0.$$

We use (6.24) to check that (6.23) is valid and, in this situation, does not depend on the particular selection from $\mathbb{Q}$ of the constants $A_{\mu,\nu}^{i,j}$ in (6.3).

6.2. IDENTITIES FOR THE COEFFICIENTS OF RELATED CANONICAL FORMS

Suppose that the transformation of (1.1) into (5.1) is given by (6.8) of page 54. We find that (6.17) yields

(6.25) $d^{(\mu+\nu)}_{i-\mu,j-\nu}(\zeta)$
$$\equiv \left(f'(\zeta)\right)^{i+j} \sum_{p=0}^{i-\mu} \sum_{q=0}^{j-\nu} \sum_{r=0}^{\mu+\nu} D^{i-\mu,j-\nu,\mu+\nu}_{p,q,r} \left[\frac{\zeta-\zeta_0}{b}\right]^{p+q+r} c^{(\mu+\nu-r)}_{i-\mu-p,j-\nu-q}(f(\zeta)),$$

$$\text{for } 0 \le \mu \le i \text{ and } 0 \le \nu \le j.$$

Letting $I^{\sharp}_{i,j}(\zeta)$ denote the left member of (6.23), we apply (6.25) to obtain

(6.26) $$I^{\sharp}_{i,j}(\zeta) \equiv \left(f'(\zeta)\right)^{i+j} \sum_{\mu=0}^{i} \sum_{\nu=0}^{j} \sum_{p=0}^{i-\mu} \sum_{q=0}^{j-\nu} \sum_{r=0}^{\mu+\nu} T^{i,j,\mu,\nu}_{p,q,r}(\zeta),$$

where $T^{i,j,\mu,\nu}_{p,q,r}(\zeta)$ is defined in terms of (6.3) and (6.16) by

(6.27) $$T^{i,j,\mu,\nu}_{p,q,r}(\zeta) \equiv A^{i,j}_{\mu,\nu} D^{i-\mu,j-\nu,\mu+\nu}_{p,q,r} \left[\frac{\zeta-\zeta_0}{b}\right]^{p+q+r} c^{(\mu+\nu-r)}_{i-\mu-p,j-\nu-q}(f(\zeta)),$$

for $0 \le \mu \le i$, $0 \le \nu \le j$, $0 \le p \le i-\mu$, $0 \le q \le j-\nu$ and $0 \le r \le \mu+\nu$.

Using (6.26) and (6.27), we introduce $\kappa = \mu + p$ followed by $\lambda = \nu + q$ and then $s = r + p + q$ to verify that

(6.28)
$$\frac{I^{\sharp}_{i,j}(\zeta)}{(f'(\zeta))^{i+j}} \equiv \sum_{\mu=0}^{i} \sum_{p=0}^{i-\mu} \sum_{\nu=0}^{j} \sum_{q=0}^{j-\nu} \sum_{r=0}^{\mu+\nu} T^{i,j,\mu,\nu}_{p,q,r}(\zeta)$$

$$\equiv \sum_{\kappa=0}^{i} \sum_{p=0}^{\kappa} \sum_{\nu=0}^{j} \sum_{q=0}^{j-\nu} \sum_{r=0}^{\kappa-p+\nu} T^{i,j,\kappa-p,\nu}_{p,q,r}(\zeta)$$

$$\equiv \sum_{\kappa=0}^{i} \sum_{p=0}^{\kappa} \sum_{\lambda=0}^{j} \sum_{q=0}^{\lambda} \sum_{r=0}^{\kappa-p+\lambda-q} T^{i,j,\kappa-p,\lambda-q}_{p,q,r}(\zeta)$$

$$\equiv \sum_{\kappa=0}^{i} \sum_{\lambda=0}^{j} \sum_{p=0}^{\kappa} \sum_{q=0}^{\lambda} \sum_{s=p+q}^{\kappa+\lambda} T^{i,j,\kappa-p,\lambda-q}_{p,q,s-p-q}(\zeta).$$

We note that the right member of (6.23) is equal to $\left(f'(\zeta)\right)^{i+j} I_{i,j}(f(\zeta))$, where $I_{i,j}(z)$ is defined by

(6.29) $$I_{i,j}(z) \equiv \sum_{\mu=0}^{i} \sum_{\nu=0}^{j} A^{i,j}_{\mu,\nu} c^{(\mu+\nu)}_{i-\mu,j-\nu}(z), \quad \text{on } \Omega.$$

Formulas (6.27), (6.16), (6.5), and (6.29) yield

(6.30) $$\sum_{\kappa=0}^{i} \sum_{\lambda=0}^{j} T^{i,j,\kappa,\lambda}_{0,0,0}(\zeta) \equiv \sum_{\kappa=0}^{i} \sum_{\lambda=0}^{j} A^{i,j}_{\kappa,\lambda} c^{(\kappa+\lambda)}_{i-\kappa,j-\lambda}(f(\zeta)) \equiv I_{i,j}(f(\zeta)).$$

In view of (6.28) and (6.30), we have

(6.31) $$\frac{I^{\sharp}_{i,j}(\zeta)}{(f'(\zeta))^{i+j}} - I_{i,j}(f(\zeta)) \equiv \sum_{\kappa=0}^{i} \sum_{\lambda=0}^{j} \Phi_{\kappa,\lambda}(\zeta),$$

where

$$(6.32) \quad \Phi_{\kappa,\lambda}(\zeta) \equiv \left[\sum_{p=0}^{\kappa} \sum_{q=0}^{\lambda} \sum_{s=p+q}^{\kappa+\lambda} T_{p,q,s-p-q}^{i,j,\kappa-p,\lambda-q}(\zeta) \right] - T_{0,0,0}^{i,j,\kappa,\lambda}(\zeta),$$

for $0 \leq \kappa \leq i$ and $0 \leq \lambda \leq j$.

We shall establish (6.23) by proving that $\Phi_{\kappa,\lambda}(\zeta) \equiv 0$, for each κ and λ; then, the right member of (6.31) is identically zero. To accomplish this, more information is needed about $T_{p,q,s-p-q}^{i,j,\kappa-p,\lambda-q}(\zeta)$. For later reference, we define $H_{p,q,s}^{i,j,\kappa,\lambda}$ in $\mathbb{Q}$ by

$$(6.33) \quad H_{p,q,s}^{i,j,\kappa,\lambda} \equiv (-1)^{p+q} \binom{\kappa}{p} \binom{\lambda}{q} \frac{\binom{2i+2j-\kappa-\lambda-1}{s-p-q}}{\binom{2i+2j-2}{\kappa+\lambda-p-q}},$$

when $0 \leq p$, $0 \leq q$, $0 \leq \kappa \leq i$, $0 \leq \lambda \leq j$, and $p+q \leq s \leq \kappa + \lambda$.

With $i + j \geq 2$, the restrictions on κ, λ, p, q, s in (6.33) gives $2i + 2j - 2 \geq 2$ and

$$0 \leq \kappa + \lambda - s \leq \kappa + \lambda - p - q \leq \kappa + \lambda \leq i + j \leq 2i + 2j - 2$$

so that the binomial coefficient in the denominator of (6.33) is nonzero. We note that $H_{p,q,s}^{i,j,\kappa,\lambda}$ is defined by (6.33) whenever the integers κ, λ, p, q, s satisfy the more restrictive condition

$$(6.34) \quad 0 \leq p \leq \kappa \leq i, \quad 0 \leq q \leq \lambda \leq j, \quad \text{and} \quad p + q \leq s \leq \kappa + \lambda.$$

For any integers κ, λ, p, q, and s that satisfy (6.34), we find that (6.27) yields

$$(6.35) \quad T_{p,q,s-p-q}^{i,j,\kappa-p,\lambda-q}(\zeta) \equiv \left[Q_{\kappa,\lambda,s}^{i,j}(\zeta) \right] \left[A_{\kappa-p,\lambda-q}^{i,j} \, D_{p,q,s-p-q}^{i-\kappa+p,j-\lambda+q,\kappa+\lambda-p-q} \right]$$

in terms of (6.3), (6.16), and

$$(6.36) \quad Q_{\kappa,\lambda,s}^{i,j}(\zeta) \equiv \left[\frac{\zeta - \zeta_0}{b} \right]^s c_{i-\kappa,j-\lambda}^{(\kappa+\lambda-s)}(f(\zeta)).$$

For any integers κ, λ, p, q, and s that satisfy (6.34), we use the identities

$$\binom{i-1}{\kappa-p}\binom{i-\kappa+p-1}{p} \equiv \binom{i-1}{\kappa}\binom{\kappa}{p},$$

$$\binom{j-1}{\lambda-q}\binom{j-\lambda+q-1}{q} \equiv \binom{j-1}{\lambda}\binom{\lambda}{q},$$

$$\binom{m-i+\kappa}{p}\binom{m-i+\kappa-p}{\kappa-p} \equiv \binom{m-i+\kappa}{\kappa}\binom{\kappa}{p},$$

$$\binom{m-j+\lambda}{q}\binom{m-j+\lambda-q}{\lambda-q} \equiv \binom{m-j+\lambda}{\lambda}\binom{\lambda}{q},$$

6.2. IDENTITIES FOR THE COEFFICIENTS OF RELATED CANONICAL FORMS

and

$$\binom{\kappa}{p}\binom{\lambda}{q}\frac{\binom{\kappa+\lambda-p-q}{s-p-q}(s-p-q)!}{\binom{\kappa+\lambda-p-q}{\lambda-q}} p!\, q! \equiv \binom{\kappa}{p}\binom{\lambda}{q}\frac{(\kappa-p)!\,(\lambda-q)!}{(\kappa+\lambda-s)!} p!\, q!$$

$$\equiv \frac{\kappa!\,\lambda!}{(\kappa+\lambda-s)!}$$

as well as (6.3), (6.16), (6.5), and (6.33) to verify that

(6.37) $\quad A^{i,j}_{\kappa-p,\lambda-q} D^{i-\kappa+p,j-\lambda+q,\kappa+\lambda-p-q}_{p,q,s-p-q}$

$$\equiv \frac{(-1)^{\kappa+\lambda-p-q}\binom{i-1}{\kappa-p}\binom{j-1}{\lambda-q}}{\binom{\kappa+\lambda-p-q}{\lambda-q}\binom{2i+2j-2}{\kappa+\lambda-p-q}}\binom{m-i+\kappa-p}{\kappa-p}\binom{m-j+\lambda-q}{\lambda-q}$$

$$\times \binom{\kappa+\lambda-p-q}{s-p-q}(s-p-q)!\binom{2i+2j-\kappa-\lambda-1}{s-p-q}$$

$$\times p!\,q!\binom{i-\kappa+p-1}{p}\binom{j-\lambda+q-1}{q}\binom{m-i+\kappa}{p}\binom{m-j+\lambda}{q}$$

$$\equiv (-1)^{\kappa+\lambda+p+q}\binom{i-1}{\kappa}\binom{\kappa}{p}\binom{j-1}{\lambda}\binom{\lambda}{q}\binom{m-i+\kappa}{\kappa}\binom{\kappa}{p}\binom{m-j+\lambda}{\lambda}\binom{\lambda}{q}$$

$$\times \frac{\binom{\kappa+\lambda-p-q}{s-p-q}(s-p-q)!\binom{2i+2j-\kappa-\lambda-1}{s-p-q}p!\,q!}{\binom{\kappa+\lambda-p-q}{\lambda-q}\binom{2i+2j-2}{\kappa+\lambda-p-q}}$$

$$\equiv (-1)^{\kappa+\lambda} H^{i,j,\kappa,\lambda}_{p,q,s}\binom{\kappa}{p}\binom{\lambda}{q}\frac{\binom{\kappa+\lambda-p-q}{s-p-q}(s-p-q)!}{\binom{\kappa+\lambda-p-q}{\lambda-q}}p!\,q!$$

$$\times \binom{i-1}{\kappa}\binom{j-1}{\lambda}\binom{m-i+\kappa}{\kappa}\binom{m-j+\lambda}{\lambda}$$

$$\equiv \frac{(-1)^{\kappa+\lambda}\kappa!\,\lambda!}{(\kappa+\lambda-s)!}\binom{i-1}{\kappa}\binom{j-1}{\lambda}\binom{m-i+\kappa}{\kappa}\binom{m-j+\lambda}{\lambda} H^{i,j,\kappa,\lambda}_{p,q,s}.$$

After setting

(6.38) $\quad R^{i,j}_{\kappa,\lambda,s} \equiv \dfrac{(-1)^{\kappa+\lambda}\kappa!\,\lambda!}{(\kappa+\lambda-s)!}\binom{i-1}{\kappa}\binom{j-1}{\lambda}\binom{m-i+\kappa}{\kappa}\binom{m-j+\lambda}{\lambda},$

we introduce
$$P_{\kappa,\lambda,s}^{i,j}(\zeta) \equiv Q_{\kappa,\lambda,s}^{i,j}(\zeta) \, R_{\kappa,\lambda,s}^{i,j} \tag{6.39}$$

in terms of (6.36) and (6.38). Due to (6.35)–(6.39), we have
$$T_{p,q,s-p-q}^{i,j,\kappa-p,\lambda-q}(\zeta) \equiv P_{\kappa,\lambda,s}^{i,j}(\zeta) \, H_{p,q,s}^{i,j,\kappa,\lambda}, \tag{6.40}$$

for all integers i, j, κ, λ, p, q, and s that satisfy (6.34). Note that $P_{\kappa,\lambda,s}^{i,j}(\zeta)$ in (6.40) does not involve either p or q.

Let κ and λ be fixed integers subject to $0 \leq \kappa \leq i$ and $0 \leq \lambda \leq j$. Because (6.32) is symmetric with respect to an interchange of i, κ, and p with j, λ, and q, *we henceforth assume that $\kappa \leq \lambda$*. After introducing the abbreviation
$$x_{p,q,s} \equiv P_{\kappa,\lambda,s}^{i,j}(\zeta) \, H_{p,q,s}^{i,j,\kappa,\lambda}, \tag{6.41}$$

we use (6.32), (6.40), and (6.41) to obtain
$$\Phi_{\kappa,\lambda}(\zeta) \equiv \sum_{p=0}^{\kappa}\left[\sum_{s=p}^{p+\lambda}\sum_{q=0}^{s-p} x_{p,q,s} + \sum_{s=p+\lambda+1}^{\kappa+\lambda}\sum_{q=0}^{\lambda} x_{p,q,s}\right] - x_{0,0,0} \equiv S_1 + S_2, \tag{6.42}$$

where
$$S_1 \equiv \left[\sum_{p=0}^{\kappa}\sum_{s=p}^{p+\lambda}\sum_{q=0}^{s-p} x_{p,q,s}\right] - x_{0,0,0} \tag{6.43}$$
$$\equiv \sum_{s=0}^{\kappa}\sum_{p=0}^{s}\sum_{q=0}^{s-p} x_{p,q,s} + \sum_{s=\kappa+1}^{\lambda}\sum_{p=0}^{\kappa}\sum_{q=0}^{s-p} x_{p,q,s} + \sum_{s=\lambda+1}^{\kappa+\lambda}\sum_{p=s-\lambda}^{\kappa}\sum_{q=0}^{s-p} x_{p,q,s}$$
$$\quad - x_{0,0,0}$$
$$\equiv \sum_{s=1}^{\kappa}\sum_{p=0}^{s}\sum_{q=0}^{s-p} x_{p,q,s} + \sum_{s=\kappa+1}^{\lambda}\sum_{p=0}^{\kappa}\sum_{q=0}^{s-p} x_{p,q,s} + \sum_{s=\lambda+1}^{\kappa+\lambda}\sum_{p=s-\lambda}^{\kappa}\sum_{q=0}^{s-p} x_{p,q,s}$$

and
$$S_2 \equiv \sum_{p=0}^{\kappa}\sum_{s=p+\lambda+1}^{\kappa+\lambda}\sum_{q=0}^{\lambda} x_{p,q,s} \equiv \sum_{s=\lambda+1}^{\kappa+\lambda}\sum_{p=0}^{s-\lambda-1}\sum_{q=0}^{\lambda} x_{p,q,s}. \tag{6.44}$$

By employing (6.41) and (6.40), we rewrite (6.43) and (6.44) as
$$S_1 \equiv \left[\sum_{s=1}^{\kappa} P_{\kappa,\lambda,s}^{i,j}(\zeta) \sum_{p=0}^{s}\sum_{q=0}^{s-p} H_{p,q,s}^{i,j,\kappa,\lambda}\right] + \left[\sum_{s=\kappa+1}^{\lambda} P_{\kappa,\lambda,s}^{i,j}(\zeta) \sum_{p=0}^{\kappa}\sum_{q=0}^{s-p} H_{p,q,s}^{i,j,\kappa,\lambda}\right] \tag{6.45}$$
$$\quad + \left[\sum_{s=\lambda+1}^{\kappa+\lambda} P_{\kappa,\lambda,s}^{i,j}(\zeta) \sum_{p=s-\lambda}^{\kappa}\sum_{q=0}^{s-p} H_{p,q,s}^{i,j,\kappa,\lambda}\right]$$

and
$$S_2 \equiv \left[\sum_{s=\lambda+1}^{\kappa+\lambda} P_{\kappa,\lambda,s}^{i,j}(\zeta) \sum_{p=0}^{s-\lambda-1}\sum_{q=0}^{\lambda} H_{p,q,s}^{i,j,\kappa,\lambda}\right]. \tag{6.46}$$

6.2. IDENTITIES FOR THE COEFFICIENTS OF RELATED CANONICAL FORMS

We shall relate S_1 and S_2 to the expressions

$$(6.47) \qquad \mathcal{M}_s \equiv \sum_{p=0}^{s} \sum_{q=0}^{s-p} H_{p,q,s}^{i,j,\kappa,\lambda}, \quad \text{for } 1 \leq s \leq \kappa + \lambda.$$

We note that

$$(6.48) \qquad S_1 \equiv \left[\sum_{s=1}^{\kappa} P_{\kappa,\lambda,s}^{i,j} \mathcal{M}_s\right] + T_2 + T_3,$$

where T_2 and T_3 are the corresponding second and third terms in (6.45). Several short arguments are needed.

1. Suppose that $\kappa + 1 \leq s \leq \lambda$. If $\kappa < p \leq s$, then (6.33) yields $H_{p,q,s}^{i,j,\kappa,\lambda} \equiv 0$. Thus, we have

$$(6.49) \qquad T_2 \equiv \sum_{s=\kappa+1}^{\lambda} P_{\kappa,\lambda,s}^{i,j}(\zeta) \sum_{p=0}^{\kappa} \sum_{q=0}^{s-p} H_{p,q,s}^{i,j,\kappa,\lambda} \equiv \sum_{s=\kappa+1}^{\lambda} P_{\kappa,\lambda,s}^{i,j}(\zeta) \sum_{p=0}^{s} \sum_{q=0}^{s-p} H_{p,q,s}^{i,j,\kappa,\lambda}$$

$$\equiv \sum_{s=\kappa+1}^{\lambda} P_{\kappa,\lambda,s}^{i,j}(\zeta) \mathcal{M}_s.$$

2. Suppose that $\lambda + 1 \leq s \leq \kappa + \lambda$. Due to $\kappa \leq \lambda$, we have $\kappa < s$. If $\kappa < p \leq s$ and $0 \leq q \leq s - p$, then (6.33) gives $H_{p,q,s}^{i,j,\kappa,\lambda} \equiv 0$. Hence, we find that

$$(6.50) \qquad T_3 \equiv \sum_{s=\lambda+1}^{\kappa+\lambda} P_{\kappa,\lambda,s}^{i,j}(\zeta) \sum_{p=s-\lambda}^{\kappa} \sum_{q=0}^{s-p} H_{p,q,s}^{i,j,\kappa,\lambda} \equiv \sum_{s=\lambda+1}^{\kappa+\lambda} P_{\kappa,\lambda,s}^{i,j}(\zeta) \sum_{p=s-\lambda}^{s} \sum_{q=0}^{s-p} H_{p,q,s}^{i,j,\kappa,\lambda}.$$

3. Suppose that $\lambda + 1 \leq s \leq \kappa + \lambda$ and $0 \leq p \leq s - \lambda - 1$. Then, we have $\lambda + 1 \leq s - p$. If $\lambda < q \leq s - p$, then (6.33) yields $H_{p,q,s}^{i,j,\kappa,\lambda} \equiv 0$. Thus, we obtain

$$(6.51) \qquad S_2 \equiv \sum_{s=\lambda+1}^{\kappa+\lambda} P_{\kappa,\lambda,s}^{i,j}(\zeta) \sum_{p=0}^{s-\lambda-1} \sum_{q=0}^{\lambda} H_{p,q,s}^{i,j,\kappa,\lambda} \equiv \sum_{s=\lambda+1}^{\kappa+\lambda} P_{\kappa,\lambda,s}^{i,j}(\zeta) \sum_{p=0}^{s-\lambda-1} \sum_{q=0}^{s-p} H_{p,q,s}^{i,j,\kappa,\lambda}.$$

By adding the right members of (6.50) and (6.51), we observe that

$$(6.52) \qquad T_3 + S_2 \equiv \sum_{s=\lambda+1}^{\kappa+\lambda} P_{\kappa,\lambda,s}^{i,j}(\zeta) \sum_{p=0}^{s} \sum_{q=0}^{s-p} H_{p,q,s}^{i,j,\kappa,\lambda} \equiv \sum_{s=\lambda+1}^{\kappa+\lambda} P_{\kappa,\lambda,s}^{i,j}(\zeta) \mathcal{M}_s.$$

We combine (6.42), (6.48), (6.49), and (6.52) to deduce that

$$(6.53) \qquad \Phi_{\kappa,\lambda}(\zeta) \equiv \sum_{s=1}^{\kappa+\lambda} P_{\kappa,\lambda,s}^{i,j}(\zeta) \mathcal{M}_s.$$

Lemma 6.4 establishes that $\mathcal{M}_s \equiv 0$, for $1 \leq s \leq \kappa + \lambda$. Consequently, (6.53) yields $\Phi_{\kappa,\lambda}(\zeta) \equiv 0$. The supposition $\kappa \leq \lambda$ introduced just before (6.41) on page 62 can be removed by a completely symmetric argument in which the roles of i, κ, p and j, λ, q are interchanged. The validity of $\Phi_{\kappa,\lambda}(\zeta) \equiv 0$ for all κ, λ satisfying $0 \leq \kappa \leq i$ and $0 \leq \lambda \leq j$ shows that the right member of (6.31) is identically zero. Consequently, (6.23) is valid. This completes the proof of Theorem 6.3. $\square$

LEMMA 6.4. *The rational numbers $H_{p,q,s}^{i,j,\kappa,\lambda}$ in (6.33) satisfy*

$$\sum_{p=0}^{s}\sum_{q=0}^{s-p} H_{p,q,s}^{i,j,\kappa,\lambda} \equiv 0, \quad \text{for } 0 \leq \kappa \leq i,\ 0 \leq \lambda \leq j,\ \text{and } 1 \leq s \leq \kappa+\lambda. \tag{6.54}$$

PROOF. The left member $\mathcal{L}$ of (6.54) is well defined. Setting $r = p + q$ to transform $\mathcal{L}$ and using (6.33), we find that

$$\mathcal{L} \equiv \sum_{r=0}^{s}\sum_{q=0}^{r} H_{r-q,q,s}^{i,j,\kappa,\lambda} \equiv \sum_{r=0}^{s}\sum_{q=0}^{r} (-1)^r \binom{\kappa}{r-q}\binom{\lambda}{q} \frac{\binom{2i+2j-\kappa-\lambda-1}{s-r}}{\binom{2i+2j-2}{\kappa+\lambda-r}} \tag{6.55}$$

$$\equiv \sum_{r=0}^{s} (-1)^r \frac{\binom{2i+2j-\kappa-\lambda-1}{s-r}}{\binom{2i+2j-2}{\kappa+\lambda-r}} \sum_{q=0}^{r} \binom{\kappa}{r-q}\binom{\lambda}{q}.$$

Historical commentary and details about Vandermonde's convolution

$$\sum_{k} \binom{a}{k}\binom{b}{j-k} \equiv \binom{a+b}{j}, \quad \text{for any integer } j, \tag{6.56}$$

can be found in [**40**, page 59]. We apply (6.56) to rewrite (6.55) as

$$\mathcal{L} \equiv \sum_{r=0}^{s} (-1)^r \binom{\kappa+\lambda}{r} \frac{\binom{2i+2j-\kappa-\lambda-1}{s-r}}{\binom{2i+2j-2}{\kappa+\lambda-r}} \tag{6.57}$$

$$\equiv \sum_{r=0}^{s} (-1)^r \binom{\kappa+\lambda}{r} \frac{\binom{2i+2j-\kappa-\lambda-1}{s-r}\binom{2i+2j-\kappa-\lambda+r-2}{r}}{\binom{2i+2j-2}{\kappa+\lambda-r}\binom{2i+2j-\kappa-\lambda+r-2}{r}}$$

$$\equiv \sum_{r=0}^{s} (-1)^r \frac{\binom{2i+2j-\kappa-\lambda-1}{s-r}\binom{2i+2j-\kappa-\lambda+r-2}{r}}{\binom{2i+2j-2}{\kappa+\lambda}}.$$

Setting $t = 2i + 2j - \kappa - \lambda - 1$, we use (6.57), (6.56), and $s \geq 1$ to obtain

$$\mathcal{L} \equiv \frac{1}{\binom{2i+2j-2}{\kappa+\lambda}} \sum_{r=0}^{s} \binom{t}{s-r}(-1)^r \binom{t+r-1}{r}$$

$$\equiv \frac{1}{\binom{2i+2j-2}{\kappa+\lambda}} \sum_{r=0}^{s} \binom{t}{s-r}\binom{-t}{r} \equiv \frac{1}{\binom{2i+2j-2}{\kappa+\lambda}} \binom{0}{s} \equiv 0.$$

Thus, (6.54) is valid. This completes the proof of Lemma 6.4. □

6.2. IDENTITIES FOR THE COEFFICIENTS OF RELATED CANONICAL FORMS

OBSERVATIONS ABOUT THEOREM 6.3. For i_0, j_0 subject to $0 \leq i_0, j_0 \leq m$ and $2 \leq i_0 + j_0$, the relations (6.23) specified by $(i,j) = (i_0, j_0)$ and $(i,j) = (j_0, i_0)$ are equivalent due to $A_{\nu,\mu}^{j_0,i_0} = A_{\mu,\nu}^{i_0,j_0}$, $c_{\nu,\mu}(z) \equiv c_{\mu,\nu}(z)$, and $d_{\nu,\mu}(\zeta) \equiv d_{\mu,\nu}(\zeta)$. Thus, in counting the distinct relations (6.23), we can restrict the integers i, j for (6.23) to ones satisfying $0 \leq i \leq j \leq m$ and $2 \leq i+j$. Of these, the relation (6.23) specified by $(i,j) = (0,2)$ is the trivial one $0 \equiv 0$ due to $c_{0,0}(z) \equiv 1$, $d_{0,0}(\zeta) \equiv 1$, $c_{0,1}(z) \equiv c_{0,2}(z) \equiv 0$, and $d_{0,1}(\zeta) \equiv d_{0,2}(\zeta) \equiv 0$. Were we to have defined $A_{p,q}^{i,j}$ in $\mathbb{Q}$ when $i+j = 0$ and when $i+j = 1$, there would be corresponding trivial relations $A_{0,0}^{0,0} \equiv A_{0,0}^{0,0}$, for $(i,j) = (0,0)$, and $0 \equiv 0$, for $(i,j) = (0,1)$. We find that the nontrivial relations provided by (6.23) are those for which (i,j) satisfies either $1 \leq i \leq j \leq m$ or $i = 0 < 3 \leq j \leq m$. In correspondence with them are the indices (l,n) for the basic relative invariants in our Main Theorem of page 8 where either $1 \leq l \leq n \leq m$ or $l = 0 < 3 \leq n \leq m$.

For the equations $Q_1 = 0$ (of order 1), all of the relative invariants were specified in Theorem 4.10 on page 41. They have just one basic relative invariant given by $\mathcal{I}_{1,1,1} \equiv w_{1,1} - (w_{0,1})^2$. Moreover, a polynomial $\boldsymbol{P}$ in $\mathcal{S}_1$ is a relative invariant of the equations $Q_1 = 0$ if and only if $\boldsymbol{P} \equiv \gamma(w_{1,1} - (w_{0,1})^2)^\kappa$, for some nonzero rational number γ and some positive integer κ.

Throughout Part 2, as in this chapter, we consider only equations $Q_m = 0$ of order $m \geq 2$. For such equations, we shall demonstrate that the basic relative invariants are in one-to-one correspondence with the pairs (l,n) of integers l, n that satisfy either $1 \leq l \leq n \leq m$ or $l = 0 < 3 \leq n \leq m$. The form of the left member in (6.23) was originally used to formulate the definitions for $\mathcal{I}_{m,l,n}$ in (1.28) on page 8 and $\mathcal{J}_{m,l,n}$ in (1.38) on page 9. Traces of this approach are preserved in Theorem 7.10 on page 85 and in Theorem 9.10 on page 109.

Our arguments throughout Part 2 do not depend on Theorem 6.3. In fact, they can be applied to give an independent proof of Theorem 6.3 in the same manner that Corollary 27.8 on page 315 is used to establish the generalization of Theorem 6.3 presented in Theorem 23.8 on page 271.

Part 2

The Basic Relative Invariants for $Q_m = 0$ when $m \geq 2$

CHAPTER 7

Formulas That Involve $L_{i,j}(z)$

Throughout Part 2, there are no conditions placed on the coefficients of (1.1). Moreover, with $m \geq 2$, $\boldsymbol{a}_{m,2}$ and $\boldsymbol{b}_{m,2}$ are well defined in $\mathcal{S}_m$ by (1.14) and (1.15) on page 7. The other formulas of pages 7–9 are also applicable. Using the abbreviated notation of Section 1.4 on page 10, we observe that $\boldsymbol{K}_{i,j}$ and $\boldsymbol{L}_{i,j}$ are defined in $\mathcal{S}_m$ for all i and j by (1.18)–(1.22).

In this chapter, a key role is played by the functions $L_{i,j}(z)$ on Ω obtained by substituting $c_{\mu,\nu}^{(\kappa)}(z)$ from (1.1) for $w_{\mu,\nu}^{(\kappa)}$ in $\boldsymbol{L}_{i,j}$. It is convenient for us to suppose throughout that the composite of the substitutions (1.2) and (1.4) transforms (1.1) on Ω into a corresponding (5.1) that is defined on Ω^{**} and whose coefficients satisfy $d_{0,1}(\zeta) \equiv d_{0,2}(\zeta) \equiv 0$. This allows us to keep the notation simple. However, we should be aware that: when an equation (1.1) is given on a region Ω, it can happen that only various restrictions of (1.1) to proper subregions of Ω can be transformed into corresponding Laguerre-Forsyth canonical forms; namely, see Definition 5.6 on page 51 and the conditions for Propositions 5.3–5.4 on page 49.

7.1. The coefficients of (5.1) when $d_{0,1}(\zeta) \equiv d_{0,2}(\zeta) \equiv 0$

For the statement of Proposition 7.1, we introduce

(7.1) $$\xi_{i,j} \equiv \frac{\left[\prod_{\nu=1}^{j}(i-\nu)\right]}{2^j} \binom{m-i+j}{j}, \quad \text{for any integers } i \text{ and } j.$$

PROPOSITION 7.1. *For* (5.1) *on* Ω^{**} *into which* (1.1) *on* Ω *is transformed by* $y(z) = \rho(z)\, v(z)$ *and* $z = f(\zeta)$ *according to the context of page 47, suppose that* $d_{0,1}(\zeta) \equiv d_{0,2}(\zeta) \equiv 0$. *Then, the coefficients of* (5.1) *are given by*

(7.2) $$d_{i,j}(\zeta) \equiv \left(f'(\zeta)\right)^{i+j} \sum_{r=0}^{i} \sum_{s=0}^{j} \xi_{i,r}\, \xi_{j,s}\, L_{i-r,j-s}\bigl(f(\zeta)\bigr)\, \bigl(\mathcal{Q}(\zeta)\bigr)^{r+s},$$
$$\text{on } \Omega^{**} \text{ for } 0 \leq i,j \leq m,$$

where $L_{i-r,j-s}\bigl(f(\zeta)\bigr)$ *is obtained for* $z = f(\zeta)$ *by substituting* $c_{\mu,\nu}^{(\kappa)}(z)$ *from* (1.1) *for* $w_{\mu,\nu}^{(\kappa)}$ *in* $\boldsymbol{L}_{i-r,j-s}$ *of* (1.21) *and*

(7.3) $$\mathcal{Q}(\zeta) \equiv \frac{f''(\zeta)}{\bigl(f'(\zeta)\bigr)^2}, \quad \text{on } \Omega^{**}.$$

PROOF. We repeat (5.3) of Theorem 5.1 on page 47 as

(7.4) $$d_{i,j}(\zeta) \equiv \left(f'(\zeta)\right)^{i+j} \sum_{p=0}^{i} \sum_{q=0}^{j} \phi_{i-p,p}\bigl(f(\zeta)\bigr)\, \phi_{j-q,q}\bigl(f(\zeta)\bigr)\, c_{p,q}\bigl(f(\zeta)\bigr),$$

69

where $\phi_{i,j}(z)$ is defined on Ω by (5.4)–(5.5) of page 47 in terms of our particular $\rho(z)$ and the inverse $\zeta = g(z)$ to $z = f(\zeta)$. Due to $d_{0,1}(\zeta) \equiv d_{0,2}(\zeta) \equiv 0$, Proposition 5.3 on page 49 shows that $\rho(z)$ and $g(z)$ satisfy (5.16) and (5.17). Consequently, Lemma 7.2 yields

$$(7.5) \qquad \phi_{i,j}(z) \equiv \sum_{k=0}^{i} \xi_{i+j,k}\, K_{i-k,j}(z)\, \bigl(\mathcal{P}(z)\bigr)^k, \quad \text{on } \Omega \text{ for } i \geq 0 \text{ and } j \leq m,$$

where: $K_{i-k,j}(z)$ is obtained by substituting $c_{\mu,\nu}^{(\kappa)}(z)$ from (1.1) for $\boldsymbol{w}_{\mu,\nu}^{(\kappa)}$ in $\boldsymbol{K}_{i-k,j}$ of (1.18)–(1.20) on page 8; and

$$(7.6) \qquad \mathcal{P}(z) \equiv -\frac{g''(z)}{g'(z)}.$$

We use $g(f(\zeta)) \equiv \zeta$ to deduce that

$$(7.7) \qquad \mathcal{P}\bigl(f(\zeta)\bigr) \equiv -\frac{g''(f(\zeta))}{g'(f(\zeta))} \equiv \frac{f''(\zeta)}{(f'(\zeta))^2} \equiv \mathcal{Q}(\zeta), \quad \text{on } \Omega^{**}.$$

By employing (7.4), (7.5), and (7.7), we obtain

$$(7.8) \qquad d_{i,j}(\zeta) \equiv \bigl(f'(\zeta)\bigr)^{i+j} h\bigl(f(\zeta)\bigr), \quad \text{on } \Omega^{**},$$

where $h(z)$, on Ω, is given by

$$h(z) \equiv \sum_{p=0}^{i}\sum_{q=0}^{j} \phi_{i-p,p}(z)\, \phi_{j-q,q}(z)\, c_{p,q}(z)$$

$$\equiv \sum_{p=0}^{i}\sum_{q=0}^{j}\sum_{r=0}^{i-p} \xi_{i,r}\, K_{i-p-r,p}(z)\, \bigl(\mathcal{P}(z)\bigr)^r \sum_{s=0}^{j-q} \xi_{j,s}\, K_{j-q-s,q}(z)\, \bigl(\mathcal{P}(z)\bigr)^s c_{p,q}(z)$$

$$\equiv \sum_{r=0}^{i}\sum_{p=0}^{i-r}\sum_{s=0}^{j}\sum_{q=0}^{j-s} \xi_{i,r}\, \xi_{j,s}\, K_{i-p-r,p}(z)\, K_{j-q-s,q}(z)\, \bigl(\mathcal{P}(z)\bigr)^{r+s} c_{p,q}(z)$$

$$\equiv \sum_{r=0}^{i}\sum_{s=0}^{j} \xi_{i,r}\, \xi_{j,s}\, \bigl(\mathcal{P}(z)\bigr)^{r+s} \sum_{p=0}^{i-r}\sum_{q=0}^{j-s} K_{i-r-p,p}(z)\, K_{j-s-q,q}(z)\, c_{p,q}(z).$$

By substituting $c_{\mu,\nu}^{(\kappa)}(z)$ from (1.1) for $\boldsymbol{w}_{\mu,\nu}^{(\kappa)}$ in $\boldsymbol{L}_{i-r,j-s}$ of (1.21) on page 8, we combine the resulting expression with the preceding one to see that

$$(7.9) \qquad h(z) \equiv \sum_{r=0}^{i}\sum_{s=0}^{j} \xi_{i,r}\, \xi_{j,s}\, \bigl(\mathcal{P}(z)\bigr)^{r+s} L_{i-r,j-s}(z).$$

Formulas (7.8), (7.9), and (7.7) yield (7.2). This completes the proof. $\square$

The use of (7.5) in the proof of Proposition 7.1 is based on the next result.

LEMMA 7.2. *Suppose that $\rho(z)$ and $g(z)$ satisfy (5.16) and (5.17) on Ω. Then, the functions $\psi_{i,j}(z)$ defined in terms of (7.1), (1.18)–(1.20), and (7.6) by*

$$(7.10) \qquad \psi_{i,j}(z) \equiv \sum_{k=0}^{i} \xi_{i+j,k}\, K_{i-k,j}(z)\, \bigl(\mathcal{P}(z)\bigr)^k, \quad \text{on } \Omega \text{ for } i \geq 0 \text{ and any } j,$$

and $\phi_{i,j}(z)$, defined by (5.4)–(5.5) on page 47, are related through

$$(7.11) \qquad \phi_{i,j}(z) \equiv \psi_{i,j}(z), \quad \text{on } \Omega \text{ for } i \geq 0 \text{ and } j \leq m.$$

7.1. THE COEFFICIENTS OF (5.1) WHEN $d_{0,1}(\zeta) \equiv d_{0,2}(\zeta) \equiv 0$

PROOF. We first verify that (7.1) yields

(7.12) $\quad \xi_{i+1+j,k} - \xi_{i+j,k} \equiv \dfrac{m+k-2i-2j}{2} \xi_{i+j,k-1}, \quad$ for all i, j, k.

As an abbreviation, we set

(7.13) $\quad \theta_{i,j,k} \equiv -(m-i-j+k+1)(1-i-j+k)\, \xi_{i+j,k}, \quad$ for all i, j, k;

then, we check the validity of

(7.14) $\quad \theta_{i,j,k} \equiv 2(k+1)\, \xi_{i+j,k+1}, \quad$ for $k \geq 0$ and any i, j.

For $k \leq i$ and $j \leq m$, we find that (1.20) yields

(7.15) $\quad K_{i-k,j}^{(1)}(z) \equiv K_{i+1-k,j-1}(z) - K_{i+1-k,j}(z) + \dfrac{m-1}{2} b_{m,1}(z)\, K_{i-k,j}(z)$
$\qquad - (m-i-j+k+1)(1-i-j+k)\, a_{m,2}(z)\, K_{i-1-k,j}(z).$

We apply (7.6) and (5.18) on page 49 to verify that

(7.16) $\quad \mathcal{P}^{(1)}(z) \equiv -2\, a_{m,2}(z) - \dfrac{1}{2}\bigl(\mathcal{P}(z)\bigr)^2, \quad$ on Ω.

By combining (5.17) on page 49 with (7.6), we obtain

(7.17) $\quad \dfrac{m-1}{2} b_{m,1}(z) \equiv \dfrac{m-1}{2} \mathcal{P}(z) - \dfrac{\rho'(z)}{\rho(z)}, \quad$ on Ω.

Let i and j be fixed integers satisfying $i \geq 0$ and $j \leq m$. We use (7.10), (7.16), (7.15), (7.17), (7.13), and (1.18) to deduce that

(7.18) $\quad \psi_{i,j}^{(1)}(z) \equiv \displaystyle\sum_{k=0}^{i} \xi_{i+j,k}\, K_{i-k,j}^{(1)}(z)\, \bigl(\mathcal{P}(z)\bigr)^k$
$\qquad + \displaystyle\sum_{k=0}^{i} \xi_{i+j,k}\, K_{i-k,j}(z)\, k\, \bigl(\mathcal{P}(z)\bigr)^{k-1} \left[-2a_{m,2}(z) - \dfrac{1}{2}\bigl(\mathcal{P}(z)\bigr)^2\right]$
$\qquad \equiv \mathcal{D}_1(z) + \mathcal{D}_2(z) + \mathcal{D}_3(z) + \mathcal{D}_4(z) + \mathcal{D}_5(z) + \mathcal{D}_6(z),$

where

$\mathcal{D}_1(z) \equiv \displaystyle\sum_{k=0}^{i} \xi_{i+j,k}\, K_{i+1-k,j-1}(z)\, \bigl(\mathcal{P}(z)\bigr)^k,$

$\mathcal{D}_2(z) \equiv \displaystyle\sum_{k=0}^{i} -\xi_{i+j,k}\, K_{i+1-k,j}(z)\, \bigl(\mathcal{P}(z)\bigr)^k,$

$\mathcal{D}_3(z) \equiv \displaystyle\sum_{k=0}^{i} \xi_{i+j,k} \left[\dfrac{m-1}{2}\mathcal{P}(z) - \dfrac{\rho'(z)}{\rho(z)}\right] K_{i-k,j}(z)\, \bigl(\mathcal{P}(z)\bigr)^k,$

$\mathcal{D}_4(z) \equiv a_{m,2}(z) \displaystyle\sum_{k=0}^{i-1} \theta_{i,j,k}\, K_{i-1-k,j}(z)\, \bigl(\mathcal{P}(z)\bigr)^k,$

$\mathcal{D}_5(z) \equiv a_{m,2}(z) \displaystyle\sum_{k=1}^{i} -2k\, \xi_{i+j,k}\, K_{i-k,j}(z)\, \bigl(\mathcal{P}(z)\bigr)^{k-1}$
$\qquad \equiv a_{m,2}(z) \displaystyle\sum_{k=0}^{i-1} -2(k+1)\, \xi_{i+j,k+1}\, K_{i-1-k,j}(z)\, \bigl(\mathcal{P}(z)\bigr)^k,$

and
$$\mathcal{D}_6(z) \equiv \sum_{k=0}^{i} -\frac{k}{2} \mathcal{P}(z) \, \xi_{i+j,k} \, K_{i-k,j}(z) \, (\mathcal{P}(z))^k.$$

Due to (7.10), we have
$$\mathcal{D}_1(z) - \psi_{i+1,j-1}(z) \equiv -\xi_{i+j,i+1} \, K_{0,j-1}(z) \, (\mathcal{P}(z))^{i+1}.$$

Using $K_{0,j-1}(z) \equiv K_{0,j}(z)$ from (1.19) and applying (7.12), we see that
$$[\mathcal{D}_1(z) - \psi_{i+1,j-1}(z)] + \mathcal{D}_2(z) \equiv -\sum_{k=0}^{i+1} \xi_{i+j,k} \, K_{i+1-k,j}(z) \, (\mathcal{P}(z))^k$$
$$\equiv -\sum_{k=0}^{i+1} \left[\xi_{i+1+j,k} - \frac{m+k-2i-2j}{2} \xi_{i+j,k-1} \right] K_{i+1-k,j}(z) \, (\mathcal{P}(z))^k.$$

In view of (7.10) and $\xi_{i+j,-1} = 0$ from (7.1), this gives

(7.19) $\mathcal{D}_1(z) + \mathcal{D}_2(z) \equiv \psi_{i+1,j-1}(z) - \psi_{i+1,j}(z)$
$$+ \sum_{k=0}^{i} \frac{m+k+1-2i-2j}{2} \mathcal{P}(z) \, \xi_{i+j,k} \, K_{i-k,j}(z) \, (\mathcal{P}(z))^k.$$

Moreover, (7.14) yields $\mathcal{D}_4(z) + \mathcal{D}_5(z) \equiv 0$. By combining this with (7.18), (7.19), the formula for $\mathcal{D}_3(z)$, the formula for $\mathcal{D}_6(z)$, (7.6), and (7.10), we obtain

(7.20) $\psi_{i,j}^{(1)}(z) \equiv \psi_{i+1,j-1}(z) - \psi_{i+1,j}(z)$
$$+ \sum_{k=0}^{i} \left[(m-i-j) \mathcal{P}(z) - \frac{\rho'(z)}{\rho(z)} \right] \xi_{i+j,k} \, K_{i-k,j}(z) \, (\mathcal{P}(z))^k$$
$$\equiv \psi_{i+1,j-1}(z) - \psi_{i+1,j}(z) - \left[\frac{\rho'(z)}{\rho(z)} + (m-i-j) \frac{g''(z)}{g'(z)} \right] \psi_{i,j}(z),$$
for $i \geq 0$ and $j \leq m$.

We note that (1.20) yields $K_{i+1-k,m}(z) \equiv 0$, for $k \leq i$. By employing this with (7.10), (1.19), and (7.1), we verify that

(7.21) $\qquad \psi_{i+1,m}(z) \equiv (\xi_{i+1+m,i+1})(p(z))^{i+1} \equiv 0, \quad \text{for } i \geq 0.$

Due to (7.21) and (7.20), we have

(7.22) $\qquad \psi_{i+1,j}(z) \equiv \sum_{k=j+1}^{m} [\psi_{i+1,k-1}(z) - \psi_{i+1,k}(z)]$
$$\equiv \sum_{k=j+1}^{m} \left[\psi_{i,k}^{(1)}(z) + \left[\frac{\rho'(z)}{\rho(z)} + (m-i-k) \frac{g''(z)}{g'(z)} \right] \psi_{i,k}(z) \right],$$
for $i \geq 0$ and $j \leq m$.

We note that (7.10), (1.19), and (7.1) yield $\psi_{0,j}(z) \equiv \xi_{j,0} \equiv 1$, for any j. By using this with (5.4), (5.5), and (7.22), we find that $\phi_{i,j}(z) \equiv \psi_{i,j}(z)$, for $i \geq 0$ and $j \leq m$. Thus, (7.11) is valid. This completes the proof of Lemma 7.2. $\square$

7.2. Derivatives of the coefficients in (5.1) when $d_{0,1}(\zeta) \equiv d_{0,2}(\zeta) \equiv 0$

In terms of the polynomials $\boldsymbol{L}_{i,j}$ defined in $\mathcal{S}_m$ for all i and j by (1.21)–(1.22) on page 8, we introduce $\mathcal{M}_{i,j,k,t}$ in $\mathcal{S}_m$ for $k \geq 0$ and all integers i, j, t through

$$\mathcal{M}_{i,j,0,t} \equiv \sum_{r=0}^{t} \xi_{i,r}\, \xi_{j,t-r}\, \boldsymbol{L}_{i-r,j+r-t}, \quad \text{for any } i,\, j,\, t, \tag{7.23}$$

and

$$\mathcal{M}_{i,j,k+1,t} \equiv \mathcal{M}^{(1)}_{i,j,k,t} + \frac{2i+2j+2k-t+1}{2}\mathcal{M}_{i,j,k,t-1} \\ - 2(t+1)a_{m,2}\,\mathcal{M}_{i,j,k,t+1}, \quad \text{for } k \geq 0 \text{ and any } i,\, j,\, t. \tag{7.24}$$

PROPOSITION 7.3. *For (5.1) on Ω^{**} into which (1.1) on Ω is transformed by $y(z) = \rho(z)\, v(z)$ and $z = f(\zeta)$ according to the context of page 47, suppose that $d_{0,1}(\zeta) \equiv d_{0,2}(\zeta) \equiv 0$. Then, the derivatives of the coefficients of (5.1) are given by*

$$d^{(k)}_{i,j}(\zeta) \equiv \bigl(f'(\zeta)\bigr)^{i+j+k} \sum_{t=0}^{i+j+k} \mathcal{M}_{i,j,k,t}\bigl(f(\zeta)\bigr)\bigl(\mathcal{Q}(\zeta)\bigr)^{t}, \tag{7.25}$$
$$\text{for } 0 \leq i,\, j \leq m \text{ and } k \geq 0,$$

where $\mathcal{M}_{i,j,k,t}\bigl(f(\zeta)\bigr)$ is obtained for $z = f(\zeta)$ by substituting $c^{(\kappa)}_{\mu,\nu}(z)$ from (1.1) for $w^{(\kappa)}_{\mu,\nu}$ in $\mathcal{M}_{i,j,k,t}$ of (7.23)–(7.24) and $\mathcal{Q}(\zeta)$ is defined in (7.3).

PROOF. For $t < 0$, (7.23) yields $\mathcal{M}_{i,j,0,t} \equiv 0$. Suppose that $t \geq i+j+1$. Then, in view of $(i-r)+(j+r-t) = i+j-t \leq -1$, we have either $i-r < 0$ or $j+r-t < 0$ so that formulas (1.22) and (7.23) give $\mathcal{M}_{i,j,0,t} \equiv 0$. Thus, the formula $\mathcal{M}_{i,j,k,t} \equiv 0$ is valid whenever $k = 0$ and either $t < 0$ or $t > i+j+k$. Suppose that k is a nonnegative integer such that $\mathcal{M}_{i,j,k,t} \equiv 0$ is valid whenever i, j, t are integers that satisfy either $t < 0$ or $t > i+j+k$. Then, for this k, we use (7.24) to see that $\mathcal{M}_{i,j,k+1,t} \equiv 0$ is valid for any integers i, j, t that satisfy either $t < 0$ or $t > i+j+k+1$. This establishes that

$$\mathcal{M}_{i,j,k,t} \equiv 0, \quad \text{whenever } k \geq 0 \text{ and either } t < 0 \text{ or } t > i+j+k. \tag{7.26}$$

Due to (1.22), we have $\boldsymbol{L}_{i-r,j-s} \equiv 0$, for $r > i$ or $s > j$. Consequently, for $0 \leq i,\, j \leq m$, we see that (7.2) of Proposition 7.1 on page 69 and (7.23) yield

$$d_{i,j}(\zeta) \equiv \bigl(f'(\zeta)\bigr)^{i+j} \sum_{r=0}^{i} \sum_{s=0}^{j} \xi_{i,r}\, \xi_{j,s}\, \boldsymbol{L}_{i-r,j-s}\bigl(f(\zeta)\bigr)\bigl(\mathcal{Q}(\zeta)\bigr)^{r+s}$$
$$\equiv \bigl(f'(\zeta)\bigr)^{i+j} \sum_{r=0}^{i+j} \sum_{s=0}^{i+j-r} \xi_{i,r}\, \xi_{j,s}\, \boldsymbol{L}_{i-r,j-s}\bigl(f(\zeta)\bigr)\bigl(\mathcal{Q}(\zeta)\bigr)^{r+s}$$
$$\equiv \bigl(f'(\zeta)\bigr)^{i+j} \sum_{t=0}^{i+j} \sum_{r=0}^{t} \xi_{i,r}\, \xi_{j,t-r}\, \boldsymbol{L}_{i-r,j+r-t}\bigl(f(\zeta)\bigr)\bigl(\mathcal{Q}(\zeta)\bigr)^{t}$$
$$\equiv \bigl(f'(\zeta)\bigr)^{i+j} \sum_{t=0}^{i+j} \mathcal{M}_{i,j,0,t}\bigl(f(\zeta)\bigr)\bigl(\mathcal{Q}(\zeta)\bigr)^{t}.$$

Thus, (7.25) is valid for $k = 0$ and $0 \leq i,\, j \leq m$.

Suppose that k_0 is a nonnegative integer such that (7.25) is valid for $k = k_0$ and any i, j satisfying $0 \leq i, j \leq m$. For $k = k_0$, we differentiate (7.25) with respect to ζ and use $f''(\zeta)/(f'(\zeta))^2 \equiv \mathcal{Q}(\zeta)$ from (7.3) to obtain

$$(7.27) \quad d_{i,j}^{(k_0+1)}(\zeta) \equiv (i+j+k_0)(f'(\zeta))^{i+j+k_0+1} \mathcal{Q}(\zeta) \sum_{t=0}^{i+j+k_0} \mathcal{M}_{i,j,k_0,t}(f(\zeta))(\mathcal{Q}(\zeta))^t$$

$$+ (f'(\zeta))^{i+j+k_0+1} \sum_{t=0}^{i+j+k_0} \mathcal{M}^{(1)}_{i,j,k_0,t}(f(\zeta))(\mathcal{Q}(\zeta))^t$$

$$+ (f'(\zeta))^{i+j+k_0+1} \sum_{t=0}^{i+j+k_0} t\, \mathcal{M}_{i,j,k_0,t}(f(\zeta))(\mathcal{Q}(\zeta))^{t-1} \frac{\mathcal{Q}'(\zeta)}{f'(\zeta)}.$$

In terms of $\mathcal{P}(z) \equiv -g''(z)/g'(z)$ from (7.6), we repeat (7.16) as

$$(7.28) \quad \mathcal{P}^{(1)}(z) \equiv -2 a_{m,2}(z) - \frac{1}{2}(\mathcal{P}(z))^2, \quad \text{on } \Omega.$$

We note that (7.7) is the relation $\mathcal{Q}(\zeta) \equiv \mathcal{P}(f(\zeta))$, on Ω^{**}. After differentiating (7.7) with respect to ζ, we use (7.28) and (7.7) to deduce that

$$(7.29) \quad \frac{\mathcal{Q}'(\zeta)}{f'(\zeta)} \equiv \mathcal{P}^{(1)}(f(\zeta)) \equiv -2 a_{m,2}(f(\zeta)) - \frac{1}{2}(\mathcal{Q}(\zeta))^2, \quad \text{on } \Omega^{**}.$$

We employ (7.27), (7.26), and (7.29) to verify that

$$(7.30) \quad d_{i,j}^{(k_0+1)}(\zeta) \equiv (f'(\zeta))^{i+j+k_0+1} [\mathcal{E}_1(\zeta) + \mathcal{E}_2(\zeta) + \mathcal{E}_3(\zeta) + \mathcal{E}_4(\zeta)],$$

where

$$\mathcal{E}_1(\zeta) \equiv \sum_{t=0}^{i+j+k_0} (i+j+k_0) \mathcal{M}_{i,j,k_0,t}(f(\zeta))(\mathcal{Q}(\zeta))^{t+1}$$

$$\equiv \sum_{t=0}^{i+j+k_0+1} (i+j+k_0) \mathcal{M}_{i,j,k_0,t-1}(f(\zeta))(\mathcal{Q}(\zeta))^t,$$

$$\mathcal{E}_2(\zeta) \equiv \sum_{t=0}^{i+j+k_0+1} \mathcal{M}^{(1)}_{i,j,k_0,t}(f(\zeta))(\mathcal{Q}(\zeta))^t,$$

$$\mathcal{E}_3(\zeta) \equiv \sum_{t=0}^{i+j+k_0} -2t\, a_{m,2}(f(\zeta)) \mathcal{M}_{i,j,k_0,t}(f(\zeta))(\mathcal{Q}(\zeta))^{t-1}$$

$$\equiv \sum_{t=0}^{i+j+k_0+1} -2(t+1)\, a_{m,2}(f(\zeta)) \mathcal{M}_{i,j,k_0,t+1}(f(\zeta))(\mathcal{Q}(\zeta))^t,$$

$$\mathcal{E}_4(\zeta) \equiv \sum_{t=0}^{i+j+k_0} -\frac{1}{2} t\, \mathcal{M}_{i,j,k_0,t}(f(\zeta))(\mathcal{Q}(\zeta))^{t+1}$$

$$\equiv \sum_{t=0}^{i+j+k_0+1} -\frac{1}{2}(t-1)\, \mathcal{M}_{i,j,k_0,t-1}(f(\zeta))(\mathcal{Q}(\zeta))^t.$$

After adding $\mathcal{E}_1$, $\mathcal{E}_2$, $\mathcal{E}_3$, and $\mathcal{E}_4$, we apply (7.30) with a substitution of $c_{\mu,\nu}^{(\kappa)}(z)$ from (1.1) for $\boldsymbol{w}_{\mu,\nu}^{(\kappa)}$ in (7.24) to obtain

$$(7.31) \qquad d_{i,j}^{(k_0+1)}(\zeta) \equiv \bigl(f'(\zeta)\bigr)^{i+j+k_0+1} \sum_{t=0}^{i+j+k_0+1} \mathcal{M}_{i,j,k_0+1,t}\bigl(f(\zeta)\bigr)\bigl(\mathcal{Q}(\zeta)\bigr)^t.$$

In view of (7.31), (7.25) is also valid for $0 \le i, j \le m$ and $k = k_0 + 1$. Thus, (7.25) is valid for $0 \le i, j \le m$ and $k \ge 0$. This completes the proof. □

For any given (1.1) that has been transformed into an equation (5.1) having $d_{0,1}(\zeta) \equiv d_{0,2}(\zeta) \equiv 0$, we can consider the corresponding combinations of the coefficients of (5.1) and their derivatives that appear as the left members of (6.23) on page 58. Now, Proposition 7.3 enables us to express each such combination in terms of the coefficients of (1.1) and their derivatives through use of the functions $L_{i,j}(z)$ that (1.21) specifies on page 8. Our next task is to simplify such expressions.

7.3. Special combinations of the coefficients for (5.1)

For $m \ge 2$, we continue to assume that the composite of the substitutions (1.2) and (1.4) transforms (1.1) on Ω into a corresponding (5.1) on Ω^{**} for which $d_{0,1}(\zeta) \equiv d_{0,2}(\zeta) \equiv 0$. Throughout, we let l and n denote fixed integers that satisfy either $1 \le l \le n \le m$ or $l = 0 < 3 \le n \le m$. With respect to the constants $A_{p,q}^{i,j}$ of $\mathbb{Q}$ introduced in (6.3) on page 53 and the coefficients for (5.1), the function

$$(7.32) \qquad \sum_{\mu=0}^{l} \sum_{\nu=0}^{n} A_{\mu,\nu}^{l,n} \, d_{l-\mu,n-\nu}^{(\mu+\nu)}(\zeta), \quad \text{on } \Omega^{**},$$

is a specific polynomial combination over $\mathbb{Q}$ of the coefficients for (5.1) and their derivatives. Theorem 6.3 on page 58 suggests that there may exist a basic relative invariant $\mathcal{I}_{m,l,n}$ such that the corresponding sum in (7.32) is the function obtained by substituting $d_{p,q}^{(r)}(\zeta)$ from (5.1) for $\boldsymbol{w}_{p,q}^{(r)}$ in $\mathcal{I}_{m,l,n}$. The validity of this inference will be established in the proof on page 119 for Theorem 11.1. However, our goal in this section is to express (7.32) directly in terms of $\rho(z)$, $f(\zeta)$, and the derivatives $c_{\mu,\nu}^{(\kappa)}(z)$ of the coefficients from (1.1) by means of formulas based on $L_{i,j}$ from (1.21) of page 8. This will be accomplished in Theorem 7.10 on page 85. As our first step in that direction, we use (6.4) on page 54 as well as (7.23)–(7.24) on page 73 to define $\boldsymbol{N}_{i,j}$ in $\mathcal{S}_m$ (with respect to the fixed integers m, l, n) through

$$(7.33) \qquad \boldsymbol{N}_{i,j} \equiv \sum_{r=0}^{i} \sum_{p=0}^{r} A_{l-p,n-r+p}^{l,n} \, \mathcal{M}_{p,r-p,i-r,j}, \quad \text{for any integers } i, j.$$

PROPOSITION 7.4. *The function in* (7.32) *is given by*

$$(7.34) \qquad \sum_{\mu=0}^{l} \sum_{\nu=0}^{n} A_{\mu,\nu}^{l,n} \, d_{l-\mu,n-\nu}^{(\mu+\nu)}(\zeta) \equiv \bigl(f'(\zeta)\bigr)^{l+n} \sum_{t=0}^{l+n} N_{l+n,t}\bigl(f(\zeta)\bigr)\bigl(\mathcal{Q}(\zeta)\bigr)^t,$$

where: $N_{l+n,t}(z)$ *is obtained by substituting* $c_{i,j}^{(k)}(z)$ *from* (1.1) *for* $\boldsymbol{w}_{i,j}^{(k)}$ *in* $\boldsymbol{N}_{l+n,t}$; *and* $\mathcal{Q}(\zeta)$ *is defined in* (7.3) *on page 69.*

PROOF. Letting $\mathcal{L}$ denote the left member of (7.34), we use Proposition 7.3 on page 73 to obtain

$$\mathcal{L} \equiv \sum_{p=0}^{l} \sum_{q=0}^{n} A_{l-p,n-q}^{l,n} d_{p,q}^{(l+n-p-q)}(\zeta)$$

$$\equiv \sum_{p=0}^{l} \sum_{q=0}^{n} A_{l-p,n-q}^{l,n} \sum_{t=0}^{l+n} (f'(\zeta))^{l+n} \mathcal{M}_{p,q,l+n-p-q,t}(f(\zeta)) (\mathcal{Q}(\zeta))^{t}$$

$$\equiv (f'(\zeta))^{l+n} \sum_{t=0}^{l+n} (\mathcal{Q}(\zeta))^{t} \sum_{p=0}^{l} \sum_{q=0}^{n} A_{l-p,n-q}^{l,n} \mathcal{M}_{p,q,l+n-p-q,t}(f(\zeta)).$$

Since (6.4) on page 54 yields $A_{l-p,n-q}^{l,n} = 0$, for $p > l$ or $q > n$, we have

$$\mathcal{L} \equiv (f'(\zeta))^{l+n} \sum_{t=0}^{l+n} (\mathcal{Q}(\zeta))^{t} \sum_{r=0}^{l+n} \sum_{p=0}^{r} A_{l-p,n-r+p}^{l,n} \mathcal{M}_{p,r-p,l+n-r,t}(f(\zeta)).$$

In view of (7.33), this yields (7.34) and completes the proof. $\square$

PROPOSITION 7.5. *The polynomials* $\boldsymbol{N}_{i,j}$ *in* (7.33) *are given by*

(7.35) $\qquad\qquad\qquad \boldsymbol{N}_{i,j} \equiv \boldsymbol{P}_{i-j,j}, \quad \text{for all } i \text{ and } j,$

with respect to polynomials $\boldsymbol{P}_{i,j}$ *in* $\mathcal{S}_m$ *that satisfy*

(7.36) $\qquad\qquad\qquad \boldsymbol{P}_{i,j} \equiv 0, \quad \text{for } i \leq 1 \text{ and any } j,$

and

(7.37) $\qquad \boldsymbol{P}_{i+1,j} \equiv \sum_{k=0}^{j} \gamma_{i,j,k} \left[\begin{array}{l} \boldsymbol{P}_{i,k}^{(1)} - 2(k+1)\,\boldsymbol{a}_{m,2}\,\boldsymbol{P}_{i-1,k+1} \\ + \displaystyle\sum_{h=0}^{l} \sigma_{h,i+1,k}\,\boldsymbol{L}_{h,i+1-h} \end{array} \right],$

$\qquad\qquad\qquad\qquad\qquad\qquad\qquad\qquad \text{for } i \geq 0 \text{ and any } j,$

where

(7.38) $\qquad\qquad \gamma_{i,j,k} \equiv \dfrac{(j-k)!}{2^{j-k}} \binom{2i+j+1}{j-k}, \quad \text{for } k \leq j \text{ and any } i,$

(7.39) $\qquad\qquad \sigma_{h,i,j} \equiv \sum_{p=h}^{l} A_{l-p,n-i-j+p}^{l,n} \xi_{p,p-h}\,\xi_{i+j-p,j-p+h},$

$\qquad\qquad\qquad\qquad\qquad\qquad \text{for } 0 \leq h \leq l \text{ and } i+j \geq 0,$

and $\boldsymbol{L}_{i,j}$ *is defined by* (1.21)–(1.22) *on page* 8.

OBSERVATION. The requirement $(l-p) + (n-i-j+p) \leq 2l + 2n - 2$ from (6.4) on page 54 for the definition of $A_{l-p,n-i-j+p}^{l,n}$ in the right member of (7.39) is certainly satisfied when $i+j \geq 0$. It is also satisfied when $i+j \geq 2-l-n$.

PROOF OF PROPOSITION 7.5. For $i \geq 0$ and any j, we apply $'$ to (7.33) and then employ (7.24) as well as (7.33) to verify that

$$\boldsymbol{N}_{i,j}^{(1)} \equiv \sum_{r=0}^{i} \sum_{p=0}^{r} A_{l-p,n-r+p}^{l,n} \boldsymbol{\mathcal{M}}_{p,r-p,i-r,j}^{(1)}$$

$$\equiv \sum_{r=0}^{i} \sum_{p=0}^{r} A_{l-p,n-r+p}^{l,n} \left\{ \begin{array}{l} \boldsymbol{\mathcal{M}}_{p,r-p,i+1-r,j} \\ - \left[\dfrac{2i-j+1}{2}\right] \boldsymbol{\mathcal{M}}_{p,r-p,i-r,j-1} \\ + 2(j+1)\, a_{m,2}\, \boldsymbol{\mathcal{M}}_{p,r-p,i-r,j+1} \end{array} \right\}$$

and

(7.40) $\quad \boldsymbol{N}_{i,j}^{(1)} \equiv \boldsymbol{N}_{i+1,j} - \boldsymbol{F}_{i+1,j} - \dfrac{2i-j+1}{2} \boldsymbol{N}_{i,j-1} + 2(j+1)\, a_{m,2}\, \boldsymbol{N}_{i,j+1},$

where $\boldsymbol{F}_{i,j}$ is defined in $\mathcal{S}_m$ with respect to (6.4) through

(7.41) $\quad \boldsymbol{F}_{i,j} \equiv \sum_{p=0}^{i} A_{l-p,n-i+p}^{l,n} \boldsymbol{\mathcal{M}}_{p,i-p,0,j}, \quad$ for any i, j.

In particular, when the sum in (7.41) is nonvacuous, we have $i \geq 0$ with either $1 \leq l \leq n$ or $l = 0 < 3 \leq n$ so that $(l-p) + (n-i+p) \leq l + n \leq 2l + 2n - 2$. Thus, the coefficients $A_{l-p,n-i+p}^{l,n}$ for (7.41) are well defined by (6.4).

To show that (7.40) is valid for all integers i and j, we first use (7.41) and (6.4) to obtain $\boldsymbol{F}_{0,j} \equiv A_{l,n}^{l,n} \boldsymbol{\mathcal{M}}_{0,0,0,j} \equiv 0$, for any j. We apply (7.33) and (6.4) to deduce that $\boldsymbol{N}_{0,j} \equiv A_{l,n}^{l,n} \boldsymbol{\mathcal{M}}_{0,0,0,j} \equiv 0$, for any j. Also, (7.41) and (7.33) yield $\boldsymbol{F}_{i,j} \equiv 0$ and $\boldsymbol{N}_{i,j} \equiv 0$, for $i \leq -1$ and any j. Consequently, for $i \leq -1$ and any j, each term in (7.40) is equal to 0. Thus, (7.40) *is true for any i, j*.

Combining (7.41) with (7.23) on page 73, we obtain

(7.42) $\quad \boldsymbol{F}_{i,j} \equiv \sum_{p=0}^{i} A_{l-p,n-i+p}^{l,n} \left[\sum_{r=0}^{j} \xi_{p,r}\, \xi_{i-p,j-r}\, \boldsymbol{L}_{p-r,i-p+r-j} \right], \quad$ for any i, j.

If $l < p \leq i$, then $l - p < 0$ so that (6.4) gives $A_{l-p,n-i+p}^{l,n} = 0$. If $0 \leq i < p \leq l$ and $0 \leq r \leq j$, then $i - p + r - j < 0$ so that (1.22) gives $\boldsymbol{L}_{p-r,i-p+r-j} \equiv 0$. Thus, (7.42) yields

(7.43) $\quad \boldsymbol{F}_{i,j} \equiv \sum_{p=0}^{l} \sum_{r=0}^{j} A_{l-p,n-i+p}^{l,n}\, \xi_{p,r}\, \xi_{i-p,j-r}\, \boldsymbol{L}_{p-r,i-p+r-j},$

for $i \geq 0$ and any j.

If $p < r \leq j$, then $p - r < 0$ so that (1.22) gives $\boldsymbol{L}_{p-r,i-p+r-j} \equiv 0$. If $j < r \leq p$, then $j - r < 0$ and we obtain $\xi_{i-p,j-r} = 0$ from (7.1). Thus, (7.43) gives

(7.44) $\quad \boldsymbol{F}_{i,j} \equiv \sum_{p=0}^{l} \sum_{r=0}^{p} A_{l-p,n-i+p}^{l,n}\, \xi_{p,r}\, \xi_{i-p,j-r}\, \boldsymbol{L}_{p-r,i-p+r-j},$

for $i \geq 0$ and any j.

We rewrite (7.44) as

$$(7.45) \quad \boldsymbol{F}_{i,j} \equiv \sum_{p=0}^{l} \sum_{h=0}^{p} A_{l-p,n-i+p}^{l,n} \, \xi_{p,p-h} \, \xi_{i-p,j-p+h} \, \boldsymbol{L}_{h,i-j-h}$$

$$\equiv \sum_{h=0}^{l} \sum_{p=h}^{l} A_{l-p,n-i+p}^{l,n} \, \xi_{p,p-h} \, \xi_{i-p,j-p+h} \, \boldsymbol{L}_{h,i-j-h}$$

$$\equiv \sum_{h=0}^{l} \sigma_{h,i-j,j} \, \boldsymbol{L}_{h,i-j-h}, \quad \text{for } i \geq 0 \text{ and any } j,$$

where $\sigma_{h,i,j}$ is the rational number defined by (7.39) for any integers h, i, j that satisfy $0 \leq h \leq l$ and $i+j \geq 0$. Hence, in terms of (7.40) and (7.45), we have

$$(7.46) \quad \boldsymbol{N}_{i+1,j} - \left[\frac{2i-j+1}{2}\right] \boldsymbol{N}_{i,j-1}$$

$$\equiv \boldsymbol{N}_{i,j}^{(1)} - 2(j+1)\, a_{m,2} \, \boldsymbol{N}_{i,j+1} + \sum_{h=0}^{l} \sigma_{h,i-j+1,j} \, \boldsymbol{L}_{h,i+1-j-h},$$

$$\text{for } i \geq 0 \text{ and any } j.$$

In order to establish the relation

$$(7.47) \quad \boldsymbol{F}_{i,j} \equiv 0, \quad \text{for any integers } i, j \text{ that satisfy } j \geq i-1,$$

we consider four cases. For $i \leq -1$ and any j, (7.41) yields $\boldsymbol{F}_{i,j} \equiv 0$. We use (7.45), $\boldsymbol{L}_{0,1} \equiv \boldsymbol{L}_{1,0} \equiv 0$ from (1.21) of page 8, and (1.22) to obtain

$$\boldsymbol{F}_{i,i-1} \equiv \sum_{h=0}^{l} \sigma_{h,1,i-1} \, \boldsymbol{L}_{h,1-h} \equiv 0, \quad \text{for } i \geq 0.$$

We apply (7.45), $\boldsymbol{L}_{0,0} \equiv 1$ from (1.21) of page 8, (1.22), (7.39), (7.1), and (6.4) to verify that

$$\boldsymbol{F}_{i,i} \equiv \sum_{h=0}^{l} \sigma_{h,0,i} \, \boldsymbol{L}_{h,-h} \equiv \sigma_{0,0,i} \equiv \sum_{p=0}^{l} A_{l-p,n-i+p}^{l,n} \, \xi_{p,p} \, \xi_{i-p,i-p} \equiv 0, \quad \text{for } i \geq 0.$$

When $i \geq 0$ and $j \geq i+1$, we employ (7.45) with $i-j-h \leq -h-1 \leq -1$ and $\boldsymbol{L}_{h,i-j-h} \equiv 0$ from (1.22) to deduce that $\boldsymbol{F}_{i,j} \equiv 0$. By combining these results, we see that (7.47) is valid.

In view of (7.40) and (7.47), we find that

$$(7.48) \quad \boldsymbol{N}_{i+1,j} \equiv \left[\frac{2i-j+1}{2}\right] \boldsymbol{N}_{i,j-1} + \boldsymbol{N}_{i,j}^{(1)} - 2(j+1)\, a_{m,2} \, \boldsymbol{N}_{i,j+1},$$

$$\text{for any integers } i, j \text{ that satisfy } j \geq i.$$

Next, we shall use (7.48) to establish the relation

$$(7.49) \quad \boldsymbol{N}_{i_0+j,j} \equiv 0, \quad \text{for } i_0 \leq 1 \text{ and any } j.$$

To clarify the argument, we consider separately: $i_0 \leq -1$; $i_0 = 0$; and $i_0 = 1$.

1. Suppose that $i_0 \leq -1$. Then, (7.33) yields $\boldsymbol{N}_{i_0,j} \equiv 0$, for any j. Using this, we substitute $i = i_0$ and $j \geq 1 > i_0 = i$ in (7.48) to obtain $\boldsymbol{N}_{i_0+1,j} \equiv 0$, for $j \geq 1$. By substituting $i = i_0+1$ and $j \geq 2 > i_0+1 = i$ in (7.48), we deduce $\boldsymbol{N}_{i_0+2,j} \equiv 0$, for $j \geq 2$. Continuing in this manner, we obtain $\boldsymbol{N}_{i_0+j,j} \equiv 0$, for any j.

2. Suppose that $i_0 = 0$. We use (7.33) and (6.4) to verify that $\boldsymbol{N}_{j,j} \equiv 0$, for $j \leq -1$, and

(7.50) $$\boldsymbol{N}_{0,j} \equiv A_{l,n}^{l,n} \boldsymbol{\mathcal{M}}_{0,0,0,j} \equiv 0, \quad \text{for any } j.$$

Using this, we substitute $i = 0$ and $j \geq 1$ in (7.48) to deduce that $\boldsymbol{N}_{1,j} \equiv 0$, for $j \geq 1$. Substituting $i = 1$ and $j \geq 2$ in (7.48), we find that $\boldsymbol{N}_{2,j} \equiv 0$, for $j \geq 2$. Continuing in this way, we conclude that $\boldsymbol{N}_{j,j} \equiv 0$, for any j.

3. Suppose that $i_0 = 1$. Due to (7.33) and (7.50), we have $\boldsymbol{N}_{1+j,j} \equiv 0$, for $j \leq -1$. We apply (7.33) and (6.4) to observe that

$$\boldsymbol{N}_{1,j} \equiv \sum_{r=0}^{1} \sum_{p=0}^{r} A_{l-p,n-r+p}^{l,n} \boldsymbol{\mathcal{M}}_{p,r-p,1-r,j}$$
$$\equiv A_{l,n}^{l,n} \boldsymbol{\mathcal{M}}_{0,0,1,j} + A_{l,n-1}^{l,n} \boldsymbol{\mathcal{M}}_{0,1,0,j} + A_{l-1,n}^{l,n} \boldsymbol{\mathcal{M}}_{1,0,0,j}$$
$$\equiv 0, \quad \text{for any } j.$$

Using this, we substitute $i = 1$ and $j \geq 1$ in (7.48) to deduce that $\boldsymbol{N}_{2,j} \equiv 0$, for $j \geq 1$. Substituting $i = 2$ and $j \geq 2$ in (7.48), we find that $\boldsymbol{N}_{3,j} \equiv 0$, for $j \geq 2$. Continuing in this way, we conclude that $\boldsymbol{N}_{1+j,j} \equiv 0$, for any j. Consequently, (7.49) is valid.

We define $\boldsymbol{P}_{i,j}$ in $\mathcal{S}_m$ by means of

(7.51) $$\boldsymbol{P}_{i,j} \equiv \boldsymbol{N}_{i+j,j}, \quad \text{for all integers } i \text{ and } j.$$

Thus, $\boldsymbol{P}_{i,j}$ satisfies (7.35). Moreover, we use (7.51) and (7.49) to see that $\boldsymbol{P}_{i,j}$ satisfies (7.36). For any integers i, j subject to $i+j \geq 0$, we replace i in (7.46) with $i + j$ and use (7.51) to obtain

(7.52) $$\boldsymbol{P}_{i+1,j} - \left[\frac{2i+j+1}{2}\right] \boldsymbol{P}_{i+1,j-1}$$
$$\equiv \boldsymbol{P}_{i,j}^{(1)} - 2(j+1) \boldsymbol{a}_{m,2} \boldsymbol{P}_{i-1,j+1} + \sum_{h=0}^{l} \sigma_{h,i+1,j} \boldsymbol{L}_{h,i+1-h},$$
$$\text{whenever } i + j \geq 0,$$

where $\sigma_{h,i,j}$ is given in (7.39). We find that the rational numbers $\gamma_{i,j,k}$ defined in (7.38) satisfy

(7.53) $$\gamma_{i,j,j} \equiv 1, \quad \text{for any } i, j,$$

and

(7.54) $$\frac{1}{2}(2i + k + 1) \gamma_{i,j,k} \equiv \gamma_{i,j,k-1}, \quad \text{for } k \leq j \text{ and any } i.$$

We note that (7.33) and (7.26) give $\boldsymbol{N}_{i,j} \equiv 0$, for $i \geq 0$ and $j \leq -1$. Hence, we have $\boldsymbol{N}_{i,j} \equiv 0$, for $j \leq -1$ and any i. In view of (7.51), this yields

(7.55) $$\boldsymbol{P}_{i,j} \equiv 0, \quad \text{for } j \leq -1 \text{ and any } i.$$

For $i \geq 0$ and $j \geq 0$, we combine (7.53), (7.55), (7.54), and (7.52) to obtain

$$\boldsymbol{P}_{i+1,j} \equiv \sum_{k=0}^{j} \left[\gamma_{i,j,k} \, \boldsymbol{P}_{i+1,k} - \gamma_{i,j,k-1} \, \boldsymbol{P}_{i+1,k-1} \right] \tag{7.56}$$

$$\equiv \sum_{k=0}^{j} \gamma_{i,j,k} \left[\boldsymbol{P}_{i+1,k} - \frac{1}{2}(2i+k+1)\boldsymbol{P}_{i+1,k-1} \right]$$

$$\equiv \sum_{k=0}^{j} \gamma_{i,j,k} \left[\begin{array}{c} \boldsymbol{P}_{i,k}^{(1)} - 2(k+1)\boldsymbol{a}_{m,2}\,\boldsymbol{P}_{i-1,k+1} \\ + \sum_{h=0}^{l} \sigma_{h,i+1,k}\, \boldsymbol{L}_{h,i+1-h} \end{array} \right].$$

Due to (7.55), (7.56) is valid for $i \geq 0$ and any j. This yields (7.37) and completes the proof of Proposition 7.5. $\square$

PROPOSITION 7.6. *The rational numbers $\sigma_{h,i,j}$ defined in (7.39) are given in terms of*

$$\vartheta_{i,j} \equiv \frac{\dfrac{(-1)^{l+n+i+j}}{2^j} \binom{l+n-i}{j}}{\left[\displaystyle\prod_{k=1}^{l+n-i-j} (2l+2n-1-k) \right]}, \quad \text{for } i+j \geq 0, \tag{7.57}$$

by

$$\sigma_{h,i,j} \equiv \vartheta_{i,j} \binom{m-h}{l-h} \left[\prod_{r=1}^{l-h} (l-r) \right] \binom{m-i+h}{n-i+h} \left[\prod_{s=1}^{n-i+h} (n-s) \right], \tag{7.58}$$

$$\text{for } 0 \leq h \leq l \text{ and } i+j \geq 0.$$

PROOF. To facilitate reference, we repeat (7.39) as

$$\sigma_{h,i,j} \equiv \sum_{p=h}^{l} A_{l-p,n-i-j+p}^{l,n} \, \xi_{p,p-h}\, \xi_{i+j-p,j-p+h}, \tag{7.59}$$

$$\text{for } 0 \leq h \leq l \text{ and } i+j \geq 0.$$

We use (6.4) to obtain

$$A_{l-p,n-i-j+p}^{l,n} \equiv \frac{(-1)^{l+n+i+j} P_1 P_2}{Q} \binom{m-p}{l-p}\binom{m-i-j+p}{n-i-j+p}, \tag{7.60}$$

where

$$P_1 \equiv \prod_{k=1}^{l-p}(l-k), \quad P_2 \equiv \prod_{k=1}^{n-i-j+p}(n-k), \quad \text{and} \quad Q \equiv \prod_{k=1}^{l+n-i-j}(2l+2n-1-k).$$

We apply (7.1) of page 69 to see that

$$\xi_{p,p-h} \equiv \frac{P_3}{2^{p-h}} \binom{m-h}{p-h}, \tag{7.61}$$

where

$$P_3 \equiv \prod_{k=1}^{p-h}(p-k).$$

Also, (7.1) yields

(7.62) $$\xi_{i+j-p,j-p+h} \equiv \frac{P_4}{2^{j-p+h}} \binom{m-i+h}{j-p+h},$$

where

$$P_4 \equiv \prod_{k=1}^{j-p+h} (i+j-p-k).$$

For $0 \leq h \leq p \leq l$, we check that

(7.63) $$P_1 P_3 \equiv \prod_{k=1}^{l-h} (l-k).$$

For $0 \leq h \leq p \leq l$, we observe that

(7.64) $$\binom{m-i-j+p}{n-i-j+p}\binom{m-i+h}{j-p+h} P_2 P_4$$
$$\equiv \binom{m-i-j+p}{n-i-j+p}\binom{m-i+h}{j-p+h} \prod_{k=1}^{n-i+h} (n-k).$$

Namely, for either $n-i-j+p < 0$ or $j-p+h < 0$, (7.64) reduces to $0 \equiv 0$. While, for $n-i-j+p \geq 0$ and $j-p+h \geq 0$, we find that $i-h \leq i+j-p \leq n$ and $P_2 P_4 \equiv \prod_{k=1}^{n-i+h} (n-k)$. Also, we have

(7.65) $$\binom{m-h}{p-h}\binom{m-p}{l-p} \equiv \binom{m-h}{l-h}\binom{l-h}{p-h}$$

and

(7.66) $$\binom{m-i+h}{j-p+h}\binom{m-i-j+p}{n-i-j+p} \equiv \binom{m-i+h}{n-i+h}\binom{n-i+h}{j-p+h}.$$

We employ (7.59)–(7.66) to verify that

(7.67) $$\sigma_{h,i,j} \equiv \frac{(-1)^{l+n+i+j}}{2^j Q} \binom{m-h}{l-h}\left[\prod_{r=1}^{l-h}(l-r)\right]\binom{m-i+h}{n-i+h}\left[\prod_{s=1}^{n-i+h}(n-s)\right] S,$$

where

(7.68) $$S \equiv \sum_{p=h}^{l} \binom{l-h}{p-h}\binom{n-i+h}{j-p+h} \equiv \sum_{p=0}^{l-h} \binom{l-h}{p}\binom{n-i+h}{j-p}$$
$$\equiv \binom{l+n-i}{j}.$$

At the last step of (7.68), we applied Vandermonde's convolution; in this regard, see (6.56) on page 64. We combine (7.67) and (7.68) to obtain (7.58). This completes the proof of Proposition 7.6. □

Using the abbreviated notation of (1.41) on page 10, we rewrite (1.23) as

(7.69) $$\boldsymbol{M}_{h,i} \equiv \binom{m-h}{l-h}\left[\prod_{r=1}^{l-h}(l-r)\right]\binom{m-i+h}{n-i+h}\left[\prod_{s=1}^{n-i+h}(n-s)\right] \boldsymbol{L}_{h,i-h},$$

for $0 \leq h \leq l$ and any i.

COROLLARY 7.7. *The expressions $\sigma_{h,i,j}$ in (7.39), $\vartheta_{i,j}$ in (7.57), $L_{i,j}$, and $M_{h,i}$ are related through*

(7.70) $\qquad \sigma_{h,i,j}\, \boldsymbol{L}_{h,i-h} \equiv \vartheta_{i,j}\, \boldsymbol{M}_{h,i}, \quad \text{for } 0 \leq h \leq l \text{ and } i+j \geq 0.$

PROOF. We combine (7.58) and (7.69) to obtain (7.70). $\qquad \square$

For further results about $\boldsymbol{P}_{i,j}$, (7.70) plays a key role.

With respect to (1.41) and (1.24)–(1.25), we have

(7.71) $\qquad A_i \equiv \dfrac{-1}{l+n+i-1}, \quad \text{for } i \geq 1,$

and

(7.72) $\qquad B_i \equiv \dfrac{l+n-i}{l+n+i-2}, \quad \text{for } i \geq 1.$

In terms of (1.41) and (1.26)–(1.27), $\boldsymbol{I}_{h,i}$ is defined in $\mathcal{S}_m$ through

(7.73) $\qquad \boldsymbol{I}_{h,0} \equiv \boldsymbol{I}_{h,1} \equiv 0, \quad \text{for } 0 \leq h \leq l,$

and

(7.74) $\quad \boldsymbol{I}_{h,i+1} \equiv A_i\, \boldsymbol{I}_{h,i}^{(1)} + B_i\, \boldsymbol{a}_{m,2}\, \boldsymbol{I}_{h,i-1} + \boldsymbol{M}_{h,i+1},$
$\qquad\qquad\qquad\qquad\qquad \text{for } 0 \leq h \leq l \text{ and } 1 \leq i \leq l+n-1.$

We use (7.71)–(7.74) and the rational numbers

(7.75) $\quad r_{i,j} \equiv \dfrac{(l+n-i-j)(l+n+i+j-1)}{(l+n-i)(l+n+i-1)}, \quad \text{for } 0 \leq i \leq l+n-1 \text{ and any } j,$

to define $\boldsymbol{S}_{h,i,j}$ in $\mathcal{S}_m$ by

(7.76) $\qquad \boldsymbol{S}_{h,i,j} \equiv r_{i,j}\, \boldsymbol{I}_{h,i}, \quad \text{for } 0 \leq h \leq l,\ 0 \leq i \leq l+n-1 \text{ and any } j,$

and

(7.77) $\qquad \boldsymbol{S}_{h,l+n,j} \equiv \boldsymbol{I}_{h,l+n}, \quad \text{for } 0 \leq h \leq l \text{ and any } j.$

In particular, (7.77), (7.76), and (7.75) give

(7.78) $\qquad \boldsymbol{S}_{h,l+n,0} \equiv \boldsymbol{I}_{h,l+n}, \quad \text{for } 0 \leq h \leq l,$

and

(7.79) $\qquad \boldsymbol{S}_{h,l+n-s,s} \equiv 0, \quad \text{for } 0 \leq h \leq l \text{ and } 1 \leq s \leq l+n.$

Moreover, (7.76) and (7.73) yield

(7.80) $\qquad \boldsymbol{S}_{h,i,j} \equiv 0, \quad \text{for } 0 \leq h \leq l,\ 0 \leq i \leq 1, \text{ and any } j.$

We define rational numbers $e_{i,j}$, $f_{i,j}$, and $g_{i,j}$, for $1 \leq i \leq l+n-1$ and $0 \leq j \leq l+n-1-i$, by means of

(7.81) $\qquad e_{i,j} \equiv \dfrac{-2j}{(l+n-i-j)(l+n+i+j-1)},$

(7.82) $\qquad f_{i,j} \equiv \dfrac{-(l+n-i)}{(l+n-i-j)(l+n+i+j-1)},$

and

(7.83) $\qquad g_{i,j} \equiv \dfrac{-(l+n-i)(l+n-i+1)}{2\,(j+1)(l+n-i-j)(l+n+i+j-1)}.$

7.3. SPECIAL COMBINATIONS OF THE COEFFICIENTS FOR (5.1)

PROPOSITION 7.8. *In terms of (7.81)–(7.83), the polynomials $S_{h,i,j}$ satisfy*

$$(7.84) \quad S_{h,i+1,j} - \frac{2i+j+1}{2} e_{i,j} S_{h,i+1,j-1}$$
$$\equiv f_{i,j} S_{h,i,j}^{(1)} - 2(j+1) a_{m,2} g_{i,j} S_{h,i-1,j+1} + M_{h,i+1},$$
$$\text{for } 0 \leq h \leq l,\ 1 \leq i \leq l+n-1 \text{ and } 0 \leq j \leq l+n-1-i,$$

where $M_{h,i+1}$ is obtained from (7.69).

PROOF. For $1 \leq i \leq l+n-1$ and $0 \leq j \leq l+n-1-i$, we note that (7.71), (7.82), (7.75), (7.72), and (7.83) yield

$$(7.85) \qquad A_i \equiv f_{i,j} r_{i,j}$$

and

$$(7.86) \qquad B_i \equiv -2(j+1) g_{i,j} r_{i-1,j+1}.$$

For $0 \leq h \leq l$, $1 \leq i \leq l+n-1$, and $0 \leq j \leq l+n-1-i$, we employ (7.74), (7.85), (7.86), and (7.76) to verify that

$$(7.87) \quad I_{h,i+1} \equiv A_i I_{h,i}^{(1)} + B_i a_{m,2} I_{h,i-1} + M_{h,i+1}$$
$$\equiv f_{i,j}\big(r_{i,j} I_{h,i}^{(1)}\big) - 2(j+1) a_{m,2} g_{i,j}\big(r_{i-1,j+1} I_{h,i-1}\big) + M_{h,i+1}$$
$$\equiv f_{i,j} S_{h,i,j}^{(1)} - 2(j+1) a_{m,2} g_{i,j} S_{h,i-1,j+1} + M_{h,i+1}.$$

Case 1. As the first of two situations, suppose that h, i, and j satisfy $0 \leq h \leq l$, $1 \leq i \leq l+n-2$, and $0 \leq j \leq l+n-1-i$. Then, in terms of (7.76), (7.75), and (7.81), we find that

$$(7.88) \quad S_{h,i+1,j} - \frac{2i+j+1}{2} e_{i,j} S_{h,i+1,j-1}$$
$$\equiv \left[r_{i+1,j} - \frac{2i+j+1}{2} e_{i,j} r_{i+1,j-1} \right] I_{h,i+1}$$
$$\equiv \left[\frac{(l+n-i-1-j)(l+n+i+j) + j(2i+j+1)}{(l+n-i-1)(l+n+i)} \right] I_{h,i+1}$$
$$\equiv I_{h,i+1}.$$

By combining (7.88) and (7.87), we see that (7.84) is satisfied with respect to the preceding restriction.

Case 2. For the remaining situation, suppose that h, i, and j satisfy $0 \leq h \leq l$, $i = l+n-1$, and $j = 0$. Then, in view of (7.78) and (7.81), we see that the left member of (7.84) reduces to $S_{h,l+n,0} \equiv I_{h,l+n}$. We combine this with the substitution of $i = l+n-1$ and $j = 0$ in (7.87) to deduce that (7.84) is valid for these values of h, i, and j. This completes the proof. □

With reference to $S_{h,i,j}$ in (7.76)–(7.77) and $\vartheta_{i,j}$ in (7.57), we define $R_{h,i,j}$ in $\mathcal{S}_m$ through

$$(7.89) \qquad R_{h,i,j} \equiv \vartheta_{i,j} S_{h,i,j}, \quad \text{for } 0 \leq h \leq l,\ 0 \leq i \leq l+n, \text{ and } j \geq -i.$$

In particular, after observing that (7.57) gives

(7.90) $$\vartheta_{l+n,0} \equiv \frac{(-1)^{2l+2n}}{2^0}\binom{0}{0} \equiv 1,$$

we combine (7.89), (7.90), and (7.78)–(7.79) to obtain

(7.91) $$\boldsymbol{R}_{h,l+n,0} \equiv \boldsymbol{I}_{h,l+n}, \quad \text{for } 0 \le h \le l,$$

and

(7.92) $$\boldsymbol{R}_{h,l+n-s,s} \equiv 0, \quad \text{for } 0 \le h \le l \text{ and } 1 \le s \le l+n.$$

Also, (7.89) and (7.80) yield

(7.93) $$\boldsymbol{R}_{h,i,j} \equiv 0, \quad \text{for } 0 \le h \le l,\ 0 \le i \le 1, \text{ and } j \ge -i.$$

We use (7.89) to define $\boldsymbol{Q}_{i,j}$ in $\mathcal{S}_m$ by means of

(7.94) $$\boldsymbol{Q}_{i,j} \equiv \sum_{h=0}^{l} \boldsymbol{R}_{h,i,j}, \quad \text{for } 0 \le i \le l+n \text{ and } j \ge -i.$$

In particular, (7.94) and (7.91) give

(7.95) $$\boldsymbol{Q}_{l+n,0} \equiv \sum_{h=0}^{l} \boldsymbol{I}_{h,l+n}$$

while (7.94) and (7.92) yield

(7.96) $$\boldsymbol{Q}_{l+n-s,s} \equiv 0, \quad \text{for } 1 \le s \le l+n.$$

We apply (7.94) and (7.93) to see that

(7.97) $$\boldsymbol{Q}_{i,j} \equiv 0, \quad \text{for } 0 \le i \le 1 \text{ and } j \ge -i.$$

PROPOSITION 7.9. *The polynomials $\boldsymbol{Q}_{i,j}$ satisfy*

(7.98) $$\boldsymbol{Q}_{i+1,j} - \left[\frac{2i+j+1}{2}\right]\boldsymbol{Q}_{i+1,j-1}$$
$$\equiv \boldsymbol{Q}_{i,j}^{(1)} - 2(j+1)\,\boldsymbol{a}_{m,2}\,\boldsymbol{Q}_{i-1,j+1} + \sum_{h=0}^{l}\sigma_{h,i+1,j}\,\boldsymbol{L}_{h,i+1-h},$$
$$\text{for } 1 \le i \le l+n-1 \text{ and } 0 \le j \le l+n-1-i,$$

where $\sigma_{h,i,j}$ and $\boldsymbol{L}_{i,j}$ are given by (7.58) of page 80 and (1.21)–(1.22) of page 8.

PROOF. For $0 \le h \le l$, $1 \le i \le l+n-1$, and $0 \le j \le l+n-1-i$, we use (7.57) and (7.81)–(7.83) to verify that $\vartheta_{i,j}$, $e_{i,j}$, $f_{i,j}$, and $g_{i,j}$ satisfy

(7.99) $$\vartheta_{i+1,j}\,e_{i,j} \equiv \vartheta_{i+1,j-1},$$

(7.100) $$\vartheta_{i+1,j}\,f_{i,j} \equiv \vartheta_{i,j},$$

and

(7.101) $$\vartheta_{i+1,j}\,g_{i,j} \equiv \vartheta_{i-1,j+1}.$$

7.3. SPECIAL COMBINATIONS OF THE COEFFICIENTS FOR (5.1)

For $0 \leq h \leq l$, $1 \leq i \leq l+n-1$, and $0 \leq j \leq l+n-1-i$, we multiply (7.84) by $\vartheta_{i+1,j}$ and apply (7.99)–(7.101), (7.70), and (7.89) to obtain

$$0 \equiv \vartheta_{i+1,j}\, \boldsymbol{S}_{h,i+1,j} - \frac{2i+j+1}{2}\, \vartheta_{i+1,j}\, e_{i,j}\, \boldsymbol{S}_{h,i+1,j-1} - \vartheta_{i+1,j}\, f_{i,j}\, \boldsymbol{S}_{h,i,j}^{(1)}$$
$$+ 2(j+1)\, a_{m,2}\, \vartheta_{i+1,j}\, g_{i,j}\, \boldsymbol{S}_{h,i-1,j+1} - \vartheta_{i+1,j}\, \boldsymbol{M}_{h,i+1},$$

$$0 \equiv \vartheta_{i+1,j}\, \boldsymbol{S}_{h,i+1,j} - \frac{2i+j+1}{2}\, \vartheta_{i+1,j-1}\, \boldsymbol{S}_{h,i+1,j-1} - \vartheta_{i,j}\, \boldsymbol{S}_{h,i,j}^{(1)}$$
$$+ 2(j+1)\, a_{m,2}\, \vartheta_{i-1,j+1}\, \boldsymbol{S}_{h,i-1,j+1} - \sigma_{h,i+1,j}\, \boldsymbol{L}_{h,i+1-h},$$

and

(7.102) $\quad 0 \equiv \boldsymbol{R}_{h,i+1,j} - \dfrac{2i+j+1}{2}\, \boldsymbol{R}_{h,i+1,j-1} - \boldsymbol{R}_{h,i,j}^{(1)}$
$$+ 2(j+1)\, a_{m,2}\, \boldsymbol{R}_{h,i-1,j+1} - \sigma_{h,i+1,j}\, \boldsymbol{L}_{h,i+1-h}.$$

Letting $\mathcal{M}_h$ denote the right member of (7.102) for $0 \leq h \leq l$, we use (7.102) and (7.94) to deduce that

(7.103) $\quad 0 \equiv \displaystyle\sum_{h=0}^{l} \mathcal{M}_h \equiv \boldsymbol{Q}_{i+1,j} - \dfrac{2i+j+1}{2}\, \boldsymbol{Q}_{i+1,j-1} - \boldsymbol{Q}_{i,j}^{(1)}$
$$+ 2(j+1)\, a_{m,2}\, \boldsymbol{Q}_{i-1,j+1} - \sum_{h=0}^{l} \sigma_{h,i+1,j}\, \boldsymbol{L}_{h,i+1-h},$$
$$\text{for } 1 \leq i \leq l+n-1 \text{ and } 0 \leq j \leq l+n-1-i.$$

We rewrite (7.103) to obtain (7.98) and complete the proof of Proposition 7.9. $\square$

Our development throughout this chapter has been motivated by the identity (6.23) in Theorem 6.3 on page 58 and an analogy with results in [**24**]. However, unlike the situation for (6.23), no restrictions are placed on (1.1) in the next result.

THEOREM 7.10. *For (5.1) on Ω^{**} into which (1.1) on Ω is transformed by $y(z) = \rho(z)\, v(z)$ and $z = f(\zeta)$ according to the context of page 47, suppose that $d_{0,1}(\zeta) \equiv d_{0,2}(\zeta) \equiv 0$. Let l and n be integers subject to either $1 \leq l \leq n \leq m$ or $l = 0 < 3 \leq n \leq m$. Then, the coefficients $d_{i,j}(\zeta)$ of (5.1) satisfy*

(7.104) $\quad \displaystyle\sum_{\mu=0}^{l} \sum_{\nu=0}^{n} A_{\mu,\nu}^{l,n}\, d_{l-\mu,n-\nu}^{(\mu+\nu)}(\zeta) \equiv \bigl(f'(\zeta)\bigr)^{l+n} \mathcal{I}_{m,l,n}\bigl(f(\zeta)\bigr),$

where $A_{\mu,\nu}^{l,n}$ is given by (6.3) on page 53 and where $\mathcal{I}_{m,l,n}\bigl(f(\zeta)\bigr)$ is obtained for $z = f(\zeta)$ by substituting $c_{i,j}^{(k)}(z)$ from (1.1) for $\boldsymbol{w}_{i,j}^{(k)}$ in $\mathcal{I}_{m,l,n}$ of (1.28) on page 8.

PROOF. We note that (7.57) yields $\vartheta_{i+1,-1} \equiv 0$, for $0 \leq h \leq l$ and $i \geq 0$. Due to (7.89), we have $\boldsymbol{R}_{h,i+1,-1} \equiv 0$, for $0 \leq h \leq l$ and $0 \leq i \leq l+n-1$. We use (7.94) to obtain

(7.105) $\qquad \boldsymbol{Q}_{i+1,-1} \equiv 0, \quad \text{for } 0 \leq i \leq l+n-1.$

We combine (7.53), (7.105), (7.54), and (7.98) to verify that

$$(7.106) \quad \boldsymbol{Q}_{i+1,j} \equiv \sum_{k=0}^{j} \Big[\gamma_{i,j,k} \, \boldsymbol{Q}_{i+1,k} - \gamma_{i,j,k-1} \, \boldsymbol{Q}_{i+1,k-1} \Big]$$

$$\equiv \sum_{k=0}^{j} \gamma_{i,j,k} \Big[\boldsymbol{Q}_{i+1,k} - \frac{1}{2}(2i+k+1) \boldsymbol{Q}_{i+1,k-1} \Big]$$

$$\equiv \sum_{k=0}^{j} \gamma_{i,j,k} \begin{bmatrix} \boldsymbol{Q}_{i,k}^{(1)} - 2(k+1)a_{m,2} \, \boldsymbol{Q}_{i-1,k+1} \\ + \sum_{h=0}^{l} \sigma_{h,i+1,k} \, \boldsymbol{L}_{h,i+1-h} \end{bmatrix},$$

for $1 \le i \le l+n-1$ and $0 \le j \le l+n-1-i$.

In view of (7.36) on page 76 and (7.97), we see that the statement

$$(7.107) \quad \boldsymbol{P}_{i,j} \equiv \boldsymbol{Q}_{i,j}, \quad \text{for } 0 \le i \le i_0 \text{ and } 0 \le j \le l+n-i,$$

is valid for $i_0 = 1$. Suppose that i_0 is an integer satisfying $1 \le i_0 \le l+n-1$ such that (7.107) is valid. Then, we use (7.37) on page 76 as well as (7.107) and (7.106) to deduce that $\boldsymbol{P}_{i_0+1,j} \equiv \boldsymbol{Q}_{i_0+1,j}$, for $0 \le j \le l+n-1-i_0 = l+n-(i_0+1)$. Thus, we have

$$(7.108) \quad \boldsymbol{P}_{i,j} \equiv \boldsymbol{Q}_{i,j}, \quad \text{for } 0 \le i \le l+n \text{ and } 0 \le j \le l+n-i.$$

We employ (7.35) on page 76 and (7.108) to verify that

$$(7.109) \quad \boldsymbol{N}_{l+n,t} \equiv \boldsymbol{P}_{l+n-t,t} \equiv \boldsymbol{Q}_{l+n-t,t}, \quad \text{for } 0 \le t \le l+n.$$

We find that (7.109) and (7.96) yield

$$(7.110) \quad \boldsymbol{N}_{l+n,t} \equiv 0, \quad \text{for } 1 \le t \le l+n.$$

After substituting $c_{i,j}^{(k)}(z)$ from (1.1) for $\boldsymbol{w}_{i,j}^{(k)}$ in (7.110) and setting $z = f(\zeta)$, we combine the resulting expressions with (7.34) on page 75 to obtain

$$(7.111) \quad \sum_{\mu=0}^{l} \sum_{\nu=0}^{n} A_{\mu,\nu}^{l,n} \, d_{l-\mu,n-\nu}^{(\mu+\nu)}(\zeta) \equiv \big(f'(\zeta)\big)^{l+n} \boldsymbol{N}_{l+n,0}\big(f(\zeta)\big).$$

Moreover, (7.109), (7.95), and (1.28) on page 8 give

$$(7.112) \quad \boldsymbol{N}_{l+n,0} \equiv \sum_{h=0}^{l} \boldsymbol{I}_{h,l+n} \equiv \boldsymbol{\mathcal{I}}_{m,l,n}.$$

After substituting $c_{i,j}^{(k)}(z)$ from (1.1) for $\boldsymbol{w}_{i,j}^{(k)}$ in (7.112) and setting $z = f(\zeta)$, we combine the resulting expression with (7.111) to obtain (7.104). This completes the proof of Theorem 7.10. □

CHAPTER 8

Basic Semi-Invariants of the First Kind for $m \geq 2$

The polynomials $L_{i,j}$ are defined in $\mathcal{S}_m$ for all integers i and j by (1.21)–(1.22) of page 8. They satisfy $L_{j,i} \equiv L_{i,j}$, for all i and j. Here, we focus on

$$(8.1) \qquad L_{i,j} \equiv \sum_{p=0}^{i} \sum_{q=0}^{j} K_{i-p,p} K_{j-q,q} w_{p,q}, \quad \text{where } 0 \leq i \leq j \leq m,$$

for which $K_{i-p,p}$ and $K_{j-q,q}$ are defined by (1.18)–(1.20) with $m \geq 2$. In particular, these formulas yield

$$(8.2) \qquad L_{0,0} \equiv 1, \quad L_{0,1} \equiv 0, \quad L_{0,2} \equiv 0, \quad \text{and} \quad L_{1,1} \equiv w_{1,1} - (w_{0,1})^2.$$

Since $L_{0,0}$, $L_{0,1}$, and $L_{0,2}$ are constants, they are not semi-invariants.

THEOREM 8.1. *For any integers l and n that satisfy either $1 \leq l \leq n \leq m$ or $l = 0 < 3 \leq n \leq m$, $L_{l,n}$ is an isobaric semi-invariant of the first kind that is also a basic polynomial of index (l, n) according to Definition 1.5 on page 6.*

PROOF. We note that $K_{i,j} \equiv 0$, when either $i < 0$ or $i \geq 1$ and $j \geq m$. For $i = 0$ or $i \geq 1$ and $j < m$, we deduce that $K_{i,j}$ is an isobaric polynomial over $\mathbb{Q}$ of weight i in the variables

$$(8.3) \qquad w_{0,1}, w_{0,1}^{(1)}, w_{0,1}^{(2)}, \ldots \quad \text{and} \quad w_{0,2}, w_{0,2}^{(1)}, w_{0,2}^{(2)}, \ldots.$$

We use (8.1), (1.19), and the preceding observations to obtain

$$(8.4) \qquad L_{l,n} \equiv w_{l,n} + \sum_{\substack{0 \leq p \leq l \\ 0 \leq q \leq n \\ (p,q) \neq (l,n)}} K_{l-p,p} K_{n-q,q} w_{p,q}$$

and verify that the coefficient of $w_{l,n}$ in $L_{l,n}$ is 1. Moreover, (8.4) shows that $L_{l,n}$ is an isobaric polynomial in $\mathcal{S}_m$ of weight $l + n$. We use (8.4) to see that each nonzero term of $L_{l,n}$ has at most one factor $w_{i,j}^{(k)}$ from (1.6) not listed in (8.3).

Suppose that $(l, n) \neq (1, 1)$ and let $w_{i_0,j_0}^{(k_0)}$ be a variable from (1.6) that is effectively involved in $L_{l,n}$ of (8.4). Then, we have $n \geq 2$ and observe that each variable $w_{i,j}^{(k)}$ in (8.3) has $i = 0 \leq l$ and $j \leq 2 \leq n$. If $w_{i_0,j_0}^{(k_0)}$ is involved in some $K_{l-p,p}$ or $K_{n-q,q}$, then $w_{i_0,j_0}^{(k_0)}$ appears in (8.3) and therefore has $i_0 \leq l$ as well as $j_0 \leq n$. Suppose that $w_{i_0,j_0}^{(k_0)}$ is not involved in any $K_{l-p,p}$ or $K_{n-q,q}$. Then, $k_0 = 0$ and there are integers p, q satisfying $0 \leq p \leq l$ and $0 \leq q \leq n$ such that $w_{p,q} \equiv w_{i_0,j_0}$. If $w_{p,q}$ appears in (1.6), then $i_0 = p \leq l$ and $j_0 = q \leq n$. If $w_{p,q}$ is not listed in (1.6), then $w_{q,p}$ appears in (1.6) with $q < p$, (1.7) yields $w_{i_0,j_0} \equiv w_{p,q} \equiv w_{q,p}$, and we have $i_0 = q < p \leq l$ as well as $j_0 = p \leq l \leq n$.

If $(l, n) = (1, 1)$, then $\boldsymbol{L}_{l,n} \equiv \boldsymbol{L}_{1,1} \equiv \boldsymbol{w}_{1,1} - (\boldsymbol{w}_{0,1})^2$ and each variable $\boldsymbol{w}_{i,j}^{(k)}$ from (1.6) effectively involved in $\boldsymbol{L}_{l,n}$ has $i \leq 1 = l$ and $j \leq 1 = n$.

For either $1 \leq l \leq n \leq m$ or $l = 0 < 3 \leq n \leq m$, the observations of the three previous paragraphs show that $\boldsymbol{L}_{l,n}$ is a basic polynomial of index (l, n).

Substituting $c_{i,j}^{*(k)}(z)$ from (1.3) for $\boldsymbol{w}_{i,j}^{(k)}$ in (8.1), we obtain

$$(8.5) \qquad L_{l,n}^*(z) \equiv \sum_{p=0}^{l} \sum_{q=0}^{n} K_{l-p,p}^*(z)\, K_{n-q,q}^*(z)\, c_{p,q}^*(z).$$

We use (2.5) in Proposition 2.1 on page 23 to see that

$$(8.6) \qquad c_{p,q}^*(z) \equiv \sum_{\mu=0}^{p} \sum_{\nu=0}^{q} \binom{m-\mu}{p-\mu}\binom{m-\nu}{q-\nu} \frac{\rho^{(p-\mu)}(z)\,\rho^{(q-\nu)}(z)}{(\rho(z))^2} c_{\mu,\nu}(z).$$

Therefore, after introducing

$$(8.7)\quad H_{p,q,\mu,\nu}(z) \equiv K_{l-p,p}^*(z)\, K_{n-q,q}^*(z) \binom{m-\mu}{p-\mu}\binom{m-\nu}{q-\nu} \frac{\rho^{(p-\mu)}(z)\,\rho^{(q-\nu)}(z)}{(\rho(z))^2},$$

for $0 \leq \mu \leq p \leq l$ and $0 \leq \nu \leq q \leq n$,

we find that equations (8.5)–(8.7) give

$$(8.8) \qquad L_{l,n}^*(z) \equiv \sum_{p=0}^{l}\sum_{q=0}^{n}\sum_{\mu=0}^{p}\sum_{\nu=0}^{q} H_{p,q,\mu,\nu}(z)\, c_{\mu,\nu}(z)$$

$$\equiv \sum_{\mu=0}^{l}\sum_{\nu=0}^{n} c_{\mu,\nu}(z) \sum_{p=\mu}^{l}\sum_{q=\nu}^{n} H_{p,q,\mu,\nu}(z)$$

$$\equiv \sum_{\mu=0}^{l}\sum_{\nu=0}^{n} c_{\mu,\nu}(z) \sum_{p=0}^{l-\mu}\sum_{q=0}^{n-\nu} H_{p+\mu,q+\nu,\mu,\nu}(z).$$

We apply (8.7) to deduce that

$$\sum_{p=0}^{l-\mu}\sum_{q=0}^{n-\nu} H_{p+\mu,q+\nu,\mu,\nu}(z) \equiv \left[\sum_{p=0}^{l-\mu}\binom{m-\mu}{p}\frac{\rho^{(p)}(z)}{\rho(z)} K_{l-\mu-p,\mu+p}^*(z)\right] \times$$
$$\left[\sum_{q=0}^{n-\nu}\binom{m-\nu}{q}\frac{\rho^{(q)}(z)}{\rho(z)} K_{n-\nu-q,\nu+q}^*(z)\right].$$

Consequently, we can rewrite (8.8) as

$$(8.9) \qquad L_{l,n}^*(z) \equiv \sum_{\mu=0}^{l}\sum_{\nu=0}^{n} R_{l-\mu,\mu}(z)\, R_{n-\nu,\nu}(z)\, c_{\mu,\nu}(z),$$

where $R_{i,j}(z)$ is defined by

$$(8.10) \qquad R_{i,j}(z) \equiv \sum_{k=0}^{i}\binom{m-j}{k}\frac{\rho^{(k)}(z)}{\rho(z)} K_{i-k,j+k}^*(z), \quad \text{on } \Omega \text{ for all } i \text{ and } j.$$

By modifying an argument given in [**24**, pages 31–33], we shall prove in Lemma 8.2 that $R_{i,j}(z)$ is given by

$$(8.11) \qquad R_{i,j}(z) \equiv K_{i,j}(z), \quad \text{on } \Omega \text{ for } j \leq m \text{ and any } i,$$

where $K_{i,j}(z)$ on Ω is obtained by substituting $c_{r,s}^{(t)}(z)$ from (1.1) for $\boldsymbol{w}_{r,s}^{(t)}$ in $\boldsymbol{K}_{i,j}$ of (1.18)–(1.20). A substitution of $c_{r,s}^{(t)}(z)$ from (1.1) for $\boldsymbol{w}_{r,s}^{(t)}$ in $\boldsymbol{L}_{l,n}$ of (8.1) yields

$$(8.12) \qquad L_{l,n}(z) \equiv \sum_{\mu=0}^{l} \sum_{\nu=0}^{n} K_{l-\mu,\mu}(z) \, K_{n-\nu,\nu}(z) \, c_{\mu,\nu}(z).$$

We combine (8.9), (8.11), and (8.12) to verify that

$$(8.13) \qquad L_{l,n}^{*}(z) \equiv \sum_{\mu=0}^{l} \sum_{\nu=0}^{n} K_{l-\mu,\mu}(z) \, K_{n-\nu,\nu}(z) \, c_{\mu,\nu}(z) \equiv L_{l,n}(z).$$

Since $\boldsymbol{L}_{l,n}$ is not a constant, (8.13) shows that $\boldsymbol{L}_{l,n}$ is a semi-invariant of the first kind. This completes the proof of Theorem 8.1. $\square$

LEMMA 8.2. *The functions $R_{i,j}(z)$ defined in (8.10) satisfy (8.11).*

PROOF. Let i and j be fixed integers subject to $i \geq 0$ and $j \leq m$. We have

$$(8.14) \qquad R_{i,j}^{(1)}(z) \equiv -\frac{\rho^{(1)}(z)}{\rho(z)} R_{i,j}(z) + \sum_{k=0}^{i} \binom{m-j}{k} \frac{\rho^{(k+1)}(z)}{\rho(z)} K_{i-k,j+k}^{*}(z)$$

$$+ \sum_{k=0}^{i} \binom{m-j}{k} \frac{\rho^{(k)}(z)}{\rho(z)} K_{i-k,j+k}^{*(1)}(z).$$

If $0 \leq k \leq i$ and $j + k \leq m$, then (1.20) yields

$$(8.15) \qquad K_{i-k,j+k}^{*(1)}(z) \equiv K_{i+1-k,j-1+k}^{*}(z) - K_{i+1-k,j+k}^{*}(z)$$

$$+ \frac{m-1}{2} b_{m,1}^{*}(z) \, K_{i-k,j+k}^{*}(z)$$

$$- (m-i-j+1)(1-i-j) a_{m,2}^{*}(z) \, K_{i-1-k,j+k}^{*}(z).$$

However, if $0 \leq k \leq i$ and $j + k > m$, then $k > 0$ and $\binom{m-j}{k} = 0$. Thus, we can replace $K_{i-k,j+k}^{*(1)}(z)$ in (8.14) with the right member of (8.15) to obtain

$$(8.16) \qquad R_{i,j}^{(1)}(z) \equiv \mathcal{A}_1(z) + \mathcal{A}_2(z) + \mathcal{A}_3(z) + \mathcal{A}_4(z) + \mathcal{A}_5(z) + \mathcal{A}_6(z),$$

where

$$\mathcal{A}_1(z) \equiv -\frac{\rho^{(1)}(z)}{\rho(z)} R_{i,j}(z),$$

$$\mathcal{A}_2(z) \equiv \sum_{k=0}^{i} \binom{m-j}{k} \frac{\rho^{(k+1)}(z)}{\rho(z)} K_{i-k,j+k}^{*}(z),$$

$$\mathcal{A}_3(z) \equiv \sum_{k=0}^{i} \binom{m-j}{k} \frac{\rho^{(k)}(z)}{\rho(z)} K_{i+1-k,j-1+k}^{*}(z),$$

$$\mathcal{A}_4(z) \equiv -\sum_{k=0}^{i} \binom{m-j}{k} \frac{\rho^{(k)}(z)}{\rho(z)} K_{i+1-k,j+k}^{*}(z),$$

$$\mathcal{A}_5(z) \equiv \frac{m-1}{2} b_{m,1}^{*}(z) \sum_{k=0}^{i} \binom{m-j}{k} \frac{\rho^{(k)}(z)}{\rho(z)} K_{i-k,j+k}^{*}(z),$$

and
$$\mathcal{A}_6(z) \equiv -(m-i-j+1)(1-i-j)\, a_{m,2}^*(z) \sum_{k=0}^{i} \binom{m-j}{k} \frac{\rho^{(k)}(z)}{\rho(z)} K_{i-1-k,j+k}^*(z).$$

We multiply (2.21) on page 27 by $1/m$ and employ (1.16) on page 7 to verify that

(8.17) $$-\frac{\rho^{(1)}(z)}{\rho(z)} + \frac{m-1}{2} b_{m,1}^*(z) \equiv \frac{m-1}{2} b_{m,1}(z).$$

We use (8.10), (8.17), (1.19), the property $a_{m,2}^*(z) \equiv a_{m,2}(z)$ of $\boldsymbol{a}_{m,2}$ as a semi-invariant of the first kind from Corollary 4.15 on page 45, and $K_{-1,\nu}^*(z) \equiv 0$ from (1.18) to deduce that

$$\mathcal{A}_1(z) + \mathcal{A}_5(z) \equiv -\frac{\rho^{(1)}(z)}{\rho(z)} R_{i,j}(z) + \frac{m-1}{2} b_{m,1}^*(z)\, R_{i,j}(z)$$
$$\equiv \frac{m-1}{2} b_{m,1}(z)\, R_{i,j}(z),$$

$$\mathcal{A}_2(z) + \mathcal{A}_3(z) + \mathcal{A}_4(z) \equiv \sum_{k=1}^{i+1} \binom{m-j}{k-1} \frac{\rho^{(k)}(z)}{\rho(z)} K_{i+1-k,j-1+k}^*(z)$$
$$+ \sum_{k=0}^{i} \binom{m-j}{k} \frac{\rho^{(k)}(z)}{\rho(z)} K_{i+1-k,j-1+k}^*(z)$$
$$- \sum_{k=0}^{i} \binom{m-j}{k} \frac{\rho^{(k)}(z)}{\rho(z)} K_{i+1-k,j+k}^*(z)$$
$$\equiv \sum_{k=0}^{i+1} \left[\binom{m-j}{k-1} + \binom{m-j}{k} \right] \frac{\rho^{(k)}(z)}{\rho(z)} K_{i+1-k,j-1+k}^*(z)$$
$$- \sum_{k=0}^{i+1} \binom{m-j}{k} \frac{\rho^{(k)}(z)}{\rho(z)} K_{i+1-k,j+k}^*(z)$$
$$\equiv R_{i+1,j-1}(z) - R_{i+1,j}(z),$$

and
$$\mathcal{A}_6(z) \equiv -(m-i-j+1)(1-i-j)\, a_{m,2}(z)\, R_{i-1,j}(z).$$

We apply (8.16) and the preceding formulas to establish that: for $i \geq 0$ and $j \leq m$,

(8.18) $$R_{i,j}^{(1)}(z) \equiv R_{i+1,j-1}(z) - R_{i+1,j}(z) + \frac{m-1}{2} b_{m,1}(z)\, R_{i,j}(z)$$
$$- (m-i-j+1)(1-i-j)\, a_{m,2}(z)\, R_{i-1,j}(z).$$

We verify that $K_{i+1,m}^*(z) \equiv 0$, for $i \geq 0$, by substituting $c_{r,s}^{*(t)}(z)$ from (1.3) for $w_{r,s}^{(t)}$ in (1.20) with $j = m$. Thus, (8.10) shows that

$$R_{i+1,m}(z) \equiv \sum_{k=0}^{i+1} \binom{0}{k} \frac{\rho^{(k)}(z)}{\rho(z)} K_{i+1-k,m+k}^*(z) \equiv K_{i+1,m}^*(z) \equiv 0, \quad \text{for } i \geq 0.$$

This enables us to rewrite (8.18) in the form

$$(8.19) \quad R_{i+1,j}(z) \equiv \sum_{k=j+1}^{m} \left[R_{i+1,k-1}(z) - R_{i+1,k}(z) \right]$$

$$\equiv \sum_{k=j+1}^{m} \left[\begin{array}{l} R_{i,k}^{(1)}(z) - \dfrac{m-1}{2} b_{m,1}(z) R_{i,k}(z) \\ + (m-i-k+1)(1-i-k) a_{m,2}(z) R_{i-1,k}(z) \end{array} \right],$$

for $i \geq 0$ and $j \leq m$.

We see that (8.10) and (1.18) yield $R_{i,j}(z) \equiv 0 \equiv K_{i,j}(z)$, for $i \leq -1$ and any j. Also, (8.10) and (1.19) give $R_{0,j}(z) \equiv 1 \equiv K_{0,j}(z)$, for any j. Using this with (8.19) and the result of substituting $c_{r,s}^{(t)}(z)$ from (1.1) for $w_{r,s}^{(t)}$ in (1.20), we find that $R_{i,j}(z) \equiv K_{i,j}(z)$, on Ω for $j \leq m$ and any i. Thus, (8.11) is valid. This completes the proof of Lemma 8.2. $\square$

THEOREM 8.3. *For $m \geq 2$ and integers l, n that satisfy either $1 \leq l \leq n \leq m$ or $l = 0 < 3 \leq n \leq m$, the polynomial $\boldsymbol{\mathcal{I}}_{m,l,n}$ defined in $\mathcal{S}_m$ by means of (1.28) on page 8 is both a semi-invariant of the first kind for the equations (1.1) and a basic polynomial of index (l, n) according to* Definition 1.5 *on page 6.*

PROOF. To establish that $\boldsymbol{\mathcal{I}}_{m,l,n}$ satisfies (1) in Definition 1.5 on page 6, we first use (1.27) on page 8 to see that the coefficient of $\boldsymbol{w}_{l,n}$ in $\boldsymbol{I}_{l,l+n}$ is equal to the coefficient of $\boldsymbol{w}_{l,n}$ in $\boldsymbol{M}_{l,l+n}$. With $h = l$ and $i = l + n$, we find that (1.23) yields $\boldsymbol{M}_{l,l+n} \equiv \boldsymbol{L}_{l,n}$. Since Theorem 8.1 on page 87 shows that $\boldsymbol{L}_{l,n}$ is a basic polynomial of index (l, n), the coefficients of $\boldsymbol{w}_{l,n}$ in both $\boldsymbol{L}_{l,n}$ and $\boldsymbol{I}_{l,l+n}$ equal 1.

Suppose that h satisfies $0 \leq h < l$. We use (1.27) to observe that the coefficient of $\boldsymbol{w}_{l,n}$ in $\boldsymbol{I}_{h,l+n}$ is equal to the coefficient of $\boldsymbol{w}_{l,n}$ in $\boldsymbol{M}_{h,l+n}$. With $i = l + n$, $n - i + h = h - l < 0$, and $\binom{m-i+h}{n-i+h} = 0$, we find that (1.23) gives $\boldsymbol{M}_{h,l+n} \equiv 0$. Thus, the coefficient of $\boldsymbol{w}_{l,n}$ in $\boldsymbol{I}_{h,l+n}$ is 0.

We use the preceding results and (1.28) to deduce that: the coefficient of $\boldsymbol{w}_{l,n}$ in $\boldsymbol{\mathcal{I}}_{m,l,n}$ is 1 and therefore $\boldsymbol{\mathcal{I}}_{m,l,n}$ satisfies (1) in Definition 1.5.

To show that $\boldsymbol{\mathcal{I}}_{m,l,n}$ is an isobaric semi-invariant of the first kind whose weight is $l+n$, we first establish a result about $\boldsymbol{L}_{i,j}$ in (1.21) when i and j are integers such that $\boldsymbol{L}_{i,j}$ is not a constant. Namely, after setting $i_0 = \min(i, j)$ and $j_0 = \max(i, j)$, we use $\boldsymbol{L}_{i,j} \equiv \boldsymbol{L}_{j,i} \equiv \boldsymbol{L}_{i_0,j_0}$, $\boldsymbol{L}_{0,0} \equiv 1$, $\boldsymbol{L}_{0,1} \equiv 0$, $\boldsymbol{L}_{0,2} \equiv 0$, and Theorem 8.1 to deduce that $\boldsymbol{L}_{i,j}$ is an isobaric semi-invariant of the first kind that is also a basic polynomial of index (i_0, j_0) where either $1 \leq i_0 \leq j_0 \leq m$ or $i_0 = 0 < 3 \leq j_0 \leq m$. Consequently, for each $\boldsymbol{L}_{i,j}$ in (1.21) having $i + j \geq 1$, either it is identically zero or it is an isobaric semi-invariant of the first kind having weight $i + j$. Thus, in view of (1.23) and (1.22), we find that: for any $\boldsymbol{M}_{h,i}$ in (1.23) having $i \geq 1$, either it is identically zero or it is an isobaric semi-invariant of the first kind having weight i. Since Corollary 4.15 on page 45 shows that $\boldsymbol{a}_{m,2}$ in (1.14) is an isobaric semi-invariant of the first kind having weight 2, we observe that: for any $\boldsymbol{I}_{h,i}$ in (1.27) or (1.26), either it is identically zero or it is an isobaric semi-invariant of the first kind having weight i. In view of this and (1.28), we find that either $\boldsymbol{\mathcal{I}}_{m,l,n}$ is identically zero or it is an isobaric semi-invariant of the first kind with weight $l+n$. However, the coefficient of $\boldsymbol{w}_{l,n}$ in $\boldsymbol{\mathcal{I}}_{m,l,n}$ is 1. Therefore, we conclude that $\boldsymbol{\mathcal{I}}_{m,l,n}$ is an isobaric semi-invariant of the first kind having weight $l + n$. In particular, $\boldsymbol{\mathcal{I}}_{m,l,n}$ satisfies (2) in Definition 1.5.

To show that $\mathcal{I}_{m,l,n}$ satisfies (3) in Definition 1.5, suppose that $w_{i_1,i_2}^{(k)}$ is a variable from (1.6) that is effectively involved in $\mathcal{I}_{m,l,n}$. We are to establish that
(8.20) $$i_1 \leq l \quad \text{and} \quad i_2 \leq n.$$
With $n \geq 1$, we see that (8.20) is satisfied when $w_{i_1,i_2}^{(k)} = w_{0,1}^{(k)}$. If $n = 1$, then $l = 1$, $\mathcal{I}_{m,l,n} \equiv \mathcal{I}_{m,1,1} \equiv w_{1,1} - (w_{0,1})^2$, and $w_{0,2}^{(k)}$ is not involved in $\mathcal{I}_{m,l,n}$. Thus, when $w_{i_1,i_2}^{(k)} = w_{0,2}^{(k)}$, we find that $n \geq 2$ and (8.20) is satisfied. As the remaining situation, suppose that $(i_1, i_2) \neq (0, 1)$ and $(i_1, i_2) \neq (0, 2)$. Because $\boldsymbol{a}_{m,2}$ and each $\boldsymbol{K}_{i,j}$ are polynomial combinations over $\mathbb{Q}$ of the variables in (8.3) on page 87, they do not involve $w_{i_1,i_2}^{(j)}$, for any $j \geq 0$. Therefore, equations (1.21)–(1.28) show that there are integers h and i satisfying $0 \leq h \leq l$ and $1 \leq i \leq l+n$ such that $w_{i_1,i_2}^{(0)}$ is effectively involved in the corresponding left member $\boldsymbol{M}_{h,i}$ of (1.23). In view of $\boldsymbol{M}_{h,i} \not\equiv 0$, (1.23) shows that $w_{i_1,i_2}^{(0)}$ is effectively involved in $\boldsymbol{L}_{h,i-h}$ and, due to $\binom{m-i+h}{n-i+h} \neq 0$, we have $0 \leq i - h \leq n$ in addition to $0 \leq h \leq l$. We note that $\boldsymbol{L}_{h,i-h}$ is a basic polynomial whose index is $\big(\min(h, i-h), \max(h, i-h)\big)$. Consequently, as applied to $\boldsymbol{L}_{h,i-h}$ and $w_{i_1,i_2}^{(0)}$, the third part of Definition 1.5 yields
$$i_1 \leq \min(h, i-h) \leq h \leq l \quad \text{and} \quad i_2 \leq \max(h, i-h) \leq \max(l, n) = n.$$
Thus, (8.20) is valid and therefore $\mathcal{I}_{m,l,n}$ satisfies (3) in Definition 1.5.

To show that $\mathcal{I}_{m,l,n}$ satisfies (4) in Definition 1.5, let $\boldsymbol{T}$ denote a nonzero term in some polynomial $\boldsymbol{L}_{i,j}$, $\boldsymbol{M}_{h,i}$, $\boldsymbol{I}_{h,i}$, or $\mathcal{I}_{m,l,n}$ that appears in (1.21)–(1.28). Then, for some $r \geq 1$, $\boldsymbol{T}$ is a product of r factors from (1.6) and a nonzero element of $\mathbb{Q}$. Since $\boldsymbol{a}_{m,2}$ and each $\boldsymbol{K}_{i,j}$ of (1.18)–(1.20) are polynomials over $\mathbb{Q}$ in the variables of (8.3) on page 87, we use (1.21)–(1.28) to see that at least $r-1$ of the r factors from (1.6) are listed in (8.3). Thus, $\mathcal{I}_{m,l,n}$ satisfies (4) in Definition 1.5.

We combine the preceding results to conclude that $\mathcal{I}_{m,l,n}$ is a basic polynomial of index (l, n) as well as a semi-invariant of the first kind. This completes the proof. $\square$

CHAPTER 9

Formulas That Involve $V_{i,j}(z)$

With $m \geq 2$, polynomials $\boldsymbol{b}_{m,1}$, $\boldsymbol{b}_{m,2}$, $\boldsymbol{U}_{i,j}$, and $\boldsymbol{V}_{i,j}$ are defined in $\mathcal{S}_m$ for all integers i and j by means of (1.16), (1.15), and (1.30)–(1.34). In this chapter, a key role is played by the functions $V_{i,j}(z)$ on Ω obtained by substituting $c_{\mu,\nu}^{(\kappa)}(z)$ from (1.1) for $\boldsymbol{w}_{\mu,\nu}^{(\kappa)}$ in $\boldsymbol{V}_{i,j}$. As in Chapter 7, we assume throughout that the composite of the substitutions (1.2) and (1.4) transforms (1.1) on Ω into a corresponding (5.1) that is defined on Ω^{**} and whose coefficients satisfy $d_{0,1}(\zeta) \equiv d_{0,2}(\zeta) \equiv 0$.

9.1. The coefficients of (5.1) when $d_{0,1}(\zeta) \equiv d_{0,2}(\zeta) \equiv 0$

For the statement of Proposition 9.1, we introduce

$$(9.1) \qquad \eta_{i,j} \equiv \frac{\left[\prod_{\nu=1}^{j}(i-\nu)\right]}{(m-1)^j}\binom{m-i+j}{j}, \quad \text{for any integers } i \text{ and } j.$$

PROPOSITION 9.1. *For (5.1) on Ω^{**} into which (1.1) on Ω is transformed by $y(z) = \rho(z)\,v(z)$ and $z = f(\zeta)$ according to the context of page 47, suppose that $d_{0,1}(\zeta) \equiv d_{0,2}(\zeta) \equiv 0$. Then, the coefficients of (5.1) are given by*

$$(9.2) \quad d_{i,j}(\zeta) \equiv \bigl(f'(\zeta)\bigr)^{i+j} \sum_{r=0}^{i}\sum_{s=0}^{j} \eta_{i,r}\,\eta_{j,s}\,V_{i-r,j-s}\bigl(f(\zeta)\bigr)\bigl(\mathcal{S}(\zeta)\bigr)^{r+s},$$

$$\text{on } \Omega^{**} \text{ for } 0 \leq i,\,j \leq m,$$

where $V_{i-r,j-s}\bigl(f(\zeta)\bigr)$ is obtained for $z = f(\zeta)$ by substituting $c_{\mu,\nu}^{(\kappa)}(z)$ from (1.1) for $\boldsymbol{w}_{\mu,\nu}^{(\kappa)}$ in $\boldsymbol{V}_{i-r,j-s}$ of (1.33) and

$$(9.3) \qquad \mathcal{S}(\zeta) \equiv \frac{\rho^{(1)}\bigl(f(\zeta)\bigr)}{\rho\bigl(f(\zeta)\bigr)}, \quad \text{on } \Omega^{**}.$$

PROOF. We use (5.3) from Theorem 5.1 on page 47 to obtain

$$(9.4) \qquad d_{i,j}(\zeta) \equiv \bigl(f'(\zeta)\bigr)^{i+j} \sum_{p=0}^{i}\sum_{q=0}^{j} \phi_{i-p,p}\bigl(f(\zeta)\bigr)\,\phi_{j-q,q}\bigl(f(\zeta)\bigr)\,c_{p,q}\bigl(f(\zeta)\bigr),$$

where $\phi_{i,j}(z)$ is defined on Ω by (5.4)–(5.5) in terms of $\rho(z)$ and the inverse $\zeta = g(z)$ for $z = f(\zeta)$. Due to $d_{0,1}(\zeta) \equiv d_{0,2}(\zeta) \equiv 0$, Proposition 5.3 on page 49 shows that $\rho(z)$ and $g(z)$ satisfy (5.16) and (5.17). Consequently, Lemma 9.2 yields

$$(9.5) \qquad \phi_{i,j}(z) \equiv \sum_{k=0}^{i} \eta_{i+j,k}\,U_{i-k,j}(z)\bigl(\mathcal{R}(z)\bigr)^k, \quad \text{on } \Omega \text{ for } i \geq 0 \text{ and } j \leq m,$$

93

where $U_{i-k,j}(z)$ is obtained by substituting $c_{\mu,\nu}^{(\kappa)}(z)$ from (1.1) for $\boldsymbol{w}_{\mu,\nu}^{(\kappa)}$ in $\boldsymbol{U}_{i-k,j}$ of (1.30)–(1.32) on page 9 and

$$(9.6) \qquad \mathcal{R}(z) \equiv \frac{\rho^{(1)}(z)}{\rho(z)}, \quad \text{on } \Omega.$$

In terms of (9.3), (9.6) yields

$$(9.7) \qquad \mathcal{R}\bigl(f(\zeta)\bigr) \equiv \mathcal{S}(\zeta), \quad \text{on } \Omega^{**}.$$

We employ (9.4), (9.5), and (9.7) to obtain

$$(9.8) \qquad d_{i,j}(\zeta) \equiv \bigl(f'(\zeta)\bigr)^{i+j} h\bigl(f(\zeta)\bigr), \quad \text{on } \Omega^{**},$$

where $h(z)$, on Ω, is given by

$$h(z) \equiv \sum_{p=0}^{i} \sum_{q=0}^{j} \phi_{i-p,p}(z)\, \phi_{j-q,q}(z)\, c_{p,q}(z)$$

$$\equiv \sum_{p=0}^{i} \sum_{q=0}^{j} \sum_{r=0}^{i-p} \eta_{i,r}\, U_{i-p-r,p}(z)\, \bigl(\mathcal{R}(z)\bigr)^{r} \sum_{s=0}^{j-q} \eta_{j,s}\, U_{j-q-s,q}(z)\, \bigl(\mathcal{R}(z)\bigr)^{s} c_{p,q}(z)$$

$$\equiv \sum_{r=0}^{i} \sum_{p=0}^{i-r} \sum_{s=0}^{j} \sum_{q=0}^{j-s} \eta_{i,r}\, \eta_{j,s}\, U_{i-p-r,p}(z)\, U_{j-q-s,q}(z)\, \bigl(\mathcal{R}(z)\bigr)^{r+s} c_{p,q}(z)$$

$$\equiv \sum_{r=0}^{i} \sum_{s=0}^{j} \eta_{i,r}\, \eta_{j,s}\, \bigl(\mathcal{R}(z)\bigr)^{r+s} \sum_{p=0}^{i-r} \sum_{q=0}^{j-s} U_{i-r-p,p}(z)\, U_{j-s-q,q}(z)\, c_{p,q}(z).$$

By substituting $c_{\mu,\nu}^{(\kappa)}(z)$ from (1.1) for $\boldsymbol{w}_{\mu,\nu}^{(\kappa)}$ in $\boldsymbol{V}_{i-r,j-s}$ of (1.33) on page 9, we combine the resulting expression with the preceding one to see that

$$(9.9) \qquad h(z) \equiv \sum_{r=0}^{i} \sum_{s=0}^{j} \eta_{i,r}\, \eta_{j,s}\, \bigl(\mathcal{R}(z)\bigr)^{r+s} V_{i-r,j-s}(z).$$

Formulas (9.8), (9.9), and (9.7) yield (9.2). This completes the proof. $\square$

The use of (9.5) in the proof of Proposition 9.1 is based on the next result.

LEMMA 9.2. *Suppose that $\rho(z)$ and $g(z)$ satisfy (5.16) and (5.17) on Ω. Then, the functions $\chi_{i,j}(z)$ defined in terms of (9.1), (1.30)–(1.32) and (9.6) by*

$$(9.10) \qquad \chi_{i,j}(z) \equiv \sum_{k=0}^{i} \eta_{i+j,k}\, U_{i-k,j}(z)\, \bigl(\mathcal{R}(z)\bigr)^{k}, \quad \text{on } \Omega \text{ for } i \geq 0 \text{ and any } j,$$

and $\phi_{i,j}(z)$, defined on Ω by (5.4)–(5.5) of page 47, are related through

$$(9.11) \qquad \phi_{i,j}(z) \equiv \chi_{i,j}(z), \quad \text{on } \Omega \text{ for } i \geq 0 \text{ and } j \leq m.$$

PROOF. We first verify that (9.1) yields

$$(9.12) \qquad \eta_{i+1+j,k} - \eta_{i+j,k} \equiv \frac{m+k-2i-2j}{m-1}\, \eta_{i+j,k-1}, \quad \text{for all } i, j, k.$$

As an abbreviation, we set

$$(9.13) \qquad \omega_{i,j,k} \equiv -(m-i-j+k+1)(1-i-j+k)\, \eta_{i+j,k}, \quad \text{for all } i, j, k;$$

then, we check the validity of

$$(9.14) \qquad \omega_{i,j,k} \equiv (m-1)(k+1)\, \eta_{i+j,k+1}, \quad \text{for } k \geq 0 \text{ and any } i, j.$$

9.1. THE COEFFICIENTS OF (5.1) WHEN $d_{0,1}(\zeta) \equiv d_{0,2}(\zeta) \equiv 0$

For $k \leq i$ and $j \leq m$, we find that (1.32) yields

$$(9.15) \quad U^{(1)}_{i-k,j}(z) \equiv U_{i+1-k,j-1}(z) - U_{i+1-k,j}(z)$$
$$- (i + j - k - m)\, b_{m,1}(z)\, U_{i-k,j}(z)$$
$$- (m - i - j + k + 1)(1 - i - j + k)\, b_{m,2}(z)\, U_{i-1-k,j}(z).$$

We apply (9.6) and (5.16) on page 49 to observe that

$$(9.16) \quad \mathcal{R}^{(1)}(z) \equiv -\frac{1}{m-1}\left(\mathcal{R}(z)\right)^2 - b_{m,1}(z)\,\mathcal{R}(z) - (m-1)\,b_{m,2}(z).$$

By combining (5.17) on page 49 with (9.6), we obtain

$$(9.17) \quad b_{m,1}(z) \equiv -\frac{g''(z)}{g'(z)} - \frac{2\mathcal{R}(z)}{m-1}.$$

Let i and j be fixed integers satisfying $i \geq 0$ and $j \leq m$. We use (9.10), (9.15), (1.30), (9.13), and (9.16) to establish that

$$(9.18) \quad \chi^{(1)}_{i,j}(z) \equiv \sum_{k=0}^{i} \eta_{i+j,k}\left[U^{(1)}_{i-k,j}(z)\left(\mathcal{R}(z)\right)^k + U_{i-k,j}(z)\,k\,\left(\mathcal{R}(z)\right)^{k-1}\mathcal{R}^{(1)}(z)\right]$$
$$\equiv \mathcal{F}_1(z) + \mathcal{F}_2(z) + \mathcal{F}_3(z) + \mathcal{F}_4(z) + \mathcal{F}_5(z) + \mathcal{F}_6(z) + \mathcal{F}_7(z),$$

where

$$\mathcal{F}_1(z) \equiv \sum_{k=0}^{i} \eta_{i+j,k}\, U_{i+1-k,j-1}(z)\left(\mathcal{R}(z)\right)^k,$$

$$\mathcal{F}_2(z) \equiv \sum_{k=0}^{i} -\eta_{i+j,k}\, U_{i+1-k,j}(z)\left(\mathcal{R}(z)\right)^k,$$

$$\mathcal{F}_3(z) \equiv b_{m,1}(z) \sum_{k=0}^{i} \eta_{i+j,k}\,(m - i - j + k)\, U_{i-k,j}(z)\left(\mathcal{R}(z)\right)^k,$$

$$\mathcal{F}_4(z) \equiv b_{m,2}(z) \sum_{k=0}^{i-1} \omega_{i,j,k}\, U_{i-1-k,j}(z)\left(\mathcal{R}(z)\right)^k,$$

$$\mathcal{F}_5(z) \equiv \sum_{k=0}^{i} \frac{-k}{m-1}\, \eta_{i+j,k}\, U_{i-k,j}(z)\left(\mathcal{R}(z)\right)^{k+1},$$

$$\mathcal{F}_6(z) \equiv b_{m,1}(z) \sum_{k=0}^{i} -k\, \eta_{i+j,k}\, U_{i-k,j}(z)\left(\mathcal{R}(z)\right)^k,$$

$$\mathcal{F}_7(z) \equiv b_{m,2}(z) \sum_{k=0}^{i} -k(m-1)\, \eta_{i+j,k}\, U_{i-k,j}(z)\left(\mathcal{R}(z)\right)^{k-1}$$
$$\equiv b_{m,2}(z) \sum_{k=0}^{i-1} -(m-1)(k+1)\, \eta_{i+j,k+1}\, U_{i-1-k,j}(z)\left(\mathcal{R}(z)\right)^k.$$

Due to (9.10), $U_{0,j-1}(z) \equiv U_{0,j}(z)$ from (1.31), and (9.12), we have

$$\left[\mathcal{F}_1(z) - \chi_{i+1,j-1}(z)\right] + \mathcal{F}_2(z) \equiv -\sum_{k=0}^{i+1} \eta_{i+j,k}\, U_{i+1-k,j}(z)\,(\mathcal{R}(z))^k$$

$$\equiv -\sum_{k=0}^{i+1} \left[\eta_{i+1+j,k} - \frac{m+k-2i-2j}{m-1}\eta_{i+j,k-1}\right] U_{i+1-k,j}(z)\,(\mathcal{R}(z))^k.$$

In view of (9.10) and $\eta_{i+j,-1} = 0$ from (9.1), this gives

$$\mathcal{F}_1(z) + \mathcal{F}_2(z) \equiv \chi_{i+1,j-1}(z) - \chi_{i+1,j}(z)$$
$$+ \sum_{k=0}^{i} \frac{m+k+1-2i-2j}{m-1} \eta_{i+j,k}\, U_{i-k,j}(z)\,(\mathcal{R}(z))^{k+1}.$$

Thus, we see that

(9.19) $\quad \mathcal{F}_1(z) + \mathcal{F}_2(z) + \mathcal{F}_5(z) \equiv \chi_{i+1,j-1}(z) - \chi_{i+1,j}(z)$
$$+ \frac{m+1-2i-2j}{m-1}\mathcal{R}(z)\,\chi_{i,j}(z).$$

We use (9.17) to verify that

$$\mathcal{F}_3(z) + \mathcal{F}_6(z) \equiv (m-i-j)\,b_{m,1}(z)\,\chi_{i,j}(z)$$
$$\equiv (m-i-j)\left[-\frac{g''(z)}{g'(z)} - \frac{2\mathcal{R}(z)}{m-1}\right]\chi_{i,j}(z).$$

Moreover, (9.14) yields $\mathcal{F}_4(z) + \mathcal{F}_7(z) \equiv 0$. Combining (9.18), (9.19), the results following (9.19), and (9.6), we obtain

(9.20) $\quad \chi_{i,j}^{(1)}(z) \equiv \chi_{i+1,j-1}(z) - \chi_{i+1,j}(z) - \left[\frac{\rho'(z)}{\rho(z)} + (m-i-j)\frac{g''(z)}{g'(z)}\right]\chi_{i,j}(z),$
$$\text{for } i \geq 0 \text{ and } j \leq m.$$

We note that (1.32) yields $U_{i+1-k,m} \equiv 0$, for $k \leq i$. Employing this with (9.10), (1.31), and (9.1), we find that

(9.21) $\quad \chi_{i+1,m}(z) \equiv (\eta_{i+1+m,i+1})(\mathcal{R}(z))^{i+1} \equiv 0, \quad \text{for } i \geq 0.$

We apply (9.21) and (9.20) to deduce that

(9.22) $\quad \chi_{i+1,j}(z) \equiv \sum_{k=j+1}^{m} \left[\chi_{i+1,k-1}(z) - \chi_{i+1,k}(z)\right]$
$$\equiv \sum_{k=j+1}^{m} \left[\chi_{i,k}^{(1)}(z) + \left[\frac{\rho'(z)}{\rho(z)} + (m-i-k)\frac{g''(z)}{g'(z)}\right]\chi_{i,k}(z)\right],$$
$$\text{for } i \geq 0 \text{ and } j \leq m.$$

We note that (9.10), (1.31), and (9.1) yield $\chi_{0,j}(z) \equiv \eta_{j,0} \equiv 1$, for any j. By using this with (5.4), (5.5), and (9.22), we find that $\phi_{i,j}(z) \equiv \chi_{i,j}(z)$, for $i \geq 0$ and $j \leq m$. Thus, (9.11) is valid. This completes the proof of Lemma 9.2. $\square$

9.2. Derivatives of the coefficients in (5.1) when $d_{0,1}(\zeta) \equiv d_{0,2}(\zeta) \equiv 0$

In terms of the polynomials $\boldsymbol{V}_{i,j}$ defined in $\mathcal{S}_m$ for all i and j by (1.33)–(1.34) on page 9, we introduce $\boldsymbol{W}_{i,j,k,t}$ in $\mathcal{S}_m$ for $k \geq 0$ and all integers i, j, t through

$$(9.23) \qquad \boldsymbol{W}_{i,j,0,t} \equiv \sum_{r=0}^{t} \eta_{i,r}\, \eta_{j,t-r}\, \boldsymbol{V}_{i-r,j+r-t}, \quad \text{for any } i,\, j,\, t,$$

and

$$(9.24) \qquad \boldsymbol{W}_{i,j,k+1,t} \equiv \boldsymbol{W}_{i,j,k,t}^{(1)} + (i+j+k-t)\, \boldsymbol{b}_{m,1}\, \boldsymbol{W}_{i,j,k,t}$$
$$+ \frac{2i+2j+2k-t+1}{m-1}\boldsymbol{W}_{i,j,k,t-1} - (m-1)(t+1)\boldsymbol{b}_{m,2}\, \boldsymbol{W}_{i,j,k,t+1},$$

$$\text{for } k \geq 0 \text{ and any } i,\, j,\, t.$$

PROPOSITION 9.3. *For (5.1) on Ω^{**} into which (1.1) on Ω is transformed by $y(z) = \rho(z)\, v(z)$ and $z = f(\zeta)$ according to the context of page 47, suppose that $d_{0,1}(\zeta) \equiv d_{0,2}(\zeta) \equiv 0$. Then, the derivatives of the coefficients of (5.1) are given by*

$$(9.25) \qquad d_{i,j}^{(k)}(\zeta) \equiv \bigl(f'(\zeta)\bigr)^{i+j+k} \sum_{t=0}^{i+j+k} \mathcal{W}_{i,j,k,t}\bigl(f(\zeta)\bigr)\bigl(\mathcal{S}(\zeta)\bigr)^{t},$$

$$\text{for } 0 \leq i,\, j \leq m \text{ and } k \geq 0,$$

where $\mathcal{W}_{i,j,k,t}\bigl(f(\zeta)\bigr)$ is obtained for $z = f(\zeta)$ by substituting $c_{\mu,\nu}^{(\kappa)}(z)$ from (1.1) for $w_{\mu,\nu}^{(\kappa)}$ in $\boldsymbol{W}_{i,j,k,t}$ of (9.23)–(9.24) and $\mathcal{S}(\zeta)$ is defined in (9.3).

PROOF. For $t < 0$, (9.23) yields $\boldsymbol{W}_{i,j,0,t} \equiv 0$. Suppose that $t \geq i+j+1$. Then, in view of $(i-r)+(j+r-t) = i+j-t \leq -1$, we have either $i-r < 0$ or $j+r-t < 0$ so that formulas (1.34) and (9.23) give $\boldsymbol{W}_{i,j,0,t} \equiv 0$. Thus, the formula $\boldsymbol{W}_{i,j,k,t} \equiv 0$ is valid whenever $k = 0$ and either $t < 0$ or $t > i+j+k$. Suppose that k is a nonnegative integer such that $\boldsymbol{W}_{i,j,k,t} \equiv 0$ is valid whenever i, j, t are integers that satisfy either $t < 0$ or $t > i+j+k$. Then, for this k, we use (9.24) to see that $\boldsymbol{W}_{i,j,k+1,t} \equiv 0$ is valid for any integers i, j, t that satisfy either $t < 0$ or $t > i+j+k+1$. This establishes that

$$(9.26) \qquad \boldsymbol{W}_{i,j,k,t} \equiv 0, \quad \text{whenever } k \geq 0 \text{ and either } t < 0 \text{ or } t > i+j+k.$$

Due to (1.34), we have $\boldsymbol{V}_{i-r,j-s} \equiv 0$, for $r > i$ or $s > j$. Consequently, for $0 \leq i,\, j \leq m$, we see that (9.2) and (9.23) yield

$$d_{i,j}(\zeta) \equiv \bigl(f'(\zeta)\bigr)^{i+j} \sum_{r=0}^{i} \sum_{s=0}^{j} \eta_{i,r}\, \eta_{j,s}\, \boldsymbol{V}_{i-r,j-s}\bigl(f(\zeta)\bigr)\bigl(\mathcal{S}(\zeta)\bigr)^{r+s}$$
$$\equiv \bigl(f'(\zeta)\bigr)^{i+j} \sum_{r=0}^{i+j} \sum_{s=0}^{i+j-r} \eta_{i,r}\, \eta_{j,s}\, \boldsymbol{V}_{i-r,j-s}\bigl(f(\zeta)\bigr)\bigl(\mathcal{S}(\zeta)\bigr)^{r+s}$$
$$\equiv \bigl(f'(\zeta)\bigr)^{i+j} \sum_{t=0}^{i+j} \sum_{r=0}^{t} \eta_{i,r}\, \eta_{j,t-r}\, \boldsymbol{V}_{i-r,j+r-t}\bigl(f(\zeta)\bigr)\bigl(\mathcal{S}(\zeta)\bigr)^{t}$$
$$\equiv \bigl(f'(\zeta)\bigr)^{i+j} \sum_{t=0}^{i+j} \mathcal{W}_{i,j,0,t}\bigl(f(\zeta)\bigr)\bigl(\mathcal{S}(\zeta)\bigr)^{t}.$$

Thus, (9.25) is valid for $k = 0$ and $0 \leq i,\, j \leq m$.

Suppose that k_0 is a nonnegative integer such that (9.25) is valid for $k = k_0$ and any i, j satisfying $0 \leq i, j \leq m$. For $k = k_0$, we differentiate (9.25) with respect to ζ and use $f''(\zeta)/(f'(\zeta))^2 \equiv \mathcal{Q}(\zeta)$ from (7.3) of page 69 to obtain

$$(9.27) \quad d_{i,j}^{(k_0+1)}(\zeta) \equiv (i+j+k_0)(f'(\zeta))^{i+j+k_0+1} \mathcal{Q}(\zeta) \sum_{t=0}^{i+j+k_0} \mathcal{W}_{i,j,k_0,t}(f(\zeta))(\mathcal{S}(\zeta))^t$$

$$+ (f'(\zeta))^{i+j+k_0+1} \sum_{t=0}^{i+j+k_0} \mathcal{W}_{i,j,k_0,t}^{(1)}(f(\zeta))(\mathcal{S}(\zeta))^t$$

$$+ (f'(\zeta))^{i+j+k_0+1} \sum_{t=0}^{i+j+k_0} t\,\mathcal{W}_{i,j,k_0,t}(f(\zeta))(\mathcal{S}(\zeta))^{t-1} \frac{\mathcal{S}'(\zeta)}{f'(\zeta)}.$$

After differentiating $g(f(\zeta)) \equiv \zeta$ twice with respect to ζ, we combine the resulting expression with (5.17), on page 49, and (9.3) to verify that

$$(9.28) \quad \mathcal{Q}(\zeta) \equiv \frac{f''(\zeta)}{(f'(\zeta))^2} \equiv -\frac{g''(f(\zeta))}{g'(f(\zeta))} \equiv b_{m,1}(f(\zeta)) + \frac{2}{m-1}\mathcal{S}(\zeta).$$

We note that (9.7) is the relation $\mathcal{S}(\zeta) \equiv \mathcal{R}(f(\zeta))$, on Ω^{**}. After differentiating (9.7) with respect to ζ, we use (9.16) and (9.7) to deduce on Ω^{**} that

$$(9.29) \quad \frac{\mathcal{S}'(\zeta)}{f'(\zeta)} \equiv \mathcal{R}^{(1)}(f(\zeta)) \equiv -\frac{(\mathcal{S}(\zeta))^2}{m-1} - b_{m,1}(f(\zeta))\,\mathcal{S}(\zeta) - (m-1)\,b_{m,2}(f(\zeta)).$$

We apply (9.27), (9.28), (9.26), and (9.29) to establish that

$$(9.30) \quad d_{i,j}^{(k_0+1)}(\zeta) \equiv (f'(\zeta))^{i+j+k_0+1}\left[(\mathcal{G}_1 + \mathcal{G}_2 + \mathcal{G}_3 + \mathcal{G}_4 + \mathcal{G}_5 + \mathcal{G}_6)(\zeta)\right],$$

where

$$\mathcal{G}_1(\zeta) \equiv \sum_{t=0}^{i+j+k_0+1} (i+j+k_0)\,b_{m,1}(f(\zeta))\,\mathcal{W}_{i,j,k_0,t}(f(\zeta))(\mathcal{S}(\zeta))^t,$$

$$\mathcal{G}_2(\zeta) \equiv \sum_{t=0}^{i+j+k_0} \frac{2(i+j+k_0)}{m-1} \mathcal{W}_{i,j,k_0,t}(f(\zeta))(\mathcal{S}(\zeta))^{t+1}$$

$$\equiv \sum_{t=0}^{i+j+k_0+1} \frac{2(i+j+k_0)}{m-1} \mathcal{W}_{i,j,k_0,t-1}(f(\zeta))(\mathcal{S}(\zeta))^t,$$

$$\mathcal{G}_3(\zeta) \equiv \sum_{t=0}^{i+j+k_0+1} \mathcal{W}_{i,j,k_0,t}^{(1)}(f(\zeta))(\mathcal{S}(\zeta))^t,$$

$$\mathcal{G}_4(\zeta) \equiv \sum_{t=0}^{i+j+k_0} \frac{-t}{m-1} \mathcal{W}_{i,j,k_0,t}(f(\zeta))(\mathcal{S}(\zeta))^{t+1}$$

$$\equiv \sum_{t=0}^{i+j+k_0+1} -\frac{t-1}{m-1} \mathcal{W}_{i,j,k_0,t-1}(f(\zeta))(\mathcal{S}(\zeta))^t,$$

$$\mathcal{G}_5(\zeta) \equiv \sum_{t=0}^{i+j+k_0+1} -t\, b_{m,1}\big(f(\zeta)\big)\, \mathcal{W}_{i,j,k_0,t}\big(f(\zeta)\big)\, \big(\mathcal{S}(\zeta)\big)^t,$$

and

$$\mathcal{G}_6(\zeta) \equiv \sum_{t=1}^{i+j+k_0+2} -(m-1)\,t\, b_{m,2}\big(f(\zeta)\big)\, \mathcal{W}_{i,j,k_0,t}\big(f(\zeta)\big)\, \big(\mathcal{S}(\zeta)\big)^{t-1}$$
$$\equiv \sum_{t=0}^{i+j+k_0+1} -(m-1)(t+1)\, b_{m,2}\big(f(\zeta)\big)\, \mathcal{W}_{i,j,k_0,t+1}\big(f(\zeta)\big)\, \big(\mathcal{S}(\zeta)\big)^t.$$

After adding $\mathcal{G}_1(\zeta), \ldots, \mathcal{G}_6(\zeta)$, we employ (9.30) and (9.24) to obtain

$$(9.31) \qquad d_{i,j}^{(k_0+1)}(\zeta) \equiv \big(f'(\zeta)\big)^{i+j+k_0+1} \sum_{t=0}^{i+j+k_0+1} \mathcal{W}_{i,j,k_0+1,t}\big(f(\zeta)\big)\, \big(\mathcal{S}(\zeta)\big)^t.$$

In view of (9.31), (9.25) is also valid for $0 \leq i, j \leq m$ and $k = k_0 + 1$. Thus, (9.25) is valid for $0 \leq i, j \leq m$ and $k \geq 0$. This completes the proof. □

9.3. Special combinations of the coefficients for (5.1)

For $m \geq 2$, we continue to assume that the composite of the substitutions (1.2) and (1.4) transforms (1.1) on Ω into a corresponding (5.1) on Ω^{**} for which $d_{0,1}(\zeta) \equiv d_{0,2}(\zeta) \equiv 0$. Throughout, we let l and n denote fixed integers that satisfy either $1 \leq l \leq n \leq m$ or $l = 0 < 3 \leq n \leq m$. With reference to $A^{i,j}_{p,q}$ in (6.3) on page 53 and the coefficients for (5.1), Theorem 7.10 on page 85 expresses the function

$$(9.32) \qquad \sum_{\mu=0}^{l} \sum_{\nu=0}^{n} A^{l,n}_{\mu,\nu}\, d_{l-\mu,n-\nu}^{(\mu+\nu)}(\zeta), \quad \text{on } \Omega^{**},$$

directly in terms of $\rho(z)$, $f(\zeta)$, and the derivatives $c_{\mu,\nu}^{(\kappa)}$ of the coefficients for (1.1) by means of formulas based on $\boldsymbol{L}_{i,j}$ from (1.21) of page 8. A similar result for (9.32) based on $\boldsymbol{V}_{i,j}$ from (1.33) of page 8 will be developed in this section and presented in Theorem 9.10 on page 109. As our first step in this plan, we use (6.4) on page 54 as well as (9.23)–(9.24) on page 97 to define $\boldsymbol{X}_{i,j}$ in $\mathcal{S}_m$ through

$$(9.33) \qquad \boldsymbol{X}_{i,j} \equiv \sum_{r=0}^{i} \sum_{p=0}^{r} A^{l,n}_{l-p,n-r+p}\, \boldsymbol{W}_{p,r-p,i-r,j}, \quad \text{for any integers } i, j.$$

PROPOSITION 9.4. *The function in (9.32) is given by*

$$(9.34) \qquad \sum_{\mu=0}^{l} \sum_{\nu=0}^{n} A^{l,n}_{\mu,\nu}\, d_{l-\mu,n-\nu}^{(\mu+\nu)}(\zeta) \equiv \big(f'(\zeta)\big)^{l+n} \sum_{t=0}^{l+n} X_{l+n,t}\big(f(\zeta)\big)\, \big(\mathcal{S}(\zeta)\big)^t,$$

where $X_{l+n,t}(z)$ is obtained by substituting $c_{i,j}^{(k)}(z)$ from (1.1) for $\boldsymbol{w}_{i,j}^{(k)}$ in $\boldsymbol{X}_{l+n,t}$ and $\mathcal{S}(\zeta)$ is defined in (9.3) on page 93.

PROOF. Letting $\mathcal{L}$ denote the left member of (9.34), we use Proposition 9.3 on page 97 to obtain

$$\mathcal{L} \equiv \sum_{p=0}^{l} \sum_{q=0}^{n} A_{l-p,n-q}^{l,n} d_{p,q}^{(l+n-p-q)}(\zeta)$$

$$\equiv \sum_{p=0}^{l} \sum_{q=0}^{n} A_{l-p,n-q}^{l,n} \sum_{t=0}^{l+n} (f'(\zeta))^{l+n} \mathcal{W}_{p,q,l+n-p-q,t}(f(\zeta)) (\mathcal{S}(\zeta))^{t}$$

$$\equiv (f'(\zeta))^{l+n} \sum_{t=0}^{l+n} (\mathcal{S}(\zeta))^{t} \sum_{p=0}^{l} \sum_{q=0}^{n} A_{l-p,n-q}^{l,n} \mathcal{W}_{p,q,l+n-p-q,t}(f(\zeta)).$$

Since (6.4) on page 54 yields $A_{l-p,n-q}^{l,n} = 0$ for $p > l$ or $q > n$, we have

$$\mathcal{L} \equiv (f'(\zeta))^{l+n} \sum_{t=0}^{l+n} (\mathcal{S}(\zeta))^{t} \sum_{r=0}^{l+n} \sum_{p=0}^{r} A_{l-p,n-r+p}^{l,n} \mathcal{W}_{p,r-p,l+n-r,t}(f(\zeta)).$$

In view of (9.33), this yields (9.34) and completes the proof. □

PROPOSITION 9.5. *The polynomials $\boldsymbol{X}_{i,j}$ in (9.33) are given by*

(9.35) $$\boldsymbol{X}_{i,j} \equiv \boldsymbol{Y}_{i-j,j}, \quad \text{for all } i \text{ and } j,$$

with respect to polynomials $\boldsymbol{Y}_{i,j}$ in $\mathcal{S}_m$ that satisfy

(9.36) $$\boldsymbol{Y}_{i,j} \equiv 0, \quad \text{for } i \leq 1 \text{ and any } j,$$

and

(9.37) $$\boldsymbol{Y}_{i+1,j} \equiv \sum_{k=0}^{j} \epsilon_{i,j,k} \left[\begin{array}{l} \boldsymbol{Y}_{i,k}^{(1)} + i\,\boldsymbol{b}_{m,1}\boldsymbol{Y}_{i,k} + \sum_{h=0}^{l} \tau_{h,i+1,k}\,\boldsymbol{V}_{h,i+1-h} \\ - (m-1)(k+1)\,\boldsymbol{b}_{m,2}\,\boldsymbol{Y}_{i-1,k+1} \end{array} \right],$$

$$\text{for } i \geq 0 \text{ and any } j,$$

where

(9.38) $$\epsilon_{i,j,k} \equiv \frac{(j-k)!}{(m-1)^{j-k}} \binom{2i+j+1}{j-k}, \quad \text{for } k \leq j \text{ and any } i,$$

(9.39) $$\tau_{h,i,j} \equiv \sum_{p=h}^{l} A_{l-p,n-i-j+p}^{l,n}\, \eta_{p,p-h}\, \eta_{i+j-p,j-p+h},$$

$$\text{for } 0 \leq h \leq l \text{ and } i+j \geq 0,$$

and $\boldsymbol{V}_{i,j}$ is defined by (1.33)–(1.34) on page 9.

OBSERVATION. The requirement $(l-p)+(n-i-j+p) \leq 2l+2n-2$ from (6.4) on page 54 for the definition of $A_{l-p,n-i-j+p}^{l,n}$ in the right member of (9.39) is certainly satisfied when $i+j \geq 0$. It is also satisfied when $i+j \geq 2-l-n$.

9.3. SPECIAL COMBINATIONS OF THE COEFFICIENTS FOR (5.1)

PROOF OF PROPOSITION 9.5. For $i \geq 0$ and any j, we apply $'$ to (9.33) and then employ (9.24) as well as (9.33) to verify that

$$X_{i,j}^{(1)} \equiv \sum_{r=0}^{i}\sum_{p=0}^{r} A_{l-p,n-r+p}^{l,n} \, \mathcal{W}_{p,r-p,i-r,j}^{(1)}$$

$$\equiv \sum_{r=0}^{i}\sum_{p=0}^{r} A_{l-p,n-r+p}^{l,n} \left\{ \begin{array}{l} \mathcal{W}_{p,r-p,i+1-r,j} - (i-j)b_{m,1}\,\mathcal{W}_{p,r-p,i-r,j} \\ - \left[\dfrac{2i-j+1}{m-1}\right]\mathcal{W}_{p,r-p,i-r,j-1} \\ + (m-1)(j+1)\,b_{m,2}\,\mathcal{W}_{p,r-p,i-r,j+1} \end{array} \right\}$$

and

(9.40) $\qquad X_{i,j}^{(1)} \equiv X_{i+1,j} - G_{i+1,j} - (i-j)b_{m,1}\,X_{i,j} - \dfrac{2i-j+1}{m-1}X_{i,j-1}$
$\qquad\qquad\qquad + (m-1)(j+1)\,b_{m,2}\,X_{i,j+1},$

where $G_{i,j}$ is defined in $\mathcal{S}_m$ with respect to (6.4) through

(9.41) $\qquad G_{i,j} \equiv \sum_{p=0}^{i} A_{l-p,n-i+p}^{l,n} \, \mathcal{W}_{p,i-p,0,j}, \quad \text{for any } i,\,j.$

In particular, when the sum in (9.41) is nonvacuous, we have $i \geq 0$ with either $1 \leq l \leq n$ or $l = 0 < 3 \leq n$ so that $(l-p) + (n-i+p) \leq l + n \leq 2l + 2n - 2$ and the coefficients $A_{l-p,n-i+p}^{l,n}$ for (7.41) are well defined by (6.4).

To show that (9.40) is valid for all integers i and j, we first use (9.41) and (6.4) to obtain $G_{0,j} \equiv A_{l,n}^{l,n}\,\mathcal{W}_{0,0,0,j} \equiv 0$, for any j. We apply (9.33) and (6.4) to deduce that $X_{0,j} \equiv A_{l,n}^{l,n}\,\mathcal{W}_{0,0,0,j} \equiv 0$, for any j. Also, (9.41) and (9.33) yield $G_{i,j} \equiv 0$ and $X_{i,j} \equiv 0$, for $i \leq -1$ and any j. Consequently, for $i \leq -1$ and any j, each term in (9.40) is equal to 0. Thus, (9.40) *is true for any* $i,\,j$.

Combining (9.41) with (9.23) on page 97, we obtain

(9.42) $\quad G_{i,j} \equiv \sum_{p=0}^{i} A_{l-p,n-i+p}^{l,n} \left[\sum_{r=0}^{j} \eta_{p,r}\,\eta_{i-p,j-r}\,V_{p-r,i-p+r-j}\right], \quad \text{for all } i \text{ and } j.$

If $l < p \leq i$, then $l - p < 0$ so that (6.4) gives $A_{l-p,n-i+p}^{l,n} = 0$. If $0 \leq i < p \leq l$ and $0 \leq r \leq j$, then $i - p + r - j < 0$ so that (1.34) gives $V_{p-r,i-p+r-j} \equiv 0$. Thus, (9.42) yields

(9.43) $\quad G_{i,j} \equiv \sum_{p=0}^{l}\sum_{r=0}^{j} A_{l-p,n-i+p}^{l,n}\,\eta_{p,r}\,\eta_{i-p,j-r}\,V_{p-r,i-p+r-j},$

$\qquad\qquad\qquad\qquad\qquad\qquad\qquad\qquad\qquad\qquad \text{for } i \geq 0 \text{ and any } j.$

If $p < r \leq j$, then $p - r < 0$ so that (1.34) gives $V_{p-r,i-p+r-j} \equiv 0$. If $j < r \leq p$, then $j - r < 0$ and we obtain $\eta_{i-p,j-r} = 0$ from (9.1). Thus, (9.43) yields

(9.44) $\quad G_{i,j} \equiv \sum_{p=0}^{l}\sum_{r=0}^{p} A_{l-p,n-i+p}^{l,n}\,\eta_{p,r}\,\eta_{i-p,j-r}\,V_{p-r,i-p+r-j},$

$\qquad\qquad\qquad\qquad\qquad\qquad\qquad\qquad\qquad\qquad \text{for } i \geq 0 \text{ and any } j.$

We rewrite (9.44) as

$$(9.45) \quad \boldsymbol{G}_{i,j} \equiv \sum_{p=0}^{l} \sum_{h=0}^{p} A_{l-p,n-i+p}^{l,n} \, \eta_{p,p-h} \, \eta_{i-p,j-p+h} \, \boldsymbol{V}_{h,i-j-h}$$

$$\equiv \sum_{h=0}^{l} \sum_{p=h}^{l} A_{l-p,n-i+p}^{l,n} \, \eta_{p,p-h} \, \eta_{i-p,j-p+h} \, \boldsymbol{V}_{h,i-j-h}$$

$$\equiv \sum_{h=0}^{l} \tau_{h,i-j,j} \, \boldsymbol{V}_{h,i-j-h}, \quad \text{for } i \geq 0 \text{ and any } j,$$

where $\tau_{h,i,j}$ is the rational number defined by (9.39) for any integers h, i, j that satisfy $0 \leq h \leq l$ and $i+j \geq 0$. Hence, in terms of (9.40) and (9.45), we have

$$(9.46) \quad \boldsymbol{X}_{i+1,j} - \left[\frac{2i-j+1}{m-1}\right]\boldsymbol{X}_{i,j-1}$$

$$\equiv \boldsymbol{X}_{i,j}^{(1)} + (i-j)\,\boldsymbol{b}_{m,1}\,\boldsymbol{X}_{i,j} - (m-1)(j+1)\,\boldsymbol{b}_{m,2}\,\boldsymbol{X}_{i,j+1}$$

$$+ \sum_{h=0}^{l} \tau_{h,i-j+1,j} \, \boldsymbol{V}_{h,i+1-j-h}, \quad \text{for } i \geq 0 \text{ and any } j.$$

In order to establish the relation

$$(9.47) \quad \boldsymbol{G}_{i,j} \equiv 0, \quad \text{for any integers } i, j \text{ that satisfy } j \geq i-1,$$

we consider four cases. For $i \leq -1$ and any j, (9.41) yields $\boldsymbol{G}_{i,j} \equiv 0$. We use (9.45), $\boldsymbol{V}_{0,1} \equiv \boldsymbol{V}_{1,0} \equiv 0$ from (1.33) of page 9, and (1.34) to obtain

$$\boldsymbol{G}_{i,i-1} \equiv \sum_{h=0}^{l} \tau_{h,1,i-1} \, \boldsymbol{V}_{h,1-h} \equiv 0, \quad \text{for } i \geq 0.$$

We apply (9.45), $\boldsymbol{V}_{0,0} \equiv 1$ from (1.33) of page 9, (1.34), (9.39), (9.1), and (6.4) to verify that

$$\boldsymbol{G}_{i,i} \equiv \sum_{h=0}^{l} \tau_{h,0,i} \, \boldsymbol{V}_{h,-h} \equiv \tau_{0,0,i} \equiv \sum_{p=0}^{l} A_{l-p,n-i+p}^{l,n} \, \eta_{p,p} \, \eta_{i-p,i-p} \equiv 0, \quad \text{for } i \geq 0.$$

When $i \geq 0$ and $j \geq i+1$, we employ (9.45) with $i-j-h \leq -h-1 \leq -1$ and $\boldsymbol{V}_{h,i-j-h} \equiv 0$ from (1.34) to deduce that $\boldsymbol{G}_{i,j} \equiv 0$. By combining these results, we find that (9.47) is valid.

In view of (9.40) and (9.47), we find that

$$(9.48) \quad \boldsymbol{X}_{i+1,j} \equiv \left[\frac{2i-j+1}{m-1}\right]\boldsymbol{X}_{i,j-1} + \boldsymbol{X}_{i,j}^{(1)} + (i-j)\,\boldsymbol{b}_{m,1}\,\boldsymbol{X}_{i,j}$$

$$- (m-1)(j+1)\,\boldsymbol{b}_{m,2}\,\boldsymbol{X}_{i,j+1}, \quad \text{whenever } j \geq i.$$

Next, we shall use (9.48) to establish the relation

$$(9.49) \quad \boldsymbol{X}_{i_0+j,j} \equiv 0, \quad \text{for } i_0 \leq 1 \text{ and any } j.$$

To clarify the argument, we consider separately: $i_0 \leq -1$; $i_0 = 0$; and $i_0 = 1$.

1. Suppose that $i_0 \leq -1$. Then, (9.33) yields $\boldsymbol{X}_{i_0,j} \equiv 0$, for any j. Using this, we substitute $i = i_0$ and $j \geq 1 > i_0 = i$ in (9.48) to obtain $\boldsymbol{X}_{i_0+1,j} \equiv 0$, for $j \geq 1$. By substituting $i = i_0+1$ and $j \geq 2 > i_0+1 = i$ in (9.48), we deduce $\boldsymbol{X}_{i_0+2,j} \equiv 0$, for $j \geq 2$. Continuing in this manner, we obtain $\boldsymbol{X}_{i_0+j,j} \equiv 0$, for any j.

9.3. SPECIAL COMBINATIONS OF THE COEFFICIENTS FOR (5.1)

2. Suppose that $i_0 = 0$. We use (9.33) and (6.4) to verify that $\boldsymbol{X}_{j,j} \equiv 0$, for $j \leq -1$, and

(9.50) $$\boldsymbol{X}_{0,j} \equiv A^{l,n}_{l,n} \boldsymbol{W}_{0,0,0,j} \equiv 0, \quad \text{for any } j.$$

Using this, we substitute $i = 0$ and $j \geq 1$ in (9.48) to deduce that $\boldsymbol{X}_{1,j} \equiv 0$, for $j \geq 1$. Substituting $i = 1$ and $j \geq 2$ in (9.48), we find that $\boldsymbol{X}_{2,j} \equiv 0$, for $j \geq 2$. Continuing in this way, we conclude that $\boldsymbol{X}_{j,j} \equiv 0$, for any j.

3. Suppose that $i_0 = 1$. Due to (9.33) and (9.50), we have $\boldsymbol{X}_{1+j,j} \equiv 0$, for $j \leq -1$. We apply (9.33) and (6.4) to observe that

$$\boldsymbol{X}_{1,j} \equiv \sum_{r=0}^{1} \sum_{p=0}^{r} A^{l,n}_{l-p,n-r+p} \boldsymbol{W}_{p,r-p,1-r,j}$$
$$\equiv A^{l,n}_{l,n} \boldsymbol{W}_{0,0,1,j} + A^{l,n}_{l,n-1} \boldsymbol{W}_{0,1,0,j} + A^{l,n}_{l-1,n} \boldsymbol{W}_{1,0,0,j}$$
$$\equiv 0, \quad \text{for any } j.$$

Using this, we substitute $i = 1$ and $j \geq 1$ in (9.48) to deduce that $\boldsymbol{X}_{2,j} \equiv 0$, for $j \geq 1$. Substituting $i = 2$ and $j \geq 2$ in (9.48), we find that $\boldsymbol{X}_{3,j} \equiv 0$, for $j \geq 2$. Continuing in this way, we conclude that $\boldsymbol{X}_{1+j,j} \equiv 0$, for any j. Consequently, (9.49) is valid.

We define $\boldsymbol{Y}_{i,j}$ in $\mathcal{S}_m$ by means of

(9.51) $$\boldsymbol{Y}_{i,j} \equiv \boldsymbol{X}_{i+j,j}, \quad \text{for all integers } i \text{ and } j.$$

Thus, $\boldsymbol{Y}_{i,j}$ satisfies (9.35). Moreover, we use (9.51) and (9.49) to see that $\boldsymbol{Y}_{i,j}$ satisfies (9.36). For any integers i, j subject to $i + j \geq 0$, we replace i in (9.46) with $i + j$ and use (9.51) to obtain

(9.52) $$\boldsymbol{Y}_{i+1,j} - \left[\frac{2i+j+1}{m-1}\right] \boldsymbol{Y}_{i+1,j-1}$$
$$\equiv \boldsymbol{Y}^{(1)}_{i,j} + i\,\boldsymbol{b}_{m,1}\,\boldsymbol{Y}_{i,j} - (m-1)(j+1)\,\boldsymbol{b}_{m,2}\,\boldsymbol{Y}_{i-1,j+1}$$
$$+ \sum_{h=0}^{l} \tau_{h,i+1,j}\,\boldsymbol{V}_{h,i+1-h}, \quad \text{whenever } i+j \geq 0,$$

where $\tau_{h,i,j}$ is given in (9.39). We find that the rational numbers $\epsilon_{i,j,k}$ defined in (9.38) satisfy

(9.53) $$\epsilon_{i,j,j} \equiv 1, \quad \text{for any } i, j,$$

and

(9.54) $$\frac{(2i+k+1)}{m-1} \epsilon_{i,j,k} \equiv \epsilon_{i,j,k-1}, \quad \text{for } k \leq j \text{ and any } i.$$

We note that (9.33) and (9.26) give $\boldsymbol{X}_{i,j} \equiv 0$, for $i \geq 0$ and $j \leq -1$. Hence, we have $\boldsymbol{X}_{i,j} \equiv 0$, for $j \leq -1$ and any i. In view of (9.51), this yields

(9.55) $$\boldsymbol{Y}_{i,j} \equiv 0, \quad \text{for } j \leq -1 \text{ and any } i.$$

For $i \geq 0$ and $j \geq 0$, we combine (9.53), (9.55), (9.54), and (9.52) to obtain

$$(9.56) \quad Y_{i+1,j} \equiv \sum_{k=0}^{j} \left[\epsilon_{i,j,k} \, Y_{i+1,k} - \epsilon_{i,j,k-1} \, Y_{i+1,k-1} \right]$$

$$\equiv \sum_{k=0}^{j} \epsilon_{i,j,k} \left[Y_{i+1,k} - \frac{(2i+k+1)}{m-1} Y_{i+1,k-1} \right]$$

$$\equiv \sum_{k=0}^{j} \epsilon_{i,j,k} \left[\begin{array}{l} Y_{i,k}^{(1)} + i \, \boldsymbol{b}_{m,1} \, Y_{i,k} + \sum_{h=0}^{l} \tau_{h,i+1,k} \, V_{h,i+1-h} \\ - (m-1)(k+1) \boldsymbol{b}_{m,2} \, Y_{i-1,k+1} \end{array} \right].$$

Due to (9.55), (9.56) is valid for $i \geq 0$ and any j. This yields (9.37) and completes the proof of Proposition 9.5. $\square$

PROPOSITION 9.6. *The rational numbers $\tau_{h,i,j}$ defined in (9.39) are given in terms of*

$$(9.57) \quad v_{i,j} \equiv \frac{\dfrac{(-1)^{l+n+i+j}}{(m-1)^j} \dbinom{l+n-i}{j}}{\left[\displaystyle\prod_{k=1}^{l+n-i-j} (2l+2n-1-k) \right]}, \quad \text{for } 0 \leq h \leq l \text{ and } i+j \geq 0,$$

by

$$(9.58) \quad \tau_{h,i,j} \equiv v_{i,j} \binom{m-h}{l-h} \left[\prod_{r=1}^{l-h} (l-r) \right] \binom{m-i+h}{n-i+h} \left[\prod_{s=1}^{n-i+h} (n-s) \right],$$

for $0 \leq h \leq l$ and $i+j \geq 0$.

PROOF. To facilitate reference, we repeat (9.39) as

$$(9.59) \quad \tau_{h,i,j} \equiv \sum_{p=h}^{l} A_{l-p,n-i-j+p}^{l,n} \, \eta_{p,p-h} \, \eta_{i+j-p,j-p+h},$$

for $0 \leq h \leq l$ and $i+j \geq 0$.
We use (6.4) of page 54 to obtain

$$(9.60) \quad A_{l-p,n-i-j+p}^{l,n} \equiv \frac{(-1)^{l+n+i+j} P_1 P_2}{Q} \binom{m-p}{l-p} \binom{m-i-j+p}{n-i-j+p},$$

where

$$P_1 \equiv \prod_{k=1}^{l-p}(l-k), \quad P_2 \equiv \prod_{k=1}^{n-i-j+p}(n-k), \quad \text{and} \quad Q \equiv \prod_{k=1}^{l+n-i-j}(2l+2n-1-k).$$

We apply (9.1) of page 93 to see that

$$(9.61) \quad \eta_{p,p-h} \equiv \frac{P_3}{(m-1)^{p-h}} \binom{m-h}{p-h},$$

where

$$P_3 \equiv \prod_{k=1}^{p-h}(p-k).$$

Also, (9.1) yields

$$\eta_{i+j-p,j-p+h} \equiv \frac{P_4}{(m-1)^{j-p+h}} \binom{m-i+h}{j-p+h}, \tag{9.62}$$

where

$$P_4 \equiv \prod_{k=1}^{j-p+h} (i+j-p-k).$$

For $0 \le h \le p \le l$, we check that

$$P_1 P_3 \equiv \prod_{k=1}^{l-h} (l-k). \tag{9.63}$$

For $0 \le h \le p \le l$, we observe that

$$\binom{m-i-j+p}{n-i-j+p}\binom{m-i+h}{j-p+h} P_2 P_4 \tag{9.64}$$

$$\equiv \binom{m-i-j+p}{n-i-j+p}\binom{m-i+h}{j-p+h} \prod_{k=1}^{n-i+h} (n-k).$$

Namely, for either $n-i-j+p < 0$ or $j-p+h < 0$, (7.64) reduces to $0 \equiv 0$. While, for $n-i-j+p \ge 0$ and $j-p+h \ge 0$, we find that $i-h \le i+j-p \le n$ and $P_2 P_4 \equiv \prod_{k=1}^{n-i+h}(n-k)$. Also, to maintain the correspondence with Chapter 7, we repeat (7.65) and (7.66) as

$$\binom{m-h}{p-h}\binom{m-p}{l-p} \equiv \binom{m-h}{l-h}\binom{l-h}{p-h} \tag{9.65}$$

and

$$\binom{m-i+h}{j-p+h}\binom{m-i-j+p}{n-i-j+p} \equiv \binom{m-i+h}{n-i+h}\binom{n-i+h}{j-p+h}. \tag{9.66}$$

We use (9.59)–(9.66) to verify that

$$\tau_{h,i,j} \equiv \frac{(-1)^{l+n+i+j}}{(m-1)^j Q} \binom{m-h}{l-h} \left[\prod_{r=1}^{l-h}(l-r)\right]\binom{m-i+h}{n-i+h}\left[\prod_{s=1}^{n-i+h}(n-s)\right] S, \tag{9.67}$$

where, as in (7.68),

$$S \equiv \sum_{p=h}^{l} \binom{l-h}{p-h}\binom{n-i+h}{j-p+h} \equiv \binom{l+n-i}{j}. \tag{9.68}$$

By combining (9.67) and (9.68), we obtain (9.58). This completes the proof. $\square$

Using the abbreviated notation of (1.41), we rewrite (1.35) as

$$\boldsymbol{W}_{h,i} \equiv \binom{m-h}{l-h}\left[\prod_{r=1}^{l-h}(l-r)\right]\binom{m-i+h}{n-i+h}\left[\prod_{s=1}^{n-i+h}(n-s)\right] \boldsymbol{V}_{h,i-h}, \tag{9.69}$$

for $0 \le h \le l$ and any i.

COROLLARY 9.7. *The expressions* $\tau_{h,i,j}$ *in* (9.39), $v_{i,j}$ *in* (9.57), $\boldsymbol{V}_{i,j}$, *and* $\boldsymbol{W}_{h,i}$ *are related through*

(9.70) $\qquad \tau_{h,i,j}\, \boldsymbol{V}_{h,i-h} \equiv v_{i,j}\, \boldsymbol{W}_{h,i}, \quad \text{for } 0 \le h \le l \text{ and } i+j \ge 0.$

PROOF. We combine (9.58) and (9.69) to obtain (9.70). $\qquad \square$

For further results about $\boldsymbol{Y}_{i,j}$, (9.70) plays a key role.

With respect to (1.41) and (1.24)–(1.25), we have

(9.71) $\qquad A_i \equiv \dfrac{-1}{l+n+i-1}, \quad \text{for } 0 \le h \le l \text{ and } i \ge 1,$

and

(9.72) $\qquad B_i \equiv \dfrac{l+n-i}{l+n+i-2}, \quad \text{for } 0 \le h \le l \text{ and } i \ge 1.$

In terms of (1.41) and (1.36)–(1.37), $\boldsymbol{J}_{h,i}$ is defined in $\mathcal{S}_m$ through

(9.73) $\qquad \boldsymbol{J}_{h,0} \equiv \boldsymbol{J}_{h,1} \equiv 0, \quad \text{for } 0 \le h \le l,$

and

(9.74) $\quad \boldsymbol{J}_{h,i+1} \equiv A_i\bigl(\boldsymbol{J}_{h,i}^{(1)} + i\, \boldsymbol{b}_{m,1}\, \boldsymbol{J}_{h,i}\bigr) + B_i\, \boldsymbol{b}_{m,2}\, \boldsymbol{J}_{h,i-1} + \boldsymbol{W}_{h,i+1},$
$\qquad\qquad\qquad\qquad\qquad \text{for } 0 \le h \le l \text{ and } 1 \le i \le l+n-1.$

We use (9.71)–(9.74) and the rational numbers

(9.75) $\quad r_{i,j} \equiv \dfrac{(l+n-i-j)(l+n+i+j-1)}{(l+n-i)(l+n+i-1)}, \quad \text{for } 0 \le i \le l+n-1 \text{ and any } j,$

to define $\mathfrak{S}_{h,i,j}$ in $\mathcal{S}_m$ by

(9.76) $\qquad \mathfrak{S}_{h,i,j} \equiv r_{i,j}\, \boldsymbol{J}_{h,i}, \quad \text{for } 0 \le h \le l,\ 0 \le i \le l+n-1 \text{ and any } j,$

and

(9.77) $\qquad \mathfrak{S}_{h,l+n,j} \equiv \boldsymbol{J}_{h,l+n}, \quad \text{for } 0 \le h \le l \text{ and any } j.$

In particular, (9.77), (9.76), and (9.75) give

(9.78) $\qquad \mathfrak{S}_{h,l+n,0} \equiv \boldsymbol{J}_{h,l+n}, \quad \text{for } 0 \le h \le l,$

and

(9.79) $\qquad \mathfrak{S}_{h,l+n-s,s} \equiv 0, \quad \text{for } 0 \le h \le l \text{ and } 1 \le s \le l+n.$

Moreover, (9.76) and (9.73) yield

(9.80) $\qquad \mathfrak{S}_{h,i,j} \equiv 0, \quad \text{for } 0 \le h \le l,\ 0 \le i \le 1 \text{ and any } j.$

For $1 \le i \le l+n-1$ and $0 \le j \le l+n-1-i$, we define rational numbers $E_{i,j}$, $F_{i,j}$, $G_{i,j}$ by means of

(9.81) $\qquad E_{i,j} \equiv \dfrac{-(m-1)j}{(l+n-i-j)(l+n+i+j-1)},$

(9.82) $\qquad F_{i,j} \equiv \dfrac{-(l+n-i)}{(l+n-i-j)(l+n+i+j-1)},$

and

(9.83) $\qquad G_{i,j} \equiv \dfrac{-(l+n-i)(l+n-i+1)}{(m-1)(j+1)(l+n-i-j)(l+n+i+j-1)}.$

9.3. SPECIAL COMBINATIONS OF THE COEFFICIENTS FOR (5.1)

PROPOSITION 9.8. *In terms of* (9.81)–(9.83), *the polynomials* $\mathfrak{S}_{h,i,j}$ *satisfy*

$$(9.84) \quad \mathfrak{S}_{h,i+1,j} - \frac{2i+j+1}{m-1} E_{i,j} \, \mathfrak{S}_{h,i+1,j-1}$$
$$\equiv F_{i,j}\bigl(\mathfrak{S}_{h,i,j}^{(1)} + i\, \boldsymbol{b}_{m,1}\, \mathfrak{S}_{h,i,j}\bigr)$$
$$- (m-1)(j+1)\, \boldsymbol{b}_{m,2}\, G_{i,j}\, \mathfrak{S}_{h,i-1,j+1} + \boldsymbol{W}_{h,i+1},$$

for $0 \le h \le l$, $1 \le i \le l+n-1$, *and* $0 \le j \le l+n-1-i$,

where $\boldsymbol{W}_{h,i+1}$ *is obtained from* (9.69).

PROOF. For $1 \le i \le l+n-1$ and $0 \le j \le l+n-1-i$, we note that (9.71), (9.82), (9.75), (9.72), and (9.83) yield

$$(9.85) \quad A_i \equiv F_{i,j}\, r_{i,j},$$

and

$$(9.86) \quad B_i \equiv -(m-1)(j+1)\, G_{i,j}\, r_{i-1,j+1}.$$

For $0 \le h \le l$, $1 \le i \le l+n-1$, and $0 \le j \le l+n-1-i$, we employ (9.74), (9.85), (9.86), and (9.76) to verify that

$$(9.87) \quad \boldsymbol{J}_{h,i+1} \equiv A_i\bigl(\boldsymbol{J}_{h,i}^{(1)} + i\, \boldsymbol{b}_{m,1}\, \boldsymbol{J}_{h,i}\bigr) + B_i\, \boldsymbol{b}_{m,2}\, \boldsymbol{J}_{h,i-1} + \boldsymbol{W}_{h,i+1}$$
$$\equiv F_{i,j}\bigl(r_{i,j}\, \boldsymbol{J}_{h,i}^{(1)} + i\, \boldsymbol{b}_{m,1}\, r_{i,j}\, \boldsymbol{J}_{h,i}\bigr)$$
$$- (m-1)(j+1)\, \boldsymbol{b}_{m,2}\, G_{i,j}\bigl(r_{i-1,j+1}\, \boldsymbol{J}_{h,i-1}\bigr) + \boldsymbol{W}_{h,i+1}$$
$$\equiv F_{i,j}\bigl(\mathfrak{S}_{h,i,j}^{(1)} + i\boldsymbol{b}_{m,1}\, \mathfrak{S}_{h,i,j}\bigr)$$
$$- (m-1)(j+1)\, \boldsymbol{b}_{m,2}\, G_{i,j}\, \mathfrak{S}_{h,i-1,j+1} + \boldsymbol{W}_{h,i+1}.$$

Case 1. As the first of two situations, suppose now that h, i, and j satisfy $0 \le h \le l$, $1 \le i \le l+n-2$, and $0 \le j \le l+n-1-i$. Then, in terms of (9.76), (9.75), and (9.81), we find that

$$(9.88) \quad \mathfrak{S}_{h,i+1,j} - \frac{2i+j+1}{m-1} E_{i,j}\, \mathfrak{S}_{h,i+1,j-1}$$
$$\equiv \left[r_{i+1,j} - \frac{2i+j+1}{m-1} E_{i,j}\, r_{i+1,j-1}\right] \boldsymbol{J}_{h,i+1}$$
$$\equiv \left[\frac{(l+n-i-1-j)(l+n+i+j) + j(2i+j+1)}{(l+n-i-1)(l+n+i)}\right] \boldsymbol{J}_{h,i+1}$$
$$\equiv \boldsymbol{J}_{h,i+1}.$$

By combining (9.88) and (9.87), we conclude that (9.84) is satisfied with respect to the preceding restriction.

Case 2. For the remaining situation, suppose that h, i, and j satisfy $0 \le h \le l$, $i = l+n-1$, and $j = 0$. Then, in view of (9.78) and (9.81), we see that the left member of (9.84) reduces to $\mathfrak{S}_{h,l+n,0} \equiv \boldsymbol{J}_{h,l+n}$. We combine this with the substitution of $i = l+n-1$ and $j = 0$ in (9.87) to deduce that (9.84) is valid for these values of h, i, and j. This completes the proof. $\square$

With reference to $\mathfrak{S}_{h,i,j}$ in (9.76)–(9.77) and $v_{i,j}$ in (9.57), we define $\mathfrak{R}_{h,i,j}$ in $\mathcal{S}_m$ through

(9.89) $\quad \mathfrak{R}_{h,i,j} \equiv v_{i,j}\, \mathfrak{S}_{h,i,j}, \quad$ for $0 \leq h \leq l$, $0 \leq i \leq l+n$, and $j \geq -i$.

In particular, after observing that (9.57) gives

$$\text{(9.90)} \qquad v_{l+n,0} \equiv \frac{(-1)^{2l+2n}}{(m-1)^0} \binom{0}{0} \equiv 1,$$

we combine (9.89), (9.90), and (9.78)–(9.79) to obtain

(9.91) $\quad \mathfrak{R}_{h,l+n,0} \equiv J_{h,l+n}, \quad$ for $0 \leq h \leq l$,

and

(9.92) $\quad \mathfrak{R}_{h,l+n-s,s} \equiv 0, \quad$ for $0 \leq h \leq l$ and $1 \leq s \leq l+n$.

Also, (9.89) and (9.80) yield

(9.93) $\quad \mathfrak{R}_{h,i,j} \equiv 0, \quad$ for $0 \leq h \leq l$, $0 \leq i \leq 1$, and $j \geq -i$.

We use (9.89) to define $Z_{i,j}$ in $\mathcal{S}_m$ by means of

$$\text{(9.94)} \qquad Z_{i,j} \equiv \sum_{h=0}^{l} \mathfrak{R}_{h,i,j}, \quad \text{for } 0 \leq i \leq l+n \text{ and } j \geq -i.$$

In particular, (9.94) and (9.91) give

$$\text{(9.95)} \qquad Z_{l+n,0} \equiv \sum_{h=0}^{l} J_{h,l+n},$$

while (9.94) and (9.92) yield

(9.96) $\quad Z_{l+n-s,s} \equiv 0, \quad$ for $1 \leq s \leq l+n$.

We apply (9.94) and (9.93) to see that

(9.97) $\quad Z_{i,j} \equiv 0, \quad$ for $0 \leq i \leq 1$, and $j \geq -i$.

PROPOSITION 9.9. *The polynomials $Z_{i,j}$ satisfy*

$$\text{(9.98)} \quad Z_{i+1,j} - \frac{2i+j+1}{m-1} Z_{i+1,j-1} \equiv Z_{i,j}^{(1)} + i\, \boldsymbol{b}_{m,1}\, Z_{i,j} + \sum_{h=0}^{l} \tau_{h,i+1,j}\, \boldsymbol{V}_{h,i+1-h}$$
$$- (m-1)(j+1)\, \boldsymbol{b}_{m,2}\, Z_{i-1,j+1},$$

for $1 \leq i \leq l+n-1$ and $0 \leq j \leq l+n-1-i$,

where $\tau_{h,i,j}$ and $\boldsymbol{V}_{i,j}$ are given by (9.58) of page 104 and (1.33)–(1.34) of page 9.

PROOF. For $0 \leq h \leq l$, $1 \leq i \leq l+n-1$, and $0 \leq j \leq l+n-1-i$, we use (9.57) and (9.81)–(9.83) to verify that $v_{i,j}$, $E_{i,j}$, $F_{i,j}$, and $G_{i,j}$ satisfy

(9.99) $\qquad\qquad\qquad v_{i+1,j}\, E_{i,j} \equiv v_{i+1,j-1}$,

(9.100) $\qquad\qquad\qquad v_{i+1,j}\, F_{i,j} \equiv v_{i,j}$,

and

(9.101) $\qquad\qquad\qquad v_{i+1,j}\, G_{i,j} \equiv v_{i-1,j+1}$.

For $0 \leq h \leq l$, $1 \leq i \leq l+n-1$, and $0 \leq j \leq l+n-1-i$, we multiply (9.84) by $v_{i+1,j}$ and apply (9.99)–(9.101), (9.70), and (9.89) to obtain

$$0 \equiv v_{i+1,j}\, \mathfrak{S}_{h,i+1,j} - \frac{2i+j+1}{m-1}\, v_{i+1,j}\, E_{i,j}\, \mathfrak{S}_{h,i+1,j-1}$$
$$- v_{i+1,j}\, F_{i,j}\big(\mathfrak{S}^{(1)}_{h,i,j} + i\, \boldsymbol{b}_{m,1}\, \mathfrak{S}_{h,i,j}\big)$$
$$+ (m-1)(j+1)\, \boldsymbol{b}_{m,2}\, v_{i+1,j}\, G_{i,j}\, \mathfrak{S}_{h,i-1,j+1} - v_{i+1,j}\, W_{h,i+1},$$

$$0 \equiv v_{i+1,j}\, \mathfrak{S}_{h,i+1,j} - \frac{2i+j+1}{m-1}\, v_{i+1,j-1}\, \mathfrak{S}_{h,i+1,j-1}$$
$$- v_{i,j}\big(\mathfrak{S}^{(1)}_{h,i,j} + i\, \boldsymbol{b}_{m,1}\, \mathfrak{S}_{h,i,j}\big)$$
$$+ (m-1)(j+1)\, \boldsymbol{b}_{m,2}\, v_{i-1,j+1}\, \mathfrak{S}_{h,i-1,j+1} - \tau_{h,i+1,j}\, V_{h,i+1-h},$$

and

(9.102) $\quad 0 \equiv \mathfrak{R}_{h,i+1,j} - \dfrac{2i+j+1}{m-1}\, \mathfrak{R}_{h,i+1,j-1} - \mathfrak{R}^{(1)}_{h,i,j} - i\, \boldsymbol{b}_{m,1}\, \mathfrak{R}_{h,i,j}$
$$+ (m-1)(j+1)\, \boldsymbol{b}_{m,2}\, \mathfrak{R}_{h,i-1,j+1} - \tau_{h,i+1,j}\, V_{h,i+1-h}.$$

Letting $\mathcal{N}_h$ denote the right member of (9.102) for $0 \leq h \leq l$, we use (9.102) and (9.94) to deduce, for $1 \leq i \leq l+n-1$ and $0 \leq j \leq l+n-1-i$, that

(9.103) $\quad 0 \equiv \displaystyle\sum_{h=0}^{l} \mathcal{N}_h \equiv Z_{i+1,j} - \frac{2i+j+1}{m-1} Z_{i+1,j-1} - Z^{(1)}_{i,j} - i\, \boldsymbol{b}_{m,1}\, Z_{i,j}$
$$+ (m+1)(j+1)\, \boldsymbol{b}_{m,2}\, Z_{i-1,j+1} - \sum_{h=0}^{l} \tau_{h,i+1,j}\, V_{h,i+1-h}.$$

We rewrite (9.103) to obtain (9.98) and complete the proof of Proposition 9.9. $\square$

Our development throughout this chapter has been motivated by the identity (6.23) in Theorem 6.3 on page 58 and an analogy with results in [24]. However, unlike the situation for (6.23), no restrictions are placed on (1.1) in the next result.

THEOREM 9.10. *For (5.1) on Ω^{**} into which (1.1) on Ω is transformed by $y(z) = \rho(z)\, v(z)$ and $z = f(\zeta)$ according to the context of page 47, suppose that $d_{0,1}(\zeta) \equiv d_{0,2}(\zeta) \equiv 0$. Let l and n be integers subject to either $1 \leq l \leq n \leq m$ or $l = 0 < 3 \leq n \leq m$. Then, the coefficients $d_{i,j}(\zeta)$ of (5.1) satisfy*

(9.104) $\quad \displaystyle\sum_{\mu=0}^{l} \sum_{\nu=0}^{n} A^{l,n}_{\mu,\nu}\, d^{(\mu+\nu)}_{l-\mu,n-\nu}(\zeta) \equiv \big(f'(\zeta)\big)^{l+n} \mathcal{J}_{m,l,n}\big(f(\zeta)\big),$

where $A^{l,n}_{\mu,\nu}$ is given by (6.3) on page 53 and where $\mathcal{J}_{m,l,n}\big(f(\zeta)\big)$ is obtained for $z = f(\zeta)$ by substituting $c^{(k)}_{i,j}(z)$ from (1.1) for $w^{(k)}_{i,j}$ in $\mathcal{J}_{m,l,n}$ of (1.38) on page 9.

PROOF. We note that (9.57) yields $v_{i+1,-1} \equiv 0$, for $0 \leq h \leq l$ and $i \geq 0$. Due to (9.89), we have $\mathfrak{R}_{h,i+1,-1} \equiv 0$, for $0 \leq h \leq l$ and $0 \leq i \leq l+n-1$. We use (9.94) to obtain

(9.105) $\quad\quad\quad Z_{i+1,-1} \equiv 0, \quad \text{for } 0 \leq i \leq l+n-1.$

We combine (9.53), (9.105), (9.54), and (9.98) to check that

$$(9.106) \quad \boldsymbol{Z}_{i+1,j} \equiv \sum_{k=0}^{j} \left[\epsilon_{i,j,k} \, \boldsymbol{Z}_{i+1,k} - \epsilon_{i,j,k-1} \, \boldsymbol{Z}_{i+1,k-1} \right]$$

$$\equiv \sum_{k=0}^{j} \epsilon_{i,j,k} \left[\boldsymbol{Z}_{i+1,k} - \frac{(2i+k+1)}{m-1} \boldsymbol{Z}_{i+1,k-1} \right]$$

$$\equiv \sum_{k=0}^{j} \epsilon_{i,j,k} \left[\begin{array}{c} \boldsymbol{Z}_{i,k}^{(1)} + i\,\boldsymbol{b}_{m,1}\,\boldsymbol{Z}_{i,k} + \sum_{h=0}^{l} \tau_{h,i+1,k}\,\boldsymbol{V}_{h,i+1-h} \\ - (m-1)(k+1)\boldsymbol{b}_{m,2}\,\boldsymbol{Z}_{i-1,k+1} \end{array} \right],$$

for $1 \leq i \leq l+n-1$ and $0 \leq j \leq l+n-1-i$.

In view of (9.36) on page 100 and (9.97), we see that the statement

$$(9.107) \quad \boldsymbol{Y}_{i,j} \equiv \boldsymbol{Z}_{i,j}, \quad \text{for } 0 \leq i \leq i_0 \text{ and } 0 \leq j \leq l+n-i,$$

is valid for $i_0 = 1$. Suppose that i_0 is an integer satisfying $1 \leq i_0 \leq l+n-1$ such that (9.107) is valid. Then, we use (9.37) on page 100 as well as (9.107) and (9.106) to deduce that $\boldsymbol{Y}_{i_0+1,j} \equiv \boldsymbol{Z}_{i_0+1,j}$, for $0 \leq j \leq l+n-1-i_0 = l+n-(i_0+1)$. Consequently, we have

$$(9.108) \quad \boldsymbol{Y}_{i,j} \equiv \boldsymbol{Z}_{i,j}, \quad \text{for } 0 \leq i \leq l+n \text{ and } 0 \leq j \leq l+n-i.$$

We employ (9.35) on page 100 and (9.108) to verify that

$$(9.109) \quad \boldsymbol{X}_{l+n,t} \equiv \boldsymbol{Y}_{l+n-t,t} \equiv \boldsymbol{Z}_{l+n-t,t}, \quad \text{for } 0 \leq t \leq l+n.$$

We find that (9.109) and (9.96) yield

$$(9.110) \quad \boldsymbol{X}_{l+n,t} \equiv 0, \quad \text{for } 1 \leq t \leq l+n.$$

After substituting $c_{i,j}^{(k)}(z)$ from (1.1) for $w_{i,j}^{(k)}$ in (9.110) and setting $z = f(\zeta)$, we combine the resulting expressions with (9.34) on page 99 to conclude that

$$(9.111) \quad \sum_{\mu=0}^{l} \sum_{\nu=0}^{n} A_{\mu,\nu}^{l,n} \, d_{l-\mu,n-\nu}^{(\mu+\nu)}(\zeta) \equiv \left(f'(\zeta)\right)^{l+n} X_{l+n,0}\bigl(f(\zeta)\bigr).$$

Moreover, (9.109), (9.95), and (1.38) on page 9 give

$$(9.112) \quad \boldsymbol{X}_{l+n,0} \equiv \sum_{h=0}^{l} \boldsymbol{J}_{h,l+n} \equiv \boldsymbol{\mathcal{J}}_{m,l,n}.$$

After substituting $c_{i,j}^{(k)}(z)$ from (1.1) for $w_{i,j}^{(k)}$ in (9.112) and setting $z = f(\zeta)$, we combine the resulting expression with (9.111) to obtain (9.104). This completes the proof of Theorem 9.10. □

CHAPTER 10

Basic Semi-Invariants of the Second Kind for $m \geq 2$

The polynomials $V_{i,j}$ are defined in S_m for all integers i and j by (1.33)–(1.34) of page 9. They satisfy $V_{j,i} \equiv V_{i,j}$, for all i and j. Here, we focus on

(10.1) $$V_{i,j} \equiv \sum_{p=0}^{i} \sum_{q=0}^{j} U_{i-p,p}\, U_{j-q,q}\, w_{p,q}, \quad \text{where } 0 \leq i \leq j \leq m,$$

for which $U_{i-p,p}$ and $U_{j-q,q}$ are defined by (1.30)–(1.32) with $m \geq 2$. In particular, these formulas yield

(10.2) $\quad V_{0,0} \equiv 1, \quad V_{0,1} \equiv 0, \quad V_{0,2} \equiv 0, \quad \text{and} \quad V_{1,1} \equiv w_{1,1} - (w_{0,1})^2.$

Since $V_{0,0}$, $V_{0,1}$, and $V_{0,2}$ are constants, they are not semi-invariants.

THEOREM 10.1. *For any integers l and n that satisfy either $1 \leq l \leq n \leq m$ or $l = 0 < 3 \leq n \leq m$, $V_{l,n}$ is a semi-invariant of the second kind that is also a basic polynomial of index (l, n) according to* Definition 1.5 *on page* 6.

PROOF. We note that $U_{i,j} \equiv 0$, when either $i < 0$ or $i \geq 1$ and $j \geq m$. For $i = 0$ or $i \geq 1$ and $j < m$, we check that $U_{i,j}$ is an isobaric polynomial over $\mathbb{Q}$ of weight i in the variables

(10.3) $\quad w_{0,1},\, w_{0,1}^{(1)},\, w_{0,1}^{(2)},\, \ldots \quad \text{and} \quad w_{0,2},\, w_{0,2}^{(1)},\, w_{0,2}^{(2)},\, \ldots .$

We use (10.1), (1.31), and the preceding observations to obtain

(10.4) $$V_{l,n} \equiv w_{l,n} + \sum_{\substack{0 \leq p \leq l \\ 0 \leq q \leq n \\ (p,q) \neq (l,n)}} U_{l-p,p}\, U_{n-q,q}\, w_{p,q}$$

and verify that the coefficient of $w_{l,n}$ in $V_{l,n}$ is 1. Moreover, (10.4) shows that $V_{l,n}$ is an isobaric polynomial in S_m of weight $l + n$. We use (10.4) to see that each nonzero term of $V_{l,n}$ has at most one factor $w_{i,j}^{(k)}$ from (1.6) not listed in (10.3).

Suppose that $(l, n) \neq (1, 1)$ and let $w_{i_0,j_0}^{(k_0)}$ be a variable from (1.6) that is effectively involved in $V_{l,n}$ of (10.4). Then, we have $n \geq 2$ and observe that each variable $w_{i,j}^{(k)}$ in (10.3) has $i = 0 \leq l$ and $j \leq 2 \leq n$. If $w_{i_0,j_0}^{(k_0)}$ is involved in some $U_{l-p,p}$ or $U_{n-q,q}$, then $w_{i_0,j_0}^{(k_0)}$ appears in (10.3) and therefore has $i_0 \leq l$ as well as $j_0 \leq n$. Suppose that $w_{i_0,j_0}^{(k_0)}$ is not involved in any $U_{l-p,p}$ or $U_{n-q,q}$. Then, $k_0 = 0$ and there are integers p, q satisfying $0 \leq p \leq l$ and $0 \leq q \leq n$ such that $w_{p,q} \equiv w_{i_0,j_0}$. If $w_{p,q}$ appears in (1.6), then $i_0 = p \leq l$ and $j_0 = q \leq n$. If $w_{p,q}$ is not listed in (1.6), then $w_{q,p}$ appears in (1.6) with $q < p$, (1.7) yields $w_{i_0,j_0} \equiv w_{p,q} \equiv w_{q,p}$, and we have $i_0 = q < p \leq l$ as well as $j_0 = p \leq l \leq n$.

111

If $(l, n) = (1, 1)$, then $\boldsymbol{V}_{l,n} \equiv \boldsymbol{V}_{1,1} \equiv \boldsymbol{w}_{1,1} - (\boldsymbol{w}_{0,1})^2$ and each variable $\boldsymbol{w}_{i,j}^{(k)}$ from (1.6) effectively involved in $\boldsymbol{V}_{l,n}$ has $i \leq 1 = l$ and $j \leq 1 = n$.

For either $1 \leq l \leq n \leq m$ or $l = 0 < 3 \leq n \leq m$, the observations of the three preceding paragraphs show that $\boldsymbol{V}_{l,n}$ is a basic polynomial of index (l, n).

Substituting $c_{i,j}^{**(k)}(\zeta)$ from (1.5) for $\boldsymbol{w}_{i,j}^{(k)}$ in (10.1), we obtain

$$(10.5) \qquad V_{l,n}^{**}(\zeta) \equiv \sum_{p=0}^{l} \sum_{q=0}^{n} U_{l-p,p}^{**}(\zeta)\, U_{n-q,q}^{**}(\zeta)\, c_{p,q}^{**}(\zeta), \quad \text{on } \Omega^{**}.$$

For ζ in Ω^{**}, we use (3.9) of Theorem 3.2 on page 30 to see that

$$(10.6) \qquad c_{p,q}^{**}(\zeta) \equiv \sum_{\mu=0}^{p} \sum_{\nu=0}^{q} (f'(\zeta))^{\mu+\nu} \alpha_{p-\mu,m-p}(\zeta)\, \alpha_{q-\nu,m-q}(\zeta)\, c_{\mu,\nu}\big(f(\zeta)\big),$$

where $\alpha_{i,j}(\zeta)$ is defined on Ω^{**}, for $i \geq 0$ and all j, by (3.1)–(3.2) of page 29. Therefore, after introducing

$$(10.7) \quad \mathcal{H}_{p,q,\mu,\nu}(\zeta) \equiv \alpha_{p-\mu,m-p}(\zeta)\, \alpha_{q-\nu,m-q}(\zeta)\, U_{l-p,p}^{**}(\zeta)\, U_{n-q,q}^{**}(\zeta),$$

$$\text{on } \Omega^{**} \text{ for } 0 \leq \mu \leq p \leq l \text{ and } 0 \leq \nu \leq q \leq n,$$

we find that equations (10.5)–(10.7) give

$$(10.8) \qquad V_{l,n}^{**}(\zeta) \equiv \sum_{p=0}^{l} \sum_{q=0}^{n} \sum_{\mu=0}^{p} \sum_{\nu=0}^{q} \mathcal{H}_{p,q,\mu,\nu}(\zeta)\, (f'(\zeta))^{\mu+\nu} c_{\mu,\nu}\big(f(\zeta)\big)$$

$$\equiv \sum_{\mu=0}^{l} \sum_{\nu=0}^{n} (f'(\zeta))^{\mu+\nu} c_{\mu,\nu}\big(f(\zeta)\big) \sum_{p=\mu}^{l} \sum_{q=\nu}^{n} \mathcal{H}_{p,q,\mu,\nu}(\zeta)$$

$$\equiv \sum_{\mu=0}^{l} \sum_{\nu=0}^{n} (f'(\zeta))^{\mu+\nu} c_{\mu,\nu}\big(f(\zeta)\big) \sum_{p=0}^{l-\mu} \sum_{q=0}^{n-\nu} \mathcal{H}_{p+\mu,q+\nu,\mu,\nu}(\zeta).$$

We apply (10.7) to deduce that

$$\sum_{p=0}^{l-\mu} \sum_{q=0}^{n-\nu} \mathcal{H}_{p+\mu,q+\nu,\mu,\nu}(\zeta) \equiv \left[\sum_{p=0}^{l-\mu} \alpha_{p,m-\mu-p}(\zeta)\, U_{l-\mu-p,\mu+p}^{**}(\zeta)\right] \times$$

$$\left[\sum_{q=0}^{n-\nu} \alpha_{q,m-\nu-q}(\zeta)\, U_{n-\nu-q,\nu+q}^{**}(\zeta)\right], \quad \text{on } \Omega^{**}.$$

Consequently, we can rewrite (10.8) as

$$(10.9) \qquad V_{l,n}^{**}(\zeta) \equiv \sum_{\mu=0}^{l} \sum_{\nu=0}^{n} (f'(\zeta))^{\mu+\nu} S_{l-\mu,\mu}(\zeta)\, S_{n-\nu,\nu}(\zeta)\, c_{\mu,\nu}\big(f(\zeta)\big),$$

where $S_{i,j}(\zeta)$ is defined by

$$(10.10) \qquad S_{i,j}(\zeta) \equiv \sum_{k=0}^{i} \alpha_{k,m-j-k}(\zeta)\, U_{i-k,j+k}^{**}(\zeta), \quad \text{on } \Omega^{**} \text{ for all } i \text{ and } j.$$

By modifying an argument given in [**24**, pages 34–37], we shall prove in Lemma 10.2 that $S_{i,j}(\zeta)$ is given by

$$(10.11) \qquad S_{i,j}(\zeta) \equiv (f'(\zeta))^{i}\, U_{i,j}\big(f(\zeta)\big), \quad \text{on } \Omega^{**} \text{ for } j \leq m \text{ and any } i,$$

where $U_{i,j}(z)$ on Ω is obtained by substituting $c_{r,s}^{(t)}(z)$ from (1.1) for $\boldsymbol{w}_{r,s}^{(t)}$ in $\boldsymbol{U}_{i,j}$ of (1.30)–(1.32). A substitution of $c_{r,s}^{(t)}(z)$ from (1.1) for $\boldsymbol{w}_{r,s}^{(t)}$ in $\boldsymbol{V}_{l,n}$ of (10.1) yields

$$(10.12) \qquad V_{l,n}(z) \equiv \sum_{\mu=0}^{l} \sum_{\nu=0}^{n} U_{l-\mu,\mu}(z)\, U_{n-\nu,\nu}(z)\, c_{\mu,\nu}(z), \quad \text{on } \Omega.$$

We combine (10.9), (10.11), and (10.12) to verify that

$$(10.13) \qquad V_{l,n}^{**}(\zeta) \equiv \bigl(f'(\zeta)\bigr)^{l+n} \sum_{\mu=0}^{l} \sum_{\nu=0}^{n} U_{l-\mu,\mu}\bigl(f(\zeta)\bigr)\, U_{n-\nu,\nu}\bigl(f(\zeta)\bigr)\, c_{\mu,\nu}\bigl(f(\zeta)\bigr)$$

$$\equiv \bigl(f'(\zeta)\bigr)^{l+n} V_{l,n}\bigl(f(\zeta)\bigr), \quad \text{on } \Omega^{**}.$$

Since $\boldsymbol{V}_{l,n}$ is not a constant, (10.13) shows that $\boldsymbol{V}_{l,n}$ is a semi-invariant of the second kind. Thus, $\boldsymbol{V}_{l,n}$ is a basic semi-invariant of the second kind whose index is (l, n). This completes the proof of Theorem 10.1. □

LEMMA 10.2. *The functions $S_{i,j}(\zeta)$ defined in (10.10) satisfy (10.11).*

PROOF. We define $T_{i,j}(\zeta)$ by means of

$$(10.14) \qquad T_{i,j}(\zeta) \equiv \bigl(f'(\zeta)\bigr)^{i} U_{i,j}\bigl(f(\zeta)\bigr), \quad \text{on } \Omega^{**} \text{ for any } i \text{ and } j.$$

To establish Lemma 10.2, we shall prove that $S_{i,j}(\zeta) \equiv T_{i,j}(\zeta)$, on Ω^{**} for $j \leq m$ and any i.

Let i and j be fixed integers subject to $i \geq 0$ and $j \leq m$. Then, (10.10) yields

$$(10.15) \quad S_{i,j}^{(1)}(\zeta) \equiv \sum_{k=0}^{i} \alpha_{k,m-j-k}^{(1)}(\zeta)\, U_{i-k,j+k}^{**}(\zeta) + \sum_{k=0}^{i} \alpha_{k,m-j-k}(\zeta)\, U_{i-k,j+k}^{**(1)}(\zeta).$$

If $0 \leq k \leq i$ and $m - j - k \geq 0$, then (3.2) on page 29 shows that

$$(10.16) \qquad \alpha_{k,m-j-k}^{(1)}(\zeta) \equiv \alpha_{k+1,m-j-k}(\zeta) - \alpha_{k+1,m-j-k-1}(\zeta)$$
$$- (j-m)\frac{f''(\zeta)}{f'(\zeta)}\, \alpha_{k,m-j-k}(\zeta).$$

If $0 \leq k \leq i$ and $m - j - k < 0$, then $k > 0$ and we apply (3.2) to see that each term in (10.16) is zero. If $0 \leq k \leq i$ and $j + k \leq m$, then (1.32) on page 9 gives

$$(10.17) \qquad U_{i-k,j+k}^{**(1)}(\zeta) \equiv U_{i+1-k,j-1+k}^{**}(\zeta) - U_{i+1-k,j+k}^{**}(\zeta)$$
$$- (i+j-m)\, b_{m,1}^{**}(\zeta)\, U_{i-k,j+k}^{**}(\zeta)$$
$$- (m-i-j+1)(1-i-j)\, b_{m,2}^{**}(\zeta)\, U_{i-1-k,j+k}^{**}(\zeta).$$

And, if $0 \leq k \leq i$ and $j + k > m$, then $k > 0$, $m - j - k < 0$, and (3.2) yield $\alpha_{k,m-j-k}(\zeta) \equiv 0$. Consequently, we can replace $\alpha_{k,m-j-k}^{(1)}(\zeta)$ and $U_{i-k,j+k}^{**(1)}(\zeta)$ in (10.15) with the corresponding right members of (10.16) and (10.17) to obtain

(10.18) $$S_{i,j}^{(1)}(\zeta) \equiv (\mathcal{B}_1 + \mathcal{B}_2 + \mathcal{B}_3 + \mathcal{B}_4 + \mathcal{B}_5 + \mathcal{B}_6 + \mathcal{B}_7)(\zeta),$$

where

$$\mathcal{B}_1(\zeta) \equiv \sum_{k=0}^{i} \alpha_{k+1,m-j-k}(\zeta)\, U_{i-k,j+k}^{**}(\zeta),$$

$$\mathcal{B}_2(\zeta) \equiv -\sum_{k=0}^{i} \alpha_{k+1,m-j-k-1}(\zeta)\, U_{i-k,j+k}^{**}(\zeta),$$

$$\mathcal{B}_3(\zeta) \equiv -(j-m)\frac{f''(\zeta)}{f'(\zeta)} \sum_{k=0}^{i} \alpha_{k,m-j-k}(\zeta)\, U_{i-k,j+k}^{**}(\zeta),$$

$$\mathcal{B}_4(\zeta) \equiv \sum_{k=0}^{i} \alpha_{k,m-j-k}(\zeta)\, U_{i+1-k,j-1+k}^{**}(\zeta),$$

$$\mathcal{B}_5(\zeta) \equiv -\sum_{k=0}^{i} \alpha_{k,m-j-k}(\zeta)\, U_{i+1-k,j+k}^{**}(\zeta),$$

$$\mathcal{B}_6(\zeta) \equiv -(i+j-m)\, b_{m,1}^{**}(\zeta) \sum_{k=0}^{i} \alpha_{k,m-j-k}(\zeta)\, U_{i-k,j+k}^{**}(\zeta),$$

and

$$\mathcal{B}_7(\zeta) \equiv -(m-i-j+1)(1-i-j)\, b_{m,2}^{**}(\zeta) \sum_{k=0}^{i} \alpha_{k,m-j-k}(\zeta)\, U_{i-1-k,j+k}^{**}(\zeta).$$

By using (10.10), $U_{0,\nu}^{**}(\zeta) \equiv 1$ from (1.31), $\alpha_{0,\nu}(\zeta) \equiv 1$ from (3.1), and $U_{-1,\nu}^{**}(\zeta) \equiv 0$ from (1.30), we can rewrite these expressions to deduce that

$$\mathcal{B}_1 + \mathcal{B}_5 \equiv \left[\sum_{k=0}^{i} \alpha_{k+1,m-j-k}\, U_{i-k,j+k}^{**}\right] - \left[\sum_{k=0}^{i} \alpha_{k,m-j-k}\, U_{i+1-k,j+k}^{**}\right]$$

$$\equiv \left[\sum_{k=1}^{i+1} \alpha_{k,m-j+1-k}\, U_{i+1-k,j-1+k}^{**}\right] - S_{i+1,j} + \alpha_{i+1,m-i-j-1}$$

$$\equiv S_{i+1,j-1} - U_{i+1,j-1}^{**} - S_{i+1,j} + \alpha_{i+1,m-i-j-1},$$

$$\mathcal{B}_4 + \mathcal{B}_2 \equiv \sum_{k=0}^{i} \alpha_{k,m-j-k}\, U_{i+1-k,j-1+k}^{**} - \sum_{k=1}^{i+1} \alpha_{k,m-j-k}\, U_{i+1-k,j-1+k}^{**}$$

$$\equiv U_{i+1,j-1}^{**} - \alpha_{i+1,m-i-j-1},$$

$$\mathcal{B}_6 + \mathcal{B}_3 \equiv -\left[(i+j-m)\, b_{m,1}^{**} + (j-m)\frac{f''}{f'}\right] S_{i,j},$$

and

$$\mathcal{B}_7 \equiv -(m-i-j+1)(1-i-j)\, b_{m,2}^{**}\, S_{i-1,j}.$$

We substitute these expressions in (10.18) to verify that: for $i \geq 0$ and $j \leq m$,

$$(10.19) \quad S_{i,j}^{(1)}(\zeta) \equiv S_{i+1,j-1}(\zeta) - S_{i+1,j}(\zeta)$$
$$- \left[(i+j-m) b_{m,1}^{**}(\zeta) + (j-m) \frac{f''(\zeta)}{f'(\zeta)} \right] S_{i,j}(\zeta)$$
$$- (m-i-j+1)(1-i-j) b_{m,2}^{**}(\zeta) S_{i-1,j}(\zeta).$$

Since (3.2) yields $\alpha_{k,-k}(\zeta) \equiv 0$, for $k \geq 1$, while (3.1) gives $\alpha_{0,0}(\zeta) \equiv 1$ and (1.32) establishes $U_{i+1,m}^{**}(\zeta) \equiv 0$, for $i \geq 0$, we use (10.10) to check that

$$S_{i+1,m}(\zeta) \equiv \sum_{k=0}^{i+1} \alpha_{k,-k}(\zeta) U_{i+1-k,m+k}^{**}(\zeta) \equiv U_{i+1,m}^{**}(\zeta) \equiv 0, \quad \text{for } i \geq 0.$$

Thus, we have

$$S_{i+1,j}(\zeta) \equiv \sum_{k=j+1}^{m} \left[S_{i+1,k-1}(\zeta) - S_{i+1,k}(\zeta) \right], \quad \text{for } i \geq 0 \text{ and } j \leq m.$$

We employ this and (10.19) to conclude that

$$(10.20) \quad S_{i+1,j}(\zeta) \equiv \sum_{k=j+1}^{m} \left[\begin{array}{l} S_{i,k}^{(1)}(\zeta) \\ + \left[(i+k-m) b_{m,1}^{**}(\zeta) + (k-m) \frac{f''(\zeta)}{f'(\zeta)} \right] S_{i,k}(\zeta) \\ + (m-i-k+1)(1-i-k) b_{m,2}^{**}(\zeta) S_{i-1,k}(\zeta) \end{array} \right],$$

for $i \geq 0$ and $j \leq m$.

We differentiate (10.14) with respect to ζ to see that

$$(10.21) \quad \left(f'(\zeta) \right)^{i+1} U_{i,j}^{(1)}\left(f(\zeta)\right) \equiv T_{i,j}^{(1)}(\zeta) - i \frac{f''(\zeta)}{f'(\zeta)} T_{i,j}(\zeta).$$

By combining (1.16) with (3.19), we obtain

$$(10.22) \quad f'(\zeta) b_{m,1}\left(f(\zeta)\right) \equiv b_{m,1}^{**}(\zeta) + \frac{f''(\zeta)}{f'(\zeta)}.$$

The property that $b_{m,2}$ in (1.15) is a semi-invariant of the second kind of weight 2 was established in Proposition 4.14 on page 43. It yields

$$(10.23) \quad \left(f'(\zeta) \right)^2 b_{m,2}\left(f(\zeta)\right) \equiv b_{m,2}^{**}(\zeta).$$

After substituting $c_{r,s}^{(t)}(z)$ from (1.1) for $w_{r,s}^{(t)}$ in (1.32) on page 9, we replace z with $f(\zeta)$ and multiply by $\left(f'(\zeta)\right)^{i+1}$; then, (10.14) and (10.21)–(10.23) give

$$(10.24) \quad T_{i+1,j}(\zeta) \equiv \sum_{k=j+1}^{m} \left[\begin{array}{l} T_{i,k}^{(1)}(\zeta) \\ + \left[(i+k-m) b_{m,1}^{**}(\zeta) + (k-m) \frac{f''(\zeta)}{f'(\zeta)} \right] T_{i,k}(\zeta) \\ + (m-i-k+1)(1-i-k) b_{m,2}^{**}(\zeta) T_{i-1,k}(\zeta) \end{array} \right],$$

for $i \geq 0$ and any j.

We see that (10.10), (10.14), and (1.30) yield $S_{i,j}(\zeta) \equiv 0 \equiv T_{i,j}(\zeta)$, for $i \leq -1$ and any j. Also, (10.10), (10.14), and (1.31) give $S_{0,j}(\zeta) \equiv 1 \equiv T_{0,j}(\zeta)$, for any j. Using this with (10.20) and (10.24), we deduce that $S_{i,j}(\zeta) \equiv T_{i,j}(\zeta)$, for $j \leq m$ and any i. Thus, $S_{i,j}(\zeta)$ in (10.10) satisfies (10.11). This completes the proof. □

THEOREM 10.3. *For $m \geq 2$ and integers l, n that satisfy either $1 \leq l \leq n \leq m$ or $l = 0 < 3 \leq n \leq m$, the polynomial $\mathcal{J}_{m,l,n}$ defined in $\mathcal{S}_m$ by means of (1.38) on page 9 is both a semi-invariant of the second kind for the equations (1.1) and a basic polynomial of index (l, n) according to Definition 1.5 on page 6.*

PROOF. To establish that $\mathcal{J}_{m,l,n}$ satisfies (1) in Definition 1.5 on page 6, we first use (1.37) on page 9 to see that the coefficient of $\boldsymbol{w}_{l,n}$ in $\boldsymbol{J}_{l,l+n}$ is equal to the coefficient of $\boldsymbol{w}_{l,n}$ in $\boldsymbol{W}_{l,l+n}$. With $h = l$ and $i = l + n$, we find that (1.35) yields $\boldsymbol{W}_{l,l+n} \equiv \boldsymbol{V}_{l,n}$. Since Theorem 10.1 on page 111 shows that $\boldsymbol{V}_{l,n}$ is a basic polynomial of index (l, n), the coefficients of $\boldsymbol{w}_{l,n}$ in both $\boldsymbol{V}_{l,n}$ and $\boldsymbol{J}_{l,l+n}$ equal 1.

Suppose that h satisfies $0 \leq h < l$. We use (1.37) to observe that the coefficient of $\boldsymbol{w}_{l,n}$ in $\boldsymbol{J}_{h,l+n}$ is equal to the coefficient of $\boldsymbol{w}_{l,n}$ in $\boldsymbol{W}_{h,l+n}$. With $i = l + n$, $n - i + h = h - l < 0$, and $\binom{m-i+h}{n-i+h} = 0$, we find that (1.35) gives $\boldsymbol{W}_{h,l+n} \equiv 0$. Thus, the coefficient of $\boldsymbol{w}_{l,n}$ in $\boldsymbol{J}_{h,l+n}$ is 0.

We use the preceding results and (1.38) to deduce that: the coefficient of $\boldsymbol{w}_{l,n}$ in $\mathcal{J}_{m,l,n}$ is 1 and therefore $\mathcal{J}_{m,l,n}$ satisfies (1) in Definition 1.5.

To show that $\mathcal{J}_{m,l,n}$ is a semi-invariant of the second kind having weight $l + n$, we first establish a result about $\boldsymbol{V}_{i,j}$ in (1.33) when i and j are integers such that $\boldsymbol{V}_{i,j}$ is not a constant. Namely, after setting $i_0 = \min(i, j)$ and $j_0 = \max(i, j)$, we use $\boldsymbol{V}_{i,j} \equiv \boldsymbol{V}_{j,i} \equiv \boldsymbol{V}_{i_0,j_0}$, $\boldsymbol{V}_{0,0} \equiv 1$, $\boldsymbol{V}_{0,1} \equiv 0$, $\boldsymbol{V}_{0,2} \equiv 0$, and Theorem 10.1 to deduce that $\boldsymbol{V}_{i,j}$ is a semi-invariant of the second kind that is also a basic polynomial of index (i_0, j_0) where either $1 \leq i_0 \leq j_0 \leq m$ or $i_0 = 0 < 3 \leq j_0 \leq m$. Consequently, for each $\boldsymbol{V}_{i,j}$ in (1.33) having $i + j \geq 1$, either it is identically zero or it is a semi-invariant of the second kind having weight $i + j$. Thus, in view of (1.35) and (1.34), we find that: for any $\boldsymbol{W}_{h,i}$ in (1.35), either it is identically zero or it is a semi-invariant of the second kind having weight i. Corollary 4.15 on page 45 shows that $\boldsymbol{b}_{m,2}$ in (1.15) is a semi-invariant of the second kind having weight 2. We employ this and Proposition 4.17 on page 45 to verify that: for any $\boldsymbol{J}_{h,i}$ in (1.37) or (1.36), either it is identically zero or it is a semi-invariant of the second kind having weight i. In view of this and (1.38), we find that either $\mathcal{J}_{m,l,n}$ is identically zero or it is a semi-invariant of the second kind having weight $l + n$. Since the coefficient of $\boldsymbol{w}_{l,n}$ in $\mathcal{J}_{m,l,n}$ is 1, we conclude that $\mathcal{J}_{m,l,n}$ is a semi-invariant of the second kind having weight $l + n$. In particular, $\mathcal{J}_{m,l,n}$ satisfies (2) in Definition 1.5.

To show that $\mathcal{J}_{m,l,n}$ satisfies (3) in Definition 1.5, suppose that $\boldsymbol{w}^{(k)}_{i_1,i_2}$ is a variable from (1.6) that is effectively involved in $\mathcal{J}_{m,l,n}$. We are to establish that

(10.25) $$i_1 \leq l \quad \text{and} \quad i_2 \leq n.$$

With $n \geq 1$, we see that (10.25) is satisfied when $\boldsymbol{w}^{(k)}_{i_1,i_2} = \boldsymbol{w}^{(k)}_{0,1}$. If $n = 1$, then $l = 1$, $\mathcal{J}_{m,l,n} \equiv \mathcal{J}_{m,1,1} \equiv \boldsymbol{w}_{1,1} - (\boldsymbol{w}_{0,1})^2$, and $\boldsymbol{w}^{(k)}_{0,2}$ is not involved in $\mathcal{J}_{m,l,n}$. Thus, when $\boldsymbol{w}^{(k)}_{i_1,i_2} = \boldsymbol{w}^{(k)}_{0,2}$, we find that $n \geq 2$ and (10.25) is satisfied. As the remaining situation, suppose that $(i_1, i_2) \neq (0, 1)$ and $(i_1, i_2) \neq (0, 2)$. Because $\boldsymbol{b}_{m,1}$, $\boldsymbol{b}_{m,2}$, and each $\boldsymbol{U}_{i,j}$ are polynomial combinations over $\mathbb{Q}$ of the variables in (10.3) on page 111, they do not involve $\boldsymbol{w}^{(j)}_{i_1,i_2}$, for any $j \geq 0$. Therefore, equations (1.33)–(1.38) show that there are integers h and i satisfying $0 \leq h \leq l$ and $1 \leq i \leq l + n$ such that $\boldsymbol{w}^{(0)}_{i_1,i_2}$ is effectively involved in the corresponding left member $\boldsymbol{W}_{h,i}$ of (1.35). In view of $\boldsymbol{W}_{h,i} \not\equiv 0$, (1.35) shows that $\boldsymbol{w}^{(0)}_{i_1,i_2}$ is effectively involved in $\boldsymbol{V}_{h,i-h}$ and, due to $\binom{m-i+h}{n-i+h} \neq 0$, we have $0 \leq i - h \leq n$ in addition to $0 \leq h \leq l$. We note that $V_{h,i-h}$ is a basic polynomial whose index is $\bigl(\min(h,\, i-h),\, \max(h,\, i-h)\bigr)$. Consequently, as applied to $V_{h,i-h}$ and $w^{(0)}_{i_1,i_2}$, the third part of Definition 1.5 yields

$$i_1 \leq \min(h,\, i-h) \leq h \leq l \quad \text{and} \quad i_2 \leq \max(h,\, i-h) \leq \max(l,\, n) = n.$$

Thus, (10.25) is valid and therefore $\mathcal{J}_{m,l,n}$ satisfies (3) in Definition 1.5.

To show that $\mathcal{J}_{m,l,n}$ satisfies (4) in Definition 1.5, let T denote a nonzero term in some polynomial $V_{i,j}$, $W_{h,i}$, $J_{h,i}$, or $\mathcal{J}_{m,l,n}$ that appears in (1.33)–(1.38). Then, for some $r \geq 1$, T is a product of r factors from (1.6) and a nonzero element of $\mathbb{Q}$. Since $b_{m,1}$, $b_{m,2}$, and each $U_{i,j}$ of (1.30)–(1.32) are polynomials over $\mathbb{Q}$ in the variables of (10.3) on page 111, we use (1.33)–(1.38) to see that at least $r-1$ of the r factors from (1.6) are listed in (10.3). Thus, $\mathcal{J}_{m,l,n}$ satisfies (4) in Definition 1.5.

We combine the preceding results to conclude that $\mathcal{J}_{m,l,n}$ is a basic polynomial of index $(l,\, n)$ as well as a semi-invariant of the second kind. This completes the proof. □

CHAPTER 11

The Existence of Basic Relative Invariants

THEOREM 11.1. *For $m \geq 2$, the polynomials $\mathcal{I}_{m,l,n}$ and $\mathcal{J}_{m,l,n}$ defined in $\mathcal{S}_m$ by (1.28) and (1.38) on pages 8–9 satisfy*

(11.1) $\mathcal{I}_{m,l,n} \equiv \mathcal{J}_{m,l,n}$,   *when* $1 \leq l \leq n \leq m$ *or* $l = 0 < 3 \leq n \leq m$.

PROOF. Let l and n be fixed integers subject to either $1 \leq l \leq n \leq m$ or $l = 0 < 3 \leq n \leq m$. For any given (1.1) on a region Ω, Proposition 5.4 on page 49 yields a subregion $\mathcal{U}_0$ of Ω on which there are defined a meromorphic function $\rho_0(z) \not\equiv 0$ and a univalent analytic function $g_0(z)$ such that $\rho_0(z)$ and $g_0(z)$ are solutions on $\mathcal{U}_0$ of (5.16)–(5.17). Let $z = f_0(\zeta)$ on $\mathcal{V}_0 = g_0(\mathcal{U}_0)$ denote the inverse function for $\zeta = g_0(z)$ on $\mathcal{U}_0$. Theorem 5.1 on page 47 shows that the restriction of (1.1) to $\mathcal{U}_0$ is transformed by

$$y(z) = \rho_0(z)\, v(z), \quad z = f_0(\zeta), \quad \text{and} \quad w(\zeta) = (v \circ f_0)(\zeta)$$

into a unique differential equation

$$\left(w^{(m)}(\zeta)\right)^2 + \sum_{\substack{0 \leq i,j \leq m \\ (i,j) \neq (0,0)}} d_{i,j}(\zeta)\, w^{(m-i)}(\zeta)\, w^{(m-j)}(\zeta) = 0, \quad \text{on } \mathcal{V}_0,$$

where $d_{j,i}(\zeta) \equiv d_{i,j}(\zeta)$, for $0 \leq i < j \leq m$.

Proposition 5.3 on page 49 yields $d_{0,1}(\zeta) \equiv d_{0,2}(\zeta) \equiv 0$. Consequently, with respect to $A^{i,j}_{p,q}$ in (6.3) on page 53, we apply Theorem 7.10 on page 85 to obtain

(11.2) $\displaystyle\sum_{\mu=0}^{l} \sum_{\nu=0}^{n} A^{l,n}_{\mu,\nu}\, d^{(\mu+\nu)}_{l-\mu,n-\nu}(\zeta) \equiv \left(f'_0(\zeta)\right)^{l+n} \mathcal{I}_{m,l,n}\!\left(f_0(\zeta)\right)$,   on $\mathcal{V}_0$,

where $\mathcal{I}_{m,l,n}(z)$ is the function on Ω obtained by substituting $c^{(k)}_{i,j}(z)$ from (1.1) for $w^{(k)}_{i,j}$ in $\mathcal{I}_{m,l,n}$ of (1.28) on page 8. Also, Theorem 9.10 on page 109 yields

(11.3) $\displaystyle\sum_{\mu=0}^{l} \sum_{\nu=0}^{n} A^{l,n}_{\mu,\nu}\, d^{(\mu+\nu)}_{l-\mu,n-\nu}(\zeta) \equiv \left(f'_0(\zeta)\right)^{l+n} \mathcal{J}_{m,l,n}\!\left(f_0(\zeta)\right)$,   on $\mathcal{V}_0$,

where $\mathcal{J}_{m,l,n}(z)$ on Ω is obtained by substituting $c^{(k)}_{i,j}(z)$ from (1.1) for $w^{(k)}_{i,j}$ in $\mathcal{J}_{m,l,n}$ of (1.38) on page 9. We equate the right members of (11.2) and (11.3), cancel the nonzero factor $\left(f'_0(\zeta)\right)^{l+n}$, set $\zeta = g_0(z)$, and use $f_0\!\left(g_0(z)\right) \equiv z$ to deduce that

(11.4) $\mathcal{I}_{m,l,n}(z) \equiv \mathcal{J}_{m,l,n}(z)$,

for each z in $\mathcal{U}_0$. Since the coefficients $c_{i,j}(z)$ are meromorphic on Ω, (11.4) is valid for each z in Ω. Thus, for the polynomial $\boldsymbol{P}$ defined in $\mathcal{S}_m$ by $\boldsymbol{P} \equiv \mathcal{I}_{m,l,n} - \mathcal{J}_{m,l,n}$, the substitution of $c^{(k)}_{i,j}(z)$ from (1.1) for $w^{(k)}_{i,j}$ in $\boldsymbol{P}$ yields $P(z) \equiv 0$, for any (1.1). We apply Corollary 4.3 on page 37 to conclude that $\boldsymbol{P} \equiv 0$ and, therefore, (11.1) is valid. This completes the proof.   □

THEOREM 11.2. *For $m \geq 2$ and integers l, n that satisfy either $1 \leq l \leq n \leq m$ or $l = 0 < 3 \leq n \leq m$, the polynomial $\mathcal{I}_{m,l,n}$ defined in $\mathcal{S}_m$ by (1.28) on page 8 is a basic relative invariant of index (l, n) for the equations (1.1).*

PROOF. Theorem 8.3 on page 91 establishes that: $\mathcal{I}_{m,l,n}$ is a semi-invariant of the first kind for the equations (1.1); and $\mathcal{I}_{m,l,n}$ is a basic polynomial of index (l, n). Theorem 10.3 on page 116 proves that $\mathcal{J}_{m,l,n}$ is a semi-invariant of the second kind for the equations (1.1). The identity (11.1) of Theorem 11.1 shows that $\mathcal{I}_{m,l,n}$ is also a semi-invariant of the second kind. Thus, $\mathcal{I}_{m,l,n}$ is a basic relative invariant of index (l, n) for the equations (1.1). This completes the proof. □

CHAPTER 12

The Uniqueness of Basic Relative Invariants

Each basic relative invariant for the equations (1.1) has an index (l, n) given by Definition 1.5 on page 6. The index is a pair (l, n) of integers l, n that satisfy either $1 \leq l \leq n \leq m$ or $l = 0 < 3 \leq n \leq m$. For $m = 1$, Theorem 4.10 on page 41 shows that the only basic relative invariant is $\mathcal{I}_{1,1,1} \equiv \boldsymbol{w}_{1,1} - (\boldsymbol{w}_{0,1})^2$ and its index is $(1, 1)$. For $m \geq 2$ and any integers l, n subject to either $1 \leq l \leq n \leq m$ or $l = 0 < 3 \leq n \leq m$, a corresponding basic relative invariant $\mathcal{I}_{m,l,n}$ of index (l, n) was explicitly constructed for Theorem 11.2. To establish for the situation $m \geq 2$ in Theorem 12.2 on page 130 that $\mathcal{I}_{m,l,n}$ is the only basic relative invariant having index (l, n), we shall base the argument on a more general result presented in Theorem 12.1.

12.1. Some polynomials that are not relative invariants

We begin by recalling the result of Corollary 5.8 on page 52 that: if $\boldsymbol{R}$ in $\mathcal{S}_m$ is a relative invariant for the equations (1.1), then $\boldsymbol{R}$ must effectively involve some variable $\boldsymbol{w}_{p,q}^{(r)}$ having $r \geq 0$ and either $1 \leq p \leq q \leq m$ or $p = 0 < 3 \leq q \leq m$. For Proposition 5.7 on page 52, we introduced the symbolism $\mathbb{Q}\{\boldsymbol{w}_{0,1}, \boldsymbol{w}_{0,2}\}$ to designate the subring of $\mathcal{S}_m$ that consists of the polynomials over $\mathbb{Q}$ in the variables $\boldsymbol{w}_{0,1}^{(\kappa)}$, $\boldsymbol{w}_{0,2}^{(\lambda)}$, for $\kappa, \lambda \geq 0$. In particular, no polynomial in $\mathbb{Q}\{\boldsymbol{w}_{0,1}, \boldsymbol{w}_{0,2}\}$ can be a relative invariant.

THEOREM 12.1. *In terms of fixed nonnegative integers m, p, q, r that satisfy $m \geq 2$ and either $1 \leq p \leq q \leq m$ or $p = 0 < 3 \leq q \leq m$, suppose that $\boldsymbol{R}$ in $\mathcal{S}_m$ is an isobaric polynomial of weight $s > p + q$ that is expressible as*

(12.1) $$\boldsymbol{R} \equiv \boldsymbol{S}\, \boldsymbol{w}_{p,q}^{(r)} + \boldsymbol{T},$$

where $\boldsymbol{S}$ is a nonzero polynomial in $\mathbb{Q}\{\boldsymbol{w}_{0,1}, \boldsymbol{w}_{0,2}\}$ and $\boldsymbol{T}$ is a polynomial in $\mathcal{S}_m$ such that any variable $\boldsymbol{w}_{i,j}^{(k)}$ from (1.6) that is effectively involved in $\boldsymbol{T}$ has either (1) $i < p$; or (2) $i = p$ and $j < q$; or (3) $i = p$, $j = q$, and $k < r$. Then, $\boldsymbol{R}$ is not a relative invariant for the equations (1.1).

PROOF. For an indirect argument (terminating on page 130), suppose that $\boldsymbol{R}$ is a relative invariant for the equations (1.1). Since Proposition 5.7 on page 52 shows that $\boldsymbol{S}$ is not a relative invariant, there are three cases according to whether: (i) $\boldsymbol{S}$ is a constant; or (ii) $\boldsymbol{S}' \not\equiv 0$ and $\boldsymbol{S}$ is not a semi-invariant of the first kind; or (iii) $\boldsymbol{S}' \not\equiv 0$ and $\boldsymbol{S}$ is not a semi-invariant of the second kind.

12.1.1. Case 1. Suppose $\boldsymbol{S}$ is a constant. Since $\boldsymbol{R}$ is isobaric of weight s, we have $r = s - p - q \geq 1$. Thus, there are rational numbers $\gamma_0 \neq 0$ and γ_1 such that

(12.2) $$\boldsymbol{R} \equiv \gamma_0\, \boldsymbol{w}_{p,q}^{(r)} + \gamma_1\, \boldsymbol{w}_{0,1}^{(0)}\, \boldsymbol{w}_{p,q}^{(r-1)} + \boldsymbol{T}_0,$$

where T_0 is a polynomial over $\mathbb{Q}\{\boldsymbol{w}_{0,1}, \boldsymbol{w}_{0,2}\}$ in variables $w_{i,j}^{(k)}$ from (1.6) having $0 \leq i \leq p$, $\max(1, i) \leq j \leq m$, $k \geq 0$, and either (1) $i < p$; or (2) $i = p$ and $j < q$; or (3) $i = p$, $j = q$, $r \geq 2$, and $k \leq r - 2$. We shall show that the requirement for $\boldsymbol{R}$ to be a semi-invariant of the first kind imposes a condition on γ_1 that is inconsistent with the condition imposed on γ_1 by the requirement for $\boldsymbol{R}$ to be a semi-invariant of the second kind. Thus, we consider two subcases.

12.1.1.1. *Consequences of $\boldsymbol{R}$ as a semi-invariant of the first kind.* For any (1.1) and any transformation (1.2) of (1.1) into a corresponding (1.3), the property of $\boldsymbol{R}$ as a semi-invariant of the first kind in combination with (12.2) yields

(12.3) $\quad 0 \equiv R^*(z) - R(z)$
$$\equiv \gamma_0 \left[c_{p,q}^{*(r)}(z) - c_{p,q}^{(r)}(z) \right] + \gamma_1 \left[c_{0,1}^*(z) c_{p,q}^{*(r-1)} - c_{0,1}(z) c_{p,q}^{(r-1)}(z) \right]$$
$$+ T_0^*(z) - T_0(z).$$

To eliminate each $c_{i,j}^{*(k)}(z)$ from the right member of (12.3), we first use (2.5) on page 23 to obtain

(12.4) $\quad c_{i,j}^*(z) \equiv c_{i,j}(z) + \displaystyle\sum_{\substack{0 \leq \mu \leq i \\ 0 \leq \nu \leq j \\ (\mu,\nu) \neq (i,j)}} \binom{m-\mu}{i-\mu}\binom{m-\nu}{j-\nu} \frac{\rho^{(i-\mu)}(z)\rho^{(j-\nu)}(z)}{(\rho(z))^2} c_{\mu,\nu}(z),$

on Ω for $0 \leq i \leq j \leq m$.

If $i \leq j$, $0 \leq \mu \leq i$, $0 \leq \nu \leq j$, $(\mu, \nu) \neq (0, 0)$, $(\mu, \nu) \neq (i, j)$, and $\mu \leq \nu$, then $0 \leq \mu \leq i$, $\max(1, \mu) \leq \nu \leq j$, and the condition $\nu < j$ is satisfied whenever $\mu = i$. If $i \leq j$, $0 \leq \mu \leq i$, $0 \leq \nu \leq j$, $(\mu, \nu) \neq (0, 0)$, $(\mu, \nu) \neq (i, j)$, and $\mu > \nu$, then $c_{\mu,\nu}(z) \equiv c_{\nu,\mu}(z)$, $0 \leq \nu < \mu \leq i$, $\max(1, \nu) \leq \mu \leq i \leq j$, and we must have $\nu < i$. In view of this and $c_{0,0}(z) \equiv 1$, we can rewrite (12.4) as

(12.5) $\quad c_{i,j}^*(z) \equiv c_{i,j}(z) + \Upsilon_{i,j}(z), \quad$ on Ω for $0 \leq i \leq j \leq m$,

where $\Upsilon_{i,j}(z)$ is a polynomial combination over $\mathbb{Q}$ of the expressions $\rho^{(\lambda)}(z)/\rho(z)$ and $c_{\mu,\nu}(z)$ having $\lambda \geq 1$, $0 \leq \mu \leq i$, $\max(1, \mu) \leq \nu \leq j$, and $(\mu, \nu) \neq (i, j)$. In particular, (12.4) and (12.5) respectively give

(12.6) $\quad c_{0,1}^*(z) \equiv c_{0,1}(z) + m \dfrac{\rho^{(1)}(z)}{\rho(z)}$

and

(12.7) $\quad c_{p,q}^*(z) \equiv c_{p,q}(z) + \Phi_0(z),$

where $\Phi_0(z)$ is a polynomial combination over $\mathbb{Q}$ of the expressions $\rho^{(\lambda)}(z)/\rho(z)$ and $c_{\mu,\nu}(z)$ having $\lambda \geq 1$, $0 \leq \mu \leq p$, $\max(1, \mu) \leq \nu \leq q$, and $(\mu, \nu) \neq (p, q)$. In view of (12.7), we find that

(12.8) $\quad c_{p,q}^{*(r)}(z) - c_{p,q}^{(r)}(z) \equiv \Phi_1(z),$

where $\Phi_1(z)$, is a polynomial combination over $\mathbb{Q}$ of the expressions $\rho^{(\lambda)}(z)/\rho(z)$ and $c_{\mu,\nu}^{(\kappa)}(z)$ having $\lambda \geq 1$, $0 \leq \mu \leq p$, $\max(1, \mu) \leq \nu \leq q$, $0 \leq \kappa \leq r$, and $(\mu, \nu) \neq (p, q)$. We employ (12.7) and (12.6) to verify that

(12.9) $\quad c_{0,1}^*(z) c_{p,q}^{*(r-1)}(z) - c_{0,1}(z) c_{p,q}^{(r-1)}(z) \equiv m \dfrac{\rho^{(1)}(z)}{\rho(z)} c_{p,q}^{(r-1)}(z) + \Phi_2(z),$

12.1. SOME POLYNOMIALS THAT ARE NOT RELATIVE INVARIANTS

where $\Phi_2(z)$, is a polynomial combination over $\mathbb{Q}$ of the expressions $\rho^{(\lambda)}(z)/\rho(z)$ and $c_{\mu,\nu}^{(\kappa)}(z)$ having $\lambda \geq 1$, $0 \leq \mu \leq p$, $\max(1, \mu) \leq \nu \leq q$, $0 \leq \kappa \leq r-1$, and $(\mu, \nu) \neq (p, q)$.

Suppose that $w_{i,j}^{(k)}$ is a variable from (1.6) that is effectively involved in T_0. Then, (12.5) yields

(12.10) $$c_{i,j}^{*(k)}(z) \equiv c_{i,j}^{(k)}(z) + \Psi(z),$$

where $\Psi(z)$ is a polynomial combination over $\mathbb{Q}$ of the expressions $\rho^{(\lambda)}(z)/\rho(z)$ and $c_{\mu,\nu}^{(\kappa)}(z)$, for $\lambda \geq 1$ and various integers μ, ν, κ. If $i < p$, then $0 \leq \mu \leq i < p$, $\max(1, \mu) \leq \nu \leq j \leq m$, and $\kappa \geq 0$. If $i = p$ and $j < q$, then $0 \leq \mu \leq p$, $\max(1, \mu) \leq \nu \leq j < q$, and $\kappa \geq 0$. If $i = p$, $j = q$, $r \geq 2$, and $k \leq r-2$, then $0 \leq \mu \leq p$, $\max(1, \mu) \leq \nu \leq q$, and $0 \leq \kappa \leq k \leq r-2$. Thus, $\Psi(z)$ does not involve any $c_{p,q}^{(\kappa)}(z)$ having $\kappa \geq r-1$. Since each term of $T_0^*(z)$ is a product of a rational number and expressions like $c_{i,j}^{*(k)}(z)$ in (12.10), we see that

(12.11) $$T_0^*(z) - T_0(z) \equiv \Phi_3(z),$$

where $\Phi_3(z)$, is a polynomial combination over $\mathbb{Q}$ of various $\rho^{(\lambda)}(z)/\rho(z)$ and $c_{\mu,\nu}^{(\kappa)}(z)$ having $\lambda \geq 1$, $0 \leq \mu \leq p$, $\max(1, \mu) \leq \nu \leq m$, and $\kappa \geq 0$ such that $\Phi_3(z)$ does not involve any $c_{p,q}^{(\kappa)}(z)$ with $\kappa \geq r-1$.

For any (1.1) and (1.2), we combine (12.3), (12.8), (12.9), and (12.11) to obtain

(12.12) $$0 \equiv \gamma_1 m \frac{\rho^{(1)}(z)}{\rho(z)} c_{p,q}^{(r-1)}(z) + \Phi_4(z),$$

where $\Phi_4(z)$, is a polynomial combination over $\mathbb{Q}$ of $\rho^{(\lambda)}(z)/\rho(z)$ and $c_{\mu,\nu}^{(\kappa)}(z)$ having $\lambda \geq 1$, $0 \leq \mu \leq p$, $\max(1, \mu) \leq \nu \leq m$, and $\kappa \geq 0$ such that $\Phi_4(z)$ does not involve any $c_{p,q}^{(\kappa)}(z)$ with $\kappa \geq r-1$. Thus, if $\gamma_1 \neq 0$, then the right member of (12.12) effectively involves $c_{p,q}^{(r-1)}(z)$.

Let t_1 be an integer such that: for any (1.1) and any (1.2), the product of $(\rho(z))^{t_1}$ and the right member of (12.12) is a polynomial combination over $\mathbb{Q}$ of the expressions $\rho^{(\lambda)}(z)$ and $c_{\mu,\nu}^{(\kappa)}(z)$ having $\lambda \geq 0$, $0 \leq \mu \leq p$, $\max(1, \mu) \leq \nu \leq m$, and $\kappa \geq 0$. We consider the ring $\mathfrak{S}_{\chi(m,m)+1}$ of polynomials over $\mathbb{Q}$ specified for Proposition 4.2 on page 36. With respect to (4.1) on page 35, $\mathfrak{S}_{\chi(m,m)+1}$ is the ring of polynomials over $\mathbb{Q}$ in the variables

$$w_{0,1}^{(k)}, \; w_{1,1}^{(k)}, \; w_{0,2}^{(k)}, \; w_{1,2}^{(k)}, \; w_{2,2}^{(k)}, \; \ldots, \; w_{m,m}^{(k)}, \; w_{0,m+1}^{(k)}, \quad \text{for } k = 0, 1, 2, 3, \ldots.$$

Let Q_1 be the polynomial in $\mathfrak{S}_{\chi(m,m)+1}$ such that: for any (1.1) and any (1.2), the function obtained by substituting $c_{\mu,\nu}^{(\kappa)}(z)$ for $w_{\mu,\nu}^{(\kappa)}$ and $\rho^{(\lambda)}(z)$ for $w_{0,m+1}^{(\lambda)}$ in Q_1 is equal to the product of $(\rho(z))^{t_1+1}$ and the right member of (12.12). In order to establish the relation

(12.13) $$\gamma_1 = 0,$$

suppose that (12.13) is not satisfied. Then, the coefficient of $c_{p,q}^{(r-1)}(z)$ in (12.12) is not identically zero for some transformations (1.2). Therefore, Q_1 satisfies $Q_1 \not\equiv 0$. We apply Proposition 4.2 on page 36 to obtain meromorphic functions $\rho(z)$ and $c_{i,j}(z)$ on a region Ω, for $0 \leq i \leq m$ and $\max(1, i) \leq j \leq m$, such that the substitution of $c_{\mu,\nu}^{(\kappa)}(z)$ for $w_{\mu,\nu}^{(\kappa)}$ and $\rho^{(\lambda)}(z)$ for $w_{0,m+1}^{(\lambda)}$ in Q_1 yields a function that is not identically zero on Ω. Since the (t_1+1)th power of $\rho(z)$ was used in defining $\boldsymbol{Q}_1$, we have $\rho(z) \not\equiv 0$. Then, for the particular (1.1) and (1.2) that this specifies, we obtain a contradiction to (12.12). Thus, (12.2) must satisfy (12.13).

12.1.1.2. *Consequences of $\boldsymbol{R}$ as a semi-invariant of the second kind.* For any (1.1) and any transformation (1.4) of (1.1) into a corresponding (1.5), the property of $\boldsymbol{R}$ in (12.2) as a semi-invariant of the second kind of weight $s = p+q+r$ requires

$$
\begin{aligned}
(12.14) \quad 0 &\equiv R^{**}(\zeta) - \bigl(f'(\zeta)\bigr)^s R\bigl(f(\zeta)\bigr) \\
&\equiv \gamma_0 \left[c_{p,q}^{**(r)}(\zeta) - \bigl(f'(\zeta)\bigr)^s c_{p,q}^{(r)}\bigl(f(\zeta)\bigr) \right] \\
&\quad + \gamma_1 \left[c_{0,1}^{**}(\zeta)\, c_{p,q}^{**(r-1)}(\zeta) - \bigl(f'(\zeta)\bigr)^s c_{0,1}\bigl(f(\zeta)\bigr) c_{p,q}^{(r-1)}\bigl(f(\zeta)\bigr) \right] \\
&\quad + T_0^{**}(\zeta) - \bigl(f'(\zeta)\bigr)^s T_0\bigl(f(\zeta)\bigr).
\end{aligned}
$$

As a first step in using (3.9) on page 30 to eliminate each expression $c_{\mu,\nu}^{**(\kappa)}(\zeta)$ from the right member of (12.14), we apply the definitions (3.1)–(3.2) for $\alpha_{i,j}(\zeta)$ in (3.9) to see that $\alpha_{r,s}(\zeta)$ consists of terms having the form

$$
v \prod_{k=1}^{r} \frac{f^{(\nu_k)}(\zeta)}{f'(\zeta)}
$$

where: v is an integer; $1 \leq \nu_k \leq r+1$, for $1 \leq k \leq r$; and $\sum_{k=1}^{r}(\nu_k - 1) = r$. Thus, (3.9) on page 30 yields

$$
(12.15) \quad c_{i,j}^{**}(\zeta) \equiv \bigl(f'(\zeta)\bigr)^{i+j} c_{i,j}\bigl(f(\zeta)\bigr) + \sum_{\substack{0 \leq \mu \leq i \\ 0 \leq \nu \leq j \\ (\mu,\nu) \neq (i,j)}} \bigl(f'(\zeta)\bigr)^{\mu+\nu} \Gamma_{\mu,\nu}^{i,j}(\zeta)\, c_{\mu,\nu}\bigl(f(\zeta)\bigr),
$$

on Ω^{**} for $0 \leq i \leq j \leq m$,

where each $\Gamma_{\mu,\nu}^{i,j}(\zeta)$ is a polynomial combination over $\mathbb{Q}$ of $\bigl(f'(\zeta)\bigr)^{-1}$ and $f^{(k)}(\zeta)$, for $2 \leq k \leq j+1$. The argument applied to deduce (12.5) from (12.4) enables us to rewrite (12.15) as

$$
(12.16) \quad c_{i,j}^{**}(\zeta) \equiv \bigl(f'(\zeta)\bigr)^{i+j} c_{i,j}\bigl(f(\zeta)\bigr) + E_{i,j}(\zeta), \quad \text{on } \Omega^{**} \text{ for } 0 \leq i \leq j \leq m,
$$

where $E_{i,j}(\zeta)$ is a polynomial combination over $\mathbb{Q}$ of $\bigl(f'(\zeta)\bigr)^{-1}$, $f'(\zeta)$, ..., $f^{(j+1)}(\zeta)$ and various $c_{\mu,\nu}\bigl(f(\zeta)\bigr)$ having $0 \leq \mu \leq i$, $\max(1,\mu) \leq \nu \leq j$, and $(\mu,\nu) \neq (i,j)$. In particular, (3.19) of page 34 and (12.16) respectively give

$$
(12.17) \quad c_{0,1}^{**}(\zeta) \equiv f'(\zeta)\, c_{0,1}\bigl(f(\zeta)\bigr) - \binom{m}{2} \frac{f''(\zeta)}{f'(\zeta)}
$$

and

$$
(12.18) \quad c_{p,q}^{**}(\zeta) \equiv \bigl(f'(\zeta)\bigr)^{p+q} c_{p,q}\bigl(f(\zeta)\bigr) + F_0(\zeta),
$$

where $F_0(\zeta)$ is a polynomial combination over $\mathbb{Q}$ of $\bigl(f'(\zeta)\bigr)^{-1}$, $f'(\zeta)$, ..., $f^{(q+1)}(\zeta)$ and various $c_{\mu,\nu}\bigl(f(\zeta)\bigr)$ having $0 \leq \mu \leq p$, $\max(1,\mu) \leq \nu \leq q$, and $(\mu,\nu) \neq (p,q)$. We set $c_{p,q}^{(-1)}(z) \equiv 0$ and then use mathematical induction to check that, for $k \geq 0$,

$$
\frac{d^k}{d\zeta^k}\bigl[c_{p,q}\bigl(f(\zeta)\bigr)\bigr] \equiv \bigl(f'(\zeta)\bigr)^k c_{p,q}^{(k)}\bigl(f(\zeta)\bigr) + \binom{k}{2}\bigl(f'(\zeta)\bigr)^{k-2} f''(\zeta)\, c_{p,q}^{(k-1)}\bigl(f(\zeta)\bigr) + \dots,
$$

12.1. SOME POLYNOMIALS THAT ARE NOT RELATIVE INVARIANTS

where the ellipsis signifies a sum of terms that do not involve any $c_{p,q}^{(\kappa)}(f(\zeta))$ having $\kappa \geq \max(k-1, 0)$. By applying this to (12.18), we find that

$$c_{p,q}^{**(r)}(\zeta) \equiv \frac{d^r}{d\zeta^r}\left[(f'(\zeta))^{p+q} c_{p,q}(f(\zeta))\right] + \cdots$$

$$\equiv (f'(\zeta))^{p+q} \frac{d^r}{d\zeta^r} c_{p,q}(f(\zeta)) + \binom{r}{1}\left[\frac{d}{d\zeta}(f'(\zeta))^{p+q}\right]\left[\frac{d^{r-1}}{d\zeta^{r-1}} c_{p,q}(f(\zeta))\right] + \cdots$$

$$\equiv (f'(\zeta))^{p+q}\left[(f'(\zeta))^r c_{p,q}^{(r)}(f(\zeta)) + \binom{r}{2}(f'(\zeta))^{r-2} f''(\zeta) c_{p,q}^{(r-1)}(f(\zeta)) + \cdots\right]$$

$$+ \binom{r}{1}(p+q)(f'(\zeta))^{p+q-1} f''(\zeta)\left[(f'(\zeta))^{r-1} c_{p,q}^{(r-1)}(f(\zeta)) + \cdots\right] + \cdots$$

$$\equiv (f'(\zeta))^{p+q+r} c_{p,q}^{(r)}(f(\zeta))$$

$$+ \frac{r(2p+2q+r-1)}{2}(f'(\zeta))^{p+q+r-2} f''(\zeta) c_{p,q}^{(r-1)}(f(\zeta)) + \cdots$$

and

(12.19) $\quad c_{p,q}^{**(r)}(\zeta) - (f'(\zeta))^{p+q+r} c_{p,q}^{(r)}(f(\zeta))$

$$\equiv \frac{r(2p+2q+r-1)}{2}(f'(\zeta))^{p+q+r-2} f''(\zeta) c_{p,q}^{(r-1)}(f(\zeta)) + \cdots,$$

where the various ellipses indicate sums of terms not involving any $c_{p,q}^{(\kappa)}(f(\zeta))$ having $\kappa \geq r - 1$. Interpreting each ellipsis in this manner, we apply (12.17)–(12.18) and $p + q + r = s$ to verify that

$$c_{0,1}^{**}(\zeta) c_{p,q}^{**(r-1)}(\zeta)$$

$$\equiv \left[f'(\zeta) c_{0,1}(f(\zeta)) - \binom{m}{2}\frac{f''(\zeta)}{f'(\zeta)}\right]\left[(f'(\zeta))^{p+q+r-1} c_{p,q}^{(r-1)}(f(\zeta)) + \cdots\right]$$

$$\equiv (f'(\zeta))^s c_{0,1}(f(\zeta)) c_{p,q}^{(r-1)}(f(\zeta)) - \binom{m}{2}(f'(\zeta))^{s-2} f''(\zeta) c_{p,q}^{(r-1)}(f(\zeta)) + \cdots$$

and

(12.20) $\quad c_{0,1}^{**}(\zeta) c_{p,q}^{**(r-1)}(\zeta) - (f'(\zeta))^s c_{0,1}(f(\zeta)) c_{p,q}^{(r-1)}(f(\zeta))$

$$\equiv -\binom{m}{2}(f'(\zeta))^{s-2} f''(\zeta) c_{p,q}^{(r-1)}(f(\zeta)) + \cdots.$$

Suppose that $w_{i,j}^{(k)}$ is a variable from (1.6) that is effectively involved in T_0. Then, (12.16) yields

(12.21) $\quad c_{i,j}^{**(k)}(\zeta) \equiv (f'(\zeta))^{i+j+k} c_{i,j}^{(k)}(f(\zeta)) + H(\zeta),$

where $H(\zeta)$ is a polynomial combination over $\mathbb{Q}$ of $(f'(\zeta))^{-1}, f'(\zeta), \ldots, f^{(j+k+1)}(\zeta)$ and expressions $c_{\mu,\nu}^{(\kappa)}(f(\zeta))$ for various integers μ, ν, κ. If $i < p$, then $\mu \leq i < p$. If $i = p$ and $j < q$, then $\nu \leq j < q$. If $i = p$, $j = q$, $r \geq 2$, and $k \leq r - 2$, then $\kappa \leq k \leq r - 2$. Thus, the right member of (12.21) does not involve any $c_{p,q}^{(\kappa)}(f(\zeta))$ having $\kappa \geq r - 1$. Since each term of $T_0^{**}(\zeta)$ is a product of a rational number and expressions like $c_{i,j}^{**(k)}(\zeta)$ in (12.21), we see that

(12.22) $\quad T_0^{**}(\zeta) \equiv (f'(\zeta))^s T_0(f(\zeta)) + F_1(\zeta),$

where $F_1(\zeta)$ is a product of $\bigl(f'(\zeta)\bigr)^{n_1}$, for some integer n_1, and a polynomial combination over $\mathbb{Q}$ of $f^{(\lambda)}(\zeta)$ and $c_{\mu,\nu}^{(\kappa)}(f(\zeta))$, for $\lambda \geq 2$, $0 \leq \mu \leq p$, $\max(1,\mu) \leq \nu \leq m$, and $\kappa \geq 0$, such that $F_1(\zeta)$ does not involve any $c_{p,q}^{(\kappa)}(f(\zeta))$ having $\kappa \geq r-1$.

For any (1.1) and (1.4), we use (12.14), (12.19), (12.20), and (12.22) to obtain

$$(12.23) \quad 0 \equiv \left[\frac{r(2p+2q+r-1)}{2}\gamma_0 - \binom{m}{2}\gamma_1\right]\bigl(f'(\zeta)\bigr)^{s-2} f''(\zeta)\, c_{p,q}^{(r-1)}(f(\zeta)) + F_2,$$

where $F_2 \equiv F_2(\zeta)$ is a product of $\bigl(f'(\zeta)\bigr)^{n_2}$, for some integer n_2, and a polynomial combination over $\mathbb{Q}$ of various $f^{(\lambda)}(\zeta)$ and $c_{\mu,\nu}^{(\kappa)}(f(\zeta))$ having $\lambda \geq 2$, $0 \leq \mu \leq p$, $\max(1,\mu) \leq \nu \leq m$, and $\kappa \geq 0$ such that $F_2(\zeta)$ does not involve any $c_{p,q}^{(\kappa)}(f(\zeta))$ with $\kappa \geq r-1$. By setting $\zeta = g(z)$ in (12.23) and applying

$$(12.24) \quad f'\bigl(g(z)\bigr) \equiv \frac{1}{g'(z)}, \quad f''\bigl(g(z)\bigr) \equiv \frac{-g''(z)}{\bigl(g'(z)\bigr)^3}, \quad \ldots,$$

we find that, for any (1.1) and any univalent analytic function $g(z)$ on Ω,

$$(12.25) \quad 0 \equiv \left[\frac{r(2p+2q+r-1)}{2}\gamma_0 - \binom{m}{2}\gamma_1\right]\frac{\bigl(-g''(z)\bigr)}{\bigl(g'(z)\bigr)^{s+1}} c_{p,q}^{(r-1)}(z) + G_2(z),$$

where $G_2(z)$ is a product of $\bigl(g'(z)\bigr)^{n_0}$, for some integer n_0, and a polynomial combination over $\mathbb{Q}$ of $g^{(\lambda)}(z)$ and $c_{\mu,\nu}^{(\kappa)}(z)$, for $\lambda \geq 2$, $0 \leq \mu \leq p$, $\max(1,\mu) \leq \nu \leq m$, and $\kappa \geq 0$, such that $G_2(z)$ does not involve any $c_{p,q}^{(\kappa)}(z)$ having $\kappa \geq r-1$.

For any (1.1) and any univalent analytic function $\zeta = g(z)$ on the domain Ω of (1.1), there is a corresponding (12.25) whose deduction involves the inverse $z = f(\zeta)$ of $\zeta = g(z)$ as the transformation (1.4) of (1.1) into (1.5).

Let t_2 be an integer such that: for any (1.1) and any $\zeta = g(z)$, the product of $\bigl(g'(z)\bigr)^{t_2}$ and the right member of (12.25) is a polynomial combination over $\mathbb{Q}$ of $g^{(\lambda)}(z)$ and $c_{\mu,\nu}^{(\kappa)}(z)$, for $\lambda \geq 1$, $0 \leq \mu \leq p$, $\max(1,\mu) \leq \nu \leq m$, and $\kappa \geq 0$. Again, as in the argument following (12.12), we consider the particular ring $\mathfrak{S}_{\chi(m,m)+1}$ of polynomials over $\mathbb{Q}$ specified by the context for Proposition 4.2 on page 36. Let $\boldsymbol{Q}_2$ be the polynomial in $\mathfrak{S}_{\chi(m,m)+1}$ such that: for any (1.1) and any $\zeta = g(z)$, the function obtained by substituting $c_{\mu,\nu}^{(\kappa)}(z)$ for $\boldsymbol{w}_{\mu,\nu}^{(\kappa)}$ and substituting $g^{(\lambda)}(z)$ for $\boldsymbol{w}_{0,q+1}^{(\lambda)}$ in $\boldsymbol{Q}_2$ is equal to the product of $\bigl(g'(z)\bigr)^{t_2+1}$ and the right member of (12.25). In order to establish the relation

$$(12.26) \quad \gamma_1 \equiv \frac{r(2p+2q+r-1)}{m(m-1)}\gamma_0,$$

suppose that (12.26) is not satisfied. Then, there are situations where the coefficient of $c_{p,q}^{(r-1)}(z)$ in (12.25) is not identically zero. Consequently, $\boldsymbol{Q}_2$ satisfies $\boldsymbol{Q}_2 \neq 0$. We apply Proposition 4.2 on page 36 to obtain meromorphic functions $h(z)$ and $c_{i,j}(z)$, for $0 \leq i \leq m$ and $\max(1,i) \leq j \leq m$, on a region Ω_0, such that the substitution of $c_{\mu,\nu}^{(\kappa)}(z)$ for $\boldsymbol{w}_{\mu,\nu}^{(\kappa)}$ and the substitution of $h^{(\lambda)}(z)$ for $\boldsymbol{w}_{0,q+1}^{(\lambda)}$ in $\boldsymbol{Q}_2$ yields a function not identically zero on Ω_0. Since the (t_2+1)th power of $g'(z)$ was used in defining $\boldsymbol{Q}_2$, we have $h(z) \not\equiv 0$. Let Ω be a subregion of Ω_0 on which the restriction of $h(z)$ is a univalent analytic function; and set $g(z) \equiv h(z)$ on Ω. Then, for the particular (1.1) and (1.4) that this specifies, we obtain a contradiction to (12.25). Thus, (12.2) must satisfy (12.26).

12.1. SOME POLYNOMIALS THAT ARE NOT RELATIVE INVARIANTS

12.1.1.3. *Conclusion of* Case 1. In view of $\gamma_0 \neq 0$, the deductions (12.13) and (12.26) in Subsections 12.1.1.1 and 12.1.1.2 yield $0 = \gamma_1 \neq 0$. Thus, Case 1 *leads to a contradiction*.

12.1.2. Case 2. Suppose that $S' \neq 0$ and S is not a semi-invariant of the first kind. For any (1.1) and any transformation (1.2) of (1.1) into a corresponding (1.3), the property of $\boldsymbol{R}$ as a semi-invariant of the first kind requires

(12.27)
$$0 \equiv R^*(z) - R(z)$$
$$\equiv S^*(z)\, c^{*(r)}_{p,q}(z) - S(z)\, c^{(r)}_{p,q}(z) + T^*(z) - T(z).$$

Observing that $\boldsymbol{S}$ is a nonconstant isobaric polynomial over $\mathbb{Q}$ in the variables $\boldsymbol{w}^{(\kappa_1)}_{0,1}$ and $\boldsymbol{w}^{(\kappa_2)}_{0,2}$, we apply (12.4) to obtain

(12.28) $$S^*(z) \equiv S(z) + \Phi_5(z),$$

where $\Phi_5(z)$ is a polynomial combination over $\mathbb{Q}$ of $\rho^{(\lambda)}(z)/\rho(z)$, $c^{(\kappa_1)}_{0,1}(z)$, and $c^{(\kappa_2)}_{0,2}(z)$, for $\lambda \geq 1$, $\kappa_1 \geq 0$, and $\kappa_2 \geq 0$. The reasoning for (12.8) gives

(12.29) $$c^{*(r)}_{p,q}(z) \equiv c^{(r)}_{p,q}(z) + \Phi_1(z),$$

where $\Phi_1(z)$ is a polynomial combination over $\mathbb{Q}$ of expressions $\rho^{(\lambda)}(z)/\rho(z)$ and $c^{(\kappa)}_{\mu,\nu}(z)$ having $\lambda \geq 1$, $0 \leq \mu \leq p$, $\max(1, \mu) \leq \nu \leq q$, $\kappa \geq 0$, and $(\mu, \nu) \neq (p, q)$.

Suppose that $\boldsymbol{w}^{(k)}_{i,j}$ is a variable from (1.6) that is effectively involved in $\boldsymbol{T}$. Then, (12.5) yields

(12.30) $$c^{*(k)}_{i,j}(z) \equiv c^{(k)}_{i,j}(z) + \Psi(z),$$

where $\Psi(z)$ is a polynomial combination over $\mathbb{Q}$ of expressions $\rho^{(\lambda)}(z)/\rho(z)$ and $c^{(\kappa)}_{\mu,\nu}(z)$, for $\lambda \geq 1$ and various integers μ, ν, κ. If $i < p$, then $\mu \leq i < p$. If $i = p$ and $j < q$, then $\nu \leq j < q$. If $i = p$, $j = q$, and $k < r$, then $0 \leq \mu \leq p$, $\max(1, \mu) \leq \nu \leq q$, and $0 \leq \kappa \leq k < r$. Thus, the right member of (12.30) does not involve any $c^{(\kappa)}_{p,q}(z)$ having $\kappa \geq r$. Since each term of $T^*(z)$ is a product of a rational number and expressions like $c^{*(k)}_{i,j}(z)$ in (12.30), we see that

(12.31) $$T^*(z) \equiv T(z) + \Phi_6(z),$$

where $\Phi_6(z)$, is a polynomial combination over $\mathbb{Q}$ of $\rho^{(\lambda)}(z)/\rho(z)$ and $c^{(\kappa)}_{\mu,\nu}(z)$ having $\lambda \geq 1$, $0 \leq \mu \leq p$, $\max(1, \mu) \leq \nu \leq m$, and $\kappa \geq 0$ such that $\Phi_6(z)$ does not involve any $c^{(\kappa)}_{p,q}(z)$ with $\kappa \geq r$.

For any (1.1) and (1.2), we combine (12.27)–(12.29) and (12.31) to obtain

(12.32) $$0 \equiv \Phi_5(z)\, c^{(r)}_{p,q}(z) + \Phi_7(z),$$

where $\Phi_7(z)$ is a polynomial combination over $\mathbb{Q}$ of $\rho^{(\lambda)}(z)/\rho(z)$ and $c^{(\kappa)}_{\mu,\nu}(z)$, for $\lambda \geq 1$, $0 \leq \mu \leq p$, $\max(1, \mu) \leq \nu \leq m$, and $\kappa \geq 0$, that does not involve $c^{(\kappa)}_{p,q}(z)$ with $\kappa \geq r$.

Let t_3 be an integer such that: for any (1.1) and (1.2), the product of $(\rho(z))^{t_3}$ and the right member of (12.32) is a polynomial combination over $\mathbb{Q}$ of $\rho^{(\lambda)}(z)$ and $c^{(\kappa)}_{\mu,\nu}(z)$, for $\lambda \geq 0$, $0 \leq \mu \leq p$, $\max(1, \mu) \leq \nu \leq m$, and $\kappa \geq 0$. Again, as in the argument following (12.12), we consider the particular ring $\mathfrak{S}_{\chi(m,m)+1}$ of polynomials over $\mathbb{Q}$ specified by the context for Proposition 4.2 on page 36. Let $\boldsymbol{Q}_3$ be the polynomial in $\mathfrak{S}_{\chi(m,m)+1}$ such that: for any (1.1) and any transformation

(1.2) of (1.1) into a corresponding (1.3), the function obtained by substituting $c_{\mu,\nu}^{(\kappa)}(z)$ from (1.1) for $\boldsymbol{w}_{\mu,\nu}^{(\kappa)}$ and substituting $\rho^{(\lambda)}(z)$ for $\boldsymbol{w}_{0,q+1}^{(\lambda)}$ in $\boldsymbol{Q}_3$ is equal to the product of $(\rho(z))^{t_3+1}$ and the right member of (12.32). Since $\boldsymbol{S}$ is not a semi-invariant of the first kind, there is a particular (1.1) and a particular (1.2) for which (12.28) has $\Phi_5(z) \not\equiv 0$. Moreover, neither $\Phi_5(z)$ nor $\Phi_7(z)$ involves $c_{p,q}^{(r)}(z)$. Thus, we have $\boldsymbol{Q}_3 \not\equiv 0$. We apply Proposition 4.2 on page 36 to obtain meromorphic functions $\rho(z)$ and $c_{i,j}(z)$, for $0 \leq i \leq m$ and $\max(1, i) \leq j \leq m$, on a region Ω, such that the substitution of $c_{\mu,\nu}^{(\kappa)}(z)$ for $\boldsymbol{w}_{\mu,\nu}^{(\kappa)}$ and the substitution of $\rho^{(\lambda)}(z)$ for $\boldsymbol{w}_{0,m+1}^{(\lambda)}$ in $\boldsymbol{Q}_3$ yields a function not identically zero on Ω. The (t_3+1)th power of $\rho(z)$ was used in defining $\boldsymbol{Q}_3$; thus, we have $\rho(z) \not\equiv 0$. Consequently, for the particular (1.1) and (1.2) that this specifies, we obtain a contradiction to (12.32). Hence, Case 2 *leads to a contradiction.*

12.1.3. Case 3. Suppose that $S' \not\equiv 0$ and S is not a semi-invariant of the second kind. For any (1.1) and any transformation (1.4) of (1.1) into a corresponding (1.5), the property of $\boldsymbol{R}$ as a semi-invariant of the second kind having weight s requires that

$$(12.33) \quad 0 \equiv R^{**}(\zeta) - (f'(\zeta))^s R(f(\zeta))$$
$$\equiv S^{**}(\zeta)\, c_{p,q}^{**(r)}(\zeta) - (f'(\zeta))^s S(f(\zeta))\, c_{p,q}^{(r)}(f(\zeta))$$
$$+ T^{**}(\zeta) - (f'(\zeta))^s T(f(\zeta)).$$

Since $\boldsymbol{S}$ is a nonconstant isobaric polynomial over $\mathbb{Q}$ in the variables $\boldsymbol{w}_{0,1}^{(\kappa_1)}$ and $\boldsymbol{w}_{0,2}^{(\kappa_2)}$ of weight $s-p-q-r > 0$, we apply (12.15) to obtain

$$(12.34) \qquad S^{**}(\zeta) \equiv (f'(\zeta))^{s-p-q-r} S(f(\zeta)) + F_3(\zeta),$$

where $F_3(\zeta)$ is a product of $(f'(\zeta))^{n_3}$, for some integer n_3, and a polynomial combination over $\mathbb{Q}$ of $f^{(\lambda)}(\zeta)$ and $c_{0,\nu}^{(\kappa)}(f(\zeta))$, for $\lambda \geq 2$, $1 \leq \nu \leq 2$, and $\kappa \geq 0$. Moreover, for $\kappa \geq 0$, $F_3(\zeta)$ does not involve $c_{p,q}^{(\kappa)}(f(\zeta))$. From (12.19), we obtain

$$(12.35) \qquad c_{p,q}^{**(r)}(\zeta) \equiv (f'(\zeta))^{p+q+r} c_{p,q}^{(r)}(f(\zeta)) + F_4(\zeta),$$

where $F_4(\zeta)$ is a product of $(f'(\zeta))^{n_4}$, for some integer n_4, and a polynomial combination over $\mathbb{Q}$ of expressions $f^{(\lambda)}(\zeta)$ and $c_{\mu,\nu}^{(\kappa)}(f(\zeta))$ having $\lambda \geq 2$, $0 \leq \mu \leq p$, $\max(1, \mu) \leq \nu \leq q$, and $\kappa \geq 0$ such that $F_4(\zeta)$ does not involve any $c_{p,q}^{(\kappa)}(f(\zeta))$ having $\kappa \geq r$. We use (12.34) and (12.35) to obtain

$$(12.36) \quad S^{**}(\zeta)\, c_{p,q}^{**(r)}(\zeta) \equiv (f'(\zeta))^s S(f(\zeta))\, c_{p,q}^{(r)}(f(\zeta))$$
$$+ F_3(\zeta)(f'(\zeta))^{p+q+r} c_{p,q}^{(r)}(f(\zeta)) + F_5(\zeta),$$

where $F_5(\zeta)$ is a product of $(f'(\zeta))^{n_5}$, for some integer n_5, and a polynomial combination over $\mathbb{Q}$ of expressions $f^{(\lambda)}(\zeta)$ and $c_{\mu,\nu}^{(\kappa)}(f(\zeta))$, for $\lambda \geq 2$, $0 \leq \mu \leq p$, $\max(1, \mu) \leq \nu \leq m$, and $\kappa \geq 0$ such that $F_5(\zeta)$ does not involve any $c_{p,q}^{(\kappa)}(f(\zeta))$ having $\kappa \geq r$.

Suppose that $\boldsymbol{w}_{i,j}^{(k)}$ is a variable from (1.6) that is effectively involved in $\boldsymbol{T}$. Then, (12.16) yields

$$(12.37) \qquad c_{i,j}^{**(k)}(\zeta) \equiv (f'(\zeta))^{i+j+k} c_{i,j}^{(k)}(f(\zeta)) + H(\zeta),$$

where $H(\zeta)$ is a polynomial combination over $\mathbb{Q}$ of $(f'(\zeta))^{-1}$, $f'(\zeta)$, ..., $f^{(j+k+1)}(\zeta)$ and expressions $c_{\mu,\nu}^{(\kappa)}(f(\zeta))$ for various integers μ, ν, κ. If $i < p$, then $\mu \leq i < p$. If $i = p$ and $j < q$, then $\nu \leq j < q$. If $i = p$, $j = q$, and $k < r$, then $\mu \leq p$, $\nu \leq q$, and $\kappa \leq k < r$. Thus, the right member of (12.37) does not involve any $c_{p,q}^{(\kappa)}(f(\zeta))$ having $\kappa \geq r$. Since $\boldsymbol{T}$ is isobaric and $T^{**}(\zeta)$ is a sum of terms consisting of a rational number times a product of expressions like $c_{i,j}^{**(k)}(\zeta)$ in (12.37), we see that

$$(12.38) \qquad T^{**}(\zeta) \equiv (f'(\zeta))^s T(f(\zeta)) + F_6(\zeta),$$

where s is the weight of $\boldsymbol{T}$ and $F_6(\zeta)$ is a product of $(f'(\zeta))^{n_6}$, for some integer n_6, and a polynomial combination over $\mathbb{Q}$ of various $f^{(\lambda)}(\zeta)$ and $c_{\mu,\nu}^{(\kappa)}(f(\zeta))$ having $\lambda \geq 2$, $0 \leq \mu \leq p$, $\max(1, \mu) \leq \nu \leq m$, and $\kappa \geq 0$ such that $F_6(\zeta)$ does not involve any $c_{p,q}^{(\kappa)}(f(\zeta))$ having $\kappa \geq r$.

For any (1.1) and (1.4), we use (12.33), (12.36), and (12.38) to deduce that

$$(12.39) \qquad 0 \equiv F_3(\zeta)(f'(\zeta))^{p+q+r} c_{p,q}^{(r)}(f(\zeta)) + F_7(\zeta),$$

where $F_7(\zeta)$ is a product of $(f'(\zeta))^{n_7}$, for some integer n_7, and a polynomial combination over $\mathbb{Q}$ of expressions $f^{(\lambda)}(\zeta)$ and $c_{\mu,\nu}^{(\kappa)}(f(\zeta))$ having $\lambda \geq 2$, $0 \leq \mu \leq p$, $\max(1, \mu) \leq \nu \leq m$, and $\kappa \geq 0$ such that $F_7(\zeta)$ does not involve any $c_{p,q}^{(\kappa)}(f(\zeta))$ having $\kappa \geq r$. By setting $\zeta = g(z)$ in (12.39) and using (12.24), we find that, for any (1.1) and any univalent analytic function $g(z)$ on Ω,

$$(12.40) \qquad 0 \equiv G_3(z) c_{p,q}^{(r)}(z) + G_7(z),$$

where each of $G_3(z) \equiv F_3(g(z)) (g'(z))^{-p-q-r}$ and $G_7(z) \equiv F_7(g(z))$ is a product of some integral power of $g'(z)$ and a polynomial combination over $\mathbb{Q}$ of expressions $g^{(\lambda)}(z)$ and $c_{\mu,\nu}^{(\kappa)}(z)$ having $\lambda \geq 2$, $0 \leq \mu \leq p$, $\max(1, \mu) \leq \nu \leq m$, $\kappa \geq 0$, and not involving $c_{p,q}^{(\kappa)}(f(\zeta))$, for $\kappa \geq r$.

Let t_4 be an integer such that: for any (1.1) and any $\zeta = g(z)$ for (1.4), the product of $(g'(z))^{t_4}$ and the right member of (12.40) is a polynomial combination over $\mathbb{Q}$ of $g^{(\lambda)}(z)$ and $c_{\mu,\nu}^{(\kappa)}(z)$, for $\lambda \geq 2$, $0 \leq \mu \leq p$, $\max(1, \mu) \leq \nu \leq m$, and $\kappa \geq 0$. As in the argument following (12.12), we consider the ring $\mathfrak{S}_{\chi(m,m)+1}$ of polynomials over $\mathbb{Q}$ specified by the context for Proposition 4.2 on page 36. Let $\boldsymbol{Q}_4$ be the polynomial in $\mathfrak{S}_{\chi(m,m)+1}$ such that: for any (1.1) and any $\zeta = g(z)$ for (1.4), the function obtained by substituting $c_{\mu,\nu}^{(\kappa)}(z)$ for $\boldsymbol{w}_{\mu,\nu}^{(\kappa)}$ and $g^{(\lambda)}(z)$ for $\boldsymbol{w}_{0,q+1}^{(\lambda)}$ in $\boldsymbol{Q}_4$ is equal to the product of $(g'(z))^{t_4+1}$ and the right member of (12.40). Since $\boldsymbol{S}$ is not a semi-invariant of the second kind, there is a (1.1) and a $\zeta = g(z)$ for (1.4) such that $F_3(\zeta)$ in (12.34) and $G_3(z)$ in (12.40) satisfy $F_3(\zeta) \not\equiv 0$ and $G_3(z) \not\equiv 0$. Hence, we must have $\boldsymbol{Q}_4 \neq 0$. We apply Proposition 4.2 on page 36 to obtain meromorphic functions $h(z)$ and $c_{i,j}(z)$, for $0 \leq i \leq m$ and $\max(1, i) \leq j \leq m$, on a region Ω_0, such that the substitution of $c_{\mu,\nu}^{(\kappa)}(z)$ for $\boldsymbol{w}_{\mu,\nu}^{(\kappa)}$ and $h^{(\lambda)}(z)$ for $\boldsymbol{w}_{0,m+1}^{(\lambda)}$ in $\boldsymbol{Q}_4$ yield a function not identically zero on Ω_0. Since the (t_4+1)th power of $g'(z)$ was used in defining $\boldsymbol{Q}_4$, we have $h(z) \not\equiv 0$. Let Ω be a subregion of Ω_0 on which the restriction of $h(z)$ is a univalent analytic function; and set $g(z) \equiv h(z)$ on Ω. Then, for the particular (1.1) on Ω and (1.4) that this specifies, (12.40) is not satisfied. Thus, Case 3 *leads to a contradiction*.

Since each of the three possible cases leads to a contradiction, we conclude that the supposition "R is a relative invariant" is false. Thus, R is not a relative invariant. This completes the proof of Theorem 12.1 on page 121. $\square$

12.2. The uniqueness of basic relative invariants

Each basic relative invariant for the equations (1.1) has a corresponding index (l, n) according to Definition 1.5 on page 6. The index is a pair (l, n) of integers that satisfy either $1 \leq l \leq n \leq m$ or $l = 0 < 3 \leq n \leq m$. That each basic relative invariant for $m \geq 2$ is uniquely specified by its index is established next.

THEOREM 12.2. *For $m \geq 2$ and integers l, n subject to either $1 \leq l \leq n \leq m$ or $l = 0 < 3 \leq n \leq m$, the basic relative invariant $\boldsymbol{\mathcal{I}}_{m,l,n}$ constructed on page 8 for Theorem 11.2 is the only basic relative invariant of the equations (1.1) that has the index (l, n).*

PROOF. Let $\boldsymbol{P}_1$ and $\boldsymbol{P}_2$ in $\mathcal{S}_m$ be two basic relative invariants having the same index (l, n). To prove that $\boldsymbol{P}_1 \equiv \boldsymbol{P}_2$, suppose that $\boldsymbol{P}_1 \not\equiv \boldsymbol{P}_2$. Then the polynomial

$$(12.41) \qquad \boldsymbol{R} \equiv \boldsymbol{P}_1 - \boldsymbol{P}_2, \quad \text{in } \mathcal{S}_m,$$

is a relative invariant of weight $l + n$ for the equations (1.1). Since $\boldsymbol{P}_1$ and $\boldsymbol{P}_2$ are basic relative invariants of index (l, n), the coefficients of $\boldsymbol{w}_{l,n}$ in both $\boldsymbol{P}_1$ and $\boldsymbol{P}_2$ are equal to 1. Hence, (12.41) shows that the coefficient of $\boldsymbol{w}_{l,n}$ in $\boldsymbol{R}$ is 0. Also, (12.41) shows that any variable $\boldsymbol{w}_{i,j}^{(k)}$ from (1.6) that is effectively involved in $\boldsymbol{R}$ must be effectively involved in at least one of $\boldsymbol{P}_1$, $\boldsymbol{P}_2$. Consequently, because $\boldsymbol{P}_1$ and $\boldsymbol{P}_2$ are basic relative invariants of index (l, n), any variable $\boldsymbol{w}_{i,j}^{(k)}$ from (1.6) that is effectively involved in $\boldsymbol{R}$ must have $i \leq l$, $j \leq n$, $i+j \leq l+n$, $(i, j) \neq (l, n)$, and $i + j < l + n$. Let p be the greatest integer such that $\boldsymbol{w}_{p,j}^{(k)}$ from (1.6) is effectively involved in $\boldsymbol{R}$ for some integers j and k. Let q be the greatest integer such that $\boldsymbol{w}_{p,q}^{(k)}$ from (1.6) is effectively involved in $\boldsymbol{R}$ for some integer k. Let r be the greatest integer such that $\boldsymbol{w}_{p,q}^{(r)}$ from (1.6) is effectively involved in $\boldsymbol{R}$. Since $\boldsymbol{R}$ is a relative invariant, Corollary 5.8 of page 52 shows that $\boldsymbol{R}$ must effectively involve at least one variable $\boldsymbol{w}_{i,j}^{(k)}$ from (1.6) having $k \geq 0$ and either $1 \leq i \leq j \leq m$ or $i = 0 < 3 \leq j \leq m$. Thus, we find that $\boldsymbol{w}_{p,q}^{(r)}$ is not an element of $\mathbb{Q}\{\boldsymbol{w}_{0,1}, \boldsymbol{w}_{0,2}\}$. Since $\boldsymbol{P}_1$ and $\boldsymbol{P}_2$ are basic polynomials, Condition 4 of Definition 1.5 on page 6 shows that the coefficient of $\boldsymbol{w}_{p,q}^{(r)}$ in $\boldsymbol{P}_1$ and the coefficient of $\boldsymbol{w}_{p,q}^{(r)}$ in $\boldsymbol{P}_2$ are polynomials in $\mathbb{Q}\{\boldsymbol{w}_{0,1}, \boldsymbol{w}_{0,2}\}$. Consequently, there is a nonzero polynomial $\boldsymbol{S}$ in $\mathbb{Q}\{\boldsymbol{w}_{0,1}, \boldsymbol{w}_{0,2}\}$ and there is a polynomial $\boldsymbol{T}$ in $\mathcal{S}_m$ in which $\boldsymbol{w}_{p,q}^{(r)}$ is not effectively involved such that

$$(12.42) \qquad \boldsymbol{R} \equiv \boldsymbol{S}\,\boldsymbol{w}_{p,q}^{(r)} + \boldsymbol{T}.$$

We have $p + q < s$, where $s = l + n$ is the weight of $\boldsymbol{R}$. The definition of $\boldsymbol{w}_{p,q}^{(r)}$ shows that: if $\boldsymbol{w}_{i,j}^{(k)}$ is any variable that is effectively involved in $\boldsymbol{T}$, then $0 \leq i \leq p$, $\max(1, i) \leq j \leq n \leq m$, $k \geq 0$, and either (1) $i < p$; or (2) $i = p$ and $j < q$; or (3) $i = p$, $j = q$, and $k < r$. Thus, $\boldsymbol{R}$ satisfies the hypotheses of Theorem 12.1 on page 121. We use that result to conclude that $\boldsymbol{R}$ is not a relative invariant. This contradiction establishes that $\boldsymbol{P}_1 \equiv \boldsymbol{P}_2$ and completes the proof. $\square$

OBSERVATION 12.3. We proved for Theorem 1.13 of page 18 that each relative invariant for homogeneous linear differential equations of order $m \geq 3$ is specified by a corresponding polynomial over $\mathbb{Q}$ in variables $\boldsymbol{w}_{0,j}^{(k)}$ from (1.6) having $1 \leq j \leq m$ and $k \geq 0$. The latter is expressible as a polynomial over $\mathbb{Q}\{\boldsymbol{w}_{0,1}, \boldsymbol{w}_{0,2}\}$ in the variables $\boldsymbol{w}_{0,j}^{(k)}$ having $3 \leq j \leq m$ and $k \geq 0$. With this identification, we proved in [**24**, Proposition C.7 of page 164] that: if a relative invariant $\boldsymbol{I}$ for homogeneous linear differential equations of order $m \geq 3$ is expressible as a *linear polynomial* over $\mathbb{Q}\{\boldsymbol{w}_{0,1}, \boldsymbol{w}_{0,2}\}$ in the variables $\boldsymbol{w}_{0,j}^{(k)}$, for $3 \leq j \leq m$, then there is a nonzero rational number γ and an integer s satisfying $3 \leq s \leq m$ such that $\boldsymbol{I} \equiv \gamma \boldsymbol{\mathcal{I}}_{m,0,s}$, where $\boldsymbol{\mathcal{I}}_{m,0,s}$ from (1.28) of page 8 is identified with the basic relative invariant of weight s for homogeneous linear differential equations of order m.

For equations (1.1) of order $m \geq 3$, an analogously stated result is not valid. In particular, the relative invariant $(\boldsymbol{\mathcal{I}}_{m,0,3} + \boldsymbol{\mathcal{I}}_{m,1,2})$ of weight 3 for equations (1.1) of order $m \geq 3$ is linear over $\mathbb{Q}\{\boldsymbol{w}_{0,1}, \boldsymbol{w}_{0,2}\}$ in the variables $\boldsymbol{w}_{0,3}, \boldsymbol{w}_{1,2}, \boldsymbol{w}_{1,1}^{(1)}, \boldsymbol{w}_{1,1}$; however, for any nonzero γ in $\mathbb{Q}$, $(1/\gamma)(\boldsymbol{\mathcal{I}}_{m,0,3} + \boldsymbol{\mathcal{I}}_{m,1,2})$ is not a basic polynomial.

12.3. Algebraic independence

The set $\mathcal{B}_{m,0}$ of basic relative invariants is given by (1.29) on page 8. It yields the corresponding sets of polynomials in $\mathcal{S}_m$ defined by

$$(12.43) \quad \mathcal{B}_{m,k} = \left\{ \boldsymbol{\mathcal{I}}_{m,l,n}^{(k)} \mid \text{ for } 1 \leq l \leq n \leq m \right\} \bigcup \left\{ \boldsymbol{\mathcal{I}}_{m,0,n}^{(k)} \mid \text{ for } 3 \leq n \leq m \right\},$$

for $k \geq 0$.

For $m \geq 2$, we use (1.14) and (12.43) to introduce

$$(12.44) \quad \mathcal{A}_m = \left\{ \boldsymbol{w}_{0,1}^{(k)} \mid \text{ for } k \geq 0 \right\} \bigcup \left\{ \boldsymbol{a}_{m,2}^{(k)} \mid \text{ for } k \geq 0 \right\} \bigcup \left[\bigcup_{k \geq 0} \mathcal{B}_{m,k} \right].$$

PROPOSITION 12.4. *For $m \geq 2$, the set $\mathcal{A}_m$ is algebraically independent over $\mathbb{Q}$.*

PROOF. Let $\mathcal{X}_m = \left\{ \boldsymbol{X}_{i,j}^{(k)} \mid \text{ for } 1 \leq j \leq m, 0 \leq i \leq j, \text{ and } k \geq 0 \right\}$ be an algebraically independent set of variables over $\mathbb{Q}$; let $\mathcal{P}_m$ denote the ring of polynomials over $\mathbb{Q}$ in the variables of $\mathcal{X}_m$; let $\mathfrak{P}$ denote a nonzero polynomial in $\mathcal{P}_m$; and let P denote the polynomial in $\mathcal{S}_m$ obtained from $\mathfrak{P}$ by: replacing $\boldsymbol{X}_{0,1}^{(k)}$ with $\boldsymbol{w}_{0,1}^{(k)}$; replacing $\boldsymbol{X}_{0,2}^{(k)}$ with $\boldsymbol{a}_{m,2}^{(k)}$; and, for either $1 \leq i \leq j \leq m$ or $i = 0 < 3 \leq j \leq m$, replacing $\boldsymbol{X}_{i,j}^{(k)}$ with $\boldsymbol{\mathcal{I}}_{m,i,j}^{(k)}$. To complete the proof, we shall show that $P \not\equiv 0$ in $\mathcal{S}_m$. Clearly, we may assume that $\mathfrak{P}$ is a nonconstant polynomial in $\mathcal{P}_m$.

For $\boldsymbol{X}_{a,b}^{(d)}, \boldsymbol{X}_{\alpha,\beta}^{(\delta)}$ in $\mathcal{X}_m$, we define $\boldsymbol{X}_{a,b}^{(d)}$ *precedes* $\boldsymbol{X}_{\alpha,\beta}^{(\delta)}$ in $\mathcal{X}_m$ to mean that either (1) $a > \alpha$; or (2) $a = \alpha$ and $b > \beta$; or (3) $a = \alpha$, $b = \beta$, and $d > \delta$. The usual *equality* of $\boldsymbol{X}_{a,b}^{(d)}$ and $\boldsymbol{X}_{\alpha,\beta}^{(\delta)}$ in $\mathcal{X}_m$ occurs if and only if $a = \alpha$, $b = \beta$, and $d = \delta$. For $\boldsymbol{w}_{a,b}^{(d)}, \boldsymbol{w}_{\alpha,\beta}^{(\delta)}$ in the set $\mathcal{W}_m = \left\{ \boldsymbol{w}_{i,j}^{(k)} \mid \text{ for } 1 \leq j \leq m, 0 \leq i \leq j, \text{ and } k \geq 0 \right\}$ of variables for (1.6), we similarly define the phrase $\boldsymbol{w}_{a,b}^{(d)}$ *precedes* $\boldsymbol{w}_{\alpha,\beta}^{(\delta)}$ in $\mathcal{W}_m$ to mean that either (1) $a > \alpha$; or (2) $a = \alpha$ and $b > \beta$; or (3) $a = \alpha$, $b = \beta$, and $d > \delta$. Here, precedence specifies a total ordering for $\mathcal{X}_m$ and one for $\mathcal{W}_m$. With respect to the one-to-one correspondence $\boldsymbol{X}_{i,j}^{(k)} \longleftrightarrow \boldsymbol{w}_{i,j}^{(k)}$ between $\mathcal{X}_m$ and $\mathcal{W}_m$, we see that $\boldsymbol{w}_{a,b}^{(d)}$ precedes $\boldsymbol{w}_{\alpha,\beta}^{(\delta)}$ in $\mathcal{W}_m$ if and only if $\boldsymbol{X}_{a,b}^{(d)}$ precedes $\boldsymbol{X}_{\alpha,\beta}^{(\delta)}$ in $\mathcal{X}_m$.

Let $\mathfrak{M}(\mathcal{X}_m)$ and $\mathfrak{M}(\mathcal{W}_m)$ be the sets of monomials in $\mathcal{P}_m$ and $\mathcal{S}_m$, respectively. The elements of $\mathbb{Q}$ are the constant monomials of $\mathfrak{M}(\mathcal{X}_m)$ and $\mathfrak{M}(\mathcal{W}_m)$. Any two nonconstant monomials $\boldsymbol{M}_1$ and $\boldsymbol{M}_2$ in $\mathfrak{M}(\mathcal{X}_m)$ can be represented by

$$(12.45) \qquad \boldsymbol{M}_1 \equiv \gamma_1 \prod_{\lambda=1}^{\sigma} \boldsymbol{X}_{a_\lambda,b_\lambda}^{(d_\lambda)} \quad \text{and} \quad \boldsymbol{M}_2 \equiv \gamma_2 \prod_{\lambda=1}^{\tau} \boldsymbol{X}_{\alpha_\lambda,\beta_\lambda}^{(\delta_\lambda)},$$

where: γ_1, γ_2 are nonzero rational numbers; σ, τ are positive integers; $\boldsymbol{X}_{a_\lambda,b_\lambda}^{(d_\lambda)}$ precedes or equals $\boldsymbol{X}_{a_{\lambda+1},b_{\lambda+1}}^{(d_{\lambda+1})}$ in $\mathcal{X}_m$ for $1 \leq \lambda \leq \sigma - 1$; and, $\boldsymbol{X}_{\alpha_\lambda,\beta_\lambda}^{(\delta_\lambda)}$ precedes or equals $\boldsymbol{X}_{\alpha_{\lambda+1},\beta_{\lambda+1}}^{(\delta_{\lambda+1})}$ in $\mathcal{X}_m$, for $1 \leq \lambda \leq \tau - 1$. We define $\boldsymbol{M}_1$ *precedes* $\boldsymbol{M}_2$ *in* $\mathfrak{M}(\mathcal{X}_m)$ to mean there is an integer κ for which: $1 \leq \kappa \leq \min(\sigma,\tau)$; $\boldsymbol{X}_{a_\kappa,b_\kappa}^{(d_\kappa)}$ precedes $\boldsymbol{X}_{\alpha_\kappa,\beta_\kappa}^{(\delta_\kappa)}$ in $\mathfrak{X}_m$; while $a_\lambda = \alpha_\lambda$, $b_\lambda = \beta_\lambda$, and $d_\lambda = \delta_\lambda$, for $1 \leq \lambda \leq \kappa - 1$. Also, we agree that any nonconstant monomial in $\mathfrak{M}(\mathcal{X}_m)$ precedes any constant monomial in $\mathfrak{M}(\mathcal{X}_m)$. For any $\boldsymbol{M}_1$, $\boldsymbol{M}_2$, $\boldsymbol{M}_3$ in $\mathfrak{M}(\mathcal{X}_m)$, we see that: if $\boldsymbol{M}_1$ precedes $\boldsymbol{M}_2$ in $\mathfrak{M}(\mathcal{X}_m)$ and $\boldsymbol{M}_2$ precedes $\boldsymbol{M}_3$ in $\mathfrak{M}(\mathcal{X}_m)$, then $\boldsymbol{M}_1$ precedes $\boldsymbol{M}_3$ in $\mathfrak{M}(\mathcal{X}_m)$. Thus, *precedence* is a partial ordering for $\mathfrak{M}(\mathfrak{X}_m)$. It is not a total ordering for $\mathfrak{M}(\mathfrak{X}_m)$; for example, $2\boldsymbol{X}_{0,1}$ and $3\boldsymbol{X}_{0,1}$ are unequal yet neither $2\boldsymbol{X}_{0,1}$ nor $3\boldsymbol{X}_{0,1}$ precedes the other. Also, for any $\boldsymbol{M}_1$, $\boldsymbol{M}_2$, $\boldsymbol{N}_1$, $\boldsymbol{N}_2$ in $\mathfrak{M}(\mathcal{X}_m)$, we find that: if $\boldsymbol{M}_1$ precedes $\boldsymbol{M}_2$ in $\mathfrak{M}(\mathcal{X}_m)$ and $\boldsymbol{N}_1$ precedes $\boldsymbol{N}_2$ in $\mathfrak{M}(\mathcal{X}_m)$, then $\boldsymbol{M}_1\boldsymbol{N}_1$ precedes $\boldsymbol{M}_2\boldsymbol{N}_2$ in $\mathfrak{M}(\mathcal{X}_m)$. By extending the bijection $\boldsymbol{X}_{i,j}^{(k)} \longleftrightarrow \boldsymbol{w}_{i,j}^{(k)}$ between $\mathfrak{X}_m$ and $\mathcal{W}_m$ to one between $\mathfrak{M}(\mathcal{X}_m)$ and $\mathfrak{M}(\mathcal{W}_m)$, we similarly define a partial ordering of *precedence* for $\mathfrak{M}(\mathcal{W}_m)$.

We write the nonconstant polynomial $\mathfrak{P}$ in $\mathcal{P}_m$ as a sum of terms that are minimal in number. Thus, if $\mathfrak{T}_1$ and $\mathfrak{T}_2$ are terms of this expansion subject to $\mathfrak{T}_1 \neq \mathfrak{T}_2$, then either $\mathfrak{T}_1$ precedes $\mathfrak{T}_2$ or $\mathfrak{T}_2$ precedes $\mathfrak{T}_1$. Consequently, $\mathfrak{P}$ contains a term $\mathfrak{T}_0$ that precedes any other term. Hence, we can express $\mathfrak{P}$ as

$$(12.46) \qquad \mathfrak{P} \equiv \mathfrak{T}_0 + \mathfrak{R}_0, \quad \text{where} \quad \mathfrak{T}_0 \equiv \gamma \prod_{\lambda=1}^{\upsilon} \boldsymbol{X}_{p_\lambda,q_\lambda}^{(r_\lambda)},$$

γ is a nonzero rational number, υ is a positive integer, and $\mathfrak{R}_0$ is a polynomial in $\mathcal{P}_m$ that can be written with a minimal number of terms each of which is preceded in $\mathfrak{M}(\mathcal{X}_m)$ by $\mathfrak{T}_0$. We use (1.14) on page 7 to obtain

$$(12.47) \qquad \boldsymbol{a}_{m,2}^{(k)} \equiv \boldsymbol{w}_{0,2}^{(k)} + \boldsymbol{R}_{0,2,k}, \quad \text{for } k \geq 0,$$

where $\boldsymbol{R}_{0,2,k}$ is a polynomial in $\mathcal{S}_m$ that can be written with a minimal number of terms each of which is preceded in $\mathfrak{M}(\mathcal{W}_m)$ by $\boldsymbol{w}_{0,2}^{(k)}$. Our Main Theorem on page 8 shows that $\boldsymbol{\mathcal{I}}_{m,i,j}$ is a basic polynomial of index (i,j). This enables us to write

$$(12.48) \qquad \boldsymbol{\mathcal{I}}_{m,i,j}^{(k)} \equiv \boldsymbol{w}_{i,j}^{(k)} + \boldsymbol{R}_{i,j,k},$$

for $k \geq 0$ and either $1 \leq i \leq j \leq m$ or $i = 0 < 3 \leq j \leq m$,

where $\boldsymbol{R}_{i,j,k}$ is a polynomial in $\mathcal{S}_m$ that can be written with a minimal number of terms each of which is preceded in $\mathfrak{M}(\mathcal{W}_m)$ by $\boldsymbol{w}_{i,j}^{(k)}$. In view of (12.46)–(12.48), we find that: when the substitution described earlier is made in $\mathfrak{P}$ to obtain $\boldsymbol{P}$, the same substitution in $\mathfrak{T}_0$ and $\mathfrak{R}_0$ yields polynomials $(\boldsymbol{T}_0 + \boldsymbol{Q}_0)$ and $\boldsymbol{R}_0$ in $\mathcal{S}_m$ such that

$$\boldsymbol{P} \equiv (\boldsymbol{T}_0 + \boldsymbol{Q}_0) + \boldsymbol{R}_0, \quad \text{where} \quad \boldsymbol{T}_0 \equiv \gamma \prod_{\lambda=1}^{\upsilon} \boldsymbol{w}_{p_\lambda,q_\lambda}^{(r_\lambda)},$$

and $\boldsymbol{Q}_0$, $\boldsymbol{R}_0$ are polynomials in $\mathcal{S}_m$ that can both be written with a minimal number of terms each of which is preceded in $\mathfrak{M}(\mathcal{W}_m)$ by $\boldsymbol{T}_0$. Since $\boldsymbol{P}$ is a polynomial in the algebraically independent variables of $\mathcal{W}_m$ over $\mathbb{Q}$ that effectively involves the nonzero term $\boldsymbol{T}_0$, this establishes that $\boldsymbol{P} \not\equiv 0$. Thus, $\mathcal{A}_m$ is algebraically independent over $\mathbb{Q}$. This completes the proof. □

COROLLARY 12.5. *The set $\mathcal{B}_{m,0}$ of basic relative invariants is algebraically independent over $\mathbb{Q}$.*

PROOF. For $m = 1$, we have $\mathcal{B}_{1,0} = \left\{\boldsymbol{\mathcal{I}}_{1,1,1}\right\}$ and $\boldsymbol{\mathcal{I}}_{1,1,1} \not\equiv 0$. For $m \geq 2$, $\mathcal{B}_{m,0}$ is a nonvacuous subset of the algebraically independent set $\mathcal{A}_m$. □

CHAPTER 13

Real-Valued Functions of a Real Variable

Thus far, our results have been presented in the context of Section 1.2. However, other conditions can be placed on (1.1), (1.2), and (1.4) so that each $\mathcal{I}_{m,l,n}$ of page 8 yields corresponding identities like (1.9) and (1.10). For real-valued functions of a real variable, we place suitable restrictions on (1.1), (1.2), and (1.4) in this chapter.

In and only in this chapter, we restrict Ω to be a nonvacuous open subinterval of the real line and we regard z as a real variable. For any nonvacuous open subinterval $\mathcal{U}$ of the real line and any integer $k \geq 0$, we let $C^k(\mathcal{U})$ denote the class of real-valued functions on $\mathcal{U}$ that have continuous kth derivatives on $\mathcal{U}$.

13.1. A suitable context for the evaluations when $m \geq 2$

For $m \geq 2$, we consider the equations having the form

$$(13.1) \qquad \left(y^{(m)}(z)\right)^2 + \sum_{\substack{0 \leq i,j \leq m \\ (i,j) \neq (0,0)}} c_{i,j}(z)\, y^{(m-i)}(z)\, y^{(m-j)}(z) = 0, \quad \text{on } \Omega,$$

where $c_{i,j}$ belongs to $C^{2m-i-j}(\Omega)$ and satisfies $c_{j,i}(z) \equiv c_{i,j}(z)$, on Ω, whenever $0 \leq i, j \leq m$ and $(i,j) \neq (0,0)$. For (13.2-A) and (13.4-A), we set $c_{0,0}(z) \equiv 1$ on Ω.

PROPOSITION 13.1. *With respect to (13.1) on Ω, suppose that ρ for*

$$(13.2) \qquad y(z) = \rho(z)\, v(z)$$

is a function in $C^{2m}(\Omega)$ that satisfies $\rho(z) \neq 0$, for each z in Ω. Then, the formula

$$(13.2\text{-A}) \quad c_{i,j}^*(z) \equiv \sum_{\mu=0}^{i} \sum_{\nu=0}^{j} \binom{m-\mu}{i-\mu}\binom{m-\nu}{j-\nu} \frac{\rho^{(i-\mu)}(z)\, \rho^{(j-\nu)}(z)}{\left(\rho(z)\right)^2} c_{\mu,\nu}(z), \quad \text{on } \Omega,$$
$$\text{for } 0 \leq i, j \leq m,$$

defines functions $c_{i,j}^*$ in $C^{2m-i-j}(\Omega)$ with $c_{0,0}^*(z) \equiv 1$ and $c_{j,i}^*(z) \equiv c_{i,j}^*(z)$, when $0 \leq i, j \leq m$, in terms of which the differential equation

$$(13.3) \qquad \left(v^{(m)}(z)\right)^2 + \sum_{\substack{0 \leq i,j \leq m \\ (i,j) \neq (0,0)}} c_{i,j}^*(z)\, v^{(m-i)}(z)\, v^{(m-j)}(z) = 0, \quad \text{on } \Omega,$$

is such that: for any real-valued functions y and v that are defined on an open subinterval $\mathcal{U}$ of Ω and related on $\mathcal{U}$ by (13.2), y belongs to $C^m(\mathcal{U})$ and is a solution of (13.1) on $\mathcal{U}$ if and only if v belongs to $C^m(\mathcal{U})$ and is a solution of (13.3) on $\mathcal{U}$.

OBSERVATION. Distinct equations of the form (13.3) may possess the same local solutions. For instance, each of the differential equations given by

$$\left(v^{(m)}(z)\right)^2 + p^2\left(v(z)\right)^2 = 0 \quad \text{or} \quad \left(v^{(m)}(z)\right)^2 + q^2\left(v(z)\right)^2 = 0, \quad \text{for } 1 \le p < q,$$

has only the real-valued solutions that are identically zero on their domains. Thus, Proposition 13.1 is not a direct restatement of Proposition 2.1 on page 23.

PROOF OF PROPOSITION 13.1. Let i, j, μ, and ν satisfy $0 \le \mu \le i \le m$ and $0 \le \nu \le j \le m$. In view of $2m - (i - \mu) \ge 2m - i \ge 2m - i - j$, $\rho \in C^{2m}$, and $\rho(z) \ne 0$, for each z in Ω, we have

(a) $$\frac{\rho^{(i-\mu)}}{\rho} \in C^{2m-(i-\mu)}(\Omega) \subseteq C^{2m-i-j}(\Omega).$$

Due to $2m - (j - \nu) \ge 2m - j \ge 2m - i - j$, $\rho \in C^{2m}$, and $\rho(z) \ne 0$, for each z in Ω, we find that

(b) $$\frac{\rho^{(j-\nu)}}{\rho} \in C^{2m-(j-\nu)}(\Omega) \subseteq C^{2m-i-j}(\Omega).$$

We use $2m - \mu - \nu \ge 2m - i - j$ and $c_{\mu,\nu} \in C^{2m-\mu-\nu}(\Omega)$ to obtain

(c) $$c_{\mu,\nu} \in C^{2m-i-j}(\Omega).$$

Since the product of the functions in the left members of (a), (b), and (c) is an element of $C^{2m-i-j}(\Omega)$, the right member of (13.2-A) defines $c^*_{i,j}$ as an element of $C^{2m-i-j}(\Omega)$. Moreover, in view of $c_{\nu,\mu} = c_{\mu,\nu}$, for $0 \le \mu, \nu \le m$, and $c_{0,0}(z) \equiv 1$, (13.2-A) yields $c^*_{j,i} = c^*_{i,j}$, for $0 \le i, j \le m$, as well as $c^*_{0,0}(z) \equiv 1$. Thus, (13.1) and ρ for (13.2) uniquely specify coefficients of the indicated type for (13.3).

In terms of the coefficients $c_{i,j}$ for (13.1) and $c^*_{i,j}$ for (13.3), we set

$$Q\left[z; Y^{(0)}, \ldots, Y^{(m)}\right] \equiv \left(Y^{(m)}\right)^2 + \sum_{\substack{0 \le i,j \le m \\ (i,j) \ne (0,0)}} c_{i,j}(z)\, Y^{(m-i)}\, Y^{(m-j)}, \quad \text{on } \Omega,$$

and

$$Q^*\left[z; V^{(0)}, \ldots, V^{(m)}\right] \equiv \left(V^{(m)}\right)^2 + \sum_{\substack{0 \le i,j \le m \\ (i,j) \ne (0,0)}} c^*_{i,j}(z)\, V^{(m-i)}\, V^{(m-j)}, \quad \text{on } \Omega.$$

Let $\mathcal{U}$ be a nonvacuous open subinterval of Ω. When real-valued functions y and v on $\mathcal{U}$ satisfy (13.2) on $\mathcal{U}$, y belongs to $C^m(\mathcal{U})$ if and only if v belongs to $C^m(\mathcal{U})$. For any functions y and v in $C^m(\mathcal{U})$ that are related by (13.2) on $\mathcal{U}$, we have

$$Q\left[z; y^{(0)}(z), \ldots, y^{(m)}(z)\right] \equiv \sum_{0 \le \mu,\nu \le m} c_{\mu,\nu}(z)\, y^{(m-\mu)}(z)\, y^{(m-\nu)}(z)$$

$$\equiv \sum_{\mu=0}^{m} \sum_{\nu=0}^{m} c_{\mu,\nu}(z) \sum_{r=0}^{m-\mu} \binom{m-\mu}{r} \rho^{(r)}(z)\, v^{(m-\mu-r)}(z) \sum_{s=0}^{m-\nu} \binom{m-\nu}{s} \rho^{(s)}(z)\, v^{(m-\nu-s)}(z)$$

$$\equiv \sum_{\mu=0}^{m} \sum_{\nu=0}^{m} \sum_{i=\mu}^{m} \sum_{j=\nu}^{m} \binom{m-\mu}{i-\mu}\binom{m-\nu}{j-\nu} \rho^{(i-\mu)}(z)\, \rho^{(j-\nu)}(z)\, c_{\mu,\nu}(z)\, v^{(m-i)}(z)\, v^{(m-j)}(z)$$

$$\equiv \sum_{i=0}^{m} \sum_{j=0}^{m} \left[\sum_{\mu=0}^{i} \sum_{\nu=0}^{j} \binom{m-\mu}{i-\mu}\binom{m-\nu}{j-\nu} \rho^{(i-\mu)}(z)\, \rho^{(j-\nu)}(z)\, c_{\mu,\nu}(z)\right] v^{(m-i)}(z)\, v^{(m-j)}(z)$$

13.1. A SUITABLE CONTEXT FOR THE EVALUATIONS WHEN $m \geq 2$ 137

and therefore

$$Q\Big[z;\, y^{(0)}(z),\, \ldots,\, y^{(m)}(z)\Big] \equiv \big(\rho(z)\big)^2 \sum_{0 \leq i,j \leq m} c_{i,j}^*(z)\, v^{(m-i)}(z)\, v^{(m-j)}(z)$$

$$\equiv \big(\rho(z)\big)^2 Q^*\Big[z;\, v^{(0)}(z),\, \ldots,\, v^{(m)}(z)\Big].$$

Consequently, y is a local solution of (13.1) if and only if v is a local solution of (13.3). This completes the proof. □

PROPOSITION 13.2. *With respect to* (13.1) *on* Ω, *suppose that g is a function in* $C^{2m+1}(\Omega)$ *that satisfies $g'(z) \neq 0$, for each z in Ω. Then, $\zeta = g(z)$ is a univalent function on Ω that has a univalent inverse function*

$$(13.4) \qquad z = f(\zeta), \quad \text{on } \Omega^{**} = g(\Omega),$$

such that f belongs to $C^{2m+1}(\Omega^{**})$ and, for each ζ in Ω^{**}, $f'(\zeta) \neq 0$. Moreover, in terms of (13.4) *and* (3.1)–(3.2) *on page 29, the formula*

$$(13.4\text{-A}) \quad c_{i,j}^{**}(\zeta) \equiv \sum_{\mu=0}^{i} \sum_{\nu=0}^{j} \big(f'(\zeta)\big)^{\mu+\nu}\, \alpha_{i-\mu, m-i}(\zeta)\, \alpha_{j-\nu, m-j}(\zeta)\, c_{\mu,\nu}\big(f(\zeta)\big),$$

$$\text{on } \Omega^{**} \text{ for } 0 \leq i,\, j \leq m,$$

defines functions $c_{i,j}^{**}$ in $C^{2m-i-j}(\Omega^{**})$ with $c_{0,0}^{**}(\zeta) \equiv 1$ and $c_{j,i}^{**}(\zeta) \equiv c_{i,j}^{**}(\zeta)$, for $0 \leq i,\, j \leq m$, in terms of which the differential equation

$$(13.5) \qquad \big(u^{(m)}(\zeta)\big)^2 + \sum_{\substack{0 \leq i,j \leq m \\ (i,j) \neq (0,0)}} c_{i,j}^{**}(\zeta)\, u^{(m-i)}(\zeta)\, u^{(m-j)}(\zeta) = 0, \quad \text{on } \Omega^{**},$$

is such that: for any nonvacuous open subinterval $\mathcal{U}$ of Ω with corresponding open subinterval $\mathcal{U}^{**} = g(\mathcal{U})$ of Ω^{**}, a real-valued function y on $\mathcal{U}$ belongs to $C^m(\mathcal{U})$ and is a solution of (13.1) *on $\mathcal{U}$ if and only if the real-valued function $u = y \circ f$ on $\mathcal{U}^{**}$ belongs to $C^m(\mathcal{U}^{**})$ and is a solution of* (13.5) *on $\mathcal{U}^{**}$.*

PROOF. Since g' is a continuous function on the open interval Ω such that $g'(z) \neq 0$, for each z in Ω, we have either $g'(z) > 0$, for each z in Ω, or $g'(z) < 0$, for each z in Ω. Therefore, g is a univalent function on Ω; and g possesses a univalent inverse function f on the corresponding open interval $\Omega^{**} = g(\Omega)$ such that $f'(\zeta) = 1/g'\big(f(\zeta)\big)$, on Ω^{**}, and $f'(\zeta) \neq 0$, for each ζ in Ω^{**}. We combine this with $g \in C^{2m+1}(\Omega)$ to conclude that $f \in C^{2m+1}(\Omega)$ and $f''/f' \in C^{2m-1}(\Omega)$. Thus, (3.1)–(3.2) on page 29 show that $\alpha_{i,j}$ belongs to $C^{2m-i}(\Omega^{**})$, for $0 \leq i \leq m$ and any j.

Let i, j, μ, and ν satisfy $0 \leq \mu \leq i \leq m$ and $0 \leq \nu \leq j \leq m$. In view of $2m - (i - \mu) \geq 2m - i \geq 2m - i - j$, we have

$$(13.6) \qquad \alpha_{i-\mu, m-i} \in C^{2m-(i-\mu)}(\Omega^{**}) \subseteq C^{2m-i-j}(\Omega^{**}).$$

Due to $2m - (j - \nu) \geq 2m - j \geq 2m - i - j$, we find that

$$(13.7) \qquad \alpha_{j-\nu, m-j} \in C^{2m-(j-\nu)}(\Omega^{**}) \subseteq C^{2m-i-j}(\Omega^{**}).$$

We use $c_{\mu,\nu} \in C^{2m-\mu-\nu}(\Omega)$ and $2m - \mu - \nu \geq 2m - i - j$ to obtain

$$(13.8) \qquad c_{\mu,\nu} \circ f \in C^{2m-\mu-\nu}(\Omega^{**}) \subseteq C^{2m-i-j}(\Omega^{**}).$$

The product of the functions in the left members of (13.6)–(13.8) is an element of $C^{2m-i-j}(\Omega^{**})$; hence, the right member of (13.4-A) defines $c_{i,j}^{**}$ as an element of $C^{2m-i-j}(\Omega^{**})$. We apply (13.4-A) with $\alpha_{0,j}(\zeta) \equiv 1$ from (3.1) and $c_{0,0}(z) \equiv 1$ to obtain $c_{0,0}^{**}(\zeta) \equiv 1$ on Ω^{**}. We deduce $c_{j,i}^{**}(\zeta) = c_{i,j}^{**}(\zeta)$, on Ω^{**} for $0 \leq i, j \leq m$, from (13.4-A) and $c_{\nu,\mu}(z) = c_{\mu,\nu}(z)$, on Ω for $0 \leq \mu, \nu \leq m$. Thus, (13.1) and (13.4-A) uniquely specify coefficients of the indicated type for (13.5).

Let y be a real-valued function on an nonvacuous open subinterval $\mathcal{U}$ of Ω and set $u = y \circ f$ on the corresponding open subinterval $\mathcal{U}^{**} = g(\mathcal{U})$ of Ω^{**}. Since we also have $y = u \circ g$ on $\mathcal{U}$, we see that: y belongs to $C^m(\mathcal{U})$ if and only if u belongs to $C^m(\mathcal{U}^{**})$. Suppose that y belongs to $C^m(\mathcal{U})$. Then, with u in $C^m(\mathcal{U}^{**})$, we use

$$Q\Big[z; Y^{(0)}, \ldots, Y^{(m)}\Big] \equiv \big(Y^{(m)}\big)^2 + \sum_{\substack{0 \leq i,j \leq m \\ (i,j) \neq (0,0)}} c_{i,j}(z)\, Y^{(m-i)}\, Y^{(m-j)}, \quad \text{on } \Omega,$$

a natural modification for Proposition 3.1 of page 29, (13.4-A), and

$$Q^{**}\Big[\zeta; U^{(0)}, \ldots, U^{(m)}\Big] \equiv \big(U^{(m)}\big)^2 + \sum_{\substack{0 \leq i,j \leq m \\ (i,j) \neq (0,0)}} c_{i,j}^{**}(\zeta)\, U^{(m-i)}\, U^{(m-j)}, \quad \text{on } \Omega^{**},$$

to verify that

$$\big(f'(\zeta)\big)^{2m} \left\{ Q\Big[z; y^{(0)}(z), \ldots, y^{(m)}(z)\Big] \right\}_{z=f(\zeta)}$$

$$\equiv \big(f'(\zeta)\big)^{2m} Q\Big[f(\zeta); y^{(0)}\big(f(\zeta)\big), \ldots, y^{(m)}\big(f(\zeta)\big)\Big]$$

$$\equiv \big(f'(\zeta)\big)^{2m} \sum_{0 \leq \mu,\nu \leq m} c_{\mu,\nu}\big(f(\zeta)\big)\, y^{(m-\mu)}\big(f(\zeta)\big)\, y^{(m-\nu)}\big(f(\zeta)\big)$$

$$\equiv \sum_{\mu=0}^{m} \sum_{\nu=0}^{m} \big(f'(\zeta)\big)^{\mu+\nu} c_{\mu,\nu}\big(f(\zeta)\big) \sum_{r=0}^{m-\mu} \alpha_{m-\mu-r,r}(\zeta)\, u^{(r)}(\zeta) \sum_{s=0}^{m-\nu} \alpha_{m-\nu-s,s}(\zeta)\, u^{(s)}(\zeta)$$

$$\equiv \sum_{\mu=0}^{m} \sum_{\nu=0}^{m} \sum_{i=\mu}^{m} \sum_{j=\nu}^{m} \frac{\alpha_{i-\mu,m-i}(\zeta)\, \alpha_{j-\nu,m-j}(\zeta)}{\big(f'(\zeta)\big)^{-\mu-\nu}}\, c_{\mu,\nu}\big(f(\zeta)\big)\, u^{(m-i)}(\zeta)\, u^{(m-j)}(\zeta)$$

$$\equiv \sum_{i=0}^{m} \sum_{j=0}^{m} \sum_{\mu=0}^{i} \sum_{\nu=0}^{j} \frac{\alpha_{i-\mu,m-i}(\zeta)\, \alpha_{j-\nu,m-j}(\zeta)}{\big(f'(\zeta)\big)^{-\mu-\nu}}\, c_{\mu,\nu}\big(f(\zeta)\big)\, u^{(m-i)}(\zeta)\, u^{(m-j)}(\zeta)$$

$$\equiv \sum_{0 \leq i,j \leq m} c_{i,j}^{**}(\zeta)\, u^{(m-i)}(\zeta)\, u^{(m-j)}(\zeta) \equiv Q^{**}\Big[\zeta; u^{(0)}(\zeta), \ldots, u^{(m)}(\zeta)\Big], \quad \text{on } \mathcal{U}^{**}.$$

Thus, y is a local solution of (13.1) on $\mathcal{U}$ if and only if u is a local solution of (13.5) on $\mathcal{U}^{**}$. This completes the proof. $\square$

We set $\mathfrak{A}_0 = \mathbb{Q}$ and, for $t = 1, 2, \ldots, 2m-1, 2m$, we introduce $\mathfrak{A}_t$ as the subset of $\mathcal{S}_m$ that consists of the polynomials over $\mathbb{Q}$ in the particular variables from (1.6) given by

(13.9) $\quad w_{i,j}^{(k)}, \quad \text{for } 1 \leq j \leq m,\ 0 \leq i \leq j,\ k \geq 0,\ \text{and } i+j+k \leq t.$

We continue to employ the notation $w_{j,i} \equiv w_{i,j}$, for $0 \leq i < j \leq m$, and $w_{0,0} \equiv 1$.

13.1. A SUITABLE CONTEXT FOR THE EVALUATIONS WHEN $m \geq 2$

PROPOSITION 13.3. *For either $1 \leq l \leq n \leq m$ or $l = 0 < 3 \leq n \leq m$, let $\boldsymbol{P}$ denote any polynomial that appears in at least one of the formulas (1.14)–(1.28) or (1.30)–(1.38) on pages 7–9. Then, $\boldsymbol{P}$ is an element of $\mathfrak{A}_{2m}$ and, for any variable $\boldsymbol{w}_{i,j}^{(k)}$ that appears effectively in $\boldsymbol{P}$, the corresponding coefficients $c_{i,j}(z)$ of (13.1), $c_{i,j}^*(z)$ of (13.3), and $c_{i,j}^{**}(\zeta)$ of (13.5) specify continuous functions $c_{i,j}^{(k)}(z)$ on Ω, $c_{i,j}^{*(k)}(z)$ on Ω, and $c_{i,j}^{**(k)}(\zeta)$ on Ω^{**}.*

PROOF. In view of (1.16) and (1.14)–(1.15) on page 7, we see that $\boldsymbol{b}_{m,1}$ is a polynomial in $\mathfrak{A}_1$ while $\boldsymbol{a}_{m,2}$ and $\boldsymbol{b}_{m,2}$ are polynomials in $\mathfrak{A}_2$. Combining this with (1.18)–(1.20), and (1.30)–(1.32), we use mathematical induction to deduce that: $\boldsymbol{K}_{m,i,j} \in \mathfrak{A}_i$, for $0 \leq i \leq m$ and any j; and $\boldsymbol{U}_{m,i,j} \in \mathfrak{A}_i$, for $0 \leq i \leq m$ and any j. Next, we find that (1.21) and (1.33) yield $\boldsymbol{L}_{m,i,j} \in \mathfrak{A}_{i+j}$ and $\boldsymbol{V}_{m,i,j} \in \mathfrak{A}_{i+j}$, for $0 \leq i, j \leq m$. By employing this with (1.22)–(1.23) and (1.34)–(1.35), we obtain $\boldsymbol{M}_{m,l,n,h,i} \in \mathfrak{A}_i$ and $\boldsymbol{W}_{m,l,n,h,i} \in \mathfrak{A}_i$, for $0 \leq h \leq l$ and $0 \leq i \leq l+n$. Using this with (1.26)–(1.27) and (1.36)–(1.37), we verify that $\boldsymbol{I}_{m,l,n,h,i} \in \mathfrak{A}_i$ and $\boldsymbol{J}_{m,l,n,h,i} \in \mathfrak{A}_i$, for $0 \leq h \leq l$ and $0 \leq i \leq l+n$. Consequently, (1.28) and (1.38) show that $\boldsymbol{\mathcal{I}}_{m,l,n} \in \mathfrak{A}_{l+n}$ and $\boldsymbol{\mathcal{J}}_{m,l,n} \in \mathfrak{A}_{l+n}$. In view of $l+n \leq 2m$ and

$$\mathfrak{A}_0 \subset \mathfrak{A}_1 \subset \cdots \subset \mathfrak{A}_{2m-1} \subset \mathfrak{A}_{2m},$$

we have just shown that each polynomial that appears in any of the formulas (1.14)–(1.28) or (1.30)–(1.38) belongs to $\mathfrak{A}_{2m}$. Thus, we obtain $\boldsymbol{P} \in \mathfrak{A}_{2m}$.

For any variable $\boldsymbol{w}_{i,j}^{(k)}$ that $\boldsymbol{P}$ effectively involves, we have $0 \leq k \leq 2m - i - j$. Since $c_{i,j}(z)$ in (13.1) belongs to $C^{2m-i-j}(\Omega)$, its kth derivative $c_{i,j}^{(k)}(z)$ exists as a continuous function in $C^{2m-i-j-k}(\Omega) \subseteq C^0(\Omega)$. Similarly, we find that $c_{i,j}^{*(k)}(z)$ and $c_{i,j}^{**(k)}(\zeta)$ exist as continuous functions. This completes the proof. □

CONTEXT FOR THEOREM 13.4. Let an equation (13.1) be given on Ω; let (13.2) transform (13.1) on Ω into (13.3) on Ω as indicated in Proposition 13.1; and let (13.4) transform (13.1) on Ω into (13.5) on Ω^{**} according to Proposition 13.2. For any variable $\boldsymbol{w}_{i,j}^{(k)}$ in $\mathfrak{A}_{2m}$, Proposition 13.3 shows that: $c_{i,j}(z)$ for (13.1), $c_{i,j}^*(z)$ for (13.3), and $c_{i,j}^{**}(\zeta)$ for (13.5) specify corresponding continuous functions $c_{i,j}^{(k)}(z)$ on Ω, $c_{i,j}^{*(k)}(z)$ on Ω, and $c_{i,j}^{**(k)}(\zeta)$ on Ω^{**}. This enables us to introduce the following notation. For any polynomial $\boldsymbol{P}$ in $\mathfrak{A}_{2m}$, let $P(z)$ denote the continuous function on Ω obtained by replacing each $\boldsymbol{w}_{i,j}^{(k)}$ in $\boldsymbol{P}$ with the corresponding $c_{i,j}^{(k)}(z)$ from (13.1), let $P^*(z)$ denote the continuous function on Ω obtained by replacing each $\boldsymbol{w}_{i,j}^{(k)}$ in $\boldsymbol{P}$ with the corresponding $c_{i,j}^{*(k)}(z)$ from (13.3), and let $P^{**}(\zeta)$ denote the continuous function on Ω^{**} obtained by replacing each $\boldsymbol{w}_{i,j}^{(k)}$ in $\boldsymbol{P}$ with the corresponding $c_{i,j}^{**(k)}(\zeta)$ from (13.5).

THEOREM 13.4. *For $m \geq 2$ and integers l, n that satisfy either $1 \leq l \leq n \leq m$ or $l = 0 < 3 \leq n \leq m$, the polynomial $\boldsymbol{\mathcal{I}}_{m,l,n}$ defined in (1.28) of page 8 yields*

(13.10) $$\mathcal{I}_{m,l,n}^*(z) \equiv \mathcal{I}_{m,l,n}(z), \quad \text{on } \Omega,$$

and

(13.11) $$\mathcal{I}_{m,l,n}^{**}(\zeta) \equiv \bigl(f'(\zeta)\bigr)^{l+n} \mathcal{I}_{m,l,n}\bigl(f(\zeta)\bigr), \quad \text{on } \Omega^{**}.$$

PROOF. To verify at each step that the arguments for Theorem 8.1, Lemma 8.2, and Theorem 8.3 can be repeated in our present context, we check that substitutions are made only into polynomials defined in $\mathfrak{A}_{2m}$ and, with $\rho \in C^{2m}(\Omega)$, we use (13.2-A) in place of (2.5). Thus, substitutions into $\mathcal{I}_{m,l,n}$ yield (13.10). To establish at each step that the proofs of Theorem 10.1, Lemma 10.2, and Theorem 10.3 are applicable for our present context, we observe that substitutions are made only into polynomials belonging to $\mathfrak{A}_{2m}$ and, with $f \in C^{2m+1}(\Omega^{**})$, we apply (13.4-A) in place of (3.9). Consequently, substitutions into $\mathcal{J}_{m,l,n}$ of (1.38) on page 9 yield

$$(13.12) \qquad \mathcal{J}^{**}_{m,l,n}(\zeta) \equiv \bigl(f'(\zeta)\bigr)^{l+n} \mathcal{J}_{m,l,n}\bigl(f(\zeta)\bigr), \quad \text{on } \Omega^{**}.$$

The polynomial equality $\mathcal{J}_{m,l,n} \equiv \mathcal{I}_{m,l,n}$ in $\mathfrak{A}_{2m}$ was established for Theorem 11.1 on page 119. Hence, (13.12) yields (13.11). This completes the proof. $\square$

When Definitions 1.1–1.5 of page 6 are restated in our present context for polynomials of $\mathfrak{A}_{2m}$, we conclude that: *for $m \geq 2$ and either $1 \leq l \leq n \leq m$ or $l = 0 < 3 \leq n \leq m$, the polynomial $\mathcal{I}_{m,l,n}$ defined in $\mathfrak{A}_{2m}$ by (1.28) on page 8 is a basic relative invariant of weight $l + n$ for the equations* (13.1).

Through analogy with [**24**, Sections 12.3–12.4, pages 96–103], one could develop results in the present context about equations having fundamental systems of local solutions like those of (17.2) on page 179. Then, linear independence would be with respect to the field $\mathbb{R}$ of real numbers. In the next chapter, we consider polynomials in $\mathcal{S}_m$ that may not belong to $\mathfrak{A}_{2m}$. By returning there to the context of Section 1.2 where the functions are meromorphic, we can perform substitutions for $w_{i,j}^{(k)}$ without restriction on the nonnegative integer k.

13.2. Appropriate hypotheses when $m = 1$

For $m = 1$ and transformations of the equations

$$(13.13) \qquad \bigl(y'(z)\bigr)^2 + 2c_{0,1}(z)\, y'(z)\, y(z) + c_{1,1}(z)\bigl(y(z)\bigr)^2 = 0$$

by means of substitutions $y(z) = \rho(z) v(z)$ and $z = f(\zeta)$, we assume that: *the coefficients $c_{0,1}(z)$, $c_{1,1}(z)$ of (13.13) are continuous real-valued functions of a real variable z on an open subinterval Ω of the real line; $\rho \in C^1(\Omega)$; $\rho(z) \neq 0$, for each z in Ω; g is a function belonging to $C^2(\Omega)$ such that $g'(z) \neq 0$, for each z in Ω; and f is the inverse function for g on the open interval $\Omega^{**} = g(\Omega)$.*

The case $m = 1$ of (2.21) and (2.22) on page 27 yields continuous real-valued functions $c_{0,1}^*(z)$, $c_{1,1}^*(z)$ on Ω such that: (13.13) is transformed into

$$(13.14) \qquad \bigl(v'(z)\bigr)^2 + 2c_{0,1}^*(z)\, v'(z)\, v(z) + c_{1,1}^*(z)\bigl(v(z)\bigr)^2 = 0, \quad \text{on } \Omega,$$

by the substitution $y(z) = \rho(z) v(z)$; and, (2.23) on page 27 is satisfied on Ω.

After deducing that $z = f(\zeta)$ is a univalent function in $C^2(\Omega^{**})$ that has $f'(\zeta) \neq 0$, for each ζ in Ω^{**}, we apply the case $m = 1$ of (3.19) and (3.20) on page 34 to obtain $c_{0,1}^{**}(\zeta)$, $c_{1,1}^{**}(\zeta)$ in $C^0(\Omega^{**})$ such that: (13.13) is transformed into

$$(13.15) \qquad \bigl(u'(z)\bigr)^2 + 2c_{0,1}^{**}(z)\, u'(z)\, u(z) + c_{1,1}^{**}(z)\bigl(u(z)\bigr)^2 = 0, \quad \text{on } \Omega^{**},$$

by $z = f(\zeta)$ and $u(\zeta) = (y \circ f)(\zeta)$. Then, (3.21) on page 34 is satisfied on Ω^{**}.

Substitutions from (13.13)–(13.15) in $\mathcal{I}_{1,1,1} \equiv \boldsymbol{w}_{1,1} - (\boldsymbol{w}_{0,1})^2$ therefore give

$$\mathcal{I}_{1,1,1}^*(z) \equiv \mathcal{I}_{1,1,1}(z), \quad \text{on } \Omega, \quad \text{and} \quad \mathcal{I}_{1,1,1}^{**}(\zeta) \equiv \bigl(f'(\zeta)\bigr)^2 \mathcal{I}_{1,1,1}\bigl(f(\zeta)\bigr), \quad \text{on } \Omega^{**}.$$

Thus, for the context about (13.13), $\mathcal{I}_{1,1,1}$ has the properties of a relative invariant.

Part 3

Supplementary Results

CHAPTER 14

Relative Invariants via Basic Ones for $m \geq 2$

14.1. Relative invariants in terms of basic ones and $a_{m,2}$

For the equations (1.1) having $m = 1$, each relative invariant is expressed in terms of the basic relative invariant $\mathcal{I}_{1,1,1} \equiv w_{1,1} - (w_{0,1})^2$ by the explicit formula (4.23) on page 41. For $m \geq 2$, we shall establish in Corollary 14.4 the result that any relative invariant for the equations (1.1) is expressible as a differential-polynomial combination over $\mathbb{Q}$ of $a_{m,2}$ and the basic relative invariants for (1.1). First, we explain the terminology *differential-polynomial combination*.

DEFINITION 14.1. For any P in $\mathcal{S}_m$, we set $P^{(0)} \equiv P$ and $P^{(k+1)} \equiv (P^{(k)})'$, for $k = 0, 1, 2, \ldots$. Let $\mathcal{A}$ denote any subset of $\mathcal{S}_m$. Then, there is a corresponding subset $\mathcal{A}'$ defined by

$$\mathcal{A}' \equiv \left\{ P^{(k)} \mid \text{ for some } P \text{ in } \mathcal{A} \text{ and some } k \geq 0 \right\}.$$

To say that P in $\mathcal{S}_m$ is a *differential-polynomial combination over $\mathbb{Q}$ of the elements in $\mathcal{A}$* means that P is a polynomial combination over $\mathbb{Q}$ of the elements in $\mathcal{A}'$.

We let $\mathbb{Q}[\mathcal{A}]$ denote the set of polynomial combinations over $\mathbb{Q}$ of the elements in $\mathcal{A}$. It is a subring of $\mathcal{S}_m$. Then, $\mathbb{Q}[\mathcal{A}']$ is the set of polynomial combinations over $\mathbb{Q}$ of the elements in $\mathcal{A}'$; and we use the notation $\mathbb{Q}\{\mathcal{A}\} \equiv \mathbb{Q}[\mathcal{A}']$ for it. Thus, the elements of $\mathbb{Q}\{\mathcal{A}\}$ are the *differential-polynomial combinations over $\mathbb{Q}$ of the elements in $\mathcal{A}$*. In particular, $\mathbb{Q}\{\mathcal{A}\}$ is closed under the derivation ' for $\mathcal{S}_m$. It is the differential subring of $\mathcal{S}_m$ that is generated by $\mathcal{A}$. Our terminology is consistent with [41].

EXAMPLE 14.2. For $\mathcal{A} = \{w_{0,1}, w_{0,2}\}$, we have

(14.1) $\quad \mathcal{A}' = \left\{ w_{0,1}^{(k)} \mid \text{ for } k = 0, 1, \ldots \right\} \bigcup \left\{ w_{0,2}^{(k)} \mid \text{ for } k = 0, 1, \ldots \right\}.$

Here, the elements of $\mathbb{Q}\{\mathcal{A}\}$ are the polynomial combinations over $\mathbb{Q}$ of the elements in $\mathcal{A}'$ of (14.1). The notation $\mathbb{Q}\{w_{0,1}, w_{0,2}\}$ for $\mathbb{Q}\{\mathcal{A}\}$ was introduced in Section 5.5 on page 52 and was employed throughout Chapter 12.

With reference to (1.14) on page 7 and Corollary 4.15 on page 45, we define

(14.2) $\quad G_{m,2} \equiv \binom{m+1}{3} a_{m,2} \equiv w_{0,2} - \frac{m-1}{2} w_{0,1}^{(1)} - \frac{m-1}{2m}(w_{0,1})^2$

as the semi-invariant of the first kind that is also a basic polynomial of index $(0, 2)$.

THEOREM 14.3. *Each relative invariant for the equations* (1.1) *of order $m \geq 2$ is expressible as a differential-polynomial combination over $\mathbb{Q}$ of $G_{m,2}$ and the basic relative invariants for such equations.*

PROOF. Let $\boldsymbol{P}$ be a relative invariant for equations (1.1) of order $m \geq 2$. Then, $\boldsymbol{P}$ is expressible as a differential-polynomial combination over $\mathbb{Q}$ of the variables from (1.6) given by

(14.3) $\qquad \boldsymbol{w}_{i,j}, \quad$ for $1 \leq j \leq m$ and $0 \leq i \leq j$.

We order these variables according to the scheme for (4.1) on page 35 where $\boldsymbol{w}_{i,j}$ precedes $\boldsymbol{w}_{p,q}$ if and only if either $j < q$ or $j = q$ and $i < p$. We use (14.2) as well as the general structure of the basic relative invariants to verify that

(14.4) $\qquad \boldsymbol{w}_{0,2} \equiv \boldsymbol{G}_{m,2} + \dfrac{m-1}{2} \boldsymbol{w}_{0,1}^{(1)} + \dfrac{m-1}{2m}(\boldsymbol{w}_{0,1})^2 \equiv \boldsymbol{G}_{m,2} + \boldsymbol{R}_{m,0,2,0}$

and

(14.5) $\qquad \boldsymbol{w}_{i,j} \equiv \boldsymbol{\mathcal{I}}_{m,i,j} + \boldsymbol{R}_{m,i,j,0}, \quad$ for $1 \leq i \leq j \leq m$ or $i = 0 < 3 \leq j \leq m$,

where each $\boldsymbol{R}_{m,i,j,0}$ is a differential-polynomial combination over $\mathbb{Q}$ of the variables that precede $\boldsymbol{w}_{i,j}$ in the ordering for (14.3). By repeatedly applying the derivation $'$ of $\mathcal{S}_m$ to each of (14.4)–(14.5), we find that: for $k \geq 0$,

(14.6) $\qquad \boldsymbol{w}_{0,2}^{(k)} \equiv \boldsymbol{G}_{m,2}^{(k)} + \boldsymbol{R}_{m,0,2,k}$

and

(14.7) $\qquad \boldsymbol{w}_{i,j}^{(k)} \equiv \boldsymbol{\mathcal{I}}_{m,i,j}^{(k)} + \boldsymbol{R}_{m,i,j,k}, \quad$ when $1 \leq i \leq j \leq m$ or $i = 0 < 3 \leq j \leq m$,

where each $\boldsymbol{R}_{m,i,j,k}$ is a differential-polynomial combination over $\mathbb{Q}$ of the variables that precede $\boldsymbol{w}_{i,j}$ in the ordering for (14.3).

Suppose that $m = 2$. Then, the ordering for (14.3) is given by

$$\boldsymbol{w}_{0,1},\ \boldsymbol{w}_{1,1},\ \boldsymbol{w}_{0,2},\ \boldsymbol{w}_{1,2},\ \boldsymbol{w}_{2,2}$$

and $\boldsymbol{P}$ is a differential-polynomial combination over $\mathbb{Q}$ of these variables. For $k \geq 0$, we replace each $\boldsymbol{w}_{2,2}^{(k)}$ in $\boldsymbol{P}$ with $\boldsymbol{\mathcal{I}}_{2,2,2}^{(k)} + \boldsymbol{R}_{2,2,2,k}$ from (14.7) to express $\boldsymbol{P}$ as a differential-polynomial combination over $\mathbb{Q}$ of

$$\boldsymbol{w}_{0,1},\ \boldsymbol{w}_{1,1},\ \boldsymbol{w}_{0,2},\ \boldsymbol{w}_{1,2},\ \boldsymbol{\mathcal{I}}_{2,2,2}.$$

Then, for $k \geq 0$, we replace each $\boldsymbol{w}_{1,2}^{(k)}$ in $\boldsymbol{P}$ with $\boldsymbol{\mathcal{I}}_{2,1,2}^{(k)} + \boldsymbol{R}_{2,1,2,k}$ from (14.7) to express $\boldsymbol{P}$ as a differential-polynomial combination over $\mathbb{Q}$ of

$$\boldsymbol{w}_{0,1},\ \boldsymbol{w}_{1,1},\ \boldsymbol{w}_{0,2},\ \boldsymbol{\mathcal{I}}_{2,1,2},\ \boldsymbol{\mathcal{I}}_{2,2,2}.$$

Then, for $k \geq 0$, we replace each $\boldsymbol{w}_{0,2}^{(k)}$ in $\boldsymbol{P}$ with $\boldsymbol{G}_{2,2}^{(k)} + \boldsymbol{R}_{2,0,2,k}$ from (14.6) to express $\boldsymbol{P}$ as a differential-polynomial combination over $\mathbb{Q}$ of

$$\boldsymbol{w}_{0,1},\ \boldsymbol{w}_{1,1},\ \boldsymbol{G}_{2,2},\ \boldsymbol{\mathcal{I}}_{2,1,2},\ \boldsymbol{\mathcal{I}}_{2,2,2}.$$

Finally, for $k \geq 0$, we replace each $\boldsymbol{w}_{1,1}^{(k)}$ in $\boldsymbol{P}$ with $\boldsymbol{\mathcal{I}}_{2,1,1}^{(k)} + \boldsymbol{R}_{2,1,1,k}$ from (14.7) to express $\boldsymbol{P}$ as a differential-polynomial combination over $\mathbb{Q}$ of

$$\boldsymbol{w}_{0,1},\ \boldsymbol{\mathcal{I}}_{2,1,1},\ \boldsymbol{G}_{2,2},\ \boldsymbol{\mathcal{I}}_{2,1,2},\ \boldsymbol{\mathcal{I}}_{2,2,2}.$$

Hence, we can represent $\boldsymbol{P}$ in the form

(14.8) $\qquad \boldsymbol{P} \equiv \boldsymbol{P}_0 + \boldsymbol{Z}_0,$

where: $\boldsymbol{P}_0$ is a differential-polynomial combination over $\mathbb{Q}$ of $\boldsymbol{G}_{2,2}$ and the basic relative invariants $\boldsymbol{\mathcal{I}}_{2,1,1},\ \boldsymbol{\mathcal{I}}_{2,1,2},\ \boldsymbol{\mathcal{I}}_{2,2,2}$ for (1.1) when $m = 2$; and where $\boldsymbol{Z}_0$ is a differential-polynomial combination over $\mathbb{Q}$ of $\boldsymbol{G}_{2,2},\ \boldsymbol{\mathcal{I}}_{2,1,1},\ \boldsymbol{\mathcal{I}}_{2,1,2},\ \boldsymbol{\mathcal{I}}_{2,2,2}$, and $\boldsymbol{w}_{0,1}$

such that each term of $\boldsymbol{Z}_0$ has at least one factor of the form $\boldsymbol{w}_{0,1}^{(r)}$, for some $r \geq 0$. Next, we shall show that $\boldsymbol{P}$ has a similar representation when $m \geq 3$.

Suppose that $m \geq 3$. Then, the ordering for (14.3) is indicated by

$$\boldsymbol{w}_{0,1},\ \boldsymbol{w}_{1,1},\ \boldsymbol{w}_{0,2},\ \boldsymbol{w}_{1,2},\ \boldsymbol{w}_{2,2},\ \boldsymbol{w}_{0,3},\ \boldsymbol{w}_{1,3},\ \ldots,\ \boldsymbol{w}_{m-2,m},\ \boldsymbol{w}_{m-1,m},\ \boldsymbol{w}_{m,m}.$$

and $\boldsymbol{P}$ is a differential-polynomial combination over $\mathbb{Q}$ of these variables. For $k \geq 0$, we replace each $\boldsymbol{w}_{m,m}^{(k)}$ in $\boldsymbol{P}$ with $\boldsymbol{\mathcal{I}}_{m,m,m}^{(k)} + \boldsymbol{R}_{m,m,m,k}$ from (14.7) to express $\boldsymbol{P}$ as a differential-polynomial combination over $\mathbb{Q}$ of

$$\boldsymbol{w}_{0,1},\ \boldsymbol{w}_{1,1},\ \boldsymbol{w}_{0,2},\ \boldsymbol{w}_{1,2},\ \boldsymbol{w}_{2,2},\ \boldsymbol{w}_{0,3},\ \ldots,\ \boldsymbol{w}_{m-2,m},\ \boldsymbol{w}_{m-1,m},\ \boldsymbol{\mathcal{I}}_{m,m,m}.$$

Then, for $k \geq 0$, we replace each $\boldsymbol{w}_{m-1,m}^{(k)}$ in $\boldsymbol{P}$ with $\boldsymbol{\mathcal{I}}_{m,m-1,m}^{(k)} + \boldsymbol{R}_{m,m-1,m,k}$ from (14.7) to express $\boldsymbol{P}$ as a differential-polynomial combination over $\mathbb{Q}$ of

$$\boldsymbol{w}_{0,1},\ \boldsymbol{w}_{1,1},\ \boldsymbol{w}_{0,2},\ \boldsymbol{w}_{1,2},\ \boldsymbol{w}_{2,2},\ \boldsymbol{w}_{0,3},\ \ldots,\ \boldsymbol{w}_{m-2,m},\ \boldsymbol{\mathcal{I}}_{m,m-1,m},\ \boldsymbol{\mathcal{I}}_{m,m,m}.$$

Continuing in this manner, we successively eliminate: $\boldsymbol{w}_{m-2,m}^{(k)}$, for $k \geq 0$; $\ldots$; $\boldsymbol{w}_{2,2}^{(k)}$, for $k \geq 0$; $\boldsymbol{w}_{1,2}^{(k)}$, for $k \geq 0$; $\boldsymbol{w}_{0,2}^{(k)}$, for $k \geq 0$; and $\boldsymbol{w}_{1,1}^{(k)}$, for $k \geq 0$. This yields an expression for $\boldsymbol{P}$ as a differential-polynomial combination over $\mathbb{Q}$ of

$$\boldsymbol{w}_{0,1},\ \boldsymbol{\mathcal{I}}_{m,1,1},\ \boldsymbol{G}_{m,2},\ \boldsymbol{\mathcal{I}}_{m,1,2},\ \boldsymbol{\mathcal{I}}_{m,2,2},\ \boldsymbol{\mathcal{I}}_{m,0,3},\ \ldots,\ \boldsymbol{\mathcal{I}}_{m,m-2,m},\ \boldsymbol{\mathcal{I}}_{m,m-1,m},\ \boldsymbol{\mathcal{I}}_{m,m,m}.$$

Consequently, we can represent $\boldsymbol{P}$ in the form

$$(14.9) \qquad \boldsymbol{P} \equiv \boldsymbol{P}_0 + \boldsymbol{Z}_0,$$

where, for $m \geq 3$, $\boldsymbol{P}_0$ is a differential-polynomial combination over $\mathbb{Q}$ of $\boldsymbol{G}_{m,2}$ and the basic relative invariants for the equations (1.1) while $\boldsymbol{Z}_0$ is a differential-polynomial combination over $\mathbb{Q}$ of $\boldsymbol{w}_{0,1}$, $\boldsymbol{G}_{m,2}$, and the basic relative invariants for the equations (1.1) such that any nonzero term of $\boldsymbol{Z}_0$ has at least one factor of the form $\boldsymbol{w}_{0,1}^{(r)}$, for some $r \geq 0$.

In either (14.8) or (14.9), $\boldsymbol{P}_0$ is a differential-polynomial combination over $\mathbb{Q}$ of semi-invariants of the first kind. Therefore, in view of Proposition 4.7 on page 39, we see that $\boldsymbol{P}_0$ is either a constant (namely 0) or a semi-invariant of the first kind. For any given (1.1) on a region Ω, let $\rho_0(z)$ be a meromorphic function on a subregion $\mathcal{U}$ of Ω such that $\rho_0(z) \not\equiv 0$ and

$$(14.10) \qquad \rho_0^{(1)}(z) + \frac{c_{0,1}(z)}{m}\rho_0(z) \equiv 0, \quad \text{on } \mathcal{U}.$$

Due to (14.10), an application of (2.21) on page 27 shows that the substitution $y(z) = \rho_0(z)\,v(z)$ transforms the restriction of (1.1) to $\mathcal{U}$ into a corresponding equation (1.3) on $\mathcal{U}$ having $c_{0,1}^*(z) \equiv 0$. Hence, the substitution of $c_{i,j}^{*(k)}(z)$ from that (1.3) for $\boldsymbol{w}_{i,j}^{(k)}$ in $\boldsymbol{Z}_0$ yields $Z_0^*(z) \equiv 0$, on $\mathcal{U}$. We combine this with properties of $\boldsymbol{P}$ and $\boldsymbol{P}_0$ to obtain

$$P(z) \equiv P^*(z) \equiv P_0^*(z) + Z_0^*(z) \equiv P_0^*(z) \equiv P_0(z), \quad \text{on } \mathcal{U}.$$

Since $P(z)$ and $P_0(z)$ are meromorphic on Ω, we have $P(z) \equiv P_0(z)$, on Ω. Thus, for any given (1.1) on Ω, the substitution of $c_{i,j}^{(k)}(z)$ from (1.1) for $\boldsymbol{w}_{i,j}^{(k)}$ in $\boldsymbol{P} - \boldsymbol{P}_0$ yields $P(z) - P_0(z) \equiv 0$. We use Corollary 4.3 on page 37 to conclude that $\boldsymbol{P} - \boldsymbol{P}_0 \equiv 0$ and $\boldsymbol{P} \equiv \boldsymbol{P}_0$. Thus, for $m \geq 2$, $\boldsymbol{P}$ is expressible as a differential-polynomial combination over $\mathbb{Q}$ of $\boldsymbol{G}_{m,2}$ and the basic relative invariants for (1.1). This completes the proof. $\square$

COROLLARY 14.4. *Any relative invariant for the equations* (1.1) *of order* $m \geq 2$ *is expressible as a differential-polynomial combination over* $\mathbb{Q}$ *of* $\boldsymbol{a}_{m,2}$ *and the basic relative invariants for such equations.*

PROOF. In view of $\boldsymbol{G}_{m,2} \equiv \binom{m+1}{3} \boldsymbol{a}_{m,2}$, each differential-polynomial combination over $\mathbb{Q}$ of $\boldsymbol{G}_{m,2}$ and the basic relative invariants is a differential-polynomial combination over $\mathbb{Q}$ of $\boldsymbol{a}_{m,2}$ and the basic relative invariants; and vice versa. We invoke Theorem 14.3 to complete the proof. □

REMARK 14.5. The context for (1.83)–(1.86) of Section 1.7 shows how $\boldsymbol{G}_{m,2}$ and $\boldsymbol{a}_{m,2}$ can be interpreted as semi-invariants of the first kind for homogeneous linear differential equations of order $m \geq 2$. Moreover, the relative invariants for homogeneous linear differential equations of order $m \geq 3$ are given by those relative invariants for the equations (1.1) of order $m \geq 3$ that belong to the subring $\mathcal{R}_m$ of $\mathcal{S}_m$ consisting of the polynomials over $\mathbb{Q}$ in the variables $w_j^{(k)} = w_{0,j}^{(k)}$, for $1 \leq j \leq m$ and $k \geq 0$. In particular, the basic relative invariants for homogeneous linear differential equations of order $m \geq 3$ are given by $\mathcal{I}_{m,0,n}$, for $3 \leq n \leq m$. Consequently, for the context of Section 1.7, Theorem 14.3 shows that: *each relative invariant for homogeneous linear differential equations of order* $m \geq 3$ *is expressible as a differential-polynomial combination over* $\mathbb{Q}$ *of* $\boldsymbol{G}_{m,2}$, *or* $\boldsymbol{a}_{m,2}$, *and the basic relative invariants for such equations.*

The preceding result was stated without proof in [**24**, Section E.3, page 195]. For the context used there, a proof can be given by a natural modification of the argument for Theorem 14.3.

14.2. Combinations of invariants that yield other invariants

The product $\boldsymbol{P}_1 \boldsymbol{P}_2$ of two relative invariants $\boldsymbol{P}_1$ and $\boldsymbol{P}_2$ of respective weights n_1 and n_2 is a relative invariant of weight $n_1 + n_2$. A nonzero scalar multiple from $\mathbb{Q}$ of a relative invariant is a relative invariant. And, a nonzero sum of two relative invariants having the same weight is a relative invariant. However, Proposition 4.7 on page 39 shows that a direct application of the derivation ′ for $\mathcal{S}_m$ to a relative invariant does not yield a relative invariant. Various procedures to involve ′ in the construction of relative invariants are presented in this section. Proposition 14.6 and Corollary 14.9 are generalizations to equations having the form (1.1) of results known to G.-H. Halphen and A. R. Forsyth. The other results are new. All of them will be employed in Section 14.3.

For homogeneous linear differential equations of order m, G.-H. Halphen and A. R. Forsyth used expressions like (14.11) to construct relative invariants from given ones. In [**33**, pages 339–340] or [**39**, page 476], G.-H. Halphen restricted his attention to $m = 4$ and $p = 3$. In [**31**, page 409, Formula (vii.)], A. R. Forsyth presented the general situation for any $m \geq 3$. As we shall see next, similar arguments are applicable to the equations (1.1).

PROPOSITION 14.6 (G.-H. Halphen and A. R. Forsyth). *For the equations* (1.1), *suppose that* $\boldsymbol{P}_p$ *and* $\boldsymbol{Q}_q$ *in* $\mathcal{S}_m$ *are relative invariants of respective weights* p *and* q *such that the polynomial* $\boldsymbol{R}$ *defined in* $\mathcal{S}_m$ *by*

(14.11) $$\boldsymbol{R} \equiv \boldsymbol{P}_p \boldsymbol{Q}_q^{(1)} - \frac{q}{p} \boldsymbol{P}_p^{(1)} \boldsymbol{Q}_q$$

satisfies $\boldsymbol{R} \not\equiv 0$. *Then,* $\boldsymbol{R}$ *is a relative invariant of weight* $p + q + 1$.

PROOF. For any transformation (1.4) of (1.1) on Ω into (1.5) on Ω^{**}, we have
$$P_p^{**}(\zeta) \equiv (f'(\zeta))^p P_p(f(\zeta)) \tag{14.12}$$
and
$$P_p^{**(1)}(\zeta) \equiv (f'(\zeta))^{p+1} P_p^{(1)}(f(\zeta)) + p(f'(\zeta))^{p-1} f''(\zeta) P_p(f(\zeta)) \tag{14.13}$$
as well as
$$Q_q^{**}(\zeta) \equiv (f'(\zeta))^q Q_q(f(\zeta)) \tag{14.14}$$
and
$$Q_q^{**(1)}(\zeta) \equiv (f'(\zeta))^{q+1} Q_q^{(1)}(f(\zeta)) + q(f'(\zeta))^{q-1} f''(\zeta) Q_q(f(\zeta)). \tag{14.15}$$
We use (14.11)–(14.15) to deduce that
$$\begin{aligned}R^{**}(\zeta) &\equiv P_p^{**}(\zeta) Q_q^{**(1)}(\zeta) - \frac{q}{p} P_p^{**(1)}(\zeta) Q_q^{**}(\zeta) \\ &\equiv (f'(\zeta))^{p+q+1} \left[P_p(f(\zeta)) Q_q^{(1)}(f(\zeta)) - \frac{q}{p} P_p^{(1)}(f(\zeta)) Q_q(f(\zeta)) \right] \\ &\equiv (f'(\zeta))^{p+q+1} R(f(\zeta)).\end{aligned} \tag{14.16}$$

Since $P_p Q_q^{(1)}$ and $P_p^{(1)} Q_q$ in (14.11) are isobaric polynomials of the same positive weight $p+q+1$, R is not a nonzero rational number. In view of (14.16), the nonconstant R is a semi-invariant of the second kind of weight $p+q+1$. Due to Proposition 4.7 on page 39, each of P_p, $P_p^{(1)}$, Q_q, $Q_q^{(1)}$, and R is a semi-invariant of the first kind. This shows that R is a relative invariant and completes the proof. □

EXAMPLE 14.7. The first terms in the right members of (1.65), (1.81), and (1.82) are relative invariants because they are nonzero isobaric polynomial combinations over $\mathbb{Q}$ of relative invariants. Proposition 14.6 shows that the second terms in the right members of (1.65), (1.81), and (1.82) are relative invariants. That the third terms in the right members of (1.65), (1.81), and (1.82) are relative invariants is a consequence of the following new result.

THEOREM 14.8. *Suppose that P_p and Q_q in S_m are any two relative invariants of respective weights p and q for equations of the form (1.1). In terms of rational numbers α, β, and γ, let S be defined in S_m by*
$$S \equiv P_p Q_q^{(2)} + \alpha P_p^{(1)} Q_q^{(1)} + \beta P_p^{(2)} Q_q + \gamma a_{m,2} P_p Q_q. \tag{14.17}$$
Then, for S to be a relative invariant of the equations (1.1) having weight $p+q+2$, it is necessary and sufficient that
$$\alpha = \frac{-(2q+1)}{p}, \quad \beta = \frac{q(2q+1)}{p(2p+1)}, \quad \text{and} \quad \gamma = \frac{-4q(p+q+1)}{2p+1}. \tag{14.18}$$

PROOF. For a transformation (1.4) of (1.1) on Ω into (1.5) on Ω^{**}, Lemma 4.13 of page 43 gives
$$a_{m,2}^{**}(\zeta) \equiv (f'(\zeta))^2 a_{m,2}(f(\zeta)) + \frac{f'''(\zeta)}{2f'(\zeta)} - \frac{3}{4}\left[\frac{f''(\zeta)}{f'(\zeta)}\right]^2, \quad \text{on } \Omega^{**}. \tag{14.19}$$
Since P_p is a relative invariant of weight p, we have
$$P_p^{**}(\zeta) \equiv (f'(\zeta))^p P_p(f(\zeta)), \quad \text{for } \zeta \text{ in } \Omega^{**}. \tag{14.20}$$

For ζ in Ω^{**}, we differentiate (14.20) with respect to ζ to deduce

$$(14.21) \quad P_p^{**(1)}(\zeta) \equiv \left(f'(\zeta)\right)^{p+1} P_p^{(1)}\left(f(\zeta)\right) + p\left(f'(\zeta)\right)^{p-1} f''(\zeta)\, P_p\left(f(\zeta)\right)$$

and

$$(14.22) \quad P_p^{**(2)}(\zeta) \equiv \left(f'(\zeta)\right)^{p+2} P_p^{(2)}\left(f(\zeta)\right) + (2p+1)\left(f'(\zeta)\right)^p f''(\zeta)\, P_p^{(1)}\left(f(\zeta)\right)$$
$$+ \left[p\left(f'(\zeta)\right)^{p-1} f'''(\zeta) + p(p-1)\left(f'(\zeta)\right)^{p-2}\left(f''(\zeta)\right)^2\right] P_p\left(f(\zeta)\right).$$

Moreover, there are formulas similar to (14.20), (14.21), and (14.22) for $Q_q^{**}(\zeta)$, $Q_q^{**(1)}(\zeta)$, and $Q_q^{**(2)}(\zeta)$. We use them with (14.17) and (14.19)–(14.22) to obtain

$$(14.23) \quad S^{**}(\zeta) - \left(f'(\zeta)\right)^{p+q+2} S\left(f(\zeta)\right)$$
$$\equiv \left(f'(\zeta)\right)^{p+q} \left[A_1 \frac{f'''(\zeta)}{f'(\zeta)} + A_2 \left[\frac{f''(\zeta)}{f'(\zeta)}\right]^2\right] P_p\left(f(\zeta)\right) Q_q\left(f(\zeta)\right)$$
$$+ \left(f'(\zeta)\right)^{p+q} f''(\zeta) \left[A_3 P_p^{(1)}\left(f(\zeta)\right) Q_q\left(f(\zeta)\right) + A_4 P_p\left(f(\zeta)\right) Q_q^{(1)}\left(f(\zeta)\right)\right],$$

where

$$A_1 \equiv p\beta + (1/2)\gamma + q, \qquad A_2 \equiv pq\alpha + p(p-1)\beta - (3/4)\gamma + q(q-1),$$
$$A_3 \equiv q\alpha + (2p+1)\beta, \qquad A_4 \equiv p\alpha + (2q+1).$$

Thus, the condition $S^{**}(\zeta) \equiv \left(f'(\zeta)\right)^{p+q+2} S\left(f(\zeta)\right)$ is satisfied if and only if α, β, and γ can be selected so that $A_1 \equiv A_2 \equiv A_3 \equiv A_4 \equiv 0$. This condition specifies a system of four linear equations in α, β, γ that has the unique solution given by (14.18). In particular, (14.18) yields $\gamma \neq 0$. Each nonzero monomial of the nonzero polynomial $\gamma\, a_{m,2}\, P_p Q_q$ is expressible in the form

$$(14.24) \quad M \equiv \Gamma \prod_{\lambda=1}^{v} w_{i_\lambda, j_\lambda}^{(k_\lambda)},$$

where Γ is a nonzero rational number and, for $1 \leq \lambda \leq v$, the variables $w_{i_\lambda,j_\lambda}^{(k_\lambda)}$ satisfy $k_\lambda \geq 0$, $1 \leq j_\lambda \leq m$, and $0 \leq i_\lambda \leq j_\lambda$. We apply (14.17) to see that the M for which $(k_1 + k_2 + \cdots + k_v)$ is minimal are also monomials of S. Thus, with the selection (14.18), S is a nonconstant polynomial in S_m; and therefore S is a semi-invariant of the second kind having weight $p+q+2$.

For S defined by (14.17) and (14.18), we use Corollary 4.15 on page 45 and Proposition 4.7 on page 39 to see that S is a semi-invariant of the first kind. Thus, S is a relative invariant of weight $p+q+2$. This completes the proof. $\square$

COROLLARY 14.9 (G.-H. Halphen and A. R. Forsyth). *Suppose that P_p in S_m is a relative invariant of weight p for the equations (1.1). Then, the polynomial T defined in S_m by*

$$(14.25) \quad T \equiv P_p P_p^{(2)} - \frac{2p+1}{2p}\left(P_p^{(1)}\right)^2 - 2p\, a_{m,2}\, (P_p)^2$$

is a relative invariant of weight $2p+2$ for the equations (1.1).

PROOF. For the special case of S in (14.17) where $q = p$ and $Q_q \equiv P_p$, we see that T in (14.25) is given by $T \equiv \frac{1}{2} S$. Thus, Theorem 14.8 shows that T is a relative invariant of weight $2p+2$. This completes the proof. $\square$

14.2. COMBINATIONS OF INVARIANTS THAT YIELD OTHER INVARIANTS

OBSERVATION. For homogeneous linear differential equations of order $m = 4$, G.-H. Halphen considered a result analogous to the special case of (14.25) having $p = 3$ in [**33**, page 335, (11)] or [**39**, page 472].

Let $\boldsymbol{P}$ be a relative invariant in the variables of (1.83) over $\mathbb{Q}$ for the monic homogeneous linear differential equations (5.21) of order $m \geq 3$; and, let (5.26) be a Laguerre-Forsyth canonical form for a particular (5.21). A. R. Forsyth presented in [**31**, page 408, Formula (vi.)] an expression like

$$T^\sharp(\zeta) \equiv P_p^{\sharp(0)}(\zeta)\, P_p^{\sharp(2)}(\zeta) - \frac{2p+1}{2p}\left(P_p^{\sharp(1)}(\zeta)\right)^2$$

as the evaluated form of $\boldsymbol{T}$ in (14.25) at (5.26). Here, for $0 \leq r \leq 2$, we let $P_p^{\sharp(r)}(\zeta)$ designate the function obtained by substituting $d_j^{(k)}(\zeta)$ from (5.26) for $w_j^{(k)}$ in $\boldsymbol{P}_p^{(r)}$. Due to $d_1(\zeta) \equiv d_2(\zeta) \equiv 0$ for (5.26), such a substitution in $\boldsymbol{a}_{m,2}$ of (1.14) on page 7 yields $a_{m,2}^\sharp(\zeta) \equiv 0$.

EXAMPLE 14.10. For $m \geq 4$, $p = 3$, and $q = 4$, we apply Theorem 14.8 as well as our Main Theorem on page 8 to verify that the polynomial

(14.26) $$\boldsymbol{U} \equiv \boldsymbol{\mathcal{I}}_{m,0,3}\, \boldsymbol{\mathcal{I}}_{m,0,4}^{(2)} - 3 \boldsymbol{\mathcal{I}}_{m,0,3}^{(1)}\, \boldsymbol{\mathcal{I}}_{m,0,4}^{(1)} + \frac{12}{7} \boldsymbol{\mathcal{I}}_{m,0,3}^{(2)}\, \boldsymbol{\mathcal{I}}_{m,0,4}$$
$$- \frac{128}{7}\, \boldsymbol{a}_{m,2}\, \boldsymbol{\mathcal{I}}_{m,0,3}\, \boldsymbol{\mathcal{I}}_{m,0,4}$$

is a relative invariant of weight 9 for the equations (1.1). Moreover, with the context for Section 1.7 on pages 17–18, we see that $\boldsymbol{U}$ specifies a relative invariant of weight 9 for homogeneous linear differential equations of order $m \geq 4$.

OBSERVATION 14.11. A nonzero linear combination over $\mathbb{Q}$ of relative invariants having the same weight is a relative invariant; and the product of two relative invariants is a relative invariant. Proposition 14.6 and Corollary 14.9 yield additional relative invariants. For homogeneous linear differential equations, A. R. Forsyth asked in [**31**, page 418] whether any relative invariant could be obtained by starting with various basic relative invariants and using the preceding operations repeatedly. In [**24**, Section E.4, pages 195–196], we asked whether such could merely be done for those relative invariants of total degree 2 over $\mathbb{Q}\{\boldsymbol{w}_{0,1}, \boldsymbol{w}_{0,2}\}$ in the variables $\boldsymbol{w}_{0,j}^{(k)}$ having $j \geq 3$ and $k \geq 0$. Example 14.10 shows that "No" is the answer to both questions. Namely, the total degree of $\boldsymbol{U}$ over $\mathbb{Q}\{\boldsymbol{w}_{0,1}, \boldsymbol{w}_{0,2}\}$ in the variables $\boldsymbol{w}_{0,j}^{(k)}$, for $j \geq 3$ and $k \geq 0$, is 2. Yet, $\boldsymbol{U}$ can not be deduced from the basic relative invariants by using only those operations considered by Forsyth.

Propositions 14.13–14.14 and Formulations 14.15–14.16 provide new techniques for constructing relative invariants from given ones. The arguments needed for Proposition 14.13 and Formulation 14.15 use the case $n = 2$ of the following result.

LEMMA 14.12. *Polynomials $\boldsymbol{P_1}, \boldsymbol{P_2}, \ldots, \boldsymbol{P_n}$ in $\mathcal{S}_m$ are linearly independent over $\mathbb{Q}$ if and only if their Wronskian W satisfies $W \not\equiv 0$.*

PROOF. The derivation $'$ for $\mathcal{S}_m$ has a unique extension to a derivation $'$ for the quotient field $\mathcal{Q}_m$ of $\mathcal{S}_m$ and the constants in $\mathcal{Q}_m$ are the elements of $\mathbb{Q}$; e.g., see page 219. In particular, $\boldsymbol{P_1}, \boldsymbol{P_2}, \ldots, \boldsymbol{P_n}$ are elements of the ordinary differential field $\mathcal{Q}_m$. It was established in [**47**, pages 34–35] that n elements in an ordinary differential field are linearly independent over the subfield of constant elements if and only if their Wronskian is nonzero. This yields the desired conclusion. □

PROPOSITION 14.13. *Suppose that P_p and Q_q are relative invariants of respective weights p and q for the equations (1.1) such that P_p and Q_q are linearly independent over $\mathbb{Q}$. Let $x_1, x_2, x_3, x_4, x_5, x_6$ denote rational numbers. Then, the polynomial X defined in S_m by*

(14.27) $$X \equiv P_p\,Q_q^{(3)} + x_1\,P_p^{(1)}\,Q_q^{(2)} + x_2\,P_p^{(2)}\,Q_q^{(1)} + x_3\,P_p^{(3)}\,Q_q$$
$$+ x_4\,a_{m,2}\,P_p\,Q_q^{(1)} + x_5\,a_{m,2}\,P_p^{(1)}\,Q_q + x_6\,a_{m,2}^{(1)}\,P_p\,Q_q$$

is a relative invariant of weight $p+q+3$ for the equations (1.1) if and only if

(14.28) $\quad x_1 = \dfrac{-3(q+1)}{p}, \qquad x_4 = \dfrac{-4(3pq + 3q^2 + p + 6q + 2)}{2p+1},$

(14.29) $\quad x_2 = \dfrac{3(q+1)(2q+1)}{p(2p+1)}, \qquad x_5 = \dfrac{4q(q+1)(3p^2 + 3pq + 6p + q + 2)}{p(p+1)(2p+1)},$

(14.30) $\quad x_3 = \dfrac{-q(q+1)(2q+1)}{p(p+1)(2p+1)}, \qquad x_6 = \dfrac{2q(q-p)(2p+2q+3)}{(p+1)(2p+1)}.$

PROOF. We can use computer algebra throughout. Starting with

$$P_p^{**}(\zeta) \equiv (f'(\zeta))^p P_p(f(\zeta)) \quad \text{and} \quad Q_q^{**}(\zeta) \equiv (f'(\zeta))^q Q_q(f(\zeta)),$$

we obtain $P_p^{**(k)}(\zeta)$ and $Q_q^{**(k)}(\zeta)$, for $0 \le k \le 3$. We then substitute these along with $a_{m,2}^{**}(\zeta)$, $a_{m,2}^{**(1)}(\zeta)$ from (14.19) and $z = f(\zeta)$ into the right member of

(14.31) $\quad X^{**}(\zeta) - (f'(\zeta))^{p+q+3} X(f(\zeta))$
$$\equiv P_p^{**}(\zeta)\,Q_q^{**(3)}(\zeta) + \cdots + x_6\,a_{m,2}^{**(1)}(\zeta)\,P_p^{**}(\zeta)\,Q_q^{**}(\zeta)$$
$$- (f'(\zeta))^{p+q+3} \left[P_p(z)\,Q_q^{(3)}(z) + \cdots + x_6\,a_{m,2}^{(1)}(z)\,P_p(z)\,Q_q(z)\right]$$

and expand the result to write it as a linear combination over $\mathbb{Q}$ of eleven distinct expressions having the form

$$P^{(n_1)}(f(\zeta))\,Q^{(n_2)}(f(\zeta))\,a_{m,2}^{(n_3)}(f(\zeta))\,\prod_{j=1}^{4}\left(f^{(j)}(\zeta)\right)^{k_j},$$

for particular values of nonnegative integers n_1, n_2, n_3 and k_1, k_2, k_3, k_4. Then, we equate to zero the eleven coefficients of these eleven expressions and obtain a system of eleven linear equations over $\mathbb{Q}$ in $x_1, x_2, \ldots, x_6$. This system has the unique solution given by (14.28)–(14.30). Thus, (14.31) shows that (14.27) satisfies

(14.32) $$X^{**}(\zeta) \equiv (f'(\zeta))^{p+q+3} X(f(\zeta))$$

if and only if $x_1, x_2, \ldots, x_6$ are given by (14.28)–(14.30).

Suppose that $x_1, x_2, \ldots, x_6$ satisfy (14.28)–(14.30). (i) When $p \ne q$, we use (14.27) so see that the nonzero polynomial $x_6\,a_{m,2}^{(1)}\,P_p\,Q_q$ has a nonzero monomial M of the form (14.24) with $(k_1 + k_2 + \cdots + k_v)$ minimal such that M is a monomial of X. This gives $X \not\equiv 0$. (ii) When $p = q$, (14.28)–(14.30) and Lemma 14.12 yield $x_6\,a_{m,2}^{(1)}\,P_p\,Q_q \equiv 0$ and

$$\left\{ \begin{array}{l} x_4\,a_{m,2}\,P_p\,Q_q^{(1)} \\ + x_5\,a_{m,2}\,P_p^{(1)}\,Q_q \end{array} \right\} \equiv \dfrac{-4(6q^2 + 7q + 2)}{2q+1} a_{m,2}\big(P_p\,Q_q^{(1)} - P_p^{(1)}\,Q_q\big) \not\equiv 0.$$

14.2. COMBINATIONS OF INVARIANTS THAT YIELD OTHER INVARIANTS

We use (14.27) to see that the polynomial $x_4\,\boldsymbol{a}_{m,2}\,\boldsymbol{P}_p\,\boldsymbol{Q}_q^{(1)} + x_5\,\boldsymbol{a}_{m,2}\,\boldsymbol{P}_p^{(1)}\,\boldsymbol{Q}_q$ has a nonzero monomial $\boldsymbol{M}$ of the form (14.24) having minimal $(k_1 + k_2 + \cdots + k_v)$ such that $\boldsymbol{M}$ is a monomial in $\boldsymbol{X}$. This yields $\boldsymbol{X} \not\equiv 0$. In each case, $\boldsymbol{X}$ is a nonconstant polynomial in $\mathcal{S}_m$ and (14.32) shows that $\boldsymbol{X}$ is therefore a semi-invariant of the second kind having weight $p + q + 3$. We apply Corollary 4.15 on page 45 and Proposition 4.7 on page 39 to deduce that $\boldsymbol{X}$ is a semi-invariant of the first kind. Thus, $\boldsymbol{X}$ is a relative invariant of weight $p + q + 3$. This completes the proof. □

Both Proposition 14.13 and the following result will be applied in Section 14.3.

PROPOSITION 14.14. *Suppose that $\boldsymbol{P}_p$ and $\boldsymbol{Q}_q$ in $\mathcal{S}_m$ are relative invariants of respective weights p and q for the equations (1.1). Then, in terms of rational numbers $y_1, y_2, \ldots, y_{11}$, the polynomial $\boldsymbol{Y}$ defined in $\mathcal{S}_m$ by*

$$(14.33)\quad \boldsymbol{Y} \equiv \boldsymbol{P}_p\,\boldsymbol{Q}_q^{(4)} + y_1\,\boldsymbol{P}_p^{(1)}\,\boldsymbol{Q}_q^{(3)} + y_2\,\boldsymbol{P}_p^{(2)}\,\boldsymbol{Q}_q^{(2)} + y_3\,\boldsymbol{P}_p^{(3)}\,\boldsymbol{Q}_q^{(1)} + y_4\,\boldsymbol{P}_p^{(4)}\,\boldsymbol{Q}_q$$
$$+\,\boldsymbol{a}_{m,2}\Big[y_5\,\boldsymbol{P}_p\,\boldsymbol{Q}_q^{(2)} + y_6\,\boldsymbol{P}_p^{(1)}\,\boldsymbol{Q}_q^{(1)} + y_7\,\boldsymbol{P}_p^{(2)}\,\boldsymbol{Q}_q\Big]$$
$$+\,\boldsymbol{a}_{m,2}^{(1)}\Big[y_8\,\boldsymbol{P}_p\,\boldsymbol{Q}_q^{(1)} + y_9\,\boldsymbol{P}_p^{(1)}\,\boldsymbol{Q}_q\Big] + \Big[y_{10}\,\boldsymbol{a}_{m,2}^{(2)} + y_{11}\,(\boldsymbol{a}_{m,2})^2\Big]\boldsymbol{P}_p\,\boldsymbol{Q}_q$$

is a relative invariant of weight $p + q + 4$ for the equations (1.1) if and only if

$$(14.34)\quad y_1 = \frac{-2(2q+3)}{p},$$

$$(14.35)\quad y_2 = \frac{6(q+1)(2q+3)}{p(2p+1)},$$

$$(14.36)\quad y_3 = \frac{-2(q+1)(2q+1)(2q+3)}{p(p+1)(2p+1)},$$

$$(14.37)\quad y_4 = \frac{q(q+1)(2q+1)(2q+3)}{p(p+1)(2p+1)(2p+3)},$$

$$(14.38)\quad y_5 = \frac{-4(6pq + 6q^2 + 4p + 18q + 11)}{2p+1},$$

$$(14.39)\quad y_6 = \frac{8(2q+3)(3p^2q + 3pq^2 + p^2 + 9pq + q^2 + 3p + 3q + 1)}{p(p+1)(2p+1)},$$

$$(14.40)\quad y_7 = \frac{-4q(q+1)(2q+3)(6p^2 + 6pq + 18p + 4q + 11)}{p(p+1)(2p+1)(2p+3)},$$

$$(14.41)\quad y_8 = \frac{-2(8p^2q - 8q^3 + 2p^2 + 12pq - 24q^2 + 3p - 18q - 5)}{(p+1)(2p+1)},$$

$$(14.42)\quad y_9 = \frac{2q(2q+3)(8p^3 - 8pq^2 + 24p^2 - 12pq - 2q^2 + 18p - 3q + 5)}{p(p+1)(2p+1)(2p+3)},$$

$$(14.43)\quad y_{10} = \frac{-2q(p+q+2)(4p^2 - 4pq + 4q^2 + 4p + 4q + 3)}{(p+1)(2p+1)(2p+3)},$$

and

$$(14.44)\quad y_{11} = \frac{48q(q+1)(p+q+2)(p+q+3)}{(2p+1)(2p+3)}.$$

PROOF. We use computer algebra throughout. Starting with
$$P_p^{**}(\zeta) \equiv (f'(\zeta))^p P_p(f(\zeta)) \quad \text{and} \quad Q_q^{**}(\zeta) \equiv (f'(\zeta))^q Q_q(f(\zeta)),$$
we obtain $P_p^{**(k)}(\zeta)$ and $Q_q^{**(k)}(\zeta)$, for $0 \leq k \leq 4$. We substitute these along with $a_{m,2}^{**}(\zeta)$, $a_{m,2}^{**(1)}(\zeta)$, $a_{m,2}^{**(2)}(\zeta)$ from (14.19) and $z = f(\zeta)$ into the right member of

(14.45) $\quad Y^{**}(\zeta) - (f'(\zeta))^{p+q+4} Y(f(\zeta))$
$$\equiv P_p^{**}(\zeta) Q_q^{**(4)}(\zeta) + \cdots + y_{11} (a_{m,2}^{**}(\zeta))^2 P_p^{**}(\zeta) Q_q^{**}(\zeta)$$
$$- (f'(\zeta))^{p+q+4} \left[P_p(z) Q_q^{(4)}(z) + \cdots + y_{11} (a_{m,2}(z))^2 P_p(z) Q_q(z) \right]$$

and expand the result to write it as a linear combination over $\mathbb{Q}$ of twenty-six distinct expressions having the form
$$P^{(n_1)}(f(\zeta)) \, Q^{(n_2)}(f(\zeta)) \, \left(a_{m,2}^{(n_3)}(f(\zeta))\right)^{n_4} \prod_{j=1}^{5} (f^{(j)}(\zeta))^{k_j},$$
for particular nonnegative integers n_1, n_2, n_3, n_4 and k_1, k_2, k_3, k_4, k_5. Then, we equate to zero the twenty-six coefficients of these twenty-six expressions and obtain a system of twenty-six linear equations over $\mathbb{Q}$ in y_1, y_2, ..., y_{11}. This system has the unique solution of (14.34)–(14.44). Thus, (14.45) shows that (14.33) satisfies

(14.46) $\qquad\qquad Y^{**}(\zeta) \equiv (f'(\zeta))^{p+q+4} Y(f(\zeta))$

if and only if y_1, y_2, ..., y_{11} are given by (14.34)–(14.44).

Suppose that y_1, y_2, ..., y_{11} for (14.33) are given by (14.34)–(14.44). Then, we have $y_{11} \neq 0$ and see that the polynomial $\boldsymbol{T} \equiv y_{11}(\boldsymbol{a}_{m,2})^2$ is a nonzero element of $\mathbb{Q}\{\boldsymbol{w}_{0,1}, \boldsymbol{w}_{0,2}\}$. Each nonzero monomial of the nonzero polynomial $\boldsymbol{T} \boldsymbol{P}_p \boldsymbol{Q}_q$ is expressible in the form described for (14.24). We use (14.33) to see that each such monomial having minimal $(k_1 + k_2 + \cdots + k_v)$ is a nonzero monomial of $\boldsymbol{Y}$ and yields $\boldsymbol{Y} \not\equiv 0$. Thus, $\boldsymbol{Y}$ is a nonconstant polynomial in $\mathcal{S}_m$. In view of (14.46), $\boldsymbol{Y}$ is therefore a semi-invariant of the second kind having weight $p + q + 4$. We apply Corollary 4.15 on page 45 and Proposition 4.7 on page 39 to deduce that $\boldsymbol{Y}$ is a semi-invariant of the first kind. Thus, $\boldsymbol{Y}$ is a relative invariant of weight $p + q + 4$. This completes the proof. $\square$

FORMULATION 14.15. *Suppose that $\boldsymbol{P}_p$ and $\boldsymbol{Q}_q$ are linearly independent relative invariants of respective weights p and q for the equations (1.1). Then, there are unique rational numbers z_1, z_2, ..., z_{18} such that the expression*

(14.47) $\quad \boldsymbol{Z} \equiv \boldsymbol{P}_p \boldsymbol{Q}_q^{(5)} + \sum_{i=1}^{5} z_i \, \boldsymbol{P}_p^{(i)} \boldsymbol{Q}_q^{(5-i)} + \boldsymbol{a}_{m,2} \sum_{i=0}^{3} z_{6+i} \, \boldsymbol{P}_p^{(i)} \boldsymbol{Q}_q^{(3-i)}$
$$+ \boldsymbol{a}_{m,2}^{(1)} \sum_{i=0}^{2} z_{10+i} \boldsymbol{P}_p^{(i)} \boldsymbol{Q}_q^{(2-i)} + \boldsymbol{a}_{m,2}^{(2)} \sum_{i=0}^{1} z_{13+i} \, \boldsymbol{P}_p^{(i)} \boldsymbol{Q}_q^{(1-i)}$$
$$+ (\boldsymbol{a}_{m,2})^2 \sum_{i=0}^{1} z_{15+i} \, \boldsymbol{P}_p^{(i)} \boldsymbol{Q}_q^{(1-i)} + \left[z_{17} \, \boldsymbol{a}_{m,2}^{(3)} + z_{18} \, \boldsymbol{a}_{m,2} \, \boldsymbol{a}_{m,2}^{(1)} \right] \boldsymbol{P}_p \boldsymbol{Q}_q$$

is a relative invariant of weight $p + q + 5$ for the equations (1.1).

Modern personal computers have the capacity to easily verify Formulation 14.15 as well as further results having similar assertions of which we include the following.

FORMULATION 14.16. *Suppose that P_p and Q_q are any two relative invariants of respective weights p and q for the equations* (1.1). *Then, there are unique rational numbers v_1, v_2, ..., v_{29} such that the expression*

$$(14.48) \quad V \equiv P_p Q_q^{(6)} + \sum_{i=1}^{6} v_i P_p^{(i)} Q_q^{(6-i)}$$

$$+ a_{m,2} \sum_{i=0}^{4} v_{7+i} P_p^{(i)} Q_q^{(4-i)} + a_{m,2}^{(1)} \sum_{i=0}^{3} v_{12+i} P_p^{(i)} Q_q^{(3-i)}$$

$$+ a_{m,2}^{(2)} \sum_{i=0}^{2} v_{16+i} P_p^{(i)} Q_q^{(2-i)} + (a_{m,2})^2 \sum_{i=0}^{2} v_{19+i} P_p^{(i)} Q_q^{(2-i)}$$

$$+ a_{m,2}^{(3)} \sum_{i=0}^{1} v_{22+i} P_p^{(i)} Q_q^{(1-i)} + a_{m,2} a_{m,2}^{(1)} \sum_{i=0}^{1} v_{24+i} P_p^{(i)} Q_q^{(1-i)}$$

$$+ \left[v_{26} a_{m,2}^{(4)} + v_{27} a_{m,2} a_{m,2}^{(2)} + v_{28} \left(a_{m,2}^{(1)} \right)^2 + v_{29} (a_{m,2})^3 \right] P_p Q_q$$

is a relative invariant of weight $p+q+6$ for the equations (1.1).

We have suspected the existence of a general result for which (14.11), (14.17), (14.27), (14.33), (14.47), and (14.48) are particular instances. However, our efforts in that direction have not been productive.

14.3. The relative invariants of weight ≤ 9 for the equations $Q_2 = 0$

Theorem 1.8 on page 13 shows that the differential equations

$$(14.49) \quad \left(y^{(2)}(z)\right)^2 + \sum_{\substack{0 \leq i,j \leq 2 \\ (i,j) \neq (0,0)}} c_{i,j}(z)\, y^{(m-i)}(z)\, y^{(m-j)}(z) = 0, \quad \text{where} \quad c_{j,i}(z) \equiv c_{i,j}(z),$$

have three basic relative invariants $\mathcal{I}_{2,1,1}$, $\mathcal{I}_{2,1,2}$, and $\mathcal{I}_{2,2,2}$ given by (1.60), (1.61), and (1.62). For $s = 2, 3, \ldots$, there are linearly independent relative invariants $F_{s,1}, \ldots, F_{s,n(s)}$ over $\mathbb{Q}$ of weight s for the equations (14.49) such that the relative invariants of weight s for the equations (14.49) are given by the linear combinations

$$(14.50) \quad \sum_{i=1}^{n(s)} K_i\, F_{s,i}, \quad \text{with } K_1, \ldots, K_{n(s)} \text{ in } \mathbb{Q} \text{ and some } K_i \neq 0.$$

Here, we present particular selections for various $F_{s,i}$ as differential-polynomial combinations over $\mathbb{Q}$ of $\mathcal{I}_{2,1,1}$, $\mathcal{I}_{2,1,2}$, $\mathcal{I}_{2,2,2}$, and $a_{m,2}$. That the corresponding linear combinations in (14.50) yield all the relative invariants of weight $s \leq 9$ for the equations (14.49) will be established by computations presented in Section 16.4.

There are no relative invariants of weight 1; in this regard, see page 45.
For $s = 2$, we select $F_{2,1} \equiv \mathcal{I}_{2,1,1}$ and have $n(2) = 1$.
For $s = 3$, we select $F_{3,1} \equiv \mathcal{I}_{2,1,2}$ and have $n(3) = 1$.
For $s = 4$, we select $F_{4,1} \equiv (\mathcal{I}_{2,1,1})^2$, $F_{4,2} \equiv \mathcal{I}_{2,2,2}$, and have $n(4) = 2$.
For $s = 5$, we select $F_{5,1} \equiv \mathcal{I}_{2,1,1} \mathcal{I}_{2,1,2}$ and have $n(5) = 1$.

For $s = 6$, we use (14.11) and (14.25) to see that the polynomials

$$F_{6,1} \equiv (\mathcal{I}_{2,1,1})^3,$$
$$F_{6,2} \equiv (\mathcal{I}_{2,1,2})^2,$$
$$F_{6,3} \equiv \mathcal{I}_{2,1,1}\,\mathcal{I}_{2,2,2},$$
$$F_{6,4} \equiv \mathcal{I}_{2,1,1}\,\mathcal{I}_{2,1,2}^{(1)} - \frac{3}{2}\mathcal{I}_{2,1,1}^{(1)}\,\mathcal{I}_{2,1,2},$$
$$F_{6,5} \equiv \mathcal{I}_{2,1,1}\,\mathcal{I}_{2,1,1}^{(2)} - \frac{5}{4}(\mathcal{I}_{2,1,1}^{(1)})^2 - 4a_{2,2}(\mathcal{I}_{2,1,1})^2,$$

are $n(6) = 5$ linearly independent relative invariants for the equations (14.49).

For $s = 7$, we use (14.11) and (14.17) to verify that the polynomials

$$F_{7,1} \equiv (\mathcal{I}_{2,1,1})^2\,\mathcal{I}_{2,1,2},$$
$$F_{7,2} \equiv \mathcal{I}_{2,1,2}\,\mathcal{I}_{2,2,2},$$
$$F_{7,3} \equiv \mathcal{I}_{2,1,1}\,\mathcal{I}_{2,2,2}^{(1)} - 2\mathcal{I}_{2,1,1}^{(1)}\,\mathcal{I}_{2,2,2},$$
$$F_{7,4} \equiv \mathcal{I}_{2,1,1}\,\mathcal{I}_{2,1,2}^{(2)} - \frac{7}{2}\mathcal{I}_{2,1,1}^{(1)}\,\mathcal{I}_{2,1,2}^{(1)} + \frac{21}{10}\mathcal{I}_{2,1,1}^{(2)}\,\mathcal{I}_{2,1,2} - \frac{72}{5}a_{2,2}\,\mathcal{I}_{2,1,1}\,\mathcal{I}_{2,1,2},$$

are $n(7) = 4$ linearly independent relative invariants for the equations (14.49).

For $s = 8$, we use (14.11), (14.25), (14.17), (14.27), and (14.33) to deduce that the polynomials

$$F_{8,1} \equiv (\mathcal{I}_{2,1,1})^4,$$
$$F_{8,2} \equiv \mathcal{I}_{2,1,1}\,(\mathcal{I}_{2,1,2})^2,$$
$$F_{8,3} \equiv (\mathcal{I}_{2,1,1})^2\,\mathcal{I}_{2,2,2},$$
$$F_{8,4} \equiv (\mathcal{I}_{2,2,2})^2,$$
$$F_{8,5} \equiv \mathcal{I}_{2,1,1}\,F_{6,4}$$
$$F_{8,6} \equiv \mathcal{I}_{2,1,1}\,F_{6,5}$$
$$F_{8,7} \equiv \mathcal{I}_{2,1,2}\,\mathcal{I}_{2,2,2}^{(1)} - \frac{4}{3}\mathcal{I}_{2,1,2}^{(1)}\,\mathcal{I}_{2,2,2}$$
$$F_{8,8} \equiv \mathcal{I}_{2,1,2}\,\mathcal{I}_{2,1,2}^{(2)} - \frac{7}{6}(\mathcal{I}_{2,1,2}^{(1)})^2 - 6a_{2,2}\,(\mathcal{I}_{2,1,2})^2,$$
$$F_{8,9} \equiv \mathcal{I}_{2,1,1}\,\mathcal{I}_{2,2,2}^{(2)} - \frac{9}{2}\mathcal{I}_{2,1,1}^{(1)}\,\mathcal{I}_{2,2,2}^{(1)} + \frac{18}{5}\mathcal{I}_{2,1,1}^{(2)}\,\mathcal{I}_{2,2,2} - \frac{112}{5}a_{2,2}\,\mathcal{I}_{2,1,1}\,\mathcal{I}_{2,2,2},$$
$$F_{8,10} \equiv \mathcal{I}_{2,1,1}\,\mathcal{I}_{2,1,2}^{(3)} - 6\mathcal{I}_{2,1,1}^{(1)}\,\mathcal{I}_{2,1,2}^{(2)} + \frac{42}{5}\mathcal{I}_{2,1,1}^{(2)}\,\mathcal{I}_{2,1,2}^{(1)} - \frac{14}{5}\mathcal{I}_{2,1,1}^{(3)}\,\mathcal{I}_{2,1,2}$$
$$- \frac{268}{5}a_{2,2}\,\mathcal{I}_{2,1,1}\,\mathcal{I}_{2,1,2}^{(1)} + \frac{376}{5}a_{2,2}\,\mathcal{I}_{2,1,1}^{(1)}\,\mathcal{I}_{2,1,2} + \frac{26}{5}a_{2,2}^{(1)}\,\mathcal{I}_{2,1,1}\,\mathcal{I}_{2,1,2},$$
$$F_{8,11} \equiv \mathcal{I}_{2,1,1}\,\mathcal{I}_{2,1,1}^{(4)} - 7\mathcal{I}_{2,1,1}^{(1)}\,\mathcal{I}_{2,1,1}^{(3)} + \frac{63}{10}(\mathcal{I}_{2,1,1}^{(2)})^2$$
$$- \frac{412}{5}a_{2,2}\,\mathcal{I}_{2,1,1}\,\mathcal{I}_{2,1,1}^{(2)} + 98a_{2,2}\,(\mathcal{I}_{2,1,1}^{(1)})^2 + 10a_{2,2}^{(1)}\,\mathcal{I}_{2,1,1}\,\mathcal{I}_{2,1,1}^{(1)}$$
$$- 4a_{2,2}^{(2)}\,(\mathcal{I}_{2,1,1})^2 + \frac{864}{5}(a_{2,2})^2(\mathcal{I}_{2,1,1})^2,$$

are $n(8) = 11$ linearly independent relative invariants for the equations (14.49).

14.3. THE RELATIVE INVARIANTS OF WEIGHT ≤ 9 FOR THE EQUATIONS $Q_2 = 0$

For $s = 9$, we use (14.17), (14.27), and (14.33) to check that the polynomials

$\boldsymbol{F}_{9,1} \equiv (\boldsymbol{\mathcal{I}}_{2,1,1})^3 \boldsymbol{\mathcal{I}}_{2,1,2},$

$\boldsymbol{F}_{9,2} \equiv (\boldsymbol{\mathcal{I}}_{2,1,2})^3,$

$\boldsymbol{F}_{9,3} \equiv \boldsymbol{\mathcal{I}}_{2,1,1} \boldsymbol{\mathcal{I}}_{2,1,2} \boldsymbol{\mathcal{I}}_{2,2,2},$

$\boldsymbol{F}_{9,4} \equiv \boldsymbol{\mathcal{I}}_{2,1,2} \boldsymbol{F}_{6,4},$

$\boldsymbol{F}_{9,5} \equiv \boldsymbol{\mathcal{I}}_{2,1,2} \boldsymbol{F}_{6,5},$

$\boldsymbol{F}_{9,6} \equiv \boldsymbol{\mathcal{I}}_{2,1,1} \boldsymbol{F}_{7,3},$

$\boldsymbol{F}_{9,7} \equiv \boldsymbol{\mathcal{I}}_{2,1,1} \boldsymbol{F}_{7,4},$

$\boldsymbol{F}_{9,8} \equiv \boldsymbol{\mathcal{I}}_{2,1,2} \boldsymbol{\mathcal{I}}^{(2)}_{2,2,2} - 3\boldsymbol{\mathcal{I}}^{(1)}_{2,1,2} \boldsymbol{\mathcal{I}}^{(1)}_{2,2,2} + \frac{12}{7} \boldsymbol{\mathcal{I}}^{(2)}_{2,1,2} \boldsymbol{\mathcal{I}}_{2,2,2} - \frac{128}{7} a_{2,2} \boldsymbol{\mathcal{I}}_{2,1,2} \boldsymbol{\mathcal{I}}_{2,2,2},$

$\boldsymbol{F}_{9,9} \equiv (\boldsymbol{\mathcal{I}}_{2,1,1})^2 \boldsymbol{\mathcal{I}}^{(3)}_{2,1,1} - \frac{9}{2} \boldsymbol{\mathcal{I}}_{2,1,1} \boldsymbol{\mathcal{I}}^{(1)}_{2,1,1} \boldsymbol{\mathcal{I}}^{(2)}_{2,1,1} + \frac{15}{4} \left(\boldsymbol{\mathcal{I}}^{(1)}_{2,1,1}\right)^3$
$\quad + 4a_{2,2} (\boldsymbol{\mathcal{I}}_{2,1,1})^2 \boldsymbol{\mathcal{I}}^{(1)}_{2,1,1} - 4a^{(1)}_{2,2} (\boldsymbol{\mathcal{I}}_{2,1,1})^3,$

$\boldsymbol{F}_{9,10} \equiv \boldsymbol{\mathcal{I}}_{2,1,1} \boldsymbol{\mathcal{I}}^{(3)}_{2,2,2} - \frac{15}{2} \boldsymbol{\mathcal{I}}^{(1)}_{2,1,1} \boldsymbol{\mathcal{I}}^{(2)}_{2,2,2} + \frac{27}{2} \boldsymbol{\mathcal{I}}^{(2)}_{2,1,1} \boldsymbol{\mathcal{I}}^{(1)}_{2,2,2} - 6 \boldsymbol{\mathcal{I}}^{(3)}_{2,1,1} \boldsymbol{\mathcal{I}}_{2,2,2}$
$\quad + a_{2,2} \left[-80 \boldsymbol{\mathcal{I}}_{2,1,1} \boldsymbol{\mathcal{I}}^{(1)}_{2,2,2} + 144 \boldsymbol{\mathcal{I}}^{(1)}_{2,1,1} \boldsymbol{\mathcal{I}}_{2,2,2}\right] + 16 a^{(1)}_{2,2} \boldsymbol{\mathcal{I}}_{2,1,1} \boldsymbol{\mathcal{I}}_{2,2,2},$

$\boldsymbol{F}_{9,11} \equiv \boldsymbol{\mathcal{I}}_{2,1,1} \boldsymbol{\mathcal{I}}^{(4)}_{2,1,2} - 9 \boldsymbol{\mathcal{I}}^{(1)}_{2,1,1} \boldsymbol{\mathcal{I}}^{(3)}_{2,1,2} + \frac{108}{5} \boldsymbol{\mathcal{I}}^{(2)}_{2,1,1} \boldsymbol{\mathcal{I}}^{(2)}_{2,1,2}$
$\quad - \frac{84}{5} \boldsymbol{\mathcal{I}}^{(3)}_{2,1,1} \boldsymbol{\mathcal{I}}^{(1)}_{2,1,2} + \frac{18}{5} \boldsymbol{\mathcal{I}}^{(4)}_{2,1,1} \boldsymbol{\mathcal{I}}_{2,1,2}$
$\quad + a_{2,2} \left[-\frac{652}{5} \boldsymbol{\mathcal{I}}_{2,1,1} \boldsymbol{\mathcal{I}}^{(2)}_{2,1,2} + \frac{2076}{5} \boldsymbol{\mathcal{I}}^{(1)}_{2,1,1} \boldsymbol{\mathcal{I}}^{(1)}_{2,1,2} - \frac{1224}{5} \boldsymbol{\mathcal{I}}^{(2)}_{2,1,1} \boldsymbol{\mathcal{I}}_{2,1,2}\right]$
$\quad + a^{(1)}_{2,2} \left[+\frac{206}{5} \boldsymbol{\mathcal{I}}_{2,1,1} \boldsymbol{\mathcal{I}}^{(1)}_{2,1,2} - \frac{54}{5} \boldsymbol{\mathcal{I}}^{(1)}_{2,1,1} \boldsymbol{\mathcal{I}}_{2,1,2}\right]$
$\quad + \left[-\frac{102}{5} a^{(2)}_{2,2} + \frac{4608}{5} (a_{2,2})^2\right] \boldsymbol{\mathcal{I}}_{2,1,1} \boldsymbol{\mathcal{I}}_{2,1,2}.$

are $n(9) = 11$ linearly independent relative invariants for the equations (14.49).

The various $\boldsymbol{F}_{s,i}$ were obtained from machine evaluations of Inv[s,i] described in Section 16.4. To check their character as relative invariants, we note the following.

Selection of $\boldsymbol{P}_2 \equiv \boldsymbol{\mathcal{I}}_{2,1,1}$ and $\boldsymbol{Q}_3 \equiv \boldsymbol{\mathcal{I}}_{2,1,2}$ in (14.11) yields $\boldsymbol{F}_{6,4}$.
Selection of $\boldsymbol{P}_2 \equiv \boldsymbol{\mathcal{I}}_{2,1,1}$ in (14.25) yields $\boldsymbol{F}_{6,5}$.
Selection of $\boldsymbol{P}_2 \equiv \boldsymbol{\mathcal{I}}_{2,1,1}$ and $\boldsymbol{Q}_4 \equiv \boldsymbol{\mathcal{I}}_{2,2,2}$ in (14.11) yields $\boldsymbol{F}_{7,3}$.
Selection of $\boldsymbol{P}_2 \equiv \boldsymbol{\mathcal{I}}_{2,1,1}$ and $\boldsymbol{Q}_3 \equiv \boldsymbol{\mathcal{I}}_{2,1,2}$ in (14.17) yields $\boldsymbol{F}_{7,4}$.
Selection of $\boldsymbol{P}_3 \equiv \boldsymbol{\mathcal{I}}_{2,1,2}$ and $\boldsymbol{Q}_4 \equiv \boldsymbol{\mathcal{I}}_{2,2,2}$ in (14.11) yields $\boldsymbol{F}_{8,7}$.
Selection of $\boldsymbol{P}_3 \equiv \boldsymbol{\mathcal{I}}_{2,1,2}$ in (14.25) yields $\boldsymbol{F}_{8,8}$.
Selection of $\boldsymbol{P}_2 \equiv \boldsymbol{\mathcal{I}}_{2,1,1}$ and $\boldsymbol{Q}_4 \equiv \boldsymbol{\mathcal{I}}_{2,2,2}$ in (14.17) yields $\boldsymbol{F}_{8,9}$.
Selection of $\boldsymbol{P}_2 \equiv \boldsymbol{\mathcal{I}}_{2,1,1}$ and $\boldsymbol{Q}_3 \equiv \boldsymbol{\mathcal{I}}_{2,1,2}$ in (14.27) yields $\boldsymbol{F}_{8,10}$.
Selection of $\boldsymbol{P}_2 \equiv \boldsymbol{\mathcal{I}}_{2,1,1}$ and $\boldsymbol{Q}_2 \equiv \boldsymbol{\mathcal{I}}_{2,1,1}$ in (14.33) yields $\boldsymbol{F}_{8,11}$.
Selection of $\boldsymbol{P}_3 \equiv \boldsymbol{\mathcal{I}}_{2,1,2}$ and $\boldsymbol{Q}_4 \equiv \boldsymbol{\mathcal{I}}_{2,2,2}$ in (14.17) yields $\boldsymbol{F}_{9,8}$.

For $\boldsymbol{F}_{9,9}$, we first note that $\boldsymbol{P}_4 \equiv (\boldsymbol{\mathcal{I}}_{2,1,1})^2$ and $\boldsymbol{Q}_2 \equiv \boldsymbol{\mathcal{I}}_{2,1,1}$ are linearly independent over $\mathbb{Q}$. Then, after defining $\boldsymbol{X}$ in terms of this $\boldsymbol{P}_4$ and $\boldsymbol{Q}_2$ by means of (14.27), we find that $\boldsymbol{F}_{9,9} \equiv \frac{3}{2} \boldsymbol{X}$.

Selection of $\boldsymbol{P}_2 \equiv \boldsymbol{\mathcal{I}}_{2,1,1}$ and $\boldsymbol{Q}_4 \equiv \boldsymbol{\mathcal{I}}_{2,2,2}$ in (14.27) yields $\boldsymbol{F}_{9,10}$.

Selection of $\boldsymbol{P}_2 \equiv \boldsymbol{\mathcal{I}}_{2,1,1}$ and $\boldsymbol{Q}_3 \equiv \boldsymbol{\mathcal{I}}_{2,1,2}$ in (14.33) yields $\boldsymbol{F}_{9,11}$.

This technique can be extended to greater values of s. When $s = 10$, there are $n(10) = 22$ linearly independent relative invariants of weight 10 such that their nonzero linear combinations over $\mathbb{Q}$ yield all the relative invariants of weight 10 for the equations (14.49). To select and represent them as simple differential-polynomial combinations over $\mathbb{Q}$ of $\boldsymbol{\mathcal{I}}_{2,1,1}$, $\boldsymbol{\mathcal{I}}_{2,1,2}$, $\boldsymbol{\mathcal{I}}_{2,2,2}$, and $\boldsymbol{a}_{2,2}$, we would use (14.47) and (14.48) of Formulations 14.15 and 14.16 in addition to the other results in Section 14.2.

CHAPTER 15

Results about Q_m as a Quadratic Form

15.1. For Q_m to have a nontrivial factorization

The coefficients $c_{i,j}(z)$ of the left member

$$(15.1) \qquad Q_m \equiv \left(y^{(m)}\right)^2 + \sum_{\substack{0 \le i,j \le m \\ (i,j) \ne 0}} c_{i,j}(z)\, y^{(m-i)}\, y^{(m-j)}$$

of (1.1) are meromorphic functions on a region Ω of the complex plane and satisfy $c_{j,i}(z) \equiv c_{i,j}(z)$, for $0 \le i < j \le m$. We shall establish in Theorem 15.3 a necessary and sufficient condition on these coefficients in order that: for any subregion $\mathcal{U}$ of Ω, there is a subregion $\mathcal{V}$ of $\mathcal{U}$ such that the restriction of Q_m to $\mathcal{V}$ has a factorization

$$(15.2) \qquad Q_m \equiv \left[y^{(m)} + \sum_{i=1}^{m} d_i(z)\, y^{(m-i)}\right]\left[y^{(m)} + \sum_{j=0}^{m} e_j(z)\, y^{(m-j)}\right],$$

for some meromorphic functions on $d_i(z)$, $e_j(z)$ on $\mathcal{V}$. To do that, we introduce the polynomials $\boldsymbol{B}_{i,j}$ that are defined in $\mathcal{S}_m$ with respect to (1.6)–(1.8) by

$$(15.3) \qquad \boldsymbol{B}_{i,j} \equiv \boldsymbol{w}_{i,j} - \boldsymbol{w}_{0,i}\, \boldsymbol{w}_{0,j}, \quad \text{for } 1 \le i,\, j \le m;$$

and, in $\mathcal{S}_m$, we set

$$(15.4) \qquad \boldsymbol{\Delta}_{i,j,k} \equiv \boldsymbol{B}_{i,j} \boldsymbol{B}_{k,k} - \boldsymbol{B}_{i,k} \boldsymbol{B}_{j,k}, \quad \text{for } 1 \le i,\, j,\, k \le m.$$

In terms of the meromorphic functions $\Delta_{i,j,k}(z)$ on Ω obtained by replacing each $w_{\mu,\nu}$ in $\boldsymbol{\Delta}_{i,j,k}$ with the corresponding $c_{\mu,\nu}(z)$ from (15.1), the proof of Theorem 15.3 shows that: Q_m has a factorization of the form (15.2) if and only if

$$\Delta_{i,j,k}(z) \equiv 0, \quad \text{for } 1 \le i,\, j,\, k \le m.$$

By removing various redundancies of the form $0 \equiv 0$ from the preceding conditions, we deduce the equivalent requirements of (15.14).

PROPOSITION 15.1. *With respect to the meromorphic functions $B_{i,j}(z)$ on Ω that are obtained by replacing each $w_{\mu,\nu}$ in $\boldsymbol{B}_{i,j}$ with the corresponding $c_{\mu,\nu}(z)$ from (15.1), the polynomial Q_m in (15.1) is expressible as*

$$(15.5) \qquad Q_m \equiv \left(S_m\right)^2 + \sum_{1 \le i,j \le m} B_{i,j}(z)\, y^{(m-i)}\, y^{(m-j)},$$

where

$$(15.6) \qquad S_m \equiv y^{(m)} + \sum_{j=1}^{m} c_{0,j}(z)\, y^{(m-j)}.$$

PROOF. In terms of
$$R_m \equiv \sum_{1 \leq i,j \leq m} c_{i,j}(z) \, y^{(m-i)} \, y^{(m-j)},$$
we use (15.1), (15.6), and (15.3) to obtain
$$Q_m \equiv \left(y^{(m)}\right)^2 + \sum_{j=1}^{m} c_{0,j}(z) \, y^{(m)} \, y^{(m-j)} + \sum_{i=1}^{m} c_{i,0}(z) \, y^{(m-i)} \, y^{(m)} + R_m$$
$$\equiv \left(y^{(m)}\right)^2 + 2 y^{(m)} \sum_{j=1}^{m} c_{0,j}(z) \, y^{(m-j)} + R_m$$
$$\equiv \left[y^{(m)} + \sum_{j=1}^{m} c_{0,j}(z) \, y^{(m-j)} \right]^2 + R_m - \left[\sum_{j=1}^{m} c_{0,j}(z) \, y^{(m-j)} \right]^2$$
$$\equiv (S_m)^2 + \sum_{1 \leq i,j \leq m} c_{i,j}(z) \, y^{(m-i)} \, y^{(m-j)} - \sum_{1 \leq i,j \leq m} c_{0,i}(z) \, c_{0,j}(z) \, y^{(m-i)} \, y^{(m-j)}$$
$$\equiv (S_m)^2 + \sum_{1 \leq i,j \leq m} B_{i,j}(z) \, y^{(m-i)} \, y^{(m-j)}.$$

This yields (15.5) and completes the proof. $\square$

PROPOSITION 15.2. *For any subregion $\mathcal{V}$ of Ω, there is a factorization of Q_m having the form (15.2) on $\mathcal{V}$ if and only if meromorphic functions $\gamma_1(z), \ldots, \gamma_m(z)$ on $\mathcal{V}$ exist such that*

(15.7) $$\sum_{1 \leq i,j \leq m} B_{i,j}(z) \, y^{(m-i)} \, y^{(m-j)} \equiv -\left[\sum_{i=1}^{m} \gamma_i(z) \, y^{(m-i)} \right]^2, \quad \text{on } \mathcal{V}.$$

Moreover, Q_m has a factorization of the form (15.2) in which the factors are equal if and only if

(15.8) $$B_{i,j}(z) \equiv 0, \quad \text{on } \Omega \text{ for } 1 \leq i, j \leq m;$$

and, when (15.8) is satisfied, (15.5) yields $Q_m \equiv (S_m)^2$, on Ω.

PROOF. (i) Suppose there are meromorphic functions $\gamma_1(z), \ldots, \gamma_m(z)$ on $\mathcal{V}$ that satisfy (15.7). We combine the restriction of (15.5) to $\mathcal{V}$ with (15.7) to verify that

(15.9) $$Q_m \equiv (S_m)^2 - \left[\sum_{i=1}^{m} \gamma_i(z) \, y^{(m-i)} \right]^2$$
$$\equiv \left[S_m + \sum_{i=1}^{m} \gamma_i(z) \, y^{(m-i)} \right] \left[S_m - \sum_{i=1}^{m} \gamma_i(z) \, y^{(m-i)} \right].$$

By applying (15.6) to rewrite (15.9), we obtain a factorization for Q_m of the form (15.2) on $\mathcal{V}$.

(ii) Suppose the restriction of Q_m to $\mathcal{V}$ has a factorization of the form (15.2) on $\mathcal{V}$. Regarding $y^{(0)}$, ..., $y^{(m)}$ as algebraically independent variables over the field $\mathfrak{F}_\mathcal{V}$ of meromorphic functions on $\mathcal{V}$, we rewrite (15.6) to obtain

$$y^{(m)} \equiv S_m - \sum_{j=1}^{m} c_{0,j}(z)\, y^{(m-j)}. \tag{15.10}$$

We employ (15.10) to eliminate $y^{(m)}$ from the factors in the right member of (15.2) and obtain

$$Q_m \equiv (S_m + F)(S_m + G) \equiv (S_m)^2 + (F+G)S_m + FG, \tag{15.11}$$

where F and G are linear forms in the symbols $y^{(0)}$, ..., $y^{(m-1)}$ over $\mathfrak{F}_\mathcal{V}$ and do not involve $y^{(m)}$. By combining (15.11) with (15.5), we find that

$$(F+G)S_m + FG \equiv \sum_{1 \le i,j \le m} B_{i,j}(z)\, y^{(m-i)}\, y^{(m-j)}. \tag{15.12}$$

Since only S_m in (15.12) involves the variable $y^{(m)}$, we must have $F + G \equiv 0$ and $F \equiv -G$. Thus, (15.12) yields

$$-G^2 \equiv \sum_{1 \le i,j \le m} B_{i,j}(z)\, y^{(m-i)}\, y^{(m-j)}. \tag{15.13}$$

We use (15.13) and the property that G is a linear form in $y^{(0)}$, ..., $y^{(m-1)}$ over $\mathfrak{F}_\mathcal{V}$ to establish the existence of functions $\gamma_1(z)$, ..., $\gamma_m(z)$ on $\mathcal{V}$ that satisfy (15.7).

Suppose that Q_m has a factorization (15.2) in which the factors are equal. Then, we have $F \equiv G$ for (15.11) in addition to $F \equiv -G$. This gives $F \equiv G \equiv 0$. We apply (15.13) with $G \equiv 0$ to deduce that $B_{i,j}(z) \equiv 0$, on $\mathcal{V}$, for $1 \le i, j \le m$. Meromorphic continuation shows that (15.8) is also satisfied.

Suppose that (15.8) is satisfied. Then, (15.5) yields $Q_m \equiv (S_m)^2$, over Ω, as a factorization with equal factors. This completes the proof. □

THEOREM 15.3. *With respect to Q_m in (15.1) and the functions $\Delta_{i,j,k}(z)$ on Ω obtained by substituting $c_{\mu,\nu}(z)$ from (15.1) for $w_{\mu,\nu}$ in (15.4), the condition*

$$\Delta_{i,j,k}(z) \equiv 0, \quad \text{for } 1 \le i, j, k \le m \text{ and either } i = j < k \text{ or } k \ne i < j \ne k, \tag{15.14}$$

is necessary and sufficient in order that: for any subregion $\mathcal{U}$ of Ω, there is a subregion $\mathcal{V}$ of $\mathcal{U}$ on which meromorphic functions $d_i(z)$, $e_j(z)$ are defined such that Q_m has the factorization (15.2) on $\mathcal{V}$. Moreover, Q_m has such a factorization in which the factors are equal if and only if $B_{i,j}(z) \equiv 0$, for $1 \le i \le j \le m$.

PROOF. (i) Suppose that Q_m possesses the factorization (15.2) on a subregion $\mathcal{V}$ of Ω. Then, in view of Proposition 15.2, meromorphic functions $\gamma_1(z)$, ..., $\gamma_m(z)$ on $\mathcal{V}$ exists such that

$$\sum_{1 \le i,j \le m} B_{i,j}(z)\, y^{(m-i)}\, y^{(m-j)} \equiv \sum_{1 \le i,j \le m} -\gamma_i(z)\, \gamma_j(z)\, y^{(m-i)}\, y^{(m-j)}, \quad \text{on } \mathcal{V}. \tag{15.15}$$

Regarding $y^{(0)}$, ..., $y^{(m-1)}$ as algebraically independent variables over the field $\mathfrak{F}_\mathcal{V}$ of meromorphic functions on $\mathcal{V}$, we use (15.15) and $B_{j,i}(z) \equiv B_{i,j}(z)$ to deduce that

$$B_{i,j}(z) \equiv -\gamma_i(z)\, \gamma_j(z), \quad \text{on } \mathcal{V}, \text{ for } 1 \le i, j \le m. \tag{15.16}$$

We employ (15.4) and (15.16) to verify that: for $1 \leq i, j, k \leq m$ and z in $\mathcal{V}$,
$$\Delta_{i,j,k}(z) \equiv B_{i,j}(z)\, B_{k,k}(z) - B_{i,k}(z)\, B_{j,k}(z)$$
$$\equiv \gamma_i(z)\, \gamma_j(z)\, \gamma_k(z)\, \gamma_k(z) - \gamma_i(z)\, \gamma_k(z)\, \gamma_j(z)\, \gamma_k(z) \equiv 0.$$

Thus, (15.14) is satisfied. Moreover, if the factors for Q_m in (15.2) are equal, then Proposition 15.2 shows that (15.8) is satisfied.

(ii) Suppose that (15.14) is satisfied and let $\mathcal{U}$ be a subregion of Ω. In terms of integers i, j, k subject to $1 \leq i, j, k \leq m$, we use (15.4) to establish that
$$\Delta_{i,i,k} \equiv \Delta_{k,k,i}, \quad \Delta_{i,j,k} \equiv \Delta_{j,i,k}, \quad \Delta_{i,j,i} \equiv 0, \quad \text{and} \quad \Delta_{i,j,j} \equiv 0.$$
Consequently, (15.14) yields

(15.17) $\qquad \Delta_{i,j,k}(z) \equiv 0, \quad \text{on } \Omega \text{ for } 1 \leq i, j, k \leq m.$

If we have $B_{k,k}(z) \equiv 0$, for $1 \leq k \leq m$, then (15.17) and (15.4) give
$$0 \equiv \Delta_{i,i,j}(z) \equiv -\big(B_{i,j}(z)\big)^2 \quad \text{and} \quad B_{i,j}(z) \equiv 0, \quad \text{for } 1 \leq i, j \leq m.$$
In this situation, (15.5) and (15.6) yield the factorization $Q_m \equiv (S_m)^2$ on Ω and therefore also on $\mathcal{V} = \mathcal{U}$. Next, suppose there is an integer k_0 such that $1 \leq k_0 \leq m$ and $B_{k_0,k_0}(z) \not\equiv 0$. Then, there is a meromorphic function $\gamma_{k_0}(z)$ on a subregion $\mathcal{V}$ of $\mathcal{U}$ such that $\big(\gamma_{k_0}(z)\big)^2 \equiv -B_{k_0,k_0}(z)$, on $\mathcal{V}$. After observing that $\gamma_{k_0}(z) \not\equiv 0$, we set $\gamma_i(z) \equiv -B_{i,k_0}(z)/\gamma_{k_0}(z)$, on $\mathcal{V}$ for $1 \leq i \leq m$ and $i \neq k_0$. Hence, we have

(15.18) $\qquad \gamma_i(z) \equiv -\dfrac{B_{i,k_0}(z)}{\gamma_{k_0}(z)}, \quad \text{on } \mathcal{V} \text{ for } 1 \leq i \leq m.$

We apply (15.17), (15.4), and (15.18) to deduce that: for $1 \leq i, j \leq m$,
$$B_{i,j}(z) \equiv \frac{B_{i,k_0}(z)\, B_{j,k_0}(z)}{B_{k_0,k_0}(z)} \equiv -\left[\frac{B_{i,k_0}(z)}{\gamma_{k_0}(z)}\right]\left[\frac{B_{j,k_0}(z)}{\gamma_{k_0}(z)}\right] \equiv -\gamma_i(z)\, \gamma_j(z), \quad \text{on } \mathcal{V}.$$
This yields

(15.19) $\displaystyle\sum_{1 \leq i,j \leq m} B_{i,j}(z)\, y^{(m-i)}\, y^{(m-j)} \equiv \sum_{1 \leq i,j \leq m} -\gamma_i(z)\, \gamma_j(z)\, y^{(m-i)}\, y^{(m-j)}$
$$\equiv -\left[\sum_{i=1}^{m} \gamma_i(z)\, y^{(m-i)}\right]^2, \quad \text{on } \mathcal{V}.$$

In view of (15.19), Proposition 15.2 shows that there is a factorization of the form (15.2) on $\mathcal{V}$ for Q_m. Also, if $B_{i,j}(z) \equiv 0$, on Ω for $1 \leq i \leq j \leq m$, then we use $B_{j,i}(z) \equiv B_{i,j}(z)$, for $1 \leq i, j \leq m$, and (15.5) to obtain the factorization $Q_m \equiv (S_m)^2$ with equal factors on Ω. This completes the proof. $\square$

REMARK 15.4. Let the symbols $\mathfrak{F}_\Omega$ and $\mathfrak{F}_\mathcal{V}$ denote the fields of meromorphic functions defined respectively on Ω and on a subregion $\mathcal{V}$ of Ω. For any h in $\mathfrak{F}_\Omega$, let $h_\mathcal{V}$ denote the restriction of h to $\mathcal{V}$. A one-to-one mapping σ of $\mathfrak{F}_\Omega$ into $\mathfrak{F}_\mathcal{V}$ is defined by $\sigma(h) = h_\mathcal{V}$, for each h in $\mathfrak{F}_\Omega$. We find that σ is an isomorphism of $\mathfrak{F}_\Omega$ onto a corresponding subfield of $\mathfrak{F}_\mathcal{V}$.

Throughout, the various factorizations for Q_m could be formulated as ones over a quadratic extension field of $\mathfrak{F}_\Omega$. For a suitable subregion $\mathcal{V}$ of Ω, our employment of $\mathfrak{F}_\mathcal{V}$ as an extension field of a field isomorphic to $\mathfrak{F}_\Omega$ enables us to avoid the use of objects that correspond to multiple-valued functions.

For a given (15.1) of order m, let $\mathfrak{B}_m(z)$ denote the symmetric $m \times m$ matrix over $\mathfrak{F}_\Omega$ whose (i,j)-component $[\mathfrak{B}_m(z)]_{i,j}$ is specified in terms of (15.3) by

(15.20) $\quad [\mathfrak{B}_m(z)]_{i,j} \equiv B_{i,j}(z) \equiv c_{i,j}(z) - c_{0,i}(z)\, c_{0,j}(z), \quad \text{for } 1 \leq i, j \leq m.$

PROPOSITION 15.5. *For Q_m in (1.1), the condition (15.14) is satisfied if and only if the rank of $\mathfrak{B}_m(z)$ over $\mathfrak{F}_\Omega$ is 1 or 0. If the rank of $\mathfrak{B}_m(z)$ is 1, then any factorization for Q_m of the form (15.2) has unequal factors; and, if the rank of $\mathfrak{B}_m(z)$ is 0, then Q_m has the factorization $Q_m \equiv (S_m)^2$ with S_m in (15.6).*

PROOF. If $\operatorname{rank}(\mathfrak{B}_m(z)) = 0$, then (15.20) gives $B_{i,j}(z) \equiv 0$, for $1 \leq i, j \leq m$. Hence, (15.4) shows that (15.14) is satisfied; and (15.5) gives $Q_m \equiv (S_m)^2$ on Ω.

Suppose that the rank of $\mathfrak{B}_m(z)$ is 1. Then, any two rows of $\mathfrak{B}_m(z)$ are linearly dependent over $\mathfrak{F}_\Omega$. Since the determinant of each 2×2 submatrix of $\mathfrak{B}_m(z)$ is identically zero on Ω, we find that

$$0 \equiv B_{i,j}(z)\, B_{k,k}(z) - B_{i,k}(z)\, B_{j,k}(z) \equiv \Delta_{i,j,k}(z), \quad \text{for } 1 \leq i, j, k \leq m,$$

and (15.14) is satisfied. However, not all the components of $\mathfrak{B}_m(z)$ are 0. Hence, Theorem 15.3 yields a factorization for Q_m of the form (15.2); and, for any such factorization, the factors are unequal.

Suppose that (15.14) is satisfied. We apply Theorem 15.3 to see that Q_m has a factorization (15.2) on some subregion $\mathcal{V}$ of Ω. Consequently, Proposition 15.2 yields meromorphic functions $\gamma_1(z), \ldots, \gamma_m(z)$ on a subregion $\mathcal{V}$ of Ω such that

(15.21) $\quad \displaystyle\sum_{1 \leq i,j \leq m} B_{i,j}(z)\, X_i\, X_j \equiv -\left[\sum_{i=1}^m \gamma_i(z)\, X_i\right]^2, \quad \text{on } \mathcal{V}.$

If $\gamma_i(z) \equiv 0$ for $1 \leq i \leq m$, then $B_{i,j}(z) \equiv 0$, for $1 \leq i, j \leq m$, and the rank of $\mathfrak{B}_m(z)$ is 0. Suppose that $\gamma_{k_0}(z) \not\equiv 0$ for some integer k_0 that satisfies $1 \leq k_0 \leq m$. Then, the linear substitution specified by

(15.22) $\quad \widehat{X}_{k_0} = \displaystyle\sum_{i=1}^m \gamma_i(z)\, X_i \quad \text{and} \quad \widehat{X}_i = X_i, \quad \text{for } 1 \leq i \leq m \text{ and } i \neq k_0,$

is nonsingular. Let N denote the inverse matrix for the transformation (15.22) so that, in terms of $X = \begin{bmatrix} X_1, X_2, \ldots, X_m \end{bmatrix}$ and $\widehat{X} = \begin{bmatrix} \widehat{X}_1, \widehat{X}_2, \ldots, \widehat{X}_m \end{bmatrix}$, we have $X^T = N\widehat{X}^T$, where T denotes the operation of transposition. In particular, N is a nonsingular $m \times m$ matrix and we find that

(15.23) $\quad X\, \mathfrak{B}_m(z)\, X^T \equiv \widehat{X}\, (N^T \mathfrak{B}_m(z)\, N)\, \widehat{X}^T.$

The substitutional inverse for (15.22) transforms the quadratic form in the left member of (15.21) into the quadratic form $-\widehat{X}_{k_0}^2$ whose matrix has rank 1. We use (15.23) to see that the matrix of the quadratic form $-\widehat{X}_{k_0}^2$ is $N^T \mathfrak{B}_m(z)\, N$. Since N is nonsingular, the rank of $\mathfrak{B}_m(z)$ is 1. This completes the proof. □

EXAMPLE 15.6. For $m = 2$, (15.14) is the condition $\Delta_{1,1,2}(z) \equiv 0$. We use (15.4) and (1.70) on page 14 to obtain $\boldsymbol{\Delta}_{1,1,2} \equiv \boldsymbol{B}_{1,1}\boldsymbol{B}_{2,2} - (\boldsymbol{B}_{1,2})^2 \equiv \boldsymbol{D}_2$, where $\boldsymbol{D}_2$ is the particular relative invariant for the equations (1.0–A) on page 3 that was studied in Example 1.9 on page 14. In particular, (1.0–A) has a factorization of the type (1.0-B) on page 3 if and only if the function $D_2(z)$ on Ω obtained by replacing $\boldsymbol{w}_{i,j}$ in $\boldsymbol{D}_2$ with $c_{i,j}(z)$ from (1.0–A) satisfies $D_2(z) \equiv 0$.

EXAMPLE 15.7. For $m = 3$, (15.14) is the condition
$$\Delta_{1,1,2}(z) \equiv \Delta_{1,1,3}(z) \equiv \Delta_{2,2,3}(z) \equiv \Delta_{1,2,3}(z) \equiv \Delta_{1,3,2}(z) \equiv \Delta_{2,3,1}(z) \equiv 0,$$
where
$$\Delta_{1,1,2} \equiv B_{1,1}B_{2,2} - B_{1,2}B_{1,2}, \qquad \Delta_{1,1,3} \equiv B_{1,1}B_{3,3} - B_{1,3}B_{1,3},$$
$$\Delta_{2,2,3} \equiv B_{2,2}B_{3,3} - B_{2,3}B_{2,3}, \qquad \Delta_{1,2,3} \equiv B_{1,2}B_{3,3} - B_{1,3}B_{2,3},$$
$$\Delta_{1,3,2} \equiv B_{1,3}B_{2,2} - B_{1,2}B_{2,3}, \qquad \Delta_{2,3,1} \equiv B_{2,3}B_{1,1} - B_{1,2}B_{1,3}.$$

Machine computations show that: $\Delta_{1,1,2}$ is a relative invariant for the equations (1.1) having $m = 3$; however, $\Delta_{1,1,3}$ and $\Delta_{2,3,1}$ are merely semi-invariants of the second kind for such equations; and each of $\Delta_{2,2,3}$, $\Delta_{1,2,3}$, $\Delta_{1,3,2}$ is not a semi-invariant of either kind. We note that $\Delta_{1,1,2} \equiv D_2$, where D_2 is defined by (1.64) on page 14.

15.2. Relative invariants defined by determinants

For $1 \leq k \leq m$, we use (1.6)–(1.7) on page 5 to define D_k in $\mathcal{S}_m$ through

(15.24) $$D_k \equiv \begin{vmatrix} 1 & w_{0,1} & w_{0,2} & \cdots & w_{0,k} \\ w_{1,0} & w_{1,1} & w_{1,2} & \cdots & w_{1,k} \\ w_{2,0} & w_{2,1} & w_{2,2} & \cdots & w_{2,k} \\ \vdots & \vdots & \vdots & \ddots & \vdots \\ w_{k,0} & w_{k,1} & w_{k,2} & \cdots & w_{k,k} \end{vmatrix}.$$

In other words, with $w_{0,0} \equiv 1$ from (1.8), D_k is defined by $D_k \equiv det(\mathfrak{A}_k)$, where $\mathfrak{A}_k$ is the $(k+1) \times (k+1)$ matrix that has $[\mathfrak{A}_k]_{i,j} \equiv w_{i,j}$, for $0 \leq i, j \leq k$. We shall prove in Theorem 15.10 that D_k *is a relative invariant of weight* $k(k+1)$ *for the equations* (1.1) *of order* $m \geq k$. For $k = 1$ and any fixed integer $m \geq 1$, D_1 is the basic relative invariant $\mathcal{I}_{m,1,1}$ for the equations (1.1) of order m. For $k = 2$ and any fixed integer $m \geq 2$, D_2 is a relative invariant for the equations (1.1) of order m. This generalizes the restriction $m = 2$ for D_2 in Example 1.9 on page 14.

By using elementary row operations and the definition of $B_{i,j}$ in (15.3) along with $w_{j,i} \equiv w_{i,j}$, we find that (15.24) yields

(15.25) $$D_k \equiv \begin{vmatrix} 1 & w_{0,1} & w_{0,2} & \cdots & w_{0,k} \\ 0 & B_{1,1} & B_{1,2} & \cdots & B_{1,k} \\ 0 & B_{2,1} & B_{2,2} & \cdots & B_{2,k} \\ \vdots & \vdots & \vdots & \ddots & \vdots \\ 0 & B_{k,1} & B_{k,2} & \cdots & B_{k,k} \end{vmatrix} \equiv \begin{vmatrix} B_{1,1} & B_{1,2} & \cdots & B_{1,k} \\ B_{2,1} & B_{2,2} & \cdots & B_{2,k} \\ \vdots & \vdots & \ddots & \vdots \\ B_{k,1} & B_{k,2} & \cdots & B_{k,k} \end{vmatrix}.$$

Thus, we have $D_k \equiv det(\mathfrak{B}_k)$, where $\mathfrak{B}_k$ is the $k \times k$ matrix with $[\mathfrak{B}_k]_{i,j} \equiv B_{i,j}$, for $1 \leq i, j \leq k$. When $c_{i,j}(z)$ from (1.1) is substituted for $w_{i,j}$ in $\mathfrak{A}_k$ and $\mathfrak{B}_k$, we obtain matrices $\mathfrak{A}_k(z)$ and $\mathfrak{B}_k(z)$ over $\mathfrak{F}_\Omega$ of respective sizes $(k+1) \times (k+1)$ and $k \times k$. Then, the row operations employed to obtain (15.25) show that

(15.26) $$rank(\mathfrak{A}_k(z)) = rank(\mathfrak{B}_k(z)) + 1, \quad \text{for } 1 \leq k \leq m.$$

For $k = m$, (15.26) gives $rank(\mathfrak{B}_m(z)) = i$ if and only if $rank(\mathfrak{A}_m(z)) = i + 1$. However, we prefer the present formulation of Proposition 15.5 with respect to the rank of $\mathfrak{B}_m(z)$ rather than one in terms of the rank of $\mathfrak{A}_m(z)$. Also, the notation $D_k \equiv det(\mathfrak{B}_k)$ leads naturally to Propositions 15.8 and 15.9.

15.2. RELATIVE INVARIANTS DEFINED BY DETERMINANTS

PROPOSITION 15.8. *When (1.2) transforms (1.1) into (1.3), the functions*

(15.27) $\quad B_{i,j}(z) \equiv c_{i,j}(z) - c_{0,i}(z)\, c_{0,j}(z) \quad \text{and} \quad B^*_{i,j}(z) \equiv c^*_{i,j}(z) - c^*_{0,i}(z)\, c^*_{0,j}(z)$

satisfy

(15.28) $\quad B^*_{i,j}(z) \equiv \sum_{\mu=1}^{i} \sum_{\nu=1}^{j} \binom{m-\mu}{i-\mu} \binom{m-\nu}{j-\nu} \frac{\rho^{(i-\mu)}(z)\, \rho^{(j-\nu)}(z)}{(\rho(z))^2}\, B_{\mu,\nu}(z),$

$$\text{on } \Omega \text{ for } 1 \le i,\, j \le m.$$

PROOF. For $1 \le i, j \le m$, (15.27), (2.5) on page 23, and $c_{0,0}(z) \equiv 1$ give

$$B^*_{i,j}(z) \equiv \sum_{\mu=0}^{i} \sum_{\nu=0}^{j} \binom{m-\mu}{i-\mu} \binom{m-\nu}{j-\nu} \frac{\rho^{(i-\mu)}(z)\, \rho^{(j-\nu)}(z)}{(\rho(z))^2}\, c_{\mu,\nu}(z)$$

$$- \left[\sum_{\mu=0}^{i} \binom{m-\mu}{i-\mu} \frac{\rho^{(i-\mu)}(z)}{\rho(z)}\, c_{0,\mu}(z) \right] \left[\sum_{\nu=0}^{j} \binom{m-\nu}{j-\nu} \frac{\rho^{(j-\nu)}(z)}{\rho(z)}\, c_{0,\nu}(z) \right]$$

$$\equiv \sum_{\mu=0}^{i} \sum_{\nu=0}^{j} \binom{m-\mu}{i-\mu} \binom{m-\nu}{j-\nu} \frac{\rho^{(i-\mu)}(z)\, \rho^{(j-\nu)}(z)}{(\rho(z))^2} \left[c_{\mu,\nu}(z) - c_{0,\mu}(z)\, c_{0,\nu}(z) \right]$$

$$\equiv \sum_{\mu=1}^{i} \sum_{\nu=1}^{j} \binom{m-\mu}{i-\mu} \binom{m-\nu}{j-\nu} \frac{\rho^{(i-\mu)}(z)\, \rho^{(j-\nu)}(z)}{(\rho(z))^2}\, B_{\mu,\nu}(z).$$

This yields (15.28) and completes the proof. □

PROPOSITION 15.9. *When (1.4) transforms (1.1) into (1.5), the functions*

(15.29) $\quad B_{i,j}(z) \equiv c_{i,j}(z) - c_{0,i}(z)\, c_{0,j}(z) \quad \text{and} \quad B^{**}_{i,j}(\zeta) \equiv c^{**}_{i,j}(\zeta) - c^{**}_{0,i}(\zeta)\, c^{**}_{0,j}(\zeta)$

satisfy

(15.30) $\quad B^{**}_{i,j}(\zeta) \equiv \sum_{\mu=1}^{i} \sum_{\nu=1}^{j} (f'(\zeta))^{\mu+\nu} \alpha_{i-\mu,m-i}(\zeta)\, \alpha_{j-\nu,m-j}(\zeta)\, B_{\mu,\nu}(f(\zeta)),$

$$\text{on } \Omega^{**} \text{ for } 1 \le i,\, j \le m,$$

where $\alpha_{i,j}(\zeta)$ is defined on Ω^{**} through (3.1)–(3.2) of page 29.

PROOF. For $1 \le i, j \le m$, (15.29), (3.9) on page 30, (3.1), and $c_{0,0}(z) \equiv 1$ give

$$B^{**}_{i,j}(\zeta) \equiv \sum_{\mu=0}^{i} \sum_{\nu=0}^{j} (f'(\zeta))^{\mu+\nu} \alpha_{i-\mu,m-i}(\zeta)\, \alpha_{j-\nu,m-j}(\zeta)\, c_{\mu,\nu}(f(\zeta))$$

$$- \left[\sum_{\mu=0}^{i} (f'(\zeta))^{\mu} \alpha_{i-\mu,m-i}(\zeta)\, c_{0,\mu}(f(\zeta)) \right] \left[\sum_{\nu=0}^{j} (f'(\zeta))^{\nu} \alpha_{j-\nu,m-j}(\zeta)\, c_{0,\nu}(f(\zeta)) \right]$$

$$\equiv \sum_{\mu=0}^{i} \sum_{\nu=0}^{j} (f'(\zeta))^{\mu+\nu} \alpha_{i-\mu,m-i}(\zeta)\, \alpha_{j-\nu,m-j}(\zeta) \left[c_{\mu,\nu}(f(\zeta)) - c_{0,\mu}(f(\zeta))\, c_{0,\nu}(f(\zeta)) \right]$$

$$\equiv \sum_{\mu=1}^{i} \sum_{\nu=1}^{j} (f'(\zeta))^{\mu+\nu} \alpha_{i-\mu,m-i}(\zeta)\, \alpha_{j-\nu,m-j}(\zeta)\, B_{\mu,\nu}(f(\zeta)).$$

This yields (15.30) and completes the proof. □

THEOREM 15.10. *For $1 \leq k \leq m$, the polynomial $\boldsymbol{D}_k$ defined in $\mathcal{S}_m$ by (15.24) is a relative invariant of weight $k(k+1)$ for the equations (1.1) of order m.*

PROOF. Suppose that (1.2) transforms (1.1) on Ω into (1.3) on $\tilde{\Omega}$. In terms of the field $\mathfrak{F}_\Omega$ of meromorphic functions on Ω and the function $\rho(z) \not\equiv 0$ in $\mathfrak{F}_\Omega$ for (1.2), let L and U denote the $k \times k$ matrices over $\mathfrak{F}_\Omega$ whose (i,j)-components are given by

$$(15.31) \qquad [L]_{i,j} \equiv \begin{cases} \binom{m-j}{i-j} \dfrac{\rho^{(i-j)}(z)}{\rho(z)}, & \text{for } 1 \leq j \leq i \leq k, \\ 0, & \text{for } 1 \leq i < j \leq k, \end{cases}$$

and

$$(15.32) \qquad [U]_{i,j} \equiv \begin{cases} \binom{m-i}{j-i} \dfrac{\rho^{(j-i)}(z)}{\rho(z)}, & \text{for } 1 \leq i \leq j \leq k, \\ 0, & \text{for } 0 \leq j < i \leq k. \end{cases}$$

In particular, L is a lower-triangular matrix with 1's along its main diagonal and $det(L) = 1$. Furthermore, U is an upper-triangular matrix having 1's along its main diagonal and $det(U) = 1$. For the function $D_k(z)$ on Ω obtained by replacing each $w_{i,j}$ in $\boldsymbol{D}_k$ with the corresponding $c_{i,j}(z)$ from (1.1), we use (15.25) to obtain

$$(15.33) \qquad D_k(z) \equiv det(\mathfrak{B}_k(z)),$$

where $\mathfrak{B}_k(z)$ is the $k \times k$ matrix whose components are given by

$$(15.34) \quad [\mathfrak{B}_k(z)]_{i,j} \equiv B_{i,j}(z) \equiv c_{i,j}(z) - c_{0,i}(z)\, c_{0,j}(z), \quad \text{in } \mathfrak{F}_\Omega \text{ for } 1 \leq i,j \leq m,$$

We introduce the $k \times k$ matrix

$$(15.35) \qquad \mathfrak{X}_k(z) = L\, \mathfrak{B}_k(z)\, U, \quad \text{over } \mathfrak{F}_\Omega.$$

For $1 \leq i,j \leq k$, we employ (15.35), (15.31), (15.32), and (15.28) to verify that the (i,j)-component of $\mathfrak{X}_k(z)$ is given by

$$(15.36) \qquad [\mathfrak{X}_k(z)]_{i,j} \equiv \sum_{\mu=1}^{k} [L]_{i,\mu} \sum_{\nu=1}^{k} [\mathfrak{B}_k(z)]_{\mu,\nu}\, [U]_{\nu,j}$$

$$\equiv \sum_{\mu=1}^{i} \sum_{\nu=1}^{j} [L]_{i,\mu}\, [\mathfrak{B}_k(z)]_{\mu,\nu}\, [U]_{\nu,j}$$

$$\equiv \sum_{\mu=1}^{i} \sum_{\nu=1}^{j} \binom{m-\mu}{i-\mu}\binom{m-\nu}{j-\nu} \frac{\rho^{(i-\mu)}(z)\, \rho^{(j-\nu)}(z)}{(\rho(z))^2} B_{\mu,\nu}(z)$$

$$\equiv B_{i,j}^*(z).$$

After replacing each $w_{i,j}$ in (15.25) with the corresponding $c_{i,j}^*(z)$ from (1.3), we apply (15.36) to obtain $D_k^*(z) \equiv det(\mathfrak{X}_k(z))$. We combine this with (15.35) and (15.33) to deduce that

$$(15.37) \quad D_k^*(z) \equiv det(\mathfrak{X}_k(z)) \equiv det(L)\, det(\mathfrak{B}_k(z))\, det(U) \equiv det(\mathfrak{B}_k(z)) \equiv D_k(z).$$

Thus, $\boldsymbol{D}_k$ is a semi-invariant of the first kind for the equations (1.1).

15.2. RELATIVE INVARIANTS DEFINED BY DETERMINANTS

Suppose that (1.4) transforms (1.1) on Ω into (1.5) on Ω^{**}. In terms of the field $\mathfrak{F}_{\Omega^{**}}$ of meromorphic functions on Ω^{**}, the function $f(\zeta)$ for (1.4), and the analytic functions $\alpha_{i,j}(\zeta)$ defined in $\mathfrak{F}_{\Omega^{**}}$ by means of (3.1)–(3.2) on page 29, let P and Q denote the $k \times k$ matrices over $\mathfrak{F}_{\Omega^{**}}$ whose (i, j)-components are given by

(15.38) $$[P]_{i,j} \equiv \begin{cases} (f'(\zeta))^j \, \alpha_{i-j,m-i}(\zeta), & \text{for } 1 \leq j \leq i \leq k, \\ 0, & \text{for } 1 \leq i < j \leq k, \end{cases}$$

and

(15.39) $$[Q]_{i,j} \equiv \begin{cases} (f'(\zeta))^i \, \alpha_{j-i,m-j}(\zeta), & \text{for } 1 \leq i \leq j \leq k, \\ 0, & \text{for } 0 \leq j < i \leq k. \end{cases}$$

In particular, P is a lower-triangular matrix. In view of (15.38) and $\alpha_{0,j}(\zeta) \equiv 1$ from (3.1) on page 29, we find that $[P]_{i,i} \equiv (f'(\zeta))^i$, for $1 \leq i \leq k$. Hence, we have

(15.40) $$\det(P) \equiv \prod_{i=1}^{k} (f'(\zeta))^i \equiv (f'(\zeta))^{k(k+1)/2}.$$

Moreover, Q is an upper-triangular matrix. Since (15.39) and $\alpha_{0,j}(\zeta) \equiv 1$ from (3.1) yield $[Q]_{i,i} \equiv (f'(\zeta))^i$, for $1 \leq i \leq k$, we find that

(15.41) $$\det(Q) \equiv \prod_{i=1}^{k} (f'(\zeta))^i \equiv (f'(\zeta))^{k(k+1)/2}.$$

We set $z = f(\zeta)$ in (15.33) and (15.34) to obtain

(15.42) $$D_k(f(\zeta)) \equiv \det\bigl(\mathfrak{B}_k(f(\zeta))\bigr), \quad \text{over } \mathfrak{F}_{\Omega^{**}},$$

where $\mathfrak{B}_k(f(\zeta))$ is the $k \times k$ matrix having

(15.43) $$[\mathfrak{B}_k(f(\zeta))]_{i,j} \equiv B_{i,j}(f(\zeta)) \equiv c_{i,j}(f(\zeta)) - c_{0,i}(f(\zeta)) \, c_{0,j}(f(\zeta)),$$
$$\text{for } 1 \leq i, j \leq k.$$

We introduce the $k \times k$ matrix

(15.44) $$\mathfrak{Y}_k(\zeta) = P \, \mathfrak{B}_k(f(\zeta)) \, Q, \quad \text{over } \mathfrak{F}_{\Omega^{**}}.$$

For $1 \leq i, j \leq k$, we employ (15.44), (15.38), (15.39), (15.43), and (15.30) to verify that the (i, j)-component of $\mathfrak{Y}_k(\zeta)$ is given by

(15.45) $$[\mathfrak{Y}_k(\zeta)]_{i,j} \equiv \sum_{\mu=1}^{k} [P]_{i,\mu} \sum_{\nu=1}^{k} [\mathfrak{B}_k(f(\zeta))]_{\mu,\nu} \, [Q]_{\nu,j}$$
$$\equiv \sum_{\mu=1}^{i} \sum_{\nu=1}^{j} [P]_{i,\mu} \, [\mathfrak{B}_k(f(\zeta))]_{\mu,\nu} \, [Q]_{\nu,j}$$
$$\equiv \sum_{\mu=1}^{i} \sum_{\nu=1}^{j} (f'(\zeta))^{\mu+\nu} \alpha_{i-\mu,m-i}(\zeta) \, \alpha_{j-\nu,m-j}(\zeta) \, B_{\mu,\nu}(f(\zeta))$$
$$\equiv B_{i,j}^{**}(\zeta).$$

After replacing each $\boldsymbol{w}_{i,j}$ in (15.25) with the corresponding $c_{i,j}^{**}(\zeta)$ from (1.5), we apply (15.45) to obtain $D_k^{**}(\zeta) \equiv det(\mathfrak{Y}_k(\zeta))$. We combine this with (15.44), (15.40), (15.41), and (15.42) to deduce that

$$(15.46) \quad D_k^{**}(\zeta) \equiv det(\mathfrak{Y}_k(\zeta)) \equiv det(P)\, det(\mathfrak{B}_k(f(\zeta)))\, det(Q)$$
$$\equiv (f'(\zeta))^{k(k+1)} det(\mathfrak{B}_k(f(\zeta))) \equiv (f'(\zeta))^{k(k+1)} D_k(f(\zeta)).$$

Thus, $\boldsymbol{D}_k$ is a semi-invariant of the second kind of weight $k(k+1)$ for the equations (1.1) having order m.

Consequently, in view of (15.37) and (15.46), $\boldsymbol{D}_k$ is a relative invariant of weight $k(k+1)$ for the equations (1.1). This completes the proof of Theorem 15.10. □

Theorem 15.10 and our Main Theorem, page 8, show that: for $1 \leq k \leq m$, the polynomial $\boldsymbol{D}_k$ in (15.24) is expressible as a differential-polynomial combination over $\mathbb{Q}$ of $\boldsymbol{a}_{m,2}$ and the polynomials in

$$\mathcal{B}_{m,0} = \{\boldsymbol{\mathcal{I}}_{m,l,n} | \text{ for } 1 \leq l \leq n \leq m\} \bigcup \{\boldsymbol{\mathcal{I}}_{m,0,n} | \text{ for } 3 \leq n \leq m\}.$$

We recall that $\mathcal{B}_{m,0}$ is the set of basic relative invariants for the equations (1.1) of order m.

EXAMPLE 15.11. For $k = 1$ and $m \geq 1$, we have $\boldsymbol{D}_1 \equiv \boldsymbol{w}_{1,1} - (\boldsymbol{w}_{0,1})^2 \equiv \boldsymbol{\mathcal{I}}_{m,1,1}$. For $k = m = 2$, $\boldsymbol{D}_2$ was expressed as a differential-polynomial combination of $\boldsymbol{a}_{2,2}$, $\boldsymbol{\mathcal{I}}_{2,1,1}$, $\boldsymbol{\mathcal{I}}_{2,1,2}$, and $\boldsymbol{\mathcal{I}}_{2,2,2}$ over $\mathbb{Q}$ in (1.65) of Example 1.9 on page 14.

For $m \geq 2$, $\boldsymbol{D}_2$ is a relative invariant for equations (1.1) of order m. Moreover, $\boldsymbol{a}_{m,2}$, $\boldsymbol{\mathcal{I}}_{m,1,1}$, $\boldsymbol{\mathcal{I}}_{m,1,2}$, and $\boldsymbol{\mathcal{I}}_{m,2,2}$ are explicitly given in (1.14), (1.49), (1.53), and (1.59). Section 16.1 shows how machine computations and the algorithmic proof on pages 144–145 for Theorem 14.3 enabled us to discover the identity

$$(15.47) \quad \boldsymbol{D}_2 \equiv \left[\boldsymbol{\mathcal{I}}_{m,2,2}\,\boldsymbol{\mathcal{I}}_{m,1,1} - (\boldsymbol{\mathcal{I}}_{m,1,2})^2\right]$$
$$+ \frac{(m-1)}{3}\left[\boldsymbol{\mathcal{I}}_{m,1,1}\,\boldsymbol{\mathcal{I}}_{m,1,2}^{(1)} - \frac{3}{2}\boldsymbol{\mathcal{I}}_{m,1,1}^{(1)}\,\boldsymbol{\mathcal{I}}_{m,1,2}\right]$$
$$+ \frac{(m-1)^2}{20}\left[\boldsymbol{\mathcal{I}}_{m,1,1}\,\boldsymbol{\mathcal{I}}_{m,1,1}^{(2)} - \frac{5}{4}(\boldsymbol{\mathcal{I}}_{m,1,1}^{(1)})^2 - 4\boldsymbol{a}_{m,2}(\boldsymbol{\mathcal{I}}_{m,1,1})^2\right],$$

which is valid for any $m \geq 2$. Proposition 14.6 on page 146 and Corollary 14.9 on page 148 show that each of the three bracketed expressions in the right member of (15.47) is a relative invariant of weight 6 for the equations (1.1) of order m. For $m = 2$, (15.47) yields (1.65) on page 14.

CHAPTER 16

Machine Computations

Here, we indicate our use of MATHEMATICA from [51] to deduce (15.47), to generate basic relative invariants, and to check various formulas. The commands used can be easily modified for other systems of computer algebra. To represent the derivation $'$ for $\mathcal{S}_m$ we use a fictitious independent variable that we choose to write as z. Then, $w_{i,j}^{(k)}$ can be interpreted as the kth derivative of $w_{i,j}$ with respect to z. We represent it in the system as the evaluation of D[w[i,j][z],{z,k}].

16.1. Expansion of D_2 in terms of $a_{m,2}$, $\mathcal{I}_{m,1,1}$, $\mathcal{I}_{m,1,2}$, and $\mathcal{I}_{m,2,2}$

The identity $w_{1,1} \equiv \mathcal{I}_{m,1,1} + (w_{0,1})^2$ is obtained by solving for $w_{1,1}$ in (1.49) on page 11. Its right member in the machine commands listed below is represented by the evaluation of x[1,1][z]. The evaluations of x[0,2][z], x[1,2][z], and x[2,2][z] represent the right members of the expressions obtained when (1.14), (1.53), and (1.59) are respectively solved for $w_{0,2}$, $w_{1,2}$ and $w_{2,2}$. We have found that: the successive evaluations with MATHEMATICA of the instructions

```
B[1,1][z_] = w[1,1][z] - (w[0,1][z])^2;

B[1,2][z_] = w[1,2][z] - w[0,1][z]*w[0,2][z];

B[2,2][z_] = w[2,2][z] - (w[0,2][z])^2;

    D2[z_] = B[1,1][z]*B[2,2][z] - (B[1,2][z])^2;

x[2,2][z_] = ( Inv[m,2,2][z] + (2(m-1)/m)*w[0,1][z]*w[1,2][z]
             - (6(m-1)/(5(m+1)m))*w[0,2][z]*w[1,1][z]
             + (w[0,2][z])^2
             - ((5m+2)(m-1)^2/(5(m+1)m^2))*(w[0,1][z])^2*w[1,1][z]
             - (2(5m+2)(m-1)/(5(m+1)m))*(w[0,1][z])^2*w[0,2][z]
             + ((5m+2)(m-1)^2/(5(m+1)m^2))*(w[0,1][z])^4
             + ((m-1)/3)*D[w[1,2][z], z]
             - ((m-1)^2/30)*D[w[1,1][z],{z,2}]
             - ((m-1)^2/(3m))*w[0,1][z]*D[w[1,1][z], z]
             - ((5m-4)(m-1)^2/(15(m+1)m))*D[w[0,1][z], z]*w[1,1][z]
          - ((m-1)/3)*w[0,1][z]*D[w[0,2][z],z]
         - ((m-1)/3)*D[w[0,1][z],z]*w[0,2][z]
        + ((m-1)^2/15)*w[0,1][z]*D[w[0,1][z],{z,2}]
        + ((m-1)^2/15)*(D[w[0,1][z],z])^2
        + ((5m+2)(m-1)^2/(5(m+1)m))*(w[0,1][z])^2*D[w[0,1][z],z] );
```

```
x[1,2][z_] = ( Inv[m,1,2][z] + ((m-1)/m)*w[0,1][z]*w[1,1][z]
                + w[0,1][z]*w[0,2][z]
                - ((m-1)/m)(w[0,1][z])^3
                + ((m-1)/4)*D[w[1,1][z],z]
                - ((m-1)/2)w[0,1][z]*D[w[0,1][z],z] );

x[1,1][z_] = Inv[m,1,1][z] + (w[0,1][z])^2;

x[0,2][z_] = ( Binomial[m+1,3]*a[m,0,2][z]
                + ((m-1)/2)*D[w[0,1][z],z]
                + ((m-1)/(2m))(w[0,1][z])^2 );

A1 = Expand[ D2[z] /. w[2,2][z] -> x[2,2][z] ];

A2 = Expand[ A1 /. w[1,2]'[z] -> x[1,2]'[z] ];

A3 = Expand[ A2 /. w[1,2][z] -> x[1,2][z] ];

A4 = Expand[ A3 /. w[1,1]''[z] -> x[1,1]''[z] ];

A5 = Expand[ A4 /. w[1,1]'[z] -> x[1,1]'[z] ];

A6 = Expand[ A5 /. w[1,1][z] -> x[1,1][z] ];

A7 = Expand[ FullSimplify[ A6 /. w[0,2][z] -> x[0,2][z] ] ]
```

yield the evaluation A7 for D_2 from which the representations of $w_{2,2}$, $w_{1,2}^{(1)}$, $w_{1,2}$, $w_{1,1}^{(2)}$, $w_{1,1}^{(1)}$, $w_{1,1}$, and $w_{0.2}$ have been eliminated. In fact, this expression A7 for D_2 involves only the symbols representing $\mathcal{I}_{m,2,2}$, $\mathcal{I}_{m,1,2}^{(1)}$, $\mathcal{I}_{m,1,2}$, $\mathcal{I}_{m,1,1}^{(2)}$, $\mathcal{I}_{m,1,1}^{(1)}$, $\mathcal{I}_{m,1,1}$, and $a_{m,2}$. We rewrite it to easily obtain the expansion for D_2 in (15.47) on page 166. The special case $m = 2$ yields (1.65) on page 14.

16.2. The expansions for E_6 and E_7 in (1.81) and (1.82)

After MATHEMATICA has evaluated the expressions given in Section 16.1, we have found that the additional sequential evaluation of

```
    m = 2;

E6[z_] = ( B[1,1][z]*B[2,2][z] - 2(B[1,2][z])^2
            + w[0,1][z]*B[1,1][z]*B[1,2][z] - w[0,2][z](B[1,1][z])^2
            + B[1,1][z]*D[B[1,2][z],z] - D[B[1,1][z],z]*B[1,2][z] );

H7[z_] = ( B[1,1][z]*D[B[2,2][z],z] - D[B[1,1][z],z]*B[2,2][z]
            - 2B[1,2][z]*B[2,2][z] + 2w[0,1][z]*B[1,1][z]*B[2,2][z]
            - 2w[0,2][z]*B[1,1][z]*B[1,2][z] );
```

```
E7[z_] = ( H7[z] - (1/6)(D[E6[z],z] + 6w[0,1][z]*E6[z]) );

{B1, C1} = Expand[ {E6[z], E7[z]} /. w[2,2]'[z] -> x[2,2]'[z] ];

{B2, C2} = Expand[ {B1, C1} /. w[2,2][z] -> x[2,2][z] ];

{B3, C3} = Expand[ {B2, C2} /. w[1,2]''[z] -> x[1,2]''[z] ];

{B4, C4} = Expand[ {B3, C3} /. w[1,2]'[z] -> x[1,2]'[z] ];

{B5, C5} = Expand[ {B4, C4} /. w[1,2][z] -> x[1,2][z] ];

{B6, C6} = Expand[ {B5, C5} /. w[1,1]''[z] -> x[1,1]''[z] ];

{B7, C7} = Expand[ {B6, C6} /. w[1,1]'[z] -> x[1,1]'[z] ];

{B8, C8} = Expand[ {B7, C7} /. w[1,1][z] -> x[1,1][z] ];

{B9, C9} = Expand[ {B8, C8} /. w[0,2][z] -> x[0,2][z] ];
```

yields useful expressions for B9 and C9. Namely, B9 represents E_6 in terms $\mathcal{I}_{2,2,2}$, $\mathcal{I}^{(1)}_{2,1,2}$, $\mathcal{I}_{2,1,2}$, $\mathcal{I}^{(2)}_{2,1,1}$, $\mathcal{I}^{(1)}_{2,1,1}$, $\mathcal{I}_{2,1,1}$, and $a_{2,0,2}$; it immediately gives (1.81). And, C9 represents E_7 in terms of $\mathcal{I}^{(1)}_{2,2,2}$, $\mathcal{I}_{2,2,2}$, $\mathcal{I}^{(2)}_{2,1,2}$, $\mathcal{I}^{(1)}_{2,1,2}$, $\mathcal{I}_{2,1,2}$, $\mathcal{I}^{(2)}_{2,1,1}$, $\mathcal{I}^{(1)}_{2,1,1}$, $\mathcal{I}_{2,1,1}$, and $a_{2,0,2}$; it easily establishes (1.82).

16.3. A comprehensive check on the consistency of (1.14)–(1.38)

With respect to $\mathcal{I}_{m,l,n}$ in (1.28) and $\mathcal{J}_{m,l,n}$ in (1.38), the equality

$$\mathcal{I}_{m,l,n} \equiv \mathcal{J}_{m,l,n}, \quad \text{for } 1 \leq l \leq n \leq m \quad \text{or} \quad l = 0 < 3 \leq n \leq m,$$

was established in Theorem 11.1 on page 119. It is the principal result needed for our Main Theorem on page 8. Computer algebra provides an excellent check on the correctness of this relation. Regarding m as a symbol for any integer $\geq n$, we shall evaluate $\mathcal{I}_{m,l,n} - \mathcal{J}_{m,l,n}$, for various specific integers l and n. The technique presented next also provides a useful method to represent the basic relative invariants for use in other computations.

Using MATHEMATICA (Version 3.0 or later) without previous assignments, we find that the successive evaluations of the following twenty-five commands

```
p = 7;   Date[]

TrueQ[ $VersionNumber >= 3.0]

a[2][z_]:= b[2][z] - D[b[1][z],z]/2 -(b[1][z])^2/4

w[0,0][z_] = 1;

w[0,1][z_] = Binomial[m,2]*b[1][z];
```

```
w[0,2][z_] = ( Binomial[m+1,3]*a[2][z] +
    ((m-1)/2)*D[w[0,1][z],z] + ((m-1)/(2m))*(w[0,1][z])^2 );

w[j_,i_][z_] := w[i,j][z] /; j > i

Do[ SumOfPowers[n,j_] = Sum[k^n,{k,j+1,m}], {n, 0, 2*p+1} ]

K[-1,j_] = 0;

K[0,j_] = 1;

summandK[i_,j_] := Expand[ D[K[i,j],z]
    - ((m-1)*b[1][z]*K[i,j])/2
    + (m-i-j+1)*(1-i-j)*a[2][z]*K[i-1,j] ]

Do[ K[i,j_] = Module[{t,k},
    t = Expand[summandK[i-1,k]];
    Sum[Coefficient[t,k,n]*SumOfPowers[n,j],
    {n,0,2*i-1}]], {i,1,p} ]

L[i_,j_] := Sum[ K[i-p,p]*K[j-q,q]*w[p,q][z],
    {p,0,i},{q,0,j} ]

M[l_,n_,h_,i_] := Binomial[m-h,l-h]*
    (Product[(l-r),{r,1,l-h}])*Binomial[m-i+h,n-i+h]*
    (Product[(n-s),{s,1,n-i+h}])*L[h,i-h]

U[-1,j_] = 0;

U[0,j_] = 1;

summandForU[i_,j_] := Expand[
    D[U[i,j],z] +
    (i+j-m)*b[1][z]*U[i,j] +
    (m-i-j+1)*(1-i-j)*b[2][z]*U[i-1,j] ]

Do[ U[i,j_] = Module[{u,r},
    u = Expand[summandForU[i-1,r]];
    Sum[Coefficient[u,r,n]*SumOfPowers[n,j],
    {n,0,2*i-1}]], {i,1,p} ]

V[i_,j_] := Sum[ U[i-p,p]*U[j-q,q]*w[p,q][z],
    {p,0,i}, {q,0,j} ]

W[l_,n_,h_,i_] := Binomial[m-h,l-h]*
    (Product[(l-r),{r,1,l-h}])*Binomial[m-i+h,n-i+h]*
    (Product[(n-s),{s,1,n-i+h}])*V[h,i-h]
```

16.3. A COMPREHENSIVE CHECK ON THE CONSISTENCY OF (1.14)–(1.38)

```
A[l_,n_,i_] := -1/(1+n+i-1)

B[l_,n_,i_] := (1+n-i)/(1+n+i-2)

difference[l_,n_] := Module[{t,u},
    Do[ t[h,0] = 0, {h,0,l} ];
    Do[ t[h,1] = 0, {h,0,l} ];
    Do[ t[h,i] = M[l,n,h,i]
        + A[l,n,i-1]*D[ t[h,i-1],z]
        + B[l,n,i-1]*a[2][z]*t[h,i-2], {i,2,1+n}, {h,0,l} ];
    Inv[l,n] = Sum[ t[h,1+n], {h,0,l} ];
    Do[ u[h,0] = 0, {h,0,l} ];
    Do[ u[h,1] = 0, {h,0,l} ];
    Do[ u[h,i] = W[l,n,h,i]
        + A[l,n,i-1]*(D[u[h,i-1],z] + (i-1)*b[1][z]*u[h,i-1])
        + B[l,n,i-1]*b[2][z]*u[h,i-2], {i,2,1+n}, {h,0,l} ];
    J[l,n] = Sum[ u[h,1+n], {h,0,l} ];
    Print["The number of terms in Inv[", l, ",", n, "] is ",
        Length[ Expand[ Inv[l,n] ] ]];
    Print["The number of terms in J[", l, ",", n, "] is ",
        Length[ Expand[ J[l,n] ] ]];
    Print["The memory currently used (in bytes) is ",
        MemoryInUse[] ];
    Print["Inv[", l, ",", n, "] - J[", l, ",", n, "] = ",
        Expand[ Inv[l,n] - J[l,n] ]];
    Print["The maximum memory used (in bytes) is ",
        MaxMemoryUsed[] ];
    Print[ Date[] ];  Print["      "]; Clear[t,u,Inv,J] ]

Do[ difference[0,n], {n,3,p} ]

Do[ difference[l,n], {l,1,p}, {n,1,p-1} ]
```

verify $\mathcal{I}_{m,0,n} \equiv \mathcal{J}_{m,0,n}$, for $3 \leq n \leq p = 7$ and $n \leq m$, as well as $\mathcal{I}_{m,l,n} \equiv \mathcal{J}_{m,l,n}$, for $1 \leq l \leq n \leq p = 7$ and $n \leq m$. The initial assignment p = 7 was made so that all twenty-five of the preceding commands can be evaluated in less than three minutes on typical personal computers. The assignment p = 7 can be increased to p = 8, p = 9, p = 10, etc. until the capacity of the computer is exceeded or the computation times become inconveniently long. This procedure is a natural extension of the one presented in [**24**, pages 26–30]. Here, the main advance is the introduction of the function difference[l,n]. By conserving memory requirements, it improves the efficiency of the earlier computation. Also, elimination of the command Together[ ] used in [**24**, page 29] is recommended. Otherwise, some issues of MATHEMATICA after Version 3.0 will complete the computations too slowly.

A similar check on the formulas in Chapter 24 and the identity (27.34) in Theorem 27.7 on page 314 can be obtained by making suitable alterations in the preceding computation.

16.4. The relative invariants of weight ≤ 9 for the equations $Q_2 = 0$

For a fixed integer $s \geq 1$, let a nonnegative integer $e_{s,i,j,k}$ be given for each triple (i, j, k) of integers that satisfy

$$0 \leq i \leq 2, \quad \max(1, i) \leq j \leq 2, \quad \text{and} \quad 0 \leq k \leq s - i - j.$$

Then, the expression

(16.1)
$$M = \prod_{\substack{0 \leq i \leq 2 \\ \max(1, i) \leq j \leq 2 \\ 0 \leq k \leq s-i-j}} \left[w_{i,j}^{(k)} \right]^{e_{s,i,j,k}}$$

is a monic monomial of weight s in $\mathcal{S}_2$ from page 5 if and only if the condition

(16.2)
$$\sum_{\substack{0 \leq i \leq 2 \\ \max(1, i) \leq j \leq 2 \\ 0 \leq k \leq s-i-j}} (i + j + k)\, e_{s,i,j,k} = s$$

is satisfied. Moreover, each monic monomial of weight s in $\mathcal{S}_2$ is expressible in the form (16.1) for some nonnegative integers $e_{s,\,i,\,j,\,k}$ that satisfy (16.2).

The isobaric polynomials of weight s in $\mathcal{S}_2$ are the nonzero linear combinations over $\mathbb{Q}$ of the monic monomials of weight s in $\mathcal{S}_2$. Thus, our main task is to acquire an efficient means of computing and enumerating the monic monomials of weight s in $\mathcal{S}_2$. This requires that the trial-and-error assignment of integers $e_{s,\,i,\,j,\,k}$ be efficiently restricted to reduce computation times. The first two of the following three MATHEMATICA commands

```
u[s_,p_,q_,r_] := Floor[ (1/(p+q+r))*( s
    -Sum[(p+q+k)*e[s,p,q,k],{k,r+1,s-p-q}]
    -Sum[(p+j+k)*e[s,p,j,k],{j,q+1,2},{k,0,s-p-j}]
    -Sum[(i+j+k)*e[s,i,j,k],{i,p+1,2},{j,Max[1,i],2},{k,0,s-i-j}])]

nM[s_] := Module[ {counter, seq}, counter = 0;
    seq = Apply[ Sequence, Reverse[ Flatten[
        Table[ {e[s,i,j,k], 0, u[s,i,j,k]},
        {i,0,2},{j,Max[1,i],2},{k,0,s-i-j}], 2 ] ] ];
    Do[ If[ (Sum[(i+j+k)*e[s,i,j,k],
        {i,0,2},{j,Max[1,i],2},{k,0,s-i-j}]) == s,
        ( counter = counter + 1;
        monM[s,counter] =
        Product[ D[w[i,j][z],{z,k}]^e[s,i,j,k],
        {i,0,2},{j,Max[1,i],2},{k,0,s-i-j}] ),
        counter = counter ],
    Evaluate[seq]]; counter ]

printM[s_] := Do[ Print["monM[",s,", ",i,"] = ",monM[s,i]],
        {i,1,nM[s]}]
```

accomplish that purpose. For instance, after all three are entered, the evaluation of printM[2] computes and prints the four monic monomials in $\mathcal{S}_2$ of weight 2.

16.4. THE RELATIVE INVARIANTS OF WEIGHT ≤ 9 FOR THE EQUATIONS $Q_2 = 0$

Any isobaric polynomial $\boldsymbol{P}$ of weight s in $\mathcal{S}_2$ is representable by the input

```
wPolynomial[s_] := Sum[ x[s,i]*monM[s,i], {i,1,nM[s]} ]
```

for some rational numbers `x[s,i]`. In view of (1.9)–(1.10), the expressions

$$(16.3) \qquad \rho(z)^s\bigl[P^*(z) - P(z)\bigr] \quad \text{and} \quad \bigl(f'(\zeta)\bigr)^s\bigl[P^{**}(\zeta) - (f'(\zeta))^s P(f(\zeta))\bigr]$$

are identically zero if and only if $\boldsymbol{P}$ is a relative invariant. We use (2.5) and (3.9) to represent $P(z)$, $P^*(z)$, $P^{**}(\zeta)$, as well as the expressions in (16.3) through

```
sub[s_] := Table[ D[w[i,j][z],{z,k}]->D[c[i,j][z],{z,k}],
               {i,0,2},{j,Max[1,i],2},{k,0,s-i-j}] //Flatten;

cPoly[s_] := wPolynomial[s] /. sub[s]

m = 2;   c[0,0][z_] = 1;   c[j_,i_][z_] := c[i,j][z] /; j > i;

cOneStar[i_,j_][z_] := Sum[ Binomial[m-mu,i-mu]*
   Binomial[m-nu,j-nu]*D[ rho[z], {z,i-mu}]*
   D[ rho[z], {z,j-nu}]/rho[z]^2*c[mu,nu][z], {mu,0,i},{nu,0,j}];

sub1[s_] := Flatten[ Table[
             D[w[i,j][z],{z,k}]->D[cOneStar[i,j][z],{z,k}],
             {i,0,2},{j,Max[1,i],2},{k,0,s-i-j} ] ]

cOneStarPoly[s_] := wPolynomial[s] /. sub1[s]

alpha[0][j_,zet_] = 1;   g[zet_] = f''[zet]/f'[zet];

alpha[1][j_,zet_] = g[zet]*Sum[-k, {k,1,j}];

alpha[2][j_,zet_] = ( D[g[zet],zet]*Sum[-k*(k+1)/2, {k,1,j}]
            + g[zet]^2*Sum[(1+k)^2*k/2,{k,1,j}] )//Simplify;

cTwoStar[i_,j_][zet_] := Sum[ f'[zet]^(mu+nu)*
   alpha[i-mu][m-i,zet]*alpha[j-nu][m-j,zet]*
   c[mu,nu][f[zet]],{mu,0,i},{nu,0,j}]

sub2[s_] := Flatten[ Table[
             D[ w[i,j][z],{z,k}]->D[ cTwoStar[i,j][zet],{zet,k}],
             {i,0,2},{j,Max[1,i],2},{k,0,s-i-j} ] ];

cTwoStarPoly[s_] := wPolynomial[s] /. sub2[s]

firstCondition[s_] := firstCondition[s] = Expand[
                 rho[z]^s*(cOneStarPoly[s] - cPoly[s]) ]

secondCondition[s_] := secondCondition[s] = Expand[ f'[zet]^s*
     ( cTwoStarPoly[s] - f'[zet]^s*(cPoly[s] /. z->f[zet]) )]
```

and note that conditions (1.9)–(1.10) of page 6 require the coefficients x[s,i] of wPolynomial[s] to be such that firstCondition[s] and secondCondition[s] are identically zero.

We can represent $\rho(z)^s [P^*(z) - P(z)]$ as a linear combination over $\mathbb{Q}$ of the various expressions

$$(16.4) \qquad \prod_{\substack{0 \leq i \leq 2 \\ \max(1,i) \leq j \leq 2 \\ 0 \leq k \leq s-i-j}} \left[c_{i,j}^{(k)}(z) \right]^{e_{s,i,j,k}} \prod_{0 \leq t \leq s} \left[\rho^{(t)}(z) \right]^{h_{s,t}}$$

that correspond to nonnegative integers $e_{s,i,j,k}$, $h_{s,t}$ satisfying

$$(16.5) \qquad \sum_{\substack{0 \leq i \leq 2 \\ \max(1,i) \leq j \leq 2 \\ 0 \leq k \leq s-i-j}} (i+j+k) \, e_{s,i,j,k} + \sum_{t=1}^{s} t \, h_{s,t} = s \quad \text{and} \quad \sum_{t=0}^{s} h_{s,t} = s.$$

After the evaluation of the input statement beginning with u[s_,i_,j_,k_] on page 172, the expressions corresponding to (16.4) needed for firstCondition[s] can be obtained from the evaluation of the first two of the following input commands

```
v[s_,p_] := Floor[(1/p)*( s
   - Sum[ (i+j+k)*e[s,i,j,k],{i,0,2},{j,Max[1,i],2},{k,0,s-i-j}]
   - Sum[ t*h[s,t], {t,p+1,s}] ) ]

nA[s_] := Module[ {counter, seq1, seq2}, counter = 0;
   seq1 = Apply[ Sequence, Reverse[ Flatten[
       Table[ {e[s,i,j,k], 0, u[s,i,j,k]},
       {i,0,2},{j,Max[1,i],2},{k,0,s-i-j}], 2] ] ];
   seq2 = Apply[ Sequence, Reverse[
       Table[ {h[s,t],0,v[s,t]},{t,1,s}]]];
   Do[ If[(Sum[(i+j+k)*e[s,i,j,k],{i,0,2},
       {j,Max[1,i],2},{k,0,s-i-j}]
       + Sum[t*h[s,t],{t,1,s}]) == s,
       ( counter = counter + 1;
       monA[s,counter] =
       Product[ D[c[i,j][z],{z,k}]^e[s,i,j,k],
       {i,0,2},{j,Max[1,i],2},{k,0,s-i-j}]*
       rho[z]^(s - Sum[h[s,t],{t,1,s}])*
       Product[ D[ rho[z],{z,t}]^h[s,t], {t,1,s}] ),
       counter = counter ],
       Evaluate[seq1], Evaluate[seq2] ]; counter ]

printA[s_] := Do[ Print["monA[",s,", ",i,"] = ",monA[s,i]],
       {i,1,nA[s]}]
```

via MATHEMATICA. For example, after the three preceding input statements and the one beginning with u[s_,p_,q_,r_] are evaluated, the evaluation of printA[2] prints the seven expressions of the form (16.4) that satisfy (16.5) for $s = 2$.

16.4. THE RELATIVE INVARIANTS OF WEIGHT ≤ 9 FOR THE EQUATIONS $Q_2 = 0$ 175

We can represent $(f'(\zeta))^s [P^{**}(\zeta) - (f'(\zeta))^s P(f(\zeta))]$ as a linear combination over $\mathbb{Q}$ of the various expressions

(16.6) $$\prod_{\substack{0 \leq i \leq 2 \\ \max(1,i) \leq j \leq 2 \\ 0 \leq k \leq s-i-j}} \left[c_{i,j}^{(k)}(f(\zeta)) \right]^{e_{s,i,j,k}} \prod_{0 \leq t \leq s} \left[f^{(t+1)}(z) \right]^{h_{s,t}}$$

that correspond to nonnegative integers $e_{s,i,j,k}$, $h_{s,t}$ satisfying

(16.7) $$\sum_{\substack{0 \leq i \leq 2 \\ \max(1,i) \leq j \leq 2 \\ 0 \leq k \leq s-i-j}} (i+j+k) \, e_{s,i,j,k} + \sum_{t=1}^{s} t \, h_{s,t} = s \quad \text{and} \quad \sum_{t=0}^{s} (t+1) \, h_{s,t} = 2s.$$

After the two input statements beginning with u[s_,i_,j_,k_] and v[s_,p_] on pages 172 and 174 are evaluated, the expressions corresponding to (16.6) needed for secondCondition[s] can be obtained from the evaluation of the first of the following two input commands

```
nB[s_] := Module[ {counter, seq1, seq2}, counter = 0;
    seq1 = Apply[ Sequence, Reverse[ Flatten[
        Table[ {e[s,i,j,k], 0, u[s,i,j,k]},
        {i,0,2},{j,Max[1,i],2},{k,0,s-i-j}], 2] ] ];
    seq2 = Apply[ Sequence, Reverse[
        Table[ {h[s,t],0,v[s,t]},{t,1,s}]]];
    Do[ If[ (Sum[(i+j+k)*e[s,i,j,k],
        {i,0,2},{j,Max[1,i],2},{k,0,s-i-j}]
        + Sum[ t*h[s,t],{t,1,s}]) == s,
        ( counter = counter + 1;
        monB[s,counter] = Product[
        (D[c[i,j][z],{z,k}] /. z->f[zet] )^e[s,i,j,k],
        {i,0,2},{j,Max[1,i],2},{k,0,s-i-j}]*
        f'[zet]^(2s - Sum[(t+1)*h[s,t],{t,1,s}])*
        Product[ D[f[zet],{zet,t+1}]^h[s,t], {t,1,s}] ),
        counter = counter ],
        Evaluate[seq1], Evaluate[seq2] ]; counter ]

printB[s_] := Do[ Print["monB[",s,", ",i,"] = ",monB[s,i]],
    {i,1,nB[s]}]
```

via MATHEMATICA. For example, after the two preceding input statements as well as those beginning with u[s_,p_,q_,r_] and v[s_,p_] are evaluated, the evaluation of printB[2] prints the seven expressions of the form (16.6) that satisfy (16.7) for $s = 2$.

For i = 1, ..., nA[s] and for j = 1, ..., nB[s], the coefficient a[s,i] of monA[s,i] in first[s] and the coefficient b[s,j] of monB[s,i] in second[s] are linear combinations over $\mathbb{Q}$ of the coefficients x[s,1], ..., x[s,nM[s]] for wPoly[s]. The nonzero solutions for x[s,1], ..., x[s,nM[s]] of the system of linear equations obtained by equating to zero each a[s,i] and b[s,j] yield the polynomials wPoly[s] that are relative invariants for the equations $Q_2 = 0$. As corresponding input statements, we have

```
firstCoefficients[s_] := Do[ a[s,i] =
   Coefficient[firstCondition[s], monA[s,i]], {i,1,nA[s]}]

firstZeroCheck[s_] := Expand[ firstCondition[s]
                 - Sum[ a[s,i]*monA[s,i], {i,1,nA[s]}] ]

secondCoefficients[s_] := Do[ b[s,j] =
   Coefficient[ secondCondition[s], monB[s,j]], {j,1,nB[s]}]

secondZeroCheck[s_] := Expand[ secondCondition[s]
                 - Sum[ b[s,j]*monB[s,j], {j,1,nB[s]}] ]

eqs[s_] := Join[ Table[ a[s,i] == 0, {i,1,nA[s]}],
              Table[ b[s,j] == 0, {j,1,nB[s]}] ]

rules[s_] := Flatten[
           Solve[ eqs[s], Table[ x[s,i], {i,1,nM[s]}] ] ]

invariants[s_] :=
   Module[ {expr, counter}, counter = 0;
   Print["  "];
   Print["  Computation of invariants[",s,"]"];
   nM[s]; Print["Number of Monic Monomials = ", nM[s]];
   nA[s]; Print["Number of firstCondition Terms = ", nA[s]];
   nB[s]; Print["Number of secondCondition Terms = ", nB[s]];
   firstCondition[s];  firstCoefficients[s];
   Print["firstZeroCheck[",s,"] = ", firstZeroCheck[s]];
   secondCondition[s]; secondCoefficients[s];
   Print["secondZeroCheck[",s,"] = ", secondZeroCheck[s]];
   expr = Expand[
         ( Sum[ x[s,i]*monM[s,i], {i,1,nM[s]} ]) /. rules[s] ];
   Do[ If[ Length[ Coefficient[ expr, x[s,i]]]! = 0,
       ( counter = counter + 1;
       Inv[s,counter] = Coefficient[expr,x[s,i]];
       Print["Inv[", s, ", ", counter, "] = ", Inv[s,counter]] ),
       counter = counter],
   {i,1,nM[s]} ]; n[s] = counter; Print["n[",s,"] = ", n[s]] ]
```

for MATHEMATICA to evaluate. After all of the preceding input commands of this section are successively entered and evaluated, the entry and evaluation of

```
Do[ invariants[s], {s,2,8} ]
```

can be done by MATHEMATICA in less than 15 minutes on a modern personal computer. In terms of $n(2) = 1$, $n(3) = 1$, $n(4) = 2$, $n(5) = 1$, $n(6) = 5$, $n(7) = 4$, and $n(8) = 11$, we find that: for $2 \leq s \leq 8$, it yields $n(s)$ linearly independent relative invariants whose nonzero linear combinations over $\mathbb{Q}$ give all of the relative invariants of weight s for the equations $Q_2 = 0$. The evaluations of `invariants[9]`, `invariants[10]`, and `invariants[11]` require longer times and yield $n(9) = 11$, $n(10) = 22$, and $n(11) = 26$ respectively. Each of the relative invariants that these evaluations produce can be expressed as a differential-polynomial combination over $\mathbb{Q}$ of $a_{2,2}$ and the basic relative invariants $\mathcal{I}_{2,1,1}$, $\mathcal{I}_{2,1,2}$, and $\mathcal{I}_{2,2,2}$ by the technique of Sections 16.1–16.2, pages 167–169. When this is done for $2 \leq s \leq 9$ and $1 \leq i \leq n(s)$, the resulting expression for `Inv[s,i]` enables it to be recognized as a linear combination over $\mathbb{Q}$ of the relative invariants $\boldsymbol{F}_{s,1}, \ldots, \boldsymbol{F}_{s,n(s)}$ from Section 14.3, pages 153–156. Except for `Inv[9,6]`, the linear combinations that give `Inv[s,i]` are easily recognized. Writing $\boldsymbol{I}_{9,6}$ in $\mathcal{S}_2$ for `Inv[9,6]`, we have

$$\boldsymbol{I}_{9,6} \equiv (\mathcal{I}_{2,1,1})^2 \, \mathcal{I}_{2,1,2}^{(2)} - \frac{7}{2} \mathcal{I}_{2,1,1} \, \mathcal{I}_{2,1,1}^{(1)} \, \mathcal{I}_{2,1,2}^{(1)} + \frac{21}{8} (\mathcal{I}_{2,1,1}^{(1)})^2 \, \mathcal{I}_{2,1,2}$$

$$- 6 a_{2,2} (\mathcal{I}_{2,1,1})^2 \, \mathcal{I}_{2,1,2}) + \frac{1}{4} \boldsymbol{F}_{9,9}$$

$$\equiv \mathcal{I}_{2,1,1} \, \boldsymbol{F}_{7,4} - \frac{21}{10} \mathcal{I}_{2,1,2} \, \boldsymbol{F}_{6,5} + \frac{1}{4} \boldsymbol{F}_{9,9}$$

$$\equiv \boldsymbol{F}_{9,7} - \frac{21}{10} \boldsymbol{F}_{9,5} + \frac{1}{4} \boldsymbol{F}_{9,9}.$$

Thus, *for the relative invariants $\boldsymbol{F}_{s,i}$ of pages 153–156, the italicized assertion about (14.50) on page 153 is valid.*

16.5. Entry of keyboard instructions

To make the MATHEMATICA input statements of this chapter easy to read and recognize, we have introduced some spaces and invisible return characters. When we check for accuracy by using copy-and-paste techniques to transfer the input commands from the Latex files for this Memoir into cells of MATHEMATICA notebooks, we remove all spaces and hidden end-of-line characters before evaluating the notebooks. We have found that Versions 3.0, 4.1, 5.0, and 5.1 produce the desired output for the input of this chapter as well as for Sections 23.5 and 28.4

We can not be sure that versions of MATHEMATICA to be introduced later will yield satisfactory output for the input presented here. For instance, we used Version 3.0 to discover the identities for Theorem 14.8 on page 147 and those in the remainder of Section 14.2. The corresponding notebooks also evaluate correctly in Version 4.1 but fail to do so in Version 5.0. In particular, Version 5.0 returns 0 as the evaluation of the input statement

`Coefficient[ x^(p+q-1),  x^(p+q-1) ]`

whereas Versions 3.0 and 4.1 correctly give 1 as the coefficient of `x^(p+q-1)` in the expression `x^(p+q-1)`. In the recently released Version 5.1, the preceding error has been corrected. However, for our research, we have found that the implementation of the `Coefficient` command in Version 5.1 is much less efficient than it was in either Versions 3.0 or 4.1. This is particularly noticeable when one applies it to obtain explicit expressions for the unique coefficients of (14.47) and (14.48) on pages 152–153.

CHAPTER 17

The Simplest of the Fano-Type Problems for (1.1)

17.1. Historical motivation

In [30], G. Fano studied algebraic relations that may be satisfied by various solutions of mth-order homogeneous linear differential equations having the form

$$(17.1) \qquad y^m(z) + \sum_{j=1}^{m} c_{0,j}(z)\, y^{(m-j)}(z) = 0,$$

where the coefficients $c_{0,j}(z)$ are given meromorphic functions on a region Ω of the complex plane. As an instance when $m \geq 3$, there may exist meromorphic functions $\phi(z)$, $\psi(z)$ on a subregion of Ω such that the functions

$$(17.2) \qquad y_k(z) \equiv \big(\phi(z)\big)^{m-k} \big(\psi(z)\big)^{k-1}, \quad \text{for } 1 \leq k \leq m.$$

form a fundamental system over $\mathbb{C}$ of local solutions for (17.1). Then, we have $y_k(z)\, y_{k+2}(z) - \big(y_{k+1}(z)\big)^2 \equiv 0$, when $1 \leq k \leq m-2$. In general, the identity

$$W[y_1, y_2, \ldots, y_m](z) \equiv \left[\prod_{k=1}^{m-1}(k!)\right]\!\left[W[\phi,\psi](z)\right]^{\frac{(m-1)m}{2}}$$

relates the Wronskians of $y_1(z), \ldots, y_m(z)$ and $\phi(z)$, $\psi(z)$; see [19, Corollary 5.2].

In constructive supplements to Fano's research, one seeks explicit conditions on the coefficients of (17.1) that are necessary and sufficient for the existence of a fundamental system of local solutions of (17.1) that satisfy algebraic relations of a given type. One can also seek explicit procedures to specify all the differential equations (17.1) for which such solutions exist.

EXAMPLE 17.1. For $m \geq 3$, we introduce the following restriction on (17.1).

Condition A is the requirement that: for any subregion $\mathcal{U}$ of Ω, there are linearly independent meromorphic functions $\phi(z)$, $\psi(z)$ on a subregion $\mathcal{U}_0$ of $\mathcal{U}$ in terms of which the formulas of (17.2) specify $\{y_1(z), y_2(z), \ldots, y_m(z)\}$ as a fundamental system of local solutions on $\mathcal{U}_0$ for (17.1).

In Section 1.7, we observed that: for $3 \leq n \leq m$, the basic relative invariant $\mathcal{I}_{m,n}$ of weight n for the equations (17.1) coincides with the polynomial formally obtained by replacing each $w_{0,j}^{(k)}$ in $\mathcal{I}_{m,0,n}$ of (1.28) on page 8 with $w_j^{(k)}$. Thus, the value $\mathcal{I}_{m,n}(z)$ of $\mathcal{I}_{m,n}$ at (17.1) is the function on Ω obtained from $\mathcal{I}_{m,0,n}$ of (1.28) by replacing each $w_{0,j}^{(k)}$ in $\mathcal{I}_{m,0,n}$ with $c_{0,j}^{(k)}(z)$ from (17.1).

We established in [24, page 71, Theorem 9.1] that: *a given differential equation* (17.1) *satisfies* Condition A *if and only if*

$$(17.3) \qquad \mathcal{I}_{m,n}(z) \equiv 0, \quad \text{for } 3 \leq n \leq m.$$

Other conditions that are equivalent to (17.3) were also presented in [24].

We showed in [24, pages 13–17] that: *a given equation* (17.1) *having* $m \geq 3$ *satisfies Condition A if and only the coefficients of* (17.1) *and the meromorphic functions* $\alpha(z)$, $\beta(z)$ *defined on* Ω *through*

(17.4) $\qquad \alpha(z) \equiv \dfrac{1}{\binom{m}{2}} c_{0,1}(z)$

and

(17.5) $\qquad \beta(z) \equiv \dfrac{1}{\binom{m+1}{3}} \left[c_{0,2}(z) - \dfrac{m-2}{3} c_{0,1}^{(1)}(z) - \dfrac{(3m-1)(m-2)}{6m(m-1)} \left(c_{0,1}(z)\right)^2 \right]$

satisfy

(17.6) $\qquad c_{0,j}(z) \equiv \Gamma_{j,m-j}(z), \quad$ for $3 \leq j \leq m$,

where $\Gamma_{i,j}(z)$ is the meromorphic function defined on Ω through

(17.7) $\qquad \Gamma_{-1,j}(z) \equiv 0, \quad$ for any j,

(17.8) $\qquad \Gamma_{0,j}(z) \equiv 1, \quad$ for any j,

and

(17.9) $\qquad \Gamma_{i+1,j}(z) \equiv \sum_{k=0}^{j} \left[\begin{array}{l} \Gamma_{i,k}^{(1)}(z) + (i+k)\,\alpha(z)\,\Gamma_{i,k}(z) \\ + (i+k)(m-i-k)\,\beta(z)\,\Gamma_{i-1,k}(z) \end{array} \right], \quad \begin{array}{l} \text{for } i \geq 0 \\ \text{and any } j. \end{array}$

Also, when (17.6) is true, $\{\phi(z), \psi(z)\}$ for (17.2) may be any fundamental system of solutions on $\mathcal{U}_0$ for $y'' + \alpha(z)\,y' + \beta(z)\,y = 0$; see Theorem 19.22 on page 210.

For any (17.1) having $m = 2$, Condition A is satisfied. Then, the corresponding (17.6) is valid with $\alpha(z) \equiv c_{0,1}(z)$ and $\beta(z) \equiv c_{0,2}(z)$ for (17.7)–(17.9).

17.2. Results for (1.1) analogous to those in Example 17.1 for (17.1)

For convenience, we repeat (1.1) of page 4 as

(17.10) $\qquad \left(y^{(m)}(z)\right)^2 + \sum_{\substack{0 \leq i,j \leq m \\ (i,j) \neq (0,0)}} c_{i,j}(z)\, y^{(m-i)}(z)\, y^{(m-j)}(z) = 0, \quad$ on Ω,

where $c_{j,i}(z) \equiv c_{i,j}(z)$, for $0 \leq i < j \leq m$.

Throughout this section, we assume that $m \geq 2$.

Condition B is the requirement that: for each subregion $\mathcal{U}$ of Ω, there are linearly independent meromorphic functions $\phi(z)$, $\psi(z)$ on a subregion $\mathcal{U}_0$ of $\mathcal{U}$ such that each of the functions

(17.11) $\qquad y(z) \equiv \sum_{k=1}^{m} C_k \left(\phi(z)\right)^{m-k} \left(\psi(z)\right)^{k-1}, \quad$ for $C_1, \ldots, C_m$ in $\mathbb{C}$,

is a local solution of (17.10) on $\mathcal{U}_0$ and each local solution of (17.10) on any subregion of $\mathcal{U}_0$ is expressible in the form (17.11).

In view of $m \geq 2$, (17.10) specifies corresponding functions $\alpha(z)$ and $\beta(z)$ by means of (17.4) and (17.5). In particular, $\alpha(z)$ and $\beta(z)$ are the functions obtained by substituting $c_{i,j}^{(k)}(z)$ from (17.10) for $w_{i,j}^{(k)}$ in $\boldsymbol{b}_{m,1}$ and $\boldsymbol{b}_{m,2}$ of (1.16) and (1.15). The basic relative invariants for the equations (17.10) are given by (1.28) of page 8 as $\mathcal{I}_{m,l,n}$, for either $1 \leq l \leq n \leq m$ or $l = 0 < 3 \leq n \leq m$. Here, for a given (17.10), the symbolism $\mathcal{I}_{m,l,n}(z)$ denotes the function on Ω obtained by substituting $c_{i,j}^{(k)}(z)$ from (17.10) for $w_{i,j}^{(k)}$ in $\mathcal{I}_{m,l,n}$.

17.2. RESULTS FOR (1.1) ANALOGOUS TO THOSE IN EXAMPLE 17.1 FOR (17.1)

THEOREM 17.2. *For $m \geq 2$, (17.10) satisfies* Condition B *if and only if*

(17.12) $\quad \mathcal{I}_{m,l,n}(z) \equiv 0, \quad$ *on Ω for each (l, n) in $\mathcal{S}$,*

$$\text{where } \mathcal{S} = \big\{(i, j) \mid 1 \leq i \leq j \leq m \text{ or } i = 0 < 3 \leq j \leq m\big\}.$$

Moreover, (17.10) satisfies Condition B *if and only if its left member is given by*

(17.13) $$Q_m \equiv \left[y^{(m)}(z) + \sum_{j=1}^{m} c_{0,j}(z)\, y^{(m-j)}(z) \right]^2$$

and the coefficients $c_{0,3}(z), \ldots, c_{0,m}(z)$ satisfy (17.6) in terms of $c_{0,1}(z)$, $c_{0,2}(z)$, (17.4), (17.5), and (17.7)–(17.9).

PROOF. Suppose that (17.10) satisfies Condition B and let $\phi(z)$, $\psi(z)$ denote linearly independent meromorphic functions on a subregion $\mathcal{U}_0$ of Ω such that all of the solutions for (17.10) on subregions of $\mathcal{U}_0$ are given by (17.11). By replacing $\mathcal{U}_0$ with a suitable subregion of $\mathcal{U}_0$, we may assume that $\phi(z)$, $\psi(z)$ are analytic functions without zeros on $\mathcal{U}_0$ such that $g(z) \equiv \psi(z)/\phi(z)$ is a univalent analytic function on $\mathcal{U}_0$. After setting $\rho(z) \equiv \big(\phi(z)\big)^{m-1}$ on $\mathcal{U}_0$ and observing that the substitution $y(z) = \rho(z)\,v(z)$ transforms the restriction of (17.10) to $\mathcal{U}_0$ into an equation

(17.14) $\quad \big(v^{(m)}(z)\big)^2 + \sum_{\substack{0 \leq i,j \leq m \\ (i,j) \neq (0,0)}} c^*_{i,j}(z)\, v^{(m-i)}(z)\, v^{(m-j)}(z) = 0, \quad$ on $\mathcal{U}_0$,

$\quad\quad\quad$ where $c^*_{j,i}(z) \equiv c^*_{i,j}(z)$, for $0 \leq i < j \leq m$,

we find that all of the solutions for (17.14) on subregions of $\mathcal{U}_0$ are given by

$$v(z) \equiv \sum_{k=1}^{m} C_k\, \big(g(z)\big)^{k-1}, \quad \text{for } C_1, \ldots, C_m \text{ in } \mathbb{C}.$$

Let $z = f(\zeta)$ on $g(\mathcal{U}_0)$ denote the inverse function for $\zeta = g(z)$ on $\mathcal{U}_0$. Then, the substitutions $z = f(\zeta)$, $t(\zeta) = (v \circ f)(\zeta)$ transform (17.14) on $\mathcal{U}_0$ into an equation

(17.15) $\quad \big(t^{(m)}(\zeta)\big)^2 + \sum_{\substack{0 \leq i,j \leq m \\ (i,j) \neq (0,0)}} d_{i,j}(\zeta)\, t^{(m-i)}(\zeta)\, t^{(m-j)}(\zeta) = 0, \quad$ on $g(\mathcal{U}_0)$,

$\quad\quad\quad$ where $d_{j,i}(\zeta) \equiv d_{i,j}(\zeta)$, for $0 \leq i < j \leq m$,

such that all of the solutions for (17.15) on subregions of $g(\mathcal{U}_0)$ are given by

$$t(\zeta) \equiv \sum_{k=1}^{m} C_k\, \zeta^{k-1}, \quad \text{for } C_1, \ldots, C_m \text{ in } \mathbb{C}.$$

Through application of Proposition 17.3 on page 183 to (17.15) on $g(\mathcal{U}_0)$ in place of (17.10) on Ω, we obtain $d_{i,j}(\zeta) \equiv 0$, for $0 \leq i, j \leq m$ and $(i, j) \neq (0,0)$. Therefore the functions $\mathcal{I}^{\sharp}_{m,l,n}(\zeta)$ on $g(\mathcal{U}_0)$ obtained by substituting $d^{(k)}_{i,j}(\zeta)$ from (17.15) for $w^{(k)}_{i,j}$ in $\mathcal{I}_{m,l,n}$ of (1.28) satisfy

(17.16) $\quad \mathcal{I}^{\sharp}_{m,l,n}(\zeta) \equiv 0, \quad$ on $g(\mathcal{U}_0)$ for each (l, n) in $\mathcal{S}$ of (17.12).

We employ (17.16) and the properties (1.9)–(1.10) of relative invariants to deduce that (17.10) satisfies (17.12) when Ω is replaced with $\mathcal{U}_0$. Since each $\mathcal{I}_{m,l,n}(z)$ for (17.12) is a meromorphic function on Ω, we find see that (17.12) is satisfied without restriction. Moreover, by using $\zeta = g(z)$ to transform $\big(t^{(m)}(\zeta)\big)^2 = 0$ for (17.15) on $g(\mathcal{U}_0)$ into (17.14) on $\mathcal{U}_0$ and then using $v(z) = \bigl(1/\rho(z)\bigr)y(z)$ to transform (17.14) on $\mathcal{U}_0$ into (17.10) on $\mathcal{U}_0$, we see that the left member Q_m of (17.10) has a factorization of the form (15.2) on page 157 in which the two factors are equal. Thus, for S_m in (15.6), Proposition 15.2 on page 158 shows that $Q_m \equiv (S_m)^2$ on Ω. This gives (17.13). Since the corresponding equation (17.1) on Ω satisfies Condition A, its coefficients are related to $\alpha(z)$ and $\beta(z)$ through (17.6)–(17.9). Hence, when (17.10) satisfies Condition B, (17.10) satisfies (17.12) and Q_m is specified by (17.13) together with (17.6) in terms of $c_{0,1}(z)$, $c_{0,2}(z)$, (17.4)–(17.5), and (17.7)–(17.9).

Suppose that Q_m is specified by (17.13) together with (17.6) in terms of $c_{0,1}(z)$, $c_{0,2}(z)$, (17.4)–(17.5), and (17.7)–(17.9). Then, we have $Q_m \equiv (S_m)^2$ where S_m is the left member of an equation (17.1) given by (17.6) with respect to $c_{0,1}(z)$, $c_{0,2}(z)$, (17.4)–(17.5), and (17.7)–(17.9). Consequently, $S_m = 0$ satisfies Condition A and therefore (17.10) satisfies Condition B. It remains to be shown that: if (17.10) satisfies (17.12), then (17.10) satisfies Condition B.

Suppose that (17.10) satisfies (17.12) and let $\mathcal{U}$ be a subregion of Ω. We apply Propositions 5.3 and 5.4 of page 49 to obtain a subregion $\mathcal{U}_0$ of $\mathcal{U}$ on which are defined meromorphic functions $\phi(z) \not\equiv 0$ and $\rho(z) \equiv \bigl(\phi(z)\bigr)^{m-1}$ as well as a univalent analytic function $\zeta = g(z)$ with inverse designated by $z = f(\zeta)$ on $g(\mathcal{U}_0)$ such that: the substitutions $y(z) = \rho(z)v(z)$, $z = f(\zeta)$ with $t(\zeta) = (v \circ f)(\zeta)$ transform the restriction of (17.10) to $\mathcal{U}_0$ into a Laguerre-Forsyth canonical form on $g(\mathcal{U}_0)$. Thus, $y(z) = \rho(z)v(z)$ transforms (17.10) on $\mathcal{U}_0$ into an equation (17.14) on $\mathcal{U}_0$ and $z = f(\zeta)$ with $t(\zeta) = (v \circ f)(\zeta)$ transforms (17.14) on $\mathcal{U}_0$ into an equation (17.15) on $g(\mathcal{U}_0)$ having $d_{0,1}(\zeta) \equiv d_{1,2}(\zeta) \equiv 0$. We use (17.12) and the properties (1.9)–(1.10) of relative invariants to verify that the functions $\mathcal{I}^{\sharp}_{m,l,n}(\zeta)$ on $g(\mathcal{U}_0)$ obtained by substituting $d^{(k)}_{i,j}(\zeta)$ from (17.15) for $w^{(k)}_{i,j}$ in $\mathcal{I}_{m,l,n}$ of (1.28) satisfy

(17.17) $\qquad \mathcal{I}^{\sharp}_{m,l,n}(\zeta) \equiv 0, \quad \text{on } g(\mathcal{U}_0) \text{ for each } (l,n) \text{ in } \mathcal{S} \text{ of } (17.12).$

We have $d_{0,1}(\zeta) \equiv d_{0,2}(\zeta) \equiv 0$. Moreover, in view of this and (17.17), a substitution of $d_{i,j}(\zeta)$ from (17.15) for $w_{i,j}$ in $\mathcal{I}_{m,1,1}$ of (1.49) on page 11 yields

$$d_{1,1}(\zeta) \equiv \mathcal{I}^{\sharp}_{m,1,1}(\zeta) + \bigl(d_{0,1}(\zeta)\bigr)^2 \equiv 0.$$

Consequently, for (17.15), the condition

(17.18) $\qquad d_{i,j}(\zeta) \equiv 0, \quad \text{for } 0 \leq i \leq j, \ i+j = p, \text{ and } (i,j) \neq (0,0),$

is valid for $1 \leq p \leq 2$. Suppose that (17.18) is valid for some integer $p = p_0 \geq 2$ such that $p_0 < 2m$. Let i_0 and j_0 be integers subject to $0 \leq i_0 \leq j_0$ and $i_0 + j_0 = p_0 + 1$. Then, there is a basic relative invariant $\mathcal{I}_{m,i_0,j_0}$ of index (i_0, j_0) and it is given by

(17.19) $\qquad \mathcal{I}_{i_0,j_0} \equiv w_{i_0,j_0} + R_{i_0,j_0},$

where R_{i_0,j_0} is a polynomial over $\mathbb{Q}$ in variables $w^{(k)}_{i,j}$ from (1.6) for which

$$1 \leq j \leq j_0, \quad 0 \leq i \leq \min(i_0, j), \quad k \geq 0, \quad i+j < i_0 + j_0,$$

and $i + j + k \leq i_0 + j_0$. In particular, each $w^{(k)}_{i,j}$ effectively involved in R_{i_0,j_0} has $0 \leq i \leq j$, $i + j \leq p_0$, and $(i,j) \neq (0,0)$. Thus, by using (17.17) and (17.18), we see that the substitution of $d^{(k)}_{i,j}(\zeta)$ from (17.15) for $w^{(k)}_{i,j}$ in (17.19) yields

$$d_{i_0,j_0}(\zeta) \equiv \mathcal{I}^{\sharp}_{i_0,j_0}(\zeta) - R^{\sharp}_{i_0,j_0}(\zeta) \equiv -R^{\sharp}_{i_0,j_0}(\zeta) \equiv 0.$$

This shows that (17.18) is valid for $p = p_0 + 1$. Now, mathematical induction establishes that (17.18) is valid for $1 \leq p \leq 2m$. In view of $d_{\nu,\mu}(\zeta) \equiv d_{\mu,\nu}(\zeta)$, we therefore have $d_{\mu,\nu}(\zeta) \equiv 0$, for $0 \leq \mu, \nu \leq m$ and $(\mu, \nu) \neq (0,0)$. Hence, (17.15) is the equation $\left(t^{(m)}(\zeta)\right)^2 = 0$ and its solutions have the form

$$(17.20) \qquad t(\zeta) \equiv \sum_{k=1}^{m} C_k \zeta^{k-1}, \quad \text{for } C_1, \ldots, C_m \text{ in } \mathbb{C},$$

on any subregion of $g(\mathcal{U}_0)$. We use (17.20) and the transformation $\zeta = g(z)$ of (17.15) on $g(\mathcal{U}_0)$ into (17.14) on $\mathcal{U}_0$ to see that the formula

$$(17.21) \qquad v(z) \equiv \sum_{k=1}^{m} C_k\, g(z)^{k-1}, \quad \text{for } C_1, \ldots, C_m \text{ in } \mathbb{C},$$

yields the solutions of (17.14) on any subregion of $\mathcal{U}_0$. We set $\psi(z) \equiv \phi(z)\,g(z)$ and employ (17.21) as well as the transformation $v(z) = \left(1/\phi(z)\right)^{m-1} y(z)$ of (17.14) on $\mathcal{U}_0$ into (17.10) on $\mathcal{U}_0$ to verify that the solutions of (17.10) on any subregion of $\mathcal{U}_0$ are given by

$$\begin{aligned}y(z) &\equiv \sum_{k=1}^{m} C_k \left(\phi(z)\right)^{m-1} \left(g(z)\right)^{k-1} \\ &\equiv \sum_{k=1}^{m} C_k \left(\phi(z)\right)^{m-k} \left(\psi(z)\right)^{k-1}, \quad \text{for } C_1, \ldots, C_m \text{ in } \mathbb{C}.\end{aligned}$$

Thus, (17.10) satisfies Condition B. This completes the proof. $\square$

17.3. An equation (1.1) that has special polynomial solutions

We seek information about the left member Q_m of (17.10) on page 180 when $Q_m = 0$ has solutions of the type

$$(17.22) \qquad y(z) \equiv \sum_{k=1}^{m} C_k\, z^{k-1}, \quad \text{for } C_1, \ldots, C_m \text{ in } \mathbb{C}.$$

PROPOSITION 17.3. *For $m \geq 1$, suppose that each $y(z)$ in (17.22) is a solution of $Q_m = 0$. Then, Q_m has the factorization*

$$(17.23) \qquad Q_m \equiv y^{(m)}(z) \left[y^{(m)}(z) + \sum_{j=1}^{m} 2c_{0,j}(z)\, y^{(m-j)}(z) \right], \quad \text{on } \Omega.$$

Moreover, if each solution of $Q_m = 0$ is given by (17.22), then $Q_m \equiv \left(y^{(m)}(z)\right)^2$.

PROOF. Each function in (17.22) is a solution of both

$$(17.24) \qquad 0 = Q_m - \left(y^{(m)}(z)\right)^2 \equiv y^{(m)}(z) \left[\sum_{j=1}^{m} 2c_{0,j}(z)\, y^{(m-j)}(z) \right] + R_m$$

and $R_m = 0$, where

$$(17.25) \qquad R_m \equiv \sum_{1 \leq i,j \leq m} c_{i,j}(z)\, y^{(m-i)}(z)\, y^{(m-j)}(z).$$

Since 1 is a solution of $Q_m = 0$, we have $c_{m,m}(z) \equiv 0$. In particular, for $m = 1$, we find that $R_1 \equiv 0$.

To prove that $R_m \equiv 0$ for $m \geq 2$, suppose that $m \geq 2$ and $R_m \not\equiv 0$. Then, we use $c_{m,m}(z) \equiv 0$ and (17.25) to see that the order r of R_m satisfies $1 \leq r \leq m-1$. Hence, we can write $R_m = 0$ as

$$\left[\sum_{i=0}^{r} p_i(z)\, y^{(i)}(z)\right] y^{(r)}(z) + \sum_{0 \leq i \leq j \leq r-1} q_{i,j}(z)\, y^{(i)}(z)\, y^{(j)}(z) = 0, \tag{17.26}$$

for some meromorphic coefficients $p_i(z)$, $q_{i,j}(z)$ on Ω such that at least one of $p_0(z), p_1(z), \ldots, p_r(z)$ is not identically zero. Let i_0 be the greatest integer $\leq r$ such that $p_{i_0}(z) \not\equiv 0$. In particular, we have $0 \leq i_0 \leq r$. Let $\mathcal{U}$ be a subregion of Ω on which the coefficients of (17.26) are analytic and $p_{i_0}(z) \neq 0$, for each z in $\mathcal{U}$. Let z_0 be a complex number in $\mathcal{U}$. To obtain a contradiction, we consider two cases.

Case 1. Suppose that $i_0 = r$. Any solution $y(z)$ of (17.26) that also satisfies

$$y^{(i)}(z_0) = 0, \quad \text{for } 0 \leq i \leq r-1, \tag{17.27}$$

can have no more than two possible values for $y^{(r)}(z_0)$. However, for any complex number α, (17.22) yields the contradiction that (17.26) has a solution $y(z)$ that satisfies (17.27) and $y^{(r)}(z_0) = \alpha$.

Case 2. Suppose that $0 \leq i_0 < r$. Then, we use (17.26) and $p_i(z_0) = 0$, for $i_0 < i \leq r$, to see that any local solution $y(z)$ of (17.26) that also satisfies

$$y^{(i)}(z_0) = 0, \quad \text{for } 0 \leq i \leq i_0 - 1, \quad \text{and} \quad y^{(i)}(z_0) = 1, \quad \text{for } i_0 \leq i \leq r-1, \tag{17.28}$$

has a unique value for $y^{(r)}(z_0)$. However, for any complex number α, (17.22) gives the contradiction that (17.26) has solutions $y(z)$ that satisfies (17.28) and $y^{(r)}(z_0) = \alpha$.

Consequently, we have $R_m \equiv 0$. Thus, (17.24) yields (17.23).

Next, assume that each solution of $Q_m = 0$ is given by (17.22). To prove that $c_{0,j}(z) \equiv 0$, for $1 \leq j \leq m$, suppose that this is not the situation. Then, there is an open disk D_1 centered at z_1 and contained in Ω such that the coefficients of

$$y^{(m)}(z) + \sum_{j=1}^{m} 2c_{0,j}(z)\, y^{(m-j)}(z) = 0 \tag{17.29}$$

are analytic on D_1 and at least one of the complex numbers $c_{0,1}(z_1), \ldots, c_{0,m}(z_1)$ is nonzero. In terms of the complex conjugate $\overline{c_{0,j}(z_1)}$ of $c_{0,j}(z_1)$, for $1 \leq j \leq m$, let $y_0(z)$ denote the unique solution of (17.29) on D_1 that satisfies

$$y_0^{(j)}(z_1) = -\,\overline{c_{0,m-j}(z_1)}, \quad \text{for } 0 \leq j \leq m-1.$$

Then, (17.23) shows that $y_0(z)$ is a solution of $Q_m = 0$ on D_1. Thus, $y_0(z)$ is a function given by (17.22); and $y_0(z)$ is therefore a solution of $y^{(m)}(z) = 0$. Hence, we obtain the contradiction

$$0 = y_0^{(m)}(z_1) = -\sum_{j=1}^{m} 2c_{0,j}(z_1)\, y_0^{(m-j)}(z_1) = \sum_{j=1}^{m} 2|c_{0,j}(z_1)|^2 \neq 0.$$

Consequently, we must have $c_{0,j}(z) \equiv 0$, for $1 \leq j \leq m$. Now, (17.23) yields $Q_m \equiv \left(y^{(m)}(z)\right)^2$, on Ω. This completes the proof. $\square$

CHAPTER 18

Paul Appell's Condition of Solvability for $Q_m = 0$

18.1. Context and historical motivation

Let $\{y^{(k)} \mid \text{for } k \geq 0\}$ be an algebraically independent set of variables over the field $\mathfrak{F}_\Omega$ of meromorphic functions defined on a region Ω of the complex plane and let $\mathfrak{D}$ denote the ring of polynomials in those variables over $\mathfrak{F}_\Omega$. In terms of the derivation $'$ for $\mathfrak{F}_\Omega$ that assigns to each $f(z)$ in $\mathfrak{F}_\Omega$ its first derivative $f'(z)$, there is a unique extension of $'$ to a derivation $'$ for $\mathfrak{D}$ such that

$$\left(y^{(k)}\right)' \equiv y^{(k+1)}, \quad \text{for } k \geq 0;$$

e.g., see [10, page 139] or [11, page A.V.130]. The left member of (1.1) specifies

$$(18.1) \qquad Q_m \equiv \left(y^{(m)}\right)^2 + \sum_{\substack{0 \leq i,j \leq m \\ (i,j) \neq (0,0)}} c_{i,j}(z)\, y^{(m-i)}\, y^{(m-j)}, \quad \text{in } \mathfrak{D},$$

where $c_{j,i}(z) \equiv c_{i,j}(z)$, on Ω for $0 \leq i < j \leq m$. We note that

$$(Q_m)' \equiv 2\, y^{(m+1)}\, y^{(m)} + \sum_{\substack{0 \leq i,j \leq m \\ (i,j) \neq (0,0)}} \begin{bmatrix} c_{i,j}^{(1)}(z)\, y^{(m-i)}\, y^{(m-j)} \\ +\, c_{i,j}(z)\, y^{(m+1-i)}\, y^{(m-j)} \\ +\, c_{i,j}(z)\, y^{(m-i)}\, y^{(m+1-j)} \end{bmatrix}, \quad \text{in } \mathfrak{D}.$$

Paul Appell introduced in [5] of 1888 the condition on the coefficients of (18.1) that there exist a meromorphic function $\lambda(z)$ in $\mathfrak{F}_\Omega$ such that $(Q_m)' + \lambda(z)\, Q_m$ has a nontrivial factorization in $\mathfrak{D}$; and he observed in [5] that a nontrivial factorization for $(Q_m)' + \lambda(z)\, Q_m$ must have the form

$$(18.2) \quad (Q_m)' + \lambda Q_m \equiv L_m[y^{(0)}, y^{(1)}, \ldots, y^{(m)}]\, L_{m+1}[y^{(0)}, y^{(1)}, \ldots, y^{(m+1)}],$$

where L_m and L_{m+1} are homogeneous linear polynomials in the indicated variables over $\mathfrak{F}_\Omega$ of respective orders m and $m+1$. He then proved in [5] that: *when Q_m has a representation (18.2), the problem of finding the solutions of $Q_m = 0$ can be reduced to the problem of separately solving two homogeneous linear differential equations of respective orders m and $m+1$.* The main idea is the observation that: if Q_m has a representation (18.2) and if $y_0(z)$ is a local solution of (1.1), then $y_0(z)$ is a solution of both $Q_m = 0$ and $(Q_m)' = 0$ so that (since all functions are meromorphic) $y_0(z)$ is either a solution of $L_m = 0$ or a solution of $L_{m+1} = 0$. For any $m \geq 1$, Theorem 18.1 of page 186 provides a constructive method of deciding whether a given (18.1) satisfies (18.2) for some λ. Previously, only the special case $m = 2$ had been studied further; namely, see the work of Paul Appell in [7] and our completion of that research in Theorem 1.10 of page 15 by means of [18]. We can free Theorem 1.10 from its dependence on [18] by basing it on the consequence of Theorem 18.1 presented in Proposition 19.1 of page 194.

18.2. The equivalent condition of Theorem 18.1

For a fixed integer $m \geq 1$ and a given Q_m, we continue to use the notation

$$(18.3) \qquad Q_m \equiv (S_m)^2 + \sum_{1 \leq i,j \leq m} B_{i,j}(z)\, y^{(m-i)}\, y^{(m-j)}$$

of (15.5) where

$$(18.4) \qquad S_m \equiv y^{(m)} + \sum_{j=1}^{m} c_{0,j}(z)\, y^{(m-j)}$$

and

$$(18.5) \qquad B_{i,j}(z) \equiv c_{i,j}(z) - c_{0,i}(z)\, c_{0,j}(z), \quad \text{on } \Omega \text{ for } 1 \leq i,j \leq m.$$

Here, we supplement (18.5) with the additional definition

$$(18.6) \qquad B_{i,m+1}(z) \equiv B_{m+1,j}(z) \equiv 0, \quad \text{on } \Omega \text{ for } 1 \leq i,j \leq m+1.$$

In terms of (18.5) and (18.6), we introduce

$$(18.7) \qquad A_{i,j}(z) \equiv B_{i,j}^{(1)}(z) + B_{i+1,j}(z) + B_{i,j+1}(z), \quad \text{on } \Omega \text{ for } 1 \leq i,j \leq m.$$

We set

$$(18.8) \qquad \Theta_{i,j}^{r,s}(z) \equiv B_{r,s}(z)\left[A_{i,j}(z) - B_{1,i}(z)\, c_{0,j}(z) - B_{1,j}(z)\, c_{0,i}(z)\right]$$
$$- B_{i,j}(z)\left[A_{r,s}(z) - B_{1,r}(z)\, c_{0,s}(z) - B_{1,s}(z)\, c_{0,r}(z)\right],$$
$$\text{on } \Omega \text{ for } 1 \leq i,j,r,s \leq m.$$

In view of Proposition 15.2 on page 158, we see that: if $B_{i,j}(z) \equiv 0$, for $1 \leq i,j \leq m$, then $Q_m \equiv (S_m)^2$ and we obtain

$$(18.9) \qquad (Q_m)' + \lambda\, Q_m \equiv S_m\bigl(2\,(S_m)' + \lambda\, S_m\bigr)$$

as a nontrivial factorization (18.2) for any selection of $\lambda(z)$ in $\mathfrak{F}_\Omega$.

THEOREM 18.1. *For the left member Q_m of a given (1.1), suppose that p, q are integers such that $1 \leq p \leq q \leq m$ and $B_{p,q}(z) \not\equiv 0$. Then, Q_m has a representation (18.2) for some $\lambda(z)$ in $\mathfrak{F}_\Omega$ if and only if the functions $\Theta_{i,j}^{r,s}$ defined in $\mathfrak{F}_\Omega$ by (18.8) satisfy*

$$(18.10) \qquad \Theta_{i,j}^{p,q}(z) \equiv 0, \quad \text{on } \Omega \text{ for } 1 \leq i \leq j \leq m,$$

and $\lambda(z)$ is given by

$$(18.11) \qquad \lambda(z) \equiv \frac{B_{1,p}(z)\, c_{0,q}(z) + B_{1,q}(z)\, c_{0,p}(z) - A_{p,q}(z)}{B_{p,q}(z)}.$$

Moreover, when (18.10) is satisfied, $\lambda(z)$ in (18.11) yields

$$(18.12) \qquad (Q_m)' + \lambda(z)\, Q_m \equiv 2 S_m\left[(S_m)' + \frac{\lambda(z)}{2} S_m + \sum_{i=1}^{m} B_{1,i}(z)\, y^{(m-i)}\right].$$

18.2. THE EQUIVALENT CONDITION OF THEOREM 18.1

PROOF. For any given Q_m and $\lambda(z)$ in $\mathfrak{F}_\Omega$, we use (18.3)–(18.7) to obtain

$$(Q_m)' \equiv 2\,S_m\,(S_m)' + \sum_{i=1}^{m}\sum_{j=1}^{m} B_{i,j}^{(1)}(z)\, y^{(m-i)}\, y^{(m-j)}$$

$$+ \sum_{i=1}^{m+1}\sum_{j=1}^{m} B_{i,j}(z)\, y^{(m-i+1)}\, y^{(m-j)}$$

$$+ \sum_{i=1}^{m}\sum_{j=1}^{m+1} B_{i,j}(z)\, y^{(m-i)}\, y^{(m-j+1)}$$

$$\equiv 2\,S_m\,(S_m)' + \sum_{i=1}^{m}\sum_{j=1}^{m} B_{i,j}^{(1)}(z)\, y^{(m-i)}\, y^{(m-j)}$$

$$+ \sum_{i=1}^{m}\sum_{j=1}^{m} B_{i+1,j}(z)\, y^{(m-i)}\, y^{(m-j)} + \sum_{j=1}^{m} B_{1,j}(z)\, y^{(m)}\, y^{(m-j)}$$

$$+ \sum_{i=1}^{m}\sum_{j=1}^{m} B_{i,j+1}(z)\, y^{(m-i)}\, y^{(m-j)} + \sum_{i=1}^{m} B_{i,1}(z)\, y^{(m-i)}\, y^{(m)}$$

$$\equiv 2\,S_m\,(S_m)' + \sum_{i=1}^{m}\sum_{j=1}^{m} A_{i,j}(z)\, y^{(m-i)}\, y^{(m-j)} + \sum_{i=1}^{m} 2\,B_{1,i}(z)\, y^{(m)}\, y^{(m-i)}$$

and

(18.13) $$(Q_m)' + \lambda(z)\, Q_m \equiv S_m\bigl[2(S_m)' + \lambda(z)\, S_m\bigr] + \Phi_m,$$

where

(18.14) $$\Phi_m \equiv \sum_{i=1}^{m}\sum_{j=1}^{m} \bigl[A_{i,j}(z) + \lambda(z)\, B_{i,j}(z)\bigr] y^{(m-i)}\, y^{(m-j)}$$

$$+ \sum_{i=1}^{m} 2\,B_{1,i}(z)\, y^{(m)}\, y^{(m-i)}.$$

Since (18.4) and (18.13) show that the coefficient of $y^{(m+1)}$ in $(Q_m)' + \lambda(z)\, Q_m$ is $2S_m$, we see that $(Q_m)' + \lambda(z)\, Q_m$ has a nontrivial factorization if and only if it is divisible by S_m. Therefore, in view of (18.13), there is a nontrivial factorization for $(Q_m)' + \lambda(z)\, Q_m$ if and only if S_m divides Φ_m in $\mathfrak{D}$. By observing that the coefficient of $y^{(m)}$ in Φ_m is $\sum_{i=1}^{m} 2\,B_{1,i}(z)\, y^{(m-i)}$, we use the form of S_m in (18.4) to deduce that: S_m divides Φ_m in $\mathfrak{D}$ if and only if

(18.15) $$\Phi_m \equiv \left[\sum_{i=1}^{m} 2\,B_{1,i}(z) y^{(m-i)}\right] S_m.$$

We use (18.4) and the definition of Φ_m to see that (18.15) is satisfied if and only if

(18.16) $$\sum_{i=1}^{m}\sum_{j=1}^{m}\left[A_{i,j}(z)+\lambda(z)\,B_{i,j}(z)\right]y^{(m-i)}\,y^{(m-j)}$$
$$\equiv \left[\sum_{i=1}^{m}2B_{1,i}(z)\,y^{(m-i)}\right]\left[\sum_{j=1}^{m}c_{0,j}(z)\,y^{(m-j)}\right]$$
$$\equiv \sum_{i=1}^{m}\sum_{j=1}^{m}2B_{1,i}(z)\,c_{0,j}(z)\,y^{(m-i)}\,y^{(m-j)}.$$

We equate the sum of the coefficients of $y^{(m-i)}\,y^{(m-j)}$ and $y^{(m-j)}\,y^{(m-i)}$ in the left member of (18.16) with the sum of the coefficients of the same expressions in the right member of (18.16) to obtain

(18.17) $\left[A_{i,j}(z)+\lambda(z)\,B_{i,j}(z)\right]+\left[A_{j,i}(z)+\lambda(z)\,B_{j,i}(z)\right]$
$$\equiv 2B_{1,i}(z)\,c_{0,j}(z)+2B_{1,j}(z)\,c_{0,i}(z),\quad\text{for } 1\le i\le j\le m.$$

Thus, for the given Q_m and $\lambda(z)$, we combine (18.17) with $B_{j,i}(z)\equiv B_{i,j}(z)$ and $A_{j,i}(z)\equiv A_{i,j}(z)$ to deduce that $(Q_m)'+\lambda(z)\,Q_m$ has a nontrivial factorization in $\mathfrak{D}$ if and only if

(18.18) $A_{i,j}(z)+\lambda(z)\,B_{i,j}(z)\equiv B_{1,i}(z)\,c_{0,j}(z)+B_{1,j}(z)\,c_{0,i}(z),$
$$\text{on } \Omega \text{ for } 1\le i\le j\le m.$$

The particular condition

(18.19) $A_{p,q}(z)+\lambda(z)\,B_{p,q}(z)\equiv B_{1,p}(z)\,c_{0,q}(z)+B_{1,q}(z)\,c_{0,p}(z),\quad\text{on } \Omega,$

in (18.18) has $B_{p,q}(z)\not\equiv 0$. It shows that: if a suitable $\lambda(z)$ exists, then $\lambda(z)$ is unique and given by (18.11). Moreover, it enables us to eliminate $\lambda(z)$ from (18.18) and see that: (18.18) is equivalent to the requirement that the functions $\Theta_{i,j}^{r,s}(z)$ in (18.8) satisfy (18.10). When (18.10) is valid and $\lambda(z)$ is defined by (18.11), we apply (18.13)–(18.15) to obtain the factorization (18.12). This completes the proof. □

REMARK 18.2. Suppose that $\lambda(z)$ is a meromorphic function on a subregion $\mathcal{U}$ of Ω such that $(Q_m)'+\lambda Q_m$ has a factorization (18.2) in which all the functions belong to the field $\mathfrak{F}_\mathcal{U}$ of meromorphic functions defined on $\mathcal{U}$. Then, with $\mathfrak{F}_\Omega$ replaced throughout by $\mathfrak{F}_\mathcal{U}$, the argument for Theorem 18.1 is valid and shows that: when $B_{p,q}(z)\not\equiv 0$ for some integers p, q satisfying $1\le p\le q\le m$, the functions $\Theta_{i,j}^{r,s}(z)$ defined by (18.8) satisfy $\Theta_{i,j}^{p,q}(z)\equiv 0$ on $\mathcal{U}$, for $1\le i\le j\le m$; and $\lambda(z)$ is given on $\mathcal{U}$ by (18.11). However, in that situation, (18.8) is satisfied on Ω and (18.11) extends the definition of $\lambda(z)$ to a meromorphic function defined on Ω. Hence, Theorem 18.1 yields a factorization (18.2) over $\mathfrak{F}_\Omega$. Thus, *less restrictive conditions are not gained by considering factorizations* (18.2) *over subregions of* Ω.

EXAMPLE 18.3. For $m=1$ and $B_{1,1}(z)\equiv 0$, we have $Q_1=(S_m)^2$ as in (4.29) of page 42 where $I_{1,1,1}(z)\equiv B_{1,1}(z)\equiv 0$. Then, $(Q_1)'+\lambda Q_1$ has the nontrivial factorization (18.9) on page 186 for each meromorphic function λ on Ω. When $m=1$ and $B_{1,1}(z)\not\equiv 0$, Theorem 18.1 is applicable with $p=q=1$ and shows that $(Q_1)'+\lambda Q_1$ has the factorization (18.12) in which $m=1$ and λ is given by (18.11). Thus, any equation $Q_1=0$ satisfies Appell's condition.

18.3. Solutions of $Q_m = 0$ when Appell's condition is satisfied

When Q_m satisfies (18.2), any solution of $Q_m = 0$ is either a solution of $L_m = 0$ or a solution of $L_{m+1} = 0$. Then, the solutions of $Q_m = 0$ can be obtained from the solutions of $L_{m+1} = 0$ and the solutions of $L_m = 0$ by a technique that Paul Appell presented in [**5**]. It is explained by the proof of Proposition 18.4.

Let M be the symmetric $(m+1) \times (m+1)$ matrix whose (i,j)-component $[M]_{i,j}$ is defined in terms of $c_{0,0}(z) \equiv 1$ and the coefficients $c_{i,j}(z)$ of (1.1) by

$$[M]_{i,j} \equiv c_{i,j}(z), \quad \text{for } 0 \leq i,j \leq m.$$

Then, as an equality of two 1×1 matrices, (1.1) can be rewritten as

$$(18.20) \quad [y^{(m)}(z), y^{(m-1)}(z), \ldots, y(z)] \, M \, [y^{(m)}(z), y^{(m-1)}(z), \ldots, y(z)]^T = [0].$$

Regarding Q_m as a quadratic form, we note that its determinant is $det(M)$.

PROPOSITION 18.4. *For a given (1.1) of order $m \geq 1$, suppose that $\lambda(z)$ and the coefficients of a homogeneous linear differential equation $L = 0$ are meromorphic functions on Ω such that: the order r of $L = 0$ satisfies $1 \leq r \leq m+1$; and each local solution of $L = 0$ is a solution of $(Q_m)' + \lambda Q_m = 0$. Let $\{\phi_1(z), \phi_2(z), \ldots, \phi_r(z)\}$ be a fundamental system of local solutions of $L = 0$ on a subregion $\mathcal{U}$ of Ω. Then, there is a symmetric $r \times r$ matrix B over $\mathbb{C}$ in terms of which the solutions of both $Q_m = 0$ and $L = 0$ on $\mathcal{U}$ are given by*

$$(18.21) \quad y(z) \equiv K_1 \phi_1(z) + K_2 \phi_2(z) + \cdots + K_r \phi_r(z)$$

as $K_1, K_2, \ldots, K_r$ range through the elements of $\mathbb{C}$ that satisfy

$$(18.22) \quad [K_1, K_2, \ldots, K_r] \, B \, [K_1, K_2, \ldots, K_r]^T = [0].$$

Moreover, when $det(M) \neq 0$ and $r = m+1$, B is nonsingular.

PROOF. The solutions of $L = 0$ on $\mathcal{U}$ are provided by

$$(18.23) \quad y_0(z) \equiv \sum_{j=1}^{r} K_j \phi_j(z), \quad \text{for } K_1, K_2, \ldots, K_r \text{ in } \mathbb{C}.$$

The equations $y_0^{(i)}(z) \equiv \sum_{j=1}^{r} K_j \phi_j^{(i)}(z)$, derived from (18.23) for $0 \leq i \leq m$, enable us to construct an $r \times (m+1)$ matrix N over the field $\mathfrak{F}_\mathcal{U}$ of meromorphic functions on $\mathcal{U}$ such that

$$(18.24) \quad [y_0^{(m)}(z), y_0^{(m-1)}(z), \ldots, y_0^{(0)}(z)] = [K_1, K_2, \ldots, K_r] \, N.$$

For the function $h_{K_1, \ldots, K_r}(z)$ on $\mathcal{U}$ that is obtained by substituting $y_0^{(i)}(z)$ for $y^{(i)}$ in Q_m, we use (18.20) and (18.24) to see that $h_{K_1, \ldots, K_r}(z)$ is the $(1,1)$-component of the 1×1 matrix

$$(18.25) \quad [h_{K_1, \ldots, K_r}(z)] = [K_1, K_2, \ldots, K_r] \, (NMN^T) \, [K_1, K_2, \ldots, K_r]^T.$$

Let $\Lambda(z)$ be a meromorphic function on a subregion $\mathcal{U}_0$ of $\mathcal{U}$ such that $\Lambda'(z) \equiv \lambda(z)$ on $\mathcal{U}_0$. Because the function on $\mathcal{U}$ obtained by replacing each $y^{(i)}$ in $(Q_m)'$ with $y_0^{(i)}(z)$ from (18.23) is the same as the one obtained by differentiating $h_{K_1,\ldots,K_r}(z)$

with respect to z, we apply the property that each solution (18.23) of $L = 0$ is a solution of $(Q_m)' + \lambda Q_m = 0$ to deduce on $\mathcal{U}_0$ that: for any $K_1, \ldots, K_r$ in $\mathbb{C}$,

$$(18.26) \quad \frac{d}{dz}\left[e^{\Lambda(z)} h_{K_1,\ldots,K_r}(z)\right] \equiv e^{\Lambda(z)}\left[\frac{d}{dz} h_{K_1,\ldots,K_r}(z) + \lambda(z) h_{K_1,\ldots,K_r}(z)\right] \equiv 0.$$

We restrict the meromorphic components of NMN^T to $\mathcal{U}_0$ and scalarly multiply by $e^{\Lambda(z)}$ to obtain the $r \times r$ matrix $B = e^{\Lambda(z)} NMN^T$ with meromorphic components on $\mathcal{U}_0$. This matrix B satisfies $B^T \equiv B$. After multiplying both members of (18.25) with $e^{\Lambda(z)}$ and assigning particular values from $\{0,1\}$ to the constants $K_1, \ldots, K_r$, we use (18.26) to conclude that each component of B is a constant function on $\mathcal{U}_0$. Consequently, we can regard B as a symmetric $r \times r$ matrix whose components belong to $\mathbb{C}$. We apply (18.25) to verify that a solution (18.23) of $L = 0$ is a solution of $Q_m = 0$ if and only if the corresponding constants $K_1, \ldots, K_r$ in $\mathbb{C}$ for (18.23) satisfy (18.22).

Finally, when $det(M) \neq 0$ and $r = m + 1$, we find that the matrix N has size $(m + 1) \times (m + 1)$ and its determinant is given by $det(N) = \varepsilon W$, where $\varepsilon = \pm 1$ and W is the Wronskian on $\mathcal{U}$ of the linearly independent meromorphic functions $\phi_1(z), \phi_2(z), \ldots, \phi_{m+1}(z)$. Thus, we have $det(N) \neq 0$. Hence, by viewing the $(m + 1) \times (m + 1)$ matrix B over $\mathbb{C}$ as a matrix of constant functions on $\mathcal{U}_0$, we obtain

$$det(B) \equiv det(e^{\Lambda(z)} NMN^T) \equiv \left[e^{\Lambda(z)}\right]^{m+1} det(N)\, det(M)\, det(N^T) \not\equiv 0.$$

Consequently, B is nonsingular. This completes the proof. $\square$

EXAMPLE 18.5. Computer algebra enables us to easily check the various details of Theorem 18.1 and Proposition 18.4 for the differential equation

$$(18.27) \quad (y^{(3)})^2 - \frac{18}{z} y^{(3)} y^{(2)} + \frac{288}{5z^2} y^{(3)} y^{(1)} - \frac{312}{5z^3} y^{(3)} y^{(0)} + \frac{396}{5z^2} (y^{(2)})^2$$
$$- \frac{2466}{5z^3} y^{(2)} y^{(1)} + \frac{522}{z^4} y^{(2)} y^{(0)} + \frac{747}{z^4} (y^{(1)})^2 - \frac{1566}{z^5} y^{(1)} y^{(0)} + \frac{819}{z^6} (y^{(0)})^2 = 0,$$

where Ω is the complex plane. The determinant for the left member Q_3 of (18.27) is equal to $\frac{186624}{625 z^{12}} \neq 0$. We use (18.5)–(18.8) of page 186 to obtain

$$B_{1,1}(z) \equiv -\frac{9}{5z^2} \not\equiv 0 \quad \text{and} \quad \Theta^{1,1}_{i,j}(z) \equiv 0, \quad \text{for } 1 \leq i \leq j \leq 3.$$

Thus, (18.10) of page 186 is satisfied with $p = q = 1$. Hence, Theorem 18.1 yields

$$\lambda(z) \equiv -\frac{2}{z} \quad \text{and} \quad (Q_3)' + \lambda Q_3 \equiv L_3 L_4,$$

where

$$(18.28) \quad L_3 \equiv 2\left[y^{(3)} - \frac{9}{z} y^{(2)} + \frac{144}{5z^2} y^{(1)} - \frac{156}{5z^3} y^{(0)}\right]$$

and

$$(18.29) \quad L_4 \equiv y^{(4)} - \frac{10}{z} y^{(3)} + \frac{45}{z^2} y^{(2)} - \frac{105}{z^3} y^{(1)} + \frac{105}{z^4} y^{(0)}.$$

Each solution of $Q_3 = 0$ is either a solution of $L_3 = 0$ or a solution of $L_4 = 0$. On any simply-connected region $\mathcal{U}$ of not containing 0, there are analytic branches $s_1(z)$ of $z^{(20-\sqrt{205})/5}$, $s_2(z)$ of z^4, and $s_3(z)$ of $z^{(20+\sqrt{205})/5}$. The solutions of $L_3 = 0$

on $\mathcal{U}$ are given by $y_0(z) \equiv K_1 s_1(z) + K_2 s_2(z) + K_3 s_3(z)$, for K_1, K_2, K_3 in $\mathbb{C}$. The substitution of $y_0^{(i)}(z)$ for $y^{(i)}$ in Q_3 yields $(-369/125)(25(K_2)^2 + 64 K_1 K_3) z^2$. Hence, with $i^2 = -1$, the solutions on $\mathcal{U}$ of $L_3 = 0$ and $Q_3 = 0$ are

$$(18.30) \quad y_0(z) \equiv C^2 \big(5 s_1(z)\big) + CK \big(8i\, s_2(z)\big) + K^2 \big(5 s_3(z)\big), \quad \text{for } C, K \text{ in } \mathbb{C}.$$

We find that the formula

$$(18.31) \quad y(z) \equiv C_1 z + C_2 z^3 + C_3 z^5 + C_4 z^7, \quad \text{for } C_1, C_2, C_3, C_4 \text{ in } \mathbb{C},$$

gives the solutions of $L_4 = 0$ on any subregion of Ω. The result of substituting $y(z)$ from (18.31) and its derivatives into Q_3 is equal to $(-1152/5)(C_1 C_4 + C_2 C_3) z^2$. Thus, the solutions of $L_4 = 0$ that are also solutions of $Q_3 = 0$ are

$$(18.32) \quad y(z) \equiv \sum_{i=1}^{4} C_i z^{2i-1}, \quad \text{for } C_1, C_2, C_3, C_4 \text{ in } \mathbb{C} \text{ and } C_1 C_4 + C_2 C_3 = 0.$$

In terms of meromorphic continuation, each nonzero function given by (18.30) has $z = 0$ as a singular point. Thus, $y(z) \equiv 0$ is the only solution of $L_4 = 0$ that is also a solution of $L_3 = 0$. Consequently, each nonzero solution of $Q_3 = 0$ on $\mathcal{U}$ is given by either (18.30) or (18.32) but not by both.

DEFINITION 18.6. According to [47, page 5] and [41, page 75], the *separant* of Q_m in (18.1) is the formal partial derivative of Q_m with respect to $y^{(m)}$. It is equal to $2 S_m$, where S_m is defined in (18.4). Consequently, in the terminology of [47, page 32] and [41, page 155], a local solution of $Q_m = 0$ is a *singular solution* of $Q_m = 0$ if and only if it is also a solution of $S_m = 0$.

When Paul Appell's condition (18.2) on page 185 is satisfied for Q_m of (1.1), the factorization given by (18.12) in Theorem 18.1 shows the existence of a meromorphic function $\chi(z) \not\equiv 0$ on Ω such that $L_m \equiv \chi(z) S_m$. Then, the singular solutions of $Q_m = 0$ are the solutions of $Q_m = 0$ that are also solutions of $L_m = 0$. As applied to $Q_3 = 0$ in (18.27) of Example 18.5, the singular solutions on $\mathcal{U}$ of (18.27) are given by (18.30); and the nonsingular solutions of (18.27) on Ω or on any subregion of Ω are the functions of (18.32) that are not identically zero.

PROPOSITION 18.7. *Let $\phi_1(z), \phi_2(z) \ldots, \phi_r(z)$ be meromorphic functions on a region $\mathcal{U}$ that are linearly independent over $\mathbb{C}$, let B be a nonsingular symmetric $r \times r$ matrix over $\mathbb{C}$, let $\mathcal{S}$ be the set of functions given by*

$$(18.33) \quad y(z) = \sum_{j=1}^{r} K_j\, \phi_j(z),$$

for $K_1, \ldots, K_r$ in $\mathbb{C}$ and $[K_1, \ldots, K_r] B [K_1, \ldots, K_r]^T = [0]$,

and let A be a nonsingular symmetric $r \times r$ matrix over $\mathbb{C}$. Then, there are linearly independent functions $\psi_1(z), \psi_2(z) \ldots, \psi_r(z)$ over $\mathbb{C}$ in $\mathcal{S}$ such that the elements of $\mathcal{S}$ are given by

$$(18.34) \quad y(z) = \sum_{j=1}^{r} C_j\, \psi_j(z),$$

for $C_1, \ldots, C_r$ in $\mathbb{C}$ and $[C_1, \ldots, C_r] A [C_1, \ldots, C_r]^T = [0]$.

OBSERVATION. When (18.33) occurs through application of Proposition 18.4, a simpler relation (18.34) may be desired. Convenient selections for A are available. For instance, we may choose A so that the condition for (18.34) is given by

$$\text{(18.35)} \qquad C_1 C_r - \sum_{2 \leq i \leq (r+1)/2} C_i C_{r+1-i} = 0.$$

For $r = 2q - 1$, (18.35) has the form $C_1 C_r - (C_2 C_{r-1} + \cdots + (C_q)^2) = 0$. When $r = 2q$, (18.35) can be written as $C_1 C_r - (C_2 C_{r-1} + \cdots + C_q C_{q+1}) = 0$.

PROOF OF PROPOSITION 18.7. Let U be a nonsingular $r \times r$ matrix over $\mathbb{C}$ such that $U^T B U$ is the $r \times r$ identity matrix. For instance, an algorithm presented in [**9**, pages 131–136] yields a linear substitution over $\mathbb{C}$ that transforms the quadratic form specified by $[K_1, \ldots, K_r] B [K_1, \ldots, K_r]^T$ to a sum of r squares; then, a suitable U is provided by the linear substitution. Similarly, we obtain a nonsingular $r \times r$ matrix V over $\mathbb{C}$ such that $V^T A V$ is the $r \times r$ identity matrix. In terms of the $r \times r$ nonsingular matrix $R = U V^{-1}$ over $\mathbb{C}$, we have $A = R^T B R$ and we define $\psi_1(z), \psi_2(z), \ldots, \psi_r(z)$ on $\mathcal{U}$ through

$$[\psi_1(z), \psi_2(z), \ldots, \psi_r(z)]^T \equiv R^T [\phi_1(z), \phi_2(z), \ldots, \phi_r(z)]^T.$$

As $[C_1, C_2, \ldots, C_r]$ ranges through $\mathbb{C}^r$, the r-tuple $[K_1, K_2, \ldots, K_r]$ defined by

$$[K_1, K_2, \ldots, K_r]^T \equiv R [C_1, C_2, \ldots, C_r]^T$$

ranges through $\mathbb{C}^r$. For this situation, we find that

$$[K_1, \ldots, K_r][\phi_1(z), \ldots, \phi_r(z)]^T \equiv [C_1, \ldots, C_r] R^T (R^T)^{-1} [\psi_1(z), \ldots, \psi_r(z)]^T$$

$$\equiv [C_1, \ldots, C_r][\psi_1(z), \ldots, \psi_r(z)]^T$$

as well as

$$[K_1, \ldots, K_r] B [K_1, \ldots, K_r]^T \equiv [C_1, \ldots, C_r] R^T B R [C_1, \ldots, C_r]^T$$

$$\equiv [C_1, \ldots, C_r] A [C_1, \ldots, C_r]^T.$$

Thus, (18.33) and (18.34) specify the same set of functions. This completes the proof. □

REMARK 18.8. If Q_m in (18.1) has a nontrivial factorization, then the solutions of $Q_m = 0$ are solutions of homogeneous linear differential equations and are therefore free of movable branch points in the precise sense of Definition 19.14 on page 206.

Suppose that Q_m has no nontrivial factorization. In that situation, we have $B_{p,q}(z) \not\equiv 0$, for some p, q satisfying $1 \leq p, q \leq m$. If Q_m satisfies (18.10), then the factorization (18.12) for $(Q_m)' + \lambda Q_m$ shows that the solutions of $Q_m = 0$ are solutions of homogeneous linear differential equations and are therefore free of movable branch points. For $m = 2$, the converse of the preceding assertion is also true; it is an immediate consequence of Theorem 19.16, Corollary 19.2, and Proposition 19.1. For $m = 1$, Example 18.3 shows that the solutions of $Q_1 = 0$ are always free of movable branch points. However, for $m \geq 3$, convenient conditions based on Definition 19.14 of page 206 that are necessary as well as sufficient for $Q_m = 0$ to have its solutions free of movable branch points are not known. For this, one would need new techniques having far greater complexity than those presented on page 207 for $m = 2$.

CHAPTER 19

Appell's Condition for $Q_2 = 0$ and Related Topics

Substituting $m = 2$ in (18.1) on page 185, we see that Q_2 is the left member of (1.0–A) on page 3. It has the form

$$(19.1) \qquad Q_2 \equiv \left(y^{(2)}\right)^2 + \sum_{\substack{0 \le i,j \le 2 \\ (i,j) \ne (0,0)}} c_{i,j}(z)\, y^{(2-i)}\, y^{(2-j)}, \quad \text{on } \Omega,$$

where $c_{j,i}(z) \equiv c_{i,j}(z)$, on Ω for $0 \le i < j \le 2$. The determinant of the matrix M for (19.1) is given by (15.24)–(15.25) as

$$(19.2) \qquad D_2(z) \equiv B_{1,1}(z)\, B_{2,2}(z) - \left(B_{1,2}(z)\right)^2,$$

where $B_{i,j}(z) \equiv c_{i,j}(z) - c_{0,i}(z)\, c_{0,j}(z)$, for $1 \le i,j \le 2$. Example 15.6 on page 161 shows that the condition $D_2(z) \not\equiv 0$ is satisfied if and only if Q_2 has no nontrivial factorization. When Q_2 does have a factorization, one can equate the factors to zero and solve the corresponding second-order homogeneous linear differential equations to obtain local solutions of $Q_2 = 0$.

Section 19.1 gives details about Appell's condition for $Q_2 = 0$ and his solution procedure based on solving a third-order homogeneous linear differential equation. However, when $Q_2 = 0$ satisfies Appell's condition, its solutions can be obtained by solving homogeneous linear differential equations of orders ≤ 2. This improvement of his technique for the situation $D_2(z) \not\equiv 0$ is presented in Theorem 19.7 of page 197 and illustrated in Example 19.8 of page 198. In [**19**, page 106, Corollary 6.5], we proved that: when $D_2(z) \not\equiv 0$, Appell's condition for $Q_2 = 0$ is satisfied if and only if *the solutions of $Q_2 = 0$ are free of movable branch points* in the precise sense of Definition 19.14 on page 206. Our consideration of $Q_2 = 0$ in this chapter leads naturally to other equations

$$a(z)\, y''^2 + b(z)\, y''y' + c(z)\, y''y + d(z)\, y'^2 + e(z)\, y'y + f(z)\, y^2 = 0$$

having meromorphic coefficients without the restriction $a(z) \not\equiv 0$. Conditions for their solutions to be free of movable branch points are recalled in Section 19.7.

19.1. Consequences of Chapter 18 for the equations $Q_2 = 0$

For immediate reference, we note that (18.8) yields

$$(19.3) \qquad \Theta^{1,1}_{1,2}(z) \equiv B_{1,1}(z)\, B_{2,2}(z) - 2(B_{1,2}(z))^2 + c_{0,1}(z)\, B_{1,1}(z)\, B_{1,2}(z)$$
$$ - c_{0,2}(z)\, (B_{1,1}(z))^2 + B_{1,1}(z)\, B^{(1)}_{1,2}(z) - B^{(1)}_{1,1}(z)\, B_{1,2}(z)$$

and

$$(19.4) \qquad \Theta^{1,1}_{2,2}(z) \equiv B_{1,1}(z)\, B^{(1)}_{2,2}(z) - B^{(1)}_{1,1}(z)\, B_{2,2}(z) - 2\, B_{1,2}(z)\, B_{2,2}(z)$$
$$ + 2\, c_{0,1}(z)\, B_{1,1}(z)\, B_{2,2}(z) - 2\, c_{0,2}(z)\, B_{1,1}(z)\, B_{1,2}(z).$$

PROPOSITION 19.1. *For Q_2 in (19.1), suppose that at least one of $B_{1,1}(z)$, $B_{1,2}(z)$, and $B_{2,2}(z)$ is not identically zero. Then, with reference to $\mathfrak{F}_\Omega$ and $\mathfrak{D}$ in Section 18.1, there is a function $\lambda(z)$ in $\mathfrak{F}_\Omega$ such that $(Q_2)' + \lambda Q_2$ has a nontrivial factorization in $\mathfrak{D}$ if and only if*

(19.5) $$B_{1,1}(z) \not\equiv 0, \quad \Theta_{1,2}^{1,1}(z) \equiv 0, \quad \text{and} \quad \Theta_{2,2}^{1,1}(z) \equiv 0.$$

Moreover, when (19.5) is satisfied, there is only one such $\lambda(z)$; it is given by

(19.6) $$\lambda(z) \equiv 2c_{0,1}(z) - \frac{B_{1,1}^{(1)}(z) + 2B_{1,2}(z)}{B_{1,1}(z)}, \quad \text{on } \Omega;$$

and it yields

(19.7) $$(Q_2)' + \lambda(z)\, Q_2 \equiv 2 S_2\, \mathcal{L}_3, \quad \text{on } \Omega,$$

where

(19.8) $$S_2 \equiv y''(z) + c_{0,1}(z)\, y'(z) + c_{0,2}(z)\, y(z)$$

and

(19.9) $$\mathcal{L}_3 \equiv (S_2)' + \frac{\lambda(z)}{2} S_2 + B_{1,1}(z)\, y' + B_{1,2}(z)\, y.$$

PROOF. For a given function $\lambda(z)$ in $\mathfrak{F}_\Omega$, suppose that $(Q_2)' + \lambda(z)\, Q_2$ has a nontrivial factorization. We note that (18.8) yields

(19.10) $$\Theta_{1,2}^{2,2}(z) \equiv (B_{2,2}(z))^2 - c_{0,1}(z)\, B_{1,2}(z)\, B_{2,2}(z) - c_{0,2}(z) B_{1,1}(z)\, B_{2,2}(z)$$
$$+ 2c_{0,2}(z)(B_{1,2}(z))^2 + B_{1,2}^{(1)}(z)\, B_{2,2}(z) - B_{1,2}(z)\, B_{2,2}^{(1)}(z).$$

If $B_{2,2}(z) \not\equiv 0$ and $B_{1,2}(z) \equiv B_{1,1}(z) \equiv 0$, then (19.10) and Theorem 18.1 (with $m = p = q = 2$) give the contradiction
$$0 \equiv \Theta_{1,2}^{2,2}(z) \equiv (B_{2,2}(z))^2 \not\equiv 0.$$
Thus, either $B_{1,2}(z) \not\equiv 0$ or $B_{1,1}(z) \not\equiv 0$. If $B_{1,2}(z) \not\equiv 0$ and $B_{1,1}(z) \equiv 0$, then Theorem 18.1 (with $m = q = 2$ and $p = 1$), (18.8), and (19.3) yield the contradiction
$$0 \equiv \Theta_{1,1}^{1,2}(z) \equiv -\Theta_{1,2}^{1,1}(z) \equiv 2(B_{1,2}(z))^2 \not\equiv 0.$$
Consequently, we must have $B_{1,1}(z) \not\equiv 0$. Now, Theorem 18.1 (with $m = 2$ and $p = q = 1$) yields $\Theta_{1,2}^{1,1}(z) \equiv 0$ and $\Theta_{2,2}^{1,1}(z) \equiv 0$. Thus, (19.5) is satisfied.

Suppose that Q_2 satisfies (19.5). We always have $\Theta_{1,1}^{1,1}(z) \equiv 0$. Thus, for $m = 2$, $p = 1$, and $q = 1$, we see that: Q_2 satisfies (18.10) of Theorem 18.1; and (18.11)–(18.12) give (19.6)–(19.9). This completes the proof. □

COROLLARY 19.2. *Suppose Q_2 satisfies $D_2(z) \not\equiv 0$. Then, the condition*

(19.11) $$\Theta_{1,2}^{1,1}(z) \equiv 0 \quad \text{and} \quad \Theta_{2,2}^{1,1}(z) \equiv 0$$

for Q_2 is equivalent to (19.5).

PROOF. Suppose that (19.11) is valid. To establish $B_{1,1}(z) \not\equiv 0$, assume that $B_{1,1}(z) \equiv 0$. Then, the hypothesis $D_2(z) \not\equiv 0$ yields $B_{1,2}(z) \not\equiv 0$ while (19.11) and (19.3) give $0 \equiv \Theta_{1,2}^{1,1}(z) \equiv -2(B_{1,2}(z))^2 \not\equiv 0$. This contradiction shows that (19.5) is satisfied. Of course, (19.5) implies (19.11). This completes the proof. □

REMARK 19.3. The functions $E_6(z)$ and $H_7(z)$ obtained by substituting $c_{i,j}^{(k)}(z)$ from Q_2 for $w_{i,j}^{(k)}$ in $\boldsymbol{E}_6$ and $\boldsymbol{H}_7$ of (1.72) and (1.71) on page 15 and the functions $\Theta_{1,2}^{1,1}(z)$ and $\Theta_{2,2}^{1,1}(z)$ in (19.3) and (19.4) are related by

(19.12) $\qquad E_6(z) \equiv \Theta_{1,2}^{1,1}(z) \quad \text{and} \quad H_7(z) \equiv \Theta_{2,2}^{1,1}(z), \quad \text{on } \Omega.$

Suppose that $D_2(z) \not\equiv 0$. Then, we see that the condition $E_6(z) \equiv E_7(z) \equiv 0$ of (1.74) is equivalent to the condition $\Theta_{1,2}^{1,1}(z) \equiv \Theta_{2,2}^{1,1}(z) \equiv 0$ of (19.11). Thus, Proposition 19.1 and Corollary 19.2 can be used to make the proof on page 16 for Theorem 1.10 independent of the research in [**18**]. In particular, if both $D_2(z) \not\equiv 0$ and (19.11) are satisfied, then (19.6)–(19.9) yield (1.75)–(1.77) of page 15.

PROPOSITION 19.4. *For a given Q_2, suppose that $D_2(z) \not\equiv 0$. Then, Q_2 satisfies (19.11) if and only if $B_{1,1}(z) \not\equiv 0$ and the coefficients $c_{0,1}(z), c_{0,2}(z)$ are given by*

(19.13) $\qquad c_{0,1}(z) \equiv \dfrac{B_{1,1}^{(1)}(z) + 2 B_{1,2}(z)}{B_{1,1}(z)} - \dfrac{D_2^{(1)}(z)}{2 D_2(z)}$

and

(19.14) $\qquad c_{0,2}(z) \equiv \dfrac{B_{1,2}^{(1)}(z) + B_{2,2}(z)}{B_{1,1}(z)} - \dfrac{B_{1,2}(z) D_2^{(1)}(z)}{2 B_{1,1}(z) D_2(z)}.$

Moreover, as $B_{1,1}(z), B_{1,2}(z)$, and $D_2(z)$ range over the meromorphic functions on Ω subject to $B_{1,1}(z) \not\equiv 0$ and $D_2(z) \not\equiv 0$, all of the differential equations $Q_2 = 0$ on Ω that satisfy $D_2(z) \not\equiv 0$ and (19.11) are specified parametrically by

(19.15) $Q_2 \equiv \left[y'' + c_{0,1}(z)\, y' + c_{0,2}(z)\, y \right]^2 + B_{1,1}(z)\, (y')^2 + B_{1,2}(z)\, y'\, y + B_{2,2}(z)\, y^2$

when $B_{2,2}(z), c_{0,1}(z)$, and $c_{0,2}(z)$ are given by

(19.16) $\qquad B_{2,2}(z) \equiv \dfrac{D_2(z) + \bigl(B_{1,2}(z)\bigr)^2}{B_{1,1}(z)}$

and (19.13)–(19.14) after the substitution for $B_{2,2}(z)$ has been made.

PROOF. When (19.11) is satisfied, Corollary 19.2 shows that $B_{1,1}(z) \not\equiv 0$. Then, we can rewrite $\Theta_{1,2}^{1,1}(z) \equiv 0$ and $\Theta_{2,2}^{1,1}(z) \equiv 0$ in the form of (19.13) and (19.14). Conversely, when $B_{1,1}(z) \not\equiv 0$ and (19.13)–(19.14) are satisfied, we can rewrite (19.13) and (19.14) in the form $\Theta_{1,2}^{1,1}(z) \equiv 0$ and $\Theta_{2,2}^{1,1}(z) \equiv 0$ to conclude that (19.11) is satisfied.

To complete the proof, we combine the preceding deduction with the case $m = 2$ of (18.3)–(18.5) on page 186. $\square$

PROPOSITION 19.5. *Suppose that Q_2 in (19.1) satisfies $D_2(z) \not\equiv 0$ and (19.11). Then, for the corresponding factorization (19.7), a local solution of $Q_2 = 0$ is a nonsingular solution of $Q_2 = 0$ if and only if it is a nonzero solution of $\mathcal{L}_3 = 0$.*

PROOF. Because (19.7) shows that each local solution of $Q_2 = 0$ is either a solution of $S_2 = 0$ or a solution of $\mathcal{L}_3 = 0$, we are to prove that: *each nonzero local solution of both $Q_2 = 0$ and $\mathcal{L}_3 = 0$ is not a solution of $S_2 = 0$*. To do this, suppose that $y_0(z)$ is a solution of both $S_2 = 0$ and $\mathcal{L}_3 = 0$. Then, (19.8) and (19.9) yield

$$B_{1,1}(z)\, y_0'(z) + B_{1,2}(z)\, y_0(z) \equiv 0.$$

With $B_{1,1}(z) \not\equiv 0$, we therefore have

(19.17) $\quad y_0'(z) \equiv \left[\dfrac{-B_{1,2}(z)}{B_{1,1}(z)} \right] y_0(z)$

and

(19.18) $\quad y_0''(z) \equiv \left[\dfrac{-B_{1,1}(z)\, B_{1,2}^{(1)}(z) + B_{1,2}(z)\, B_{1,1}^{(1)}(z) + \left(B_{1,2}(z)\right)^2}{\left(B_{1,1}(z)\right)^2} \right] y_0(z).$

We use (19.17), (19.18), (19.3), $\Theta_{1,2}^{1,1}(z) \equiv 0$, and (19.2) to deduce that

$$0 \equiv \left(B_{1,1}(z)\right)^2 \left[y_0''(z) + c_{0,1}(z)\, y_0'(z) + c_{0,2}(z)\, y_0(z) \right]$$

$$\equiv \left[\begin{array}{l} -B_{1,1}(z)\, B_{1,2}^{(1)}(z) + B_{1,2}(z)\, B_{1,1}^{(1)}(z) + \left(B_{1,2}(z)\right)^2 \\ -c_{0,1}(z)\, B_{1,1}(z)\, B_{1,2}(z) + c_{0,2}(z)\, \left(B_{1,1}(z)\right)^2 \end{array} \right] y_0(z)$$

$$\equiv \left[B_{1,1}(z)\, B_{2,2}(z) - \left(B_{1,2}(z)\right)^2 - \Theta_{1,2}^{1,1}(z) \right] y_0(z) \equiv D_2(z)\, y_0(z)$$

and therefore $y_0(z) \equiv 0$. Thus, the nonzero solutions of $Q_2 = 0$ and $\mathcal{L}_3 = 0$ are the nonsingular solutions of $Q_2 = 0$. This completes the proof. $\square$

PROPOSITION 19.6. *Suppose that Q_2 in (19.1) satisfies $D_2(z) \not\equiv 0$ and (19.11) on Ω; and, with respect to $\mathcal{L}_3$ in the corresponding factorization (19.7), suppose there are three linearly independent solutions of $\mathcal{L}_3 = 0$ on a subregion $\mathcal{U}$ of Ω. Then, there are linearly independent solutions $u_0(z)$, $u_1(z)$, $u_2(z)$ of $\mathcal{L}_3 = 0$ on $\mathcal{U}$ such that the nonsingular solutions of $Q_2 = 0$ on $\mathcal{U}$ are given by*

(19.19) $\quad y(z) \equiv C^2 u_0(z) + CK\, u_1(z) + K^2 u_2(z),$

for C, K in $\mathbb{C}$ and either $C \neq 0$ or $K \neq 0$.

PROOF. Let $\phi_1(z)$, $\phi_2(z)$, $\phi_3(z)$ be linearly independent solutions of $\mathcal{L}_3 = 0$ on $\mathcal{U}$. For $m = 2$ and $r = m+1$, Proposition 18.4 on page 189 yields a nonsingular symmetric 3×3 matrix B such that the solutions of both $Q_2 = 0$ and $\mathcal{L}_3 = 0$ on $\mathcal{U}$ are given by $y(z) \equiv K_1 \phi_1(z) + K_2 \phi_2(z) + K_3 \phi_3(z)$ as K_1, K_2, K_3 range over the elements of $\mathbb{C}$ that satisfy $[K_1, K_2, K_3]\, B\, [K_1, K_2, K_3]^T = [0]$. The matrix

$$A = \begin{bmatrix} 0 & 0 & 1/2 \\ 0 & -1 & 0 \\ 1/2 & 0 & 0 \end{bmatrix}$$

yields the relation $C_1 C_3 - (C_2)^2 = 0$ for $r = 3$ in (18.35). In terms of it, we apply Proposition 18.7 of page 191 to obtain three linearly independent solutions $u_0(z)$, $u_1(z)$, $u_2(z)$ of $\mathcal{L}_3 = 0$ on $\mathcal{U}$ such that the solutions of $Q_2 = 0$ and $\mathcal{L}_3 = 0$ on $\mathcal{U}$ are given by $y(z) \equiv C_1 u_0(z) + C_2 u_1(z) + C_3 u_2(z)$ as C_1, C_2, C_3 range over the elements of $\mathbb{C}$ that satisfy $C_1 C_3 - (C_2)^2 = 0$. These constants are given by $C_1 = C^2$, $C_2 = CK$, and $C_3 = K^2$ as C and K range through $\mathbb{C}$. Proposition 19.5 shows that the nonsingular solutions of $Q_2 = 0$ are the solutions of $Q_2 = 0$ and $\mathcal{L}_3 = 0$ that are not identically zero. Hence, we have the restriction $C \neq 0$ or $K \neq 0$ for (19.19). This completes the proof. $\square$

19.2. An improvement for Proposition 19.6

Rather than solve a third-order homogeneous linear differential equation and then make a linear substitution to obtain suitable functions $u_0(z)$, $u_1(z)$, $u_2(z)$ for (19.19) in Proposition 19.6, we noticed in [16] of 1960 that it is sufficient to solve a Riccati differential equation and a first-order homogeneous linear differential equation. Complete details were given in [18] for [18, Corollary 4.9, page 100]. Since the solving of a Riccati differential equation is equivalent to the solving of a corresponding second-order homogeneous linear differential equation, we were led to the following formulation in [20, Theorem 2.1].

THEOREM 19.7. *For Q_2 in (19.1) and $D_2(z)$ in (19.2), suppose that Q_2 satisfies $D_2(z) \not\equiv 0$ as well as (19.11) and therefore has $B_{1,1}(z) \not\equiv 0$. Let $\tau(z)$, $\eta(z)$, $\phi(z)$, $\psi(z)$ be meromorphic functions on a subregion $\mathcal{V}$ of Ω such that:*

$$(19.20) \qquad \bigl(\tau(z)\bigr)^2 \equiv \frac{-D_2(z)}{\bigl(B_{1,1}(z)\bigr)^2}, \quad \text{on } \mathcal{V};$$

$$(19.21) \qquad \eta^{(1)}(z) + \left[\frac{B_{1,1}^{(1)}(z) + B_{1,2}(z)}{B_{1,1}(z)} - \tau(z)\right]\eta(z) \equiv 0, \quad \text{on } \mathcal{V};$$

$\eta(z) \not\equiv 0$; and $\phi(z)$, $\psi(z)$ are linearly independent solutions on $\mathcal{V}$ of

$$(19.22) \qquad w'' + \left[\tau(z) - \frac{B_{1,1}^{(1)}(z)}{2B_{1,1}(z)}\right] w' + \frac{B_{1,1}(z)}{4} w = 0.$$

Then, the nonsingular solutions on $\mathcal{V}$ of $Q_2 = 0$ are given by

$$(19.23) \quad y(z) \equiv B_{1,1}(z)\,\eta(z)\Bigl[C_1\phi(z) + C_2\psi(z)\Bigr]^2 + 4\eta(z)\Bigl[C_1\phi^{(1)}(z) + C_2\psi^{(1)}(z)\Bigr]^2,$$

for C_1, C_2 in $\mathbb{C}$ and either $C_1 \neq 0$ or $C_2 \neq 0$.

Moreover, when (19.23) is written as

$$(19.24) \qquad y(z) \equiv (C_1)^2\, v_0(z) + C_1 C_2\, v_1(z) + (C_2)^2\, v_2(z),$$

the functions $v_0(z)$, $v_1(z)$, $v_2(z)$ on $\mathcal{V}$ are linearly independent over $\mathbb{C}$ and satisfy

$$(19.25) \qquad \bigl(v_1(z)\bigr)^2 - 4 v_0(z)\, v_2(z) \not\equiv 0.$$

And, the singular solutions of $Q_2 = 0$ on subregions of $\mathcal{V}$ are the local solutions of

$$(19.26) \qquad y' + \left[\frac{B_{1,2}(z)}{B_{1,1}(z)} - \tau(z)\right] y = 0 \quad \text{or} \quad y' + \left[\frac{B_{1,2}(z)}{B_{1,1}(z)} + \tau(z)\right] y = 0.$$

PROOF. In [20, Theorem 2.1], we set $a \equiv 1$, $b \equiv 2c_{0,1}$, $c \equiv 2c_{0,2}$, $d \equiv c_{1,1}$, $e \equiv 2c_{1,2}$, $f \equiv c_{2,2}$, $A_2 \equiv 4B_{1,1}$, $A_3 \equiv 8B_{1,2}$, $A_4 \equiv 4B_{2,2}$, $D \equiv -64D_2$ as well as replace C_1 with $C_1/2$ and C_2 with $C_2/2$. This yields our present formulation. □

The following example illustrates the use of Theorem 19.7 and its advantages over Proposition 19.6. That (19.19) and any such expression of that type must satisfy a relation like (19.25) is established by means of Proposition 19.9.

19.3. An example to illustrate Theorem 19.7

EXAMPLE 19.8. Here, we use MATHEMATICA to demonstrate the advantage of applying Theorem 19.7 rather than Proposition 19.6 to the differential equation

$$(19.27) \quad (y'')^2 - \frac{10z^2+2}{z(z^2+1)}y''y' + \frac{2(23z^4-2z^2-1)}{3z^2(z^2+1)^2}y''y + \frac{4(19z^4+8z^2+1)}{3z^2(z^2+1)^2}(y')^2$$
$$- \frac{4(59z^4+10z^2-1)}{3z(z^2+1)^3}y'y + \frac{4(46z^6-3z^4+1)}{3z^2(z^2+1)^4}y^2 = 0,$$

where Ω is the complex plane. With $c_{0,1}(z) \equiv -(5z^2+1)/(z(z^2+1))$, ..., we have

$$B_{1,1}(z) \equiv \frac{1}{3z^2}, \quad B_{1,2}(z) \equiv \frac{-3z^2-1}{3z^3(z^2+1)}, \quad B_{2,2}(z) \equiv \frac{23z^4+10z^2-1}{9z^4(z^2+1)^3},$$

as well as

$$D_2(z) \equiv \frac{-4}{27z^6}, \quad \Theta^{1,1}_{1,2}(z) \equiv 0, \quad \Theta^{1,1}_{2,2}(z) \equiv 0, \quad \text{and} \quad \lambda(z) \equiv -\frac{2(z^2-1)}{z(z^2+1)}.$$

The left member Q_2 of (19.27) satisfies $(Q_2)' + \lambda Q_2 \equiv 2 S_2 \mathcal{L}_3$, where

$$(19.28) \qquad 2S_2 \equiv 2y'' - \frac{10z^2+2}{z(z^2+1)}y' + \frac{2(23z^4-2z^2-1)}{3z^2(z^2+1)^2}y$$

and

$$(19.29) \qquad \mathcal{L}_3 \equiv y''' - \frac{6z}{z^2+1}y'' + \frac{6(3z^2-1)}{(z^2+1)^2}y' - \frac{24z(z^2-1)}{(z^2+1)^3}y.$$

For Version 3.0 of MATHEMATICA, the DSolve command does not directly yield solutions for $\mathcal{L}_3 = 0$. However, Theorem 19.7 is applicable and, as we shall see, it gives three linearly independent solutions $v_0(z)$, $v_1(z)$, $v_2(z)$ of $\mathcal{L}_3 = 0$.

Given a subregion $\mathcal{U}$ of Ω, let $\mathcal{V}$ be a simply-connected subregion of $\mathcal{U}$ and select a fixed logarithm on $\mathcal{V}$ in terms of which z^r is defined on $\mathcal{V}$, for each r in $\mathbb{C}$. With $\frac{-D_2(z)}{(B_{1,1}(z))^2} \equiv \frac{4}{3z^2}$, we select $\tau(z) \equiv \frac{2}{\sqrt{3}z}$ for (19.20). When the MATHEMATICA commands DSolve and FullSimplify are used with (19.21), we find that $\eta(z)$ may be selected as $\eta(z) \equiv z^{(3+2/\sqrt{3})}(z^2+1)$. We apply DSolve and Simplify to (19.22) in order to obtain $\phi(z) \equiv z^{(-3-2\sqrt{3})/6}$ and $\psi(z) \equiv z^{(3-2\sqrt{3})/6}$ as linearly independent solutions of (19.22). Next, we employ FullSimplify with (19.23) to conclude that the nonsingular solutions on $\mathcal{V}$ of (19.27) are given by

$$(19.30) \qquad y(z) \equiv \frac{4}{3}(z^2+1)\Big[(2+\sqrt{3})(C_1)^2 + C_1 C_2 z + (2-\sqrt{3})(C_2)^2 z^2\Big],$$

for C_1, C_2 in $\mathbb{C}$ and either $C_1 \neq 0$ or $C_2 \neq 0$.

We can simplify (19.30) by substituting $C_1 = C/h_0$ and $C_2 = K/k_0$, where

$$h_0 = 2\sqrt{\frac{2}{3}+\frac{1}{\sqrt{3}}} \quad \text{and} \quad k_0 = 2\sqrt{\frac{2}{3}-\frac{1}{\sqrt{3}}}.$$

We do this with FullSimplify. Thus, the nonsingular solutions of (19.27) are

$$(19.31) \qquad y(z) \equiv (z^2+1)\big[C^2 + CKz + K^2 z^2\big], \quad \text{for } C, K \text{ in } \mathbb{C} \text{ not both zero.}$$

We use (19.26) with DSolve and FullSimplify to deduce that the singular solutions on $\mathcal{V}$ of (19.27) are given by

$$y(z) \equiv C_1(z^3+z)z^{2\sqrt{3}/3} \quad \text{or} \quad y(z) \equiv C_2(z^3+z)z^{-2\sqrt{3}/3}, \quad \text{for } C_1, C_2 \text{ in } \mathbb{C}.$$

19.4. Other forms for the nonsingular solutions in Theorem 19.7

PROPOSITION 19.9. *Let $\{v_0(z), v_1(z), v_2(z)\}$ be a linearly independent set over $\mathbb{C}$ of meromorphic functions on a region $\mathcal{V}$; let A be a nonsingular 3×3 matrix over $\mathbb{C}$, let meromorphic functions $u_0(z)$, $u_1(z)$, $u_2(z)$ be defined on $\mathcal{V}$ by*

$$(19.32) \qquad [u_0(z), u_1(z), u_2(z)] \equiv [v_0(z), v_1(z), v_2(z)] A;$$

let $\mathcal{S}_1$ be the set of functions $y_1(z)$ defined on $\mathcal{V}$ by

$$(19.33) \qquad y_1(z) \equiv C^2 u_0(z) + CK u_1(z) + K^2 u_2(z), \quad \text{for } C, K \text{ in } \mathbb{C};$$

and let $\mathcal{S}_2$ be the set of functions $y_2(z)$ defined on $\mathcal{V}$ by

$$(19.34) \qquad y_2(z) \equiv (C_1)^2 v_0(z) + C_1 C_2 v_1(z) + (C_2)^2 v_2(z), \quad \text{for } C_1, C_2 \text{ in } \mathbb{C}.$$

Then, $\mathcal{S}_1$ is contained in $\mathcal{S}_2$ if and only if $\mathcal{S}_1 \equiv \mathcal{S}_2$; and the condition $\mathcal{S}_1 \equiv \mathcal{S}_2$ is satisfied if and only if there are complex numbers $\alpha, \beta, \gamma, \delta$ such that

$$(19.35) \qquad A = \begin{bmatrix} \alpha^2 & 2\alpha\beta & \beta^2 \\ \alpha\gamma & \alpha\delta + \beta\gamma & \beta\delta \\ \gamma^2 & 2\gamma\delta & \delta^2 \end{bmatrix} \quad \text{and} \quad \det(A) \equiv (\alpha\delta - \beta\gamma)^3 \neq 0.$$

Moreover, whenever A is given by (19.35), (19.32) yields

$$(19.36) \qquad \bigl(u_1(z)\bigr)^2 - 4 u_0(z) u_2(z) \equiv (\alpha\delta - \beta\gamma)^2 \Bigl[\bigl(v_1(z)\bigr)^2 - 4 v_0(z) v_2(z)\Bigr]$$

and

$$(19.37) \qquad \begin{vmatrix} u_0(z) & u_2(z) \\ u_0^{(1)}(z) & u_2^{(1)}(z) \end{vmatrix}^2 - \begin{vmatrix} u_0(z) & u_1(z) \\ u_0^{(1)}(z) & u_1^{(1)}(z) \end{vmatrix} \begin{vmatrix} u_1(z) & u_2(z) \\ u_1^{(1)}(z) & u_2^{(1)}(z) \end{vmatrix}$$

$$\equiv (\alpha\delta - \beta\gamma)^4 \left\{ \begin{vmatrix} v_0(z) & v_2(z) \\ v_0^{(1)}(z) & v_2^{(1)}(z) \end{vmatrix}^2 - \begin{vmatrix} v_0(z) & v_1(z) \\ v_0^{(1)}(z) & v_1^{(1)}(z) \end{vmatrix} \begin{vmatrix} v_1(z) & v_2(z) \\ v_1^{(1)}(z) & v_2^{(1)}(z) \end{vmatrix} \right\}.$$

OBSERVATION. Matrices having the form of A in (19.35) occur when binary forms of degree 2 are subjected to linear substitutions as in [**48**, page 16]. Complete details are given by the special case $n = 2$ of Proposition 19.26 on page 213.

Using (19.36) and Notation 19.29 of page 215 with $n = 2$, we see that the polynomial $(x_1)^2 - 4 x_0 x_2$ is a type of invariant under substitutions of $u_i(z)$ for x_i vis-à-vis substitutions of $v_i(z)$ for x_i. In view of (19.37), the polynomial

$$(19.38) \qquad \bigl(x_0 x_2^{(1)} - x_0^{(1)} x_2\bigr)^2 - \bigl(x_0 x_1^{(1)} - x_0^{(1)} x_1\bigr)\bigl(x_1 x_2^{(1)} - x_1^{(1)} x_2\bigr)$$

is also a type of invariant under substitutions of $u_i^{(j)}(z)$ for $x_i^{(j)}$ in relation to substitutions of $v_i^{(j)}(z)$ for $x_i^{(j)}$. Similarly, for Γ_n in (19.120) on page 215, we have (19.125) on page 217. Thus, by permitting the coefficients of binary forms to be elements of an ordinary differential field (such as the field $\mathfrak{F}_\mathcal{V}$ of meromorphic functions on $\mathcal{V}$), *new invariants are deduced for an old subject.*

PROOF OF PROPOSITION 19.9. For $A = \begin{bmatrix} a_{0,0} & a_{0,1} & a_{0,2} \\ a_{1,0} & a_{1,1} & a_{1,2} \\ a_{2,0} & a_{2,1} & a_{2,2} \end{bmatrix}$ with $a_{i,j}$ in $\mathbb{C}$, we use (19.32) to rewrite (19.33) as $y_1(z) \equiv R_0 v_0(z) + R_1 v_1(z) + R_2 v_2(z)$, where $R_i \equiv C^2 a_{i,0} + CK a_{i,1} + K^2 a_{i,2}$, for $0 \leq i \leq 2$ and C, K in $\mathbb{C}$.

Suppose that $\mathcal{S}_1$ is contained in $\mathcal{S}_2$. Then, each $y_1(z)$ given by (19.33) is also given by (19.34). This requires $R_0 R_2 - (R_1)^2 \equiv 0$, for each C, K in $\mathbb{C}$. We have

(19.39) $\qquad R_0 R_2 - (R_1)^2 \equiv C^4 L_0 + C^3 K L_1 + C^2 K^2 L_2 + C K^3 L_3 + K^4 L_4,$

where L_0, L_4, L_1, L_3, L_2 are the respective left members of

(19.40) $\qquad\qquad\qquad\qquad a_{0,0}\, a_{2,0} - (a_{1,0})^2 = 0,$

(19.41) $\qquad\qquad\qquad\qquad a_{0,2}\, a_{2,2} - (a_{1,2})^2 = 0,$

(19.42) $\qquad\qquad\qquad a_{0,0}\, a_{2,1} + a_{0,1}\, a_{2,0} - 2\, a_{1,0}\, a_{1,1} = 0,$

(19.43) $\qquad\qquad\qquad a_{0,1}\, a_{2,2} + a_{0,2}\, a_{2,1} - 2\, a_{1,1}\, a_{1,2} = 0,$

(19.44) $\qquad a_{0,0}\, a_{2,2} + a_{0,1}\, a_{2,1} + a_{0,2}\, a_{2,0} - 2\, a_{1,0}\, a_{1,2} - (a_{1,1})^2 = 0.$

Thus, when $\mathcal{S}_1 \subseteq \mathcal{S}_2$, the components $a_{i,j}$ of A satisfy (19.40)–(19.44).

Suppose that (19.40)–(19.44) are satisfied. Let $\alpha, \beta, \gamma, \delta$ be complex numbers such that $\alpha^2 = a_{0,0}$, $\beta^2 = a_{0,2}$, $\gamma^2 = a_{2,0}$, and $\delta^2 = a_{2,2}$. Then, (19.40) and (19.41) require $a_{1,0} = \pm\alpha\gamma$ and $a_{1,2} = \pm\beta\delta$. By replacing γ with $-\gamma$ or δ with $-\delta$ whenever necessary, we obtain complex numbers $\alpha, \beta, \gamma, \delta$ such that

(19.45) $\qquad a_{0,0} = \alpha^2,\; a_{0,2} = \beta^2,\; a_{2,0} = \gamma^2,\; a_{2,2} = \delta^2,\; a_{1,0} = \alpha\gamma,\; a_{1,2} = \beta\delta.$

We use (19.42)–(19.43) and (19.45) to obtain

(19.46) $\qquad \alpha^2 a_{2,1} + \gamma^2 a_{0,1} = 2\alpha\gamma a_{1,1} \quad \text{and} \quad \beta^2 a_{2,1} + \delta^2 a_{0,1} = 2\beta\delta a_{1,1}.$

In view of $\det(A) \neq 0$, the first column $[\alpha^2, \alpha\gamma, \gamma^2]^T$ of A and the third column $[\beta^2, \beta\delta, \delta^2]^T$ of A are linearly independent over $\mathbb{C}$. Thus, we have $\alpha\delta - \beta\gamma \neq 0$. Using this, we apply Cramer's rules to the linear system (19.46) in $a_{2,1}$ and $a_{0,1}$ to verify that

(19.47) $\qquad (\alpha\delta + \beta\gamma) a_{0,1} = 2\alpha\beta a_{1,1} \quad \text{and} \quad (\alpha\delta + \beta\gamma) a_{2,1} = 2\gamma\delta a_{1,1}.$

Moreover, (19.44) and (19.45) yield

(19.48) $\qquad\qquad\qquad (a_{1,1})^2 - a_{0,1}\, a_{2,1} = (\alpha\delta - \beta\gamma)^2.$

There are two cases depending on whether $\alpha\delta + \beta\gamma \neq 0$ or $\alpha\delta + \beta\gamma = 0$.

(i) Suppose that $\alpha\delta + \beta\gamma \neq 0$. Then, (19.47) and (19.48) yield

$$a_{0,1} = \frac{2\alpha\beta}{\alpha\delta + \beta\gamma} a_{1,1}, \quad a_{2,1} = \frac{2\gamma\delta}{\alpha\delta + \beta\gamma} a_{1,1},$$

$$(\alpha\delta - \beta\gamma)^2 = (a_{1,1})^2 - \frac{4\alpha\beta\gamma\delta}{(\alpha\delta + \beta\gamma)^2} (a_{1,1})^2 = \frac{(\alpha\delta - \beta\gamma)^2}{(\alpha\delta + \beta\gamma)^2} (a_{1,1})^2,$$

and $(a_{1,1})^2 = (\alpha\delta + \beta\gamma)^2$. Consequently, for $\epsilon = \pm 1$, we have $a_{1,1} = \epsilon(\alpha\delta + \beta\gamma)$, $a_{0,1} = 2\epsilon\alpha\beta$, and $a_{2,1} = 2\epsilon\gamma\delta$. When $\epsilon = 1$, (19.35) is valid. When $\epsilon = -1$, we can replace β and δ throughout with $-\beta$ and $-\delta$ to satisfy (19.35).

(ii) Suppose that $\alpha\delta + \beta\gamma = 0$. Then, we use $\alpha\delta - \beta\gamma \neq 0$, (19.47), (19.46), and (19.48) to deduce that $\alpha\delta \neq 0$, $\beta\gamma \neq 0$, $\alpha \neq 0$, $\beta \neq 0$, $\gamma \neq 0$, $\delta \neq 0$, $a_{1,1} = 0$,

$$a_{0,1} = -\frac{\alpha^2}{\gamma^2} a_{2,1}, \quad (a_{0,1})^2 = -\frac{\alpha^2}{\gamma^2} a_{0,1}\, a_{2,1} = \frac{\alpha^2}{\gamma^2}(\alpha\delta - \beta\gamma)^2 = (2\alpha\beta)^2,$$

and $a_{0,1} = 2\epsilon\alpha\beta$, for $\epsilon = \pm 1$. We also have
$$a_{2,1} = -\frac{\delta^2}{\beta^2} a_{0,1} = \frac{-2\epsilon(\alpha\delta)(\beta\delta)}{\beta^2} = \frac{2\epsilon(\beta\gamma)(\beta\delta)}{\beta^2} = 2\epsilon\gamma\delta.$$

When $\epsilon = 1$, (19.35) is valid. When $\epsilon = -1$, we can replace β and δ throughout with $-\beta$ and $-\delta$ to satisfy (19.35).

Thus, if $\mathcal{S}_1 \subseteq \mathcal{S}_2$, then the matrix A for (19.32) has the structure provided by (19.35). In this situation, (19.32) and (19.35) yield (19.36) and (19.37).

Suppose that A for (19.32) is given by (19.35). Due to $\alpha\delta - \beta\gamma \neq 0$, we see that: as (C, K) ranges through $\mathbb{C} \times \mathbb{C}$, the corresponding (C_1, C_2) specified by

(19.49) $\qquad C_1 = \alpha C + \beta K \quad \text{and} \quad C_2 = \gamma C + \delta K$

ranges through $\mathbb{C} \times \mathbb{C}$. After substituting (19.49) into (19.34), we use (19.32) and (19.35) to deduce that $y_2(z) \equiv y_1(z)$, for $y_2(z)$ in (19.34) and $y_1(z)$ in (19.33). This yields $\mathcal{S}_2 \equiv \mathcal{S}_1$ and completes the proof. $\square$

19.5. Conditions of the type $\bigl(u_1(z)\bigr)^2 - 4 u_0(z)\, u_2(z) \not\equiv 0$

As a supplement to Theorem 19.7 of page 197, we have the following result.

PROPOSITION 19.10. *For Q_2 in (19.1) and $D_2(z)$ in (19.2), suppose that Q_2 satisfies $D_2(z) \not\equiv 0$ and (19.11). Let $u_0(z), u_1(z), u_2(z)$ be linearly independent meromorphic functions on a subregion $\mathcal{U}$ of Ω such that each of the functions*

(19.50) $\qquad y(z) \equiv C^2 u_0(z) + CK\, u_1(z) + K^2 u_2(z), \quad \text{for } C,\, K \text{ in } \mathbb{C},$

is a solution of $Q_2 = 0$. Then, $u_0(z), u_1(z)$, and $u_2(z)$ satisfy

(19.51) $\qquad \bigl(u_1(z)\bigr)^2 - 4 u_0(z)\, u_2(z) \not\equiv 0, \quad \text{on } \mathcal{U};$

the nonzero functions $y(z)$ given by (19.50) are the nonsingular solutions of $Q_2 = 0$ on $\mathcal{U}$; and each nonsingular solution of $Q_2 = 0$ on a subregion $\mathcal{U}_0$ of $\mathcal{U}$ is the restriction to $\mathcal{U}_0$ of some nonzero function given by (19.50).

PROOF. Let $y_0(z)$ be a nonsingular solution of $Q_2 = 0$ on a subregion $\mathcal{U}_0$ of $\mathcal{U}$. We apply Theorem 19.7 with $\mathcal{U}_0$ in place of Ω to obtain meromorphic functions $v_0(z), v_1(z), v_2(z)$ on a subregion $\mathcal{V}$ of $\mathcal{U}_0$ such that (19.25) is satisfied and the zero function as well as all of the nonsingular solutions of $Q_2 = 0$ on $\mathcal{V}$ are given by

(19.52) $\qquad y(z) \equiv (C_1)^2 v_0(z) + C_1 C_2\, v_1(z) + (C_2)^2 v_2(z), \quad \text{for } C_1,\, C_2 \text{ in } \mathbb{C}.$

We let $\bar{u}_0(z), \bar{u}_1(z), \bar{u}_2(z)$ denote the restrictions of $u_0(z), u_1(z), u_2(z)$ to $\mathcal{V}$ and use (19.50) to see that each of the five functions

(19.53) $\qquad \bar{u}_0, \quad \bar{u}_2, \quad \bar{u}_0 + \bar{u}_1 + \bar{u}_2, \quad 4\bar{u}_0 + 2\bar{u}_1 + \bar{u}_2, \quad \bar{u}_0 + 2\bar{u}_1 + 4\bar{u}_2$

is a solutions of $Q_2 = 0$ on $\mathcal{V}$. The factorization $(Q_2)' + \lambda(z)\, Q_2 \equiv 2 S_2\, \mathcal{L}_3$ in (19.7) on page 194 shows that each of the five functions in (19.53) is a solution on $\mathcal{V}$ of either $S_2 = 0$ or $\mathcal{L}_3 = 0$. Since any three of the five functions in (19.53) are linearly independent over $\mathbb{C}$ and the order of $S_2 = 0$ is 2, we deduce that $\bar{u}_0(z), \bar{u}_1(z), \bar{u}_2(z)$ are solutions on $\mathcal{V}$ of $\mathcal{L}_3 = 0$. Thus, each of the functions

(19.54) $\qquad y(z) \equiv C^2 \bar{u}_0(z) + CK\, \bar{u}_1(z) + K^2 \bar{u}_2(z), \quad \text{for } C,\, K \text{ in } \mathbb{C},$

is a solution of both $Q_2 = 0$ and $\mathcal{L}_3 = 0$. We apply Proposition 19.5 on page 195 to see that each nonzero function of (19.54) is a nonsingular solution on $\mathcal{V}$ of $Q_2 = 0$. Since each function specified by (19.54) is therefore included among those given by (19.52), Proposition 19.9 shows that (19.54) and (19.52) define the same set of functions on $\mathcal{V}$. Thus, the nonzero functions of (19.54) are the nonsingular solutions on $\mathcal{V}$ of $Q_2 = 0$. Meromorphic continuation of them to $\mathcal{U}$ shows that each nonzero function of (19.50) is a nonsingular solution on $\mathcal{U}$ of $Q_2 = 0$. For any nonsingular solution of $Q_2 = 0$ on $\mathcal{U}$, its restriction to $\mathcal{V}$ must be given by (19.54) for some C, K in $\mathbb{C}$ not both zero. Hence, the nonsingular solutions of $Q_2 = 0$ on $\mathcal{U}$ are the nonzero functions of (19.50). Since the restriction to $\mathcal{V}$ of $y_0(z)$ on $\mathcal{U}_0$ is a nonsingular solution, it is given by (19.54), for some C, K in $\mathbb{C}$ not both zero. Thus, $y_0(z)$ is the restriction to $\mathcal{U}_0$ of the nonsingular solution on $\mathcal{U}$ that is defined by (19.50) for the same C and K. Finally, we apply (19.36) and (19.25) to obtain

$$\left(\bar{u}_1(z)\right)^2 - 4\bar{u}_0(z)\,\bar{u}_2(z) \equiv (\alpha\delta - \beta\gamma)^2 \left[\left(v_1(z)\right)^2 - 4v_0(z)\,v_2(z)\right] \not\equiv 0, \quad \text{on } \mathcal{V}.$$

This yields (19.51) and completes the proof. $\square$

PROPOSITION 19.11. *Suppose that meromorphic functions $u_0(z)$, $u_1(z)$, $u_2(z)$ on a region Ω are linearly independent over $\mathbb{C}$ and set*

$$(19.55) \qquad \Delta(z) \equiv \left(u_1(z)\right)^2 - 4u_0(z)\,u_2(z), \quad \text{on } \Omega,$$

as well as

$$(19.56)\quad a(z) \equiv \begin{vmatrix} u_0(z) & u_2(z) \\ u_0^{(1)}(z) & u_2^{(1)}(z) \end{vmatrix}^2 - \begin{vmatrix} u_0(z) & u_1(z) \\ u_0^{(1)}(z) & u_1^{(1)}(z) \end{vmatrix} \begin{vmatrix} u_1(z) & u_2(z) \\ u_1^{(1)}(z) & u_2^{(1)}(z) \end{vmatrix}, \quad \text{on } \Omega.$$

Then, the condition $\Delta(z) \not\equiv 0$ is satisfied if and only if $a(z) \not\equiv 0$.

PROOF. We shall establish that $\Delta(z) \equiv 0$ is equivalent to $a(z) \equiv 0$.

(i) Suppose that $\Delta(z) \equiv 0$. Then, there are meromorphic functions $\phi(z)$, $\psi(z)$ on a subregion $\mathcal{U}$ of Ω such that

$$(19.57) \quad u_0(z) \equiv \left(\phi(z)\right)^2, \quad u_2(z) \equiv \left(\psi(z)\right)^2, \quad \text{and} \quad u_1(z) \equiv 2\phi(z)\,\psi(z), \quad \text{on } \mathcal{U}.$$

We use (19.56)–(19.57) and the notation $\chi(z) \equiv \phi(z)\,\psi^{(1)}(z) - \phi^{(1)}(z)\,\psi(z)$ to obtain

$$a(z) \equiv \left[2\phi(z)\,\psi(z)\,\chi(z)\right]^2 - \left[2\left(\phi(z)\right)^2 \chi(z)\right]\left[2\left(\psi(z)\right)^2 \chi(z)\right] \equiv 0, \quad \text{on } \mathcal{U}.$$

Since $a(z)$ is meromorphic on Ω, this yields the conclusion $a(z) \equiv 0$ on Ω.

(ii) Suppose that $a(z) \equiv 0$ on Ω. Then, there are meromorphic functions $\phi(z)$, $\psi(z)$ on a subregion $\mathcal{U}$ of Ω such that $u_0(z) \equiv \left(\phi(z)\right)^2$ and $u_2(z) \equiv \left(\psi(z)\right)^2$ on $\mathcal{U}$. We substitute these expressions for $u_0(z)$ and $u_2(z)$ into $a(z)$ and see that the restriction of $u_1(z)$ to $\mathcal{U}$ gives a solution $w(z) \equiv u_1(z)$ of the differential equation

$$(19.58) \quad \left[2\phi(z)\,\psi(z)\left[\phi(z)\,\psi^{(1)}(z) - \phi^{(1)}(z)\,\psi(z)\right]\right]^2$$
$$- \left[\left(\phi(z)\right)^2 w' - 2\phi(z)\,\phi^{(1)}(z)\,w\right]\left[2\psi(z)\,\psi^{(1)}(z)\,w - \left(\psi(z)\right)^2 w'\right] = 0.$$

In view of

$$0 \not\equiv u_0(z)\,u_2^{(1)}(z) - u_0^{(1)}(z)\,u_2(z) \equiv 2\phi(z)\,\psi(z)\left(\phi(z)\,\psi^{(1)}(z) - \phi^{(1)}(z)\,\psi(z)\right),$$

we can write (19.58) in the equivalent form

$$(19.59) \qquad \left[w' - \left(\frac{\phi^{(1)}(z)\,\psi(z) + \phi(z)\,\psi^{(1)}(z)}{\phi(z)\,\psi(z)}\right)w\right]^2 + F(w) = 0,$$

where

$$F(w) \equiv -\left[\frac{\phi^{(1)}(z)}{\phi(z)} - \frac{\psi^{(1)}(z)}{\psi(z)}\right]^2 \left[w - 2\phi(z)\,\psi(z)\right]\left[w + 2\phi(z)\,\psi(z)\right].$$

Regarding $F(w)$ as a polynomial in w over the field $\mathfrak{F}_\mathcal{U}$ of meromorphic functions on $\mathcal{U}$, we see that each of the two roots $r_1(z) \equiv 2\phi(z)\,\psi(z)$ and $r_2(z) \equiv -2\phi(z)\,\psi(z)$ of $F(w)$ is a solution of (19.59). Hence, the solution procedure of [**17**, Theorem 1] is applicable to (19.59) and shows that, for nonzero C in $\mathbb{C}$, the meromorphic solutions $w(z)$ on $\mathcal{U}$ of (19.59) subject to $w(z) \not\equiv r_1(z)$ and $w(z) \not\equiv r_2(z)$ are given by

$$(19.60) \qquad w(z) \equiv \left[\frac{r_1(z) - r_2(z)}{4}\right]\left(C\,\Theta(z) + \frac{1}{C\,\Theta(z)}\right) + \frac{r_1(z) + r_2(z)}{2},$$

where $\Theta(z)$ is a meromorphic solution on $\mathcal{U}$ of $\Theta' - \eta(z)\,\Theta = 0$ subject to $\Theta(z) \not\equiv 0$, $\eta(z)$ is a meromorphic solution on $\mathcal{U}$ of $(\eta(z))^2 = -A_2(z)/4$, and $A_2(z)/4$ is the coefficient of w^2 in $F(w)$. After selecting

$$\eta(z) \equiv \frac{\phi^{(1)}(z)}{\phi(z)} - \frac{\psi^{(1)}(z)}{\psi(z)} \quad \text{and} \quad \Theta(z) \equiv \frac{\phi(z)}{\psi(z)}, \quad \text{on } \mathcal{U},$$

we observe that (19.60) yields the meromorphic solutions $w(z)$ on $\mathcal{U}$ of (19.59) subject to $w(z) \not\equiv r_1(z)$ and $w(z) \not\equiv r_2(z)$ as

$$(19.61) \qquad w(z) \equiv C(\phi(z))^2 + \frac{1}{C}(\psi(z))^2 \equiv C\,u_0(z) + \frac{1}{C}\,u_2(z),$$

$$\text{for each nonzero } C \text{ in } \mathbb{C}.$$

The restriction of $u_1(z)$ to $\mathcal{U}$ is a solution of (19.58) and (19.59) on $\mathcal{U}$. However, it is not given by (19.61) because $u_1(z)$ is linearly independent of $u_0(z)$ and $u_2(z)$ over $\mathbb{C}$. Thus, the restriction to $\mathcal{U}$ of $u_1(z)$ is either $r_1(z)$ or $r_2(z)$. Consequently, we have $\Delta(z) \equiv 0$ on $\mathcal{U}$ and therefore also on Ω. This completes the proof. $\square$

For a proof of Proposition 19.11 that is independent of [**17**, Theorem 1], see the case $n = 2$ of Proposition 19.30 on page 216.

19.6. Equations constructed to have given nonsingular solutions

The next result shows how equations like (19.27) in Example 19.8 on page 198 can be constructed. Its computational proof provides an excellent check on earlier formulas of this chapter.

PROPOSITION 19.12. *Suppose that meromorphic functions $u_0(z)$, $u_1(z)$, $u_2(z)$ on Ω are linearly independent over $\mathbb{C}$ and satisfy $\Delta(z) \not\equiv 0$, for $\Delta(z)$ in (19.55). Then, there are unique meromorphic coefficients $c_{i,j}(z)$ on Ω for Q_2 in (19.1) on page 193 such that the nonzero functions of*

$$(19.62) \qquad y(z) \equiv C^2 u_0(z) + CK\,u_1(z) + K^2 u_2(z), \quad \text{for } C,\,K \text{ in } \mathbb{C},$$

are the nonsingular solutions on Ω of $Q_2 = 0$; and each local nonsingular solution of $Q_2 = 0$ on a subregion $\mathcal{U}$ of Ω is the restriction to $\mathcal{U}$ of some function in (19.62). Moreover, Q_2 satisfies $D_2(z) \not\equiv 0$, $B_{1,1}(z) \not\equiv 0$, and $\Theta_{1,2}^{1,1}(z) \equiv \Theta_{2,2}^{1,1}(z) \equiv 0$.

PROOF. The Wronskian $W(z)$ of $u_0(z)$, $u_1(z)$, $u_2(z)$ is defined on Ω by

$$(19.63) \quad W(z) \equiv det(\mathcal{M}(z)), \quad \text{where} \quad \mathcal{M}(z) \equiv \begin{bmatrix} u_0(z) & u_1(z) & u_2(z) \\ u_0^{(1)}(z) & u_1^{(1)}(z) & u_2^{(1)}(z) \\ u_0^{(2)}(z) & u_1^{(2)}(z) & u_2^{(2)}(z) \end{bmatrix}.$$

Since the meromorphic functions $u_0(z)$, $u_1(z)$, $u_2(z)$ are linearly independent over $\mathbb{C}$, we have $W(z) \not\equiv 0$. The functions specified on Ω by (19.62) are solutions of

$$y^{(i)} = C^2 u_0^{(i)}(z) + CK u_1^{(i)}(z) + K^2 u_2^{(i)}(z), \quad \text{for } i = 0,\ 1,\ 2.$$

Therefore, with respect to $K_0 = C^2$, $K_1 = CK$, and $K_2 = K^2$, the functions $y(z)$ in (19.62) are solutions of

$$W(z)\, K_j = det(U_j(z)), \quad \text{for } j = 0,\ 1,\ 2,$$

where: for $0 \leq j \leq 2$, $U_j(z)$ is the 3×3 matrix obtained from $\mathcal{M}(z)$ by replacing the $(j+1)$th column $[u_j(z), u_j^{(1)}(z), u_j^{(2)}(z)]^T$ of $\mathcal{M}(z)$ with $[y^{(0)}, y^{(1)}, y^{(2)}]^T$. In view of $(W(z))^2 ((K_1)^2 - K_0 K_2) \equiv 0$, the functions $y(z)$ in (19.62) are also solutions of

$$(19.64) \quad \left(det(U_1(z))\right)^2 - det(U_0(z))\, det(U_2(z)) = 0.$$

After writing y for $y^{(0)}$, y' for $y^{(1)}$, and y'' for $y^{(2)}$ in (19.64), let the left member of

$$(19.65) \quad a(z)(y'')^2 + b(z)y''y' + c(z)y''y + d(z)(y')^2 + e(z)y'y + f(z)y^2 = 0$$

be obtained by expanding the left member of (19.64). Hence, each $y(z)$ in (19.62) is a solution of (19.65); and each of $a(z), \dots f(z)$ is a homogeneous quadratic polynomial combination of $u_i^{(j)}(z)$, for $0 \leq i \leq 2$ and $0 \leq j \leq 2$. We see that

$$(19.66) \quad a(z) \equiv \begin{vmatrix} u_0(z) & u_2(z) \\ u_0^{(1)}(z) & u_2^{(1)}(z) \end{vmatrix}^2 - \begin{vmatrix} u_0(z) & u_1(z) \\ u_0^{(1)}(z) & u_1^{(1)}(z) \end{vmatrix} \begin{vmatrix} u_1(z) & u_2(z) \\ u_1^{(1)}(z) & u_2^{(1)}(z) \end{vmatrix}, \quad \text{on } \Omega.$$

Due to $\Delta(z) \not\equiv 0$, Proposition 19.11 gives $a(z) \not\equiv 0$. We multiply each coefficient in (19.65) by $1/a(z)$ to obtain the equivalent differential equation

$$(19.67) \quad \left(y^{(2)}\right)^2 + \sum_{\substack{0 \leq i,j \leq 2 \\ (i,j) \neq (0,0)}} c_{i,j}(z)\, y^{(2-i)} y^{(2-j)} = 0, \quad \text{on } \Omega,$$

$$\text{where } c_{j,i}(z) \equiv c_{i,j}(z), \text{ for } 0 \leq i < j \leq 2.$$

Thus, we have $c_{0,1}(z) \equiv b(z)/(2a(z))$, $c_{0,2}(z) \equiv c(z)/(2a(z))$, $c_{1,1}(z) \equiv d(z)/a(z)$, $c_{1,2}(z) \equiv e(z)/(2a(z))$, and $c_{2,2}(z) \equiv f(z)/a(z)$. Let Q_2 denote the left member of (19.67). We note that each $y(z)$ in (19.62) is a solution of $Q_2 = 0$.

After computer algebra yields explicit expressions for the coefficients of (19.65) and (19.67), machine computations for (19.67) in combination with (18.5), (19.55), (19.63), (19.2)–(19.4), and (19.6)–(19.9) give

$$B_{1,1}(z) \equiv \frac{-\Delta(z)(W(z))^2}{4(a(z))^2} \not\equiv 0, \quad D_2(z) \equiv -\frac{(W(z))^4}{4(a(z))^3} \not\equiv 0,$$

$$\Theta_{1,2}^{1,1}(z) \equiv 0, \quad \Theta_{2,2}^{1,1}(z) \equiv 0, \quad \lambda(z) \equiv 2\left[c_{0,1}(z) + \frac{a^{(1)}(z)}{a(z)} - \frac{W^{(1)}(z)}{W(z)}\right],$$

as well as $(Q_2)' + \lambda Q_2 \equiv 2 S_2 \mathcal{L}_3$, where $2 S_2 \equiv \dfrac{\partial Q_2}{\partial y''}$ and

$$\mathcal{L}_3 \equiv \frac{1}{W(z)} \begin{vmatrix} u_0^{(0)}(z) & u_1^{(0)}(z) & u_2^{(0)}(z) & y \\ u_0^{(1)}(z) & u_1^{(1)}(z) & u_2^{(1)}(z) & y' \\ u_0^{(2)}(z) & u_1^{(2)}(z) & u_2^{(2)}(z) & y'' \\ u_0^{(3)}(z) & u_1^{(3)}(z) & u_2^{(3)}(z) & y''' \end{vmatrix}.$$

In particular, the equation $Q_2 = 0$ satisfies $D_2(z) \not\equiv 0$ as well as (19.11) on page 194. Consequently, Proposition 19.10 on page 201 shows that the nonzero $y(z)$ given by (19.62) are the nonsingular solutions of $Q_2 = 0$ on Ω.

To establish the uniqueness of $Q_2 = 0$ in (19.67), suppose that $P_2 = 0$ is also an equation of the form (19.67) for which each $y(z)$ given by (19.62) is a solution. Then, $P_2 - Q_2$ is the left member of an equation having the form

(19.68) $\qquad p(z)\, y''y' + q(z)\, y''y + r(z)\, (y')^2 + s(z)\, y'y + t(z)\, (y)^2 = 0,$

for which each $y(z)$ given by (19.62) is a solution. In particular, the five functions

$y_1 \equiv u_0, \quad y_2 \equiv u_2, \quad y_3 \equiv u_0 + u_1 + u_2, \quad y_4 \equiv 4u_0 + 2u_1 + u_2, \quad y_5 \equiv u_0 + 2u_1 + 4u_2$

from (19.62) are solutions of (19.68). Let Y be the 5×5 matrix whose ith row is

$$\left[y_i^{(2)} y_i^{(1)},\ y_i^{(2)} y_i,\ \left(y_i^{(1)}\right)^2,\ y_i^{(1)} y_i,\ (y_i)^2 \right], \quad \text{for } i = 1, 2, 3, 4, 5.$$

We have $Y\, [p(z), q(z), r(z), s(z), t(z)]^T \equiv [0, 0, 0, 0, 0]^T$. Machine computations yield $det(Y) \equiv -12\, a(z) \left(W(z)\right)^2 \not\equiv 0$. Consequently, we find that

$$p(z) \equiv q(z) \equiv r(z) \equiv s(z) \equiv t(z) \equiv 0 \quad \text{and} \quad P_2 \equiv Q_2.$$

This shows that the equation (19.67) is unique and completes the proof. $\square$

For reference in Sections 19.7 and 19.8, we include Proposition 19.13. It is obtained by applying the previous construction to the situation where $\Delta(z) \equiv 0$.

PROPOSITION 19.13. *Suppose that meromorphic functions* $u_0(z), u_1(z), u_2(z)$ *on Ω are linearly independent over $\mathbb{C}$ and satisfy $\Delta(z) \equiv 0$, for $\Delta(z)$ in (19.55). Then, unique meromorphic coefficients $\alpha(z), \beta(z)$ exist on Ω such that the functions*

(19.69) $\qquad y(z) \equiv C^2 u_0(z) + C K\, u_1(z) + K^2 u_2(z), \quad \text{for } C, K \text{ in } \mathbb{C},$

are the solutions on Ω of the differential equation

(19.70) $\qquad y''y - \tfrac{1}{2} y'^2 + \alpha(z)\, y'y + 2\beta(z)\, y^2 = 0.$

Moreover, each local solution of (19.70) *on a subregion $\mathcal{U}$ of Ω is the restriction to $\mathcal{U}$ of some function in* (19.69). *Also, the left member R_3 of* (19.70) *satisfies*

(19.71) $\qquad (R_3)' + 2\alpha\, R_3 \equiv y\, \mathcal{L}_3,$

where

(19.72) $\qquad \mathcal{L}_3 \equiv y''' + 3\alpha\, y'' + \left(\alpha' + 2\alpha^2 + 4\beta\right) y' + \left(2\beta' + 4\alpha\, \beta\right) y.$

PROOF. We repeat the reasoning for (19.63), (19.69), and (19.64) that leads to (19.65). However, due to $\Delta(z) \equiv 0$, Proposition 19.11 on page 202 yields $a(z) \equiv 0$. This gives $b(z) \equiv -a'(z) \equiv 0$. To establish $c(z) \equiv -2\, d(z) \not\equiv 0$, we use $\Delta(z) \equiv 0$ to obtain meromorphic functions $\sigma(z), \tau(z)$ on a subregion $\mathcal{V}$ of Ω such that

$$\left(\sigma(z)\right)^2 \equiv u_0(z), \quad \left(\tau(z)\right)^2 \equiv u_2(z), \quad \text{and} \quad 2\sigma(z)\, \tau(z) \equiv u_1(z), \quad \text{on } \mathcal{V}.$$

For the Wronskian $W(z)$ in (19.63), we have $0 \not\equiv W(z) \equiv 4\bigl(W[\sigma,\tau](z)\bigr)^3$, on $\mathcal{V}$, where $W[\sigma,\tau](z) \equiv \sigma(z)\,\tau'(z) - \sigma'(z)\,\tau(z)$, on $\mathcal{V}$. Machine computations give

$$c(z) \equiv 8\bigl(W[\sigma,\tau](z)\bigr)^4 \not\equiv 0 \quad \text{and} \quad d(z) \equiv -4\bigl(W[\sigma,\tau](z)\bigr)^4 \not\equiv 0, \quad \text{on } \mathcal{V}.$$

Thus, we have $c(z) \equiv -2d(z) \not\equiv 0$, on $\mathcal{V}$, and therefore also on Ω. Consequently, (19.65) can be rewritten as (19.70) and, for it, each $y(z)$ in (19.69) is a solution. The factorization (19.71) involving $\mathfrak{L}_3$ in (19.72) is easily verified. When two equations of the form (19.70) have two common linearly independent solutions, their difference is identically zero. Hence, (19.70) is unique for the properties mentioned.

Let $y_0(z)$ be a meromorphic solution of (19.70) on a subregion $\mathcal{U}$ of Ω. Let $t_0(z)$, $t_1(z)$, and $t_2(z)$ be meromorphic functions on a subregion $\mathcal{U}_0$ of $\mathcal{U}$ such that

$$\bigl(t_0(z)\bigr)^2 \equiv y_0(z), \quad \bigl(t_1(z)\bigr)^2 \equiv u_0(z), \quad \text{and} \quad \bigl(t_2(z)\bigr)^2 \equiv u_2(z), \quad \text{on } \mathcal{U}_0.$$

Due to $\Delta(z) \equiv 0$, we select $t_1(z)$ and $t_2(z)$ so as to also have $2t_1(z)\,t_2(z) \equiv u_1(z)$. The substitution $y = t^2$ transforms (19.70) into the differential equation

(19.73) $$t'' + \alpha(z)\,t' + \beta(z)\,t = 0.$$

Since $y_0(z)$, $u_0(z)$, $u_2(z)$ are solutions of (19.70) on Ω, we see that $t_0(z)$, $t_1(z)$, $t_2(z)$ are solutions of (19.73) on $\mathcal{U}_0$. Our previous argument involving $\sigma(z)$ and $\tau(z)$ shows that $t_1(z)$ and $t_2(z)$ are linearly independent over $\mathbb{C}$. Hence, there are elements C, K in $\mathbb{C}$ such that $t_0(z) \equiv C\,t_1(z) + K\,t_2(z)$, on $\mathcal{U}_0$, and

$$y_0(z) \equiv \bigl(t_0(z)\bigr)^2 \equiv \bigl(C\,t_1(z) + K\,t_2(z)\bigr)^2 \equiv C^2 u_0(z) + CK u_1(z) + K^2 u_2(z), \quad \text{on } \mathcal{U}_0.$$

Thus, we have $y_0(z) \equiv C^2 u_0(z) + CK u_1(z) + K^2 u_2(z)$, on $\mathcal{U}$. This shows that each local solution of (19.70) on a subregion $\mathcal{U}$ of Ω is the restriction to $\mathcal{U}$ of some function in (19.69). In particular, all of the solutions on Ω of (19.70) are those given by (19.69). This completes the proof. $\square$

19.7. Absence of movable branch points

The differential equations constructed by means of Propositions 19.12 and 19.13 have their solutions free of movable branch points. Here, we shall clarify this concept and place these results in better perspective

For the subject of algebraic differential equations having given meromorphic coefficients on a region Ω, L. Bieberbach pointed out in [**8**, page 91] of 1965 that, prior to 1965, all of the considerable research about movable branch points, movable poles, and various other types of movable singularities for such equations was based on imprecise definitions. For a given algebraic differential equation $P = 0$ on a region Ω, he proposed the following definition.

DEFINITION 19.14. *Let S_{br} denote the subset of Ω such that: for each z_0 in Ω, z_0 belongs to S_{br} if and only if z_0 is a branch point for some local solution of $P = 0$ with respect to meromorphic continuation. Then, a movable branch point for $P = 0$ is an interior point of S_{br}. And, the solutions of $P = 0$ are free of movable branch points when the interior of S_{br} is vacuous.*

Other types of movable singularities are similarly defined.

DEFINITION 19.15. *Let S_{po} denote the subset of Ω such that: for each z_0 in Ω, z_0 belongs to S_{po} if and only if z_0 is a pole for some local solution of $P = 0$. Then, a movable pole for $P = 0$ is an interior point of S_{po}. And, the solutions of $P = 0$ are free of movable poles when the interior of S_{po} is vacuous.*

19.7. ABSENCE OF MOVABLE BRANCH POINTS

When $P = 0$ is a first-order algebraic differential equation, important work about necessary and sufficient conditions on the coefficients of P in order for the solutions of $P = 0$ to be free of movable branch points was presented by L. Fuchs in [32] of 1884. After his work was completed with the supplements of M. J. M. Hill and A. Berry in [36], the form E. L. Ince presented it in [38, pages 304–311] of 1927 enables one to make minor changes of wording in order for it to be consistent with Definition 19.14. For details about this, see [18, pages 77–81].

In the particular case where $P = 0$ is an equation of the form

$$(19.74) \quad a_1(z)\,w'^2 + \big(a_2(z)\,w^2 + a_3(z)\,w + a_4(z)\big)\,w' \\ + a_5(z)\,w^4 + a_6(z)\,w^3 + a_7(z)\,w^2 + a_8(z)\,w + a_9(z) = 0$$

having meromorphic coefficients on a region Ω, the results in [38, pages 304–311] can be specialized to obtain convenient explicit conditions on the coefficients of (19.74) that are necessary and sufficient for the solutions of (19.74) to be free of movable branch points. Complete results about this are presented in [18].

Both of the equations $Q_2 = 0$ and (19.70) have the form

$$(19.75) \quad c_{0,0}(z)\,y''^2 + 2\,c_{0,1}(z)\,y''\,y' + 2\,c_{0,2}(z)\,y''\,y \\ + c_{1,1}(z)\,y'^2 + 2\,c_{1,2}(z)\,y'\,y + c_{2,2}(z)\,y^2 = 0,$$

where the coefficients $c_{i,j}(z)$ are meromorphic functions on a region Ω and at least one of the coefficients is not identically zero. In [18, 19], we presented simple explicit conditions on the coefficients of (19.75) that are necessary and sufficient for the solutions of (19.75) to be free of movable branch points in the precise sense of Definition 19.14. The results were derived by using the substitution

$$\frac{y'}{y} = w, \quad \frac{y''}{y} = w' + w^2$$

to relate the solutions of (19.75) to those of a special type of equation (19.74). For our summary of them next, we introduce

$$M(z) \equiv \begin{bmatrix} c_{0,0}(z) & c_{0,1}(z) & c_{0,2}(z) \\ c_{0,1}(z) & c_{1,1}(z) & c_{1,2}(z) \\ c_{0,2}(z) & c_{1,2}(z) & c_{2,2}(z) \end{bmatrix}$$

and we assume that (19.75) is rewritten to have $c_{0,0}(z) \equiv 1$ whenever $c_{0,0}(z) \not\equiv 0$. In particular, if $c_{0,0}(z) \equiv 1$, then we obtain $\det\big(M(z)\big) \equiv D_2(z)$, for $D_2(z)$ in (19.2).

THEOREM 19.16. *When $\det\big(M(z)\big) \equiv 0$, the solutions of (19.75) are free of movable branch points. When $\det\big(M(z)\big) \not\equiv 0$ and $c_{0,0}(z) \equiv 1$, the solutions of (19.75) are free of movable branch points if and only if the left member Q_2 of (19.75) satisfies (19.11) on page 194. When $\det\big(M(z)\big) \not\equiv 0$ and $c_{0,0}(z) \equiv 0$, the solutions of (19.75) are free of movable branch points if and only if the left member Q_2 of (19.75) satisfies either*

$$(19.76) \quad c_{0,1}(z) \equiv 0 \quad \text{and} \quad c_{1,1}(z) \equiv -2\,c_{0,2}(z) \not\equiv 0$$

or

$$(19.77) \quad c_{0,1}(z) \equiv 0 \quad \text{and} \quad c_{1,1}(z) \equiv \left[\frac{2-n}{n-1}\right](2\,c_{0,2}(z)) \not\equiv 0,$$

for some integer n subject to $n \leq 0$ or $n \geq 3$.

PROOF. Letting $I_0(z)$ and $I_1(z)$ denote the left members of (1.79) and (1.80) on page 16 where $A_2(z) \equiv 4B_{1,1}(z)$, we apply [**18**, page 105, Proposition 6.4] to see that: when $c_{0,0}(z) \equiv 1$, the solutions of (19.75) are free of movable branch points if and only if either $det(M(z)) \equiv 0$ or $I_0(z) \equiv I_1(z) \equiv 0 \not\equiv A_2(z)$. In view of (1.79), (1.80), Remark 19.3, and Corollary 19.2, we conclude that: when $c_{0,0}(z) \equiv 1$, the solutions of (19.75) are free of movable branch points if and only if either $det(M(z)) \equiv 0$ or (19.11) is satisfied with $det(M(z)) \not\equiv 0$.

When $c_{0,0}(z) \equiv 0$, we apply [**19**, page 132, Theorem 3.1] to see that the solutions of (19.75) are free of movable branch points if and only if either $det(M(z)) \equiv 0$, or (19.76) is satisfied, or (19.77) is satisfied. This completes the proof. □

REMARK 19.17. If $c_{0,0}(z) \equiv 1$ and $det(M(z)) \not\equiv 0$, then the equations $Q_2 = 0$ whose solutions are free of movable branch points are the ones for which Q_2 is given by Proposition 19.4 on page 195. They are the ones for which $(Q_2)' + \lambda Q_2$ has a nontrivial factorization for some meromorphic $\lambda(z)$; and for them (19.7) is valid.

When (19.77) is satisfied, (19.75) can be rewritten as $R_n = 0$, where

$$(19.78) \quad R_n \equiv y''y + \left[\frac{2-n}{n-1}\right] y'^2 + \alpha(z)\, y'y + (n-1)\beta(z)\, y^2, \quad \text{for } n \leq 0 \text{ or } n \geq 3,$$

and $\alpha(z)$, $\beta(z)$ are meromorphic functions on Ω. We combine (19.78) with the substitution $y = u^{n-1}$ to obtain

$$(19.79) \quad (R_n)_{y=u^{n-1}} \equiv (n-1)\, u^{2n-3} [u'' + \alpha(z)\, u' + \beta(z)\, u].$$

Thus, each nonzero local solution $u(z)$ of $u'' + \alpha(z)\, u' + \beta(z)\, u = 0$ specifies a local solution $y(z) \equiv (u(z))^{n-1}$ of $R_n = 0$. For $n = 3$, we see that R_3 satisfies the identity (19.71). Analogous identities for each R_n having $n \geq 3$ are presented in Section 19.9.

19.8. Two results for third-order linear equations

PROPOSITION 19.18. *In terms of given meromorphic functions $a_1(z)$, $a_2(z)$, $a_3(z)$ on a region Ω_0, let L_3 denote the left member of*

$$(19.80) \quad y''' + a_1(z)\, y'' + a_2(z)\, y' + a_3(z)\, y = 0.$$

Then, there are meromorphic functions $c_{i,j}(z)$ on a subregion Ω of Ω_0 such that the corresponding Q_2 specified on Ω by (19.1) satisfies $D_2(z) \not\equiv 0$, (19.11), and

$$(19.81) \quad (Q_2)' + \lambda(z)\, Q_2 \equiv 2S_2 L_3, \quad \text{on } \Omega,$$

where $\lambda(z)$ and S_2 are defined on Ω by (19.6) and (19.8).

PROOF. Let $u_0(z)$, $u_1(z)$, $u_2(z)$ be linearly independent solutions of $L_3 = 0$ on a subregion Ω of Ω_0. When necessary, we can replace $u_0(z)$ with $2u_0(z)$ to ensure that $\Delta(z) \not\equiv 0$ for $\Delta(z)$ in (19.55). Then, Proposition 19.12 yields an equation $Q_2 = 0$ on Ω that satisfies $D_2(z) \not\equiv 0$ as well as (19.11) for which each of the functions $y(z)$ given by (19.62) is a solution. Moreover, in terms of the factorization $(Q_2)' + \lambda(z)\, Q_2 \equiv 2S_2 \mathcal{L}_3$ provided by (19.6)–(19.9), the equation $\mathcal{L}_3 = 0$ is a monic third-order homogeneous linear differential equation on Ω that has $u_0(z)$, $u_1(z)$, $u_2(z)$ as three linearly independent solutions. Thus, the restriction of L_3 to Ω coincides with $\mathcal{L}_3$. This yields (19.81) and completes the proof. □

19.8. TWO RESULTS FOR THIRD-ORDER LINEAR EQUATIONS

REMARK 19.19. If the left member L_3 for (19.80) is given and an explicit Q_2 can be found to satisfy (19.81), then Theorem 19.7 on page 197 reduces the problem of solving $L_3 = 0$ to that of solving a second-order homogeneous linear differential equation. For instance, when L_3 is $\mathcal{L}_3$ in (19.29) on page 198 and Q_2 is the left member of (19.27), Example 19.8 show that the solutions of $\mathcal{L}_3 = 0$ are
$$y(z) \equiv (z^2 + 1)(K_1 + K_2 z + K_3 z^2), \quad \text{for } K_1, K_2, K_3 \text{ in } \mathbb{C}.$$
But, *a useful method of discovering a suitable Q_2 for any given L_3 is not known.*

For the monic third-order homogeneous linear differential equations
$$(19.82) \qquad y''' + c_1(z)\, y'' + c_2(z)\, y' + c_3(z)\, y = 0$$
having meromorphic coefficients on a region Ω, the basic relative invariant $\boldsymbol{\mathcal{I}}_{3,3}$ of weight 3 is given by [**24**, page 7, (1.33)] in notation that is consistent with (24.30) on page 278. It yields the value $\mathcal{I}_{3,3}(z)$ of $\boldsymbol{\mathcal{I}}_{3,3}$ at (19.82) as

$$(19.83) \qquad \mathcal{I}_{3,3}(z) \equiv c_3(z) - \frac{1}{3} c_1(z)\, c_2(z) + \frac{2}{27}\bigl(c_1(z)\bigr)^3$$
$$- \frac{1}{2} c_2^{(1)}(z) + \frac{1}{3} c_1(z)\, c_1^{(1)}(z) + \frac{1}{6} c_1^{(2)}(z), \quad \text{on } \Omega.$$

PROPOSITION 19.20. *The condition $\mathcal{I}_{3,3}(z) \equiv 0$ is necessary and sufficient for the existence of meromorphic functions $\alpha(z)$ and $\beta(z)$ on Ω in terms of which the corresponding differential polynomial*
$$(19.84) \qquad R_3 \equiv y'' y - \tfrac{1}{2} y'^2 + \alpha(z)\, y' y + 2 \beta(z)\, y^2.$$
and the left member L_3 of (19.82) satisfy
$$(19.85) \qquad (R_3)' + \alpha(z)\, R_3 \equiv y\, L_3.$$

PROOF. For any meromorphic functions $\alpha(z)$ and $\beta(z)$ on Ω, the corresponding polynomial R_3 in (19.84) gives $(R_3)' + \alpha R_3 \equiv y\, \mathcal{L}_3$, where
$$(19.86) \qquad \mathcal{L}_3 \equiv y''' + 3\alpha\, y'' + \bigl(\alpha' + 2\alpha^2 + 4\beta\bigr) y' + \bigl(2\beta' + 4\alpha\, \beta\bigr) y.$$
In view of (19.83), we have $L_3 \equiv \mathcal{L}_3$ if and only if
$$(19.87) \qquad \alpha(z) \equiv \frac{c_1(z)}{3}, \quad \beta(z) \equiv \frac{c_2(z)}{4} - \frac{c_1'(z)}{12} - \frac{\bigl(c_1(z)\bigr)^2}{18},$$
and
$$0 \equiv c_3(z) - \bigl(2\beta'(z) + 4\alpha(z)\, \beta(z)\bigr) \equiv \mathcal{I}_{3,3}(z).$$
This yields the assertion of Proposition 19.20 and completes its proof. $\square$

REMARK 19.21. If (19.82) satisfies $\mathcal{I}_{3,3}(z) \equiv 0$, then the solutions of (19.82) can be obtained by solving a second-order homogeneous linear differential equation. Paul Appell presented this result in [**1**, page 213] of 1880 and in [**2**, page 414]. To explain, let $\alpha(z)$ and $\beta(z)$ be defined on Ω by (19.87). Then, with $n = 3$, we use (19.79), (19.78), and (19.85) to see that: if $u_0(z)$ is a local solution of
$$(19.88) \qquad u'' + \alpha(z)\, u' + \beta(z)\, u = 0,$$
then $y_0(z) \equiv \bigl(u_0(z)\bigr)^2$ is a local solution of $L_3 = 0$. Thus, for any subregion $\mathcal{U}$ of Ω, the deduction on page 211 about (19.97) shows that: if $\phi(z)$ and $\psi(z)$ are linearly independent solutions of (19.88) on $\mathcal{U}$, then $\bigl\{\bigl(\phi(z)\bigr)^2,\ \phi(z)\,\psi(z),\ \bigl(\psi(z)\bigr)^2\bigr\}$ is a fundamental system of local solutions on $\mathcal{U}$ for (19.82).

19.9. Extensions to linear equations of higher order

19.9.1. The basic identity that solved a problem of J. Liouville. *Let meromorphic functions $\alpha(z)$, $\beta(z)$ be given on Ω and suppose that $m \geq 2$. Then, there are unique meromorphic functions $\eta_{m,i}(z)$ on Ω, for $1 \leq i \leq m$, such that the differential polynomial defined by*

$$(19.89) \qquad R_m \equiv y''y + \left[\frac{2-m}{m-1}\right] y'^2 + \alpha(z)\, y'y + (m-1)\beta(z)\, y^2$$

satisfies

$$(19.90) \qquad \sum_{j=0}^{m-2} P_{m-1,j}\, R_m^{(j)} \equiv y^{m-1}\, \mathfrak{L}_m,$$

where

$$(19.91) \qquad \mathfrak{L}_m \equiv y^{(m)} + \sum_{i=1}^{m} \eta_{m,i}(z)\, y^{(m-i)}$$

and, for $0 \leq j \leq m-2$, $P_{m-1,j}$ is a polynomial combination of $y, y', \ldots, y^{(m-2-j)}$ over the field $\mathfrak{F}_\Omega$ of meromorphic functions on Ω. Our proof of the preceeding result in [**19**, Theorems 4.1 and 4.2] gives explicit formulas for the functions $\eta_{m,i}(z)$ and for suitable polynomials $P_{m-1,j}$. Simpler expressions for $\eta_{m,i}(z)$ were presented in [**24**, page 14, Equations (2.6)–(2.9) via Proposition B.1 of page 154]. They are

$$(19.92) \qquad \eta_{m,i}(z) \equiv \Gamma_{i,m-i}(z), \quad \text{for } i = 1, 2, \ldots, m,$$

where, for convenience, we repeat the definition of $\Gamma_{i,j}(z)$ on Ω in (17.7)–(17.9) as

$$(19.93) \qquad \Gamma_{-1,j}(z) \equiv 0, \quad \text{for any } j,$$

$$(19.94) \qquad \Gamma_{0,j}(z) \equiv 1, \quad \text{for any } j,$$

and

$$(19.95) \qquad \Gamma_{i+1,j}(z) \equiv \sum_{k=0}^{j} \left[\begin{array}{l} \Gamma_{i,k}^{(1)}(z) + (i+k)\,\alpha(z)\,\Gamma_{i,k}(z) \\ + (i+k)(m-i-k)\,\beta(z)\,\Gamma_{i-1,k}(z) \end{array} \right],$$

$$\text{for } i \geq 0 \text{ and any } j.$$

THEOREM 19.22. *Let given meromorphic functions $\alpha(z)$, $\beta(z)$ on Ω specify (19.89)–(19.91) according to the preceding context. Then, the $(m-1)$th power of each local solution of the differential equation*

$$(19.96) \qquad u'' + \alpha(z)\, u' + \beta(z)\, u = 0$$

is a local solution of $\mathfrak{L}_m = 0$. Moreover, if $L_m = 0$ is a monic homogeneous linear differential equation in $y(z)$ of order m having meromorphic coefficients on Ω such that the $(m-1)$th power of each local solution of (19.96) is a local solution of $L_m = 0$, then $L_m \equiv \mathfrak{L}_m$.

OBSERVATION. For a given (19.96) and a fixed $m \geq 2$, Joseph Liouville posed in [**45**, Section 4, pages 429–431] of 1839 the problem of finding a monic mth-order homogeneous linear differential equation $\mathfrak{L}_m = 0$ such that the $(m-1)$th power of each solution of (19.96) is a solution of $\mathfrak{L}_m = 0$. We explained in [**24**, pages 13–14] how the construction of $\mathfrak{L}_m$ by means of (19.91)–(19.95) solves this problem. Here, we also show that, apart from the names of the variables, $\mathfrak{L}_m = 0$ is the only solution of Liouville's problem.

19.9. EXTENSIONS TO LINEAR EQUATIONS OF HIGHER ORDER

PROOF OF THEOREM 19.22. For any local solution $u_0(z)$ of (19.96), we use (19.79) and (19.78) to see that $y_0(z) \equiv (u_0(z))^{m-1}$ is a local solution of $R_m = 0$; then, we apply (19.90) to deduce that $y_0(z)$ is a local solution of $\mathfrak{L}_m = 0$.

To show the uniqueness of $\mathfrak{L}_m = 0$, suppose that $L_m = 0$ is a monic mth-order homogeneous linear differential equation in y, $y^{(1)}$, ..., $y^{(m)}$ over $\mathfrak{F}_\Omega$ such that: for any solution $u_0(z)$ of (19.96), the corresponding function $y_0(z) \equiv (u_0(z))^{m-1}$ is a solution of $L_m = 0$. Let $\phi(z)$, $\psi(z)$ be linearly independent solutions of (19.96) on a subregion $\mathcal{U}$ of Ω. The Wronskian $W(z)$ of the m functions in

$$(19.97) \qquad \left\{ (\phi(z))^{m-i} (\psi(z))^{i-1} \mid \text{ for } i = 1, 2, \ldots, m \right\}$$

is given by [**19**, page 149, Proposition 5.2] as

$$(19.98) \qquad W(z) \equiv \left[\prod_{k=1}^{m-1} (k!) \right] \left(\phi(z)\,\psi^{(1)}(z) - \phi^{(1)}(z)\,\psi(z) \right)^{(m-1)m/2} \not\equiv 0;$$

and, for each C_1, C_2 in $\mathbb{C}$, the function $(C_1\,\phi(z) + C_2\,\psi(z))^{m-1}$ is a solution of both $\mathfrak{L}_m = 0$ and $L_m = 0$. This enables us to deduce as in [**19**, pages 144–145] that (19.97) is a fundamental system of local solutions on $\mathcal{U}$ for both $\mathfrak{L}_m = 0$ and $L_m = 0$. Consequently, we obtain $L_m \equiv \mathfrak{L}_m$ and conclude the proof. □

19.9.2. Solution of a related problem.

THEOREM 19.23. *In terms of a given monic homogeneous linear equation*

$$(19.99) \qquad y^{(m)} + \sum_{i=1}^{m} c_i(z)\, y^{(m-i)} = 0$$

of order $m \geq 3$ with meromorphic coefficients $c_i(z)$ on a region Ω, let $\alpha(z)$, $\beta(z)$, $\eta_{m,i}(z)$, and $B_{m,i}(z)$ be the meromorphic functions defined on Ω through

$$(19.100) \quad \alpha(z) \equiv \frac{1}{\binom{m}{2}}\, c_1(z),$$

$$(19.101) \quad \beta(z) \equiv \frac{1}{\binom{m+1}{3}} \left[c_2(z) - \frac{m-2}{3}\, c_1^{(1)}(z) - \frac{(3m-1)(m-2)}{6m(m-1)}\, (c_1(z))^2 \right],$$

(19.92)–(19.95), and

$$(19.102) \qquad B_{m,i}(z) \equiv c_i(z) - \eta_{m,i}(z), \quad \text{for } 1 \leq i \leq m.$$

Then, $B_{m,1}(z) \equiv B_{m,2}(z) \equiv 0$; and, the condition

$$(19.103) \qquad B_{m,i}(z) \equiv 0, \quad \text{on } \Omega \text{ for } 3 \leq i \leq m.$$

is necessary and sufficient for there to exist a monic second-order homogeneous linear differential equation $L_2 = 0$ on Ω such that the $(m-1)$th power of each local solution of $L_2 = 0$ is a local solution of (19.99). Moreover, when (19.103) is satisfied, there is just one such equation $L_2 = 0$ expressed in terms of $u(z)$ and it is

$$(19.104) \qquad u'' + \alpha(z)\, u' + \beta(z)\, u = 0, \quad \text{on } \Omega.$$

PROOF. We note that $c_1(z)$ and $c_2(z)$ in (19.99) specify (19.104) and

$$(19.105) \qquad y^{(m)} + \sum_{i=1}^{m} \eta_{m,i}(z)\, y^{(m-i)} = 0, \quad \text{on } \Omega,$$

by means of (19.100), (19.101), and (19.92)–(19.95). In particular, we obtain

$$(19.106) \quad \eta_{m,1}(z) \equiv \Gamma_{1,m-1}(z) \equiv \sum_{k=0}^{m-1} k\,\alpha(z) \equiv \binom{m}{2}\alpha(z) \equiv c_1(z)$$

and

$$(19.107) \quad \eta_{m,2}(z) \equiv \Gamma_{2,m-2}(z) \equiv \binom{m}{3}\alpha^{(1)}(z) + \tfrac{3m-1}{4}\binom{m}{3}(\alpha(z))^2 + \binom{m+1}{3}\beta(z) \equiv c_2(z).$$

(i) Suppose that (19.103) is satisfied. Then, we use (19.103) and (19.102) to see that $c_i(z) \equiv \eta_{m,i}(z)$, for $1 \leq i \leq m$. Since the equations (19.99) and (19.105) coincide, we obtain an equation $L_2 = 0$ having the desired properties by selecting $L_2 = 0$ to be (19.104).

(ii) For meromorphic functions $\widehat{\alpha}(z)$, $\widehat{\beta}(z)$ on Ω and $L_2 \equiv u'' + \widehat{\alpha}(z)\,u' + \widehat{\beta}(z)\,u$, suppose that the $(m-1)$th power of each solution of $L_2 = 0$ is a solution of (19.99). Let $\sigma(z)$ and $\tau(z)$ be linearly independent solutions of $L_2 = 0$ on a subregion $\mathcal{V}$ of Ω. For each C_1, C_2 in $\mathbb{C}$, the function $\bigl(C_1\,\sigma(z) + C_2\,\tau(z)\bigr)^{m-1}$ is a solution of (19.99); and a formula analogous to (19.98) shows that the Wronskian of

$$(19.108) \quad \left\{ (\sigma(z))^{m-i}(\tau(z))^{i-1} \;\middle|\; \text{for } i = 1, 2, \ldots, m \right\}$$

is not identically zero. As in as [**19**, pages 144–145], this enables us to deduce that (19.108) is a fundamental system of local solutions for (19.99). After noting that (19.99) and (19.105) have fundamental systems of local solutions of the special form in (19.108) and (19.97) while their coefficients satisfy $c_1(z) \equiv \eta_{m,1}(z)$ and $c_2(z) \equiv \eta_{m,2}(z)$, we see that [**19**, page 130, Corollary 2.19] is applicable and enables us to conclude that (19.99) and (19.105) are the same equation. Thus, (19.103) is satisfied. An application of Theorem 19.22 to $L_2 = 0$ and (19.99) yields

$$c_1(z) \equiv \binom{m}{2}\widehat{\alpha}(z) \quad \text{and} \quad c_2(z) \equiv \binom{m}{3}\widehat{\alpha}(z)^{(1)}(z) + \tfrac{3m-1}{4}\binom{m}{3}(\widehat{\alpha}(z))^2 + \binom{m+1}{3}\widehat{\beta}(z).$$

Comparing the preceding formulas for $c_1(z)$ and $c_2(z)$ with (19.106) and (19.107), we deduce that $\widehat{\alpha}(z) \equiv \alpha(z)$ and $\widehat{\beta}(z) \equiv \beta(z)$. Thus, (19.104) is unique for (19.99). This completes the proof. $\square$

REMARK 19.24. For any given (19.99), the equations (19.106) and (19.107) show that the functions $\alpha(z)$ and $\beta(z)$ defined by (19.100) and (19.101) are the only ones for which (19.102) satisfies $B_{m,1}(z) \equiv B_{m,2}(z) \equiv 0$. Thus, the left member of (19.99) is the same as $\mathfrak{L}_m$ in (19.91) if and only if (19.103) is valid. Historical information about the identity $B_{3,3}(z) \equiv \mathcal{I}_{3,3}(z)$ in regard to (19.83) is presented in [**24**, page 18, Example 2.5]. Thus, for $m = 3$, the condition (19.103) in Theorem 19.23 is the same as the condition $\mathcal{I}_{3,3}(z) \equiv 0$ in Proposition 19.20.

For conditions equivalent to (19.103) when $m \geq 3$, see *decisive sets* in [**24**].

EXAMPLE 19.25. For the fourth-order monic homogeneous linear equation

$$(19.109) \quad y^{(4)} - \frac{6z}{z-1}\,y^{(3)} + \frac{11z^2 + 10z - 6}{(z-1)^2}\,y^{(2)}$$
$$- \frac{6z^3 + 30z^2 - 13z - 8}{(z-1)^3}\,y^{(1)} + \frac{3(6z^2 + 8z + 1)}{(z-1)^3}\,y = 0,$$

where Ω is the complex plane, we use (19.100) and (19.101) to obtain

$$(19.110) \quad \alpha(z) \equiv \frac{-z}{z-1} \quad \text{and} \quad \beta(z) \equiv \frac{1}{z-1}, \quad \text{on } \Omega.$$

In terms of (19.110) and (19.92)–(19.95), we use MATHEMATICA in a manner similar to the computation for $K_{m,i,j}$ on page 170 and thereby verify that the functions $\eta_{4,1}(z)$, $\eta_{4,2}(z)$, $\eta_{4,3}(z)$, $\eta_{4,4}(z)$ for (19.109) are the respective coefficients of $y^{(3)}$, $y^{(2)}$, $y^{(1)}$, y in (19.109). Thus, (19.109) satisfies the condition $B_{4,i}(z) \equiv 0$, for $3 \leq i \leq 4$. Therefore, the third power of each solution of

$$(19.111) \qquad u'' + \frac{-z}{z-1} u' + \frac{1}{z-1} u = 0$$

is a solution of (19.109). The solutions of (19.111) are given by $u(z) \equiv K_1 z + K_2 e^z$, for K_1, K_2 in $\mathbb{C}$. Consequently, for C_1, C_2, C_3, C_4 in $\mathbb{C}$, the solutions of (19.109) are given by $y(z) \equiv C_1 z^3 + C_2 z^2 e^z + C_3 z e^{2z} + C_4 e^{3z}$.

19.10. Linear substitutions in binary forms

For a suitable context that yields $(n+1) \times (n+1)$ matrices analogous to the 3×3 matrices in (19.35), we recall how linear substitutions transform

$$(19.112) \qquad \Phi \equiv \sum_{j=0}^{n} f_j X^{n-j} Y^j,$$

where: n is a positive integer; X, Y are algebraically independent variables over a field $\mathfrak{F}$; and the coefficients f_0, f_1, ..., f_n are given elements of $\mathfrak{F}$. When the polynomial in (19.112) is nonzero, it is a *binary form of degree n in X, Y over* $\mathfrak{F}$. With respect to a fixed subfield $\mathfrak{F}_0$ of $\mathfrak{F}$, each 2×2 matrix

$$(19.113) \qquad A = \begin{bmatrix} a & b \\ c & d \end{bmatrix}, \quad \text{having } a,\, b,\, c,\, d \text{ in } \mathfrak{F}_0 \text{ and } ad - bc \neq 0,$$

defines new variables U, V by means of $\begin{bmatrix} U \\ V \end{bmatrix} = A^{-1} \begin{bmatrix} X \\ Y \end{bmatrix}$; and the linear substitution

$$(19.114) \qquad \begin{bmatrix} X \\ Y \end{bmatrix} = A \begin{bmatrix} U \\ V \end{bmatrix} = \begin{bmatrix} aU + bV \\ cU + dV \end{bmatrix}$$

transforms (19.112) into a corresponding expression

$$(19.115) \qquad \Phi \equiv \sum_{j=0}^{n} g_j U^{n-j} V^j,$$

for some elements g_0, g_1, ..., g_n in $\mathfrak{F}$. As regards (19.116), we agree that $0^0 = 1$.

PROPOSITION 19.26. *For each nonsingular 2×2 matrix A over $\mathfrak{F}_0$ specified by (19.113), the corresponding $(n+1) \times (n+1)$ matrix M_A defined over $\mathfrak{F}_0$ by*

$$(19.116) \qquad [M_A]_{i,j} = \sum_{\nu=0}^{j} \binom{n-i}{j-\nu} \binom{i}{\nu} a^{n-i-j+\nu} b^{j-\nu} c^{i-\nu} d^\nu, \quad \text{for } 0 \leq i,\, j \leq n,$$

is nonsingular and has the property that: when (19.114) transforms (19.112) into (19.115), the coefficients of (19.115) and (19.112) are related by

$$(19.117) \qquad [g_0,\, g_1,\, \ldots,\, g_n] = [f_0,\, f_1,\, \ldots,\, f_n] M_A,$$

for any f_0, f_1, ..., f_n in $\mathfrak{F}$. Moreover, if B is a nonsingular 2×2 matrix over $\mathfrak{F}_0$ and M_B as well as M_{AB} are the $(n+1) \times (n+1)$ matrices defined for B and AB in the same way that M_A is defined for A, then $M_{AB} = M_A M_B$.

PROOF. We use (19.112) and (19.114) to obtain

$$\Phi \equiv \sum_{i=0}^{n} f_i \, (aU + bV)^{n-i} \, (cU + dV)^i$$

$$\equiv \sum_{i=0}^{n} f_i \sum_{\mu=0}^{n-i} \binom{n-i}{\mu} (aU)^{n-i-\mu} (bV)^\mu \sum_{\nu=0}^{i} \binom{i}{\nu} (cU)^{i-\nu} (dV)^\nu.$$

In view of $\binom{n-i}{\mu} = 0$, for $\mu > n - i$, and $\binom{i}{\nu} = 0$, for $\nu > i$, we have

$$\Phi \equiv \sum_{i=0}^{n} f_i \sum_{\substack{0 \leq \mu,\nu \leq n \\ \mu+\nu \leq n}} \binom{n-i}{\mu} \binom{i}{\nu} (aU)^{n-i-\mu} (bV)^\mu (cU)^{i-\nu} (dV)^\nu$$

$$\equiv \sum_{i=0}^{n} f_i \sum_{j=0}^{n} \sum_{\nu=0}^{j} \binom{n-i}{j-\nu} \binom{i}{\nu} a^{n-i-j+\nu} b^{j-\nu} c^{i-\nu} d^\nu \, U^{n-j} V^j$$

$$\equiv \sum_{j=0}^{n} \left[\sum_{i=0}^{n} f_i \, [M_A]_{i,j} \right] U^{n-j} V^j \equiv \sum_{j=0}^{n} g_j \, U^{n-j} V^j,$$

where M is the $(n+1) \times (n+1)$ matrix defined over $\mathfrak{F}_0$ by (19.116) and where g_j is given by (19.117). Thus, the assertion about (19.115) and (19.112) is valid.

In terms of new variables R, S defined by $[R, S]^T = B^{-1}[U, V]^T$, we observe that the linear substitution $[U, V]^T = B[R, S]$ transforms (19.115) into

(19.118) $$\Phi \equiv \sum_{j=0}^{n} h_j \, R^{n-j} S^j,$$

for which $[h_0, h_1, \ldots, h_n] = [g_0, g_1, \ldots, g_n] M_B$. The nonsingular 2×2 matrix AB over $\mathfrak{F}_0$ specifies a linear substitution $[X, Y]^T = (AB)[R, S]^T$ that transforms (19.112) directly into (19.118). It specifies $[h_0, h_1, \ldots, h_n] = [f_0, f_1, \ldots, f_n] M_{AB}$, for any $f_0, f_1, \ldots, f_n$ in $\mathfrak{F}$. Since we also have

$$[h_0, h_1, \ldots, h_n] = [g_0, g_1, \ldots, g_n] M_B = [f_0, f_1, \ldots, f_n] M_A M_B,$$

for any $f_0, f_1, \ldots, f_n$ in $\mathfrak{F}$, we conclude that $M_{AB} = M_A M_B$. When A is the 2×2 identity matrix, we use (19.116) with $a = 1$, $b = 0$, $c = 0$, $d = 1$, and $0^0 = 1$ to deduce that M_A is the $(n+1) \times (n+1)$ identity matrix over $\mathfrak{F}_0$. Thus, for any nonsingular 2×2 matrix A over $\mathfrak{F}_0$, we set $B = A^{-1}$ and use $M_A M_{A^{-1}} = M_{AA^{-1}}$ to see that M_A is nonsingular (and $M_{A^{-1}}$ is its inverse). This completes the proof. □

To check (19.116), we found that computer algebra yields

$$det(M_A) \equiv (ad - bc)^{n(n+1)/2} \equiv \bigl(det(A)\bigr)^{n(n+1)/2}, \quad \text{for } 1 \leq n \leq 45.$$

For the groups $GL_2(\mathfrak{F}_0)$ and $GL_{n+1}(\mathfrak{F}_0)$ whose elements are the 2×2 and the $(n+1) \times (n+1)$ nonsingular matrices over $\mathfrak{F}_0$ with respect to matrix multiplication, Proposition 19.26 shows that a group homomorphism σ of $GL_2(\mathfrak{F}_0)$ into $GL_{n+1}(\mathfrak{F}_0)$ is defined for A in $GL_2(\mathfrak{F}_0)$ by $\sigma(A) = M_A$, where M_A is the $(n+1) \times (n+1)$ matrix in (19.116) that corresponds to A in (19.113). The subgroup $\sigma\bigl(GL_2(\mathfrak{F}_0)\bigr)$ of $GL_{n+1}(\mathfrak{F}_0)$ consists of the totality of $(n+1) \times (n+1)$ matrices M_A given by (19.116). It is isomorphic to the factor group $GL_2(\mathfrak{F}_0)/N_n$, where N_n is the normal subgroup of $GL_2(\mathfrak{F}_0)$ defined in terms of the $(n+1) \times (n+1)$ identity matrix I_{n+1} for $GL_{n+1}(\mathfrak{F}_0)$ by $N_n = \{A \mid A \text{ in } GL_2(\mathfrak{F}_0) \text{ and } \sigma(A) = I_{n+1}\}$. For this deduction, the particular isomorphism theorem used can be found in [**44**, page 16].

19.10. LINEAR SUBSTITUTIONS IN BINARY FORMS

PROPOSITION 19.27. *When $\mathfrak{F}_0$ contains a primitive nth root θ of unity, the normal subgroup N_n of $GL_2(\mathfrak{F}_0)$ is given by*

$$(19.119) \qquad N_n = \left\{ \begin{bmatrix} \theta^k & 0 \\ 0 & \theta^k \end{bmatrix} \mid \text{ for } k = 1, 2, \ldots, n \right\}.$$

PROOF. Starting with A in (19.113), we use (19.116) to obtain $[M_A]_{0,0} \equiv a^n$, $[M_A]_{0,1} \equiv na^{n-1}b$, $[M_A]_{1,0} \equiv a^{n-1}c$, $[M_A]_{1,1} \equiv (n-1)a^{n-2}bc + a^{n-1}d$. Since $\mathfrak{F}_0$ contains a primitive nth root θ of unity, the characteristic of $\mathfrak{F}_0$ does not divide n. Thus, when $M_A = I_{n+1}$, we see that $a^n = 1$, $b = 0$, $c = 0$, $d = a$, and A is contained in the right member of (19.119). Conversely, when A is contained in the right member of (19.119), there is an integer k such that $a = \theta^k$, $b = 0$, $c = 0$, $d = \theta^k$, and $(\theta^k)^n = 1$. Then, these formulas and (19.116) yield: $[M]_{i,j} = 0$, for $0 \leq i, j \leq n$ and $i \neq j$; $[M]_{i,i} = 1$, for $0 \leq i \leq n$; and $M_A \equiv I_{n+1}$. Consequently, N_n is given by the right member of (19.119). This completes the proof. □

EXAMPLE 19.28. Let $\mathfrak{F}$ be the field $\mathfrak{F}_\mathcal{V}$ of meromorphic functions on a region $\mathcal{V}$ of the complex plane and let $\mathfrak{F}_0$ be the field $\mathbb{C}$ of complex numbers. Then, for $n = 2$, the context becomes that employed for Proposition 19.9 on page 199 and the matrices of the type M_A are given by (19.35). They form a group under matrix multiplication that is isomorphic to $GL_2(\mathbb{C})/N_2$, where $N_2 = \{I_2, -I_2\}$ and I_2 is the 2×2 identity matrix.

NOTATION 19.29. Let $\boldsymbol{x}_j^{(k)}$, for $0 \leq j \leq n$ and $k \geq 0$, denote algebraically independent variables over the field $\mathbb{Q}$ of rational numbers, let $\mathfrak{E}$ denote the ring of polynomials in these variables over $\mathbb{Q}$, and let $'$ denote the derivation for $\mathfrak{E}$ such that $(\boldsymbol{x}_j^{(k)})' \equiv \boldsymbol{x}_j^{(k+1)}$, for $0 \leq j \leq n$ and $k \geq 0$. We set $\boldsymbol{x}_j \equiv \boldsymbol{x}_j^{(0)}$, for $0 \leq j \leq n$.

The old property of $(\boldsymbol{x}_1)^2 - 4\boldsymbol{x}_0\boldsymbol{x}_2$ as an invariant of weight 2 for binary forms of degree 2 is expressed by (19.36) on page 199. Namely, $\bigl(u_1(z)\bigr)^2 - 4u_0(z)u_2(z)$ and $\bigl(v_1(z)\bigr)^2 - 4v_0(z)v_2(z)$ are closely related to the discriminants of the quadratic forms in (19.50) and (19.52) on page 201; e.g., see [9, pages 128–129]. By considering $\mathfrak{F}_\mathcal{V}$ as an ordinary differential field in which its derivation is the usual differentiation of functions, we use (19.37) to see that an invariant of weight 4 not included in the classical theory about binary forms of degree 2 is given by the differential polynomial in (19.38). It is equal to $\boldsymbol{\Gamma}_2$ where $\boldsymbol{\Gamma}_n$ is defined in $\mathfrak{E}$, for $n \geq 1$, by

$$(19.120) \quad \boldsymbol{\Gamma}_n \equiv \begin{vmatrix} \boldsymbol{x}_0 & \boldsymbol{x}_1 & \boldsymbol{x}_2 & \cdots & \boldsymbol{x}_{n-1} & \boldsymbol{x}_n & 0 & 0 & \cdots & 0 \\ 0 & \boldsymbol{x}_0 & \boldsymbol{x}_1 & \cdots & \boldsymbol{x}_{n-2} & \boldsymbol{x}_{n-1} & \boldsymbol{x}_n & 0 & \cdots & 0 \\ 0 & 0 & \boldsymbol{x}_0 & \cdots & \boldsymbol{x}_{n-3} & \boldsymbol{x}_{n-2} & \boldsymbol{x}_{n-1} & \boldsymbol{x}_n & \cdots & 0 \\ \vdots & \vdots & \vdots & \ddots & \vdots & \vdots & \vdots & \vdots & \ddots & \vdots \\ 0 & 0 & 0 & \cdots & \boldsymbol{x}_0 & \boldsymbol{x}_1 & \boldsymbol{x}_2 & \boldsymbol{x}_3 & \cdots & \boldsymbol{x}_n \\ \boldsymbol{x}_0^{(1)} & \boldsymbol{x}_1^{(1)} & \boldsymbol{x}_2^{(1)} & \cdots & \boldsymbol{x}_{n-1}^{(1)} & \boldsymbol{x}_n^{(1)} & 0 & 0 & \cdots & 0 \\ 0 & \boldsymbol{x}_0^{(1)} & \boldsymbol{x}_1^{(1)} & \cdots & \boldsymbol{x}_{n-2}^{(1)} & \boldsymbol{x}_{n-1}^{(1)} & \boldsymbol{x}_n^{(1)} & 0 & \cdots & 0 \\ 0 & 0 & \boldsymbol{x}_0^{(1)} & \cdots & \boldsymbol{x}_{n-3}^{(1)} & \boldsymbol{x}_{n-2}^{(1)} & \boldsymbol{x}_{n-1}^{(1)} & \boldsymbol{x}_n^{(1)} & \cdots & 0 \\ \vdots & \vdots & \vdots & \ddots & \vdots & \vdots & \vdots & \vdots & \ddots & \vdots \\ 0 & 0 & 0 & \cdots & \boldsymbol{x}_0^{(1)} & \boldsymbol{x}_1^{(1)} & \boldsymbol{x}_2^{(1)} & \boldsymbol{x}_3^{(1)} & \cdots & \boldsymbol{x}_n^{(1)} \end{vmatrix}.$$

Thus, $\boldsymbol{\Gamma}_n$ is the determinant of a $2n \times 2n$ matrix and $\boldsymbol{\Gamma}_1 \equiv \boldsymbol{x}_0\boldsymbol{x}_1^{(1)} - \boldsymbol{x}_0^{(1)}\boldsymbol{x}_1$.

19.11. Properties of the polynomial Γ_n in (19.120)

Throughout this section, the field $\mathfrak{F}$ for (19.112) is specialized to be the field $\mathfrak{F}_\mathcal{V}$ of meromorphic functions defined on a region $\mathcal{V}$ of the complex plane. We have the following generalization of Proposition 19.11 on page 202.

PROPOSITION 19.30. *For $n \geq 1$, suppose that the coefficients for Φ in (19.112) are functions $f_0(z)$, $f_1(z)$, ..., $f_n(z)$ in $\mathfrak{F}_\mathcal{V}$ that are linearly independent over $\mathbb{C}$. Let $\Delta(z)$ be a discriminant for the form Φ. Then, the condition $\Delta(z) \equiv 0$ is satisfied if and only if $\Gamma_n(z) \equiv 0$, where $\Gamma_n(z)$ is the function in $\mathfrak{F}_\mathcal{V}$ obtained by substituting $f_j^{(k)}(z)$ from Φ for $x_j^{(k)}$ in Γ_n of (19.120).*

OBSERVATION. Various definitions have been proposed for the discriminant of a binary form over a field of characteristic zero. However, we shall merely use the standard property that the discriminant is identically zero if and only if the form has a factor of multiplicity ≥ 2 over some suitable extension field.

PROOF OF PROPOSITION 19.30. Due to the linearly independence of $f_0(z)$ and $f_n(z)$ over $\mathbb{C}$, we have either $f_0^{(1)}(z) \not\equiv 0$ or $f_n^{(1)}(z) \not\equiv 0$.

Suppose that $f_0^{(1)}(z) \not\equiv 0$. Let $\mathcal{V}_0$ be a subregion of $\mathcal{V}$ on which there are meromorphic functions $r_1(z)$, ..., $r_n(z)$ such that, for z in $\mathcal{V}_0$,

$$(19.121) \qquad \sum_{j=0}^{n} f_j(z) X^{n-j} \equiv f_0(z)\bigl(X - r_1(z)\bigr) \cdots \bigl(X - r_n(z)\bigr).$$

Here, we regard X as an algebraically independent variable over the field $\mathfrak{F}_{\mathcal{V}_0}$ of meromorphic functions on $\mathcal{V}_0$ and we interpret (19.121) as an identity in the ring $\mathfrak{R}$ of polynomials in X over $\mathfrak{F}_{\mathcal{V}_0}$. We have $\Delta(z) \equiv 0$ if and only if the right member of (19.121) has a factor of multiplicity ≥ 2. For each $f(z)$ in $\mathfrak{F}_\mathcal{V}$, the corresponding restriction $f_{\mathcal{V}_0}(z)$ of $f(z)$ to $\mathcal{V}_0$ specifies an embedding $f \mapsto f_{\mathcal{V}_0}$ of $\mathfrak{F}_\mathcal{V}$ in $\mathfrak{F}_{\mathcal{V}_0}$ as a subfield of $\mathfrak{F}_{\mathcal{V}_0}$. Each $\phi(X)$ in $\mathfrak{R}$ uniquely specifies a corresponding polynomial $\phi^D(X)$ in $\mathfrak{R}$ according to

$$\phi^D(X) \equiv \sum_{k=0}^{r} a_k^{(1)}(z)\, X^k \quad \text{whenever} \quad \phi(X) \equiv \sum_{k=0}^{r} a_k(z)\, X^k.$$

For any ϕ, ψ in $\mathfrak{R}$, we find that $(\phi + \psi)^D \equiv \phi^D + \psi^D$ and $(\phi\psi)^D \equiv \phi^D \psi + \phi\psi^D$. Thus, the assignment $\phi(X) \mapsto \phi^D(X)$ is a derivation for $\mathfrak{R}$. Moreover, the constants of $\mathfrak{R}$ for the derivation D are the polynomials in X over $\mathbb{C}$.

In addition to $f_0^{(1)}(z) \not\equiv 0$, suppose that $\Delta(z) \equiv 0$. Then, for some $r(z)$ in $\mathfrak{F}_{\mathcal{V}_0}$ and $Q(X)$ in $\mathfrak{R}$, we see that the polynomials

$$(19.122) \qquad P(X) \equiv \sum_{j=0}^{n} f_j(z)\, X^{n-j} \quad \text{and} \quad P^D(X) \equiv \sum_{j=0}^{n} f_j^{(1)}(z)\, X^{n-j}$$

can be written as $P(X) \equiv \bigl(X - r(z)\bigr)^2 Q(X)$ and

$$P^D(X) \equiv 2\bigl(X - r(z)\bigr)\bigl(-r^{(1)}(z)\bigr) Q(X) + \bigl(X - r(z)\bigr)^2 Q^D(X).$$

Since $P(X)$ and $P^D(X)$ have the common factor $X - r(z)$, the resultant of $P(X)$ and $P^D(X)$ must be identically zero. We use (19.122) and (19.120) as well as $f_0(z) \not\equiv 0$, $f_0^{(1)}(z) \not\equiv 0$, and the standard definition of a resultant in [**44**, page 200] to see that the resultant of $P(X)$ and $P^D(Y)$ is equal to $\Gamma_n(z)$. Hence, we have $\Gamma_n(z) \equiv 0$ on $\mathcal{V}_0$ and therefore also on $\mathcal{V}$.

In addition to $f_0^{(1)}(z) \not\equiv 0$, suppose that $\Gamma_n(z) \equiv 0$. We apply the factorization (19.121) for $P(X)$ and the observation that $\Gamma_n(z)$ is the resultant of $P(X)$ and $P^D(X)$ to deduce the existence of $r(z)$ in $\mathfrak{F}_{\mathcal{V}_0}$ and $S(X)$, $T(X)$ in $\mathfrak{R}$ such that

$$P(X) \equiv (X - r(z)) S(X) \quad \text{and} \quad P^D(X) \equiv (X - r(z)) T(X).$$

We use this and the derivation D for $\mathfrak{R}$ to obtain

(19.123) $\quad (X - r(z)) T(X) \equiv P^D(X) \equiv (-r^{(1)}(z)) S(X) + (X - r(z)) S^D(X).$

Since $f_0(z)$, ..., $f_n(z)$ are linearly independent over $\mathbb{C}$, the root $r(z)$ of $P(X)$ is not an element of $\mathbb{C}$. Thus, we have $r^{(1)}(z) \not\equiv 0$. By combining this with (19.123), we deduce that $(X - r(z))$ divides $S(X)$ and therefore $(X - r(z))^2$ divides $P(X)$. This yields $\Delta(z) \equiv 0$ on $\mathcal{V}_0$ and therefore also on $\mathcal{V}$.

Suppose that $f_0^{(1)}(z) \equiv 0$. Then, we have $f_n^{(1)}(z) \not\equiv 0$. With a context similar to that for (19.121), there are functions $s_1(z)$, ..., $s_n(z)$ in a suitable $\mathfrak{F}_{\mathcal{V}_0}$ such that

(19.124) $\quad \sum_{j=0}^{n} f_j(z) X^j \equiv f_n(z)(X - s_1(z)) \cdots (X - s_n(z)).$

Since the polynomial $P(X)$ in the left member of (19.124) has $f_n(z) \not\equiv 0$ and $f_n^{(1)}(z) \not\equiv 0$, the resultant $B_n(z)$ of $P(X)$ and $P^D(X)$ is obtained from (19.120) by replacing x_j with $f_{n-j}(z)$ and $x_j^{(1)}$ with $f_{n-j}^{(1)}(z)$. We deduce $B_n(z) \equiv (-1)^n \Gamma_n(z)$ by suitably interchanging rows and then columns of the matrix for $B_n(z)$. Now, we can repeat the earlier argument for $f_0^{(1)}(z) \not\equiv 0$ to complete the proof. $\square$

OBSERVATION 19.31. *For $n \geq 1$ and a, b, c, d in $\mathbb{C}$ satisfying $ad - bc \neq 0$, suppose that $f_0(z)$, ..., $f_n(z)$ and $g_0(z)$, ..., $g_n(z)$ in $\mathfrak{F}_{\mathcal{V}}$ are related by*

$$[g_0(z), \ldots, g_n(z)] \equiv [f_0(z), \ldots, f_n(z)] M_A,$$

where M_A is the $(n+1) \times (n+1)$ matrix over $\mathbb{C}$ defined by (19.116). Let $\Gamma_{n,f}(z)$ denote the function on $\mathcal{V}$ obtained by substituting $f_j^{(k)}(z)$ for $x_j^{(k)}$ in Γ_n of (19.120). Let $\Gamma_{n,g}(z)$ on $\mathcal{V}$ be obtained by substituting $g_j^{(k)}(z)$ for $x_j^{(k)}$ in Γ_n. We have used MATHEMATICA, Version 5.1, to verify that

(19.125) $\quad \Gamma_{n,g}(z) \equiv (ad - bc)^{n^2} \Gamma_{n,f}(z), \quad \text{for } 1 \leq n \leq 5.$

For $Q_2 = 0$ in (1.0–A) on page 3, the Appell condition that $(Q_2)' + \lambda Q_2$ have a nontrivial factorization for some $\lambda(z)$ was extended in Chapter 18 to an analogous condition involving $(Q_m)' + \lambda Q_m$ for the equations $Q_m = 0$. The form of the solutions in (1.105) on page 22 suggests another line of inquiry.

PROPOSITION 19.32. *For $n \geq 1$, let $u_0(z)$, $u_1(z)$, ..., $u_n(z)$ in $\mathfrak{F}_\Omega$ be linearly independent over $\mathbb{C}$ and suppose that the function $a(z)$ on Ω obtained from Γ_n in (19.120) by replacing $x_j^{(k)}$ with $u_j^{(k)}(z)$ satisfies $a(z) \not\equiv 0$. Then, meromorphic functions $\beta_{i_1, i_2, \ldots, i_n}(z)$ exist on Ω such that each of the functions*

(19.126) $\quad y(z) \equiv \sum_{j=0}^{n} C^{n-j} K^j u_j(z), \quad \text{for } C, K \text{ in } \mathbb{C},$

is a solution of the differential equation

(19.127) $\quad \left(y^{(2)}(z)\right)^n + \sum_{0 \leq i_1 \leq i_2 \leq \cdots \leq i_n \leq 2, \, i_n \neq 0} \beta_{i_1, i_2, \ldots, i_n}(z) \prod_{\nu=1}^{n} y^{(2-i_\nu)}(z) = 0.$

PROOF. To obtain a suitable (19.127) for which each function in (19.126) is a solution, we first note that: the $y(z)$ in (19.126) having $K \neq 0$ are given by

$$(19.128) \qquad y(z) \equiv K_0 \sum_{j=0}^{n} (C_0)^{n-j} u_j(z), \quad \text{for } C_0, K_0 \text{ in } \mathbb{C} \text{ and } K_0 \neq 0.$$

We define polynomials in the variables Y_0, Y_1, Y_2, S, T over $\mathfrak{F}_\Omega$ by means of

$$P_0 \equiv Y_0 - S \sum_{j=0}^{n} u_j^{(0)}(z) \, T^{n-j},$$

$$P_1 \equiv Y_1 - S \sum_{j=0}^{n} u_j^{(1)}(z) \, T^{n-j},$$

$$P_2 \equiv Y_2 - S \sum_{j=0}^{n} u_j^{(2)}(z) \, T^{n-j},$$

$$P_3 \equiv Resultant(P_2, P_0; S) \equiv \sum_{j=0}^{n} \left[u_j^{(0)}(z) \, Y_2 - u_j^{(2)}(z) \, Y_0 \right] T^{n-j},$$

$$P_4 \equiv Resultant(P_2, P_1; S) \equiv \sum_{j=0}^{n} \left[u_j^{(1)}(z) \, Y_2 - u_j^{(2)}(z) \, Y_1 \right] T^{n-j},$$

and

$$P_5 \equiv Resultant(P_3, P_4; T).$$

In terms of $a_j = u_j^{(0)}(z) Y_2 - u_j^{(2)}(z) Y_0$ and $b_j = u_j^{(1)}(z) Y_2 - u_j^{(2)}(z) Y_1$, for $0 \leq j \leq n$, we find that P_5 is the expression obtained by respectively replacing $\boldsymbol{x}_j$ and $\boldsymbol{x}_j^{(1)}$ in (19.120) with a_j and b_j. Let D be any one of the 2^{2n} determinants obtained from $\boldsymbol{\Gamma}_n$ in (19.120) as follows: for $1 \leq i \leq n$, replace $\boldsymbol{x}_j$ in the ith row of $\boldsymbol{\Gamma}_n$ with either $u_j^{(0)}(z) Y_2$ for $0 \leq j \leq n$ or with $-u_j^{(2)}(z) Y_0$ for $0 \leq j \leq n$; and for $n+1 \leq i \leq 2n$, replace $\boldsymbol{x}_j^{(1)}$ in the ith row of $\boldsymbol{\Gamma}_n$ with either $u_j^{(1)}(z) Y_2$ for $0 \leq j \leq n$ or with $-u_j^{(2)}(z) Y_1$ for $0 \leq j \leq n$. We observe that P_5 is expressible as a sum of determinants of the type D. In particular, one of these determinants evaluates as $a(z)(Y_2)^{2n}$ and the other nonzero ones are of degree $< 2n$ in Y_2. Also, any one of these determinants must equal 0 if, for some integer i satisfying $1 \leq i \leq n$, its ith row is obtained from $\boldsymbol{\Gamma}_n$ by replacing $\boldsymbol{x}_j$ with $-u_j^{(2)}(z) Y_0$ for $0 \leq j \leq n$ and its $(n+i)$th row is obtained by replacing $\boldsymbol{x}_j^{(1)}$ with $-u_j^{(2)}(z) Y_1$ for $0 \leq j \leq n$. Thus, if $(Y_0)^p (Y_1)^q$ is a factor of a nonzero term of P_5, then $p + q \leq n$. This shows that each nonzero term of P_5 is divisible by $(Y_2)^n$. Consequently, we have

$$P_5 \equiv (Y_2)^n \left[a(z)(Y_2)^n + \sum_{0 \leq i_1 \leq i_2 \leq \cdots \leq i_n \leq 2, \, i_n \neq 0} \alpha_{i_1, i_2, \ldots, i_n}(z) \prod_{\nu=1}^{n} Y_{2-i_\nu}(z) \right],$$

for some meromorphic functions $\alpha_{i_1, i_2, \ldots, i_n}(z)$ on Ω. We divide P_5 by the nonzero factor $a(z)(Y_2)^n$ and then replace Y_0, Y_1, Y_2 with $y(z), y^{(1)}(z), y^{(2)}(z)$ to obtain the left member of an equation (19.127) having as solutions each $y(z)$ in (19.128). To see that the functions $y(z) \equiv K_0 \sum_{k=0}^{n} (C_0)^{n-k} u_{n-k}(z)$ are also solutions, we replace $u_k^{(0)}(z), u_k^{(1)}(z), u_k^{(2)}(z)$ with $u_{n-k}^{(0)}(z), u_{n-k}^{(1)}(z), u_{n-k}^{(2)}(z)$ throughout the preceding argument and observe that P_5 is replaced with $(-1)^{(2n-1)n} P_5$. Thus, each $y(z)$ in (19.126) is a solution of the (19.127) obtained. This completes the proof. □

CHAPTER 20

Rational Semi-Invariants and Relative Invariants

20.1. Terminology for an extended context

Throughout Chapters 1–18 (except as modified in Chapter 13), we used the context of page 5 where $\mathcal{S}_m$ is the ring of polynomials over $\mathbb{Q}$ in the variables $\boldsymbol{w}_{i,j}^{(k)}$ of (1.6) and various substitutions were made for these variables. In this chapter, the semi-invariants and relative invariants introduced in Definitions 1.1–1.3 of page 6 will be referred to as *polynomial semi-invariants* and *polynomial relative invariants* to distinguish them from the more general *rational semi-invariants* and *rational relative invariants* to be introduced in Definitions 20.1–20.3. We shall find that the entire subject of rational semi-invariants and rational relative invariants developed in [**24**, Appendix D] for homogeneous linear differential equations can be extended to our study of (1.1) and its transformations (1.2), (1.4) into (1.3), (1.5).

Let $\mathcal{Q}_m$ denote the quotient field for the ring $\mathcal{S}_m$. Thus, the elements of $\mathcal{Q}_m$ can be represented as $\dfrac{\boldsymbol{P}}{\boldsymbol{Q}}$, where $\boldsymbol{P}$ and $\boldsymbol{Q}$ are polynomials in $\mathcal{S}_m$ subject to $\boldsymbol{Q} \not\equiv 0$. The derivation $'$ for $\mathcal{S}_m$ has a unique extension to a derivation $'$ for $\mathcal{Q}_m$; e.g., see [**10**, Chap. IV, page 45, Proposition 11]. For any $\boldsymbol{P}$ and $\boldsymbol{Q} \not\equiv 0$ in $\mathcal{S}_m$, it specifies

$$\left(\frac{\boldsymbol{P}}{\boldsymbol{Q}}\right)' \equiv \frac{\boldsymbol{Q}\boldsymbol{P}' - \boldsymbol{P}\boldsymbol{Q}'}{\boldsymbol{Q}^2}, \quad \text{in } \mathcal{Q}_m.$$

When $\boldsymbol{P}$ and $\boldsymbol{Q} \not\equiv 0$ are relatively prime and $(\boldsymbol{P}/\boldsymbol{Q})' \equiv 0$, we have $\boldsymbol{Q}\boldsymbol{P}' \equiv \boldsymbol{P}\boldsymbol{Q}'$. Then, $\boldsymbol{Q}$ divides $\boldsymbol{Q}'$ in $\mathcal{S}_m$. This requires $\boldsymbol{Q}' \equiv 0$ as well as $\boldsymbol{P}' \equiv 0$ and shows that the constants of $\mathcal{Q}_m$ are the constants of $\mathcal{S}_m$ — namely the rational numbers.

For any element $\boldsymbol{A}$ in $\mathcal{Q}_m$, there are relatively prime polynomials $\boldsymbol{F}$ and $\boldsymbol{G} \not\equiv 0$ in $\mathcal{S}_m$ such that $\boldsymbol{A} \equiv \boldsymbol{F}/\boldsymbol{G}$. Moreover, if $\widehat{\boldsymbol{F}}$ and $\widehat{\boldsymbol{G}} \not\equiv 0$ are also relatively prime polynomials in $\mathcal{S}_m$ subject to $\boldsymbol{A} \equiv \widehat{\boldsymbol{F}}/\widehat{\boldsymbol{G}}$, then $\widehat{\boldsymbol{F}}\boldsymbol{G} \equiv \boldsymbol{F}\widehat{\boldsymbol{G}}$, $\boldsymbol{G}$ divides $\widehat{\boldsymbol{G}}$, $\widehat{\boldsymbol{G}}$ divides $\boldsymbol{G}$, and there is a nonzero rational number γ such that $\widehat{\boldsymbol{G}} \equiv \gamma \boldsymbol{G}$ as well as $\widehat{\boldsymbol{F}} \equiv \gamma \boldsymbol{F}$. Thus, whenever (1.1)–(1.5) are such that $G(z) \not\equiv 0$, $G^*(z) \not\equiv 0$, and $G^{**}(\zeta) \not\equiv 0$, the element $\boldsymbol{A}$ uniquely specifies corresponding functions $A(z)$ on Ω, $A^*(z)$ on Ω, and $A^{**}(\zeta)$ on Ω^{**} by means of

$$A(z) \equiv \frac{F(z)}{G(z)}, \quad A^*(z) \equiv \frac{F^*(z)}{G^*(z)}, \quad \text{and} \quad A^{**}(\zeta) \equiv \frac{F^{**}(\zeta)}{G^{**}(\zeta)}.$$

We recall that $F(z)$, $G(z)$ denote the functions on Ω obtained by substituting $c_{i,j}^{(k)}(z)$ from (1.1) for $\boldsymbol{w}_{i,j}^{(k)}$ in $\boldsymbol{F}$, $\boldsymbol{G}$. Also, $F^*(z)$, $G^*(z)$ on Ω are obtained by substituting $c_{i,j}^{*(k)}(z)$ from (1.3) for $\boldsymbol{w}_{i,j}^{(k)}$ in $\boldsymbol{F}$, $\boldsymbol{G}$; and $F^{**}(\zeta)$, $G^{**}(\zeta)$ on Ω^{**} are obtained by substituting $c_{i,j}^{**(k)}(\zeta)$ from (1.5) for $\boldsymbol{w}_{i,j}^{(k)}$ in $\boldsymbol{F}$, $\boldsymbol{G}$.

DEFINITION 20.1. An element $\boldsymbol{A}$ in $\mathcal{Q}_m$ is a *rational semi-invariant of the first kind for the equations* (1.1) *having a given order* m when it is nonconstant and

$$A^*(z) \equiv A(z), \tag{20.1}$$

for each (1.1) of order m and each change (1.2) of the dependent variable for which $A(z)$ and $A^*(z)$ are both defined.

DEFINITION 20.2. An element $\boldsymbol{A}$ in $\mathcal{Q}_m$ is a *rational semi-invariant of the second kind for the equations* (1.1) *having a given order* m when it is nonconstant and there is an integer s (positive, zero, or negative) such that

$$A^{**}(\zeta) \equiv \bigl(f'(\zeta)\bigr)^s A\bigl(f(\zeta)\bigr), \tag{20.2}$$

for each (1.1) of order m and each change (1.4) of the independent variable for which $A(z)$ and $A^{**}(\zeta)$ are both defined.

DEFINITION 20.3. An element $\boldsymbol{A}$ in $\mathcal{Q}_m$ is a *rational relative invariant for the equations* (1.1) *having a given order* m when it is both a rational semi-invariant of the first kind and a rational semi-invariant of the second kind for such equations.

20.2. The integer s in Definition 20.2

THEOREM 20.4. *Suppose that $\boldsymbol{F}$ and $\boldsymbol{G}$ are relatively prime polynomials in $\mathcal{S}_m$ such that $\boldsymbol{G} \not\equiv 0$ and the element*

$$\boldsymbol{A} \equiv \frac{\boldsymbol{F}}{\boldsymbol{G}} \tag{20.3}$$

in $\mathcal{Q}_m$ is a rational semi-invariant of the second kind that satisfies (20.2) *for some fixed integer s. Then, $\boldsymbol{F}$ and $\boldsymbol{G}$ are isobaric polynomials whose weights give*

$$s = weight(\boldsymbol{F}) - weight(\boldsymbol{G}). \tag{20.4}$$

PROOF. Since $\boldsymbol{F}$ and $\boldsymbol{G}$ are both nonzero, we have

$$\boldsymbol{F} \equiv \sum_{k=0}^{p} \boldsymbol{F}_k \quad \text{and} \quad \boldsymbol{G} \equiv \sum_{k=0}^{q} \boldsymbol{G}_k, \tag{20.5}$$

where: for $0 \leq k \leq p$, either $\boldsymbol{F}_k \equiv 0$ or $\boldsymbol{F}_k$ is an isobaric polynomial in $\mathcal{S}_m$ of weight k; $\boldsymbol{F}_p \not\equiv 0$; for $0 \leq k \leq q$, either $\boldsymbol{G}_k \equiv 0$ or $\boldsymbol{G}_k$ is an isobaric polynomial in $\mathcal{S}_m$ of weight k; and $\boldsymbol{G}_q \not\equiv 0$. We set $\boldsymbol{F}_k \equiv 0$, for $k > p$, and $\boldsymbol{G}_k \equiv 0$, for $k > q$.

For any (1.1), we restrict $z = f(\zeta)$ for (1.4) to the univalent analytic functions of the type $z = f_t(\zeta) \equiv t\zeta$, where t is a fixed nonzero rational number. For each such $f_t(\zeta)$, we find that the corresponding functions $\alpha_{i,j}(\zeta)$ in (3.1)–(3.2) on page 29 are given by $\alpha_{0,j}(\zeta) \equiv 1$, for any j, and $\alpha_{i,j}(\zeta) \equiv 0$, for $1 \leq i \leq m$ and any j. Thus, (3.9) yields

$$c_{i,j}^{**}(\zeta) \equiv \bigl(f_t'(\zeta)\bigr)^{i+j} c_{i,j}\bigl(f_t(\zeta)\bigr) \equiv t^{i+j} c_{i,j}(t\zeta), \quad \text{for } 0 \leq i, j \leq m. \tag{20.6}$$

We apply (20.6) and $f_t'(\zeta) \equiv t$ to verify that

$$c_{i,j}^{**(k)}(\zeta) \equiv \bigl(t^{i+j+k}\bigr) c_{i,j}^{(k)}(t\zeta), \quad \text{for } 0 \leq i, j \leq m \text{ and } k \geq 0. \tag{20.7}$$

Any nonzero monomial in $\mathcal{S}_m$ of weight k has the form

$$\boldsymbol{P}_k \equiv \beta \prod_{j=1}^{r} \boldsymbol{w}_{\mu_j,\nu_j}^{(\kappa_j)}, \quad \text{subject to} \quad \sum_{j=1}^{r} (\mu_j + \nu_j + \kappa_j) = k, \tag{20.8}$$

where: β is a nonzero rational number; $r \geq 0$; and, for $1 \leq j \leq r$, the indices μ_j, ν_j, κ_j satisfy $1 \leq \nu_j \leq m$, $0 \leq \mu_j \leq \nu_j$, and $\kappa_j \geq 0$. Hence, (20.8) and (20.7) give

$$(20.9) \quad P_k^{**}(\zeta) \equiv \beta \prod_{j=1}^{r} c_{\mu_j,\nu_j}^{**(\kappa_j)}(\zeta) \equiv \beta \left[\prod_{j=1}^{r} t^{\mu_j+\nu_j+\kappa_j}\right]\left[\prod_{j=1}^{r} c_{\mu_j,\nu_j}^{(\kappa_j)}(t\zeta)\right] \equiv t^k P_k(t\zeta).$$

Since each of the polynomials $\boldsymbol{F}_k$ and $\boldsymbol{G}_k$ is either 0 or a sum of terms having the form (20.8) for which (20.9) is valid, we see that

$$(20.10) \quad F_k^{**}(\zeta) \equiv t^k F_k(t\zeta), \quad \text{and} \quad G_k^{**}(\zeta) \equiv t^k G_k(t\zeta), \quad \text{for } k \geq 0.$$

Applying (20.2) for $\boldsymbol{A}$ as well as (20.3), (20.5), and (20.10), we find that

$$(20.11) \quad t^s \left[\frac{F(t\zeta)}{G(t\zeta)}\right] \equiv (f_t'(\zeta))^s A(f_t(\zeta)) \equiv A^{**}(\zeta) \equiv \frac{F^{**}(\zeta)}{G^{**}(\zeta)} \equiv \frac{\sum_{k=0}^{p} t^k F_k(t\zeta)}{\sum_{k=0}^{q} t^k G_k(t\zeta)},$$

for any nonzero rational number t and any (1.1) for which the denominators in (20.11) are not identically zero. We replace $t\zeta$ with z in (20.11) and use (20.5) to obtain

$$(20.12) \quad \frac{\sum_{k=0}^{p} t^k F_k(z)}{\sum_{k=0}^{q} t^k G_k(z)} \equiv t^s \left[\frac{F(z)}{G(z)}\right] \equiv t^s \frac{\sum_{k=0}^{p} F_k(z)}{\sum_{k=0}^{q} G_k(z)},$$

for any nonzero rational number t and any (1.1) for which the denominators in (20.12) are not identically zero. Due to $\sum_{k=0}^{q} \boldsymbol{G}_k \equiv \boldsymbol{G} \not\equiv 0$, there is a real number ϵ subject to $0 < \epsilon < 1$ such that

$$(20.13) \quad \sum_{k=0}^{q} t^k \boldsymbol{G}_k \not\equiv 0, \quad \text{for each rational number } t \text{ satisfying } |t-1| < \epsilon.$$

We shall establish next that

$$(20.14) \quad \frac{\sum_{k=0}^{p} t^k \boldsymbol{F}_k}{\sum_{k=0}^{q} t^k \boldsymbol{G}_k} \equiv t^s \frac{\sum_{k=0}^{p} \boldsymbol{F}_k}{\sum_{k=0}^{q} \boldsymbol{G}_k}, \quad \text{for each } t \text{ in } \mathbb{Q} \text{ that satisfies } |t-1| < \epsilon.$$

Namely, supposes that (20.14) is not satisfied by some fixed rational number t for which $|t-1| < \epsilon$. Using that assumption and (20.13), we would obtain

$$\left[\sum_{k=0}^{q} \boldsymbol{G}_k\right]\left[\sum_{k=0}^{q} t^k \boldsymbol{G}_k\right] \left\{ \left[\sum_{k=0}^{p} t^k \boldsymbol{F}_k\right]\left[\sum_{k=0}^{q} \boldsymbol{G}_k\right] - t^s \left[\sum_{k=0}^{q} t^k \boldsymbol{G}_k\right]\left[\sum_{k=0}^{p} \boldsymbol{F}_k\right] \right\} \not\equiv 0.$$

Due to Corollary 4.3 on page 37, there is a differential equation (1.1) such that the substitution of $c_{\mu,\nu}^{(\kappa)}(z)$ from that (1.1) for $w_{\mu,\nu}^{(\kappa)}$ in the preceding expression yields

$$(20.15) \quad \sum_{k=0}^{q} t^k G_k(z) \not\equiv 0, \quad \sum_{k=0}^{q} G_k(z) \not\equiv 0,$$

and

$$(20.16) \quad \left[\sum_{k=0}^{p} t^k F_k(z)\right]\left[\sum_{k=0}^{q} G_k(z)\right] - t^s \left[\sum_{k=0}^{q} t^k G_k(z)\right]\left[\sum_{k=0}^{p} F_k(z)\right] \not\equiv 0.$$

But, (20.15) and (20.16) contradict (20.12). Thus, (20.14) is valid for some positive real number ϵ. We use (20.14), (20.5), and (20.3) to deduce in $\mathcal{Q}_m$ that

$$(20.17) \quad \sum_{k=0}^{p} \boldsymbol{F}_k t^k \equiv \sum_{k=0}^{q} \boldsymbol{A}\, \boldsymbol{G}_k t^{k+s}, \quad \text{for } t \text{ in } \mathbb{Q} \text{ and } |t-1| < \epsilon.$$

Regarding (20.17) as an equality of two polynomial functions in t over $\mathcal{Q}_m$, we note that the coefficients of like powers of t must be equal. The terms involving the greatest power of t with a nonzero coefficient in the left and right members of (20.17) are $\boldsymbol{F}_p t^p$ and $\boldsymbol{A}\,\boldsymbol{G}_q t^{q+s}$ respectively. This yields $\boldsymbol{F}_p \equiv \boldsymbol{A}\,\boldsymbol{G}_q$ and

(20.18) $$p = q + s.$$

Hence, we must have $\boldsymbol{A} \equiv \boldsymbol{F}/\boldsymbol{G} \equiv \boldsymbol{F}_p/\boldsymbol{G}_q$ and $\boldsymbol{F}\,\boldsymbol{G}_q \equiv \boldsymbol{G}\,\boldsymbol{F}_p$. Since $\boldsymbol{F}$ and $\boldsymbol{G}$ are relatively prime, we find that $\boldsymbol{F}$ divides $\boldsymbol{F}_p$ in $\mathcal{S}_m$ and $\boldsymbol{G}$ divides $\boldsymbol{G}_q$ in $\mathcal{S}_m$. In view of (20.5), there are polynomials $\boldsymbol{Q}_1$, $\boldsymbol{Q}_2$ in $\mathcal{S}_m$ such that

(20.19) $$\boldsymbol{F}_p \equiv \boldsymbol{Q}_1 \sum_{k=0}^{p} \boldsymbol{F}_k \quad \text{and} \quad \boldsymbol{G}_q \equiv \boldsymbol{Q}_2 \sum_{k=0}^{q} \boldsymbol{G}_k.$$

Due to the conditions on $\boldsymbol{F}_0$, ..., $\boldsymbol{F}_p$ and $\boldsymbol{G}_0$, ..., $\boldsymbol{G}_q$, (20.19) yields $\boldsymbol{Q}_1 \equiv \boldsymbol{Q}_2 \equiv 1$ as well as $\boldsymbol{F}_k \equiv 0$, for $0 \leq k \leq p-1$, and $\boldsymbol{G}_k \equiv 0$, for $0 \leq k \leq q-1$. This gives $\boldsymbol{F} \equiv \boldsymbol{F}_p$ and $\boldsymbol{G} \equiv \boldsymbol{G}_q$. Thus, $\boldsymbol{F}$ is isobaric of weight p, $\boldsymbol{G}$ is isobaric of weight q, and (20.18) shows that (20.4) is valid. This completes the proof. □

COROLLARY 20.5. *Suppose that $\boldsymbol{P}$ in $\mathcal{S}_m$ is a polynomial semi-invariant of the second kind for the equations (1.1) having order m. Then, $\boldsymbol{P}$ is isobaric and any integer s that is suitable for (1.10) of Definition 1.2 on page 6 is unique and given by $s = \text{weight}(\boldsymbol{P})$.*

PROOF. We set $\boldsymbol{A} \equiv \boldsymbol{P}/1$. Then, $\boldsymbol{A}$ is a rational semi-invariant of the second kind. Theorem 20.4 establishes that $\boldsymbol{P}$ is isobaric in $\mathcal{S}_m$. Moreover, for s in (1.10) on page 6, we see that (20.2) is satisfied by $\boldsymbol{A} \equiv \boldsymbol{P}/1$ for that s. Thus, (20.4) yields $s = \text{weight}(\boldsymbol{P}) - \text{weight}(1) = \text{weight}(\boldsymbol{P})$. This completes the proof. □

Suppose that $\boldsymbol{F}$, $\boldsymbol{G}$, $\boldsymbol{P}$, and $\boldsymbol{Q}$ are isobaric polynomials in $\mathcal{S}_m$ of respective weights e, f, g, and h such that $\boldsymbol{F}/\boldsymbol{G} \equiv \boldsymbol{P}/\boldsymbol{Q}$. Then, we find that $\boldsymbol{F}\boldsymbol{Q} \equiv \boldsymbol{G}\boldsymbol{P}$, $e+h = f+g$, and $e-f = g-h$. Thus, for any element $\boldsymbol{A}$ in $\mathcal{Q}_m$ that is represented as $\boldsymbol{F}/\boldsymbol{G}$ for some isobaric $\boldsymbol{F}$ and $\boldsymbol{G}$ in $\mathcal{S}_m$, the integer given by $\text{weight}(\boldsymbol{F}) - \text{weight}(\boldsymbol{G})$ does not depend on the particular representation for $\boldsymbol{A}$. And, when $\boldsymbol{G}$ is a nonzero constant, it reduces to the ordinary weight of $\boldsymbol{F}$ in $\mathcal{S}_m$.

DEFINITION 20.6. An element $\boldsymbol{A}$ in $\mathcal{Q}_m$ is said to be *rationally isobaric*, or simply *isobaric*, when there are isobaric polynomials $\boldsymbol{F}$ and $\boldsymbol{G}$ in $\mathcal{S}_m$ such that $\boldsymbol{A} \equiv \boldsymbol{F}/\boldsymbol{G}$. The *weight* of a rationally isobaric element $\boldsymbol{A}$ in $\mathcal{Q}_m$ is defined to be

(20.20) $$\text{weight}(\boldsymbol{A}) = \text{weight}(\boldsymbol{F}) - \text{weight}(\boldsymbol{G}),$$

where $\boldsymbol{A} \equiv \boldsymbol{F}/\boldsymbol{G}$ and $\boldsymbol{F}$, $\boldsymbol{G}$ are isobaric polynomials in $\mathcal{S}_m$.

As a restatement of Theorem 20.4, *a rational semi-invariant of the second kind is rationally isobaric and s for it in (20.2) is its weight given by (20.20)*.

Theorem 20.14 on page 240 provides more information about any polynomials $\boldsymbol{F}$ and $\boldsymbol{G}$ that satisfy the hypothesis of Theorem 20.4. Namely, at least one of $\boldsymbol{F}$ or $\boldsymbol{G}$ must be nonconstant; and a nonconstant $\boldsymbol{F}$ or $\boldsymbol{G}$ is a polynomial semi-invariant of the second kind. An analogous result about rational relative invariants will be presented in Theorem 20.15 on page 243.

20.3. A context for the remainder of this chapter

For $1 \leq j \leq m$, $0 \leq i \leq j$, $k \geq 0$, and $l \geq 0$, let $\boldsymbol{w}_{i,j}^{(k)}$ and $\boldsymbol{x}_0^{(l)}$ be algebraically independent variables over the field $\mathbb{Q}$ of rational numbers and let $\mathfrak{P}$ be the ring of polynomials in these variables over $\mathbb{Q}$. We also introduce $\boldsymbol{w}_{0,0}^{(0)} \equiv 1$. A derivation $'$ for $\mathbb{Q}$ must satisfy $\gamma' = 0$, for each γ in $\mathbb{Q}$. There is a unique extension of this derivation for $\mathbb{Q}$ to a derivation $'$ for $\mathfrak{P}$ such that

(20.21) $\qquad \left(\boldsymbol{w}_{i,j}^{(k)}\right)' = \boldsymbol{w}_{i,j}^{(k+1)} \quad \text{and} \quad \left(\boldsymbol{x}_0^{(l)}\right)' = \boldsymbol{x}_0^{(l+1)},$

for $1 \leq j \leq m$, $0 \leq i \leq j$, $k \geq 0$, and $l \geq 0$.

This is a consequence of [**10**, Chap. V, Sec. 9, No. 1, Proposition 4, page 139] when we identify: the field E in [**10**] with $\mathbb{Q}$; the fields F and Ω in [**10**] with the quotient field of the entire ring $\mathfrak{P}$; and the derivation D in [**10**] with the derivation $'$ for $\mathbb{Q}$. Alternatively, one can specialize [**11**, Chap. 5, Sec. 16, No. 3, Theorem 1].

Let $\mathfrak{Q}$ denote the quotient field of $\mathfrak{P}$ and let $'$ denote the unique derivation for $\mathfrak{Q}$ that is an extension of the derivation for $\mathfrak{P}$. In $\mathfrak{Q}$, we introduce

(20.22) $\qquad\qquad\qquad \boldsymbol{t}_l = \dfrac{\boldsymbol{x}_0^{(l)}}{\boldsymbol{x}_0^{(0)}}, \quad \text{for } l \geq 1.$

Since the set $\left\{\boldsymbol{w}_{i,j}^{(k)} \mid \text{for } 1 \leq j \leq m, 0 \leq i \leq j, \text{ and } k \geq 0\right\} \bigcup \left\{\boldsymbol{x}_0^{(l)} \mid \text{for } l \geq 0\right\}$ is algebraically independent over $\mathbb{Q}$, we use (20.22) to verify that the set

(20.23) $\left\{\boldsymbol{w}_{i,j}^{(k)} \mid \text{for } 1 \leq j \leq m, 0 \leq i \leq j, \text{ and } k \geq 0\right\} \bigcup \left\{\boldsymbol{t}_l \mid \text{for } l \geq 1\right\} \bigcup \left\{\boldsymbol{x}_0^{(0)}\right\}$

is algebraically independent over $\mathbb{Q}$. Due to (20.22), the derivation $'$ for $\mathfrak{Q}$ yields

(20.24) $\qquad\qquad\qquad (\boldsymbol{t}_l)' = \boldsymbol{t}_{l+1} - \boldsymbol{t}_1 \boldsymbol{t}_l, \quad \text{for } l \geq 1.$

Let $\mathfrak{S}$ denote the subring of $\mathfrak{Q}$ consisting of the polynomials over $\mathbb{Q}$ in the variables $\boldsymbol{w}_{i,j}^{(k)}$ and $\boldsymbol{t}_l$, for $1 \leq j \leq m$, $0 \leq i \leq j$, $k \geq 0$, and $l \geq 1$. Due to (20.21) and (20.24), we see that the restriction of the derivation for $\mathfrak{Q}$ to $\mathfrak{S}$ is a derivation for $\mathfrak{S}$. Moreover, for each polynomial $\boldsymbol{S}$ in $\mathfrak{S}$, there is a least nonnegative integer p such that $(\boldsymbol{x}_0^{(0)})^p \boldsymbol{S}$ is a polynomial in $\mathfrak{P}$. We also set $\boldsymbol{t}_0 = 1$ in $\mathfrak{S}$.

The ring $\mathcal{S}_m$ of polynomials over $\mathbb{Q}$ in the variables of (1.6) on page 5 can be regarded as a subring of $\mathfrak{S}$ as well as a subring of $\mathfrak{P}$ and $\mathfrak{Q}$. Then, the derivation for $\mathcal{S}_m$ specified by (1.13) on page 7 is the restriction to $\mathcal{S}_m$ of the derivation for $\mathfrak{Q}$. When $r = 0, 1, 2, \ldots$, let $\mathcal{M}_r$ denote the set of monic monomials in $\mathcal{S}_m$ of weight r. We have $\mathcal{M}_0 = \left\{\boldsymbol{w}_{0,0}^{(0)}\right\}$, $\mathcal{M}_1 = \left\{\boldsymbol{w}_{0,1}^{(0)}\right\}$, $\mathcal{M}_2 = \left\{\boldsymbol{w}_{0,2}^{(0)}, \boldsymbol{w}_{1,1}^{(0)}, \boldsymbol{w}_{0,1}^{(1)}, (\boldsymbol{w}_{0,1}^{(0)})^2\right\}, \ldots$.

Let $\mathfrak{T}$ denote the subring of $\mathfrak{S}$ that consists of the polynomials over $\mathbb{Q}$ in the variables $\boldsymbol{t}_l$, for $l \geq 1$. We use (20.24) to see that the restriction to $\mathfrak{T}$ of the derivation for $\mathfrak{S}$ is a derivation for $\mathfrak{T}$. We define the *weight* of any nonzero rational number in $\mathfrak{T}$ to be 0; we define the *weight* of $\boldsymbol{t}_l$ to be l; and we define the *weight* of any nonzero monomial in $\mathfrak{T}$ to be the sum of the weights of its factors. For instance, the weight of $3\boldsymbol{t}_1(\boldsymbol{t}_2)^5(\boldsymbol{t}_3)^2$ is 17. We define a nonzero polynomial in $\mathfrak{T}$ to be *isobaric* when any two of its nonzero terms have the same weight. And, we define the *weight* of an isobaric polynomial in $\mathfrak{T}$ to be the weight of any one of its nonzero terms. Thus, the polynomial $2\boldsymbol{t}_1(\boldsymbol{t}_2)^3 + 3(\boldsymbol{t}_1)^3(\boldsymbol{t}_2)^2$ is isobaric and its weight is 7. For $k = 0, 1, 2, \ldots$, let $\mathfrak{T}_k$ denote the subset of $\mathfrak{T}$ containing 0 and all of the isobaric polynomials in $\mathfrak{T}$ of weight k. Hence, we have $\mathfrak{T}_0 = \mathbb{Q}$, $\mathfrak{T}_1 = \left\{\alpha \boldsymbol{t}_1 \mid \text{for each } \alpha \text{ in } \mathbb{Q}\right\}$, $\mathfrak{T}_2 = \left\{\alpha \boldsymbol{t}_2 + \beta (\boldsymbol{t}_1)^2 \mid \text{for each } \alpha, \beta \text{ in } \mathbb{Q}\right\}$, and etc.

Thus, for $k \geq 0$, $\mathfrak{T}_k$ is the set of linear combinations over $\mathbb{Q}$ of the monic monomials in $\mathfrak{T}$ of weight k. In particular, $\mathfrak{T}_0$, $\mathfrak{T}_1$, $\mathfrak{T}_2$, $\mathfrak{T}_3$, ... are the linear combinations over $\mathbb{Q}$ of the respective sets $\{t_0\}$, $\{t_1\}$, $\{t_2, (t_1)^2\}$, $\{t_3, t_1 t_2, (t_1)^3\}$, ...

NOTATION 20.7. Let T be any polynomial in $\mathfrak{T}$ and let $h(z) \not\equiv 0$ be a meromorphic function on a region Ω. We introduce the notation $T[h; z]$ to denote the meromorphic function on Ω obtained from T by replacing t_l in T with $h^{(l)}(z)/h(z)$, for $l = 1, 2, \ldots$.

To establish Proposition 20.10 on page 225, we shall use the following result.

LEMMA 20.8. *Let T be a polynomial in $\mathfrak{T}$, let T' be the polynomial in $\mathfrak{T}$ obtained by applying the derivation ' for $\mathfrak{T}$ to T, and let $\rho(z) \not\equiv 0$ be a meromorphic function on a region Ω. Then the functions $T[\rho; z]$ and $(T')[\rho; z]$ on Ω obtained by replacing each t_l in T and T' with $\rho^{(l)}(z)/\rho(z)$, for $l \geq 1$, are related by*

$$\text{(20.25)} \qquad \frac{d}{dz} T[\rho; z] \equiv (T')[\rho; z].$$

PROOF. Clearly (20.25) is valid when T is a constant. For $T \equiv \gamma t_l$, γ in $\mathbb{Q}$, and $l \geq 1$, we apply (20.24) to obtain

$$\text{(20.26)} \qquad \frac{d}{dz}\left((\gamma t_l)[\rho; z]\right) \equiv \frac{d}{dz}\left[\gamma \frac{\rho^{(l)}(z)}{\rho(z)}\right] \equiv \gamma\left[\frac{\rho^{(l+1)}(z)}{\rho(z)} - \frac{\rho^{(1)}(z)}{\rho(z)}\frac{\rho^{(l)}(z)}{\rho(z)}\right]$$

$$\equiv \left(\gamma\left(t_{l+1} - t_1 t_l\right)\right)[\rho; z] \equiv (\gamma t_l')[\rho; z].$$

Thus, (20.25) is valid for $T \equiv \gamma t_l$ and $l \geq 1$. If T is a polynomial in $\mathfrak{T}$ that satisfies (20.25) and if $S \equiv t_l T$ for $l \geq 1$, then we use (20.26) and (20.25) to verify that

$$\frac{d}{dz}(S[\rho; z]) \equiv \frac{d}{dz}\left(t_l[\rho; z] T[\rho; z]\right) \equiv \left[\frac{d}{dz}t_l[\rho; z]\right] T[\rho; z] + t_l[\rho; z]\frac{d}{dz}T[\rho; z]$$

$$\equiv (t_l')[\rho; z] T[\rho; z] + t_l[\rho; z] (T')[\rho; z] \equiv (t_l' T + t_l T')[\rho; z]$$

$$\equiv ((t_l T)')[\rho; z] \equiv S'[\rho; z].$$

Thus, (20.25) is valid for any monomial in $\mathfrak{T}$. Any polynomial T in $\mathfrak{T}$ is a sum of monomials $T_1, \ldots, T_n$ in $\mathfrak{T}$. We use $\frac{d}{dz} T_k[\rho; z] \equiv (T_k')[\rho; z]$, for $1 \leq k \leq n$, as well as the linearity of both $\frac{d}{dz}$ and ' to obtain

$$\frac{d}{dz} T[\rho; z] \equiv \sum_{k=1}^{n} \frac{d}{dz} T_k[\rho; z] \equiv \sum_{k=1}^{n} T_k'[\rho; z] \equiv T'[\rho; z].$$

Thus, (20.25) is valid for any polynomial in $\mathfrak{T}$. This completes the proof. $\square$

The argument on page 235 after (20.84) will use the following result.

LEMMA 20.9. *Let T be a polynomial in $\mathfrak{T}$, let T' be the polynomial in $\mathfrak{T}$ obtained by applying the derivation ' for $\mathfrak{T}$ to T, and let $g(z)$ be a univalent analytic function on a region Ω. Then the analytic functions $T[g'; z]$ and $(T')[g'; z]$ on Ω obtained by replacing t_l in T and T' with $g^{(l+1)}(z)/g'(z)$, for $l \geq 1$, are related by*

$$\text{(20.27)} \qquad \frac{d}{dz} T[g'; z] \equiv (T')[g'; z].$$

PROOF. Since $g(z)$ is a univalent analytic function on Ω, the function $g'(z)$ is analytic on Ω and satisfies $g'(z) \neq 0$, for each z in Ω. Applying Lemma 20.8 to the situation where $\rho(z)$ is $g'(z)$, we see that (20.25) yields (20.27). □

20.4. A technical construction needed for Section 20.5

PROPOSITION 20.10. *Let $\boldsymbol{F}$ be an isobaric polynomial in $\mathcal{S}_m$ of weight $n \geq 0$. Then, for $0 \leq r \leq n$ and each monic monomial $\boldsymbol{P}$ in $\mathcal{M}_r$, there is a corresponding polynomial $\boldsymbol{T_P}$ in $\mathfrak{T}_{n-r}$ such that the following is true. For any (1.1) and any transformation (1.2) of (1.1) into a corresponding (1.3), the relations*

$$(20.28) \qquad F^*(z) \equiv \sum_{r=0}^{n} \sum_{\boldsymbol{P} \in \mathcal{M}_r} P(z) \, \boldsymbol{T_P}[\rho; z]$$

and

$$(20.29) \qquad F^*(z) \equiv F(z) + \sum_{r=0}^{n-1} \sum_{\boldsymbol{P} \in \mathcal{M}_r} P(z) \, \boldsymbol{T_P}[\rho; z]$$

are satisfied on Ω where: $F^*(z)$ on Ω is obtained by replacing each $\boldsymbol{w}_{i,j}^{(k)}$ in $\boldsymbol{F}$ with $c_{i,j}^{*(k)}(z)$ from (1.3); $F(z)$ and $P(z)$ on Ω are obtained from $\boldsymbol{F}$ and $\boldsymbol{P}$ by replacing each $\boldsymbol{w}_{i,j}^{(k)}$ in $\boldsymbol{F}$ and $\boldsymbol{P}$ with $c_{i,j}^{(k)}(z)$ from (1.1); and $\boldsymbol{T_P}[\rho; z]$ is the function of z on Ω obtained by replacing each $\boldsymbol{t}_l$ in $\boldsymbol{T_P}$ with $\rho^{(l)}(z)/\rho(z)$ from (1.2).

PROOF. For $n = 0$, $\boldsymbol{F}$ is a rational number $\gamma \neq 0$. Then, (20.28)–(20.29) are valid with $\boldsymbol{T_P} \equiv \gamma$ for $\boldsymbol{P} \equiv \boldsymbol{w}_{0,0}^{(0)} \equiv 1$ in $\mathcal{M}_0$. Thus, we may assume that $n \geq 1$.

(i) Suppose that $\boldsymbol{F} \equiv \gamma \, \boldsymbol{w}_{i,j}$, for $\gamma \neq 0$ in $\mathbb{Q}$, $1 \leq j \leq m$, and $0 \leq i \leq j$. Then, with $c_{0,0}(z) \equiv 1$ and $n = i+j$, we use (2.5) of Proposition 2.1 on page 23 to obtain

$$(20.30) \quad \gamma \, c_{i,j}^*(z) \equiv \gamma \sum_{\mu=0}^{i} \sum_{\nu=0}^{j} \binom{m-\mu}{i-\mu} \binom{m-\nu}{j-\nu} \frac{\rho^{(i-\mu)}(z) \, \rho^{(j-\nu)}(z)}{(\rho(z))^2} c_{\mu,\nu}(z)$$

$$\equiv \sum_{r=0}^{n} \left[\sum_{\substack{\mu+\nu=r \\ 0 \leq \mu < \nu \leq i}} d_{\mu,\nu}(z) + \sum_{\substack{\mu+\nu=r \\ 0 \leq \nu < \mu \leq i}} d_{\mu,\nu}(z) + \sum_{\substack{\mu+\nu=r \\ 0 \leq \mu = \nu \leq i}} d_{\mu,\nu}(z) + \sum_{\substack{\mu+\nu=r \\ 0 \leq \mu \leq i \\ i+1 \leq \nu \leq j}} d_{\mu,\nu}(z) \right]$$

in which

$$(20.31) \quad d_{\mu,\nu}(z) \equiv \gamma \binom{m-\mu}{i-\mu} \binom{m-\nu}{j-\nu} \frac{\rho^{(i-\mu)}(z) \, \rho^{(j-\nu)}(z)}{(\rho(z))^2} c_{\mu,\nu}(z)$$

$$\equiv \left[\gamma \binom{m-\mu}{i-\mu} \binom{m-\nu}{j-\nu} \boldsymbol{t}_{i-\mu}[\rho; z] \, \boldsymbol{t}_{j-\nu}[\rho; z] \right] c_{\mu,\nu}(z)$$

$$\equiv \boldsymbol{x}_{\mu,\nu}[\rho; z] \, c_{\mu,\nu}(z), \quad \text{for } 0 \leq \mu \leq i \text{ and } 0 \leq \nu \leq j,$$

where $\boldsymbol{x}_{\mu,\nu}$ is defined in $\mathfrak{T}$ by

$$\boldsymbol{x}_{\mu,\nu} \equiv \gamma \binom{m-\mu}{i-\mu} \binom{m-\nu}{j-\nu} \boldsymbol{t}_{i-\mu} \boldsymbol{t}_{j-\nu}, \quad \text{for } 0 \leq \mu \leq i \text{ and } 0 \leq \nu \leq j.$$

When $\mu + \nu = r$, the weight of $\boldsymbol{x}_{\mu,\nu}$ is $i - \mu + j - \nu = n - r$ and therefore $\boldsymbol{x}_{\mu,\nu}$ is an element of $\mathfrak{T}_{n-r}$. For $0 \le r \le n$ and $\boldsymbol{P}$ in $\mathcal{M}_r$, we define $\boldsymbol{T_P}$ in $\mathfrak{T}_{n-r}$ through

$$\boldsymbol{T_P} \equiv \begin{cases} \boldsymbol{x}_{\mu,\nu} + \boldsymbol{x}_{\nu,\mu}, & \text{if } \boldsymbol{P} \equiv \boldsymbol{w}_{\mu,\nu},\ 0 \le \mu < \nu \le i \text{ and } \mu + \nu = r; \\ \boldsymbol{x}_{\mu,\nu}, & \text{if } \boldsymbol{P} \equiv \boldsymbol{w}_{\mu,\nu},\ 0 \le \mu = \nu \le i \text{ and } \mu + \nu = r; \\ \boldsymbol{x}_{\mu,\nu}, & \text{if } \boldsymbol{P} \equiv \boldsymbol{w}_{\mu,\nu},\ 0 \le \mu \le i,\ i+1 \le \nu \le j,\text{ and } \mu + \nu = r; \\ 0, & \text{otherwise.} \end{cases}$$

For $r = n$ and $\mu + \nu = i + j$ in (20.30), there is a single corresponding term; it has $\mu = i$ and $\nu = j$. Thus, (20.30) can be rewritten as

$$\gamma c_{i,j}^*(z) \equiv \gamma c_{i,j}(z) + \sum_{r=0}^{n-1} \left[\sum_{\substack{\mu+\nu=r \\ 0 \le \mu < \nu \le i}} (d_{\mu,\nu}(z) + d_{\nu,\mu}(z)) + \sum_{\substack{\mu+\nu=r \\ 0 \le \mu = \nu \le i}} d_{\mu,\nu}(z) + \sum_{\substack{\mu+\nu=r \\ 0 \le \mu \le i \\ i+1 \le \nu \le j}} d_{\mu,\nu}(z) \right].$$

Using (20.31) and the definition of $\boldsymbol{T_P}$, we rewrite (20.30) and the preceding formula to see that (20.28) and (20.29) are valid when $\boldsymbol{F} \equiv \gamma \boldsymbol{w}_{i,j}$.

(ii) Suppose that (20.28) and (20.29) are valid when $\boldsymbol{F}$ is a particular isobaric polynomial $\boldsymbol{G}$ of weight $p \ge 1$. We shall show next that (20.28) and (20.29) are valid when $\boldsymbol{F} \equiv \boldsymbol{G}'$ and $n = p + 1$. To do so, we begin with

$$(20.32) \qquad G^*(z) \equiv \sum_{s=0}^{p} \sum_{\boldsymbol{Q} \in \mathcal{M}_s} Q(z)\, \boldsymbol{U_Q}[\rho; z]$$

and

$$(20.33) \qquad G^*(z) \equiv G(z) + \sum_{s=0}^{p-1} \sum_{\boldsymbol{Q} \in \mathcal{M}_s} Q(z)\, \boldsymbol{U_Q}[\rho; z],$$

where, for $0 \le s \le p$ and $\boldsymbol{Q}$ in $\mathcal{M}_s$, $\boldsymbol{U_Q}$ is a polynomial in $\mathcal{T}_{p-s}$. We differentiate (20.32) with respect to z and use (20.25) to obtain

$$(20.34) \quad G^{*(1)}(z) \equiv \sum_{r=0}^{p} \sum_{\boldsymbol{Q} \in \mathcal{M}_r} \left(\frac{d}{dz}Q(z)\right) \boldsymbol{U_Q}[\rho; z] + \sum_{r=0}^{p} \sum_{\boldsymbol{Q} \in \mathcal{M}_r} Q(z)\, \boldsymbol{U_Q'}[\rho; z].$$

For $r \ge 0$ and each $\boldsymbol{Q}$ in $\mathcal{M}_r$, there are unique rational numbers $\gamma_{\boldsymbol{P},\boldsymbol{Q}}$, for each $\boldsymbol{P}$ in $\mathcal{M}_{r+1}$, such that

$$(20.35) \qquad \boldsymbol{Q}' \equiv \sum_{\boldsymbol{P} \in \mathcal{M}_{r+1}} \gamma_{\boldsymbol{P},\boldsymbol{Q}}\, \boldsymbol{P}.$$

By applying (20.34), (4.13), and (20.35), we find that

$$G^{*(1)}(z) \equiv \sum_{r=0}^{p} \sum_{\boldsymbol{Q} \in \mathcal{M}_r} \sum_{\boldsymbol{P} \in \mathcal{M}_{r+1}} \gamma_{\boldsymbol{P},\boldsymbol{Q}}\, P(z)\, \boldsymbol{U_Q}[\rho; z] + \sum_{r=0}^{p} \sum_{\boldsymbol{Q} \in \mathcal{M}_r} Q(z)\, \boldsymbol{U_Q'}[\rho; z]$$

$$\equiv \sum_{r=0}^{p} \sum_{\boldsymbol{P} \in \mathcal{M}_{r+1}} P(z) \left[\sum_{\boldsymbol{Q} \in \mathcal{M}_r} \gamma_{\boldsymbol{P},\boldsymbol{Q}}\, \boldsymbol{U_Q}[\rho; z] \right] + \sum_{r=0}^{p} \sum_{\boldsymbol{P} \in \mathcal{M}_r} P(z)\, \boldsymbol{U_P'}[\rho; z]$$

$$\equiv \sum_{r=1}^{p+1} \sum_{\boldsymbol{P} \in \mathcal{M}_r} P(z) \left[\sum_{\boldsymbol{Q} \in \mathcal{M}_{r-1}} \gamma_{\boldsymbol{P},\boldsymbol{Q}}\, \boldsymbol{U_Q}[\rho; z] \right] + \sum_{r=0}^{p} \sum_{\boldsymbol{P} \in \mathcal{M}_r} P(z)\, \boldsymbol{U_P'}[\rho; z].$$

We use this formula to verify that

$$G^{*(1)}(z) \equiv \sum_{r=0}^{p+1} \sum_{\boldsymbol{P}\in\mathcal{M}_r} P(z)\, \boldsymbol{T_P}[\rho;\, z], \tag{20.36}$$

where: for $0 \leq r \leq p+1$ and each $\boldsymbol{P}$ in $\mathcal{M}_r$, $\boldsymbol{T_P}$ is defined in $\mathfrak{T}_{p+1-r}$ through

$$\boldsymbol{T_P} = \begin{cases} \boldsymbol{U'_P}, & \text{if } \boldsymbol{P} \in \mathcal{M}_0; \\ \boldsymbol{U'_P} + \displaystyle\sum_{\boldsymbol{Q}\in\mathcal{M}_{r-1}} \gamma_{\boldsymbol{P},\boldsymbol{Q}}\, \boldsymbol{U_Q}, & \text{if } \boldsymbol{P} \in \mathcal{M}_r \text{ and } 1 \leq r \leq p; \\ \displaystyle\sum_{\boldsymbol{Q}\in\mathcal{M}_p} \gamma_{\boldsymbol{P},\boldsymbol{Q}}\, \boldsymbol{U_Q}, & \text{if } \boldsymbol{P} \in \mathcal{M}_{p+1}. \end{cases} \tag{20.37}$$

We note that (20.33) and (20.32) yield

$$G(z) \equiv \sum_{\boldsymbol{Q}\in\mathcal{M}_p} Q(z)\, \boldsymbol{U_Q}[\rho;\, z]. \tag{20.38}$$

For each $\boldsymbol{Q}$ in $\mathcal{M}_p$, $\boldsymbol{U_Q}$ belongs to $\mathfrak{T}_0$ and therefore satisfies $\boldsymbol{U'_Q}[\rho;\, z] \equiv 0$. By differentiating (20.38) with respect to z and using (20.25), $\boldsymbol{U'_Q}[\rho;\, z] \equiv 0$, (20.35), (4.13), and (20.37), we obtain

$$G^{(1)}(z) \equiv \sum_{\boldsymbol{Q}\in\mathcal{M}_p} Q'(z)\, \boldsymbol{U_Q}[\rho;\, z] \equiv \sum_{\boldsymbol{Q}\in\mathcal{M}_p}\sum_{\boldsymbol{P}\in\mathcal{M}_{p+1}} \gamma_{\boldsymbol{P},\boldsymbol{Q}}\, P(z)\, \boldsymbol{U_Q}[\rho;\, z] \tag{20.39}$$

$$\equiv \sum_{\boldsymbol{P}\in\mathcal{M}_{p+1}} P(z) \sum_{\boldsymbol{Q}\in\mathcal{M}_p} \gamma_{\boldsymbol{P},\boldsymbol{Q}}\, \boldsymbol{U_Q}[\rho;\, z] \equiv \sum_{\boldsymbol{P}\in\mathcal{M}_{p+1}} P(z)\, \boldsymbol{T_P}[\rho;\, z].$$

Thus, (20.36) and (20.39) give

$$G^{*(1)}(z) \equiv G^{(1)}(z) + \sum_{r=0}^{p} \sum_{\boldsymbol{P}\in\mathcal{M}_r} P(z)\, \boldsymbol{T_P}[\rho;\, z]. \tag{20.40}$$

In view of (20.36) and (20.40), (20.28) and (20.29) are valid for $\boldsymbol{F} \equiv \boldsymbol{G}'$ with $n = p+1$.

(iii) Suppose that $\boldsymbol{G}$ and $\boldsymbol{H}$ are isobaric polynomials in $\mathcal{S}_m$ of respective weights $p \geq 1$ and $q \geq 1$ for which (20.28) and (20.29) are valid when $\boldsymbol{F} \equiv \boldsymbol{G}$ and $n = p$ as well as when $\boldsymbol{F} \equiv \boldsymbol{H}$ and $n = q$. In order to prove that there are formulas (20.28)–(20.29) for $\boldsymbol{F} \equiv \boldsymbol{GH}$ with $n = p+q$, we assume that the specific formulas for $\boldsymbol{G}$ are (20.32)–(20.33) and the specific formulas for $\boldsymbol{H}$ are

$$H^*(z) \equiv \sum_{t=0}^{q} \sum_{\boldsymbol{R}\in\mathcal{M}_t} R(z)\, \boldsymbol{V_R}[\rho;\, z] \tag{20.41}$$

and

$$H^*(z) \equiv H(z) + \sum_{t=0}^{q-1} \sum_{\boldsymbol{R}\in\mathcal{M}_t} R(z)\, \boldsymbol{V_R}[\rho;\, z], \tag{20.42}$$

where, for $0 \leq t \leq q$ and $\boldsymbol{R}$ in $\mathcal{M}_t$, $\boldsymbol{V_R}$ is a polynomial in $\mathcal{T}_{q-t}$. We extend the definitions of $\boldsymbol{U_Q}$ and $\boldsymbol{V_R}$ by setting $\boldsymbol{U_Q} \equiv 0$, for $\boldsymbol{Q}$ in $\mathcal{M}_s$ when $s > p$, and by setting $V_R \equiv 0$, for R in $\mathcal{M}_t$ when $t > q$. In view of this, (20.32) and (20.41) give

$$
(20.43) \quad G^*(z)\, H^*(z) \equiv \left[\sum_{s \geq 0} \sum_{Q \in \mathcal{M}_s} Q(z)\, U_Q[\rho;\, z] \right] \left[\sum_{t \geq 0} \sum_{R \in \mathcal{M}_t} R(z)\, V_R[\rho;\, z] \right]
$$

$$
\equiv \sum_{r=0}^{p+q} \sum_{s=0}^{r} \left[\sum_{Q \in \mathcal{M}_s} Q(z)\, U_Q[\rho;\, z] \right] \left[\sum_{R \in \mathcal{M}_{r-s}} R(z)\, V_R[\rho;\, z] \right]
$$

$$
\equiv \sum_{r=0}^{p+q} \sum_{s=0}^{r} \sum_{\substack{Q \in \mathcal{M}_s \\ R \in \mathcal{M}_{r-s}}} Q(z)\, R(z)\, U_Q[\rho;\, z]\, V_R[\rho;\, z]
$$

$$
\equiv \sum_{r=0}^{p+q} \sum_{s=0}^{r} \sum_{\substack{Q \in \mathcal{M}_s \\ R \in \mathcal{M}_{r-s} \\ P \in \mathcal{M}_r \\ P = QR}} Q(z)\, R(z)\, U_Q[\rho;\, z]\, V_R[\rho;\, z]
$$

$$
\equiv \sum_{r=0}^{p+q} \sum_{s=0}^{r} \sum_{P \in \mathcal{M}_r} \sum_{\substack{Q \in \mathcal{M}_s \\ R \in \mathcal{M}_{r-s} \\ QR = P}} P(z)\, U_Q[\rho;\, z]\, V_R[\rho;\, z]
$$

$$
\equiv \sum_{r=0}^{p+q} \sum_{P \in \mathcal{M}_r} P(z) \sum_{s=0}^{r} \sum_{\substack{Q \in \mathcal{M}_s \\ R \in \mathcal{M}_{r-s} \\ QR = P}} U_Q[\rho;\, z]\, V_R[\rho;\, z]
$$

$$
\equiv \sum_{r=0}^{p+q} \sum_{P \in \mathcal{M}_r} P(z)\, T_P[\rho;\, z],
$$

where, for $r \geq 0$ and P in $\mathcal{M}_r$, T_P is the polynomial in $\mathfrak{T}$ defined by

$$
(20.44) \quad T_P \equiv \sum_{s=0}^{r} \sum_{\substack{Q \in \mathcal{M}_s \\ R \in \mathcal{M}_{r-s} \\ QR = P}} U_Q\, V_R.
$$

For $r \geq 0$, suppose that $0 \leq s \leq r$, $Q \in \mathcal{M}_s$, $R \in \mathcal{M}_{r-s}$, and $P = QR$. If $s > p$ or $r - s > q$, then $U_Q\, V_R \equiv 0$. If $s \leq p$ and $r - s \leq q$, then $U_Q \in \mathfrak{T}_{p-s}$, $V_R \in \mathfrak{T}_{q-r+s}$, and $U_Q\, V_R \in \mathfrak{T}_{p+q-r}$. Thus, for $0 \leq r \leq p+q$ and $P \in \mathcal{M}_r$, we have $T_P \in \mathfrak{T}_{p+q-r}$.

We note that (20.44) reduces to

$$
(20.45) \quad T_P \equiv \sum_{\substack{Q \in \mathcal{M}_p \\ R \in \mathcal{M}_q \\ QR = P}} U_Q\, V_R, \quad \text{when } P \in \mathcal{M}_{p+q}.
$$

Namely, with reference to (20.44) for $r = p + q$, $0 \leq s \leq p + q$, and $s \neq p$, we see that either $s > p$ and $U_Q \equiv 0$ or $s < p$, $p + q - s > q$, and $V_R \equiv 0$.

We use (20.32)–(20.33), (20.41)–(20.42), and (20.45) to obtain

$$(20.46) \quad G(z)\,H(z) \equiv \left[\sum_{\boldsymbol{Q}\in\mathcal{M}_p} Q(z)\,\boldsymbol{U_Q}[\rho;z]\right]\left[\sum_{\boldsymbol{R}\in\mathcal{M}_q} R(z)\,\boldsymbol{V_R}[\rho;z]\right]$$

$$\equiv \sum_{\substack{\boldsymbol{Q}\in\mathcal{M}_p \\ \boldsymbol{R}\in\mathcal{M}_q \\ \boldsymbol{P}\in\mathcal{M}_{p+q} \\ \boldsymbol{P}=\boldsymbol{QR}}} Q(z)\,R(z)\,\boldsymbol{U_Q}[\rho;z]\,\boldsymbol{V_R}[\rho;z]$$

$$\equiv \sum_{\boldsymbol{P}\in\mathcal{M}_{p+q}} P(z) \sum_{\substack{\boldsymbol{Q}\in\mathcal{M}_p \\ \boldsymbol{R}\in\mathcal{M}_q \\ \boldsymbol{QR}=\boldsymbol{P}}} \boldsymbol{U_Q}[\rho;z]\,\boldsymbol{V_R}[\rho;z] \equiv \sum_{\boldsymbol{P}\in\mathcal{M}_{p+q}} P(z)\,\boldsymbol{T_P}[\rho;z].$$

We combine (20.43) and (20.46) to verify that

$$(20.47) \quad G^*(z)\,H^*(z) \equiv G(z)\,H(z) + \sum_{r=0}^{p+q-1}\sum_{\boldsymbol{P}\in\mathcal{M}_r} P(z)\,\boldsymbol{T_P}[\rho;z].$$

In view of (20.43) and (20.47), the formulas (20.28) and (20.29) are applicable for $\boldsymbol{F} \equiv \boldsymbol{GH}$ with $n = p + q$.

(iv) Parts (i), (ii), and (iii) show that (20.28) and (20.29) are valid whenever $\boldsymbol{F}$ is a monomial in $\mathcal{S}_m$. Suppose that $\boldsymbol{G}$ and $\boldsymbol{H}$ are two isobaric polynomials of the same weight $n \geq 1$ that satisfy the forms (20.32)–(20.33) and (20.41)–(20.42) of (20.28)–(20.29) with $p = q = n$. For $\boldsymbol{G}+\boldsymbol{H}$ to be isobaric of weight n, we also assume that $\boldsymbol{G}+\boldsymbol{H} \not\equiv 0$. To show that (20.28) and (20.29) are satisfied for $\boldsymbol{F} \equiv \boldsymbol{G}+\boldsymbol{H}$, we first note that: for $p = q = n$, (20.32) and (20.41) yield

$$(20.48) \quad G^*(z) + H^*(z) \equiv \sum_{r=0}^{n}\sum_{\boldsymbol{P}\in\mathcal{M}_r} P(z)\,\boldsymbol{U_P}[\rho;z] + \sum_{r=0}^{n}\sum_{\boldsymbol{P}\in\mathcal{M}_r} P(z)\,\boldsymbol{V_P}[\rho;z]$$

$$\equiv \sum_{r=0}^{n}\sum_{\boldsymbol{P}\in\mathcal{M}_r} P(z)\,\boldsymbol{T_P}[\rho;z],$$

where $\boldsymbol{T_P} \equiv \boldsymbol{U_P} + \boldsymbol{V_P}$. For $0 \leq r \leq n$ and $\boldsymbol{P}$ in $\mathcal{M}_r$, we observe that each of $\boldsymbol{U_P}$, $\boldsymbol{V_P}$, and $\boldsymbol{T_P}$ is a polynomial in $\mathfrak{T}_{n-r}$. Also, (20.32)–(20.33) and (20.41)–(20.42) give

$$(20.49) \quad G(z) + H(z) \equiv \sum_{\boldsymbol{P}\in\mathcal{M}_n} P(z)\,\boldsymbol{U_P}[\rho;z] + \sum_{\boldsymbol{P}\in\mathcal{M}_n} P(z)\,\boldsymbol{V_P}[\rho;z]$$

$$\equiv \sum_{\boldsymbol{P}\in\mathcal{M}_r} P(z)\,\boldsymbol{T_P}[\rho;z].$$

We use (20.48) and (20.49) to see that

$$(20.50) \quad G^*(z) + H^*(z) \equiv G(z) + H(z) + \sum_{r=0}^{n-1}\sum_{\boldsymbol{P}\in\mathcal{M}_r} P(z)\,\boldsymbol{T_P}[\rho;z].$$

In view of (20.48) and (20.50), (20.28) and (20.29) are satisfied by $\boldsymbol{F} \equiv \boldsymbol{G}+\boldsymbol{H}$. Consequently, (20.28) and (20.29) are valid for any isobaric polynomial $\boldsymbol{F}$ in $\mathcal{S}_m$. This completes the proof of Proposition 20.10. □

20.5. Rational semi-invariants of the first kind

For each $k \geq 0$, let $\mathcal{H}_k$ denote the subset of $\mathcal{S}_m$ consisting of 0 and all of the isobaric polynomials in $\mathcal{S}_m$ of weight k. Equivalently, we could define $\mathcal{H}_k$ as the set of linear combinations over $\mathbb{Q}$ of the elements in $\mathcal{M}_k$. For any nonzero polynomial $\boldsymbol{F}$ in $\mathcal{S}_m$, there is a unique nonnegative integer p and there are unique polynomials $\boldsymbol{F}_k$ in $\mathcal{H}_k$, for $0 \leq k \leq p$, such that $\boldsymbol{F}_p \not\equiv 0$ and

$$\boldsymbol{F} \equiv \sum_{k=0}^{p} \boldsymbol{F}_k. \tag{20.51}$$

Also, for any nonzero polynomial $\boldsymbol{G}$ in $\mathcal{S}_m$, there is a unique nonnegative integer q and there are unique polynomials $\boldsymbol{G}_k$ in $\mathcal{H}_k$, for $0 \leq k \leq q$, such that $\boldsymbol{G}_q \not\equiv 0$ and

$$\boldsymbol{G} \equiv \sum_{k=0}^{q} \boldsymbol{G}_k. \tag{20.52}$$

Thus, any nonzero element $\boldsymbol{A}$ in $\mathcal{Q}_m$ is given by

$$\boldsymbol{A} \equiv \frac{\boldsymbol{F}}{\boldsymbol{G}}, \tag{20.53}$$

for some $\boldsymbol{F}$ and $\boldsymbol{G}$ in $\mathcal{S}_m$ represented in the form (20.51) and (20.52). In particular, when $\boldsymbol{A}$ is a rational semi-invariant of the first kind, it is nonconstant and at least one of the integers p and q must be positive.

THEOREM 20.11. *Suppose that $\boldsymbol{A}$ is a rational semi-invariant of the first kind for the equations (1.1) having order m and let it be given by (20.53) in terms of relatively prime polynomials $\boldsymbol{F}$ and $\boldsymbol{G}$ from $\mathcal{S}_m$ that are represented by (20.51) and (20.52). Then, each nonconstant element of the list*

$$\boldsymbol{F}, \boldsymbol{G}, \boldsymbol{F}_0, \boldsymbol{F}_1, \ldots, \boldsymbol{F}_p, \boldsymbol{G}_0, \boldsymbol{G}_1, \ldots, \boldsymbol{G}_q \tag{20.54}$$

is a polynomial semi-invariant of the first kind for the equations (1.1).

PROOF. If $0 \leq k \leq p$ and $\boldsymbol{F}_k \not\equiv 0$, then Proposition 20.10 shows that, whenever $0 \leq r \leq k-1$ and $\boldsymbol{P} \in \mathcal{M}_r$, there is a corresponding $(\boldsymbol{T}_k)_{\boldsymbol{P}}$ in $\mathcal{T}_{k-r}$ such that: for any (1.1) and any transformation (1.2) of (1.1) into (1.3),

$$F_k^*(z) \equiv F_k(z) + \sum_{r=0}^{k-1} \sum_{\boldsymbol{P} \in \mathcal{M}_r} P(z) \, (\boldsymbol{T}_k)_{\boldsymbol{P}}[\rho; z], \quad \text{on } \Omega. \tag{20.55}$$

If $0 \leq k \leq p$ and $\boldsymbol{F}_k \equiv 0$, then we set $(\boldsymbol{T}_k)_{\boldsymbol{P}} \equiv 0$, for each $\boldsymbol{P}$, and see that (20.55) is also valid for that situation. Similarly, for $0 \leq k \leq q$, Proposition 20.10 shows that, whenever $0 \leq r \leq k-1$ and $\boldsymbol{P} \in \mathcal{M}_r$, there is a corresponding $(\boldsymbol{U}_k)_{\boldsymbol{P}}$ in $\mathcal{T}_{k-r}$ such that: for any (1.1) and any transformation (1.2) of (1.1) into (1.3),

$$G_k^*(z) \equiv G_k(z) + \sum_{r=0}^{k-1} \sum_{\boldsymbol{P} \in \mathcal{M}_r} P(z) \, (\boldsymbol{U}_k)_{\boldsymbol{P}}[\rho; z], \quad \text{on } \Omega. \tag{20.56}$$

We apply (20.51) and (20.55) to obtain

$$F^*(z) \equiv \sum_{k=0}^{p} F_k^*(z) \tag{20.57}$$

and

$$
(20.58) \quad F^*(z) \equiv \sum_{k=0}^{p}\left[F_k(z)+\sum_{r=0}^{k-1}\sum_{\boldsymbol{P}\in\mathcal{M}_r} P(z)\left(\boldsymbol{T}_k\right)_{\boldsymbol{P}}[\rho;z]\right]
$$

$$
\equiv F(z)+\sum_{k=0}^{p}\sum_{r=0}^{k-1}\sum_{\boldsymbol{P}\in\mathcal{M}_r} P(z)\left(\boldsymbol{T}_k\right)_{\boldsymbol{P}}[\rho;z]
$$

Similarly, we find that (20.52) and (20.56) yield

$$
(20.59) \quad G^*(z) \equiv \sum_{k=0}^{q} G_k^*(z)
$$

and

$$
(20.60) \quad G^*(z) \equiv G(z)+\sum_{k=0}^{q}\sum_{r=0}^{k-1}\sum_{\boldsymbol{P}\in\mathcal{M}_r} P(z)\left(\boldsymbol{U}_k\right)_{\boldsymbol{P}}[\rho;z].
$$

For any (1.1) and (1.2) such that $G(z)\not\equiv 0$ and $G^*(z)\not\equiv 0$, we use (20.58), (20.60), and the condition

$$
\frac{F^*(z)}{G^*(z)} \equiv \frac{F(z)}{G(z)}, \quad \text{on } \Omega,
$$

for $\boldsymbol{A}$ from (20.1) on page 220 to deduce that

$$
(20.61) \quad \frac{F(z)+\sum_{k=0}^{p}\sum_{r=0}^{k-1}\sum_{\boldsymbol{P}\in\mathcal{M}_r} P(z)\left(\boldsymbol{T}_k\right)_{\boldsymbol{P}}[\rho;z]}{G(z)+\sum_{k=0}^{q}\sum_{r=0}^{k-1}\sum_{\boldsymbol{P}\in\mathcal{M}_r} P(z)\left(\boldsymbol{U}_k\right)_{\boldsymbol{P}}[\rho;z]} \equiv \frac{F(z)}{G(z)}.
$$

Next, we shall employ (20.61) to prove that $\boldsymbol{M}_1\equiv 0$, where $\boldsymbol{M}_1$ is the polynomial

$$
(20.62) \quad \boldsymbol{M}_1 \equiv \left[\boldsymbol{F}+\sum_{k=0}^{p}\sum_{r=0}^{k-1}\sum_{\boldsymbol{P}\in\mathcal{M}_r} P\left(\boldsymbol{T}_k\right)_{\boldsymbol{P}}\right]\boldsymbol{G} - \left[\boldsymbol{G}+\sum_{k=0}^{q}\sum_{r=0}^{k-1}\sum_{\boldsymbol{P}\in\mathcal{M}_r} P\left(\boldsymbol{U}_k\right)_{\boldsymbol{P}}\right]\boldsymbol{F}
$$

over $\mathbb{Q}$ in the variables $\boldsymbol{w}_{i,j}^{(k)}$ and $\boldsymbol{t}_l$, for $1\leq j\leq m$, $0\leq i\leq j$, $k\geq 0$, and $l\geq 1$.

To establish that $\boldsymbol{M}_1\equiv 0$, suppose that $\boldsymbol{M}_1\not\equiv 0$. Since there are equations (1.3) satisfying $G^*(z)\not\equiv 0$, we use (20.60) to see that the polynomial

$$
(20.63) \quad \boldsymbol{D}_1 \equiv \boldsymbol{G}+\sum_{k=0}^{q}\sum_{r=0}^{k-1}\sum_{\boldsymbol{P}\in\mathcal{M}_r} P\left(\boldsymbol{U}_k\right)_{\boldsymbol{P}}
$$

satisfies $\boldsymbol{D}_1\not\equiv 0$. Also, we have $\boldsymbol{G}\not\equiv 0$. In particular, $\boldsymbol{D}_1$, $\boldsymbol{G}$, and $\boldsymbol{M}_1$ are elements of the polynomial ring $\mathfrak{S}$ introduced on page 223 just after (20.24). Let p_1, p_2, p_3 be nonnegative integers such that the elements $(\boldsymbol{x}_0^{(0)})^{p_1}\boldsymbol{D}_1$, $(\boldsymbol{x}_0^{(0)})^{p_2}\boldsymbol{G}$, and $(\boldsymbol{x}_0^{(0)})^{p_3}\boldsymbol{M}_1$ can be rewritten as polynomials over $\mathbb{Q}$ in the variables $\boldsymbol{w}_{i,j}^{(k)}$ and $\boldsymbol{x}_0^{(l)}$, for $1\leq j\leq m$, $0\leq i\leq j$, $k\geq 0$, and $l\geq 0$. Hence, a polynomial over $\mathbb{Q}$ in the variables $\boldsymbol{w}_{i,j}^{(k)}$ and $\boldsymbol{x}_0^{(l)}$, for $1\leq j\leq m$, $0\leq i\leq j$, $k\geq 0$, and $l\geq 0$ is defined by

$$
(20.64) \quad \boldsymbol{N}_1 \equiv \boldsymbol{x}_0^{(0)}\left[(\boldsymbol{x}_0^{(0)})^{p_1}\boldsymbol{D}_1\right]\left[(\boldsymbol{x}_0^{(0)})^{p_2}\boldsymbol{G}\right]\left[(\boldsymbol{x}_0^{(0)})^{p_3}\boldsymbol{M}_1\right]
$$

and it satisfies $\boldsymbol{N}_1 \not\equiv 0$. We set $\nu = \dfrac{m(m+3)}{2} + 1$ and observe that there are $\nu - 1$ variables $w_{i,j}^{(0)}$ having $1 \le j \le m$ and $0 \le i \le j$. We identify these $\nu - 1$ variables with the first $\nu-1$ variables $\mathfrak{s}_1^{(0)}, \mathfrak{s}_2^{(0)}, \ldots, \mathfrak{s}_{\nu-1}^{(0)}$ in (4.2) on page 35. And, we identify $x_0^{(0)}$ with $\mathfrak{s}_\nu^{(0)}$ in (4.2). Then, $\boldsymbol{N}_1$ belongs to the ring $\mathfrak{S}_\nu$ of polynomials over $\mathbb{Q}$ in the variables $\mathfrak{s}_h^{(k)}$ having $1 \le h \le \nu$ and $k \ge 0$. Consequently, Proposition 4.2 on page 36 is applicable and yields meromorphic functions $d_1(z), \ldots, d_\nu(z)$ on a region Ω such that the function $N_1(z)$ on Ω obtained by substituting $d_h^{(k)}(z)$ for $\mathfrak{s}_h^{(k)}$ in $\boldsymbol{N}_1$ satisfies $N_1(z) \not\equiv 0$. With a change of notation, this gives meromorphic functions $c_{i,j}(z)$ and $\rho(z)$ on Ω, for $1 \le j \le m$ and $0 \le i \le j$, such that the substitution in $\boldsymbol{N}_1$ of $c_{i,j}^{(k)}(z)$ for $w_{i,j}^{(k)}$ and $\rho^{(k)}(z)$ for $x_0^{(k)}$ yields the function $N_1(z)$ subject to $N_1(z) \not\equiv 0$. Since (20.64) shows that $x_0^{(0)}$ is a factor of $\boldsymbol{N}_1$, we have $\rho(z) \not\equiv 0$. For $1 \le j \le m$, $0 \le i \le j$, $k \ge 0$, and $l \ge 1$, we substitute $c_{i,j}^{(k)}(z)$ for $w_{i,j}^{(k)}$ and $\rho^{(l)}(z)/\rho(z)$ for t_l in the polynomials $\boldsymbol{D}_1$, $\boldsymbol{G}$, and $\boldsymbol{M}_1$ of the polynomial ring $\mathfrak{S}$; and we use (20.64), (20.63), and (20.62), as well as our usual notation to see that the preceding application of Proposition 4.2 to $\boldsymbol{N}_1$ gives

$$(20.65) \qquad G(z) + \sum_{k=0}^{q} \sum_{r=0}^{k-1} \sum_{\boldsymbol{P} \in \mathcal{M}_r} P(z)\, (U_k)_{\boldsymbol{P}}[\rho; z] \not\equiv 0, \quad G(z) \not\equiv 0,$$

and

$$(20.66) \qquad \left[F(z) + \sum_{k=0}^{p} \sum_{r=0}^{k-1} \sum_{\boldsymbol{P} \in \mathcal{M}_r} P(z)\, (T_k)_{\boldsymbol{P}}[\rho; z] \right] G(z)$$

$$- \left[G(z) + \sum_{k=0}^{q} \sum_{r=0}^{k-1} \sum_{\boldsymbol{P} \in \mathcal{M}_r} P(z)\, (U_k)_{\boldsymbol{P}}[\rho; z] \right] F(z) \not\equiv 0.$$

Let (1.1) have the coefficients $c_{i,j}(z)$ on Ω specified by Proposition 4.2 for $\boldsymbol{N}_1$ along with $c_{j,i}(z) \equiv c_{i,j}(z)$ when $0 \le j < i \le m$; and let (1.2) be the transformation for this (1.1) in which $\rho(z)$ on Ω is also given by Proposition 4.2 for $\boldsymbol{N}_1$. Due to (20.65), this (1.1) and this transformation (1.2) of this (1.1) into a corresponding (1.3) give $G(z) \not\equiv 0$ and $G^*(z) \not\equiv 0$ via (20.60). Hence, they must yield (20.61). However, (20.66) gives the contradiction that (20.61) is not satisfied. Consequently, for $\boldsymbol{M}_1$ in (20.62), we must $\boldsymbol{M}_1 \equiv 0$.

We rewrite (20.62) with $\boldsymbol{M}_1 \equiv 0$ to obtain

$$(20.67) \qquad \boldsymbol{F} \left[\sum_{k=0}^{q} \sum_{r=0}^{k-1} \sum_{\boldsymbol{P} \in \mathcal{M}_r} \boldsymbol{P}\, (U_k)_{\boldsymbol{P}} \right] \equiv \boldsymbol{G} \left[\sum_{k=0}^{p} \sum_{r=0}^{k-1} \sum_{\boldsymbol{P} \in \mathcal{M}_r} \boldsymbol{P}\, (T_k)_{\boldsymbol{P}} \right],$$

as an identity in $\mathfrak{S}$. The ring $\mathfrak{S}$ of polynomials over $\mathbb{Q}$ in the variables $w_{i,j}^{(k)}$ and t_l, for $1 \le j \le m$, $0 \le i \le j$, $k \ge 0$, and $l \ge 1$, has the subring $\mathcal{S}_m$ of polynomials over $\mathbb{Q}$ in the variables $w_{i,j}^{(k)}$, for $1 \le j \le m$, $0 \le i \le j$, and $j \ge 0$. Also, $\mathfrak{S}$ contains the subring $\mathfrak{T}$ of polynomials over $\mathbb{Q}$ in the variables t_l, for $l \ge 1$. As a polynomial ring over a field, $\mathfrak{S}$ is a unique factorization domain (i.e., $\mathfrak{S}$ is factorial); e.g., see [44]. Since $\boldsymbol{F}$ and $\boldsymbol{G}$ are relatively prime polynomials in the subring $\mathcal{S}_m$ of $\mathfrak{S}$, $\boldsymbol{F}$ and $\boldsymbol{G}$ are relatively prime in $\mathfrak{S}$. Thus, (20.67) shows that $\boldsymbol{F}$ divides the polynomial

$$(20.68) \quad K \equiv \sum_{k=0}^{p} \sum_{r=0}^{k-1} \sum_{P \in \mathcal{M}_r} P\left(T_k\right)_P \equiv \sum_{r=0}^{p-1} \sum_{P \in \mathcal{M}_r} \left[\sum_{k=r+1}^{p} \left(T_k\right)_P\right] P.$$

If $p = 0$, then the sum in (20.68) is vacuous and gives $K \equiv 0$. Suppose that $p \geq 1$. In this situation, K is a linear combination over $\mathfrak{T}$ of monic monomials P from $\mathcal{S}_m$ whose weights do not exceed $p-1$ while F effectively involves a monomial from $\mathcal{S}_m$ of weight p. Hence, the divisibility of K by F requires that $K \equiv 0$. We note that K in (20.68) is a linear combination over $\mathfrak{T}$ of the distinct monomials P in $\mathcal{M}_r$, for $0 \leq r \leq p-1$. Since the variables in (20.23) are algebraically independent over $\mathbb{Q}$, we use $K \equiv 0$ to deduce that the coefficient of each P in (20.68) is identically zero. Thus, we have

$$(20.69) \quad \sum_{k=r+1}^{p} \left(T_k\right)_P \equiv 0, \quad \text{for } 0 \leq r \leq p-1 \text{ and } P \in \mathcal{M}_r.$$

For a fixed integer r satisfying $0 \leq r \leq p-1$ and a fixed $P \in \mathcal{M}_r$, the summands in the left member of (20.69) satisfy $\left(T_k\right)_P \in \mathfrak{T}_{k-r}$, for $r+1 \leq k \leq p$. We recall from page 223 that, for $s \geq 0$, $\mathfrak{T}_s$ consists of 0 and the isobaric polynomials in $\mathfrak{T}$ of weight s. Therefore, we deduce from (20.69) that

$$(20.70) \quad \left(T_k\right)_P \equiv 0, \quad \text{for } 0 \leq r \leq p-1, \ P \in \mathcal{M}_r, \text{ and } r+1 \leq k \leq p.$$

We see that (20.70) is equivalent to

$$(20.71) \quad \left(T_k\right)_P \equiv 0, \quad \text{for } 1 \leq k \leq p, \ 0 \leq r \leq k-1, \text{ and } P \in \mathcal{M}_r.$$

A similar argument applied to G, q, and G_k in place of F, p, and F_k yields

$$(20.72) \quad \left(U_k\right)_P \equiv 0, \quad \text{for } 1 \leq k \leq q, \ 0 \leq r \leq k-1, \text{ and } P \in \mathcal{M}_r.$$

For any (1.1) on Ω and any transformation (1.2) of (1.1) into a corresponding (1.3) on Ω, we combine (20.71) with (20.55) to obtain

$$(20.73) \quad F_k^*(z) \equiv F_k(z), \quad \text{on } \Omega, \text{ for } 0 \leq k \leq p.$$

For the same situation, (20.72) and (20.56) yield

$$(20.74) \quad G_k^*(z) \equiv G_k(z), \quad \text{on } \Omega, \text{ for } 0 \leq k \leq q.$$

We use (20.57), (20.73), and (20.51) to deduce that

$$(20.75) \quad F^*(z) \equiv F(z).$$

Also, (20.59), (20.74), and (20.52) give

$$(20.76) \quad G^*(z) \equiv G(z).$$

In view of (20.75), (20.76), (20.73), and (20.74), we conclude that each nonconstant polynomial in the list (20.54) is a polynomial semi-invariant of the first kind. This completes the proof of Theorem 20.11. $\square$

We recall that any nonconstant polynomial F in $\mathcal{S}_m$ has a unique expression

$$(20.77) \quad F \equiv F_0 + F_1 + \cdots + F_p$$

as a sum of polynomials $F_0, F_1, \ldots, F_p$ in $\mathcal{S}_m$ such that: $p \geq 1$, $F_p \neq 0$, and each nonzero F_k is isobaric of weight k.

COROLLARY 20.12. *Suppose that F in $\mathcal{S}_m$ is a polynomial semi-invariant of the first kind for the equations (1.1) of order m and let F have the representation (20.77). Then, $p \geq 2$; $F_1 \equiv 0$; and, for $2 \leq k \leq p$, each nonzero F_k is an isobaric semi-invariant of the first kind of weight k for the equations (1.1).*

PROOF. Setting $G \equiv 1$ and $A \equiv F/G$, we apply Theorem 20.11 to conclude that each nonconstant element of the list $F_0, F_1, \ldots F_p$ is a semi-invariant of the first kind. Remark 3.6 on page 34 shows that there are no isobaric semi-invariants of the first kind of weight 1. This gives $F_1 \equiv 0$. Since F is nonconstant, we must have $p \geq 2$. This completes the proof. □

20.6. A technical construction needed for Section 20.7

The context employed in the formulation of the following result is explained in detail on pages 223–224 of Section 20.3.

PROPOSITION 20.13. *Let F be an isobaric polynomial in $\mathcal{S}_m$ of weight $n \geq 0$. Then, for each integer r satisfying $0 \leq r \leq n$ and each monic monomial P in $\mathcal{M}_r$, there is a corresponding polynomial X_P in $\mathfrak{T}_{n-r}$ such that the following is true. For any (1.1) on Ω and any univalent analytic function $\zeta = g(z)$ on Ω whose inverse $z = f(\zeta)$ specifies a transformation (1.4) of (1.1) on Ω into a corresponding (1.5) on $\Omega^{**} = g(\Omega)$, the relations*

$$(20.78) \qquad (g'(z))^n F^{**}(g(z)) \equiv \sum_{r=0}^{n} \sum_{P \in \mathcal{M}_r} P(z) X_P[g'; z]$$

and

$$(20.79) \qquad (g'(z))^n F^{**}(g(z)) \equiv F(z) + \sum_{r=0}^{n-1} \sum_{P \in \mathcal{M}_r} P(z) X_P[g'; z]$$

are satisfied on Ω where: $F^{**}(\zeta)$ is the function on Ω^{**} obtained by replacing each $w_{i,j}^{(k)}$ in F with $c_{i,j}^{**(k)}(\zeta)$ from (1.5); $F(z)$ and $P(z)$ on Ω are obtained from F and P by replacing each $w_{i,j}^{(k)}$ in F and P with $c_{i,j}^{(k)}(z)$ from (1.1); and $X_P[g'; z]$ is the function on Ω obtained by replacing t_l in X_P with $g^{(l+1)}(z)/g'(z)$, for $l \geq 1$.

PROOF. For $n = 0$, F is a rational number $\gamma \neq 0$. Then, (20.78)–(20.79) are valid with $X_P \equiv \gamma$ for $P \equiv w_{0,0}^{(0)} \equiv 1$ in $\mathcal{M}_0$. Thus, we may assume that $n \geq 1$.

(i) Suppose that $F \equiv \gamma w_{i,j}$, for $\gamma \neq 0$ in $\mathbb{Q}$, $1 \leq j \leq m$, and $0 \leq i \leq j$. Then, we use (3.12) on page 32 as well as $c_{0,0}(z) \equiv 1$, $n = i+j$, and $f'(g(z)) \equiv 1/g'(z)$ to obtain

$$(20.80) \quad (g'(z))^n \gamma c_{i,j}^{**}(g(z)) \equiv \gamma \sum_{\mu=0}^{i} \sum_{\nu=0}^{j} \beta_{i-\mu,\mu}(z) \beta_{j-\nu,\nu}(z) c_{\mu,\nu}(z)$$

$$\equiv \sum_{r=0}^{n} \left[\sum_{\substack{\mu+\nu=r \\ 0 \leq \mu < \nu \leq i}} e_{\mu,\nu}(z) + \sum_{\substack{\mu+\nu=r \\ 0 \leq \nu < \mu \leq i}} e_{\mu,\nu}(z) + \sum_{\substack{\mu+\nu=r \\ 0 \leq \mu = \nu \leq i}} e_{\mu,\nu}(z) + \sum_{\substack{\mu+\nu=r \\ 0 \leq \mu \leq i \\ i+1 \leq \nu \leq j}} e_{\mu,\nu}(z) \right],$$

where $\beta_{\mu,\nu}(z)$ is defined by (3.13)–(3.14) of page 32 and where

(20.81) $e_{\mu,\nu}(z) \equiv \gamma \, \beta_{i-\mu,\mu}(z) \, \beta_{j-\nu,\nu}(z) \, c_{\mu,\nu}(z)$, for $0 \le \mu \le i$ and $0 \le \nu \le j$.

We define $\boldsymbol{T}_{i,j}$ in the ring $\mathfrak{T}$ of page 223 through

(20.82) $\boldsymbol{T}_{0,j} \equiv 1$, for any j,

and

(20.83) $\boldsymbol{T}_{i+1,j} \equiv \sum_{k=j+1}^{m} \left[(\boldsymbol{T}_{i,k})' + (m-i-k) \, \boldsymbol{t}_1 \, \boldsymbol{T}_{i,k} \right]$, for $i \ge 0$ and any j,

where $'$ is the derivation for $\mathfrak{T}$. Due to (3.13) and (20.82), we see that the formula

(20.84) $\beta_{i,j}(z) \equiv \boldsymbol{T}_{i,j}[g'; z]$

is valid for $i = 0$ and any j. Suppose that i_0 is a fixed nonnegative integer such that (20.84) is valid for $i = i_0$ and any j. We use this with (3.14) on page 32, Lemma 20.9 on page 224, and (20.83) to deduce that: for any j,

$$\beta_{i_0+1,j}(z) \equiv \sum_{k=j+1}^{m} \left\{ \beta_{i_0,k}^{(1)}(z) + (m-i_0-k) \frac{g''(z)}{g'(z)} \beta_{i_0,k}(z) \right\}$$

$$\equiv \sum_{k=j+1}^{m} \left\{ (\boldsymbol{T}_{i_0,k})'[g'; z] + (m-i_0-k) \, \boldsymbol{t}_1[g'; z] \, \boldsymbol{T}_{i_0,k}[g'; z] \right\}$$

$$\equiv \boldsymbol{T}_{i_0+1,j}[g'; z].$$

Thus, (20.84) is valid for $i \ge 0$ and any j. In view of (20.81) and (20.84), we have

(20.85) $e_{\mu,\nu}(z) \equiv \boldsymbol{y}_{\mu,\nu}[g'; z] \, c_{\mu,\nu}(z)$, for $0 \le \mu \le i$ and $0 \le \nu \le j$,

where $\boldsymbol{y}_{\mu,\nu}$ is defined in $\mathfrak{T}$ through

$$\boldsymbol{y}_{\mu,\nu} \equiv \gamma \, \boldsymbol{T}_{i-\mu,\mu} \, \boldsymbol{T}_{j-\nu,\nu}, \quad \text{for } 0 \le \mu \le i \text{ and } 0 \le \nu \le j.$$

When $\mu + \nu = r$, the weight of $\boldsymbol{y}_{\mu,\nu}$ is $i - \mu + j - \nu = n - r$ and therefore $\boldsymbol{y}_{\mu,\nu}$ belongs to $\mathfrak{T}_{n-r}$. For $0 \le r \le n$ and $\boldsymbol{P}$ in $\mathcal{M}_r$, we define $\boldsymbol{X_P}$ in $\mathfrak{T}_{n-r}$ through

$$\boldsymbol{X_P} \equiv \begin{cases} \boldsymbol{y}_{\mu,\nu} + \boldsymbol{y}_{\nu,\mu}, & \text{if } \boldsymbol{P} \equiv \boldsymbol{w}_{\mu,\nu},\ 0 \le \mu < \nu \le i \text{ and } \mu + \nu = r; \\ \boldsymbol{y}_{\mu,\nu}, & \text{if } \boldsymbol{P} \equiv \boldsymbol{w}_{\mu,\nu},\ 0 \le \mu = \nu \le i \text{ and } \mu + \nu = r; \\ \boldsymbol{y}_{\mu,\nu}, & \text{if } \boldsymbol{P} \equiv \boldsymbol{w}_{\mu,\nu},\ 0 \le \mu \le i,\ i+1 \le \nu \le j, \text{ and } \mu + \nu = r; \\ 0, & \text{otherwise.} \end{cases}$$

For $r = n$ and $\mu + \nu = i + j$ in (20.80), there is a single corresponding term; it has $\mu = i$ and $\nu = j$. Thus, (20.80) can be rewritten as

$(g'(z))^n \gamma \, c_{i,j}^{**}(g(z))$

$$\equiv \gamma \, c_{i,j}(z) + \sum_{r=0}^{n-1} \left[\sum_{\substack{\mu+\nu=r \\ 0 \le \mu < \nu \le i}} \left(e_{\mu,\nu}(z) + e_{\nu,\mu}(z) \right) + \sum_{\substack{\mu+\nu=r \\ 0 \le \mu = \nu \le i}} e_{\mu,\nu}(z) + \sum_{\substack{\mu+\nu=r \\ 0 \le \mu \le i \\ i+1 \le \nu \le j}} e_{\mu,\nu}(z) \right].$$

Using (20.85) and the definition of $\boldsymbol{X_P}$, we rewrite (20.80) and the preceding formula to see that (20.78) and (20.79) are valid when $\boldsymbol{F} \equiv \gamma \, \boldsymbol{w}_{i,j}$.

(ii) Suppose that (20.78) and (20.79) are valid when $\boldsymbol{F}$ is a particular isobaric polynomial $\boldsymbol{G}$ of weight $p \geq 1$. We shall show next that (20.78) and (20.79) are valid when $\boldsymbol{F} \equiv \boldsymbol{G}'$ and $n = p+1$. To do so, we begin with

$$(20.86) \qquad (g'(z))^p \, G^{**}(g(z)) \equiv \sum_{s=0}^{p} \sum_{\boldsymbol{Q} \in \mathcal{M}_s} Q(z) \, \boldsymbol{Y_Q}[g';z]$$

and

$$(20.87) \qquad (g'(z))^p \, G^{**}(g(z)) \equiv G(z) + \sum_{s=0}^{p-1} \sum_{\boldsymbol{Q} \in \mathcal{M}_s} Q(z) \, \boldsymbol{Y_Q}[g';z].$$

Replacing s in (20.86) with r and differentiating the resulting identity with respect to z, we find that

$$(g'(z))^{p+1} \, G^{**(1)}(g(z)) + p \frac{g''(z)}{g'(z)} (g'(z))^p \, G^{**}(g(z))$$
$$\equiv \sum_{r=0}^{p} \sum_{\boldsymbol{Q} \in \mathcal{M}_r} \left(\frac{d}{dz} Q(z)\right) \boldsymbol{Y_Q}[g';z] + \sum_{r=0}^{p} \sum_{\boldsymbol{Q} \in \mathcal{M}_r} Q(z) \frac{d}{dz}\left(\boldsymbol{Y_Q}[g';z]\right).$$

To this expression we apply (20.86) and (20.27) in order to deduce that

$$(20.88) \qquad (g'(z))^{p+1} \, G^{**(1)}(g(z)) \equiv \mathcal{H}_1(z) + \mathcal{H}_2(z) + \mathcal{H}_3(z),$$

where

$$(20.89) \qquad \mathcal{H}_1(z) \equiv -p \, \frac{g''(z)}{g'(z)} \sum_{r=0}^{p} \sum_{\boldsymbol{Q} \in \mathcal{M}_r} Q(z) \, \boldsymbol{Y_Q}[g';z],$$

$$(20.90) \qquad \mathcal{H}_2(z) \equiv \sum_{r=0}^{p} \sum_{\boldsymbol{Q} \in \mathcal{M}_r} \left(\frac{d}{dz} Q(z)\right) \boldsymbol{Y_Q}[g';z],$$

and

$$(20.91) \qquad \mathcal{H}_3(z) \equiv \sum_{r=0}^{p} \sum_{\boldsymbol{Q} \in \mathcal{M}_r} Q(z) \, \boldsymbol{Y'_Q}[g';z].$$

Using $\boldsymbol{t}_1[g';z] \equiv g''(z)/g'(z)$, we rewrite (20.89) as

$$(20.92) \qquad \mathcal{H}_1(z) \equiv \sum_{r=0}^{p} \sum_{\boldsymbol{P} \in \mathcal{M}_r} P(z) \left(-p \, \boldsymbol{t}_1[g';z]\right) \boldsymbol{Y_P}[g';z].$$

In (20.90), we apply (4.13) on page 38 to interpret $\dfrac{d}{dz} Q(z)$ as the function on Ω obtained by substituting $c_{\mu,\nu}^{(\kappa)}(z)$ from (1.1) for $w_{\mu,\nu}^{(\kappa)}$ in $\boldsymbol{Q}'$. For each $r \geq 0$ and each $\boldsymbol{Q}$ in $\mathcal{M}_r$, there are unique rational numbers $\gamma_{\boldsymbol{P},\boldsymbol{Q}}$, for each $\boldsymbol{P}$ in $\mathcal{M}_{r+1}$, such that

$$(20.93) \qquad \boldsymbol{Q}' \equiv \sum_{\boldsymbol{P} \in \mathcal{M}_{r+1}} \gamma_{\boldsymbol{P},\boldsymbol{Q}} \, \boldsymbol{P}.$$

We find that (20.90) and (20.93) give

$$
(20.94) \quad \mathcal{H}_2(z) \equiv \sum_{r=0}^{p} \sum_{\boldsymbol{Q} \in \mathcal{M}_r} \sum_{\boldsymbol{P} \in \mathcal{M}_{r+1}} \gamma_{\boldsymbol{P},\boldsymbol{Q}}\, P(z)\, \boldsymbol{Y_Q}[g';z]
$$

$$
\equiv \sum_{r=0}^{p} \sum_{\boldsymbol{P} \in \mathcal{M}_{r+1}} P(z) \sum_{\boldsymbol{Q} \in \mathcal{M}_r} \gamma_{\boldsymbol{P},\boldsymbol{Q}}\, \boldsymbol{Y_Q}[g';z]
$$

$$
\equiv \sum_{r=1}^{p+1} \sum_{\boldsymbol{P} \in \mathcal{M}_r} P(z) \sum_{\boldsymbol{Q} \in \mathcal{M}_{r-1}} \gamma_{\boldsymbol{P},\boldsymbol{Q}}\, \boldsymbol{Y_Q}[g';z].
$$

A rewriting of (20.91) yields

$$
(20.95) \quad \mathcal{H}_3(z) \equiv \sum_{r=0}^{p} \sum_{\boldsymbol{P} \in \mathcal{M}_r} P(z)\, \boldsymbol{Y'_P}[g';z].
$$

We combine (20.88) with (20.92), (20.94), and (20.95) to verify that

$$
(20.96) \quad (g'(z))^{p+1}\, G^{**(1)}(g(z)) \equiv \sum_{r=0}^{p+1} \sum_{\boldsymbol{P} \in \mathcal{M}_r} P(z)\, \boldsymbol{X_P}[g';z],
$$

where: for $0 \le r \le p+1$ and each $\boldsymbol{P}$ in $\mathcal{M}_r$, $\boldsymbol{X_P}$ is defined in $\mathfrak{T}_{p+1-r}$ by

$$
(20.97) \quad \boldsymbol{X_P} = \begin{cases} \left[\boldsymbol{Y'_P} - p\, t_1 \boldsymbol{Y_P}\right], & \text{if } \boldsymbol{P} \in \mathcal{M}_0; \\[1em] \left[\begin{array}{l} \boldsymbol{Y'_P} - p\, t_1 \boldsymbol{Y_P} \\ + \displaystyle\sum_{\boldsymbol{Q} \in \mathcal{M}_{r-1}} \gamma_{\boldsymbol{P},\boldsymbol{Q}} \boldsymbol{Y_Q} \end{array}\right], & \text{if } \boldsymbol{P} \in \mathcal{M}_r \text{ and } 1 \le r \le p; \\[1em] \left[\displaystyle\sum_{\boldsymbol{Q} \in \mathcal{M}_p} \gamma_{\boldsymbol{P},\boldsymbol{Q}}\, \boldsymbol{Y_Q}\right], & \text{if } \boldsymbol{P} \in \mathcal{M}_{p+1}. \end{cases}
$$

Employing (20.86) and (20.87), we deduce that

$$
(20.98) \quad G(z) \equiv \sum_{\boldsymbol{Q} \in \mathcal{M}_p} Q(z)\, \boldsymbol{Y_Q}[g';z].
$$

For each $\boldsymbol{Q}$ in $\mathcal{M}_p$, $\boldsymbol{Y_Q}$ belongs to $\mathfrak{T}_0$ and therefore satisfies $\boldsymbol{Y'_Q} \equiv 0$. We differentiate (20.98) with respect to z and use (4.13) on page 38, (20.27), as well as the preceding observation to see that

$$
(20.99) \quad G^{(1)}(z) \equiv \sum_{\boldsymbol{Q} \in \mathcal{M}_p} Q'(z)\, \boldsymbol{Y_Q}[g';z].
$$

We combine (20.99) with (20.93) for $r = p$ and (20.97) to obtain

$$
(20.100) \quad G^{(1)}(z) \equiv \sum_{\boldsymbol{Q} \in \mathcal{M}_p} \sum_{\boldsymbol{P} \in \mathcal{M}_{p+1}} \gamma_{\boldsymbol{P},\boldsymbol{Q}}\, P(z)\, \boldsymbol{Y_Q}[g';z]
$$

$$
\equiv \sum_{\boldsymbol{P} \in \mathcal{M}_{p+1}} P(z) \sum_{\boldsymbol{Q} \in \mathcal{M}_p} \gamma_{\boldsymbol{P},\boldsymbol{Q}}\, \boldsymbol{Y_Q}[g';z]
$$

$$
\equiv \sum_{\boldsymbol{P} \in \mathcal{M}_{p+1}} P(z)\, \boldsymbol{X_P}[g';z].
$$

Thus, (20.96) and (20.100) yield

$$(20.101) \qquad \left(g'(z)\right)^{p+1} G^{**(1)}\bigl(g(z)\bigr) \equiv G^{(1)}(z) + \sum_{r=0}^{p} \sum_{\boldsymbol{P}\in\mathcal{M}_r} P(z)\, \boldsymbol{X_P}[g';z].$$

In view of (20.96) and (20.101), (20.78)–(20.79) are valid for $\boldsymbol{F} \equiv \boldsymbol{G}'$ with $n = p+1$.

(iii) Suppose that $\boldsymbol{G}$ and $\boldsymbol{H}$ are isobaric polynomials in $\mathcal{S}_m$ of respective weights $p \geq 1$ and $q \geq 1$ for which (20.78) and (20.79) are satisfied when $\boldsymbol{F} \equiv \boldsymbol{G}$ and $n = p$ as well as when $\boldsymbol{F} \equiv \boldsymbol{H}$ and $n = q$. In order to prove that (20.78) and (20.79) are valid when $\boldsymbol{F} \equiv \boldsymbol{GH}$ with $n = p+q$, we assume that the specific formulas for $\boldsymbol{G}$ are (20.86)–(20.87) and the specific formulas for $\boldsymbol{H}$ are

$$(20.102) \qquad \left(g'(z)\right)^{q} H^{**}\bigl(g(z)\bigr) \equiv \sum_{t=0}^{q} \sum_{\boldsymbol{R}\in\mathcal{M}_t} R(z)\, \boldsymbol{Z_R}[g';z]$$

and

$$(20.103) \qquad \left(g'(z)\right)^{q} H^{**}\bigl(g(z)\bigr) \equiv H(z) + \sum_{t=0}^{q-1} \sum_{\boldsymbol{R}\in\mathcal{M}_t} R(z)\, \boldsymbol{Z_R}[g';z].$$

We extend the definitions of $\boldsymbol{Y_Q}$ and $\boldsymbol{Z_R}$ by setting $\boldsymbol{Y_Q} \equiv 0$, for $\boldsymbol{Q} \in \mathcal{M}_s$ and $s > p$, as well as $\boldsymbol{Z_R} \equiv 0$, for $\boldsymbol{R} \in \mathcal{M}_t$ and $t > q$. Thus, we have

$$(20.104) \quad \left(g'(z)\right)^{p} G^{**}\bigl(g(z)\bigr) \equiv \sum_{s\geq 0} \mathfrak{A}_s(z) \quad \text{and} \quad \left(g'(z)\right)^{q} H^{**}\bigl(g(z)\bigr) \equiv \sum_{t\geq 0} \mathfrak{B}_t(z),$$

where

$$(20.105) \qquad \mathfrak{A}_s(z) \equiv \sum_{\boldsymbol{Q}\in\mathcal{M}_s} Q(z)\, \boldsymbol{Y_Q}[g';z], \quad \text{for } s \geq 0,$$

and

$$(20.106) \qquad \mathfrak{B}_t(z) \equiv \sum_{\boldsymbol{R}\in\mathcal{M}_t} R(z)\, \boldsymbol{Z_R}[g';z], \quad \text{for } t \geq 0.$$

We use (20.104)–(20.106) to obtain

$$\left(g'(z)\right)^{p+q} G^{**}\bigl(g(z)\bigr) H^{**}\bigl(g(z)\bigr) \equiv \left[\sum_{s\geq 0}\mathfrak{A}_s(z)\right]\left[\sum_{t\geq 0}\mathfrak{B}_t(z)\right]$$

$$\equiv \sum_{r=0}^{p+q}\sum_{s=0}^{r} \mathfrak{A}_s(z)\, \mathfrak{B}_{r-s}(z)$$

$$\equiv \sum_{r=0}^{p+q}\sum_{s=0}^{r} \sum_{\substack{\boldsymbol{Q}\in\mathcal{M}_s \\ \boldsymbol{R}\in\mathcal{M}_{r-s}}} Q(z)\, R(z)\, \boldsymbol{Y_Q}[g';z]\, \boldsymbol{Z_R}[g';z]$$

$$\equiv \sum_{r=0}^{p+q}\sum_{s=0}^{r} \sum_{\substack{\boldsymbol{Q}\in\mathcal{M}_s \\ \boldsymbol{R}\in\mathcal{M}_{r-s} \\ \boldsymbol{P}\in\mathcal{M}_r \\ \boldsymbol{P}=\boldsymbol{QR}}} Q(z)\, R(z)\, \boldsymbol{Y_Q}[g';z]\, \boldsymbol{Z_R}[g';z]$$

20.6. A TECHNICAL CONSTRUCTION NEEDED FOR SECTION 20.7

and

(20.107)
$$\left(g'(z)\right)^{p+q} G^{**}\!\left(g(z)\right) H^{**}\!\left(g(z)\right)$$
$$\equiv \sum_{r=0}^{p+q} \sum_{s=0}^{r} \sum_{\boldsymbol{P}\in\mathcal{M}_r} \sum_{\substack{\boldsymbol{Q}\in\mathcal{M}_s \\ \boldsymbol{R}\in\mathcal{M}_{r-s} \\ \boldsymbol{QR}=\boldsymbol{P}}} P(z)\, \boldsymbol{Y_Q}[g';z]\, \boldsymbol{Z_R}[g';z]$$
$$\equiv \sum_{r=0}^{p+q} \sum_{\boldsymbol{P}\in\mathcal{M}_r} P(z) \sum_{s=0}^{r} \sum_{\substack{\boldsymbol{Q}\in\mathcal{M}_s \\ \boldsymbol{R}\in\mathcal{M}_{r-s} \\ \boldsymbol{QR}=\boldsymbol{P}}} \boldsymbol{Y_Q}[g';z]\, \boldsymbol{Z_R}[g';z].$$
$$\equiv \sum_{r=0}^{p+q} \sum_{\boldsymbol{P}\in\mathcal{M}_r} P(z)\, \boldsymbol{X_P}[g';z],$$

where: for $\boldsymbol{P} \in \mathcal{M}_r$ and $r \geq 0$, $\boldsymbol{X_P}$ is the polynomial defined in $\mathfrak{T}$ by

(20.108)
$$\boldsymbol{X_P} \equiv \sum_{s=0}^{r} \sum_{\substack{\boldsymbol{Q}\in\mathcal{M}_s \\ \boldsymbol{R}\in\mathcal{M}_{r-s} \\ \boldsymbol{QR}=\boldsymbol{P}}} \boldsymbol{Y_Q}\, \boldsymbol{Z_R}.$$

For $r \geq 0$, suppose that $0 \leq s \leq r$, $\boldsymbol{Q} \in \mathcal{M}_s$, $\boldsymbol{R} \in \mathcal{M}_{r-s}$, and $\boldsymbol{P} = \boldsymbol{QR}$. If $s > p$ or $r - s > q$, then $\boldsymbol{Y_Q}\, \boldsymbol{Z_R} \equiv 0$. If $s \leq p$ and $r - s \leq q$, then $\boldsymbol{Y_Q} \in \mathfrak{T}_{p-s}$, $\boldsymbol{Z_R} \in \mathfrak{T}_{q-r+s}$, and $\boldsymbol{Y_Q}\, \boldsymbol{Z_R} \in \mathfrak{T}_{p+q-r}$. Thus, for $0 \leq r \leq p + q$ and $\boldsymbol{P} \in \mathcal{M}_r$, $\boldsymbol{X_P} \in \mathfrak{T}_{p+q-r}$.

We note that (20.108) reduces to

(20.109)
$$\boldsymbol{X_P} \equiv \sum_{\substack{\boldsymbol{Q}\in\mathcal{M}_p \\ \boldsymbol{R}\in\mathcal{M}_q \\ \boldsymbol{QR}=\boldsymbol{P}}} \boldsymbol{Y_Q}\, \boldsymbol{Z_R}, \quad \text{when } \boldsymbol{P} \in \mathcal{M}_{p+q}.$$

Namely, with reference to (20.108) for $r = p + q$, $0 \leq s \leq p + q$, and $s \neq p$, we see that either $s > p$ and $\boldsymbol{Y_Q} \equiv 0$ or $s < p$, $p + q - s > q$, and $\boldsymbol{Z_R} \equiv 0$.

We use (20.86)–(20.87), (20.102)–(20.103), and (20.109) to obtain

(20.110)
$$G(z)\, H(z) \equiv \left[\sum_{\boldsymbol{Q}\in\mathcal{M}_p} Q(z)\, \boldsymbol{Y_Q}[g';z]\right]\!\left[\sum_{\boldsymbol{R}\in\mathcal{M}_q} R(z)\, \boldsymbol{Z_R}[g';z]\right]$$
$$\equiv \sum_{\substack{\boldsymbol{Q}\in\mathcal{M}_p \\ \boldsymbol{R}\in\mathcal{M}_q \\ \boldsymbol{P}\in\mathcal{M}_{p+q} \\ \boldsymbol{P}=\boldsymbol{QR}}} Q(z)\, R(z)\, \boldsymbol{Y_Q}[g';z]\, \boldsymbol{Z_R}[g';z]$$
$$\equiv \sum_{\boldsymbol{P}\in\mathcal{M}_{p+q}} P(z) \sum_{\substack{\boldsymbol{Q}\in\mathcal{M}_p \\ \boldsymbol{R}\in\mathcal{M}_q \\ \boldsymbol{QR}=\boldsymbol{P}}} \boldsymbol{Y_Q}[g';z]\, \boldsymbol{Z_R}[g';z] \equiv \sum_{\boldsymbol{P}\in\mathcal{M}_{p+q}} P(z)\, \boldsymbol{X_P}[g';z].$$

We combine (20.107) and (20.110) to obtain

(20.111) $\quad (g'(z))^{p+q} G^{**}(g(z)) H^{**}(g(z)) \equiv G(z) H(z)$
$$+ \sum_{r=0}^{p+q-1} \sum_{\boldsymbol{P} \in \mathcal{M}_r} P(z) \, \boldsymbol{X_P}[g'; z].$$

In view of (20.107) and (20.111), we see that (20.78) and (20.79) are satisfied when $\boldsymbol{F} \equiv \boldsymbol{GH}$ and $n = p+q$.

(iv) Parts (i), (ii), and (iii) show that (20.78) and (20.79) are valid whenever $\boldsymbol{F}$ is a monomial in $\mathcal{S}_m$. Suppose that $\boldsymbol{G}$ and $\boldsymbol{H}$ are two isobaric polynomials of the same weight $n \geq 1$ that satisfy the specific forms of (20.78) and (20.79) given by (20.86)–(20.87) and (20.102)–(20.103) when $p = q = n$. For $\boldsymbol{G} + \boldsymbol{H}$ to be isobaric of weight n, we also assume that $\boldsymbol{G} + \boldsymbol{H} \not\equiv 0$. To show that (20.78) and (20.79) are satisfied by $\boldsymbol{F} \equiv \boldsymbol{G} + \boldsymbol{H}$, we first apply (20.86) and (20.102) to obtain

(20.112) $\quad (g'(z))^n \left(G^{**}(g(z)) + H^{**}(g(z)) \right) \equiv \sum_{r=0}^{n} \sum_{\boldsymbol{P} \in \mathcal{M}_r} P(z) \, \boldsymbol{X_P}[g'; z],$

where $\boldsymbol{X_P} \equiv \boldsymbol{Y_P} + \boldsymbol{Z_P}$. Clearly, for $0 \leq r \leq n$ and $\boldsymbol{P}$ in $\mathcal{M}_r$, each of $\boldsymbol{Y_P}$, $\boldsymbol{Z_P}$, and $\boldsymbol{X_P}$ is a polynomial in $\mathfrak{T}_{n-r}$. For $p = q = n$, we use (20.87), (20.103), and $\boldsymbol{X_P} \equiv \boldsymbol{Y_P} + \boldsymbol{Z_P}$ to verify that

(20.113) $\quad (g'(z))^n \left(G^{**}(g(z)) + H^{**}(g(z)) \right) \equiv (G(z) + H(z))$
$$+ \sum_{r=0}^{n-1} \sum_{\boldsymbol{P} \in \mathcal{M}_r} P(z) \, \boldsymbol{X_P}[g'; z].$$

In view of (20.112)–(20.113), (20.78) and (20.79) are satisfied by $\boldsymbol{F} \equiv \boldsymbol{G}+\boldsymbol{H}$. Thus, (20.78) and (20.79) are valid for any isobaric polynomial $\boldsymbol{F}$ in $\mathcal{S}_m$. This completes the proof of Proposition 20.13. $\square$

20.7. Rational semi-invariants of the second kind

THEOREM 20.14. *Suppose that $\boldsymbol{A}$ is a rational semi-invariant of the second kind of weight s for the equations (1.1) having order m and let $\boldsymbol{F}$, $\boldsymbol{G}$ be relatively prime nonzero polynomials in $\mathcal{S}_m$ such that*

(20.114) $$\boldsymbol{A} \equiv \frac{\boldsymbol{F}}{\boldsymbol{G}}.$$

Then, at least one of $\boldsymbol{F}$ and $\boldsymbol{G}$ is not a constant, each nonconstant in $\{\boldsymbol{F}, \boldsymbol{G}\}$ is a polynomial semi-invariant of the second kind for the equations (1.1) of order m, and $s = \text{weight}(\boldsymbol{F}) - \text{weight}(\boldsymbol{G})$.

PROOF. Proposition 20.4 on page 220 shows that the polynomials $\boldsymbol{F}$, $\boldsymbol{G}$ are isobaric and $s = \text{weight}(\boldsymbol{F}) - \text{weight}(\boldsymbol{G})$. We introduce $p = \text{weight}(\boldsymbol{F})$ as well as $q = \text{weight}(\boldsymbol{G})$. Since $\boldsymbol{A}$ is not a constant in $\mathcal{Q}_m$, at least one of $\boldsymbol{F}$ and $\boldsymbol{G}$ is not a constant in $\mathcal{S}_m$ and either $p \geq 1$ or $q \geq 1$. For $0 \leq r \leq p$ and each monic monomial $\boldsymbol{P}$ in $\mathcal{M}_r$, Proposition 20.13 on page 234 yields a polynomial $\boldsymbol{X_P}$ in $\mathfrak{T}_{p-r}$ for $\boldsymbol{F}$ with the following property. For any (1.1) on Ω and any univalent analytic function $\zeta = g(z)$ on Ω whose inverse on $\Omega^{**} = g(\Omega)$ is designated by $z = f(\zeta)$,

20.7. RATIONAL SEMI-INVARIANTS OF THE SECOND KIND

the substitution (1.4) specified by $z = f(\zeta)$ transforms (1.1) into a corresponding (1.5) on Ω^{**} such that

$$(20.115) \qquad \bigl(g'(z)\bigr)^p F^{**}\bigl(g(z)\bigr) \equiv F(z) + \sum_{r=0}^{p-1} \sum_{\boldsymbol{P} \in \mathcal{M}_r} P(z)\, \boldsymbol{X_P}[g'; z].$$

Similarly, for $0 \leq r \leq q$ and each monic monomial $\boldsymbol{Q}$ in $\mathcal{M}_r$, Proposition 20.13 on page 234 yields a polynomial $\boldsymbol{Y_Q}$ in $\mathfrak{T}_{q-r}$ for $\boldsymbol{G}$ such that

$$(20.116) \qquad \bigl(g'(z)\bigr)^q G^{**}\bigl(g(z)\bigr) \equiv G(z) + \sum_{s=0}^{q-1} \sum_{\boldsymbol{Q} \in \mathcal{M}_s} Q(z)\, \boldsymbol{Y_Q}[g'; z]$$

is satisfied on Ω for each transformation (1.4) of (1.1) into a corresponding (1.5). We substitute $\zeta = g(z)$ in the property $A^{**}(\zeta) \equiv \bigl(f'(\zeta)\bigr)^s A\bigl(f(\zeta)\bigr)$ of $\boldsymbol{A}$ from (20.2) of Definition 20.2 on page 220 and then use (20.114) as well as $p - q = s$ to obtain

$$(20.117) \qquad \frac{\bigl(g'(z)\bigr)^p F^{**}\bigl(g(z)\bigr)}{\bigl(g'(z)\bigr)^q G^{**}\bigl(g(z)\bigr)} \equiv \bigl(g'(z)\bigr)^s A^{**}\bigl(g(z)\bigr) \equiv A(z) \equiv \frac{F(z)}{G(z)}, \quad \text{on } \Omega,$$

whenever $G(z) \not\equiv 0$ and $G^{**}(\zeta) \not\equiv 0$. Thus, (20.115)–(20.117) yield

$$(20.118) \qquad \frac{F(z) + \sum_{r=0}^{p-1} \sum_{\boldsymbol{P} \in \mathcal{M}_r} P(z)\, \boldsymbol{X_P}[g'; z]}{G(z) + \sum_{s=0}^{q-1} \sum_{\boldsymbol{Q} \in \mathcal{M}_s} Q(z)\, \boldsymbol{Y_Q}[g'; z]} \equiv \frac{F(z)}{G(z)}, \quad \text{on } \Omega,$$

for any (1.1) on Ω satisfying $G(z) \not\equiv 0$ and any transformation (1.4) of that (1.1) on Ω into a corresponding (1.5) on Ω^{**} satisfying $G^{**}(\zeta) \not\equiv 0$. We shall use this to prove that $\boldsymbol{M_2} \equiv 0$, where $\boldsymbol{M_2}$ is the polynomial

$$(20.119) \qquad \boldsymbol{M_2} \equiv \left[\boldsymbol{F} + \sum_{r=0}^{p-1} \sum_{\boldsymbol{P} \in \mathcal{M}_r} \boldsymbol{P}\, \boldsymbol{X_P}\right] \boldsymbol{G} - \left[\boldsymbol{G} + \sum_{s=0}^{q-1} \sum_{\boldsymbol{Q} \in \mathcal{M}_s} \boldsymbol{Q}\, \boldsymbol{Y_Q}\right] \boldsymbol{F}$$

over $\mathbb{Q}$ in the variables $\boldsymbol{w}_{i,j}^{(k)}$ and $\boldsymbol{t}_l$, for $1 \leq j \leq m$, $0 \leq i \leq j$, $k \geq 0$, and $l \geq 1$.

To establish that $\boldsymbol{M_2} \equiv 0$, suppose that $\boldsymbol{M_2} \not\equiv 0$. Since there are equations (1.5) satisfying $G^{**}(\zeta) \not\equiv 0$, we use (20.116) to see that the polynomial

$$(20.120) \qquad \boldsymbol{D_2} \equiv \boldsymbol{G} + \sum_{s=0}^{q-1} \sum_{\boldsymbol{Q} \in \mathcal{M}_s} \boldsymbol{Q}\, \boldsymbol{Y_Q}$$

satisfies $\boldsymbol{D_2} \not\equiv 0$. Also, we have $\boldsymbol{G} \not\equiv 0$. In particular, $\boldsymbol{D_2}$, $\boldsymbol{G}$, and $\boldsymbol{M_2}$ are elements of the polynomial ring $\mathfrak{S}$ introduced on page 223 just after (20.24). Let q_1, q_2, q_3 be nonnegative integers such that $(\boldsymbol{x}_0^{(0)})^{q_1} \boldsymbol{D_2}$, $(\boldsymbol{x}_0^{(0)})^{q_2} \boldsymbol{G}$, and $(\boldsymbol{x}_0^{(0)})^{q_3} \boldsymbol{M_2}$ can be rewritten as polynomials over $\mathbb{Q}$ in the variables $\boldsymbol{w}_{i,j}^{(k)}$ and $\boldsymbol{x}_0^{(l)}$, for $1 \leq j \leq m$, $0 \leq i \leq j$, $k \geq 0$, and $l \geq 0$. Hence, a polynomial over $\mathbb{Q}$ in the variables $\boldsymbol{w}_{i,j}^{(k)}$ and $\boldsymbol{x}_0^{(l)}$, for $1 \leq j \leq m$, $0 \leq i \leq j$, $k \geq 0$, and $l \geq 0$, is defined by

$$(20.121) \qquad \boldsymbol{N_2} \equiv \boldsymbol{x}_0^{(0)} \left[(\boldsymbol{x}_0^{(0)})^{q_1} \boldsymbol{D_2} \right] \left[(\boldsymbol{x}_0^{(0)})^{q_2} \boldsymbol{G} \right] \left[(\boldsymbol{x}_0^{(0)})^{q_3} \boldsymbol{M_2} \right]$$

and it satisfies $\boldsymbol{N_2} \not\equiv 0$. We set $\nu = \dfrac{m(m+3)}{2} + 1$ and observe that there are $\nu - 1$ variables $\boldsymbol{w}_{i,j}^{(0)}$ having $1 \leq j \leq m$ and $0 \leq i \leq j$. We identify these $\nu - 1$ variables with the first $\nu - 1$ variables $\boldsymbol{s}_1^{(0)}, \boldsymbol{s}_2^{(0)}, \ldots, \boldsymbol{s}_{\nu-1}^{(0)}$ in (4.2) on page 35. We identify

$x_0^{(0)}$ with $\mathfrak{s}_\nu^{(0)}$ in (4.2). Then, N_2 belongs to the ring $\mathfrak{S}_\nu$ of polynomials over $\mathbb{Q}$ in the variables $\mathfrak{s}_h^{(k)}$ having $1 \leq h \leq \nu$ and $k \geq 0$. Consequently, Proposition 4.2 on page 36 is applicable and yields meromorphic functions $e_1(z)$, ..., $e_\nu(z)$ on a region Ω_0 such that the condition $N_2(z) \not\equiv 0$ is satisfied by the function $N_2(z)$ on Ω_0 obtained by substituting $e_h^{(k)}(z)$ for $\mathfrak{s}_h^{(k)}$ in N_2. With a change of notation, this gives meromorphic functions $c_{i,j}(z)$ and $h(z)$ on Ω_0, for $1 \leq j \leq m$ and $0 \leq i \leq j$, such that the meromorphic function $N_2(z)$ on Ω_0 obtained by substituting $c_{i,j}^{(k)}(z)$ for $w_{i,j}^{(k)}$ and $h^{(k)}(z)$ for $x_0^{(k)}$ in N_2 satisfies $N_2(z) \not\equiv 0$. Since (20.121) shows that $x_0^{(0)}$ is a factor of N_2, we have $h(z) \not\equiv 0$. Let Ω be a subregion of Ω_0 on which there is a univalent analytic function $g(z)$ such that $g'(z) = h(z)$, for each z in Ω. For $1 \leq j \leq m$, $0 \leq i \leq j$, $k \geq 0$, and $l \geq 1$, we substitute $c_{i,j}^{(k)}(z)$ for $w_{i,j}^{(k)}$ and $g^{(l+1)}(z)/g^{(1)}(z) \equiv h^{(l)}(z)/h(z)$ for t_l in the polynomials D_2, G, and M_2 of $\mathfrak{S}$; then we use (20.121), (20.120), and (20.119), as well as our usual notation to see that the preceding application of Proposition 4.2 to N_2 gives

$$(20.122) \quad G(z) + \sum_{s=0}^{q-1} \sum_{Q \in \mathcal{M}_s} Q(z)\, Y_Q[g'; z] \not\equiv 0, \quad G(z) \not\equiv 0,$$

and

$$(20.123) \quad \left[F(z) + \sum_{r=0}^{p-1} \sum_{P \in \mathcal{M}_r} P(z)\, X_P[g'; z] \right] G(z)$$
$$- \left[G(z) + \sum_{s=0}^{q-1} \sum_{Q \in \mathcal{M}_s} Q(z)\, Y_Q[g'; z] \right] F(z) \not\equiv 0.$$

Let (1.1) have as coefficients the functions $c_{i,j}(z)$ on Ω given by Proposition 4.2 for N_2 along with $c_{j,i}(z) \equiv c_{i,j}(z)$ when $0 \leq i < j \leq m$; and, let (1.4) be the inverse function $z = f(\zeta)$ on $\Omega^{**} = g(\Omega)$ for the univalent analytic function $g(z)$ on Ω. In view of (20.122), this (1.1) and this transformation (1.4) of it into a corresponding (1.5) yield $G(z) \not\equiv 0$ as well as $G^{**}(\zeta) \not\equiv 0$ via (20.116). Therefore, (20.118) is satisfied. However, (20.123) gives the contradiction that (20.118) is not satisfied. Consequently, for M_2 in (20.119), we conclude that $M_2 \equiv 0$.

We rewrite (20.119) with $M_2 \equiv 0$ to obtain

$$(20.124) \quad F \left[\sum_{s=0}^{q-1} \sum_{Q \in \mathcal{M}_r} Q\, Y_Q \right] \equiv G \left[\sum_{r=0}^{p-1} \sum_{P \in \mathcal{M}_r} P\, X_P \right].$$

as an identity in $\mathfrak{S}$. The polynomials F, G in $\mathcal{S}_m$ are relatively prime in $\mathcal{S}_m$. Since $\mathfrak{S}$ is a unique factorization domain that contains $\mathcal{S}_m$ as a subring, F and G are relatively prime in $\mathfrak{S}$. We use this and (20.124) to see that F divides the polynomial

$$(20.125) \quad L \equiv \sum_{r=0}^{p-1} \sum_{P \in \mathcal{M}_r} P\, X_P$$

in $\mathfrak{S}$. If $p = 0$, then F is a constant and the sum in (20.125) vacuously gives $L \equiv 0$. Suppose that $p \geq 1$. Then, L is a linear combination over $\mathfrak{T}$ of monic monomials P from $\mathcal{S}_m$ whose weights do not exceed $p-1$ while F effectively involves a monomial from $\mathcal{S}_m$ of weight p. Thus, the divisibility of $\boldsymbol{L}$ by $\boldsymbol{F}$ yields $\boldsymbol{L} \equiv 0$. We use $\boldsymbol{L} \equiv 0$ and the context for (20.115) to obtain

$$\sum_{r=0}^{p-1} \sum_{\boldsymbol{P} \in \mathcal{M}_r} P(z)\, \boldsymbol{X_P}[g'; z] \equiv 0$$

and $\bigl(g'(z)\bigr)^p F^{**}\bigl(g(z)\bigr) \equiv F(z)$, or equivalently $F^{**}(\zeta) \equiv \bigl(f'(\zeta)\bigr)^p F\bigl(f(\zeta)\bigr)$, for each (1.1) and each transformation (1.4) of (1.1) into a corresponding (1.5). Hence, when $\boldsymbol{F}$ is not a constant, it is a polynomial semi-invariant of the second kind of weight p. Since $\boldsymbol{L} \equiv 0$ and (20.124) yield

$$\sum_{s=0}^{q-1} \sum_{\boldsymbol{Q} \in \mathcal{M}_s} \boldsymbol{Q}\, \boldsymbol{Y_Q} \equiv 0,$$

we see that (20.116) reduces to $\bigl(g'(z)\bigr)^q G^{**}\bigl(g(z)\bigr) \equiv G(z)$ and gives the equivalent expression $G^{**}(\zeta) \equiv \bigl(f'(\zeta)\bigr)^q G\bigl(f(\zeta)\bigr)$. Thus, when $\boldsymbol{G}$ is not a constant, it is a polynomial semi-invariant of the second kind of weight q. This completes the proof of Theorem 20.14. □

20.8. The structure of rational relative invariants

THEOREM 20.15. *Suppose that $\boldsymbol{A}$ in $\mathcal{Q}_m$ is a rational relative invariant for the equations (1.1) of order m according to Definition 20.3 on page 220. Let $\boldsymbol{F}$ and $\boldsymbol{G}$ be relatively prime nonzero polynomials in $\mathcal{S}_m$ such that*

(20.126) $$\boldsymbol{A} \equiv \frac{\boldsymbol{F}}{\boldsymbol{G}}.$$

Then, at least one of $\boldsymbol{F}$ and $\boldsymbol{G}$ is not a constant and each nonconstant in $\{\boldsymbol{F}, \boldsymbol{G}\}$ is a polynomial relative invariant for the equations (1.1). Also, the formula

(20.127) $$s = weight(\boldsymbol{F}) - weight(\boldsymbol{G})$$

specifies the unique integer s with which $\boldsymbol{A}$ satisfies (20.2) on page 220.

PROOF. Because $\boldsymbol{A}$ is not a constant of $\mathcal{Q}_m$, at least one of $\boldsymbol{F}$, $\boldsymbol{G}$ is not a constant polynomial in $\mathcal{S}_m$. Moreover, $\boldsymbol{A}$ is a rational semi-invariant of the second kind. Consequently, Theorem 20.4 on page 220 shows that $\boldsymbol{F}$ and $\boldsymbol{G}$ are isobaric and (20.127) is valid. Because $\boldsymbol{F}$ and $\boldsymbol{G}$ for (20.126) are isobaric and $\boldsymbol{A}$ is a rational semi-invariant of the first kind, Theorem 20.11 on page 230 shows that each nonconstant element in $\{\boldsymbol{F}, \boldsymbol{G}\}$ is a polynomial semi-invariant of the first kind. Also, Theorem 20.14 on page 240 establishes that each nonconstant element in $\{\boldsymbol{F}, \boldsymbol{G}\}$ is a polynomial semi-invariant of the second kind. We combine these results to see that each nonconstant element in $\{\boldsymbol{F}, \boldsymbol{G}\}$ is a polynomial relative invariant for the equations (1.1) of order m. This completes the proof. □

For a rational relative invariant $\boldsymbol{A}$ given by (20.126), the integer s in (20.127) is the weight of $\boldsymbol{A}$ according to Definition 20.6 on page 222. The old concept of an absolute invariant for homogeneous linear differential equations was reexamined in [24, Theorem D.19 of page 188]. Here, we consider its natural extension to the equations (1.1).

DEFINITION 20.16. *An absolute invariant for the equations (1.1) of order m is a rational relative invariant of weight 0 for such equations.*

Theorem 20.15 shows that *an element A in $\mathcal{Q}_m$ is an absolute invariant for the equations* (1.1) *of order m if and only if A is expressible as the quotient of two relatively prime polynomial relative invariants having the same weight.*

EXAMPLE 20.17. For the equations (1.1) having $m = 1$, Theorem 4.10 of page 41 shows that each polynomial relative invariant has the form $\gamma(\mathcal{I}_{1,1,1})^\kappa$, for some nonzero rational number γ and some positive integer κ. Since any two relative invariants of this type are not relatively prime, there are no absolute invariants for the equations (1.1) having $m = 1$.

For any integer $m \geq 2$, three of the basic relative invariants for the equations (1.1) are $\mathcal{I}_{m,1,1}$, $\mathcal{I}_{m,1,2}$, and $\mathcal{I}_{m,2,2}$ in (1.49), (1.53), and (1.59) on pages 11, 13. They are relatively prime. Thus, each of the three elements in $\mathcal{Q}_m$ given by

$$A_1 \equiv \frac{\mathcal{I}_{m,2,2}}{(\mathcal{I}_{m,1,1})^2}, \quad A_2 \equiv \frac{(\mathcal{I}_{m,1,2})^2}{(\mathcal{I}_{m,1,1})^3}, \quad \text{and} \quad A_3 \equiv \frac{(\mathcal{I}_{m,2,2})^3}{(\mathcal{I}_{m,1,2})^4}$$

is an absolute invariant for the equations (1.1) of order m. Absolute invariants are also given by $(1/A_1)$, $(A_1 + A_2)$, $(A_1 A_2)$, etc. *Were the historic definitions of semi-invariants and relative invariants extended to include the elements of $\mathbb{Q}$, the absolute invariants would form a field under addition and multiplication.*

20.9. Substitutions into rational functions of $\mathcal{Q}_m$

PROPOSITION 20.18. *If A is a rational semi-invariant of the first kind and* (1.1) *is an equation for which $A(z)$ is defined, then the condition*

(20.128) $$A^*(z) \equiv A(z)$$

of Definition 20.1 on page 219 is valid for any transformation (1.2) *of* (1.1) *into a corresponding* (1.3). *Moreover, if A is a rational semi-invariant of the second kind of weight s and* (1.1) *is an equation for which $A(z)$ is defined, then the condition*

(20.129) $$A^{**}(\zeta) \equiv \bigl(f'(\zeta)\bigr)^s A\bigl(f(\zeta)\bigr)$$

of Definition 20.2 on page 220 is valid for any transformation (1.4) *of* (1.1) *into a corresponding* (1.5).

PROOF. Let A be a nonconstant rational function in $\mathcal{Q}_m$; let $F \not\equiv 0$ and $G \not\equiv 0$ be relatively prime polynomials in $\mathcal{S}_m$ such that $A \equiv F/G$.

Suppose that A is a rational semi-invariant of the first kind. Theorem 20.11 on page 230 shows that G is either a constant or a semi-invariant of the first kind. Consequently, for any transformation (1.2) of (1.1) into (1.3), $G^*(z) \equiv G(z)$, $G^*(z)$ satisfies $G^*(z) \not\equiv 0$ if and only if $G(z)$ satisfies $G(z) \not\equiv 0$, and $A^*(z)$ is defined if and only if $A(z)$ is defined. This yields the assertion about (20.128).

Suppose that A is a rational semi-invariant of the second kind. Theorem 20.14 of page 240 shows that G is either a constant or a semi-invariant of the second kind. Consequently, there is an integer $n \geq 0$ such that: for any transformation (1.4) of (1.1) into a corresponding (1.5), $G^{**}(\zeta) \equiv \bigl(f'(\zeta)\bigr)^n G\bigl(f(\zeta)\bigr)$, $G^{**}(\zeta)$ satisfies $G^{**}(\zeta) \not\equiv 0$ if and only if $G(z)$ satisfies $G(z) \not\equiv 0$, and $A^{**}(\zeta)$ is defined if and only if $A(z)$ is defined. This gives the assertion about (20.129) and completes the proof. □

For substitutions into rational functions of $\mathcal{Q}_m$, the next result is analogous to Proposition 4.5 on page 38 for substitutions into polynomials of $\mathcal{S}_m$.

LEMMA 20.19. *Let $\boldsymbol{A}$ be a rational function in $\mathcal{Q}_m$; let $\boldsymbol{A}'$ be the rational function in $\mathcal{Q}_m$ obtained by applying the derivation ' for $\mathcal{Q}_m$ to $\boldsymbol{A}$; set $\boldsymbol{B} \equiv \boldsymbol{A}'$; and, let meromorphic functions $c_{i,j}(z)$ on a region Ω of the complex plane be given for $1 \le j \le m$ and $0 \le i \le j$. Then, the substitution of $c_{i,j}^{(k)}(z)$ for $\boldsymbol{w}_{i,j}^{(k)}$ in $\boldsymbol{A}$ yields a well defined function $A(z)$ on Ω if and only if the substitution of $c_{i,j}^{(k)}(z)$ for $\boldsymbol{w}_{i,j}^{(k)}$ in $\boldsymbol{B}$ yields a well defined function $B(z)$ on Ω. Moreover, when $A(z)$ and $B(z)$ are defined, they are related through*

$$\tag{20.130} \frac{d}{dz}A(z) \equiv B(z), \quad \text{on } \Omega.$$

PROOF. Let $\boldsymbol{A}$ be represented as $\boldsymbol{A} \equiv \boldsymbol{F}/\boldsymbol{G}$, where $\boldsymbol{F}$ and $\boldsymbol{G} \not\equiv 0$ are relatively prime polynomials in $\mathcal{S}_m$. We set $\boldsymbol{F}_1 \equiv \boldsymbol{F}'$ and $\boldsymbol{G}_1 \equiv \boldsymbol{G}'$ in $\mathcal{S}_m$. The substitution of $c_{i,j}^{(k)}(z)$ for $\boldsymbol{w}_{i,j}^{(k)}$ in each of $\boldsymbol{F}$, $\boldsymbol{G}$, $\boldsymbol{F}_1$, and $\boldsymbol{G}_1$ yields meromorphic functions $F(z)$, $G(z)$, $F_1(z)$, and $G_1(z)$ on Ω; and Proposition 4.5 on page 38 gives

$$\frac{d}{dz}F(z) \equiv F_1(z) \quad \text{and} \quad \frac{d}{dz}G(z) \equiv G_1(z), \quad \text{on } \Omega.$$

In view of

$$\tag{20.131} \boldsymbol{A} \equiv \frac{\boldsymbol{F}}{\boldsymbol{G}} \quad \text{and} \quad \boldsymbol{B} \equiv \frac{\boldsymbol{G}\boldsymbol{F}_1 - \boldsymbol{F}\boldsymbol{G}_1}{\boldsymbol{G}^2},$$

we see that: $A(z)$ is defined on Ω if and only if $G(z) \not\equiv 0$; and, $B(z)$ is defined on Ω if and only if $G(z) \not\equiv 0$. Suppose that $G(z) \not\equiv 0$. Then, we find that

$$\frac{d}{dz}A(z) \equiv \frac{d}{dz}\left[\frac{F(z)}{G(z)}\right] \equiv \frac{G(z)\dfrac{d}{dz}F(z) - F(z)\dfrac{d}{dz}G(z)}{\left(G(z)\right)^2} \equiv B(z)$$

and (20.130) is valid. This completes the proof. □

REMARK 20.20. Whenever $\boldsymbol{A}$ in $\mathcal{Q}_m$ and (1.1) are such that $A(z)$ is defined, there are at least two interpretations for $A^{(r)}(z)$, when $r \ge 1$. We can replace $\boldsymbol{w}_{i,j}^{(k)}$ in $\boldsymbol{A}$ with $c_{i,j}^{(k)}(z)$ from (1.1) to obtain $A(z)$ and then differentiate $A(z)$ r-times with respect to z. Alternatively, we can start with $\boldsymbol{A}$ and successively apply the derivation ' for $\mathcal{Q}_m$ a total of r times to obtain $\boldsymbol{A}^{(r)}$ and then replace $\boldsymbol{w}_{i,j}^{(k)}$ in $\boldsymbol{A}^{(r)}$ with $c_{i,j}^{(k)}(z)$ from (1.1). The equivalence of the different procedures for computing $A^{(r)}(z)$ is demonstrated by applying Lemma 20.19. Similar observations can be made in regard to substitutions from either (1.3) or (1.5).

PROPOSITION 20.21. *If $\boldsymbol{A}$ in $\mathcal{Q}_m$ is a rational semi-invariant of the first kind for the equations (1.1) of order m, then $\boldsymbol{A}'$ is a rational semi-invariant of the first kind for such equations.*

PROOF. Set $\boldsymbol{B} \equiv \boldsymbol{A}'$. Since $\boldsymbol{A}$ is not a constant, $\boldsymbol{B}$ is not a constant. Let (1.1) be an equation for which $B(z)$ is defined and consider a transformation (1.2) of this (1.1) into a corresponding (1.3). Then, $A(z)$ is defined, $A^*(z)$ is defined, and $A^*(z) \equiv A(z)$. After differentiating this relation with respect to z, we apply Lemma 20.19 twice to obtain

$$B^*(z) \equiv \frac{d}{dz}A^*(z) \equiv \frac{d}{dz}A(z) \equiv B(z).$$

Thus, $\boldsymbol{A}'$ is a rational semi-invariant of the first kind. This completes the proof. □

PROPOSITION 20.22. *Suppose that A in $\mathcal{Q}_m$ is a rational relative invariant of weight n for the equations (1.1) of order m. Then, A' in $\mathcal{Q}_m$ is a rational relative invariant for such equations if and only if $n = 0$ and A is therefore an absolute invariant. Moreover, when $n = 0$, the weight of A' is 1.*

PROOF. We write $A \equiv F/G$, where F and G are relatively prime isobaric polynomials in $\mathcal{S}_m$ of respective weights n_1 and n_2 subject to $n_1 - n_2 = n$. For any (1.1) and (1.5) related by (1.4), Theorem 20.15 yields

$$(20.132) \quad F^{**}(\zeta) \equiv \bigl(f'(\zeta)\bigr)^{n_1} F\bigl(f(\zeta)\bigr) \quad \text{and} \quad G^{**}(\zeta) \equiv \bigl(f'(\zeta)\bigr)^{n_2} G\bigl(f(\zeta)\bigr),$$

on Ω^{**}. We restrict the equations (1.1) to those for which $G(z) \not\equiv 0$. Then, we also have $G^{**}(\zeta) \not\equiv 0$. With respect to the notation $B \equiv A'$, $F_1 \equiv F'$, and $G_1 \equiv G'$, we use Lemma 20.19, (20.132), Proposition 4.5 on page 38, and (20.131) to obtain

$$(20.133) \quad B^{**}(\zeta) \equiv \frac{d}{d\zeta}\left[\frac{F^{**}(\zeta)}{G^{**}(\zeta)}\right] \equiv \frac{d}{d\zeta}\left[\bigl(f'(\zeta)\bigr)^{n_1-n_2} \frac{F\bigl(f(\zeta)\bigr)}{G\bigl(f(\zeta)\bigr)}\right]$$

$$\equiv (n_1 - n_2)\bigl(f'(\zeta)\bigr)^{n_1-n_2-1} f''(\zeta) A\bigl(f(\zeta)\bigr)$$

$$+ \bigl(f'(\zeta)\bigr)^{n_1-n_2} \left[\frac{G\bigl(f(\zeta)\bigr) F_1\bigl(f(\zeta)\bigr) f'(\zeta) - F\bigl(f(\zeta)\bigr) G_1\bigl(f(\zeta)\bigr) f'(\zeta)}{\bigl(G(f(\zeta))\bigr)^2}\right]$$

$$\equiv \bigl(f'(\zeta)\bigr)^{n_1-n_2+1} \left[B\bigl(f(\zeta)\bigr) + (n_1 - n_2)\frac{f''(\zeta)}{\bigl(f'(\zeta)\bigr)^2} A\bigl(f(\zeta)\bigr)\right].$$

Also, there are particular selections of (1.1) and (1.4) such that $G(z) \not\equiv 0$ and $f''(\zeta) A\bigl(f(\zeta)\bigr) \not\equiv 0$. Since (20.131) shows that the weight of B is $n_1 - n_2 + 1$, we apply (20.133) to conclude that B is a rational semi-invariant of the second kind if and only if $n = n_1 - n_2 = 0$. Hence, in view of Proposition 20.21, A' is a rational relative invariant if and only if A is an absolute invariant. This completes the proof. $\square$

Part 4

Generalizations for $H_{m,n} = 0$

CHAPTER 21

Introduction to the Equations $H_{m,n} = 0$

For integers $m, n \geq 1$, we shall present explicit formulas for the basic relative invariants of the equations $H_{m,n} = 0$, where $H_{m,n}$ is an nth-degree homogeneous polynomial combination of $y(z), \ldots, y^{(m)}(z)$ in which the coefficient of $\left(y^{(m)}(z)\right)^n$ is 1. For $n = 1$ and $L_m \equiv H_{m,1}$, this is the subject of [24] where the equations are homogeneous linear ones of order m. For $n = 2$ and $Q_m \equiv H_{m,2}$, we have the results of Chapters 1–20. Section 1.7 shows that the semi-invariants and relative invariants for the equations $L_m = 0$ are obtained from corresponding ones for the equations $Q_m = 0$ through a minor adjustment of notation. Similarly, for any fixed integers n, q satisfying $1 \leq q \leq n$, the semi-invariants and relative invariants for the equations $H_{m,q} = 0$ are related to corresponding ones for the equations $H_{m,n} = 0$ through a minor change of notation that we present in the proof of Proposition 21.6. The main concepts will be explained in this chapter.

21.1. Transformations produced by changing the variables in $H_{m,n} = 0$

Each equation $H_{m,n} = 0$ is uniquely expressible in the form

$$(21.1) \qquad \left(y^{(m)}(z)\right)^n + \sum_{\substack{0 \leq i_1, i_2, \ldots, i_n \leq m \\ (i_1, i_2, \ldots, i_n) \neq (0, 0, \ldots, 0)}} c_{i_1, i_2, \ldots, i_n}(z) \prod_{\nu=1}^{n} y^{(m-i_\nu)}(z) = 0,$$

where $c_{0, 0, \ldots, 0}(z) \equiv 1$ and the coefficients $c_{i_1, i_2, \ldots, i_n}(z)$ are meromorphic functions on a region Ω of the complex plane such that

$$(21.1.1) \qquad c_{i_{\pi(1)}, i_{\pi(2)}, \ldots, i_{\pi(n)}}(z) \equiv c_{i_1, i_2, \ldots, i_n}(z),$$

for any permutation π of $\{1, 2, \ldots, n\}$.

For each meromorphic function $\rho(z) \not\equiv 0$ on Ω, a computation similar to that for (2.1)–(2.2) and (2.5) of page 23 shows that the change

$$(21.2) \qquad y(z) = \rho(z)\, v(z)$$

of the dependent variable in (21.1) from y to v specifies the meromorphic functions

$$(21.2.1) \qquad c^*_{i_1, i_2, \ldots, i_n}(z) \equiv \sum_{j_1=0}^{i_1} \sum_{j_2=0}^{i_2} \cdots \sum_{j_n=0}^{i_n} \mathcal{A}^{i_1, i_2, \ldots, i_n}_{j_1, j_2, \ldots, j_n}(z)\, c_{j_1, j_2, \ldots, j_n}(z),$$

on Ω for $0 \leq i_1, i_2, \ldots, i_n \leq m$,

with respect to (21.1), $c_{0, 0, \ldots, 0}(z) \equiv 1$, $\rho(z)$, and

$$(21.2.2) \qquad \mathcal{A}^{i_1, i_2, \ldots, i_n}_{j_1, j_2, \ldots, j_n}(z) \equiv \prod_{\nu=1}^{n} \left[\binom{m - j_\nu}{i_\nu - j_\nu} \frac{\rho^{(i_\nu - j_\nu)}(z)}{\rho(z)} \right]$$

in such a manner that: if meromorphic functions $y(z)$, $v(z)$ on a subregion of Ω are related by (21.2), then $y(z)$ is a solution of (21.1) if and only if $v(z)$ is a solution of

$$(21.3) \qquad \left(v^{(m)}(z)\right)^n + \sum_{\substack{0 \leq i_1, i_2, \ldots, i_n \leq m \\ (i_1, i_2, \ldots, i_n) \neq (0, 0, \ldots, 0)}} c^*_{i_1, i_2, \ldots, i_n}(z) \prod_{\nu=1}^{n} v^{(m-i_\nu)}(z) = 0.$$

We see that (21.1), (21.1.1), (21.2), (21.2.1), and (21.2.2) uniquely specify (21.3).

PROPOSITION 21.0.1. *For any permutation π of $\{1, 2, \ldots, n\}$, (21.3) has*

$$(21.3.1) \qquad c^*_{i_{\pi(1)}, i_{\pi(2)}, \ldots, i_{\pi(n)}}(z) \equiv c^*_{i_i, i_2, \ldots, i_n}(z), \quad \text{on } \Omega.$$

PROOF. We rewrite (21.2.1), substitute $k_s = j_{\pi(s)}$, for $s = 1, 2, \ldots, n$, and then employ (21.2.2), (21.1.1), and (21.2.1) to obtain

$$c^*_{i_{\pi(1)}, i_{\pi(2)}, \ldots, i_{\pi(n)}}(z) \equiv \sum_{k_1=0}^{i_{\pi(1)}} \sum_{k_2=0}^{i_{\pi(2)}} \cdots \sum_{k_n=0}^{i_{\pi(n)}} \mathcal{A}^{i_{\pi(1)}, i_{\pi(2)}, \ldots, i_{\pi(n)}}_{k_1, k_2, \ldots, k_n}(z) \, c_{k_1, k_2, \ldots, k_n}(z)$$

$$\equiv \sum_{j_{\pi(1)}=0}^{i_{\pi(1)}} \sum_{j_{\pi(2)}=0}^{i_{\pi(2)}} \cdots \sum_{j_{\pi(n)}=0}^{i_{\pi(n)}} \mathcal{A}^{i_{\pi(1)}, i_{\pi(2)}, \ldots, i_{\pi(n)}}_{j_{\pi(1)}, j_{\pi(2)}, \ldots, j_{\pi(n)}}(z) \, c_{j_{\pi(1)}, j_{\pi(2)}, \ldots, j_{\pi(n)}}(z)$$

$$\equiv \sum_{j_1=0}^{i_1} \sum_{j_2=0}^{i_2} \cdots \sum_{j_n=0}^{i_n} \mathcal{A}^{i_1, i_2, \ldots, i_n}_{j_1, j_2, \ldots, j_n}(z) \, c_{j_1, j_2, \ldots, j_n}(z) \equiv c^*_{i_i, i_2, \ldots, i_n}(z).$$

This completes the proof. $\square$

For each univalent analytic function $\zeta = g(z)$ on Ω with inverse designated by $z = f(\zeta)$ on $\Omega^{**} = g(\Omega)$, a computation similar to that for (3.5)–(3.6) and (3.9) of page 30 shows that the change

$$(21.4) \qquad z = f(\zeta)$$

of the independent variable in (21.1) from z to ζ specifies the meromorphic functions

$$(21.4.1) \qquad c^{**}_{i_1, i_2, \ldots, i_n}(\zeta) \equiv \sum_{j_1=0}^{i_1} \sum_{j_2=0}^{i_2} \cdots \sum_{j_n=0}^{i_n} \mathcal{B}^{i_1, i_2, \ldots, i_n}_{j_1, j_2, \ldots, j_n}(\zeta) \, c_{j_1, j_2, \ldots, j_n}\bigl(f(\zeta)\bigr),$$

$$\text{on } \Omega^{**} \text{ for } 0 \leq i_1, i_2, \ldots, i_n \leq m,$$

with respect to (21.1), $c_{0,0,\ldots,0}(z) \equiv 1$, $\alpha_{i,j}(\zeta)$ from (3.1)–(3.2) of page 29, and

$$(21.4.2) \qquad \mathcal{B}^{i_1, i_2, \ldots, i_n}_{j_1, j_2, \ldots, j_n}(\zeta) \equiv \bigl(f'(\zeta)\bigr)^{j_1+j_2+\cdots+j_n} \prod_{\nu=1}^{n} \alpha_{i_\nu - j_\nu, m - i_\nu}(\zeta)$$

in such a manner that: if meromorphic functions $y(z)$ on a subregion of Ω and $u(\zeta)$ on a subregion of Ω^{**} are related by $u(\zeta) \equiv (y \circ f)(\zeta)$, then $y(z)$ is a local solution of (21.1) if and only if $u(\zeta)$ is a local solution of

$$(21.5) \qquad \left(u^{(m)}(\zeta)\right)^n + \sum_{\substack{0 \leq i_1, i_2, \ldots, i_n \leq m \\ (i_1, i_2, \ldots, i_n) \neq (0, 0, \ldots, 0)}} c^{**}_{i_1, i_2, \ldots, i_n}(\zeta) \prod_{\nu=1}^{n} u^{(m-i_\nu)}(\zeta) = 0.$$

We see that (21.1), (21.1.1), (21.4), (21.4.1), and (21.4.2) uniquely specify (21.5).

PROPOSITION 21.0.2. *For any permutation π of $\{1, 2, \ldots, n\}$, (21.5) has*

(21.5.1) $\qquad c^{**}_{i_{\pi(1)}, i_{\pi(2)}, \ldots, i_{\pi(n)}}(\zeta) \equiv c^{**}_{i_i, i_2, \ldots, i_n}(\zeta), \quad \text{on } \Omega^{**}.$

PROOF. We rewrite (21.4.1), substitute $k_s = j_{\pi(s)}$, for $s = 1, 2, \ldots, n$, and then employ (21.4.2), (21.1.1), and (21.4.1) to obtain

$$c^{**}_{i_{\pi(1)}, i_{\pi(2)}, \ldots, i_{\pi(n)}}(\zeta) \equiv \sum_{k_1=0}^{i_{\pi(1)}} \sum_{k_2=0}^{i_{\pi(2)}} \cdots \sum_{k_n=0}^{i_{\pi(n)}} \mathcal{B}^{i_{\pi(1)}, i_{\pi(2)}, \ldots, i_{\pi(n)}}_{k_1, k_2, \ldots, k_n}(\zeta)\, c_{k_1, k_2, \ldots, k_n}(f(z))$$

$$\equiv \sum_{j_{\pi(1)}=0}^{i_{\pi(1)}} \sum_{j_{\pi(2)}=0}^{i_{\pi(2)}} \cdots \sum_{j_{\pi(n)}=0}^{i_{\pi(n)}} \mathcal{B}^{i_{\pi(1)}, i_{\pi(2)}, \ldots, i_{\pi(n)}}_{j_{\pi(1)}, j_{\pi(2)}, \ldots, j_{\pi(n)}}(\zeta)\, c_{j_{\pi(1)}, j_{\pi(2)}, \ldots, j_{\pi(n)}}(f(z))$$

$$\equiv \sum_{j_1=0}^{i_1} \sum_{j_2=0}^{i_2} \cdots \sum_{j_n=0}^{i_n} \mathcal{B}^{i_1, i_2, \ldots, i_n}_{j_1, j_2, \ldots, j_n}(\zeta)\, c_{j_1, j_2, \ldots, j_n}(f(z)) \equiv c^{**}_{i_i, i_2, \ldots, i_n}(\zeta).$$

This completes the proof. $\square$

21.2. Context and definitions

In order to define polynomial relative invariants for the equations (21.1), we assume that the symbols

(21.6) $\quad \boldsymbol{w}^{(k)}_{i_1, i_2, \ldots, i_n}, \quad \text{for } 1 \leq i_n \leq m,\ 0 \leq i_1 \leq i_2 \leq \cdots \leq i_n,\ \text{and}\ k \geq 0,$

are algebraically independent variables over $\mathbb{Q}$ and we let $\mathcal{S}_{m,n}$ denote the ring of polynomials in these variables over $\mathbb{Q}$. In particular, $\mathcal{S}_{m,2}$ is the ring $\mathcal{S}_m$ of page 5 and $\mathcal{S}_{m,1}$ can be identified with the ring $\mathcal{R}_m$ of page 17. To each permutation π of $\{1, 2, \ldots, n\}$, we set

(21.7) $\quad \boldsymbol{w}^{(k)}_{i_{\pi(1)}, i_{\pi(2)}, \ldots, i_{\pi(n)}} \equiv \boldsymbol{w}^{(k)}_{i_1, i_2, \ldots, i_n}, \quad \text{for } 1 \leq i_n \leq m,\ 0 \leq i_1 \leq i_2 \leq \cdots \leq i_n,$

$(i_{\pi(1)}, i_{\pi(2)}, \ldots i_{\pi(n)}) \neq (i_1, i_2, \ldots, i_n)$, and $k \geq 0$.

So that various later formulas can be written simply, we also set $\boldsymbol{w}^{(0)}_{0, 0, \ldots, 0} \equiv 1$ in $\mathcal{S}_{m,n}$ and introduce $\boldsymbol{w}_{i_1, i_2, \ldots, i_n} \equiv \boldsymbol{w}^{(0)}_{i_1, i_2, \ldots, i_n}$, when $0 \leq i_1, i_2, \ldots, i_n \leq m$.

For any polynomial $\boldsymbol{P}$ in $\mathcal{S}_{m,n}$, we let $P(z)$ denote the unique function on Ω obtained by replacing $\boldsymbol{w}^{(k)}_{i_1, i_2, \ldots, i_n}$ in $\boldsymbol{P}$ with $c^{(k)}_{i_1, i_2, \ldots, i_n}(z)$ from (21.1); we let $P^*(z)$ denote the unique function on Ω obtained by replacing $\boldsymbol{w}^{(k)}_{i_1, i_2, \ldots, i_n}$ in $\boldsymbol{P}$ with $c^{*(k)}_{i_1, i_2, \ldots, i_n}(z)$ from (21.3); and we let $P^{**}(\zeta)$ denote the unique function on Ω^{**} obtained by replacing $\boldsymbol{w}^{(k)}_{i_1, i_2, \ldots, i_n}$ in $\boldsymbol{P}$ with $c^{**(k)}_{i_1, i_2, \ldots, i_n}(\zeta)$ from (21.5).

When $\boldsymbol{P}$ is initially expressed as a polynomial combination over $\mathbb{Q}$ of the symbols in both (21.6) and (21.7), we obtain $P(z)$ according to the preceding method by first using (21.7) to express $\boldsymbol{P}$ as a polynomial in the variables (21.6) over $\mathbb{Q}$ and then substituting $c^{(k)}_{i_1, i_2, \ldots, i_n}(z)$ from (21.1) for $\boldsymbol{w}^{(k)}_{i_1, i_2, \ldots, i_n}$. However, in view of (21.1.1) and (21.7), we obtain the same function $P(z)$ by substituting $c^{(k)}_{i_1, i_2, \ldots, i_n}(z)$ from (21.1) directly for $\boldsymbol{w}^{(k)}_{i_1, i_2, \ldots, i_n}$ in the initial expression for $\boldsymbol{P}$ as a polynomial combination over $\mathbb{Q}$ of the symbols in both (21.6) and (21.7). Similar observations apply to substitutions from (21.3) and (21.5).

DEFINITION 21.1. A polynomial P of $\mathcal{S}_{m,n}$ not in $\mathbb{Q}$ is a *semi-invariant of the first kind* for the differential equations (21.1) having a given order m when
$$P^*(z) \equiv P(z), \tag{21.8}$$
for each (21.1) of order m and each change (21.2) of the dependent variable.

DEFINITION 21.2. A polynomial P of $\mathcal{S}_{m,n}$ not in $\mathbb{Q}$ is a *semi-invariant of the second kind* for the differential equations (21.1) having a given order m when there is an integer s such that
$$P^{**}(\zeta) \equiv \bigl(f'(\zeta)\bigr)^s P\bigl(f(\zeta)\bigr), \tag{21.9}$$
for each (21.1) of order m and each change (21.4) of the independent variable.

DEFINITION 21.3. A polynomial P of $\mathcal{S}_{m,n}$ not in $\mathbb{Q}$ is a *relative invariant* for the differential equations (21.1) having a given order m when it is both a semi-invariant of the first kind and a semi-invariant of the second kind for such equations.

DEFINITION 21.4. The *weight* of the variable $w^{(k)}_{i_1, i_2, \ldots, i_n}$ in (21.6) is the positive integer $i_1 + i_2 + \cdots + i_n + k$; the *weight* of a nonzero rational number is 0; and the *weight* of a nonzero monomial in $\mathcal{S}_{m,n}$ is the sum of the weights of its factors. A nonzero polynomial in $\mathcal{S}_{m,n}$ is said to be *isobaric of weight* s when each of its nonzero monomials has weight s.

MODIFICATION OF AN OLD RESULT. *If P is a semi-invariant of the second kind for the equations (21.1), then P is an isobaric polynomial, there is a unique positive integer s for P that satisfies (21.9), and s for P in (21.9) is equal to the weight of P.*

To prove this assertion, we first rewrite the argument on pages 220–222 for Theorem 20.4 so as to directly establish Corollary 20.5 without involving quotients of polynomials. Then, that reasoning can be easily adapted to the present context.

DEFINITION 21.5. A polynomial P in $\mathcal{S}_{m,n}$ is *basic* when:
(1) there are integers $l_1, l_2, \ldots, l_n$ such that $0 \leq l_1 \leq l_2 \leq \cdots \leq l_n \leq m$, $l_n \geq 1$, the coefficient in P of the variable $w_{l_1, l_2, \ldots, l_n}$ is 1, and if $l_n = 1$ or $l_n = 2$, then $n \geq 2$ and $1 \leq l_{n-1} \leq l_n$;
(2) P is isobaric of weight $l_1 + l_2 + \cdots + l_n$;
(3) if $w^{(k)}_{i_1, i_2, \ldots, i_n}$ is any variable from (21.6) that P effectively involves, then $i_\nu \leq l_\nu$, for $1 \leq \nu \leq n$ (in addition to the requirement for (21.6) that $0 \leq i_1 \leq i_2 \leq \cdots \leq i_n \leq m$ with $i_n \geq 1$); and
(4) for any variables $w^{(p)}_{i_1, i_2, \ldots, i_n}$ and $w^{(q)}_{j_1, j_2, \ldots, j_n}$ from (21.6), if the coefficient of $w^{(p)}_{i_1, i_2, \ldots, i_n} w^{(q)}_{j_1, j_2, \ldots, j_n}$ in P is nonzero, then either (i) $w^{(p)}_{i_1, i_2, \ldots, i_n}$ has $i_\nu = 0$, for $1 \leq \nu \leq n-1$, and $1 \leq i_n \leq 2$; or (ii) $w^{(q)}_{j_1, j_2, \ldots, j_n}$ has $j_\nu = 0$, for $1 \leq \nu \leq n-1$, and $1 \leq j_n \leq 2$.

A *basic polynomial P in $\mathcal{S}_{m,n}$ has index* $(l_1, l_2, \ldots, l_n)$ when $l_1, l_2, \ldots, l_n$ are the unique integers that satisfy the first three of the previous four conditions.

In particular, a *basic relative invariant* is a relative invariant that is also a basic polynomial. The terminology *basic semi-invariant* will also be employed.

For $n = 1$ and $\mathcal{S}_{m,1} = \mathcal{R}_m$ of [**24**, pages 2–3], the preceding definitions reduce to those in [**24**, pages 4–5] for homogeneous linear equations. For $n = 2$, $l_1 = l$, $l_2 = n$, and $\mathcal{S}_{m,2} = \mathcal{S}_m$ of page 5, Definitions 21.1–21.5 reduce to the corresponding Definitions 1.1–1.5 of page 6.

21.3. A summary of results and a derivation $'$ for $\mathcal{S}_{m,n}$

For m and n of (21.1), let $l_1, l_2, \ldots, l_n$ denote integers that satisfy

$$0 \leq l_1 \leq l_2 \leq \cdots \leq l_n \leq m, \quad l_n \geq 1, \quad \text{and} \quad \text{if } l_n \leq 2, \text{ then } n \geq 2 \text{ and } l_{n-1} \geq 1.$$

Then, *there is one and only one basic relative invariant of index* $(l_1, l_2, \ldots, l_n)$ *for the equations* (21.1). When $m = 1$, the existence and uniqueness are respectively established in Theorem 22.1 on page 257 and in Theorem 22.6 on page 264. When $m \geq 2$, the existence and uniqueness are respectively established in Theorem 27.7 on page 314 and in Theorem 27.13 on page 316. In particular, when $m = 1$ and therefore $n \geq 2$, we use Theorem 22.1 as well as (21.16) and (21.15) to see that the basic relative invariants can be defined without using a derivation for $\mathcal{S}_{m,n}$. However, to prove their uniqueness and to make the deductions when $m \geq 2$, we need a particular derivation for $\mathcal{S}_{m,n}$.

Henceforth, let $'$ denote the unique derivation for $\mathcal{S}_{m,n}$ such that

$$\left(\boldsymbol{w}^{(k)}_{i_1, i_2, \ldots, i_n}\right)' = \boldsymbol{w}^{(k+1)}_{i_1, i_2, \ldots, i_n}, \quad \text{for each } \boldsymbol{w}^{(k)}_{i_1, i_2, \ldots, i_n} \text{ in (21.6).}$$

By specializing [10, page 139, Proposition 4] or [11, page A.V.130, Theorem 1], we deduce the existence and uniqueness of $'$. For any $\boldsymbol{P}$ in $\mathcal{S}_{m,n}$, we agree to write $\boldsymbol{P}^{(0)} \equiv \boldsymbol{P}$ and $\boldsymbol{P}^{(r+1)} \equiv \left(\boldsymbol{P}^{(r)}\right)'$, for $r \geq 0$. The *constants* of $\mathcal{S}_{m,n}$, namely the $\boldsymbol{P}$ of $\mathcal{S}_{m,n}$ satisfying $\boldsymbol{P}' \equiv 0$, are the elements of $\mathbb{Q}$. Thus, for Definitions 21.1–21.3, the polynomials of $\mathcal{S}_{m,n}$ not in $\mathbb{Q}$ are the *nonconstant polynomials of* $\mathcal{S}_{m,n}$.

For any $\boldsymbol{P}$ in $\mathcal{S}_{m,n}$ and any (21.1), the functions $P(z)$ and $Q(z)$ obtained by substituting $c^{(k)}_{i_1, i_2, \ldots, i_n}(z)$ from (21.1) for $\boldsymbol{w}^{(k)}_{i_1, i_2, \ldots, i_n}$ in $\boldsymbol{P}$ and $\boldsymbol{Q} \equiv \boldsymbol{P}'$ satisfy $\dfrac{d}{dz}P(z) \equiv Q(z)$. This is established by an argument similar to the one given for Proposition 4.5 on page 38. Remark 4.6 and Proposition 4.7 are also applicable.

21.4. Inclusion of relative invariants for $H_{m,q} = 0$ when $1 \leq q \leq n$

PROPOSITION 21.6. *For $1 \leq q \leq n$ and the equations having the form*

$$(21.10) \qquad \left(y^{(m)}(z)\right)^q + \sum_{\substack{0 \leq k_1, k_2, \ldots, k_q \leq m \\ (k_1, k_2, \ldots, k_q) \neq (0, 0, \ldots, 0)}} c_{k_1, k_2, \ldots, k_q}(z) \prod_{\nu=1}^{q} y^{(m-k_\nu)}(z) = 0,$$

where the coefficients $c_{k_1, k_2, \ldots, k_q}(z)$ are meromorphic functions on a region Ω of the complex plane such that

$$c_{k_{\pi(1)}, k_{\pi(2)}, \ldots, k_{\pi(q)}}(z) \equiv c_{k_1, k_2, \ldots, k_q}(z), \quad \text{for any permutation } \pi \text{ of } \{1, 2, \ldots, q\},$$

each semi-invariant for the equations (21.10) *can be directly rewritten as a semi-invariant for the equations* (21.1).

PROOF. For $q = n$, the assertion is obviously true. Suppose that $1 \leq q < n$. Let $\boldsymbol{P}$ be a semi-invariant for the equations (21.10). In particular, as an element of $\mathcal{S}_{m,q}$, $\boldsymbol{P}$ is a polynomial over $\mathbb{Q}$ in the variables

$$\boldsymbol{w}^{(k)}_{k_1, k_2, \ldots, k_q}, \quad \text{for } 0 \leq k_1 \leq k_2 \leq \cdots \leq k_q \leq m, \; k_q \geq 1, \text{ and } k \geq 0.$$

We set $p = n - q$ and, with $p \geq 1$, we see that the variables of (21.6) are given by

$$\boldsymbol{w}^{(k)}_{h_1, h_2, \ldots, h_p, k_1, k_2, \ldots, k_q},$$

for $0 \leq h_1 \leq h_2 \leq \cdots \leq h_p \leq k_1 \leq k_2 \leq \cdots \leq k_q \leq m$, $k_q \geq 1$, and $k \geq 0$.

Throughout the remainder of this proof, the symbolism $\underline{0, 0, \ldots, 0}$ will designate the list consisting of p zeros. Let $\widehat{P}$ be the polynomial in $\mathcal{S}_{m,n}$ that is obtained from P by replacing $w^{(k)}_{k_1, k_2, \ldots, k_q}$ with $w^{(k)}_{\underline{0, 0, \ldots, 0}, k_1, k_2, \ldots, k_q}$. Each (21.1) satisfying (21.1.1) specifies an equation

$$(21.11) \quad \left(y^{(m)}(z)\right)^q + \sum_{\substack{0 \leq k_1, k_2, \ldots, k_q \leq m \\ (k_1, k_2, \ldots, k_q) \neq (0, 0, \ldots, 0)}} c_{\underline{0, 0, \ldots, 0}, k_1, k_2, \ldots, k_q}(z) \prod_{\nu=1}^{q} y^{(m-k_\nu)}(z) = 0$$

of the type (21.10) with $c_{\underline{0, 0, \ldots, 0}, k_{\pi(1)}, k_{\pi(2)}, \ldots, k_{\pi(q)}}(z) \equiv c_{\underline{0, 0, \ldots, 0}, k_1, k_2, \ldots, k_q}(z)$ for any permutation π of $\{1, 2, \ldots, q\}$. The functions $\widehat{P}(z)$ and $P(z)$ obtained by replacing $w^{(k)}_{i_1, i_2, \ldots, i_n}$ in $\widehat{P}$ with $c^{(k)}_{i_1, i_2, \ldots, i_n}(z)$ from (21.1) and $w^{(k)}_{k_1, k_2, \ldots, k_q}$ in P with $c^{(k)}_{\underline{0, 0, \ldots, 0}, k_1, k_2, \ldots, k_q}(z)$ from (21.11) satisfy $\widehat{P}(z) \equiv P(z)$.

Suppose that P is a semi-invariant of the first kind for the equations (21.10). The equation (21.3) into which (21.2) transforms (21.1) via (21.2.1) and (21.2.2) specifies the corresponding equation

$$(21.12) \quad \left(v^{(m)}(z)\right)^q + \sum_{\substack{0 \leq k_1, k_2, \ldots, k_q \leq m \\ (k_1, k_2, \ldots, k_q) \neq (0, 0, \ldots, 0)}} c^*_{\underline{0, 0, \ldots, 0}, k_1, k_2, \ldots, k_q}(z) \prod_{\nu=1}^{q} v^{(m-k_\nu)}(z) = 0$$

that satisfies $c^*_{\underline{0, 0, \ldots, 0}, k_{\pi(1)}, k_{\pi(2)}, \ldots, k_{\pi(q)}}(z) \equiv c^*_{\underline{0, 0, \ldots, 0}, k_1, k_2, \ldots, k_q}(z)$, where π is any permutation of $\{1, 2, \ldots, q\}$. The functions $\widehat{P}^*(z)$ and $P^*(z)$ obtained by replacing $w^{(k)}_{i_1, i_2, \ldots, i_n}$ in $\widehat{P}$ with $c^{*(k)}_{i_1, i_2, \ldots, i_n}(z)$ from (21.3) and $w^{(k)}_{k_1, k_2, \ldots, k_q}$ in P with $c^{*(k)}_{\underline{0, 0, \ldots, 0}, k_1, k_2, \ldots, k_q}(z)$ from (21.12) satisfy $\widehat{P}^*(z) \equiv P^*(z)$. We specialize (21.2.1) and (21.2.2) to see that (21.12) is the equation into which (21.2) transforms (21.11). In view of $P^*(z) \equiv P(z)$, we obtain $\widehat{P}^*(z) \equiv P^*(z) \equiv P(z) \equiv \widehat{P}(z)$ and conclude that $\widehat{P}$ is a semi-invariant of the first kind for the equations (21.1).

Suppose that P is a semi-invariant of the second kind having weight s for the equations (21.10). The equation (21.5) into which (21.4) transforms (21.1) via (21.4.1), (21.4.2), and $u(\zeta) \equiv (y \circ f)(\zeta)$ specifies the corresponding equation

$$(21.13) \quad \left(u^{(m)}(\zeta)\right)^q + \sum_{\substack{0 \leq k_1, k_2, \ldots, k_q \leq m \\ (k_1, k_2, \ldots, k_q) \neq (0, 0, \ldots, 0)}} c^{**}_{\underline{0, 0, \ldots, 0}, k_1, k_2, \ldots, k_q}(\zeta) \prod_{\nu=1}^{q} u^{(m-k_\nu)}(\zeta) = 0,$$

where $c^{**}_{\underline{0, 0, \ldots, 0}, k_{\pi(1)}, k_{\pi(2)}, \ldots, k_{\pi(q)}}(\zeta) \equiv c^{**}_{\underline{0, 0, \ldots, 0}, k_1, k_2, \ldots, k_q}(\zeta)$ is satisfied for any permutation π of $\{1, 2, \ldots, q\}$. The functions $\widehat{P}^{**}(\zeta)$ and $P^{**}(\zeta)$ obtained by replacing $w^{(k)}_{i_1, i_2, \ldots, i_n}$ in $\widehat{P}$ with $c^{**(k)}_{i_1, i_2, \ldots, i_n}(\zeta)$ from (21.5) and $w^{(k)}_{k_1, k_2, \ldots, k_q}$ in P with $c^{**(k)}_{\underline{0, 0, \ldots, 0}, k_1, k_2, \ldots, k_q}(z)$ from (21.13) satisfy $\widehat{P}^{**}(\zeta) \equiv P^{**}(\zeta)$. Specializing (21.4.1) and (21.4.2), we find that (21.13) is the equation into which (21.4) transforms (21.11). In view of $P^{**}(\zeta) \equiv \big(f'(\zeta)\big)^s P\big(f(\zeta)\big)$, we obtain

$$\widehat{P}^{**}(\zeta) \equiv P^{**}(\zeta) \equiv \big(f'(\zeta)\big)^s P\big(f(\zeta)\big) \equiv \big(f'(\zeta)\big)^s \widehat{P}\big(f(\zeta)\big)$$

and conclude that $\widehat{P}$ is a semi-invariant of the second kind having weight s for the equations (21.1). This completes the proof. $\square$

EXAMPLE 21.7. Each relative invariant for homogeneous linear differential equations of order m specifies a corresponding relative invariant for the equations (21.1) having the same order.

Proposition 3.5 on page 34 shows that the polynomial $\boldsymbol{w}_{1,1} - (\boldsymbol{w}_{0,1})^2$ in $\mathcal{S}_{m,2}$ is a basic relative invariant for the equations $Q_m = 0$. Consequently, the polynomial

$$(21.14) \qquad \boldsymbol{D}_1 \equiv \boldsymbol{w}_{\underbrace{0,\,0,\,\ldots,\,0}_{n-2},1,1} - (\boldsymbol{w}_{\underbrace{0,\,0,\,\ldots,\,0}_{n-2},0,1})^2,$$

in $\mathcal{S}_{m,n}$ for $n \geq 2$, is a basic relative invariant of index $(\overbrace{0, 0, \ldots, 0}^{n-2}, 1, 1)$ and weight 2 for the equations $H_{m,n} = 0$. In general, for $1 \leq k \leq m$ and $n \geq 2$, Theorem 15.10 on page 164 shows that a relative invariant of weight $k(k+1)$ for the equations $H_{m,n} = 0$ is specified by suitably altering (15.24) of page 162.

There are no relative invariants of weight 1 for the equations (21.1). To see this, we modify slightly the argument for Remark 3.6 on page 34.

To generalize (21.14) for any $n \geq 2$, we introduce the notation

$$(21.15) \qquad \mathfrak{w}_i \equiv \boldsymbol{w}_{\underbrace{0,\,\ldots,\,0}_{n-i},\underbrace{1,\,\ldots,\,1}_{i}}, \quad \text{for } 0 \leq i \leq n,$$

and define $\boldsymbol{B}_i$ in $\mathcal{S}_{m,n}$ by

$$(21.16) \qquad \boldsymbol{B}_i \equiv \sum_{j=0}^{i} (-1)^j \binom{i}{j} (\mathfrak{w}_1)^j \, \mathfrak{w}_{i-j}, \quad \text{for } 2 \leq i \leq n.$$

In particular, we have $\mathfrak{w}_0 \equiv \boldsymbol{w}_{\underbrace{0,\,\ldots,\,0}_{n}} \equiv 1$ and $\boldsymbol{B}_2 \equiv \boldsymbol{D}_1$, for $\boldsymbol{D}_1$ in (21.14).

We shall establish that, *when $n \geq 2$, the polynomials $\boldsymbol{B}_2, \ldots, \boldsymbol{B}_n$ are basic relative invariants for the equations* (21.1). For $m = 1$, this result is presented in Theorem 22.1 on page 257; and, for $m \geq 2$, it is the subject of Proposition 27.14 on page 317. We must distinguish between the restriction $m = 1$ in Chapter 22 and the general situation $m \geq 2$ in Chapters 23–27. Several of our principal deductions will use the results presented in the next section.

21.5. Nonsolutions of nontrivial equations

CONTEXT 21.8. In terms of a positive integer n, suppose that the symbols

$$\boldsymbol{w}^{(k)}_{i_1, i_2, \ldots, i_n}, \quad \text{for } i_n \geq 1,\ 0 \leq i_1 \leq i_2 \leq \cdots \leq i_n, \text{ and } k \geq 0,$$

are algebraically independent variables over $\mathbb{Q}$. For $\mu = 1, 2, \ldots$, let $\mathcal{S}_{\mu,n}$ denote the ring of polynomials over $\mathcal{Q}$ in those variables $\boldsymbol{w}^{(k)}_{i_1, i_2, \ldots, i_n}$ having $1 \leq i_n \leq \mu$, $0 \leq i_1 \leq i_2 \leq \cdots \leq i_\mu$, and $k \geq 0$. When $\mu = m$, this gives $\mathcal{S}_{m,n}$ of page 251.

For the variables $\boldsymbol{w}^{(0)}_{i_1, i_2, \ldots, i_n}$ having $i_n \geq 1$ and $0 \leq i_1 \leq i_2 \leq \cdots \leq i_n$, we introduce a total ordering by the requirement that $\boldsymbol{w}^{(0)}_{p_1, p_2, \ldots, p_n}$ *precedes* $\boldsymbol{w}^{(0)}_{q_1, q_2, \ldots, q_n}$ when there is an integer r such that $1 \leq r \leq n$, $p_r < q_r$, and $p_\nu = q_\nu$ for $r < \nu \leq n$. This total ordering is then exhibited by

$$(21.17) \qquad \mathfrak{s}^{(0)}_1, \mathfrak{s}^{(0)}_2, \mathfrak{s}^{(0)}_3, \mathfrak{s}^{(0)}_4, \mathfrak{s}^{(0)}_5, \mathfrak{s}^{(0)}_6, \mathfrak{s}^{(0)}_7, \mathfrak{s}^{(0)}_8, \mathfrak{s}^{(0)}_9, \mathfrak{s}^{(0)}_{10}, \mathfrak{s}^{(0)}_{11}, \mathfrak{s}^{(0)}_{12}, \ldots.$$

We note that $\mathfrak{s}_1^{(0)} = \boldsymbol{w}_{0,\ldots,0,0,1}$, $\mathfrak{s}_2^{(0)} = \boldsymbol{w}_{0,\ldots,0,1,1}$, $\ldots$, $\mathfrak{s}_n^{(0)} = \boldsymbol{w}_{1,\ldots,1,1,1}$, $\mathfrak{s}_{n+1}^{(0)} = \boldsymbol{w}_{1,\ldots,1,1,2}$, $\ldots$. There are functions $\phi_1(h), \phi_2(h), \ldots, \phi_n(h)$ such that

(21.18) $$\mathfrak{s}_h^{(0)} \equiv \boldsymbol{w}_{\phi_1(h), \phi_2(h), \ldots, \phi_n(h)}^{(0)}, \quad \text{for } h \geq 1;$$

and there is a function $\chi(i_1, i_2, \ldots, i_n)$ such that

(21.19) $\quad \boldsymbol{w}_{i_1, i_2, \ldots, i_n}^{(0)} \equiv \mathfrak{s}_{\chi(i_1, i_2, \ldots, i_n)}^{(0)}, \quad$ for $i_n \geq 1$ and $0 \leq i_1 \leq i_2 \leq \cdots \leq i_n$.

As a supplement to (21.17), we use (21.18) to introduce $\mathfrak{s}_h^{(k)} \equiv \boldsymbol{w}_{\phi_1(h), \phi_2(h), \ldots, \phi_n(h)}^{(k)}$, for $h \geq 1$ and $k \geq 1$. Now, we have

(21.20) $$\mathfrak{s}_h^{(k)} \equiv \boldsymbol{w}_{\phi_1(h), \phi_2(h), \ldots, \phi_n(h)}^{(k)}, \quad \text{for } h \geq 1 \text{ and } k \geq 0.$$

For $\nu \geq 1$, let $\mathfrak{S}_\nu$ denote the ring of polynomials over $\mathbb{Q}$ in the variables $\mathfrak{s}_h^{(k)}$ of (21.20) subject to $1 \leq h \leq \nu$ and $k \geq 0$. In particular, for Context 21.8, we note that $\mathcal{S}_{\mu,n} \equiv \mathfrak{S}_{\chi(\mu,\mu,\ldots,\mu)}$.

PROPOSITION 21.9. *For $\nu \geq 1$, let $\boldsymbol{Q}$ be a nonzero polynomial belonging to $\mathfrak{S}_\nu$ and let z_0 be a complex number. Then, there are analytic functions $d_1(z), \ldots, d_\nu(z)$ on a neighborhood Ω of z_0 such that the analytic function $Q(z)$ on Ω obtained by replacing $\mathfrak{s}_h^{(k)}$ in $\boldsymbol{Q}$ with $d_h^{(k)}(z)$ satisfies $Q(z) \neq 0$, for each z in Ω.*

PROOF. For this modification of Proposition 4.2 on page 36, the proof given there is directly applicable without alteration. $\square$

COROLLARY 21.10. *In terms of given positive integers m and n, let $\boldsymbol{P}$ be a polynomial in $\mathcal{S}_{m,n}$ subject to $\boldsymbol{P} \not\equiv 0$ and let z_0 be a complex number. Then, there is a differential equation (21.1) having analytic coefficients $c_{i_1, i_2, \ldots, i_n}(z)$ on a region Ω containing z_0 such that the analytic function $P(z)$ on Ω obtained by replacing $\boldsymbol{w}_{i_1, i_2, \ldots, i_n}^{(k)}$ in $\boldsymbol{P}$ with $c_{i_1, i_2, \ldots, i_n}^{(k)}(z)$ satisfies $P(z) \neq 0$, for each z in Ω.*

PROOF. Set $\nu = \chi(m, m, \ldots, m)$. In view of (21.19)–(21.20), $\boldsymbol{P}$ is a polynomial in $\mathfrak{S}_\nu$. Due to Proposition 21.9, there are analytic functions $d_1(z), \ldots, d_\nu(z)$ on a region Ω that contains z_0 such that the substitution of $d_h^{(k)}(z)$ for $\mathfrak{s}_h^{(k)}$ in $\boldsymbol{P}$ yields an analytic function $P(z)$ on Ω that satisfies $P(z) \neq 0$, for each z in Ω. When $1 \leq i_n \leq m$ and $0 \leq i_1 \leq i_2 \leq \cdots \leq i_n$, we set $c_{i_1, i_2, \ldots, i_n}(z) \equiv d_{\chi(i_1, i_2, \ldots, i_n)}(z)$; and we use (21.1.1) to define the remaining coefficients of a differential equation (21.1) on Ω. Then, the substitution of $c_{i_1, i_2, \ldots, i_n}^{(k)}(z)$ for $\boldsymbol{w}_{i_1, i_2, \ldots, i_n}^{(k)}$ in $\boldsymbol{P}$ yields the same function $P(z)$. This completes the proof. $\square$

CHAPTER 22

Basic Relative Invariants for $H_{1,n} = 0$ when $n \geq 2$

22.1. Existence

For $m = 1$, (21.1) shows that $H_{1,n} = 0$ is given by

$$(22.1) \qquad \left(y^{(1)}(z)\right)^n + \sum_{\substack{0 \leq i_1, i_2, \ldots, i_n \leq 1 \\ (i_1, i_2, \ldots, i_n) \neq (0, 0, \ldots, 0)}} c_{i_1, i_2, \ldots, i_n}(z) \prod_{\nu=1}^{n} y^{(1-i_\nu)}(z) = 0,$$

where $c_{0, 0, \ldots, 0}(z) \equiv 1$ and the coefficients $c_{i_1, i_2, \ldots, i_n}(z)$ are meromorphic functions on a region Ω of the complex plane such that

$$c_{i_{\pi(1)}, i_{\pi(2)}, \ldots, i_{\pi(n)}}(z) \equiv c_{i_1, i_2, \ldots, i_n}(z), \quad \text{for any permutation } \pi \text{ of } \{1, 2, \ldots, n\}.$$

When $n = 1$, (22.1) is a first-order homogeneous linear differential equation and has no relative invariants. Thus, we suppose that $n \geq 2$.

We use (3.1)–(3.2), $m = 1$, (21.4.2), and (21.4.1) to obtain $\alpha_{1,0}(\zeta) \equiv 0$ and

$$(22.2) \qquad c^{**}_{i_1, i_2, \ldots, i_n}(\zeta) \equiv \left(f'(\zeta)\right)^{i_1 + i_2 + \cdots + i_n} c_{i_1, i_2, \ldots, i_n}(f(\zeta)),$$

for any substitution $z = f(\zeta)$ that transforms $H_{1,n} = 0$ on Ω into a corresponding $H^{**}_{1,n} = 0$ on Ω^{**}. In view of (22.2), each variable $w_{i_1, i_2, \ldots, i_n}$ from (21.6) having $k = 0$ and $0 \leq i_1 \leq i_2 \leq \cdots \leq i_n = 1$ is a semi-invariant of the second kind of weight $i_1 + \cdots + i_n$ for the equations (22.1). Consequently, each nonconstant isobaric polynomial combination over $\mathbb{Q}$ of these variables is a semi-invariant of the second kind for such equations.

THEOREM 22.1. *For $2 \leq i \leq n$, the polynomial $\boldsymbol{B}_i$ defined in $\mathcal{S}_{1,n}$ by (21.16) is a basic relative invariant of index $(\underbrace{0, \ldots, 0}_{n-i}, \underbrace{1, \ldots, 1}_{i})$ for the equations (22.1).*

PROOF. The definition of $\boldsymbol{B}_i$ shows that it is a basic polynomial having the indicated index. Thus, $\boldsymbol{B}_i$ is a semi-invariant of the second kind of weight i. For $m = 1$ and any substitution $y(z) = \rho(z) v(z)$ that transforms $H_{1,n} = 0$ into a corresponding equation $H^*_{1,n} = 0$, we set $R \equiv \rho'(z)/\rho(z)$ and see that $\mathcal{A}^{i_1, i_2, \ldots, i_n}_{j_1, j_2, \ldots, j_n}(z)$ in (21.2.2) is equal to R^μ, where μ is the number of integers ν such that $1 \leq \nu \leq n$ and $0 = j_\nu < i_\nu = 1$. Hence, (21.2.1) and (21.1.1) give

$$(22.3) \quad c^*_{\underbrace{0, \ldots, 0}_{n-i}, \underbrace{1, \ldots, 1}_{i}}(z) \equiv \sum_{0 \leq j_1, \ldots, j_i \leq 1} \mathcal{A}^{\overbrace{0, \ldots, 0}^{n-i}, \overbrace{1, \ldots, 1}^{i}}_{0, \ldots, 0, j_1, \ldots, j_i}(z)\, c_{\underbrace{0, \ldots, 0}_{n-i}, j_1, \ldots, j_i}(z)$$

$$\equiv \sum_{j=0}^{i} \binom{i}{j} R^{i-j}\, c_{\underbrace{0, \ldots, 0}_{n-j}, \underbrace{1, \ldots, 1}_{j}}(z), \qquad \text{for } 0 \leq i \leq n.$$

We set $\mathfrak{c}_k(z) \equiv c_{\underbrace{0, \ldots, 0}_{n-k}, \underbrace{1, \ldots, 1}_{k}}(z)$ and $\mathfrak{c}_k^*(z) \equiv c_{\underbrace{0, \ldots, 0}_{n-k}, \underbrace{1, \ldots, 1}_{k}}^*(z)$, for $0 \leq k \leq n$.
This enables us to rewrite (22.3) as

$$\text{(22.4)} \quad \mathfrak{c}_i^*(z) \equiv \sum_{j=0}^{i} \binom{i}{j} R^{i-j} \mathfrak{c}_j(z), \quad \text{for } 0 \leq i \leq n.$$

In particular, (22.4) yields $\mathfrak{c}_1^*(z) \equiv \mathfrak{c}_1(z) + R$. For a fixed integer i satisfying $2 \leq i \leq n$, we substitute $c_{i_1, \ldots, i_n}^*(z)$ from (21.3) for $\boldsymbol{w}_{i_1, \ldots, i_n}$ in $\boldsymbol{B}_i$ of (21.16) and use (22.4) to obtain

$$\text{(22.5)} \quad B_i^*(z) \equiv \sum_{j=0}^{i} (-1)^j \binom{i}{j} \left(\mathfrak{c}_1^*(z)\right)^j \mathfrak{c}_{i-j}^*(z)$$

$$\equiv \sum_{j=0}^{i} (-1)^j \binom{i}{j} (\mathfrak{c}_1(z) + R)^j \sum_{\nu=0}^{i-j} \binom{i-j}{\nu} R^{i-j-\nu} \mathfrak{c}_\nu(z)$$

$$\equiv \sum_{j=0}^{i} (-1)^j \binom{i}{j} \sum_{\mu=0}^{j} \binom{j}{\mu} (\mathfrak{c}_1(z))^\mu R^{j-\mu} \sum_{\nu=0}^{i-j} \binom{i-j}{\nu} R^{i-j-\nu} \mathfrak{c}_\nu(z)$$

$$\equiv \sum_{\nu=0}^{i} \sum_{j=0}^{i-\nu} \sum_{\mu=0}^{j} (-1)^j \binom{i}{j} \binom{j}{\mu} \binom{i-j}{\nu} R^{i-\mu-\nu} (\mathfrak{c}_1(z))^\mu \mathfrak{c}_\nu(z)$$

$$\equiv \sum_{k=0}^{i} \sum_{j=0}^{k} \sum_{\mu=0}^{j} (-1)^j \binom{i}{j} \binom{j}{\mu} \binom{i-j}{i-k} R^{k-\mu} (\mathfrak{c}_1(z))^\mu \mathfrak{c}_{i-k}(z)$$

$$\equiv \sum_{k=0}^{i} \sum_{\mu=0}^{k} \sum_{j=\mu}^{k} (-1)^j \binom{i}{j} \binom{j}{\mu} \binom{i-j}{i-k} R^{k-\mu} (\mathfrak{c}_1(z))^\mu \mathfrak{c}_{i-k}(z)$$

$$\equiv \sum_{k=0}^{i} \sum_{\mu=0}^{k} S_{i,k,\mu} R^{k-\mu} (\mathfrak{c}_1(z))^\mu \mathfrak{c}_{i-k}(z),$$

where, for $0 \leq \mu \leq k \leq i$, $S_{i,k,\mu}$ is given by

$$\text{(22.6)} \quad S_{i,k,\mu} \equiv \sum_{j=\mu}^{k} (-1)^j \binom{j}{\mu} \binom{i}{j} \binom{i-j}{i-k} \equiv \sum_{j=\mu}^{k} (-1)^j \binom{j}{\mu} \binom{i}{k} \binom{k}{j}$$

$$\equiv \binom{i}{k} \sum_{j=\mu}^{k} (-1)^j \binom{k}{j} \binom{j}{\mu} \equiv \binom{i}{k} \sum_{j=\mu}^{k} (-1)^j \binom{k}{\mu} \binom{k-\mu}{j-\mu}$$

$$\equiv \binom{i}{k} \binom{k}{\mu} \sum_{j=0}^{k-\mu} (-1)^{j+\mu} \binom{k-\mu}{j} \equiv (-1)^\mu \binom{i}{k} \binom{k}{\mu} T$$

and T is a sum that equals 1, when $\mu = k$, and equals $(1-1)^{k-\mu} = 0$, when $\mu < k$. In terms of $\delta_{r,s} = 1$, when $r = s$, and $\delta_{r,s} = 0$, when $r \neq s$, (22.6) yields

$$\text{(22.7)} \quad S_{i,k,\mu} \equiv \delta_{\mu,k} (-1)^\mu \binom{i}{k} \binom{k}{\mu}, \quad \text{for } 0 \leq \mu \leq k \leq i.$$

We employ (22.5), (22.7), and (21.16) to deduce that

$$B_i^*(z) \equiv \sum_{k=0}^{i} \sum_{\mu=0}^{k} \delta_{\mu,k} (-1)^\mu \binom{i}{k}\binom{k}{\mu} R^{k-\mu} (\mathfrak{c}_1(z))^\mu \mathfrak{c}_{i-k}(z)$$

$$\equiv \sum_{k=0}^{i} (-1)^k \binom{i}{k} (\mathfrak{c}_1(z))^k \mathfrak{c}_{i-k}(z) \equiv B_i(z), \quad \text{on } \Omega.$$

Thus, $\boldsymbol{B}_i$ is a semi-invariant of the first kind. It is therefore a relative invariant of weight i for the equations $H_{1,n} = 0$. This completes the proof. □

The arguments for Propositions 4.5 and 4.7 on pages 38–39 show that those results are also valid for our present context. In particular, *any nonconstant differential-polynomial combination over $\mathbb{Q}$ of semi-invariants of the first kind is a semi-invariant of the first kind.*

PROPOSITION 22.2. *For $n \geq 2$, suppose that $\boldsymbol{P}$ is a nonconstant polynomial in $\mathcal{S}_{1,n}$. Then, $\boldsymbol{P}$ is a semi-invariant of the first kind for the equations (22.1) if and only if $\boldsymbol{P}$ is expressible as a differential-polynomial combination over $\mathbb{Q}$ of $\boldsymbol{B}_2, \ldots, \boldsymbol{B}_n$.*

PROOF. If $\boldsymbol{P}$ is a differential-polynomial combination over $\mathbb{Q}$ of $\boldsymbol{B}_2, \ldots, \boldsymbol{B}_n$, then $\boldsymbol{P}$ is a semi-invariant of the first kind.

Suppose that $\boldsymbol{P}$ is a semi-invariant of the first kind for the equations (22.1). We use (21.6) to see that $\boldsymbol{P}$ is a differential-polynomial combination over $\mathbb{Q}$ of $\mathfrak{w}_1, \ldots, \mathfrak{w}_n$ from (21.15). We rewrite (21.16) to obtain

(22.8) $$\mathfrak{w}_i \equiv \boldsymbol{B}_i + \boldsymbol{Q}_i, \quad \text{for } 2 \leq i \leq n,$$

where $\boldsymbol{Q}_i$ is a differential-polynomial combination over $\mathbb{Q}$ of $\mathfrak{w}_1, \ldots, \mathfrak{w}_{i-1}$. By applying the derivation $'$ for $\mathcal{S}_{1,n}$ to (22.8), we obtain

(22.9) $$\mathfrak{w}_i^{(j)} \equiv \boldsymbol{B}_i^{(j)} + \boldsymbol{R}_{i,j}, \quad \text{for } 2 \leq i \leq n \text{ and } j \geq 0,$$

where $\boldsymbol{R}_{i,j}$ is a differential-polynomial combination over $\mathbb{Q}$ of $\mathfrak{w}_1, \ldots, \mathfrak{w}_{i-1}$. For $i = n$ and $j \geq 0$, we substitute (22.9) in $\boldsymbol{P}$ to write $\boldsymbol{P}$ as a differential-polynomial combination over $\mathbb{Q}$ of $\mathfrak{w}_1, \ldots, \mathfrak{w}_{n-1}, \boldsymbol{B}_n$. Then, for $i = n-1$ and $j \geq 0$, we substitute (22.9) in that expression for $\boldsymbol{P}$ to write $\boldsymbol{P}$ as a differential-polynomial combination over $\mathbb{Q}$ of $\mathfrak{w}_1, \ldots, \mathfrak{w}_{n-2}, \boldsymbol{B}_{n-1}, \boldsymbol{B}_n$. Continuing in this manner, we express $\boldsymbol{P}$ as a differential-polynomial combination over $\mathbb{Q}$ of $\mathfrak{w}_1, \boldsymbol{B}_2, \ldots, \boldsymbol{B}_{n-1}, \boldsymbol{B}_n$. Thus, we can write $\boldsymbol{P}$ in the form

(22.10) $$\boldsymbol{P} \equiv \boldsymbol{\Phi} + \boldsymbol{\Psi},$$

where $\boldsymbol{\Phi}$ is a differential-polynomial combination over $\mathbb{Q}$ of $\boldsymbol{B}_2, \ldots, \boldsymbol{B}_n$ and $\boldsymbol{\Psi}$ is a differential-polynomial combination over $\mathbb{Q}$ of $\mathfrak{w}_1, \boldsymbol{B}_2, \ldots, \boldsymbol{B}_{n-1}, \boldsymbol{B}_n$ such that any nonzero term of $\boldsymbol{\Psi}$ has a factor of the form $\mathfrak{w}_1^{(j)}$, for some $j \geq 0$. For any given (22.1), there is a meromorphic function $\rho(z) \not\equiv 0$ on a subregion $\mathcal{U}$ of Ω such that

(22.11) $$\rho'(z) + c_{\underbrace{0, \ldots, 0}_{n-1}, 1}(z)\, \rho(z) \equiv 0, \quad \text{on } \mathcal{U}.$$

The substitution $y(z) = \rho(z)\, v(z)$ transforms the restriction of (22.1) to $\mathcal{U}$ into an equation (21.3) on $\mathcal{U}$ having $m = 1$ and satisfying (21.3.1). We use (21.2.1),

(21.2.2), and (22.11) to see that this (21.3) has

$$(22.12) \qquad c^*_{\underbrace{0,\,\ldots,\,0}_{n-1},\,1}(z) \equiv c_{\underbrace{0,\,\ldots,\,0}_{n-1},\,1}(z) + \frac{\rho'(z)}{\rho(z)} \equiv 0, \quad \text{on } \mathcal{U}.$$

Thus, the function $\Psi^*(z)$ obtained by substituting $c^*_{i_1,\ldots,i_n}(z)$ from this (21.3) for $w_{i_1,\ldots,i_n}$ in Ψ satisfies $\Psi^*(z) \equiv 0$. We have $P^*(z) \equiv P(z)$ as well as $\Phi^*(z) \equiv \Phi(z)$. Hence, we find that $\Psi(z) \equiv P(z) - \Phi(z) \equiv P^*(z) - \Phi^*(z) \equiv \Psi^*(z) \equiv 0$ on $\mathcal{U}$ and therefore also on Ω for any (21.1) having $m = 1$ and the current $n \geq 2$. Consequently, Corollary 21.10 yields $\Psi \equiv 0$. In view of (22.10), we obtain $\boldsymbol{P} \equiv \boldsymbol{\Phi}$ and see that $\boldsymbol{P}$ is a differential-polynomial combination over $\mathbb{Q}$ of $\boldsymbol{B}_2, \ldots, \boldsymbol{B}_n$. This completes the proof. □

OBSERVATION 22.3. The special case $n = 2$ of Proposition 22.2 was presented as Proposition 4.8 on page 40. It lead to the simple structural characterization of relative invariants given for $n = 2$ in Theorem 4.10 on page 41. However, when $n \geq 3$, it is not true that each relative invariant for the equations (22.1) is expressible as a polynomial combination over $\mathbb{Q}$ of $\boldsymbol{B}_2, \ldots, \boldsymbol{B}_n$. For instance, the proof on page 147 for Proposition 14.6 shows that, when $n \geq 3$, the polynomial $\boldsymbol{I} \equiv \boldsymbol{B}_2 \boldsymbol{B}_3^{(1)} - \frac{3}{2} \boldsymbol{B}_2^{(1)} \boldsymbol{B}_3$ is a relative invariant of weight 6 for the equations (22.1); but, $\boldsymbol{I}$ is not expressible as a polynomial combination over $\mathbb{Q}$ of $\boldsymbol{B}_2, \ldots, \boldsymbol{B}_n$.

The uniqueness of $\boldsymbol{B}_2, \ldots, \boldsymbol{B}_n$ as basic relative invariants will be established in Theorem 22.6. In particular, $\boldsymbol{B}_2, \ldots, \boldsymbol{B}_n$ involve only the variables $\mathfrak{w}_\mu^{(\nu)}$ having $1 \leq \mu \leq n$ and $\nu = 0$. At this point, Definition 21.5 on page 252 does not obviously rule out the existence of a basic relative invariant involving some $\mathfrak{w}_\mu^{(\nu)}$ with $\nu \geq 1$.

22.2. Some polynomials that are not relative invariants

By analogy with $\mathbb{Q}\{w_{0,1}, w_{0,2}\}$ for Theorem 12.1 on page 121, we use (21.15) to introduce $\mathbb{Q}\{\mathfrak{w}_1\}$ as the ring of polynomials over $\mathbb{Q}$ in $\mathfrak{w}_1, \mathfrak{w}_1^{(1)}, \mathfrak{w}_1^{(2)}, \ldots$.

PROPOSITION 22.4. *Suppose that $\boldsymbol{P}$ is a polynomial in $\mathbb{Q}\{\mathfrak{w}_1\}$. Then $\boldsymbol{P}$ is not a semi-invariant of the first kind for the equations* (22.1).

PROOF. For an indirect argument, assume that $\boldsymbol{P}$ in $\mathbb{Q}\{\mathfrak{w}_1\}$ is a semi-invariant of the first kind. In particular, since $\boldsymbol{P}$ is not a constant, it has at least one term of weight ≥ 1. Let $\boldsymbol{H}$ be the sum of those terms of $\boldsymbol{P}$ that have greatest weight. Then, $\boldsymbol{H}$ is an isobaric semi-invariant of the first kind for the equations (22.1). We can establish that directly with a shorter proof than the one leading to Corollary 20.12 on page 234. Each nonzero term of $\boldsymbol{H}$ has a factor $\mathfrak{w}_1^{(k)}$, for some $k \geq 0$. When any (22.1) on Ω is given, there is a corresponding meromorphic function $\rho(z) \not\equiv 0$ on a subregion $\mathcal{U}$ of Ω that satisfies (22.11). The substitution $y(z) = \rho(z) v(z)$ transforms the restriction of (22.1) to $\mathcal{U}$ into an equation (21.3) on $\mathcal{U}$ having $m = 1$ and satisfying (21.3.1). We use (22.11) as well as (21.2.1) and (21.2.2) to see that this (21.3) satisfies (22.12). Thus, for the function $H^*(z)$ obtained by substituting $c^{*(k)}_{i_1,\ldots,i_n}(z)$ from this (21.3) for $w^{(k)}_{i_1,\ldots,i_n}$ in $\boldsymbol{H}$, we have $H^*(z) \equiv 0$. We find that this yields $H(z) \equiv H^*(z) \equiv 0$ on $\mathcal{U}$ and therefore also on Ω for any (21.1) having $m = 1$ and the current $n \geq 2$. Consequently, Corollary 21.10 gives the contradiction that $\boldsymbol{H} \equiv 0$. Thus, $\boldsymbol{P}$ is not a semi-invariant of the first kind. This completes the proof. □

22.2. SOME POLYNOMIALS THAT ARE NOT RELATIVE INVARIANTS

THEOREM 22.5. *In terms of nonnegative integers n, p, r that satisfy $2 \leq p \leq n$, let $\boldsymbol{R}$ in $\mathcal{S}_{1,n}$ be an isobaric polynomial of weight $s > p$ that is expressible as*

$$(22.13) \qquad \boldsymbol{R} \equiv \boldsymbol{S}\,\mathfrak{w}_p^{(r)} + \boldsymbol{T},$$

where $\boldsymbol{S}$ is a nonzero polynomial in $\mathbb{Q}\{\mathfrak{w}_1\}$ and $\boldsymbol{T}$ is a polynomial in $\mathcal{S}_{1,n}$ such that any variable $\mathfrak{w}_\mu^{(\nu)}$ effectively involved in $\boldsymbol{T}$ has either $1 \leq \mu < p$ or $\mu = p$ and $0 \leq \nu < r$. Then, $\boldsymbol{R}$ is not a relative invariant for the equations (22.1).

PROOF. For an indirect argument, suppose that $\boldsymbol{R}$ is a relative invariant for the equations (22.1). There are two cases depending on whether $\boldsymbol{S}$ is a constant.

Case 1. Suppose that $\boldsymbol{S}$ is a constant. Since $\boldsymbol{R}$ is isobaric of weight s, we have $r = s - p \geq 1$. Thus, there are rational numbers $\gamma_0 \neq 0$ and γ_1 such that

$$(22.14) \qquad \boldsymbol{R} \equiv \gamma_0\,\mathfrak{w}_p^{(r)} + \gamma_1\,\mathfrak{w}_1\,\mathfrak{w}_p^{(r-1)} + \boldsymbol{T}_0,$$

where $\boldsymbol{T}_0$ is a polynomial over $\mathbb{Q}$ in the variables $\mathfrak{w}_i^{(k)}$ from (21.15) having either $1 \leq i < p$ or $i = p$ and $0 \leq k \leq r-2$. We use the notation

$$\mathfrak{c}_i(z) \equiv c_{\underbrace{0,\ldots,0}_{n-i},\underbrace{1,\ldots,1}_{i}}(z) \quad \text{and} \quad \mathfrak{c}_i^{**}(\zeta) \equiv c^{**}_{\underbrace{0,\ldots,0}_{n-i},\underbrace{1,\ldots,1}_{i}}(\zeta), \quad \text{for } 0 \leq i \leq n,$$

as well as the property that $\boldsymbol{R}$ in (22.14) is a semi-invariant of the second kind of weight s to obtain

$$(22.15) \quad 0 \equiv R^{**}(\zeta) - \bigl(f'(\zeta)\bigr)^s R\bigl(f(\zeta)\bigr)$$
$$\equiv \gamma_0\Bigl[\mathfrak{c}_p^{**(r)}(\zeta) - \bigl(f'(\zeta)\bigr)^s \mathfrak{c}_p^{(r)}\bigl(f(\zeta)\bigr)\Bigr] + \Bigl[T_0^{**}(\zeta) - \bigl(f'(\zeta)\bigr)^s T_0\bigl(f(\zeta)\bigr)\Bigr]$$
$$+ \gamma_1\Bigl[\mathfrak{c}_1^{**}(\zeta)\,\mathfrak{c}_p^{**(r-1)}(\zeta) - \bigl(f'(\zeta)\bigr)^s \mathfrak{c}_1\bigl(f(\zeta)\bigr)\,\mathfrak{c}_p^{(r-1)}\bigl(f(\zeta)\bigr)\Bigr],$$

for any (22.1) and $z = f(\zeta)$. We see that (22.2) yields

$$(22.16) \qquad \mathfrak{c}_i^{**}(\zeta) \equiv \bigl(f'(\zeta)\bigr)^i \mathfrak{c}_i\bigl(f(\zeta)\bigr), \quad \text{for } 0 \leq i \leq n.$$

We employ (22.16) in a computation similar to that for (12.19) on page 125 to deduce that

$$(22.17) \quad \mathfrak{c}_p^{**(r)}(\zeta) - \bigl(f'(\zeta)\bigr)^s \mathfrak{c}_p^{(r)}\bigl(f(\zeta)\bigr)$$
$$\equiv \frac{r(2p+r-1)}{2}\bigl(f'(\zeta)\bigr)^{s-2} f''(\zeta)\,\mathfrak{c}_p^{(r-1)}\bigl(f(\zeta)\bigr) + F_1(\zeta),$$

where $F_1(\zeta)$ is a polynomial combination over $\mathbb{Q}$ of $f'(\zeta)$, $f''(\zeta)$, ... and $\mathfrak{c}_p^{(\nu)}\bigl(f(\zeta)\bigr)$, for $0 \leq \nu \leq r-2$. In particular, $F_1(\zeta)$ does not involve $\mathfrak{c}_p^{(r-1)}\bigl(f(\zeta)\bigr)$. We use (22.16) to obtain

$$(22.18) \quad \mathfrak{c}_\mu^{**(\nu)}(\zeta) \equiv \bigl(f'(\zeta)\bigr)^{\mu+\nu} \mathfrak{c}_\mu^{(\nu)}\bigl(f(\zeta)\bigr) + H_{\mu,\nu}(\zeta), \quad \text{for } 1 \leq \mu \leq n \text{ and } \nu \geq 0,$$

where $H_{\mu,\nu}(\zeta)$ is a polynomial combination over $\mathbb{Q}$ of $f'(\zeta)$, $f''(\zeta)$, ... and those $\mathfrak{c}_\mu^{(k)}\bigl(f(\zeta)\bigr)$ having $0 \leq k < \nu$. Each term of $T_0^{**}(\zeta)$ consists of a rational number times a product of expressions like $\mathfrak{c}_\mu^{**(\nu)}(\zeta)$ in (22.18) for which either $1 \leq \mu < p$ or $\mu = p$ and $0 \leq \nu \leq r-2$. Consequently, we have

$$(22.19) \qquad T_0^{**}(\zeta) \equiv \bigl(f'(\zeta)\bigr)^s T_0\bigl(f(\zeta)\bigr) + F_2(\zeta),$$

where $F_2(\zeta)$ is a polynomial combination over $\mathbb{Q}$ of $f'(\zeta)$, $f''(\zeta)$, ... and expressions $\mathfrak{c}_\mu^{(\nu)}(f(\zeta))$ for which either $\mu < p$ or $\mu = p$ and $\nu \leq r - 2$. In particular, $F_2(\zeta)$ does not involve $\mathfrak{c}_p^{(r-1)}(f(\zeta))$. We verify that (22.18) yields

$$\mathfrak{c}_1^{**}(\zeta) \equiv f'(\zeta)\,\mathfrak{c}_1(f(\zeta)), \qquad \mathfrak{c}_p^{**(r-1)}(\zeta) \equiv (f'(\zeta))^{p+r-1}\mathfrak{c}_p^{(r-1)}(f(\zeta)) + \cdots$$

and

$$(22.20) \qquad \mathfrak{c}_1^{**}(\zeta)\,\mathfrak{c}_p^{**(r-1)}(\zeta) - (f'(\zeta))^s\,\mathfrak{c}_1(f(\zeta))\,\mathfrak{c}_p^{(r-1)}(f(\zeta)) \equiv F_3(\zeta),$$

where $F_3(\zeta)$ is a polynomial combination over $\mathbb{Q}$ of $f'(\zeta)$, $f''(\zeta)$, ... and expressions $\mathfrak{c}_\mu^{(\nu)}(f(\zeta))$ for which either $\mu = 1$ and $\nu = 0$ or $\mu = p$ and $\nu \leq r - 2$. We combine (22.15) with (22.17), (22.19), and (22.20) to deduce that

$$(22.21) \qquad 0 \equiv \gamma_0 \frac{r(2p + r - 1)}{2}(f'(\zeta))^{s-2}f''(\zeta)\,\mathfrak{c}_p^{(r-1)}(f(\zeta)) + F_4(\zeta)$$

where $F_4(\zeta)$ is a polynomial combination over $\mathbb{Q}$ of $f'(\zeta)$, $f''(\zeta)$, ... and expressions $\mathfrak{c}_\mu^{(\nu)}(f(\zeta))$ that do not include $\mathfrak{c}_p^{(r-1)}(f(\zeta))$. For any (22.1) on Ω and any univalent analytic function $\zeta = g(z)$ on Ω with inverse function designated by $z = f(\zeta)$, we use (12.24) on page 126 to rewrite (22.21) as

$$(22.22) \qquad 0 \equiv \gamma_0 \frac{r(2p + r - 1)}{2}\frac{(-g''(z))}{(g'(z))^{s+1}}\mathfrak{c}_p^{(r-1)}(z) + G_5(z),$$

where $G_5(z)$ is expressible as a product of an integral power of $g'(z)$ and a polynomial combination over $\mathbb{Q}$ of expressions $g^{(\lambda)}(z)$ and $\mathfrak{c}_\mu^{(\nu)}(z)$ having $\lambda \geq 1$ and either $\mu < p$ or $\mu = p$ and $\nu \leq r - 2$.

Let t_1 be an integer such that: for any (22.1) on Ω and any univalent analytic function $\zeta = g(z)$ on Ω, the product of $(g'(z))^{t_1}$ and the right member of (22.22) is a polynomial combination over $\mathbb{Q}$ of $g^{(\lambda)}(z)$ and $\mathfrak{c}_\mu^{(\nu)}(z)$, for various integers $\lambda \geq 1$, $1 \leq \mu \leq p$, and $\nu \geq 0$. We consider the ring $\mathfrak{S}_{n+1}$ that is defined on page 256 following (21.20). It consists of the polynomials over $\mathbb{Q}$ in the variables $\mathfrak{w}_i^{(k)}$ and $\mathfrak{s}_{n+1}^{(k)}$, for $1 \leq i \leq n$ and $k \geq 0$. Let Q_1 be the polynomial in $\mathfrak{S}_{n+1}$ such that: for any (22.1) and any $\zeta = g(z)$, the function obtained by substituting $\mathfrak{c}_\mu^{(\nu)}(z)$ for $\mathfrak{w}_\mu^{(\nu)}$ and $g^{(\lambda)}(z)$ for $\mathfrak{s}_{n+1}^{(\lambda)}$ in Q_1 is equal to the product of $(g'(z))^{t_1+1}$ and the right member of (22.22). We have $Q_1 \not\equiv 0$ because $\mathfrak{c}_p^{(r-1)}(z)$ is effectively involved in (22.22). Consequently, Proposition 21.9 on page 256 yields meromorphic functions $\mathfrak{c}_1(z), \ldots, \mathfrak{c}_n(z)$ and $h(z)$ on a region Ω_0 such that the function $Q_1(z)$ on Ω_0 obtained by substituting $\mathfrak{c}_\mu^{(\nu)}(z)$ for $\mathfrak{w}_\mu^{(\nu)}$ and $h^{(\lambda)}(z)$ for $\mathfrak{s}_{n+1}^{(\lambda)}$ in Q_1 satisfies $Q_1(z) \neq 0$, for each z in Ω_0. Since the $(t_1 + 1)$th power of $g'(z)$ was used in defining Q_1, we have $h(z) \not\equiv 0$. Let Ω be a subregion of Ω_0 on which $h(z)$ is a univalent analytic function and set $g(z) \equiv h(z)$ on Ω. Let (22.1) be the unique equation of that type having

$$c_{\underbrace{0,\ldots,0}_{n-i},\underbrace{1,\ldots,1}_{i}}(z) \equiv \mathfrak{c}_i(z), \quad \text{on } \Omega \text{ for } 1 \leq i \leq n.$$

Then, for this particular (22.1) and $\zeta = g(z)$, (22.22) is not satisfied. Thus, Case 1 leads to a contradiction.

22.2. SOME POLYNOMIALS THAT ARE NOT RELATIVE INVARIANTS

Case 2. Suppose that S in (22.13) is not a constant. For any (22.1) on Ω and any meromorphic function $\rho(z) \not\equiv 0$ on Ω, there is a unique equation (21.3) on Ω having $m = 1$ into which the substitution $y(z) = \rho(z)\, v(z)$ transforms (22.1). Its coefficients are given by (22.3) and (21.3.1). We use (22.13) and the property that R is a semi-invariant of the first kind to obtain

(22.23) $\quad 0 \equiv R^*(z) - R(z) \equiv \left[S^*(z)\, \mathfrak{c}_p^{*(r)}(z) - S(z)\, \mathfrak{c}_p^{(r)}(z) \right] + \left[T^*(z) - T(z) \right].$

Since S is a polynomial over $\mathbb{Q}$ in the variables $\mathfrak{w}_1^{(\nu)}$, for $\nu \geq 0$, (22.4) yields

(22.24) $\quad\quad\quad\quad\quad\quad S^*(z) \equiv S(z) + \Phi_1(z),$

where $\Phi_1(z)$ is a polynomial combination over $\mathbb{Q}$ of $\rho^\lambda(z)/\rho(z)$ and $\mathfrak{c}_1^{(\nu)}(z)$, for $\lambda \geq 1$ and $\nu \geq 0$. After setting $i = p$ in (22.4), we differentiate the resulting expression r times with respect to z and obtain

(22.25) $\quad\quad\quad\quad\quad\quad \mathfrak{c}_p^{*(r)}(z) \equiv \mathfrak{c}_p^{(r)}(z) + \Phi_2(z),$

where $\Phi_2(z)$ is a polynomial combination over $\mathbb{Q}$ of expressions $\rho^{(\lambda)}(z)/\rho(z)$ and $\mathfrak{c}_\mu^{(\nu)}(z)$, for $\lambda \geq 1$, $1 \leq \mu < p$, and $0 \leq \nu \leq r$. We combine (22.24) and (22.25) to deduce that

(22.26) $\quad S^*(z)\, \mathfrak{c}_p^{*(r)}(z) - S(z)\, \mathfrak{c}_p^{(r)}(z) \equiv \Phi_1(z)\, \mathfrak{c}_p^{*(r)}(z) + \Phi_3(z),$

where $\Phi_3(z)$ is a polynomial combination over $\mathbb{Q}$ of expressions $\rho^{(\lambda)}(z)/\rho(z)$ and $\mathfrak{c}_\mu^{(\nu)}(z)$, for $\lambda \geq 1$, $1 \leq \mu < p$, and $0 \leq \nu$. In particular, $\Phi_3(z)$ does not involve $\mathfrak{c}_p^{(r)}(z)$. We use (22.4) to verify that

(22.27) $\quad\quad \mathfrak{c}_\mu^{*(\nu)}(z) \equiv \mathfrak{c}_\mu^{(\nu)}(z) + \Psi_{\mu,\nu}(z), \quad \text{for } 1 \leq \mu \leq p \text{ and } \nu \geq 0,$

where $\Psi_{\mu,\nu}(z)$ is a polynomial combination over $\mathbb{Q}$ of expressions $\rho^{(\lambda)}(z)/\rho(z)$ and $\mathfrak{c}_i^{(k)}(z)$, for $\lambda \geq 1$, $1 \leq i < \mu$, and $k \geq 0$. Each term of $T^*(z)$ consists of a rational number times a product of expressions like $\mathfrak{c}_\mu^{*(\nu)}(z)$ in (22.27) for which either $1 \leq \mu < p$ and $\nu \geq 0$ or $\mu = p$ and $0 \leq \nu < r$. Consequently, we have

(22.28) $\quad\quad\quad\quad\quad\quad T^*(z) - T(z) \equiv \Phi_4(z),$

where $\Phi_4(z)$ is a polynomial combination over $\mathbb{Q}$ of expressions $\rho^{(\lambda)}(z)/\rho(z)$ and $\mathfrak{c}_i^{(k)}(z)$, for $\lambda \geq 1$ as well as either $0 \leq i < p$ and $k \geq 0$ or $i = p$ and $0 \leq k < r$. In particular, $\Phi_4(z)$ does not involve $\mathfrak{c}_p^{(r)}(z)$. For any (22.1) and any substitution $y(z) = \rho(z)\, v(z)$, we use (22.23), (22.26), and (22.28) to obtain

(22.29) $\quad\quad\quad\quad\quad\quad 0 \equiv \Phi_1(z)\, \mathfrak{c}_p^{(r)}(z) + \Phi_5(z),$

where $\Phi_5(z)$ is a polynomial combination over $\mathbb{Q}$ of $\rho^{(\lambda)}(z)/\rho(z)$ and $\mathfrak{c}_\mu^{(\nu)}(z)$ for $\lambda \geq 1$ and either $1 \leq \mu < p$ or $\mu = p$ and $0 \leq \nu < r$. In particular, $\Phi_5(z)$ does not involve $\mathfrak{c}_p^{(r)}(z)$.

Let t_2 be an integer such that: for any (22.1) and $y(z) = \rho(z)\, v(z)$, the product of $(\rho(z))^{t_2}$ and the right member of (22.29) is a polynomial combination over $\mathbb{Q}$ of $\rho^{(\lambda)}(z)$ and $\mathfrak{c}_\mu^{(\nu)}(z)$, for various integers $\lambda \geq 0$, $1 \leq \mu \leq p$, and $\nu \geq 0$. We consider the ring $\mathfrak{S}_{n+1}$ that is defined on page 256 following (21.20). It consists of the polynomials over $\mathbb{Q}$ in the variables $\mathfrak{w}_i^{(k)}$ and $\mathfrak{s}_{n+1}^{(k)}$, for $1 \leq i \leq n$ and $k \geq 0$. Let $\mathbf{Q}_2$ be the polynomial in $\mathfrak{S}_{n+1}$ such that: for any (22.1) and any $y(z) = \rho(z)\, v(z)$, the function obtained by substituting $\mathfrak{c}_\mu^{(\nu)}(z)$ for $\mathfrak{w}_\mu^{(\nu)}$ and $\rho^{(\lambda)}(z)$ for $\mathfrak{s}_{n+1}^{(\lambda)}$ in $\mathbf{Q}_2$ is equal to the product of $\bigl(\rho(z)\bigr)^{t_2+1}$ and the right member of (22.29). Since $\boldsymbol{S}$ is neither a constant nor a semi-invariant of the first kind, there is a particular (22.1) and a particular $y(z) = \rho(z)\,v(z)$ for which (22.24) has $\Phi_1(z) \not\equiv 0$. Moreover, $\Phi_5(z)$ does not involve $\mathfrak{c}_p^{(r)}(z)$. Consequently, we have $\boldsymbol{Q}_2 \not\equiv 0$. Now, Proposition 21.9 on page 256 yields meromorphic functions $\mathfrak{c}_1(z)$, ..., $\mathfrak{c}_n(z)$ and $\rho(z)$ on a region Ω such that the function $Q_2(z)$ on Ω obtained by substituting $\mathfrak{c}_\mu^{(\nu)}(z)$ for $\mathfrak{w}_\mu^{(\nu)}$ and $\rho^{(\lambda)}(z)$ for $\mathfrak{s}_{n+1}^{(\lambda)}$ in $\boldsymbol{Q}_2$ satisfies $Q_2(z) \neq 0$, for each z in Ω. Since the (t_2+1)th power of $\rho(z)$ was used in defining $\boldsymbol{Q}_2$, we see that $\rho(z) \not\equiv 0$. Let (22.1) be the unique equation of that type having

$$c_{\underbrace{0,\ldots,0}_{n-i},\underbrace{1,\ldots,1}_{i}}(z) \equiv \mathfrak{c}_i(z), \quad \text{on } \Omega \text{ for } 1 \le i \le n.$$

Then, for this particular (22.1) and the corresponding substitution $y(z) = \rho(z)\,v(z)$, (22.29) is not satisfied. Thus, Case 2 *leads to a contradiction*.

Since each of the two possible cases leads to a contradiction, we conclude that the supposition "$\boldsymbol{R}$ is a relative invariant" is false. Thus, $\boldsymbol{R}$ is not a relative invariant. This completes the proof of Theorem 22.5. □

22.3. Uniqueness of $\boldsymbol{B}_2, \ldots, \boldsymbol{B}_n$ as basic relative invariants

Theorem 22.1 on page 257 shows that: for $2 \le i \le n$, $\boldsymbol{B}_i$ in (21.16) is a basic relative invariant of index $(\underbrace{0,\ldots,0}_{n-i},\underbrace{1,\ldots,1}_{i})$ for the equations (22.1).

THEOREM 22.6. *Suppose that $\boldsymbol{B}$ is a basic relative invariant for the equations* (22.1). *Then, there is an integer s such that $2 \le s \le n$ and $\boldsymbol{B} \equiv \boldsymbol{B}_s$.*

PROOF. The indices for $\boldsymbol{B}_2, \ldots, \boldsymbol{B}_n$ include all possible indices that can be possessed by basic relative invariants for the equations (22.1). Let s be the integer such that $2 \le s \le n$ and the indices of $\boldsymbol{B}$ and $\boldsymbol{B}_s$ are equal.

To prove indirectly that $\boldsymbol{B} \equiv \boldsymbol{B}_s$, suppose that $\boldsymbol{B} \not\equiv \boldsymbol{B}_s$. Then, the nonconstant polynomial $\boldsymbol{R} \equiv \boldsymbol{B} - \boldsymbol{B}_s$ is a relative invariant of weight s for the equations (22.1). Since the coefficients of $\mathfrak{w}_s$ in $\boldsymbol{B}$ is 1 and the coefficient of $\mathfrak{w}_s$ in $\boldsymbol{B}_s$ is 1, the coefficient of $\mathfrak{w}_s$ in $\boldsymbol{R}$ is 0. Proposition 22.4 shows that $\boldsymbol{R}$ effectively involves some $\mathfrak{w}_\mu^{(\nu)}$ having $\mu \ge 2$ and $\nu \ge 0$. Let p be the greatest integer μ such that $\boldsymbol{R}$ effectively involves $\mathfrak{w}_\mu^{(\nu)}$, for some $\nu \ge 0$. Let r be the greatest integer ν such that $\boldsymbol{R}$ effectively involves $\mathfrak{w}_p^{(\nu)}$. The structure of basic relative invariants required by Definition 21.5 on page 252 enables us to express $\boldsymbol{R}$ in the form

$$\boldsymbol{R} \equiv \boldsymbol{S}\,\mathfrak{w}_p^{(r)} + \boldsymbol{T}, \tag{22.30}$$

where $\boldsymbol{S}$ is a nonzero polynomial in $\mathbb{Q}\{\mathfrak{w}_1\}$ and $\boldsymbol{T}$ is a polynomial in $\mathcal{S}_{1,n}$ such that any variable $\mathfrak{w}_\mu^{(\nu)}$ effectively involved in $\boldsymbol{T}$ has either $1 \le \mu < p$ or $\mu = p$ and $0 \le \nu < r$. Theorem 22.5 on page 261 is applicable to (22.30) and yields the contradiction that $\boldsymbol{R}$ is not a relative invariant for the equations (22.1). Hence, we must have $\boldsymbol{B} \equiv \boldsymbol{B}_s$. This completes the proof. □

CHAPTER 23

Laguerre-Forsyth Forms for $H_{m,n} = 0$ when $m \geq 2$

23.1. The composite of substitutions (21.2) and (21.4)

Rather than change the dependent variable alone or the independent variable alone, we now employ a composite substitution for (21.1) based on $y(z) = \rho(z)\,v(z)$ in (21.2) and $z = f(\zeta)$ in (21.4). Thus, (21.1)-(21.1.1), (21.2), and (21.4) specify unique meromorphic functions $d_{i_1, i_2, \ldots, i_n}(\zeta)$ on Ω^{**} such that: (21.1) on Ω subject to (21.1.1) is transformed by (21.2)-(21.2.1)-(21.2.2) into (21.3) on Ω; and (21.3) on Ω is transformed by (21.4)-(21.4-1)-(21.4-2) with $w(\zeta) = (v \circ f)(\zeta)$ into

$$(23.1) \qquad \left(w^{(m)}(\zeta)\right)^n + \sum_{\substack{0 \leq i_1, i_2, \ldots, i_n \leq m \\ (i_1, i_2, \ldots, i_n) \neq (0, 0, \ldots, 0)}} d_{i_1, i_2, \ldots, i_n}(\zeta) \prod_{\nu=1}^{n} w^{(m-i_\nu)}(\zeta) = 0,$$

on Ω^{**}. Propositions 21.0.1 and 21.0.2 of pages 250–251 yield

$$(23.2) \qquad d_{i_{\pi(1)}, i_{\pi(2)}, \ldots, i_{\pi(n)}}(\zeta) \equiv d_{i_1, i_2, \ldots, i_n}(\zeta),$$

for any permutation π of $\{1, 2, \ldots, n\}$.

We shall use the notation $c_{0, 0, \ldots, 0}(z) \equiv 1$ on Ω and $d_{0, 0, \ldots, 0}(\zeta) \equiv 1$ on Ω^{**}.

THEOREM 23.1. *For $m \geq 1$, the coefficients of the differential equation (23.1) into which (21.2), (21.4), and $w(\zeta) = (v \circ f)(\zeta)$ transform (21.1) are given on Ω^{**} in terms of (21.1), $\rho(z)$, $\zeta = g(z)$ on Ω, and $z = f(\zeta)$ on Ω^{**} by*

$$(23.3) \qquad d_{i_1, i_2, \ldots, i_n}(\zeta) \equiv \sum_{j_1=0}^{i_1} \sum_{j_2=0}^{i_2} \cdots \sum_{j_n=0}^{i_n} \mathcal{C}^{i_1, i_2, \ldots, i_n}_{j_1, j_2, \ldots, j_n}(\zeta)\, c_{j_1, j_2, \ldots, j_n}\bigl(f(\zeta)\bigr),$$

for $0 \leq i_1, i_2, \ldots, i_n \leq m$,

where $\mathcal{C}^{i_1, i_2, \ldots, i_n}_{j_1, j_2, \ldots, j_n}(\zeta)$ is defined on Ω^{**} in terms of

$$(23.4) \qquad \phi_{0,j}(z) \equiv 1, \quad \text{on } \Omega \text{ for any } j,$$

and

$$(23.5) \qquad \phi_{i+1, j}(z) \equiv \sum_{k=j+1}^{m} \left[\phi^{(1)}_{i,k}(z) + \left[\frac{\rho'(z)}{\rho(z)} + (m - i - k)\frac{g''(z)}{g'(z)} \right] \phi_{i,k}(z) \right],$$

on Ω for $i \geq 0$ and any j,

through

$$(23.6) \qquad \mathcal{C}^{i_1, i_2, \ldots, i_n}_{j_1, j_2, \ldots, j_n}(\zeta) \equiv \bigl(f'(\zeta)\bigr)^{i_1 + i_2 + \cdots + i_n} \prod_{\nu=1}^{n} \phi_{i_\nu - j_\nu, j_\nu}\bigl(f(\zeta)\bigr).$$

PROOF. In terms of the analytic functions $\beta_{i,j}(z)$ defined on Ω by (3.13)–(3.14) of page 32, the identity (3.15) on page 33 enables (21.4.2) to be rewritten as

$$\text{(23.7)} \quad \mathcal{B}^{i_1, i_2, \ldots, i_n}_{k_1, k_2, \ldots, k_n}(\zeta) \equiv \prod_{\nu=1}^{n} \left[(f'(\zeta))^{i_\nu} \beta_{i_\nu - k_\nu, k_\nu}(f(\zeta)) \right], \quad \text{on } \Omega^{**}.$$

For $0 \leq i_1, i_2, \ldots, i_n \leq m$, we use (21.4.1), (21.2.1), (23.7), and (21.2.2) as well as the abbreviations

$$\text{(23.8)} \quad \gamma_{i,j,k}(z) \equiv \beta_{i-k,k}(z) \binom{m-j}{k-j} \frac{\rho^{(k-j)}(z)}{\rho(z)}, \quad \text{on } \Omega \text{ for } j \leq k \leq i,$$

and $r_{i,j}(z) \equiv \sum_{k=j}^{i} \gamma_{i,j,k}(z)$, on Ω, to obtain

$$d_{i_1, i_2, \ldots, i_n}(\zeta) \equiv \sum_{k_1=0}^{i_1} \cdots \sum_{k_n=0}^{i_n} \mathcal{B}^{i_1, \ldots, i_n}_{k_1, \ldots, k_n}(\zeta) \, c^*_{k_1, \ldots, k_n}(f(\zeta))$$

$$\equiv \sum_{k_1=0}^{i_1} \cdots \sum_{k_n=0}^{i_n} \sum_{j_1=0}^{k_1} \cdots \sum_{j_n=0}^{k_n} \mathcal{B}^{i_1, \ldots, i_n}_{k_1, \ldots, k_n}(\zeta) \, \mathcal{A}^{k_1, \ldots, k_n}_{j_1, \ldots, j_n}(f(\zeta)) \, c_{j_1, \ldots, j_n}(f(\zeta))$$

$$\equiv \sum_{j_1=0}^{i_1} \cdots \sum_{j_n=0}^{i_n} \sum_{k_1=j_1}^{i_1} \cdots \sum_{k_n=j_n}^{i_n} \left[\prod_{\nu=1}^{n} (f'(\zeta))^{i_\nu} \gamma_{i_\nu, j_\nu, k_\nu}(f(\zeta)) \right] c_{j_1, \ldots, j_n}(f(\zeta))$$

$$\equiv \sum_{j_1=0}^{i_1} \cdots \sum_{j_n=0}^{i_n} (f'(\zeta))^{i_1 + \cdots + i_n} \left[c_{j_1, \ldots, j_n}(f(\zeta)) \right] \prod_{\nu=1}^{n} \left[\sum_{k_\nu=j_\nu}^{i_\nu} \gamma_{i_\nu, j_\nu, k_\nu}(f(\zeta)) \right]$$

$$\equiv \sum_{j_1=0}^{i_1} \cdots \sum_{j_n=0}^{i_n} (f'(\zeta))^{i_1 + \cdots + i_n} \left[\prod_{\nu=1}^{n} r_{i_\nu, j_\nu}(f(\zeta)) \right] c_{j_1, \ldots, j_n}(f(\zeta)).$$

We apply the identity (5.11) of page 48 to the preceding formula and deduce (23.6) for (23.3). This completes the proof. □

COROLLARY 23.2. *As the composite of (21.2) and (21.4) with $w(\zeta) = (v \circ f)(\zeta)$ transforms (21.1) on Ω into (23.1) on Ω^{**}, it transforms*

$$\text{(23.9)} \quad y^{(m)}(z) + \sum_{j=1}^{m} c_{\underbrace{0, 0, \ldots, 0}_{n-1}, j}(z) \, y^{(m-j)}(z) = 0, \quad \text{on } \Omega,$$

into

$$\text{(23.10)} \quad w^{(m)}(\zeta) + \sum_{j=1}^{m} d_{\underbrace{0, 0, \ldots, 0}_{n-1}, j}(\zeta) \, w^{(m-j)}(\zeta) = 0, \quad \text{on } \Omega^{**}.$$

PROOF. For $1 \leq i \leq m$, there are unique meromorphic coefficients $c_i^*(z)$ on Ω and $d_i(\zeta)$ on Ω^{**} such that (21.2) transforms (23.9) into

$$\text{(23.11)} \quad v^{(m)}(z) + \sum_{j=0}^{m} c_j^*(z) \, v^{(m-j)}(z) = 0, \quad \text{on } \Omega,$$

and (21.4), with $w(\zeta) = (v \circ f)(\zeta)$, transforms (23.11) into

$$(23.12) \qquad w^{(m)}(\zeta) + \sum_{j=0}^{m} d_j(\zeta)\, w^{(m-j)}(\zeta) = 0, \quad \text{on } \Omega^{**}.$$

We apply [**24**, Theorem 6.2, page 41] to obtain

$$(23.13) \qquad d_i(\zeta) \equiv \bigl(f'(\zeta)\bigr)^i \sum_{j=0}^{i} \phi_{i-j,j}\bigl(f(\zeta)\bigr)\, c_{\underbrace{0,\,0,\,\ldots,\,0}_{n-1},\,j}\bigl(f(\zeta)\bigr),$$

on Ω^{**} for $1 \leq i \leq m$. By comparing (23.13) with the formula that results by substituting $i_1 = i_2 = \cdots = i_{n-1} = 0$ in (23.3) and using (23.6), we conclude that $d_i(\zeta) \equiv d_{\underbrace{0,\,0,\,\ldots,\,0}_{n-1},\,i}(\zeta)$, for $1 \leq i \leq m$. This completes the proof. $\square$

23.2. Laguerre-Forsyth reductions when $m \geq 2$

Through analogy with (1.14), (1.16), (1.15), and (1.17), we introduce

$$(23.14) \quad \boldsymbol{a}_{m,2} \equiv \frac{1}{\binom{m+1}{3}}\Bigl[\boldsymbol{w}_{\underbrace{0,\,\ldots,\,0}_{n-1},\,2} - \tfrac{m-1}{2}\boldsymbol{w}^{(1)}_{\underbrace{0,\,\ldots,\,0}_{n-1},\,1} - \tfrac{m-1}{2m}\bigl(\boldsymbol{w}_{\underbrace{0,\,\ldots,\,0}_{n-1},\,1}\bigr)^2\Bigr],$$

$$(23.15) \quad \boldsymbol{b}_{m,1} \equiv \frac{1}{\binom{m}{2}}\boldsymbol{w}_{\underbrace{0,\,\ldots,\,0}_{n-1},\,1},$$

$$(23.16) \quad \boldsymbol{b}_{m,2} \equiv \frac{1}{\binom{m+1}{3}}\Bigl[\boldsymbol{w}_{\underbrace{0,\,\ldots,\,0}_{n-1},\,2} - \tfrac{m-2}{3}\boldsymbol{w}^{(1)}_{\underbrace{0,\,\ldots,\,0}_{n-1},\,1} - \tfrac{(3m-1)(m-2)}{6m(m-1)}\bigl(\boldsymbol{w}_{\underbrace{0,\,\ldots,\,0}_{n-1},\,1}\bigr)^2\Bigr]$$

as polynomials in $\mathcal{S}_{m,n}$ and we note the identity

$$(23.17) \quad \boldsymbol{a}_{m,2} \equiv \boldsymbol{b}_{m,2} - \tfrac{1}{2}\boldsymbol{b}^{(1)}_{m,1} - \tfrac{1}{4}(\boldsymbol{b}_{m,1})^2.$$

PROPOSITION 23.3. *For $m \geq 2$, suppose that $y(z) = \rho(z)\,v(z)$, $z = f(\zeta)$, and $w(\zeta) = (v \circ f)(\zeta)$ transform (21.1) into (23.1). Then, (23.1) satisfies*

$$(23.18) \qquad d_{i_1,\,i_2,\,\ldots,\,i_n}(\zeta) \equiv 0, \quad \text{for } i_1 = i_2 = \cdots = i_{n-1} = 0 \text{ and } 1 \leq i_n \leq 2,$$

if and only if $\rho(z)$ and the inverse $\zeta = g(z)$ of $z = f(\zeta)$ satisfy

$$(23.19) \qquad \frac{\rho''(z)}{\rho(z)} - \left[\frac{m-2}{m-1}\right]\left[\frac{\rho'(z)}{\rho(z)}\right]^2 + b_{m,1}(z)\frac{\rho'(z)}{\rho(z)} + (m-1)\,b_{m,2}(z) \equiv 0$$

and

$$(23.20) \qquad -\frac{g''(z)}{g'(z)} \equiv b_{m,1}(z) + \frac{2\rho'(z)}{(m-1)\,\rho(z)}.$$

Moreover, if $\rho(z)$ and $g(z)$ are related by (23.20), then (23.19) is equivalent to

$$(23.21) \qquad \frac{g'''(z)}{2g'(z)} - \frac{3}{4}\left[\frac{g''(z)}{g'(z)}\right]^2 \equiv a_{m,2}(z).$$

PROOF. Corollary 23.2 shows that (23.1) satisfies (23.18) if and only if (23.10) satisfies (23.18). Now, as in the argument for Proposition 5.3 on page 49, we apply [**24**, page 46, Proposition 6.7] to complete the proof. $\square$

PROPOSITION 23.4. *For any (21.1) on Ω and any subregion $\mathcal{W}$ of Ω, there are analytic function $\rho_0(z)$ and $g_0(z)$ on a subregion $\mathcal{U}$ of $\mathcal{W}$ such that: $\rho_0(z) \neq 0$, for each z in $\mathcal{U}$; $g_0(z)$ is univalent on $\mathcal{U}$; and $\rho_0(z)$, $g_0(z)$ satisfy (23.19)–(23.20) on $\mathcal{U}$.*

PROOF. The reasoning is the same as for Proposition 5.4 on page 49. □

DEFINITION 23.5. A Laguerre-Forsyth canonical form for an equation (21.1) of order $m \geq 2$ on Ω is an equation of the form (23.1) satisfying (23.2) and (23.18) on a region $\mathcal{V}$ such that the restriction of (21.1) to some subregion $\mathcal{U}$ of Ω can be transformed into this (23.1) on $\mathcal{V}$.

Propositions 23.4 and 23.3 show that each equation (21.1) having $m \geq 2$ possesses numerous Laguerre-Forsyth canonical forms.

23.3. Related Laguerre-Forsyth canonical forms

When a Laguerre-Forsyth canonical form is transformed into another, there are various identities relating their coefficients. In this section, we generalize the results of Chapter 6 to our present context. To keep the notation simple, we shall assume that one of the canonical forms is (21.1) and the other is (23.1).

Our generalization for (6.5) on page 54 is given by

$$(23.22) \quad B^{i_1, i_2, \ldots, i_n}_{j_1, j_2, \ldots, j_n} \equiv \prod_{\nu=1}^{n} \left[(j_\nu)! \binom{i_\nu - 1}{j_\nu} \binom{m - i_\nu + j_\nu}{j_\nu} \right], \quad \text{for } 0 \leq j_1, j_2, \ldots, j_n.$$

PROPOSITION 23.6. *For $m \geq 2$, suppose that (21.1) satisfies*

$$c_{i_1, i_2, \ldots, i_n}(z) \equiv 0, \quad \text{when } i_1 = i_2 = \cdots = i_{n-1} = 0 \text{ and } 1 \leq i_n \leq 2,$$

and suppose that $y(z) = \rho(z) v(z)$, $z = f(\zeta)$, and $w(\zeta) = (v \circ f)(\zeta)$ transform (21.1) on Ω into an equation (23.1) on Ω^{**} that satisfies

$$d_{i_1, i_2, \ldots, i_n}(\zeta) \equiv 0, \quad \text{when } i_1 = i_2 = \cdots = i_{n-1} = 0 \text{ and } 1 \leq i_n \leq 2.$$

Then, either there are complex numbers $a \neq 0$, $b \neq 0$, z_0 such that

$$(23.23) \qquad \rho(z) \equiv a, \quad \text{on } \Omega, \quad f(\zeta) \equiv b\zeta + z_0 \quad \text{on } \Omega^{**},$$

and

$$(23.24) \quad d_{i_1, \ldots, i_n}(\zeta) \equiv \bigl(f'(\zeta)\bigr)^{i_1 + \cdots + i_n} c_{i_1, \ldots, i_n}\bigl(f(\zeta)\bigr), \quad \text{for } 0 \leq i_1, \ldots, i_n \leq m,$$

or else there are complex numbers $a \neq 0$, $b \neq 0$, z_0, ζ_0 such that

$$(23.25) \quad \rho(z) \equiv a\,(z - z_0)^{m-1} \quad \text{on } \Omega, \quad f(\zeta) \equiv \frac{b}{\zeta - \zeta_0} + z_0, \quad \text{on } \Omega^{**},$$

and, in terms of (23.22),

$$(23.26) \quad d_{i_1, \ldots, i_n}(\zeta) \equiv$$

$$\bigl(f'(\zeta)\bigr)^{i_1 + \cdots + i_n} \sum_{j_1=0}^{i_1} \cdots \sum_{j_n=0}^{i_n} B^{i_1, \ldots, i_n}_{j_1, \ldots, j_n} \left[\frac{\zeta - \zeta_0}{b} \right]^{j_1 + \cdots + j_n} c_{i_1 - j_1, \ldots, i_n - j_n}\bigl(f(\zeta)\bigr),$$

$$\text{for } 0 \leq i_1, \ldots, i_n \leq m.$$

PROOF. Simple modifications in the proof on pages 54–56 for Proposition 6.1 suffice. The main alteration occurs at the point on page 56 where, in place of $d_{i,j}(\zeta)$, the computation of $d_{i_1,\ldots,i_n}(\zeta)$ by means of (23.3), (23.6), and (6.13) gives (23.26) with respect to (23.22). □

Through analogy with (6.16) on page 56, we define integers $D_{j_1,\ldots,j_n,r}^{i_1,\ldots,i_n,k}$ in terms of (23.22) by

$$(23.27) \quad D_{j_1,\ldots,j_n,r}^{i_1,\ldots,i_n,k} \equiv \binom{k}{r}\left[\prod_{t=1}^{r}(2i_1 + \cdots + 2i_n - j_1 - \cdots - j_n + k - t)\right] B_{j_1,\ldots,j_n}^{i_1,\ldots,i_n},$$

for $0 \le j_1 \le i_1, \ldots, 0 \le j_n \le i_n$, and any k, r.

PROPOSITION 23.7. *The kth derivative of $d_{i_1,\ldots,i_n}(\zeta)$ in (23.26) with respect to ζ is given on Ω^{**} in terms of (23.25), (23.27), and $h = i_1 + \cdots + i_n + k$ by*

$$(23.28) \quad d_{i_1,\ldots,i_n}^{(k)}(\zeta) \equiv$$

$$\left(f'(\zeta)\right)^h \sum_{j_1=0}^{i_1} \cdots \sum_{j_n=0}^{i_n} \sum_{r=0}^{k} D_{j_1,\ldots,j_n,r}^{i_1,\ldots,i_n,k} \left[\frac{\zeta - \zeta_0}{b}\right]^{j_1+\cdots+j_n+r} c_{i_1-j_1,\ldots,i_n-j_n}^{(k-r)}(f(\zeta)),$$

for $0 \le i_1, \ldots, i_n \le m$ and $k \ge 0$.

PROOF. Formula (23.26) shows that (23.28) is true for $k = 0$. Suppose that k is a positive integer such that (23.28) is valid when k is replaced with $k - 1$. In terms of $T(\zeta) \equiv (\zeta - \zeta_0)/b$ and $d = j_1 + \cdots + j_n$, we have

$$d_{i_1,\ldots,i_n}^{(k-1)}(\zeta) \equiv \left(f'(\zeta)\right)^{h-1} \sum_{j_1=0}^{i_1} \cdots \sum_{j_n=0}^{i_n} \sum_{r=0}^{k-1} D_{j_1,\ldots,j_n,r}^{i_1,\ldots,i_n,k-1} [T(\zeta)]^{d+r} c_{i_1-j_1,\ldots,i_n-j_n}^{(k-1-r)}(f(\zeta)).$$

After differentiating this expression with respect to ζ, we use $T'(\zeta) \equiv 1/b$ as well as $1/f'(\zeta) \equiv (-b)[T(\zeta)]^2$ and $f''(\zeta)/\left(f'(\zeta)\right)^2 \equiv 2T(\zeta)$ to obtain

$$d_{i_1,\ldots,i_n}^{(k)}(\zeta) \equiv \left(f'(\zeta)\right)^h \sum_{j_1=0}^{i_1} \cdots \sum_{j_n=0}^{i_n} \sum_{r=0}^{k-1} D_{j_1,\ldots,j_n,r}^{i_1,\ldots,i_n,k-1} [T(\zeta)]^{d+r} c_{i_1-j_1,\ldots,i_n-j_n}^{(k-r)}(f(\zeta))$$

$$+ \left(f'(\zeta)\right)^{h-1} \sum_{j_1=0}^{i_1} \cdots \sum_{j_n=0}^{i_n} \sum_{r=0}^{k-1} D_{j_1,\ldots,j_n,r}^{i_1,\ldots,i_n,k-1} \frac{(d+r)}{b} [T(\zeta)]^{d+r-1} c_{i_1-j_1,\ldots,i_n-j_n}^{(k-1-r)}(f(\zeta))$$

$$+ \frac{(h-1)}{\left(f'(\zeta)\right)^{-h}} \frac{f''(\zeta)}{\left(f'(\zeta)\right)^2} \sum_{j_1=0}^{i_1} \cdots \sum_{j_n=0}^{i_n} \sum_{r=0}^{k-1} D_{j_1,\ldots,j_n,r}^{i_1,\ldots,i_n,k-1} [T(\zeta)]^{d+r} c_{i_1-j_1,\ldots,i_n-j_n}^{(k-1-r)}(f(\zeta))$$

$$\equiv \left(f'(\zeta)\right)^h \sum_{j_1=0}^{i_1} \cdots \sum_{j_n=0}^{i_n} \sum_{r=0}^{k-1} D_{j_1,\ldots,j_n,r}^{i_1,\ldots,i_n,k-1} [T(\zeta)]^{d+r} c_{i_1-j_1,\ldots,i_n-j_n}^{(k-r)}(f(\zeta))$$

$$+ \left(f'(\zeta)\right)^h \sum_{j_1=0}^{i_1} \cdots \sum_{j_n=0}^{i_n} \sum_{r=0}^{k-1} (-d-r) D_{j_1,\ldots,j_n,r}^{i_1,\ldots,i_n,k-1} [T(\zeta)]^{d+r+1} c_{i_1-j_1,\ldots,i_n-j_n}^{(k-1-r)}(f(\zeta))$$

$$+ \left(f'(\zeta)\right)^h \sum_{j_1=0}^{i_1} \cdots \sum_{j_n=0}^{i_n} \sum_{r=0}^{k-1} (2h-2) D_{j_1,\ldots,j_n,r}^{i_1,\ldots,i_n,k-1} [T(\zeta)]^{d+r+1} c_{i_1-j_1,\ldots,i_n-j_n}^{(k-1-r)}(f(\zeta))$$

and

$$d_{i_1,...,i_n}^{(k)}(\zeta) \equiv (f'(\zeta))^h \sum_{j_1=0}^{i_1} \cdots \sum_{j_n=0}^{i_n} \sum_{r=0}^{k-1} D_{j_1,...,j_n,r}^{i_1,...,i_n,k-1} [T(\zeta)]^{d+r} c_{i_1-j_1,...,i_n-j_n}^{(k-r)}(f(\zeta))$$

$$+ (f'(\zeta))^h \sum_{j_1=0}^{i_1} \cdots \sum_{j_n=0}^{i_n} \sum_{r=1}^{k} \begin{bmatrix} 2h-d \\ -r-1 \end{bmatrix} D_{j_1,...,j_n,r-1}^{i_1,...,i_n,k-1} [T(\zeta)]^{d+r} c_{i_1-j_1,...,i_n-j_n}^{(k-r)}(f(\zeta)).$$

Since (23.27) yields $D_{j_1,...,j_n,r}^{i_1,...,i_n,k-1} \equiv 0$ for $r=k$ and $D_{j_1,...,j_n,r-1}^{i_1,...,i_n,k-1} \equiv 0$ for $r=0$, we see that

(23.29) $d_{i_1,...,i_n}^{(k)}(\zeta)$

$$\equiv (f'(\zeta))^h \sum_{j_1=0}^{i_1} \cdots \sum_{j_n=0}^{i_n} \sum_{r=0}^{k} E_{j_1,...,j_n,r}^{i_1,...,i_n,k} [T(\zeta)]^{d+r} c_{i_1-j_1,...,i_n-j_n}^{(k-r)}(f(\zeta)),$$

where

(23.30) $\quad E_{j_1,...,j_n,r}^{i_1,...,i_n,k} \equiv D_{j_1,...,j_n,r}^{i_1,...,i_n,k-1} + (2h-d-r-1) D_{j_1,...,j_n,r-1}^{i_1,...,i_n,k-1}$,

when $0 \le r \le k$ and $0 \le j_\nu \le i_\nu$ for $1 \le \nu \le n$.

For $1 \le r \le k$, we set $v = 2i_1 + \cdots + 2i_n - j_1 - \cdots - j_n$ and combine (23.30) with (23.27) to deduce that

(23.31) $E_{j_1,...,j_n,r}^{i_1,...,i_n,k} \equiv \binom{k-1}{r} \left[\prod_{t=1}^{r} (v+(k-1)-t) \right] B_{j_1,...,j_n}^{i_1,...,i_n}$

$$+ (v+2k-r-1) \binom{k-1}{r-1} \left[\prod_{t=1}^{r-1} (v+(k-1)-t) \right] B_{j_1,...,j_n}^{i_1,...,i_n}$$

$$\equiv R \left[\prod_{t=1}^{r-1} (v+k-1-t) \right] B_{j_1,...,j_n}^{i_1,...,i_n},$$

where

(23.32) $\quad R \equiv \binom{k-1}{r} (v+k-1-r) + (v+2k-1-r) \binom{k-1}{r-1}$

$$\equiv \binom{k}{r} (v+k-1).$$

Thus, for $1 \le r \le k$, we find that (23.31), (23.32), and (23.27) yield

(23.33) $\quad E_{j_1,...,j_n,r}^{i_1,...,i_n,k} \equiv \binom{k}{r} \left[\prod_{t=0}^{r-1} (v+k-1-t) \right] B_{j_1,...,j_n}^{i_1,...,i_n} \equiv D_{j_1,...,j_n,r}^{i_1,...,i_n,k}.$

For $r=0$, we apply (23.30) and (23.27) to verify that

$$E_{j_1,...,j_n,0}^{i_1,...,i_n,k} \equiv D_{j_1,...,j_n,0}^{i_1,...,i_n,k-1} \equiv B_{j_1,...,j_n}^{i_1,...,i_n} \equiv D_{j_1,...,j_n,0}^{i_1,...,i_n,k}.$$

Consequently, (23.33) is valid for $0 \le r \le k$. By replacing $E_{j_1,...,j_n,r}^{i_1,...,i_n,k}$ in (23.29) with $D_{j_1,...,j_n,r}^{i_1,...,i_n,k}$, we conclude that (23.28) is valid for any $k \ge 0$. This completes the proof. $\square$

23.4. Identities for the coefficients of related canonical forms

Our main developments throughout Parts 1 and 2 were based on an initial discovery of the coefficients $A^{i,j}_{p,q}$ in (6.4) of page 54 with respect to which the identities in (6.23) of Theorem 6.3 on page 58 are valid. We found those coefficients by using computer algebra with $d^{(k)}_{i,j}(\zeta)$ in (6.17) of page 56. By subtracting one member of (6.23) from the other, a trial-and-error assignment of guesses for $A^{i,j}_{p,q}$ led to (6.3) and (6.4).

For the equations $H_{m,n} = 0$ of (21.1), our generalization of (6.4) on page 54 is

$$(23.34) \quad A^{i_1, i_2, \ldots, i_n}_{j_1, j_2, \ldots, j_n} \equiv \frac{(-1)^{j_1+j_2+\cdots+j_n} \prod_{\nu=1}^{n} \left[\binom{m - i_\nu + j_\nu}{j_\nu} \prod_{k=1}^{j_\nu} (i_\nu - k) \right]}{\prod_{k=1}^{j_1+j_2+\cdots+j_n} (2i_1 + 2i_2 + \cdots + 2i_n - 1 - k)},$$

whenever $j_1 + j_2 + \cdots + j_n \leq 2i_1 + 2i_2 + \cdots + 2i_n - 2$.

These coefficients reduce to those in (6.4) when $n = 2$.

THEOREM 23.8. *For $m \geq 2$, suppose that (21.1) satisfies*

$$(23.35) \quad c_{i_1, i_2, \ldots, i_n}(z) \equiv 0, \quad \text{when } i_1 = i_2 = \cdots = i_{n-1} = 0 \text{ and } 1 \leq i_n \leq 2,$$

in addition to (21.1.1) and suppose that the substitutions $y(z) = \rho(z) v(z)$ and $z = f(\zeta)$ with $w(\zeta) = (v \circ f)(\zeta)$ transform (21.1) on Ω into an equation (23.1) on Ω^{**} that satisfies

$$(23.36) \quad d_{i_1, i_2, \ldots, i_n}(\zeta) \equiv 0, \quad \text{when } i_1 = i_2 = \cdots = i_{n-1} = 0 \text{ and } 1 \leq i_n \leq 2,$$

in addition to (23.2). Then, the coefficients of (23.1) and (21.1) are related through (23.34) and $z = f(\zeta)$ by

$$(23.37) \quad \sum_{j_1=0}^{i_1} \cdots \sum_{j_n=0}^{i_n} A^{i_1, \ldots, i_n}_{j_1, \ldots, j_n} d^{(j_1+\cdots+j_n)}_{i_1-j_1, \ldots, i_n-j_n}(\zeta)$$

$$\equiv (f'(\zeta))^{i_1+\cdots+i_n} \sum_{j_1=0}^{i_1} \cdots \sum_{j_n=0}^{i_n} A^{i_1, \ldots, i_n}_{j_1, \ldots, j_n} c^{(j_1+\cdots+j_n)}_{i_1-j_1, \ldots, i_n-j_n}(f(\zeta)),$$

for $0 \leq i_1, \ldots, i_n \leq m$ and $2 \leq i_1 + \cdots + i_n$.

COMMENT. The proofs of the various results presented from this point onward do not require Theorem 23.8. However, the coefficients $A^{i_1, i_2, \ldots, i_n}_{j_1, j_2, \ldots, j_n}$, as defined by (23.34), are essential in the remainder of Part 4. We shall establish Theorem 23.8 in Corollary 27.8 on page 315 as a simple consequence of Theorems 25.10 and 27.7.

OBSERVATION 23.9. For $n = 1$, the equations $H_{m,1} = 0$ are homogeneous linear ones. Then, (23.37) and (23.34) reduce to (5.28) and (5.27) of page 51. The relation of (5.28) and (5.27) to A. R. Forsyth's formulas in [**31**, pages 403–407] is explained in [**24**, pages 77–78 and page 49]..

For $n = 2$, the equations $H_{m,2} = 0$ are the equations $Q_m = 0$ of (1.1) on page 4. In that situation, (23.37) and (23.34) simplify to (6.23) and (6.4). Then, Theorem 23.8 reduces to Theorem 6.3 on page 58.

Theorem 23.8 provides motivation for the study of the Laguerre-Forsyth sums that we introduce in Section 24.4.

23.5. Use of computer algebra to check Theorem 23.8

Rather than establish Theorem 23.8 in Corollary 27.8 on page 315, a direct proof could be presented here. Namely, if the transformation is given by (23.23), then (23.37) is an immediate consequence of (23.24). If the transformation is given by (23.25), then we can establish (23.37) by using (23.26) and suitably modifying the proof for Theorem 6.3 that occurs on pages 58–64. However, at this point, it suffices to have a simple check based on computer algebra.

Next, for $n = 3$, we apply (23.28) on page 269 to compute the difference of the left and right members in (23.37). Using MATHEMATICA in which the kth derivative with respect to ζ of $d_{i_1,i_2,i_3}(\zeta)$ is represented by d[i1,i2,i3,k] and the kth derivative with respect to z of $c_{i_1,i_2,i_3}(z)$ at $z = f(\zeta)$ is represented by c[i1,i2,i3,k], we find that the successive evaluations of the input statements

```
p = 5;

c[i2_,i1_,i3_,k_] := c[i1,i2,i3,k] /; i2 > i1

c[i1_,i3_,i2_,k_] := c[i1,i2,i3,k] /; i3 > i2

exprD[i1_,j1_,i2_,j2_,i3_,j3_,k_,r_] :=
    Binomial[k,r]*Product[2*i1+2*i2+2*i3-j1-j2-j3+k-t,{t,1,r}]*
    (j1!)*Binomial[i1-1,j1]*Binomial[m-i1+j1,j1]*
    (j2!)*Binomial[i2-1,j2]*Binomial[m-i2+j2,j2]*
    (j3!)*Binomial[i3-1,j3]*Binomial[m-i3+j3,j3]

d[i1_,i2_,i3_,k_] := (F)^(i1+i2+i3+k)*Sum[ (T)^(j1+j2+j3+r)*
    exprD[i1,j1,i2,j2,i3,j3,k,r]*c[i1-j1,i2-j2,i3-j3,k-r],
    {j1,0,i1},{j2,0,i2},{j3,0,i3},{r,0,k}]

A[i1_,j1_,i2_,j2_,i3_,j3_]:=(-1)^(j1+j2+j3)*
    Binomial[m-i1+j1,j1]*Binomial[m-i2+j2,j2]*
    Binomial[m-i3+j3,j3]*Product[i1-k,{k,1,j1}]*
    Product[i2-k,{k,1,j2}]*Product[i3-k,{k,1,j3}]/
    Product[2i1+2i2+2i3-1-k,{k,1,j1+j2+j3}]
    /; 2i1+2i2+2i3-2>=j1+j2+j3

Inv[i1_,i2_,i3_] := Sum[ A[i1,j1,i2,j2,i3,j3]*
    c[i1-j1,i2-j2,i3-j3,j1+j2+j3],{j1,0,i1},{j2,0,i2},{j3,0,i3}]

InvSharp[i1_,i2_,i3_] := Sum[ A[i1,j1,i2,j2,i3,j3]*
    d[i1-j1,i2-j2,i3-j3,j1+j2+j3],{j1,0,i1},{j2,0,i2},{j3,0,i3}]

difference[i1_,i2_,i3_] := FullSimplify[InvSharp[i1,i2,i3]
                        - (F)^(i1+i2+i3)*Inv[i1,i2,i3]]

Do[ Print[ "difference[", i1, ", ", i2,", ", i3,  "] = ",
        Expand[ difference[i1,i2,i3] ], ",       ", Date[] ],
        {i1,0,p}, {i2,i1,p}, {i3,Max[i2,2-i1-i2],p}]
```

can be done in less than a minute. The various expressions `difference[i1,i2,i3]` evaluate to 0. Since the indices may be permuted, this checks (23.37) for $n = 3$ with $0 \leq i_1, i_2, i_3 \leq 5 \leq m$ and $i_1 + i_2 + i_3 \geq 2$. Longer computational times are required when the initial assignment `p = 5` is increased. Similar checks can be performed for $n = 4$, etc.

Formula (23.37) is the key to our discovery that the basic relative invariants in $\mathcal{S}_{m,n}$, for $m \geq 2$ and $n \geq 1$, are given by (24.14) of page 276.

23.6. Properties of $a_{m,2}$ in (23.14) and $b_{m,2}$ in (23.16)

PROPOSITION 23.10. *For $m \geq 2$ and $n \geq 1$, $a_{m,2}$ in (23.14) is an isobaric semi-invariant of the first kind having weight 2 for the equations (21.1) and $b_{m,2}$ in (23.16) is a semi-invariant of the second kind of weight 2 for the equations (21.1).*

PROOF. As an element of $\mathcal{S}_{m,1}$ for $m \geq 2$, the polynomial

$$\frac{1}{\binom{m+1}{3}}\left[w_2 - \tfrac{m-1}{2}w_1^{(1)} - \tfrac{m-1}{2m}(w_1)^2\right]$$

is a semi-invariant of the first kind for homogeneous linear differential equations of order $m \geq 2$; namely, see [**24**, page 142, Proposition A.11]. Therefore, for $m \geq 2$, $n \geq 1$, and $q = 1$, Proposition 21.6 shows that $a_{m,2}$ in (23.14) is a semi-invariant of the first kind for the equations (21.1). Clearly, $a_{m,2}$ is isobaric of weight 2. For $m \geq 2$, the polynomial

$$\frac{1}{\binom{m+1}{3}}\left[w_2 - \tfrac{m-2}{3}w_1^{(1)} - \tfrac{(3m-1)(m-2)}{6m(m-1)}(w_1)^2\right]$$

in $\mathcal{S}_{m,1}$ is a semi-invariant of the second kind for homogeneous linear differential equations of order m; namely, see [**24**, page 143, Proposition A.13]. Consequently, for $m \geq 2$, $n \geq 1$, and $q = 1$, Proposition 21.6 is applicable and establishes that $b_{m,2}$ is a semi-invariant of the second kind of weight 2 for the equations (21.1). This completes the proof. □

OBSERVATION 23.11. The property that $a_{m,2}$ is a semi-invariant of the first kind is needed to deduce on page 308 that $R_{i,j}(z)$ in (27.6) satisfies (27.7). Similarly, the property that $b_{m,2}$ is a semi-invariant of the second kind is needed on page 310 to deduce that $S_{i,j}(\zeta)$ in (27.22) satisfies (27.23).

For the following generalization of Proposition 4.14 on page 43 to the classes of equations $H_{m,n} = 0$ in (21.1), the situations $m = 1$ or $n = 1$ must be excluded.

PROPOSITION 23.12. *Suppose that m and n are fixed integers satisfying $m \geq 2$ and $n \geq 2$. Then, the isobaric semi-invariants of the first kind of weight 2 for the equations $H_{m,n} = 0$ in (21.1) are given by the nonzero linear combinations over $\mathbb{Q}$ of $a_{m,2}$ in (23.14) and D_1 in (21.14). Moreover, the semi-invariants of the second kind of weight 2 for the equations $H_{m,n} = 0$ in (1.1) are given by the nonzero linear combinations over $\mathbb{Q}$ of $b_{m,2}$ in (23.16) and D_1 in (21.14).*

PROOF. After using (21.2.1), (21.2.2), (21.4.1), (21.4.2), (3.1), (3.18), and (3.23) to obtain explicit expressions for

$$c^*_{\underbrace{0,\,0,\,\ldots,\,0}_{n-2},i,j}(z) \quad \text{and} \quad c^{**}_{\underbrace{0,\,0,\,\ldots,\,0}_{n-2},i,j}(\zeta), \quad \text{when } 0 \leq i \leq j \text{ and } 1 \leq i+j \leq 2,$$

we see that the resulting formulas are the same as those obtained formally from (2.21), (2.22), (2.24), (3.19), (3.20), and (3.24) by replacing, at each occurrence, the subscripts $_{i,j}$ with the subscripts $\underbrace{0,0,\ldots,0}_{n-2},i,j$. Similarly, this replacement of subscripts in the right members of (1.14) and (1.15) yields the right members of (23.14) and (23.16). We find that this replacement of subscripts throughout the proof on pages 43–44 for Proposition 4.14 yields a proof of Proposition 23.12. □

CHAPTER 24

Formulas for Basic Relative Invariants when $m \geq 2$

For ease of reference in Chapters 25 and 26 we place here the formulas that define the basic relative invariants for the equations (21.1) in the general case having $m \geq 2$ and $n \geq 1$. In particular, we shall include the situation of homogeneous linear differential equations where $n = 1$ along with the various nonlinear equations having $n \geq 2$. To do this, we use the convention in formulas (24.9), (24.12)–(24.14), and (24.20)–(24.23) that the subscripts $h_1, \ldots, h_{n-1}$ are absent when $n = 1$. Also, when $n = 1$, we interpret $h_1 + \cdots + h_{n-1}$ in (24.9) and (24.20) as a vacuous sum having the value 0.

24.1. Formulas for (21.1) that are analogous to (1.18)–(1.38)

With $l_0 = 0$, let m, n, and $l_1, l_2, \ldots, l_n$ be fixed integers that satisfy $m \geq 2$, $n \geq 1$, as well as

(24.1) $\qquad 0 \leq l_1 \leq l_2 \leq \cdots \leq l_n \leq m$, and either $l_{n-1} \geq 1$ or $l_n \geq 3$.

In Chapters 25 and 26, we shall use the property that

(24.2) $\qquad$ the integer $s_0 = \sum_{\nu=1}^{n} l_\nu$ satisfies $s_0 \geq 2$.

We recall that $\boldsymbol{a}_{m,2}$, $\boldsymbol{b}_{m,1}$, and $\boldsymbol{b}_{m,2}$ are defined in $\mathcal{S}_{m,n}$ by (23.14), (23.15), and (23.16) of page 267. For later reference, we introduce

(24.3) $\qquad \boldsymbol{K}_{m,i,j} \equiv 0, \quad \text{for } i \leq -1 \text{ and any } j,$

(24.4) $\qquad \boldsymbol{K}_{m,0,j} \equiv 1, \quad \text{for any } j,$

(24.5) $\qquad \boldsymbol{K}_{m,i+1,j} \equiv \sum_{k=j+1}^{m} \left[\boldsymbol{K}_{m,i,k}^{(1)} - \frac{m-1}{2} \boldsymbol{b}_{m,1} \boldsymbol{K}_{m,i,k} + (m-i-k+1)(1-i-k)\, \boldsymbol{a}_{m,2}\, \boldsymbol{K}_{m,i-1,k} \right],$
$\qquad\qquad\qquad\qquad\qquad\qquad\qquad\qquad\qquad\qquad$ for $i \geq 0$ and any j,

(24.6) $\qquad \boldsymbol{L}_{m,i_1,\ldots,i_n} \equiv \sum_{j_1=0}^{i_1} \cdots \sum_{j_n=0}^{i_n} \boldsymbol{K}_{m,i_1-j_1,j_1} \cdots \boldsymbol{K}_{m,i_n-j_n,j_n}\, \boldsymbol{w}_{j_1,\ldots,j_n},$
$\qquad\qquad\qquad\qquad\qquad\qquad\qquad\qquad$ for $0 \leq i_1, \ldots, i_n \leq m$,

(24.7) $\qquad \boldsymbol{L}_{m,i_1\ldots,i_n} \equiv 0, \quad \text{when } i_\nu < 0 \text{ or } i_\nu > m \text{ for some } \nu,$

(24.8) $\qquad H^{m,l_1,\ldots,l_n}_{h_1,\ldots,h_n} \equiv \prod_{\nu=1}^{n} \left[\binom{m-h_\nu}{l_\nu - h_\nu} \prod_{r=1}^{l_\nu - h_\nu}(l_\nu - r) \right], \quad \text{for any } h_1, \ldots, h_n,$

$$(24.9) \quad \boldsymbol{M}_{m,l_1,\ldots,l_n,h_1,\ldots,h_{n-1},i} \equiv \left[H^{m,l_1,\ldots,l_n}_{h_1,\ldots,h_n} \boldsymbol{L}_{m,h_1,\ldots,h_n} \right]_{h_n = i-(h_1+\cdots+h_{n-1})},$$

for $0 \leq h_1 \leq l_1, \ldots, 0 \leq h_{n-1} \leq l_{n-1}$, and any i,

$$(24.10) \quad A_{l_1,\ldots,l_n,i} \equiv \frac{-1}{l_1 + \cdots + l_n + i - 1}, \quad \text{for } i \geq 1,$$

$$(24.11) \quad B_{l_1,\ldots,l_n,i} \equiv \frac{l_1 + \cdots + l_n - i}{l_1 + \cdots + l_n + i - 2}, \quad \text{for } i \geq 1,$$

$$(24.12) \quad \boldsymbol{I}_{m,l_1,\ldots,l_n,h_1,\ldots,h_{n-1},0} \equiv \boldsymbol{I}_{m,l_1,\ldots,l_n,h_1,\ldots,h_{n-1},1} \equiv 0,$$

for $0 \leq h_1 \leq l_1, \ldots, 0 \leq h_{n-1} \leq l_{n-1}$,

$$(24.13) \quad \boldsymbol{I}_{m,l_1,\ldots,l_n,h_1,\ldots,h_{n-1},i+1} \equiv \boldsymbol{M}_{m,l_1,\ldots,l_n,h_1,\ldots,h_{n-1},i+1}$$
$$+ A_{l_1,\ldots,l_n,i} \boldsymbol{I}^{(1)}_{m,l_1,\ldots,l_n,h_1,\ldots,h_{n-1},i}$$
$$+ B_{l_1,\ldots,l_n,i} \boldsymbol{a}_{m,2} \boldsymbol{I}_{m,l_1,\ldots,l_n,h_1,\ldots,h_{n-1},i-1},$$

for $0 \leq h_1 \leq l_1, \ldots, 0 \leq h_{n-1} \leq l_{n-1}$, and $1 \leq i \leq l_1 + \cdots + l_n - 1$,

and

$$(24.14) \quad \boldsymbol{\mathcal{I}}_{m,l_1,\ldots,l_n} \equiv \sum_{h_0=0}^{l_0} \sum_{h_1=0}^{l_1} \cdots \sum_{h_{n-1}=0}^{l_{n-1}} \boldsymbol{I}_{m,l_1,\ldots,l_n,h_1,\ldots,h_{n-1},l_1+\cdots+l_n}.$$

We shall establish in Theorem 27.7 on page 314 that $\boldsymbol{\mathcal{I}}_{m,l_1,\ldots,l_n}$ *is a basic relative invariant of index* $(l_1, \ldots, l_n)$ *for the equations* (21.1) *on page* 249.

COMMENT. We introduce the variable h_0 that can only assume the value 0 so that the summation in (24.14) and (24.23) is not vacuous when $n = 1$. By this means, we include the basic relative invariants for homogeneous linear differential equations as the special case having $n = 1$ and $3 \leq l_1 \leq m$.

For later reference, we also introduce

$$(24.15) \quad \boldsymbol{U}_{m,i,j} \equiv 0, \quad \text{for } i \leq -1 \text{ and any } j,$$

$$(24.16) \quad \boldsymbol{U}_{m,0,j} \equiv 1, \quad \text{for any } j,$$

$$(24.17) \quad \boldsymbol{U}_{m,i+1,j} \equiv \sum_{k=j+1}^{m} \left[\begin{array}{l} \boldsymbol{U}^{(1)}_{m,i,k} + (i+k-m)\, \boldsymbol{b}_{m,1}\, \boldsymbol{U}_{m,i,k} \\ + (m-i-k+1)(1-i-k)\, \boldsymbol{b}_{m,2}\, \boldsymbol{U}_{m,i-1,k} \end{array} \right],$$

for $i \geq 0$ and any j,

$$(24.18) \quad \boldsymbol{V}_{m,i_1,\ldots,i_n} \equiv \sum_{j_1=0}^{i_1} \cdots \sum_{j_n=0}^{i_n} \boldsymbol{U}_{m,i_1-j_1,j_1} \cdots \boldsymbol{U}_{m,i_n-j_n,j_n}\, \boldsymbol{w}_{j_1,\ldots,j_n},$$

for $0 \leq i_1, \ldots, i_n \leq m$,

$$(24.19) \quad \boldsymbol{V}_{m,i_1,\ldots,i_n} \equiv 0, \quad \text{when } i_\nu < 0 \text{ or } i_\nu > m \text{ for some } \nu,$$

$$(24.20) \quad \boldsymbol{W}_{m,l_1,\ldots,l_n,h_1,\ldots,h_{n-1},i} \equiv \left[H^{m,l_1,\ldots,l_n}_{h_1,\ldots,h_n} \boldsymbol{V}_{m,h_1,\ldots,h_n} \right]_{h_n = i-(h_1+\cdots+h_{n-1})},$$

for $0 \leq h_1 \leq l_1, \ldots, 0 \leq h_{n-1} \leq l_{n-1}$, and any i,

(24.21) $$\boldsymbol{J}_{m,l_1,\ldots,l_n,h_1,\ldots,h_{n-1},0} \equiv \boldsymbol{J}_{m,l_1,\ldots,l_n,h_1,\ldots,h_{n-1},1} \equiv 0,$$

for $0 \leq h_1 \leq l_1, \ldots, 0 \leq h_{n-1} \leq l_{n-1}$,

(24.22) $$\boldsymbol{J}_{m,l_1,\ldots,l_n,h_1,\ldots,h_{n-1},i+1} \equiv \bigg[\boldsymbol{W}_{m,l_1,\ldots,l_n,h_1,\ldots,h_{n-1},i+1}$$

$$+ A_{l_1,\ldots,l_n,i} \left(\boldsymbol{J}^{(1)}_{m,l_1,\ldots,l_n,h_1,\ldots,h_{n-1},i} + i\, \boldsymbol{b}_{m,1}\, \boldsymbol{J}_{m,l_1,\ldots,l_n,h_1,\ldots,h_{n-1},i} \right)$$

$$+ B_{l_1,\ldots,l_n,i}\, \boldsymbol{b}_{m,2}\, \boldsymbol{J}_{m,l_1,\ldots,l_n,h_1,\ldots,h_{n-1},i-1} \bigg],$$

for $0 \leq h_1 \leq l_1, \ldots, 0 \leq h_{n-1} \leq l_{n-1}$, and $1 \leq i \leq l_1 + \cdots + l_n - 1$,

and

(24.23) $$\boldsymbol{\mathcal{J}}_{m,l_1,\ldots,l_n} \equiv \sum_{h_0=0}^{l_0} \sum_{h_1=0}^{l_1} \cdots \sum_{h_{n-1}=0}^{l_{n-1}} \boldsymbol{J}_{m,l_1,\ldots,l_n,h_1,\ldots,h_{n-1},l_1+\cdots+l_n}.$$

We shall establish in Theorem 27.7 on page 314 that $\boldsymbol{\mathcal{J}}_{m,l_1,\ldots,l_n} \equiv \boldsymbol{\mathcal{I}}_{m,l_1,\ldots,l_n}$ and $\boldsymbol{\mathcal{J}}_{m,l_1,\ldots,l_n}$ is therefore a basic relative invariant of index $(l_1, \ldots, l_n)$.

As a check on Theorem 27.7 and the formulas of this section, we have used computer algebra for $n = 1, 2, 3, 4, 5$ in a manner similar to that for Section 16.3 on pages 169–171.

24.2. Notational abbreviations employed in Chapters 25–26

Throughout Chapters 25 and 26, the subscripts $m, l_1, \ldots, l_n$ that appear in the expressions

$$\boldsymbol{K}_{m,i,j}, \quad \boldsymbol{L}_{m,i_1,\ldots,i_n}, \quad \boldsymbol{M}_{m,l_1,\ldots,l_n,h_1,\ldots,h_{n-1},i}, \quad \boldsymbol{I}_{m,l_1,\ldots,l_n,h_1,\ldots,h_{n-1},i},$$
$$\boldsymbol{U}_{m,i,j}, \quad \boldsymbol{V}_{m,i_1,\ldots,i_n}, \quad \boldsymbol{W}_{m,l_1,\ldots,l_n,h_1,\ldots,h_{n-1},i}, \quad \boldsymbol{J}_{m,l_1,\ldots,l_n,h_1,\ldots,h_{n-1},i}$$

will be fixed. There, we shall use the corresponding abbreviations

$$\boldsymbol{K}_{i,j}, \quad \boldsymbol{L}_{i_1,\ldots,i_n}, \quad \boldsymbol{M}_{h_1,\ldots,h_{n-1},i}, \quad \boldsymbol{I}_{h_1,\ldots,h_{n-1},i},$$
$$\boldsymbol{U}_{i,j}, \quad \boldsymbol{V}_{i_1,\ldots,i_n}, \quad \boldsymbol{W}_{h_1,\ldots,h_{n-1},i}, \quad \boldsymbol{J}_{h_1,\ldots,h_{n-1},i}.$$

Also, in place of $H^{m,l_1,\ldots,l_n}_{h_1,\ldots,h_n}$, $A_{l_1,\ldots,l_n,i}$, and $B_{l_1,\ldots,l_n,i}$, we shall write $H_{h_1,\ldots,h_n}$, A_i, and B_i.

24.3. The special situations when $n = 2$ and when $n = 1$

24.3.1. Equations $Q_m = 0$. For $n = 2$, the equations $H_{m,n} = 0$ of (21.1) are the equations $Q_m = 0$ of (1.1). Then, formulas (24.3)–(24.23) specialize to give (1.18)–(1.38) of pages 8–9 when $l_1 = l$, l_2 is the n of those formulas, and $h_1 = h$.

24.3.2. Equations $L_m = 0$. For $n = 1$, the equations $H_{m,n} = 0$ of (21.1) are monic homogeneous linear differential equations $L_m = 0$ of order m as given by (1.84) on page 17. The condition (24.1) requires $3 \leq l_1 \leq m$. To see that (24.6)–(24.14) specialize to give the basic relative invariants for homogeneous linear differential equations presented in [**24**], we note that: when $3 \leq l_1 \leq m$, they yield

$$(24.24) \qquad \boldsymbol{L}_{m,i} \equiv \sum_{j=0}^{i} \boldsymbol{K}_{m,i-j,j}\, \boldsymbol{w}_j, \quad \text{for } 0 \leq i \leq m,$$

$$(24.25) \qquad H_{h_1}^{m,l_1} \equiv \binom{m-h_1}{l_1-h_1} \prod_{r=1}^{l_1-h_1} (l_1-r), \quad \text{for any } h_1,$$

$$(24.26) \qquad \boldsymbol{M}_{m,l_1,i} \equiv \left[\binom{m-i}{l_1-i} \prod_{r=1}^{l_1-i}(l_1-r)\right] \boldsymbol{L}_{m,i}, \quad \text{for } 0 \leq i \leq m,$$

$$(24.27) \qquad A_{l_1,i} \equiv \frac{-1}{l_1+i-1} \quad \text{and} \quad B_{l_1,i} \equiv \frac{l_1-i}{l_1+i-2}, \quad \text{for } i \geq 1,$$

$$(24.28) \qquad \boldsymbol{I}_{l_1,0} \equiv \boldsymbol{I}_{l_1,1} \equiv 0,$$

$$(24.29) \qquad \boldsymbol{I}_{m,l_1,i+1} \equiv \boldsymbol{M}_{m,l_1,i+1} + A_{l_1,i}\, \boldsymbol{I}_{m,l_1,i}^{(1)} + B_{l_1,i}\, \boldsymbol{a}_{m,2}\, \boldsymbol{I}_{m,l_1,i-1},$$
$$\text{for } 1 \leq i \leq l_1-1,$$

and

$$(24.30) \qquad \boldsymbol{\mathcal{I}}_{m,l_1} \equiv \boldsymbol{I}_{m,l_1,l_1}.$$

By identifying (24.24) with (1.89) on page 18, (24.25) with (1.87), (24.26) with (1.91), and (24.27)–(24.30) with the specializations of (1.24)–(1.28) having $l = 0$ and $n = l_1$, we use the proof for Theorem 1.13 to see that the basic relative invariants in [**24**] for homogeneous linear differential equations of order $m \geq 3$ are given by $\boldsymbol{\mathcal{I}}_{m,l_1}$, when $3 \leq l_1 \leq m$.

24.4. Plan to evaluate Laguerre-Forsyth sums

Equation (23.37) on page 271 provides motivation for the following formalism.

DEFINITION 24.1. For a given equation (21.1) on page 249 that satisfies (21.1.1) but is not assumed to satisfy (23.35), suppose that $y(z) = \rho(z)\,v(z)$, $z = f(\zeta)$, and $w(\zeta) = (v \circ f)(\zeta)$ transform (21.1) on Ω into (23.1) on Ω^{**} subject to (23.2) and (23.36). Then, for this Laguerre-Forsyth canonical form (23.1) of (21.1) and the constants $A_{j_1, j_2, \ldots, j_n}^{i_1, i_2, \ldots, i_n}$ in (23.34) of page 271, we term each of the expressions

$$(24.31) \qquad \sum_{j_1=0}^{i_1} \cdots \sum_{j_n=0}^{i_n} A_{j_1, \ldots, j_n}^{i_1, \ldots, i_n}\, d_{i_1-j_1, \ldots, i_n-j_n}^{(j_1+\cdots+j_n)}(\zeta),$$
$$\text{on } \Omega^{**} \text{ for } 0 \leq i_1, \ldots, i_n \leq m \text{ and } i_1+\cdots+i_n \geq 2,$$

a *Laguerre-Forsyth sum*.

In view of (23.2) and (23.34), we may suppose that $0 \leq i_1 \leq \cdots \leq i_n \leq m$. For convenience, we set $i_0 = 0$. We may also suppose that either $i_{n-1} \geq 1$ or $i_n \geq 3$. Namely, if $i_{n-1} = 0$ and $i_n = 2$, then (23.36) and (23.34) show that the Laguerre-Forsyth sum in (24.31) equals 0. Thus, it is sufficient to study (24.31) when $i_1, \ldots, i_n$ are respectively integers $l_1, \ldots, l_n$ that satisfy (24.1). Then, we refer to $(l_1, \ldots, l_n)$ as *the index of the Laguerre-Forsyth sum*.

Theorem 25.10 on page 290 is our main goal in Chapter 25; it expresses each Laguerre-Forsyth sum of index $(l_1, \ldots, l_n)$ in terms of $z = f(\zeta)$ and the coefficients of (21.1) by means of $\boldsymbol{\mathcal{I}}_{m,l_1,\ldots,l_n}$ from (24.14).

CHAPTER 25

Extensions of Chapter 7 to $H_{m,n} = 0$, when $m \geq 2$

25.1. The coefficients of (23.1) when $d_{0,\ldots,0,1}(\zeta) \equiv d_{0,\ldots,0,2}(\zeta) \equiv 0$

For the statement of Proposition 25.1, we repeat $\xi_{i,j}$ from (7.1) as

$$(25.1) \qquad \xi_{i,j} \equiv \frac{\left[\prod_{\nu=1}^{j}(i-\nu)\right]}{2^j} \binom{m-i+j}{j}, \quad \text{for any integers } i \text{ and } j.$$

PROPOSITION 25.1. *For (23.1) on Ω^{**} into which (21.1) on Ω is transformed by $y(z) = \rho(z)\, v(z)$ and $z = f(\zeta)$ according to the context of page 265, suppose that $d_{0,\ldots,0,1}(\zeta) \equiv d_{0,\ldots,0,2}(\zeta) \equiv 0$. Then, the coefficients of (23.1) are given by*

$$(25.2) \quad d_{i_1,\ldots,i_n}(\zeta) \equiv$$

$$\bigl(f'(\zeta)\bigr)^{i_1+\cdots+i_n} \sum_{j_1=0}^{i_1}\cdots\sum_{j_n=0}^{i_n} \xi_{i_1,j_1}\cdots\xi_{i_n,j_n}\, L_{i_1-j_1,\ldots,i_n-j_n}\bigl(f(\zeta)\bigr) \bigl(\mathcal{Q}(\zeta)\bigr)^{j_1+\cdots+j_n},$$

$$\text{on } \Omega^{**} \text{ for } 0 \leq i_1, \ldots, i_n \leq m,$$

where $L_{i_1-j_1,\ldots,i_n-j_n}\bigl(f(\zeta)\bigr)$ is obtained for $z = f(\zeta)$ by substituting $c^{(\kappa)}_{\mu_1,\ldots,\mu_n}(z)$ from (21.1) for $\boldsymbol{w}^{(\kappa)}_{\mu_1,\ldots,\mu_n}$ in $\boldsymbol{L}_{i_1-j_1,\ldots,i_n-j_n}$ of (24.6) and

$$(25.3) \qquad \mathcal{Q}(\zeta) \equiv \frac{f''(\zeta)}{\bigl(f'(\zeta)\bigr)^2}, \quad \text{on } \Omega^{**}.$$

PROOF. Using (23.3) in place of (7.4), we see that it suffices to make minor changes in the proof for Proposition 7.1 of page 69. □

With respect to (25.1) and $\boldsymbol{L}_{i_1,\ldots,i_n}$ of (24.6)–(24.7) on page 275, we introduce $\boldsymbol{\mathcal{M}}_{i_1,\ldots,i_n,k,t}$ in $\mathcal{S}_{m,n}$ for $k \geq 0$ and any integers $i_1, \ldots, i_n, t$, through

$$(25.4) \qquad \boldsymbol{\mathcal{M}}_{i_1,\ldots,i_n,0,t} \equiv \sum_{\substack{0 \leq j_1,\ldots,0 \leq j_n \\ j_1+\cdots+j_n = t}} \xi_{i_1,j_1}\cdots\xi_{i_n,j_n}\, \boldsymbol{L}_{i_1-j_1,\ldots,i_n-j_n}, \quad \text{for any } i_1, \ldots, i_n, t,$$

and

$$(25.5) \quad \boldsymbol{\mathcal{M}}_{i_1,\ldots,i_n,k+1,t} \equiv \boldsymbol{\mathcal{M}}^{(1)}_{i_1,\ldots,i_n,k,t} - 2(t+1)\boldsymbol{a}_{m,2}\, \boldsymbol{\mathcal{M}}_{i_1,\ldots,i_n,k,t+1}$$

$$+ \frac{2i_1 + \cdots + 2i_n + 2k - t + 1}{2}\boldsymbol{\mathcal{M}}_{i_1,\ldots,i_n,k,t-1},$$

$$\text{for } k \geq 0 \text{ and any } i_1, \ldots, i_n, t.$$

PROPOSITION 25.2. *For (23.1) on Ω^{**} into which (21.1) on Ω is transformed by $y(z) = \rho(z)v(z)$ and $z = f(\zeta)$ according to the context of page 265, suppose that $d_{0,\ldots,0,1}(\zeta) \equiv d_{0,\ldots,0,2}(\zeta) \equiv 0$. Then, the derivatives of the coefficients of (23.1) are provided by*

$$(25.6) \quad d^{(k)}_{i_1,\ldots,i_n}(\zeta) \equiv \left(f'(\zeta)\right)^{i_1+\cdots+i_n+k} \sum_{t=0}^{i_1+\cdots+i_n+k} \mathcal{M}_{i_1,\ldots,i_n,k,t}(f(\zeta))\left(\mathcal{Q}(\zeta)\right)^t,$$

for $0 \leq i_1, \ldots, i_n \leq m$ and $k \geq 0$,

where $\mathcal{M}_{i_1,\ldots,i_n,k,t}(f(\zeta))$ is obtained for $z = f(\zeta)$ by substituting $c^{(\kappa)}_{\mu_1,\ldots,\mu_n}(z)$ from (21.1) for $w^{(\kappa)}_{\mu_1,\ldots,\mu_n}$ in $\mathcal{M}_{i_1,\ldots,i_n,k,t}$ of (25.4)–(25.5) and $\mathcal{Q}(\zeta)$ is given in (25.3).

PROOF. For $t < 0$, (25.4) yields $\mathcal{M}_{i_1,\ldots,i_n,0,t} \equiv 0$. When $t > i_1 + \cdots + i_n$, the conditions $0 \leq j_1, \ldots, j_n$ and $j_1 + \cdots + j_n = t$ in (25.4) require $j_\nu > i_\nu$, for at least one ν satisfying $1 \leq \nu \leq n$, and give $\mathcal{M}_{i_1,\ldots,i_n,0,t} \equiv 0$ because (24.7) shows that each $L_{i_1-j_1,\ldots,i_n-j_n}$ equals 0. Suppose that k is a nonnegative integer such that $\mathcal{M}_{i_1,\ldots,i_n,k,t} \equiv 0$ is valid whenever $i_1, \ldots, i_n, k, t$ are integers that satisfy either $t < 0$ or $t > i_1 + \cdots + i_n + k$. Then, for this k, we use (25.5) to see that $\mathcal{M}_{i_1,\ldots,i_n,k+1,t} \equiv 0$ is valid for any integers $i_1, \ldots, i_n, k, t$ that satisfy either $t < 0$ or $t > i_1 + \cdots + i_n + k + 1$. This establishes that

$$(25.7) \quad \mathcal{M}_{i_1,\ldots,i_n,k,t} \equiv 0, \quad \text{for } k \geq 0 \text{ and either } t < 0 \text{ or } t > i_1 + \cdots + i_n + k.$$

In view of $L_{i_1-j_1,\ldots,i_n-j_n}(f(\zeta)) \equiv 0$, whenever $j_\nu > i_\nu$ for some ν satisfying $1 \leq \nu \leq n$, we use (25.2) and (25.4) to obtain

$$\frac{d_{i_1,\ldots,i_n}(\zeta)}{\left(f'(\zeta)\right)^{i_1+\cdots+i_n}} \equiv \sum_{t=0}^{i_1+\cdots+i_n} \sum_{\substack{0 \leq j_1,\ldots,0 \leq j_n \\ j_1+\cdots+j_n=t}} \xi_{i_1,j_1} \cdots \xi_{i_n,j_n} L_{i_1-j_1,\ldots,i_n-j_n}(f(\zeta))\left(\mathcal{Q}(\zeta)\right)^t$$

$$\equiv \sum_{t=0}^{i_1+\cdots+i_n} \mathcal{M}_{i_1,\ldots,i_n,0,t}(f(\zeta))\left(\mathcal{Q}(\zeta)\right)^t.$$

This shows that (25.6) is valid for $k = 0$.

Suppose that (25.6) is valid when k is replaced with a nonnegative integer k_0. Then, by differentiating that formula with respect to ζ and using (25.5) as well as (25.7), we perform a computation similar to that for (7.27)–(7.31) to conclude that (25.6) is valid for any nonnegative integer k. □

25.2. Evaluation of Laguerre-Forsyth sums

In terms of fixed integers $m \geq 2$, $n \geq 1$, $l_0 = 0$, and $l_1, \ldots, l_n$ that satisfy $0 \leq l_1 \leq \cdots \leq l_n \leq m$ and either $l_{n-1} \geq 1$ or $l_n \geq 3$, we use (23.34) on page 271 as well as (25.4)–(25.5) to define $\mathbf{N}_{i,j}$ in $\mathcal{S}_{m,n}$ by

$$(25.8) \quad \mathbf{N}_{i,j} \equiv \sum_{r=0}^{i} \sum_{\substack{0 \leq k_1,\ldots,0 \leq k_n \\ k_1+\cdots+k_n=r}} A^{l_1,\ldots,l_n}_{l_1-k_1,\ldots,l_n-k_n} \mathcal{M}_{k_1,\ldots,k_n,i-r,j}, \quad \text{for any } i, j.$$

OBSERVATION. Whenever the sum in (25.8) is nonvacuous, we have $r \geq 0$ and

$$(2l_1 + \cdots + 2l_n - 2) - \left((l_1-k_1) + \cdots + (l_n-k_n)\right) = l_1 + \cdots + l_n - 2 + r \geq 0.$$

Thus, $A^{l_1,l_2,\ldots,l_n}_{l_1-k_1,l_2-k_2,\ldots,l_n-k_n}$ in (25.8) is well defined by (23.34) of page 271.

25.2. EVALUATION OF LAGUERRE-FORSYTH SUMS

PROPOSITION 25.3. *For (23.1) on Ω^{**} into which (21.1) on Ω is transformed by $y(z) = \rho(z) v(z)$ and $z = f(\zeta)$ according to the context of page 265, suppose that $d_{0,\ldots,0,1}(\zeta) \equiv d_{0,\ldots,0,2}(\zeta) \equiv 0$. Then,*

$$(25.9) \quad \sum_{j_1=0}^{l_1} \cdots \sum_{j_n=0}^{l_n} A_{j_1,\ldots,j_n}^{l_1,\ldots,l_n} d_{l_1-j_1,\ldots,l_n-j_n}^{(j_1+\cdots+j_n)}(\zeta)$$

$$\equiv \left(f'(\zeta)\right)^{l_1+\cdots+l_n} \sum_{t=0}^{l_1+\cdots+l_n} N_{l_1+\cdots+l_n,t}\left(f(\zeta)\right) \left(\mathcal{Q}(\zeta)\right)^t, \quad \text{on } \Omega^{**},$$

where $N_{l_1+\cdots+l_n,t}(f(\zeta))$ is obtained for $z = f(\zeta)$ by substituting $c_{\mu_1,\ldots,\mu_n}^{(\kappa)}(z)$ from (21.1) for $\boldsymbol{w}_{\mu_1,\ldots,\mu_n}^{(\kappa)}$ in $\boldsymbol{N}_{l_1+\cdots+l_n,t}$ and $\mathcal{Q}(\zeta)$ is given by (25.3).

PROOF. After setting $s_0 = l_1 + \cdots + l_n$, we employ (25.6), the property deduced from (23.34) that $A_{j_1,j_2,\ldots,j_n}^{i_1,i_2,\ldots,i_n} = 0$ when $j_\nu < 0$, for some ν, and (25.8) to obtain

$$\sum_{j_1=0}^{l_1} \cdots \sum_{j_n=0}^{l_n} A_{j_1,\ldots,j_n}^{l_1,\ldots,l_n} d_{l_1-j_1,\ldots,l_n-j_n}^{(j_1+\cdots+j_n)}(\zeta)$$

$$\equiv \sum_{k_1=0}^{l_1} \cdots \sum_{k_n=0}^{l_n} A_{l_1-k_1,\ldots,l_n-k_n}^{l_1,\ldots,l_n} d_{k_1,\ldots,k_n}^{(s_0-k_1-\cdots-k_n)}(\zeta)$$

$$\equiv \left(f'(\zeta)\right)^{s_0} \sum_{k_1=0}^{l_1} \cdots \sum_{k_n=0}^{l_n} A_{l_1-k_1,\ldots,l_n-k_n}^{l_1,\ldots,l_n} \sum_{t=0}^{s_0} \mathcal{M}_{k_1,\ldots,k_n,s_0-k_1-\cdots-k_n,t}\left(f(\zeta)\right) \left(\mathcal{Q}(\zeta)\right)^t$$

$$\equiv \left(f'(\zeta)\right)^{s_0} \sum_{t=0}^{s_0} \left(\mathcal{Q}(\zeta)\right)^t \sum_{k_1=0}^{l_1} \cdots \sum_{k_n=0}^{l_n} A_{l_1-k_1,\ldots,l_n-k_n}^{l_1,\ldots,l_n} \mathcal{M}_{k_1,\ldots,k_n,s_0-k_1-\cdots-k_n,t}\left(f(\zeta)\right)$$

$$\equiv \left(f'(\zeta)\right)^{s_0} \sum_{t=0}^{s_0} \left(\mathcal{Q}(\zeta)\right)^t \sum_{r=0}^{s_0} \sum_{\substack{0 \leq k_1,\ldots,0 \leq k_n \\ k_1+\cdots+k_n=r}} A_{l_1-k_1,\ldots,l_n-k_n}^{l_1,\ldots,l_n} \mathcal{M}_{k_1,\ldots,k_n,s_0-k_1-\cdots-k_n,t}\left(f(\zeta)\right)$$

$$\equiv \left(f'(\zeta)\right)^{l_1+\cdots+l_n} \sum_{t=0}^{l_1+\cdots+l_n} N_{l_1+\cdots+l_n,t}\left(f(\zeta)\right) \left(\mathcal{Q}(\zeta)\right)^t.$$

This yields (25.9) and completes the proof. □

We apply (23.34) on page 271 and (25.4) to define $\boldsymbol{F}_{i,j}$ in $\mathcal{S}_{m,n}$ by

$$(25.10) \quad \boldsymbol{F}_{i,j} \equiv \sum_{\substack{0 \leq k_1,\ldots,0 \leq k_n \\ k_1+\cdots+k_n=i}} A_{l_1-k_1,\ldots,l_n-k_n}^{l_1,\ldots,l_n} \mathcal{M}_{k_1,\ldots,k_n,0,j}, \quad \text{for any } i, j.$$

PROPOSITION 25.4. *The polynomials $\boldsymbol{N}_{i,j}$ and $\boldsymbol{F}_{i,j}$ satisfy*

$$(25.11) \quad \boldsymbol{N}_{i+1,j} - \frac{2i-j+1}{2} \boldsymbol{N}_{i,j-1} \equiv \boldsymbol{N}_{i,j}^{(1)} - 2(j+1)\boldsymbol{a}_{m,2} \boldsymbol{N}_{i,j+1} + \boldsymbol{F}_{i+1,j},$$

for all integers i, j.

PROOF. For $i \geq 0$ and any j, we see that (25.8), (25.5), and (25.10) give

$$\boldsymbol{N}_{i,j}^{(1)} \equiv \sum_{r=0}^{i} \sum_{\substack{0 \leq k_1, \ldots, 0 \leq k_n \\ k_1 + \cdots + k_n = r}} A_{l_1-k_1, \ldots, l_n-k_n}^{l_1, \ldots, l_n} \boldsymbol{\mathcal{M}}_{k_1, \ldots, k_n, i-r, j}^{(1)}$$

$$\equiv \sum_{r=0}^{i} \sum_{\substack{0 \leq k_1, \ldots, 0 \leq k_n \\ k_1 + \cdots + k_n = r}} A_{l_1-k_1, \ldots, l_n-k_n}^{l_1, \ldots, l_n} \left\{ \begin{array}{l} \boldsymbol{\mathcal{M}}_{k_1, \ldots, k_n, i+1-r, j} \\ - \left[\frac{2r+2(i-r)-j+1}{2} \right] \boldsymbol{\mathcal{M}}_{k_1, \ldots, k_n, i-r, j-1} \\ + 2(j+1)\, \boldsymbol{a}_{m,2}\, \boldsymbol{\mathcal{M}}_{k_1, \ldots, k_n, i-r, j+1} \end{array} \right\}$$

$$\equiv \boldsymbol{N}_{i+1,j} - \boldsymbol{F}_{i+1,j} - \frac{2i-j+1}{2} \boldsymbol{N}_{i,j-1} + 2(j+1)\boldsymbol{a}_{m,2}\, \boldsymbol{N}_{i,j+1}.$$

Thus, (25.11) is valid for $i \geq 0$ and any j.

To show that (25.11) is also valid for $i < 0$ and any j, we first use (25.10) and (23.34) to obtain $\boldsymbol{F}_{0,j} \equiv A_{l_1, \ldots, l_n}^{l_1, \ldots, l_n} \boldsymbol{\mathcal{M}}_{0, \ldots, 0, 0, j} \equiv 0$, for any j. We apply (25.8) and (23.34) to deduce that $\boldsymbol{N}_{0,j} \equiv A_{l_1, \ldots, l_n}^{l_1, \ldots, l_n} \boldsymbol{\mathcal{M}}_{0, \ldots, 0, 0, j} \equiv 0$, for any j. Also, (25.10) and (25.8) yield $\boldsymbol{F}_{i,j} \equiv 0$ and $\boldsymbol{N}_{i,j} \equiv 0$, for $i \leq -1$ and any j. Consequently, for $i \leq -1$ and any j, each term in (25.11) is equal to 0. Thus, (25.11) is valid for all integers i and j. This completes the proof. $\square$

In terms of the integer $s_0 = l_1 + \cdots + l_n$ from (24.2) that satisfies $s_0 \geq 2$, we define rational numbers $\vartheta_{i,j}$ by means of

$$(25.12) \qquad \vartheta_{i,j} \equiv \frac{(-1)^{s_0+i+j}}{2^j} \binom{s_0-i}{j} \frac{1}{\prod_{r=1}^{s_0-i-j}(2s_0-1-r)}, \quad \text{for } i+j \geq 0.$$

PROPOSITION 25.5. *The polynomial* $\boldsymbol{F}_{i,j}$ *of* (25.10) *is related to* $\vartheta_{i,j}$ *in* (25.12) *and* $\boldsymbol{M}_{h_1, \ldots, h_{n-1}, i}$ *in* (24.9) *through*

$$(25.13) \qquad \boldsymbol{F}_{i,j} \equiv \vartheta_{i-j,j} \sum_{h_0=0}^{l_0} \cdots \sum_{h_{n-1}=0}^{l_{n-1}} \boldsymbol{M}_{h_1, \ldots, h_{n-1}, i-j}, \quad \text{for } i \geq 0 \text{ and any } j.$$

PROOF. Due to (23.34), (25.1), and (24.7), we have: $A_{j_1, \ldots, j_n}^{l_1, \ldots, l_n} \equiv 0$, when $j_\nu < 0$ for some ν; $\xi_{i,j} \equiv 0$, when $j < 0$; and $\boldsymbol{L}_{j_1, \ldots, j_n} \equiv 0$, when $j_\nu < 0$ for some ν. For $i \geq 0$ and any j, we apply (25.10), (25.4), and the observations just made to obtain

$$(25.14) \quad \boldsymbol{F}_{i,j} \equiv \sum_{\substack{0 \leq k_1, \ldots, 0 \leq k_n \\ k_1 + \cdots + k_n = i}} A_{l_1-k_1, \ldots, l_n-k_n}^{l_1, \ldots, l_n} \sum_{\substack{0 \leq j_1, \ldots, 0 \leq j_n \\ j_1 + \cdots + j_n = j}} \xi_{k_1, j_1} \cdots \xi_{k_n, j_n}\, \boldsymbol{L}_{k_1-j_1, \ldots, k_n-j_n}$$

$$\equiv \sum_{\substack{0 \leq k_1, \ldots, 0 \leq k_n \\ k_1 + \cdots + k_n = i \\ 0 \leq j_1 \leq k_1, \ldots, 0 \leq j_n \leq k_n \\ j_1 + \cdots + j_n = j}} A_{l_1-k_1, \ldots, l_n-k_n}^{l_1, \ldots, l_n} \xi_{k_1, j_1} \cdots \xi_{k_n, j_n}\, \boldsymbol{L}_{k_1-j_1, \ldots, k_n-j_n}$$

$$\equiv \sum_{\substack{0 \leq k_1 \leq l_1, \ldots, 0 \leq k_n \leq l_n \\ k_1 + \cdots + k_n = i \\ 0 \leq h_1 \leq k_1, \ldots, 0 \leq h_n \leq k_n \\ h_1 + \cdots + h_n = i-j}} A_{l_1-k_1, \ldots, l_n-k_n}^{l_1, \ldots, l_n} \xi_{k_1, k_1-h_1} \cdots \xi_{k_n, k_n-h_n}\, \boldsymbol{L}_{h_1, \ldots, h_n}.$$

The restriction $k_1 + \cdots + k_n = i$ and (23.34) yield

$$A^{l_1,\ldots,l_n}_{l_1-k_1,\ldots,l_n-k_n} \equiv \frac{(-1)^{s_0-i}\prod_{\nu=1}^{n}\left[\binom{m-k_\nu}{l_\nu-k_\nu}\prod_{r=1}^{l_\nu-k_\nu}(l_\nu-r)\right]}{\prod_{r=1}^{s_0-i}(2s_0-1-r)}.$$

In view of $k_1 + \cdots + k_n = i$ and $h_1 + \cdots + h_n = i - j$, (25.1) gives

$$\prod_{\nu=1}^{n}\xi_{k_\nu,k_\nu-h_\nu} \equiv \frac{1}{2^j}\prod_{\nu=1}^{n}\left[\binom{m-h_\nu}{k_\nu-h_\nu}\prod_{r=1}^{k_\nu-h_\nu}(k_\nu-r)\right].$$

Combining the preceding expressions and setting

$$(25.15) \qquad \widehat{\vartheta}_{i,j} \equiv \frac{\dfrac{(-1)^{s_0+i}}{2^j}}{\prod_{r=1}^{s_0-i}(2s_0-1-r)},$$

we deduce that

$$A^{l_1,\ldots,l_n}_{l_1-k_1,\ldots,l_n-k_n}\prod_{\nu=1}^{n}\xi_{k_\nu,k_\nu-h_\nu}$$

$$\equiv \widehat{\vartheta}_{i,j}\prod_{\nu=1}^{n}\left\{\binom{m-h_\nu}{k_\nu-h_\nu}\binom{m-k_\nu}{l_\nu-k_\nu}\left[\prod_{r=1}^{l_\nu-k_\nu}(l_\nu-r)\right]\left[\prod_{r=1}^{k_\nu-h_\nu}(k_\nu-r)\right]\right\}$$

$$\equiv \widehat{\vartheta}_{i,j}\prod_{\nu=1}^{n}\left\{\binom{m-h_\nu}{l_\nu-h_\nu}\left[\prod_{r=1}^{l_\nu-h_\nu}(l_\nu-r)\right]\binom{l_\nu-h_\nu}{k_\nu-h_\nu}\right\}$$

$$\equiv \widehat{\vartheta}_{i,j}\,H_{h_1,\ldots,h_n}\prod_{\nu=1}^{n}\binom{l_\nu-h_\nu}{k_\nu-h_\nu},$$

where $H_{h_1,\ldots,h_n}$ is defined by (24.8). We substitute this in (25.14) to verify that

$$(25.16) \quad \boldsymbol{F}_{i,j} \equiv \sum_{\substack{0\leq h_1\leq l_1,\ldots,0\leq h_n\leq l_n \\ h_1+\cdots+h_n=i-j \\ h_1\leq k_1\leq l_1,\ldots,h_n\leq k_n\leq l_n \\ k_1+\cdots+k_n=i}} \widehat{\vartheta}_{i,j}\,H_{h_1,\ldots,h_n}\left[\prod_{\nu=1}^{n}\binom{l_\nu-h_\nu}{k_\nu-h_\nu}\right]\boldsymbol{L}_{h_1,\ldots,h_n}$$

$$\equiv \sum_{\substack{0\leq h_1\leq l_1,\ldots,0\leq h_n\leq l_n \\ h_1+\cdots+h_n=i-j}} \widehat{\vartheta}_{i,j}\,H_{h_1,\ldots,h_n}\,S_{h_1,\ldots,h_n}\,\boldsymbol{L}_{h_1,\ldots,h_n},$$

where

$$S_{h_1,\ldots,h_n} \equiv \sum_{\substack{h_1\leq k_1\leq l_1,\ldots,h_n\leq k_n\leq l_n \\ k_1+\cdots+k_n=i}}\prod_{\nu=1}^{n}\binom{l_\nu-h_\nu}{k_\nu-h_\nu}.$$

We have

$$(25.17) \qquad S_{h_1,\ldots,h_n} \equiv \sum_{\substack{0\leq i_1\leq l_1-h_1,\ldots,0\leq i_n\leq l_n-h_n \\ i_1+\cdots+i_n=j}}\prod_{\nu=1}^{n}\binom{l_\nu-h_\nu}{i_\nu},$$

when $0 \leq h_1 \leq l_1, \ldots, 0 \leq h_n \leq l_n$ and $h_1 + \cdots + h_n = i - j$.

For any nonnegative integers $a_1, a_2, \ldots, a_n$, the identity

$$(25.18) \quad \sum_{\substack{0 \le i_1 \le a_1,\, 0 \le i_2 \le a_2,\, \ldots,\, 0 \le i_n \le a_n \\ i_1 + i_2 + \cdots + i_n = j}} \binom{a_1}{i_1}\binom{a_2}{i_2}\cdots\binom{a_n}{i_n} \equiv \binom{a_1 + a_2 + \cdots + a_n}{j}$$

generalizes Vandermonde's convolution (where $n = 2$). When a set S is partitioned into n subsets $S_1, S_2, \ldots, S_n$ that contain respectively $a_1, a_2, \ldots, a_n$ elements, the left member of (25.18) counts the number of subsets of S having j elements in terms of their intersections with $S_1, S_2, \ldots, S_n$. We apply (25.17), (25.18), (25.15), and (25.12) to verify that

$$(25.19) \quad S_{h_1,\ldots,h_n} \equiv \binom{s_0 - i + j}{j} \quad \text{and} \quad \widehat{\vartheta}_{i,j}\, S_{h_1,\ldots,h_n} \equiv \vartheta_{i-j,j},$$

when $0 \le h_1 \le l_1, \ldots, 0 \le h_n \le l_n$ and $h_1 + \cdots + h_n = i - j$.

Due to (24.7) and (24.8), we have $\boldsymbol{L}_{h_1,\ldots,h_n} \equiv 0$, when $h_n < 0$, and $\boldsymbol{H}_{h_1,\ldots,h_n} \equiv 0$, when $h_n > l_n$. Thus, by employing (25.16), (25.19), $l_0 = 0$, and (24.9), we obtain

$$\boldsymbol{F}_{i,j} \equiv \sum_{\substack{0 \le h_1 \le l_1,\, \ldots,\, 0 \le h_n \le l_n \\ h_1 + \cdots + h_n = i - j}} \vartheta_{i-j,j}\, \boldsymbol{H}_{h_1,\ldots,h_n}\, \boldsymbol{L}_{h_1,\ldots,h_n}$$

$$\equiv \vartheta_{i-j,j} \sum_{h_0=0}^{l_0} \sum_{h_1=0}^{l_1} \cdots \sum_{h_{n-1}=0}^{l_{n-1}} \left[\boldsymbol{H}_{h_1,\ldots,h_n}\, \boldsymbol{L}_{h_1,\ldots,h_n} \right]_{h_n = i-j-(h_1+\cdots+h_{n-1})}$$

$$\equiv \vartheta_{i-j,j} \sum_{h_0=0}^{l_0} \sum_{h_1=0}^{l_1} \cdots \sum_{h_{n-1}=0}^{l_{n-1}} \boldsymbol{M}_{h_1,\ldots,h_{n-1},i-j}.$$

This yields (25.13) and completes the proof. $\square$

PROPOSITION 25.6. *The polynomials $\boldsymbol{N}_{i,j}$ of (25.8) satisfy*

$$(25.20) \quad \boldsymbol{N}_{i+j,j} \equiv 0, \quad \text{for } i \le 1 \text{ and any } j.$$

PROOF. We use (24.6), (24.3)–(24.5), $\underbrace{w_{0,\ldots,0,0}}_{n} \equiv 1$, and (23.15), to deduce that $\boldsymbol{L}_{\underbrace{0,\ldots,0}_{n-1},1} \equiv \boldsymbol{K}_{1,0} + \underbrace{w_{0,\ldots,0}}_{n-1},1 \equiv 0$. In view of (21.7), we obtain

$$(25.21) \quad \boldsymbol{L}_{i_1,\ldots,i_n} \equiv 0, \quad \text{for } 0 \le i_1, \ldots, 0 \le i_n, \text{ and } i_1 + \cdots + i_n = 1.$$

For $i \le -1$ and any j, (25.10) yields $\boldsymbol{F}_{i,j} \equiv 0$. We apply (25.14) and (25.21) to establish that $\boldsymbol{F}_{i,i-1} \equiv 0$, for $i \ge 0$. Since (24.6) and (24.4) yield $\boldsymbol{L}_{0,\ldots,0} \equiv 1$, we specialize (25.14) to see that

$$(25.22) \quad \boldsymbol{F}_{i,i} \equiv \sum_{\substack{0 \le k_1 \le l_1,\, \ldots,\, 0 \le k_n \le l_n \\ k_1 + \cdots + k_n = i}} A^{l_1,\ldots,l_n}_{l_1-k_1,\ldots,l_n-k_n}\, \xi_{k_1,k_1} \cdots \xi_{k_n,k_n}, \quad \text{for } i \ge 0.$$

If $i = 0$, then (25.22), (25.1), and (23.34) give $\boldsymbol{F}_{0,0} \equiv A^{l_1,\ldots,l_n}_{l_1,\ldots,l_n} \equiv 0$. Since (25.1) yields $\xi_{k_\nu,k_\nu} \equiv 0$ when $k_\nu \ge 1$, we employ (25.22) to verify that $\boldsymbol{F}_{i,i} \equiv 0$, for $i \ge 1$. For $i \ge 0$ and $j \ge i+1$, (25.14) gives $\boldsymbol{F}_{i,j} \equiv 0$. Thus, we find that

$$(25.23) \quad \boldsymbol{F}_{i,j} \equiv 0, \quad \text{for any integers } i, j \text{ that satisfy } j \ge i - 1.$$

In view of (25.11) and (25.23), we have

$$(25.24) \quad \boldsymbol{N}_{i+1,j} \equiv \left[\frac{2i-j+1}{2}\right] \boldsymbol{N}_{i,j-1} + \boldsymbol{N}_{i,j}^{(1)} - 2(j+1)\,\boldsymbol{a}_{m,2}\,\boldsymbol{N}_{i,j+1},$$

for any integers i, j that satisfy $j \geq i$.

Next, we shall apply (25.24) to establish the relation

$$(25.25) \quad \boldsymbol{N}_{i_0+j,j} \equiv 0, \quad \text{for } i_0 \leq 1 \text{ and any } j.$$

To clarify the argument, we consider separately: $i_0 \leq -1$; $i_0 = 0$; and $i_0 = 1$.

1. Suppose that $i_0 \leq -1$. Then, (25.8) yields $\boldsymbol{N}_{i_0,j} \equiv 0$, for any j. Using this, we substitute $i = i_0$ and $j \geq 1 > i_0 = i$ in (25.24) to obtain $\boldsymbol{N}_{i_0+1,j} \equiv 0$, for $j \geq 1$. By substituting $i = i_0 + 1$ and $j \geq 2 > i_0 + 1 = i$ in (25.24), we deduce $\boldsymbol{N}_{i_0+2,j} \equiv 0$, for $j \geq 2$. Continuing in this manner, we see that $\boldsymbol{N}_{i_0+j,j} \equiv 0$, for any j.

2. Suppose that $i_0 = 0$. We employ (25.8) and (23.34) to verify that $\boldsymbol{N}_{j,j} \equiv 0$, for $j \leq -1$, and

$$(25.26) \quad \boldsymbol{N}_{0,j} \equiv A^{l_1,\,\ldots,\,l_n}_{l_1,\,\ldots,\,l_n}\,\boldsymbol{M}_{0,\,\ldots,\,0,0,j} \equiv 0, \quad \text{for any } j.$$

We set $i = 0$ in (25.24) and apply (25.26) to deduce that $\boldsymbol{N}_{1,j} \equiv 0$, for $j \geq 1$. Substituting $i = 1$ in (25.24), we find that $\boldsymbol{N}_{2,j} \equiv 0$, for $j \geq 2$. Continuing in this way, we conclude that $\boldsymbol{N}_{j,j} \equiv 0$, for any j.

3. Suppose that $i_0 = 1$. Due to (25.8) and (25.26), we have $\boldsymbol{N}_{1+j,j} \equiv 0$, for $j \leq -1$. We use (25.4), (24.7), (25.21), and $\xi_{i,i} \equiv 0$ for $i \geq 1$ from (25.1) to obtain

$$(25.27) \quad \boldsymbol{M}_{k_1,\,\ldots,\,k_n,0,j} \equiv \sum_{\substack{0 \leq j_1 \leq k_1,\,\ldots,\,0 \leq j_n \leq k_n \\ j_1 + \cdots + j_n = j}} \xi_{k_1,j_1} \cdots \xi_{k_n,j_n}\,\boldsymbol{L}_{k_1-j_1,\,\ldots,\,k_n-j_n} \equiv 0,$$

for $j + 1 \geq k_1 + \cdots + k_n \geq 1$.

We employ (25.8), $A^{l_1,\,\ldots,\,l_n}_{l_1,\,\ldots,\,l_n} \equiv 0$ from (23.34), and (25.27) to see that

$$\boldsymbol{N}_{1,j} \equiv A^{l_1,\,\ldots,\,l_n}_{l_1,\,\ldots,\,l_n}\,\boldsymbol{M}_{0,\,\ldots,\,0,1,j} + \sum_{\substack{0 \leq k_1,\,\ldots,\,0 \leq k_n \\ k_1 + \cdots + k_n = 1}} A^{l_1,\,\ldots,\,l_n}_{l_1-k_1,\,\ldots,\,l_n-k_n}\,\boldsymbol{M}_{k_1,\,\ldots,\,k_n,0,j} \equiv 0,$$

for $j \geq 0$.

We set $i = 1$ in (25.24) and apply the preceding formula to deduce that $\boldsymbol{N}_{2,j} \equiv 0$, for $j \geq 1$. Substituting $i = 2$ in (25.24), we find that $\boldsymbol{N}_{3,j} \equiv 0$, for $j \geq 2$. Continuing in this manner, we conclude that $\boldsymbol{N}_{1+j,j} \equiv 0$, for any j.

Consequently, (25.25) and (25.20) are valid. This completes the proof. □

The rational numbers $\gamma_{i,j,k}$ defined by

$$(25.28) \quad \gamma_{i,j,k} \equiv \frac{(j-k)!}{2^{j-k}}\binom{2i+j+1}{j-k}, \quad \text{for } k \leq j \text{ and any } i,$$

were introduced in (7.38) on page 76. For ease of reference in the proof of the next result, we recall from (7.53) and (7.54) on page 79 that (25.28) yields

$$(25.29) \quad \gamma_{i,j,j} \equiv 1, \quad \text{for any } i, j,$$

and

$$(25.30) \quad \frac{1}{2}(2i+k+1)\,\gamma_{i,j,k} \equiv \gamma_{i,j,k-1}, \quad \text{for } k \leq j \text{ and any } i.$$

PROPOSITION 25.7. *The polynomials $N_{i,j}$ in (25.8) are given by*

(25.31) $\quad N_{i,j} \equiv P_{i-j,j}, \quad \text{for all } i, j,$

with respect to polynomials $P_{i,j}$ in $\mathcal{S}_{m,n}$ that satisfy

(25.32) $\quad P_{i,j} \equiv 0, \quad \text{for } i \leq 1 \text{ and any } j,$

and

(25.33) $\quad P_{i+1,j} \equiv \sum_{k=0}^{j} \gamma_{i,j,k} \begin{bmatrix} P_{i,k}^{(1)} - 2(k+1) \, a_{m,2} \, P_{i-1,k+1} \\ + \vartheta_{i+1,k} \sum_{h_0=0}^{l_0} \cdots \sum_{h_{n-1}=0}^{l_{n-1}} M_{h_1,\ldots,h_{n-1},i+1} \end{bmatrix},$

$\text{for } i \geq 0 \text{ and any } j.$

PROOF. We define $P_{i,j}$ in $\mathcal{S}_{m,n}$ with respect to $N_{i,j}$ in (25.8) by

(25.34) $\quad P_{i,j} \equiv N_{i+j,j}, \quad \text{for all integers } i \text{ and } j.$

Due to (25.34) and (25.20), $P_{i,j}$ satisfies (25.31) and (25.32). For any integers i, j subject to $i + j \geq -1$, we replace i in (25.11) with $i + j$, we replace i in (25.13) with $i + j + 1$, and we employ (25.34) to verify that

(25.35) $\quad P_{i+1,j} - \left[\frac{2i+j+1}{2}\right] P_{i+1,j-1}$

$\equiv \begin{bmatrix} P_{i,j}^{(1)} - 2(j+1) \, a_{m,2} \, P_{i-1,j+1} \\ + \vartheta_{i+1,j} \sum_{h_0=0}^{l_0} \cdots \sum_{h_{n-1}=0}^{l_{n-1}} M_{h_1,\ldots,h_{n-1},i+1} \end{bmatrix}, \quad \text{for } i+j \geq -1.$

We note that (25.8) and (25.7) give $N_{i,j} \equiv 0$, for $i \geq 0$ and $j \leq -1$. Hence, we have $N_{i,j} \equiv 0$, for $j \leq -1$ and any i. Due to (25.34), this yields

(25.36) $\quad P_{i,j} \equiv 0, \quad \text{for } j \leq -1 \text{ and any } i.$

For $i \geq 0$ and $j \geq 0$, we combine (25.29), (25.36), (25.30), and (25.35) to obtain

(25.37) $\quad P_{i+1,j} \equiv \sum_{k=0}^{j} \left[\gamma_{i,j,k} \, P_{i+1,k} - \gamma_{i,j,k-1} \, P_{i+1,k-1}\right]$

$\equiv \sum_{k=0}^{j} \gamma_{i,j,k} \left[P_{i+1,k} - \frac{1}{2}(2i+k+1) P_{i+1,k-1}\right]$

$\equiv \sum_{k=0}^{j} \gamma_{i,j,k} \begin{bmatrix} P_{i,k}^{(1)} - 2(k+1) \, a_{m,2} \, P_{i-1,k+1} \\ + \vartheta_{i+1,k} \sum_{h_0=0}^{l_0} \cdots \sum_{h_{n-1}=0}^{l_{n-1}} M_{h_1,\ldots,h_{n-1},i+1} \end{bmatrix}.$

In view of (25.36), (25.37) is valid for $i \geq 0$ and any j. This yields (25.33) and completes the proof of Proposition 25.7. □

25.2. EVALUATION OF LAGUERRE-FORSYTH SUMS

In terms of fixed integers $m \geq 2$, $n \geq 1$, $l_0 = 0$, as well as $l_1, \ldots, l_n$, and $s_0 = l_1 + \cdots + l_n$ that satisfy (24.1)–(24.2) of page 275, we introduce

(25.38) $\quad r_{i,j} \equiv \dfrac{(s_0 - i - j)(s_0 + i + j - 1)}{(s_0 - i)(s_0 + i - 1)}, \quad$ for $0 \leq i \leq s_0 - 1$ and any j.

For each fixed n-tuple $(h_0, \ldots, h_{n-1})$ of integers subject to

$$0 \leq h_0 \leq l_0, \ldots, 0 \leq h_{n-1} \leq l_{n-1}$$

and for each integer i satisfying $0 \leq i \leq s_0$, there is a corresponding polynomial $\boldsymbol{I}_{h_1, \ldots, h_{n-1}, i}$ given in $\mathcal{S}_{m,n}$ by (24.12)–(24.13) of page 276. We use it and (25.38) to define $\boldsymbol{S}_{h_1, \ldots, h_{n-1}, i, j}$ in $\mathcal{S}_{m,n}$ by

(25.39) $\quad \boldsymbol{S}_{h_1, \ldots, h_{n-1}, i, j} \equiv r_{i,j} \, \boldsymbol{I}_{h_1, \ldots, h_{n-1}, i}, \quad$ for $0 \leq i \leq s_0 - 1$ and any j,

and

(25.40) $\quad \boldsymbol{S}_{h_1, \ldots, h_{n-1}, s_0, j} \equiv \boldsymbol{I}_{h_1, \ldots, h_{n-1}, s_0}, \quad$ for any j.

In particular, (25.40), (25.39), and (25.38) give

(25.41) $\quad \boldsymbol{S}_{h_1, \ldots, h_{n-1}, s_0, 0} \equiv \boldsymbol{I}_{h_1, \ldots, h_{n-1}, s_0}$

and

(25.42) $\quad \boldsymbol{S}_{h_1, \ldots, h_{n-1}, s_0 - s, s} \equiv 0, \quad$ for $1 \leq s \leq s_0$.

Moreover, (25.39) and (24.12) of page 276 yield

(25.43) $\quad \boldsymbol{S}_{h_1, \ldots, h_{n-1}, i, j} \equiv 0, \quad$ for $0 \leq i \leq 1$ and any j.

For $1 \leq i \leq s_0 - 1$ and $0 \leq j \leq s_0 - 1 - i$, we introduce

(25.44) $\quad e_{i,j} \equiv \dfrac{-2j}{(s_0 - i - j)(s_0 + i + j - 1)}$,

(25.45) $\quad f_{i,j} \equiv \dfrac{-(s_0 - i)}{(s_0 - i - j)(s_0 + i + j - 1)}$,

and

(25.46) $\quad g_{i,j} \equiv \dfrac{-(s_0 - i)(s_0 - i + 1)}{2(j+1)(s_0 - i - j)(s_0 + i + j - 1)}$.

PROPOSITION 25.8. *In terms of* (25.44)–(25.46) *and* $\boldsymbol{M}_{h_1, \ldots, h_{n-1}, i+1}$ *from* (24.9) *on page 276, the identity*

(25.47) $\quad \boldsymbol{S}_{h_1, \ldots, h_{n-1}, i+1, j} - \dfrac{2i + j + 1}{2} e_{i,j} \, \boldsymbol{S}_{h_1, \ldots, h_{n-1}, i+1, j-1}$

$\equiv f_{i,j} \, \boldsymbol{S}^{(1)}_{h_1, \ldots, h_{n-1}, i, j} - 2(j+1) \, \boldsymbol{a}_{m,2} \, g_{i,j} \, \boldsymbol{S}_{h_1, \ldots, h_{n-1}, i-1, j+1}$

$\quad + \boldsymbol{M}_{h_1, \ldots, h_{n-1}, i+1}$

is satisfied for $1 \leq i \leq s_0 - 1$ *and* $0 \leq j \leq s_0 - 1 - i$.

PROOF. For $1 \leq i \leq s_0 - 1$ and $0 \leq j \leq s_0 - 1 - i$, we note that (24.10) of page 276, (25.45), (25.38), (24.11), and (25.46) yield

(25.48) $$A_i \equiv f_{i,j}\, r_{i,j}$$

and

(25.49) $$B_i \equiv -2(j+1)\, g_{i,j}\, r_{i-1,j+1}.$$

For the same i and j, we see that (24.13), (25.48), (25.49), and (25.39) give

(25.50) $\boldsymbol{I}_{h_1,\ldots,h_{n-1},i+1}$

$$\equiv A_i\, \boldsymbol{I}^{(1)}_{h_1,\ldots,h_{n-1},i} + B_i\, \boldsymbol{a}_{m,2}\, \boldsymbol{I}_{h_1,\ldots,h_{n-1},i-1} + \boldsymbol{M}_{h_1,\ldots,h_{n-1},i+1}$$

$$\equiv f_{i,j}\bigl(r_{i,j}\, \boldsymbol{I}^{(1)}_{h_1,\ldots,h_{n-1},i}\bigr) - 2(j+1)\, \boldsymbol{a}_{m,2}\, g_{i,j}\bigl(r_{i-1,j+1}\, \boldsymbol{I}_{h_1,\ldots,h_{n-1},i-1}\bigr)$$
$$+ \boldsymbol{M}_{h_1,\ldots,h_{n-1},i+1}$$

$$\equiv f_{i,j}\, \boldsymbol{S}^{(1)}_{h_1,\ldots,h_{n-1},i,j} - 2(j+1)\, \boldsymbol{a}_{m,2}\, g_{i,j}\, \boldsymbol{S}_{h_1,\ldots,h_{n-1},i-1,j+1}$$
$$+ \boldsymbol{M}_{h_1,\ldots,h_{n-1},i+1}.$$

Case 1. For $1 \leq i \leq s_0 - 2$ and $0 \leq j \leq s_0 - 1 - i$, we find that (25.39), (25.38), and (25.44) yield

(25.51) $\boldsymbol{S}_{h_1,\ldots,h_{n-1},i+1,j} - \dfrac{2i+j+1}{2}\, e_{i,j}\, \boldsymbol{S}_{h_1,\ldots,h_{n-1},i+1,j-1}$

$$\equiv \left[r_{i+1,j} - \frac{2i+j+1}{2}\, e_{i,j}\, r_{i+1,j-1} \right] \boldsymbol{I}_{h_1,\ldots,h_{n-1},i+1}$$

$$\equiv \left[\frac{(s_0 - i - 1 - j)(s_0 + i + j) + j(2i + j + 1)}{(s_0 - i - 1)(s_0 + i)} \right] \boldsymbol{I}_{h_1,\ldots,h_{n-1},i+1}$$

$$\equiv \boldsymbol{I}_{h_1,\ldots,h_{n-1},i+1}.$$

By combining (25.51) and (25.50), we see that (25.47) is satisfied with respect to the preceding restriction.

Case 2. For the remaining situation, suppose that i and j satisfy $i = s_0 - 1$ and $j = 0$. Then, using $j = 0$ in (25.44) along with (25.41), we note that the left member of (25.47) reduces to $\boldsymbol{S}_{h_1,\ldots,h_{n-1},s_0,0} \equiv \boldsymbol{I}_{h_1,\ldots,h_{n-1},s_0}$. We combine this with a substitution of $i = s_0 - 1$ and $j = 0$ in (25.50) to see that (25.47) is valid for these values of i and j. This completes the proof. □

With reference to $\boldsymbol{S}_{h_1,\ldots,h_{n-1},i,j}$ in (25.39)–(25.40) and $\vartheta_{i,j}$ in (25.12), we define $\boldsymbol{R}_{h_1,\ldots,h_{n-1},i,j}$ in $\mathcal{S}_{m,n}$ through

(25.52) $\boldsymbol{R}_{h_1,\ldots,h_{n-1},i,j} \equiv \vartheta_{i,j}\, \boldsymbol{S}_{h_1,\ldots,h_{n-1},i,j}, \quad$ for $0 \leq i \leq s_0$, and $j \geq -i$.

We employ (25.52), $\vartheta_{s_0,0} \equiv 1$ from (25.12), (25.41), and (25.42) to obtain

(25.53) $$\boldsymbol{R}_{h_1,\ldots,h_{n-1},s_0,0} \equiv \boldsymbol{I}_{h_1,\ldots,h_{n-1},s_0}$$

and

(25.54) $$\boldsymbol{R}_{h_1,\ldots,h_{n-1},s_0-s,s} \equiv 0, \quad \text{for } 1 \leq s \leq s_0.$$

Also, (25.52) and (25.43) yield

(25.55) $\quad \boldsymbol{R}_{h_1,\ldots,h_{n-1},i,j} \equiv 0, \quad \text{for } 0 \leq i \leq 1, \text{ and } j \geq -i.$

We use (25.52) to define $\boldsymbol{Q}_{i,j}$ in $\mathcal{S}_{m,n}$ by means of

(25.56) $\quad \boldsymbol{Q}_{i,j} \equiv \sum_{h_0=0}^{l_0} \cdots \sum_{h_{n-1}=0}^{l_{n-1}} \boldsymbol{R}_{h_1,\ldots,h_{n-1},i,j}, \quad \text{for } 0 \leq i \leq s_0 \text{ and } j \geq -i.$

In particular, (25.56) and (25.53) give

(25.57) $\quad \boldsymbol{Q}_{s_0,0} \equiv \sum_{h_0=0}^{l_0} \cdots \sum_{h_{n-1}=0}^{l_{n-1}} \boldsymbol{I}_{h_1,\ldots,h_{n-1},s_0}$

while (25.56) and (25.54) yield

(25.58) $\quad \boldsymbol{Q}_{s_0-s,s} \equiv 0, \quad \text{for } 1 \leq s \leq s_0.$

We apply (25.56) and (25.55) to see that

(25.59) $\quad \boldsymbol{Q}_{i,j} \equiv 0, \quad \text{for } 0 \leq i \leq 1 \text{ and } j \geq -i.$

PROPOSITION 25.9. *The identity*

(25.60) $\quad \boldsymbol{Q}_{i+1,j} - \left[\dfrac{2i+j+1}{2}\right] \boldsymbol{Q}_{i+1,j-1}$

$$\equiv \begin{bmatrix} \boldsymbol{Q}_{i,j}^{(1)} - 2(j+1)\,\boldsymbol{a}_{m,2}\,\boldsymbol{Q}_{i-1,j+1} \\ +\,\vartheta_{i+1,j} \displaystyle\sum_{h_0=0}^{l_0} \cdots \sum_{h_{n-1}=0}^{l_{n-1}} \boldsymbol{M}_{h_1,\ldots,h_{n-1},i+1} \end{bmatrix}$$

is valid for $1 \leq i \leq s_0 - 1$ *and* $0 \leq j \leq s_0 - 1 - i.$

PROOF. For $1 \leq i \leq s_0 - 1$ and $0 \leq j \leq s_0 - 1 - i$, we use (25.12) and (25.44)–(25.46) to verify that $\vartheta_{i,j}$, $e_{i,j}$, $f_{i,j}$, and $g_{i,j}$ satisfy

(25.61) $\quad\quad\quad\quad\quad\quad \vartheta_{i+1,j}\, e_{i,j} \equiv \vartheta_{i+1,j-1},$

(25.62) $\quad\quad\quad\quad\quad\quad \vartheta_{i+1,j}\, f_{i,j} \equiv \vartheta_{i,j},$

and

(25.63) $\quad\quad\quad\quad\quad\quad \vartheta_{i+1,j}\, g_{i,j} \equiv \vartheta_{i-1,j+1}.$

For the same i and j, we multiply (25.47) by $\vartheta_{i+1,j}$ and apply (25.61)–(25.63) as well as (25.52) to obtain

$$0 \equiv \vartheta_{i+1,j}\,\boldsymbol{S}_{h_1,\ldots,h_{n-1},i+1,j} - \frac{2i+j+1}{2}\,\vartheta_{i+1,j}\,e_{i,j}\,\boldsymbol{S}_{h_1,\ldots,h_{n-1},i+1,j-1}$$
$$- \vartheta_{i+1,j}\,f_{i,j}\,\boldsymbol{S}^{(1)}_{h_1,\ldots,h_{n-1},i,j} + 2(j+1)\,\boldsymbol{a}_{m,2}\,\vartheta_{i+1,j}\,g_{i,j}\,\boldsymbol{S}_{h_1,\ldots,h_{n-1},i-1,j+1}$$
$$- \vartheta_{i+1,j}\,\boldsymbol{M}_{h_1,\ldots,h_{n-1},i+1},$$

$$0 \equiv \vartheta_{i+1,j}\,\boldsymbol{S}_{h_1,\ldots,h_{n-1},i+1,j} - \frac{2i+j+1}{2}\,\vartheta_{i+1,j-1}\,\boldsymbol{S}_{h_1,\ldots,h_{n-1},i+1,j-1}$$
$$- \vartheta_{i,j}\,\boldsymbol{S}^{(1)}_{h_1,\ldots,h_{n-1},i,j} + 2(j+1)\,\boldsymbol{a}_{m,2}\,\vartheta_{i-1,j+1}\,\boldsymbol{S}_{h_1,\ldots,h_{n-1},i-1,j+1}$$
$$- \vartheta_{i+1,j}\,\boldsymbol{M}_{h_1,\ldots,h_{n-1},i+1},$$

and

(25.64) $$0 \equiv \boldsymbol{R}_{h_1,\ldots,h_{n-1},i+1,j} - \frac{2i+j+1}{2}\,\boldsymbol{R}_{h_1,\ldots,h_{n-1},i+1,j-1} - \boldsymbol{R}^{(1)}_{h_1,\ldots,h_{n-1},i,j}$$
$$+ 2(j+1)\,\boldsymbol{a}_{m,2}\,\boldsymbol{R}_{h_1,\ldots,h_{n-1},i-1,j+1} - \vartheta_{i+1,j}\,\boldsymbol{M}_{h_1,\ldots,h_{n-1},i+1}.$$

For $0 \le h_0 \le l_0,\ \ldots,\ 0 \le h_{n-1} \le l_{n-1}$, let $\mathfrak{M}_{h_1,\ldots,h_{n-1}}$ denote the right member of (25.64). We employ (25.64) and (25.56) to deduce that

(25.65) $$0 \equiv \sum_{h_0=0}^{l_0} \cdots \sum_{h_{n-1}=0}^{l_{n-1}} \mathfrak{M}_{h_1,\ldots,h_{n-1}}$$
$$\equiv \boldsymbol{Q}_{i+1,j} - \frac{2i+j+1}{2}\,\boldsymbol{Q}_{i+1,j-1} - \boldsymbol{Q}^{(1)}_{i,j}$$
$$+ 2(j+1)\,\boldsymbol{a}_{m,2}\,\boldsymbol{Q}_{i-1,j+1} - \vartheta_{i+1,j} \sum_{h_0=0}^{l_0} \cdots \sum_{h_{n-1}=0}^{l_{n-1}} \boldsymbol{M}_{h_1,\ldots,h_{n-1},i+1},$$

for $1 \le i \le s_0 - 1$ and $0 \le j \le s_0 - 1 - i$. We rewrite (25.65) to obtain (25.60) and complete the proof of Proposition 25.9. □

THEOREM 25.10. *For (23.1) on Ω^{**} into which (21.1) on Ω is transformed by $y(z) = \rho(z)\,v(z)$ and $z = f(\zeta)$ according to the context of page 265, suppose that $d_{0,\ldots,0,1}(\zeta) \equiv d_{0,\ldots,0,2}(\zeta) \equiv 0$. Let $m \ge 2$, $n \ge 1$, $l_0 = 0$, and $l_1,\ldots,l_n$ be integers subject to $0 \le l_1 \le \cdots \le l_n \le m$ and either $l_{n-1} \ge 1$ or $l_n \ge 3$. Then, the coefficients of (23.1) satisfy*

(25.66) $$\sum_{j_1=0}^{l_1} \cdots \sum_{j_n=0}^{l_n} A^{l_1,\ldots,l_n}_{j_1,\ldots,j_n}\,d^{(j_1+\cdots+j_n)}_{l_1-j_1,\ldots,l_n-j_n}(\zeta) \equiv \big(f'(\zeta)\big)^{s_0}\,\mathcal{I}_{m,l_1,\ldots,l_n}\big(f(\zeta)\big),$$

where $A^{l_1,\ldots,l_n}_{j_1,\ldots,j_n}$ is given by (23.34), $s_0 = l_1 + \cdots + l_n$, and $\mathcal{I}_{m,l_1,\ldots,l_n}\big(f(\zeta)\big)$ is obtained for $z = f(\zeta)$ by substituting $c^{(\kappa)}_{\mu_1,\ldots,\mu_n}(z)$ from (21.1) for $w^{(\kappa)}_{\mu_1,\ldots,\mu_n}$ in $\mathcal{I}_{m,l_1,\ldots,l_n}$ of (24.14) on page 276.

PROOF. We note that (25.12) yields $\vartheta_{i+1,-1} \equiv 0$, for $i \ge 0$. Due to (25.52), we have $\boldsymbol{R}_{h_1,\ldots,h_{n-1},i+1,-1} \equiv 0$, for $0 \le i \le s_0 - 1$. We use (25.56) to obtain

(25.67) $$\boldsymbol{Q}_{i+1,-1} \equiv 0, \quad \text{for } 0 \le i \le s_0 - 1.$$

25.2. EVALUATION OF LAGUERRE-FORSYTH SUMS

We combine (25.29), (25.67), (25.30), and (25.60) to verify that

$$(25.68) \quad \boldsymbol{Q}_{i+1,j} \equiv \sum_{k=0}^{j} \left[\gamma_{i,j,k}\, \boldsymbol{Q}_{i+1,k} - \gamma_{i,j,k-1}\, \boldsymbol{Q}_{i+1,k-1} \right]$$

$$\equiv \sum_{k=0}^{j} \gamma_{i,j,k}\left[\boldsymbol{Q}_{i+1,k} - \frac{1}{2}(2i+k+1)\, \boldsymbol{Q}_{i+1,k-1} \right]$$

$$\equiv \sum_{k=0}^{j} \gamma_{i,j,k} \left[\begin{array}{c} \boldsymbol{Q}_{i,k}^{(1)} - 2(k+1)\, a_{m,2}\, \boldsymbol{Q}_{i-1,k+1} \\ + \vartheta_{i+1,k} \sum_{h_0=0}^{l_0} \cdots \sum_{h_{n-1}=0}^{l_{n-1}} \boldsymbol{M}_{h_1,\ldots,h_{n-1},i+1} \end{array} \right],$$

for $1 \le i \le s_0 - 1$ and $0 \le j \le s_0 - 1 - i$.

In view of (25.32) on page 286 and (25.59), we see that the statement

$$(25.69) \quad \boldsymbol{P}_{i,j} \equiv \boldsymbol{Q}_{i,j}, \quad \text{for } 0 \le i \le i_0 \text{ and } 0 \le j \le s_0 - i,$$

is true for $i_0 = 1$. Suppose that i_0 is an integer satisfying $1 \le i_0 \le s_0 - 1$ such that (25.69) is valid. Then, we use (25.33) on page 286 as well as (25.69) and (25.68) to deduce that $\boldsymbol{P}_{i_0+1,j} \equiv \boldsymbol{Q}_{i_0+1,j}$, for $0 \le j \le s_0 - i_0 - 1$. Thus, we have

$$(25.70) \quad \boldsymbol{P}_{i,j} \equiv \boldsymbol{Q}_{i,j}, \quad \text{for } 0 \le i \le s_0 \text{ and } 0 \le j \le s_0 - i.$$

We employ (25.31) on page 286 and (25.70) to see that

$$(25.71) \quad \boldsymbol{N}_{s_0,t} \equiv \boldsymbol{P}_{s_0-t,t} \equiv \boldsymbol{Q}_{s_0-t,t}, \quad \text{for } 0 \le t \le s_0.$$

We observe that (25.71) and (25.58) yield

$$(25.72) \quad \boldsymbol{N}_{s_0,t} \equiv 0, \quad \text{for } 1 \le t \le s_0.$$

We substitute $c_{\mu_1,\ldots,\mu_n}^{(\kappa)}(z)$ from (21.1) for $\boldsymbol{w}_{\mu_1,\ldots,\mu_n}^{(\kappa)}$ in (25.72), set $z = f(\zeta)$, and combine the resulting expressions with (25.9) on page 281 to deduce that

$$(25.73) \quad \sum_{j_1=0}^{l_1} \cdots \sum_{j_n=0}^{l_n} A_{j_1,\ldots,j_n}^{l_1,\ldots,l_n}\, d_{l_1-j_1,\ldots,l_n-j_n}^{(j_1+\cdots+j_n)}(\zeta) \equiv \left(f'(\zeta)\right)^{s_0} \boldsymbol{N}_{s_0,0}\bigl(f(\zeta)\bigr).$$

Moreover, (25.71), (25.57), and (24.14) on page 276 give

$$(25.74) \quad \boldsymbol{N}_{s_0,0} \equiv \sum_{h_0=0}^{l_0} \cdots \sum_{h_{n-1}=0}^{l_{n-1}} \boldsymbol{I}_{h_1,\ldots,h_{n-1},s_0} \equiv \boldsymbol{\mathcal{I}}_{m,l_1,\ldots,l_n}.$$

After substituting $c_{\mu_1,\ldots,\mu_n}^{(\kappa)}(z)$ from (21.1) for $\boldsymbol{w}_{\mu_1,\ldots,\mu_n}^{(\kappa)}$ in (25.74), we set $z = f(\zeta)$ and combine the resulting expression with (25.73) to obtain (25.66). This completes the proof of Theorem 25.10. □

CHAPTER 26

Extensions of Chapter 9 to $H_{m,n} = 0$, when $m \geq 2$

26.1. The coefficients of (23.1) when $d_{0,\ldots,0,1}(\zeta) \equiv d_{0,\ldots,0,2}(\zeta) \equiv 0$

For the statement of Proposition 26.1, we repeat $\eta_{i,j}$ from (9.1) as

$$(26.1) \qquad \eta_{i,j} \equiv \frac{\left[\prod_{\nu=1}^{j}(i-\nu)\right]}{(m-1)^j}\binom{m-i+j}{j}, \quad \text{for any integers } i \text{ and } j.$$

PROPOSITION 26.1. *For (23.1) on Ω^{**} into which (21.1) on Ω is transformed by $y(z) = \rho(z)\,v(z)$ and $z = f(\zeta)$ according to the context of page 265, suppose that $d_{0,\ldots,0,1}(\zeta) \equiv d_{0,\ldots,0,2}(\zeta) \equiv 0$. Then, the coefficients of (23.1) are given by*

(26.2) $d_{i_1,\ldots,i_n}(\zeta) \equiv$

$$\left(f'(\zeta)\right)^{i_1+\cdots+i_n} \sum_{j_1=0}^{i_1}\cdots\sum_{j_n=0}^{i_n} \eta_{i_1,j_1}\cdots\eta_{i_n,j_n}\,V_{i_1-j_1,\ldots,i_n-j_n}\bigl(f(\zeta)\bigr)\bigl(\mathcal{S}(\zeta)\bigr)^{j_1+\cdots+j_n},$$

on Ω^{**} for $0 \leq i_1, \ldots, i_n \leq m$,

where $V_{i_1-j_1,\ldots,i_n-j_n}\bigl(f(\zeta)\bigr)$ is obtained for $z = f(\zeta)$ by substituting $c^{(\kappa)}_{\mu_1,\ldots,\mu_q}(z)$ from (21.1) for $w^{(\kappa)}_{\mu_1,\ldots,\mu_q}$ in $V_{i_1-j_1,\ldots,i_n-j_n}$ of (24.18) and

$$(26.3) \qquad \mathcal{S}(\zeta) \equiv \frac{\rho^{(1)}\bigl(f(\zeta)\bigr)}{\rho\bigl(f(\zeta)\bigr)}, \quad \text{on } \Omega^{**}.$$

PROOF. Using (23.3) in place of (9.4), we see that it suffices to make minor changes in the proof for Proposition 9.1 of page 93. □

With respect to (26.1) and $V_{i_1,\ldots,i_n}$ from (24.18)–(24.19) on page 276, we introduce $\boldsymbol{W}_{i_1,\ldots,i_n,k,t}$ in $\mathcal{S}_{m,n}$ for $k \geq 0$ and any integers $i_1, \ldots, i_n, t$, through

$$(26.4) \qquad \boldsymbol{W}_{i_1,\ldots,i_n,0,t} \equiv \sum_{\substack{0 \leq j_1,\ldots,0 \leq j_n \\ j_1+\cdots+j_n=t}} \eta_{i_1,j_1}\cdots\eta_{i_n,j_n}\,V_{i_1-j_1,\ldots,i_n-j_n},$$

for any $i_1, \ldots, i_n, t$,

and

$$(26.5) \quad \boldsymbol{W}_{i_1,\ldots,i_n,k+1,t} \equiv \boldsymbol{W}^{(1)}_{i_1,\ldots,i_n,k,t} + (i_1+\cdots+i_n+k-t)\,\boldsymbol{b}_{m,1}\boldsymbol{W}_{i_1,\ldots,i_n,k,t}$$

$$+ \frac{2i_1+\cdots+2i_n+2k-t+1}{m-1}\,\boldsymbol{W}_{i_1,\ldots,i_n,k,t-1}$$

$$- (m-1)(t+1)\,\boldsymbol{b}_{m,2}\,\boldsymbol{W}_{i_1,\ldots,i_n,k,t+1},$$

for $k \geq 0$ and any $i_1, \ldots, i_n, t$.

PROPOSITION 26.2. *For* (23.1) *on* Ω^{**} *into which* (21.1) *on* Ω *is transformed by* $y(z) = \rho(z)\,v(z)$ *and* $z = f(\zeta)$ *according to the context of page 265, suppose that* $d_{0,\ldots,0,1}(\zeta) \equiv d_{0,\ldots,0,2}(\zeta) \equiv 0$. *Then, the derivatives of the coefficients of* (23.1) *are provided by*

$$(26.6) \quad d_{i_1,\ldots,i_n}^{(k)}(\zeta) \equiv \bigl(f'(\zeta)\bigr)^{i_1+\cdots+i_n+k} \sum_{t=0}^{i_1+\cdots+i_n+k} \mathcal{W}_{i_1,\ldots,i_n,k,t}\bigl(f(\zeta)\bigr)\bigl(\mathcal{S}(\zeta)\bigr)^t,$$

for $0 \leq i_1, \ldots, i_n \leq m$ *and* $k \geq 0$,

where $\mathcal{W}_{i_1,\ldots,i_n,k,t}\bigl(f(\zeta)\bigr)$ *is obtained for* $z = f(\zeta)$ *by substituting* $c_{\mu_1,\ldots,\mu_n}^{(\kappa)}(z)$ *from* (21.1) *for* $w_{\mu_1,\ldots,\mu_n}^{(\kappa)}$ *in* $\mathcal{W}_{i_1,\ldots,i_n,k,t}$ *of* (26.4)–(26.5) *and* $\mathcal{S}(\zeta)$ *is given in* (26.3).

PROOF. For $t < 0$, (26.4) yields $\mathcal{W}_{i_1,\ldots,i_n,0,t} \equiv 0$. When $t > i_1 + \cdots + i_n$, the conditions $0 \leq j_1, \ldots, j_n$ and $j_1 + \cdots + j_n = t$ in (26.4) require $j_\nu > i_\nu$, for at least one ν satisfying $1 \leq \nu \leq n$, and give $\mathcal{W}_{i_1,\ldots,i_n,0,t} \equiv 0$ because (24.19) shows that each $\mathbf{V}_{i_1-j_1,\ldots,i_n-j_n}$ equals 0. Suppose that k is a nonnegative integer such that $\mathcal{W}_{i_1,\ldots,i_n,k,t} \equiv 0$ is valid whenever $i_1, \ldots, i_n, k, t$ are integers that satisfy either $t < 0$ or $t > i_1 + \cdots + i_n + k$. Then, for this k, we use (26.5) to see that $\mathcal{W}_{i_1,\ldots,i_n,k+1,t} \equiv 0$ is valid for any integers $i_1, \ldots, i_n, k, t$ that satisfy either $t < 0$ or $t > i_1 + \cdots + i_n + k + 1$. This establishes that

$$(26.7) \quad \mathcal{W}_{i_1,\ldots,i_n,k,t} \equiv 0, \quad \text{for } k \geq 0 \text{ and either } t < 0 \text{ or } t > i_1 + \cdots + i_n + k.$$

In view of $V_{i_1-j_1,\ldots,i_n-j_n}\bigl(f(\zeta)\bigr) \equiv 0$, whenever $j_\nu > i_\nu$ for some ν satisfying $1 \leq \nu \leq n$, we use (26.2) and (26.4) to obtain

$$\frac{d_{i_1,\ldots,i_n}(\zeta)}{\bigl(f'(\zeta)\bigr)^{i_1+\cdots+i_n}} \equiv \sum_{t=0}^{i_1+\cdots+i_n} \sum_{\substack{0\leq j_1,\ldots,0\leq j_n \\ j_1+\cdots+j_n=t}} \eta_{i_1,j_1}\cdots\eta_{i_n,j_n}\, V_{i_1-j_1,\ldots,i_n-j_n}\bigl(f(\zeta)\bigr)\bigl(\mathcal{S}(\zeta)\bigr)^t$$

$$\equiv \sum_{t=0}^{i_1+\cdots+i_n} \mathcal{W}_{i_1,\ldots,i_n,0,t}\bigl(f(\zeta)\bigr)\bigl(\mathcal{S}(\zeta)\bigr)^t.$$

This shows that (26.6) is valid for $k = 0$.

Suppose that (26.6) is valid when k is replaced with a nonnegative integer k_0. Then, by differentiating that formula with respect to ζ and using (26.5) as well as (26.7), we perform a computation similar to that for (9.27)–(9.31) to conclude that (26.6) is valid for any nonnegative integer k. □

26.2. Evaluation of Laguerre-Forsyth sums

In terms of fixed integers $m \geq 2$, $n \geq 1$, $l_0 = 0$, and $l_1, \ldots, l_n$ that satisfy $0 \leq l_1 \leq \cdots \leq l_n \leq m$ and either $l_{n-1} \geq 1$ on $l_n \geq 3$, we use (23.34) on page 271 as well as (26.4)–(26.5) to define $\boldsymbol{X}_{i,j}$ in $\mathcal{S}_{m,n}$ by

$$(26.8) \quad \boldsymbol{X}_{i,j} \equiv \sum_{r=0}^{i} \sum_{\substack{0\leq k_1,\ldots,0\leq k_n \\ k_1+\cdots+k_n=r}} A_{l_1-k_1,\ldots,l_n-k_n}^{l_1,\ldots,l_n}\, \boldsymbol{W}_{k_1,\ldots,k_n,i-r,j}, \quad \text{for any } i, j.$$

OBSERVATION. Whenever the sum in (26.8) is nonvacuous, we have $r \geq 0$ and

$$(2l_1 + \cdots + 2l_n - 2) - \bigl((l_1 - k_1) + \cdots + (l_n - k_n)\bigr) = l_1 + \cdots + l_n - 2 + r \geq 0.$$

Thus, $A_{l_1-k_1,\,l_2-k_2,\,\ldots,\,l_n-k_n}^{l_1,\,l_2,\,\ldots,\,l_n}$ in (26.8) is well defined by (23.34) of page 271.

26.2. EVALUATION OF LAGUERRE-FORSYTH SUMS

PROPOSITION 26.3. *For (23.1) on Ω^{**} into which (21.1) on Ω is transformed by $y(z) = \rho(z)\, v(z)$ and $z = f(\zeta)$ according to the context of page 265, suppose that $d_{0,\,\ldots,\,0,\,1}(\zeta) \equiv d_{0,\,\ldots,\,0,\,2}(\zeta) \equiv 0$. Then,*

$$(26.9) \quad \sum_{j_1=0}^{l_1} \cdots \sum_{j_n=0}^{l_n} A_{j_1,\,\ldots,\,j_n}^{l_1,\,\ldots,\,l_n}\, d_{l_1-j_1,\,\ldots,\,l_n-j_n}^{(j_1+\cdots+j_n)}(\zeta)$$

$$\equiv \big(f'(\zeta)\big)^{l_1+\cdots+l_n} \sum_{t=0}^{l_1+\cdots+l_n} X_{l_1+\cdots+l_n,t}\big(f(\zeta)\big) \big(\mathcal{S}(\zeta)\big)^t, \quad \text{on } \Omega^{**},$$

where $X_{l_1+\cdots+l_n,t}\big(f(\zeta)\big)$ is obtained for $z = f(\zeta)$ by substituting $c_{\mu_1,\,\ldots,\,\mu_n}^{(\kappa)}(z)$ from (21.1) for $w_{\mu_1,\,\ldots,\,\mu_n}^{(\kappa)}$ in $X_{l_1+\cdots+l_n,t}$ and $\mathcal{S}(\zeta)$ is given by (26.3).

PROOF. After setting $s_0 = l_1 + \cdots + l_n$, we employ (26.6), the property deduced from (23.34) that $A_{j_1,\,j_2,\,\ldots,\,j_n}^{i_1,\,i_2,\,\ldots,\,i_n} = 0$ when $j_\nu < 0$, for some ν, and (26.8) to obtain

$$\sum_{j_1=0}^{l_1} \cdots \sum_{j_n=0}^{l_n} A_{j_1,\,\ldots,\,j_n}^{l_1,\,\ldots,\,l_n}\, d_{l_1-j_1,\,\ldots,\,l_n-j_n}^{(j_1+\cdots+j_n)}(\zeta)$$

$$\equiv \sum_{k_1=0}^{l_1} \cdots \sum_{k_n=0}^{l_n} A_{l_1-k_1,\,\ldots,\,l_n-k_n}^{l_1,\,\ldots,\,l_n}\, d_{k_1,\,\ldots,\,k_n}^{(s_0-k_1-\cdots-k_n)}(\zeta)$$

$$\equiv \big(f'(\zeta)\big)^{s_0} \sum_{k_1=0}^{l_1} \cdots \sum_{k_n=0}^{l_n} A_{l_1-k_1,\,\ldots,\,l_n-k_n}^{l_1,\,\ldots,\,l_n} \sum_{t=0}^{s_0} \mathcal{W}_{k_1,\,\ldots,\,k_n,\,s_0-k_1-\cdots-k_n,\,t}\big(f(\zeta)\big)\big(\mathcal{S}(\zeta)\big)^t$$

$$\equiv \big(f'(\zeta)\big)^{s_0} \sum_{t=0}^{s_0} \big(\mathcal{S}(\zeta)\big)^t \sum_{k_1=0}^{l_1} \cdots \sum_{k_n=0}^{l_n} A_{l_1-k_1,\,\ldots,\,l_n-k_n}^{l_1,\,\ldots,\,l_n} \mathcal{W}_{k_1,\,\ldots,\,k_n,\,s_0-k_1-\cdots-k_n,\,t}\big(f(\zeta)\big)$$

$$\equiv \big(f'(\zeta)\big)^{s_0} \sum_{t=0}^{s_0} \big(\mathcal{S}(\zeta)\big)^t \sum_{r=0}^{s_0} \sum_{\substack{0 \leq k_1,\,\ldots,\,0 \leq k_n \\ k_1+\cdots+k_n = r}} A_{l_1-k_1,\,\ldots,\,l_n-k_n}^{l_1,\,\ldots,\,l_n} \mathcal{W}_{k_1,\,\ldots,\,k_n,\,s_0-k_1-\cdots-k_n,\,t}\big(f(\zeta)\big)$$

$$\equiv \big(f'(\zeta)\big)^{l_1+\cdots+l_n} \sum_{t=0}^{l_1+\cdots+l_n} X_{l_1+\cdots+l_n,t}\big(f(\zeta)\big)\big(\mathcal{S}(\zeta)\big)^t.$$

This yields (26.9) and completes the proof. □

In terms of (23.34) on page 271 and (26.4), we define $G_{i,j}$ in $\mathcal{S}_{m,n}$ by

$$(26.10) \qquad G_{i,j} \equiv \sum_{\substack{0 \leq k_1,\,\ldots,\,0 \leq k_n \\ k_1+\cdots+k_n = i}} A_{l_1-k_1,\,\ldots,\,l_n-k_n}^{l_1,\,\ldots,\,l_n} \mathcal{W}_{k_1,\,\ldots,\,k_n,\,0,\,j}, \quad \text{for any } i,\, j.$$

PROPOSITION 26.4. *The polynomials $X_{i,j}$ and $G_{i,j}$ satisfy*

$$(26.11) \quad X_{i+1,j} - \frac{2i-j+1}{m-1} X_{i,j-1} \equiv X_{i,j}^{(1)} + (i-j)\boldsymbol{b}_{m,1} X_{i,j}$$

$$- (m-1)(j+1)\boldsymbol{b}_{m,2}\, X_{i,j+1} + G_{i+1,j},$$

for all integers i, j.

PROOF. For $i \geq 0$ and any j, we use (26.8), (26.5), and (26.10) to verify that

$$X_{i,j}^{(1)} \equiv \sum_{r=0}^{i} \sum_{\substack{0 \leq k_1, \ldots, 0 \leq k_n \\ k_1 + \cdots + k_n = r}} A_{l_1-k_1, \ldots, l_n-k_n}^{l_1, \ldots, l_n} \mathcal{W}_{k_1, \ldots, k_n, i-r, j}^{(1)}$$

$$\equiv \sum_{r=0}^{i} \sum_{\substack{0 \leq k_1, \ldots, 0 \leq k_n \\ k_1 + \cdots + k_n = r}} A_{l_1-k_1, \ldots, l_n-k_n}^{l_1, \ldots, l_n} \left\{ \begin{array}{l} \mathcal{W}_{k_1, \ldots, k_n, i+1-r, j} \\ - (r + (i-r) - j) b_{m,1} \mathcal{W}_{k_1, \ldots, k_n, i-r, j} \\ - \left[\frac{2r + 2(i-r) - j + 1}{m-1} \right] \mathcal{W}_{k_1, \ldots, k_n, i-r, j-1} \\ + (m-1)(j+1) b_{m,2} \mathcal{W}_{k_1, \ldots, k_n, i-r, j+1} \end{array} \right\}$$

$$\equiv X_{i+1,j} - G_{i+1,j} - (i-j) b_{m,1} X_{i,j} - \frac{2i-j+1}{m-1} X_{i,j-1}$$
$$+ (m-1)(j+1) b_{m,2} X_{i,j+1}.$$

Thus, (26.11) is valid for $i \geq 0$ and any j.

To show that (26.11) is also valid for $i < 0$ and any j, we first use (26.10) and (23.34) to obtain $G_{0,j} \equiv A_{l_1, \ldots, l_n}^{l_1, \ldots, l_n} \mathcal{W}_{0, \ldots, 0, 0, j} \equiv 0$, for any j. We apply (26.8) and (23.34) to deduce that $X_{0,j} \equiv A_{l_1, \ldots, l_n}^{l_1, \ldots, l_n} \mathcal{W}_{0, \ldots, 0, 0, j} \equiv 0$, for any j. Also, (26.10) and (26.8) yield $G_{i,j} \equiv 0$ and $X_{i,j} \equiv 0$, for $i \leq -1$ and any j. Consequently, for $i \leq -1$ and any j, each term in (26.11) is equal to 0. Thus, (26.11) is valid for all integers i and j. This completes the proof. $\square$

In terms of the integer $s_0 = l_1 + \cdots + l_n$ from (24.2) that satisfies $s_0 \geq 2$, we define rational numbers $v_{i,j}$ by means of

$$(26.12) \qquad v_{i,j} \equiv \frac{(-1)^{s_0+i+j}}{(m-1)^j} \binom{s_0-i}{j} \frac{1}{\prod_{r=1}^{s_0-i-j} (2s_0 - 1 - r)}, \quad \text{for } i+j \geq 0.$$

PROPOSITION 26.5. *The polynomial $G_{i,j}$ of (26.10) is related to $v_{i,j}$ in (26.12) and $\mathcal{W}_{h_1, \ldots, h_{n-1}, i}$ in (24.20) through*

$$(26.13) \qquad G_{i,j} \equiv v_{i-j,j} \sum_{h_0=0}^{l_0} \cdots \sum_{h_{n-1}=0}^{l_{n-1}} \mathcal{W}_{h_1, \ldots, h_{n-1}, i-j}, \quad \text{for } i \geq 0 \text{ and any } j.$$

PROOF. Due to (23.34), (26.1) and (24.19), we have: $A_{j_1, \ldots, j_n}^{l_1, \ldots, l_n} \equiv 0$, when $j_\nu < 0$ for some ν; $\eta_{i,j} \equiv 0$, when $j < 0$; and $V_{j_1, \ldots, j_n} \equiv 0$, when $j_\nu < 0$ for some ν. For $i \geq 0$ and any j, we use (26.10), (26.4), and the observations just made to obtain

$$(26.14) \quad G_{i,j} \equiv \sum_{\substack{0 \leq k_1, \ldots, 0 \leq k_n \\ k_1 + \cdots + k_n = i}} A_{l_1-k_1, \ldots, l_n-k_n}^{l_1, \ldots, l_n} \sum_{\substack{0 \leq j_1, \ldots, 0 \leq j_n \\ j_1 + \cdots + j_n = j}} \eta_{k_1, j_1} \cdots \eta_{k_n, j_n} V_{k_1-j_1, \ldots, k_n-j_n}$$

$$\equiv \sum_{\substack{0 \leq k_1, \ldots, 0 \leq k_n \\ k_1 + \cdots + k_n = i \\ 0 \leq j_1 \leq k_1, \ldots, 0 \leq j_n \leq k_n \\ j_1 + \cdots + j_n = j}} A_{l_1-k_1, \ldots, l_n-k_n}^{l_1, \ldots, l_n} \eta_{k_1, j_1} \cdots \eta_{k_n, j_n} V_{k_1-j_1, \ldots, k_n-j_n}$$

$$\equiv \sum_{\substack{0 \leq k_1 \leq l_1, \ldots, 0 \leq k_n \leq l_n \\ k_1 + \cdots + k_n = i \\ 0 \leq h_1 \leq k_1, \ldots, 0 \leq h_n \leq k_n \\ h_1 + \cdots + h_n = i-j}} A_{l_1-k_1, \ldots, l_n-k_n}^{l_1, \ldots, l_n} \eta_{k_1, k_1-h_1} \cdots \eta_{k_n, k_n-h_n} V_{h_1, \ldots, h_n}.$$

26.2. EVALUATION OF LAGUERRE-FORSYTH SUMS

The restriction $k_1 + \cdots + k_n = i$ and (23.34) yield

$$A^{l_1, \ldots, l_n}_{l_1-k_1, \ldots, l_n-k_n} \equiv \frac{(-1)^{s_0-i} \prod_{\nu=1}^{n}\left[\binom{m-k_\nu}{l_\nu-k_\nu}\prod_{r=1}^{l_\nu-k_\nu}(l_\nu-r)\right]}{\prod_{r=1}^{s_0-i}(2s_0-1-r)}.$$

In view of $k_1 + \cdots + k_n = i$ and $h_1 + \cdots + h_n = i - j$, (26.1) gives

$$\prod_{\nu=1}^{n}\eta_{k_\nu, k_\nu-h_\nu} \equiv \frac{1}{(m-1)^j}\prod_{\nu=1}^{n}\left[\binom{m-h_\nu}{k_\nu-h_\nu}\prod_{r=1}^{k_\nu-h_\nu}(k_\nu-r)\right].$$

Combining the preceding expressions and setting

(26.15) $$\widehat{v}_{i,j} \equiv \frac{\frac{(-1)^{s_0+i}}{(m-1)^j}}{\prod_{r=1}^{s_0-i}(2s_0-1-r)},$$

we deduce that

$$A^{l_1, \ldots, l_n}_{l_1-k_1, \ldots, l_n-k_n}\prod_{\nu=1}^{n}\eta_{k_\nu, k_\nu-h_\nu}$$

$$\equiv \widehat{v}_{i,j}\prod_{\nu=1}^{n}\left\{\binom{m-h_\nu}{k_\nu-h_\nu}\binom{m-k_\nu}{l_\nu-k_\nu}\left[\prod_{r=1}^{l_\nu-k_\nu}(l_\nu-r)\right]\left[\prod_{r=1}^{k_\nu-h_\nu}(k_\nu-r)\right]\right\}$$

$$\equiv \widehat{v}_{i,j}\prod_{\nu=1}^{n}\left\{\binom{m-h_\nu}{l_\nu-h_\nu}\left[\prod_{r=1}^{l_\nu-h_\nu}(l_\nu-r)\right]\binom{l_\nu-h_\nu}{k_\nu-h_\nu}\right\}$$

$$\equiv \widehat{v}_{i,j}\, H_{h_1, \ldots, h_n}\prod_{\nu=1}^{n}\binom{l_\nu-h_\nu}{k_\nu-h_\nu},$$

where $H_{h_1, \ldots, h_n}$ is defined by (24.8). We substitute this in (26.14) to verify that

(26.16) $$G_{i,j} \equiv \sum_{\substack{0\leq h_1\leq l_1,\, \ldots,\, 0\leq h_n\leq l_n \\ h_1+\cdots+h_n=i-j \\ h_1\leq k_1\leq l_1,\, \ldots,\, h_n\leq k_n\leq l_n \\ k_1+\cdots+k_n=i}} \widehat{v}_{i,j}\, H_{h_1, \ldots, h_n}\left[\prod_{\nu=1}^{n}\binom{l_\nu-h_\nu}{k_\nu-h_\nu}\right]V_{h_1, \ldots, h_n}$$

$$\equiv \sum_{\substack{0\leq h_1\leq l_1,\, \ldots,\, 0\leq h_n\leq l_n \\ h_1+\cdots+h_n=i-j}} \widehat{v}_{i,j}\, H_{h_1, \ldots, h_n}\, S_{h_1, \ldots, h_n}\, V_{h_1, \ldots, h_n},$$

where

$$S_{h_1, \ldots, h_n} \equiv \sum_{\substack{h_1\leq k_1\leq l_1,\, \ldots,\, h_n\leq k_n\leq l_n \\ k_1+\cdots+k_n=i}}\prod_{\nu=1}^{n}\binom{l_\nu-h_\nu}{k_\nu-h_\nu}.$$

We have

(26.17) $$S_{h_1, \ldots, h_n} \equiv \sum_{\substack{0\leq i_1\leq l_1-h_1,\, \ldots,\, 0\leq i_n\leq l_n-h_n \\ i_1+\cdots+i_n=j}}\prod_{\nu=1}^{n}\binom{l_\nu-h_\nu}{i_\nu},$$

when $0 \leq h_1 \leq l_1, \ldots, 0 \leq h_n \leq l_n$ and $h_1 + \cdots + h_n = i - j$.

As we observed for $S_{h_1, \ldots, h_n}$ in (25.17) of page 283, the identity

(26.18) $$\sum_{\substack{0 \le i_1 \le a_1,\, 0 \le i_2 \le a_2,\, \ldots,\, 0 \le i_n \le a_n \\ i_1 + i_2 + \cdots + i_n = j}} \binom{a_1}{i_1}\binom{a_2}{i_2} \cdots \binom{a_n}{i_n} \equiv \binom{a_1 + a_2 + \cdots + a_n}{j}$$

generalizes Vandermonde's convolution. We apply (26.17), (26.18), (26.15), and (26.12) to verify that

(26.19) $$S_{h_1, \ldots, h_n} \equiv \binom{s_0 - i + j}{j} \quad \text{and} \quad \widehat{v}_{i,j}\, S_{h_1, \ldots, h_n} \equiv v_{i-j, j},$$

when $0 \le h_1 \le l_1, \ldots, 0 \le h_n \le l_n$ and $h_1 + \cdots + h_n = i - j$.

Due to (24.19) and (24.8), we have $\boldsymbol{V}_{h_1, \ldots, h_n} \equiv 0$, when $h_n < 0$, and $\boldsymbol{H}_{h_1, \ldots, h_n} \equiv 0$, when $h_n > l_n$. Thus, by employing (26.16), (26.19), $l_0 = 0$, and (24.20), we obtain

$$\boldsymbol{G}_{i,j} \equiv \sum_{\substack{0 \le h_1 \le l_1, \ldots, 0 \le h_n \le l_n \\ h_1 + \cdots + h_n = i - j}} v_{i-j, j}\, \boldsymbol{H}_{h_1, \ldots, h_n}\, \boldsymbol{V}_{h_1, \ldots, h_n}$$

$$\equiv v_{i-j,j} \sum_{h_0=0}^{l_0} \sum_{h_1=0}^{l_1} \cdots \sum_{h_{n-1}=0}^{l_{n-1}} \Big[\boldsymbol{H}_{h_1, \ldots, h_n} \boldsymbol{V}_{h_1, \ldots, h_n} \Big]_{h_n = i - j - (h_1 + \cdots + h_{n-1})}$$

$$\equiv v_{i-j,j} \sum_{h_0=0}^{l_0} \sum_{h_1=0}^{l_1} \cdots \sum_{h_{n-1}=0}^{l_{n-1}} \boldsymbol{W}_{h_1, \ldots, h_{n-1}, i-j}.$$

This yields (26.13) and completes the proof. □

PROPOSITION 26.6. *The polynomials $\boldsymbol{X}_{i,j}$ of (26.8) satisfy*

(26.20) $$\boldsymbol{X}_{i+j, j} \equiv 0, \quad \text{for } i \le 1 \text{ and any } j.$$

PROOF. We use (24.18), (24.15)–(24.17), $\underbrace{w_{0, \ldots, 0, 0}}_{n} \equiv 1$, and (23.15) to deduce that $\underbrace{\boldsymbol{V}_{0, \ldots, 0, 1}}_{n-1} \equiv \boldsymbol{U}_{1,0} + \underbrace{w_{0, \ldots, 0, 1}}_{n-1} \equiv 0$. In view of (21.7), we obtain

(26.21) $$\boldsymbol{V}_{i_1, \ldots, i_n} \equiv 0, \quad \text{for } 0 \le i_1, \ldots, 0 \le i_n \text{ and } i_1 + \cdots + i_n = 1.$$

For $i \le -1$ and any j, (26.10) yields $\boldsymbol{G}_{i,j} \equiv 0$. We apply (26.14) and (26.21) to establish that $\boldsymbol{G}_{i, i-1} \equiv 0$ for $i \ge 0$. Since (24.18) and (24.16) yield $\boldsymbol{V}_{0, \ldots, 0} \equiv 1$, we specialize (26.14) to see that

(26.22) $$\boldsymbol{G}_{i,i} \equiv \sum_{\substack{0 \le k_1 \le l_1, \ldots, 0 \le k_n \le l_n \\ k_1 + \cdots + k_n = i}} A^{l_1, \ldots, l_n}_{l_1 - k_1, \ldots, l_n - k_n}\, \eta_{k_1, k_1} \cdots \eta_{k_n, k_n}, \quad \text{for } i \ge 0.$$

If $i = 0$, then (26.22), (26.1), and (23.34) give $\boldsymbol{G}_{0,0} \equiv A^{l_1, \ldots, l_n}_{l_1, \ldots, l_n} \equiv 0$. Since (26.1) yields $\eta_{k_\nu, k_\nu} \equiv 0$ when $k_\nu \ge 1$, we employ (26.22) to verify that $\boldsymbol{G}_{i,i} \equiv 0$, for $i \ge 1$. For $i \ge 0$ and $j \ge i + 1$, (26.14) gives $\boldsymbol{G}_{i,j} \equiv 0$. Thus, we find that

(26.23) $$\boldsymbol{G}_{i,j} \equiv 0, \quad \text{for any integers } i, j \text{ that satisfy } j \ge i - 1.$$

In view of (26.11) and (26.23), we have

(26.24) $$\boldsymbol{X}_{i+1,j} \equiv \left[\frac{2i-j+1}{m-1}\right]\boldsymbol{X}_{i,j-1} + \boldsymbol{X}_{i,j}^{(1)} + (i-j)\boldsymbol{b}_{m,1}\,\boldsymbol{X}_{i,j}$$
$$- (m-1)(j+1)\,\boldsymbol{b}_{m,2}\,\boldsymbol{X}_{i,j+1},$$

for any integers i, j that satisfy $j \geq i$.

Next, we shall apply (26.24) to establish the relation

(26.25) $\quad\quad\quad\quad \boldsymbol{X}_{i_0+j,j} \equiv 0, \quad$ for $i_0 \leq 1$ and any j.

To clarify the argument, we consider separately: $i_0 \leq -1$; $i_0 = 0$; and $i_0 = 1$.

1. Suppose that $i_0 \leq -1$. Then, (26.8) yields $\boldsymbol{X}_{i_0,j} \equiv 0$, for any j. Using this, we substitute $i = i_0$ and $j \geq 1 > i_0 = i$ in (26.24) to obtain $\boldsymbol{X}_{i_0+1,j} \equiv 0$, for $j \geq 1$. By substituting $i = i_0 + 1$ and $j \geq 2 > i_0 + 1 = i$ in (26.24), we deduce $\boldsymbol{X}_{i_0+2,j} \equiv 0$, for $j \geq 2$. Continuing in this manner, we see that $\boldsymbol{X}_{i_0+j,j} \equiv 0$, for any j.

2. Suppose that $i_0 = 0$. We employ (26.8) and (23.34) to verify that $\boldsymbol{X}_{j,j} \equiv 0$, for $j \leq -1$, and

(26.26) $\quad\quad\quad \boldsymbol{X}_{0,j} \equiv A^{l_1,\ldots,l_n}_{l_1,\ldots,l_n}\,\boldsymbol{\mathcal{W}}_{0,\ldots,0,0,j} \equiv 0, \quad$ for any j.

We set $i = 0$ in (26.24) and apply (26.26) to deduce that $\boldsymbol{X}_{1,j} \equiv 0$, for $j \geq 1$. Substituting $i = 1$ in (26.24), we find that $\boldsymbol{X}_{2,j} \equiv 0$, for $j \geq 2$. Continuing in this way, we conclude that $\boldsymbol{X}_{j,j} \equiv 0$, for any j.

3. Suppose that $i_0 = 1$. Due to (26.8) and (26.26), we have $\boldsymbol{X}_{1+j,j} \equiv 0$, for $j \leq -1$. We use (26.4), (24.19), (26.21), and $\eta_{i,i} \equiv 0$ for $i \geq 1$ from (26.1) to obtain

(26.27) $\quad \boldsymbol{\mathcal{W}}_{k_1,\ldots,k_n,0,j} \equiv \sum_{\substack{0\leq j_1\leq k_1,\,\ldots,\,0\leq j_n\leq k_n \\ j_1+\cdots+j_n=j}} \eta_{k_1,j_1}\cdots\eta_{k_n,j_n}\,\boldsymbol{V}_{k_1-j_1,\ldots,k_n-j_n} \equiv 0,$

for $j + 1 \geq k_1 + \cdots + k_n \geq 1$.

We employ (26.8), $A^{l_1,\ldots,l_n}_{l_1,\ldots,l_n} \equiv 0$ from (23.34), and (26.27) to see that

$$\boldsymbol{X}_{1,j} \equiv A^{l_1,\ldots,l_n}_{l_1,\ldots,l_n}\,\boldsymbol{\mathcal{W}}_{0,\ldots,0,1,j} + \sum_{\substack{0\leq k_1,\,\ldots,\,0\leq k_n \\ k_1+\cdots+k_n=1}} A^{l_1,\ldots,l_n}_{l_1-k_1,\ldots,l_n-k_n}\,\boldsymbol{\mathcal{W}}_{k_1,\ldots,k_n,0,j} \equiv 0,$$

for $j \geq 0$.

We set $i = 1$ in (26.24) and apply the preceding formula to deduce that $\boldsymbol{X}_{2,j} \equiv 0$, for $j \geq 1$. Substituting $i = 2$ in (26.24), we find that $\boldsymbol{X}_{3,j} \equiv 0$, for $j \geq 2$. Continuing in this manner, we conclude that $\boldsymbol{X}_{1+j,j} \equiv 0$, for any j.

Consequently, (26.25) and (26.20) are valid. This completes the proof. $\quad\square$

The rational numbers $\epsilon_{i,j,k}$ defined by

(26.28) $\quad\quad\quad \epsilon_{i,j,k} \equiv \dfrac{(j-k)!}{(m-1)^{j-k}}\binom{2i+j+1}{j-k}, \quad$ for $k \leq j$ and any i,

were introduced in (9.38) on page 100. For ease of reference in the proof of the next result, we recall from (9.53) and (9.54) on page 103 that (26.28) yields

(26.29) $\quad\quad\quad\quad\quad \epsilon_{i,j,j} \equiv 1, \quad$ for any i, j,

and

(26.30) $\quad\quad\quad \dfrac{(2i+k+1)}{m-1}\epsilon_{i,j,k} \equiv \epsilon_{i,j,k-1}, \quad$ for $k \leq j$ and any i.

PROPOSITION 26.7. *The polynomials $X_{i,j}$ in (26.8) are given by*

(26.31) $\quad X_{i,j} \equiv Y_{i-j,j}, \quad \text{for all } i, j,$

with respect to polynomials $Y_{i,j}$ in $\mathcal{S}_{m,n}$ that satisfy

(26.32) $\quad Y_{i,j} \equiv 0, \quad \text{for } i \leq 1 \text{ and any } j,$

and

(26.33) $\quad Y_{i+1,j} \equiv \sum_{k=0}^{j} \epsilon_{i,j,k} \left[\begin{array}{l} Y_{i,k}^{(1)} + i\,\boldsymbol{b}_{m,1}\, Y_{i,k} - (m-1)(k+1)\,\boldsymbol{b}_{m,2}\, Y_{i-1,k+1} \\ + v_{i+1,k} \sum_{h_0=0}^{l_0} \cdots \sum_{h_{n-1}=0}^{l_{n-1}} W_{h_1,\ldots,h_{n-1},i+1} \end{array} \right],$

$\qquad\qquad\qquad\qquad\qquad\qquad\qquad\qquad\qquad\qquad$ *for $i \geq 0$ and any j.*

PROOF. We define $Y_{i,j}$ in $\mathcal{S}_{m,n}$ with respect to $X_{i,j}$ in (26.8) by

(26.34) $\qquad\qquad Y_{i,j} \equiv X_{i+j,j}, \quad \text{for all integers } i \text{ and } j.$

Due to (26.34) and (26.20), $Y_{i,j}$ satisfies (26.31) and (26.32). For any integers i, j subject to $i + j \geq -1$, we replace i in (26.11) with $i + j$, we replace i in (26.13) with $i + j + 1$, and we employ (26.34) to verify that

(26.35) $\quad Y_{i+1,j} - \left[\dfrac{2i+j+1}{m-1}\right] Y_{i+1,j-1}$

$\qquad \equiv \left[\begin{array}{l} Y_{i,j}^{(1)} + i\,\boldsymbol{b}_{m,1}\, Y_{i,j} - (m-1)(j+1)\,\boldsymbol{b}_{m,2}\, Y_{i-1,j+1} \\ + v_{i+1,j} \sum_{h_0=0}^{l_0} \cdots \sum_{h_{n-1}=0}^{l_{n-1}} W_{h_1,\ldots,h_{n-1},i+1} \end{array} \right],$

$\qquad\qquad\qquad\qquad\qquad\qquad\qquad\qquad\qquad\qquad$ for $i + j \geq -1$.

We note that (26.8) and (26.7) give $X_{i,j} \equiv 0$, for $i \geq 0$ and $j \leq -1$. Hence, we have $X_{i,j} \equiv 0$, for $j \leq -1$ and any i. Due to (26.34), this yields

(26.36) $\qquad\qquad Y_{i,j} \equiv 0, \quad \text{for } j \leq -1 \text{ and any } i.$

For $i \geq 0$ and $j \geq 0$, we combine (26.29), (26.36), (26.30), and (26.35) to obtain

(26.37) $\quad Y_{i+1,j} \equiv \sum_{k=0}^{j} \left[\epsilon_{i,j,k}\, Y_{i+1,k} - \epsilon_{i,j,k-1}\, Y_{i+1,k-1} \right]$

$\qquad \equiv \sum_{k=0}^{j} \epsilon_{i,j,k} \left[Y_{i+1,k} - \dfrac{(2i+k+1)}{m-1} Y_{i+1,k-1} \right]$

$\qquad \equiv \sum_{k=0}^{j} \epsilon_{i,j,k} \left[\begin{array}{l} Y_{i,k}^{(1)} + i\,\boldsymbol{b}_{m,1}\, Y_{i,k} - (m-1)(k+1)\,\boldsymbol{b}_{m,2}\, Y_{i-1,k+1} \\ + v_{i+1,k} \sum_{h_0=0}^{l_0} \cdots \sum_{h_{n-1}=0}^{l_{n-1}} W_{h_1,\ldots,h_{n-1},i+1} \end{array} \right].$

In view of (26.36), (26.37) is valid for $i \geq 0$ and any j. This yields (26.33) and completes the proof of Proposition 26.7. $\qquad\square$

26.2. EVALUATION OF LAGUERRE-FORSYTH SUMS

In terms of fixed integers $m \geq 2$, $n \geq 1$, $l_0 = 0$, as well as $l_1, \ldots, l_n$, and $s_0 = l_1 + \cdots + l_n$ that satisfy (24.1)–(24.2) of page 275, the rational numbers

(26.38) $\quad r_{i,j} \equiv \dfrac{(s_0 - i - j)(s_0 + i + j - 1)}{(s_0 - i)(s_0 + i - 1)}$, for $0 \leq i \leq s_0 - 1$ and any j.

were introduced in (25.38). For each fixed n-tuple $(h_0, \ldots, h_{n-1})$ of integers having

$$0 \leq h_0 \leq l_0, \ldots, 0 \leq h_{n-1} \leq l_{n-1}$$

and for each integer i satisfying $0 \leq i \leq s_0$, there is a corresponding polynomial $\boldsymbol{J}_{h_1, \ldots, h_{n-1}, i}$ given in $\mathcal{S}_{m,n}$ by (24.21)–(24.22) of page 277. We use it and (26.38) to define $\mathfrak{S}_{h_1, \ldots, h_{n-1}, i, j}$ in $\mathcal{S}_{m,n}$ by means of

(26.39) $\quad \mathfrak{S}_{h_1, \ldots, h_{n-1}, i, j} \equiv r_{i,j} \boldsymbol{J}_{h_1, \ldots, h_{n-1}, i}$, for $0 \leq i \leq s_0 - 1$ and any j,

and

(26.40) $\quad \mathfrak{S}_{h_1, \ldots, h_{n-1}, s_0, j} \equiv \boldsymbol{J}_{h_1, \ldots, h_{n-1}, s_0}$, for any j.

In particular, (26.40), (26.39), and (26.38) give

(26.41) $\quad \mathfrak{S}_{h_1, \ldots, h_{n-1}, s_0, 0} \equiv \boldsymbol{J}_{h_1, \ldots, h_{n-1}, s_0}$

and

(26.42) $\quad \mathfrak{S}_{h_1, \ldots, h_{n-1}, s_0 - s, s} \equiv 0$, for $1 \leq s \leq s_0$.

Moreover, (26.39) and (24.21) of page 277 yield

(26.43) $\quad \mathfrak{S}_{h_1, \ldots, h_{n-1}, i, j} \equiv 0$, for $0 \leq i \leq 1$ and any j.

For $1 \leq i \leq s_0 - 1$ and $0 \leq j \leq s_0 - 1 - i$, we introduce

(26.44) $\quad E_{i,j} \equiv \dfrac{-(m-1)j}{(s_0 - i - j)(s_0 + i + j - 1)}$,

(26.45) $\quad F_{i,j} \equiv \dfrac{-(s_0 - i)}{(s_0 - i - j)(s_0 + i + j - 1)}$,

and

(26.46) $\quad G_{i,j} \equiv \dfrac{-(s_0 - i)(s_0 - i + 1)}{(m-1)(j+1)(s_0 - i - j)(s_0 + i + j - 1)}$.

PROPOSITION 26.8. *In terms of (26.44)–(26.46) and $\boldsymbol{W}_{h_1, \ldots, h_{n-1}, i}$ from (24.20) on page 276, the identity*

(26.47) $\quad \mathfrak{S}_{h_1, \ldots, h_{n-1}, i+1, j} - \dfrac{2i + j + 1}{m - 1} E_{i,j} \, \mathfrak{S}_{h_1, \ldots, h_{n-1}, i+1, j-1}$

$\equiv F_{i,j} \bigl(\mathfrak{S}^{(1)}_{h_1, \ldots, h_{n-1}, i, j} + i\, \boldsymbol{b}_{m,1} \, \mathfrak{S}_{h_1, \ldots, h_{n-1}, i, j} \bigr)$

$- (m - 1)(j + 1) \boldsymbol{b}_{m,2} \, G_{i,j} \, \mathfrak{S}_{h_1, \ldots, h_{n-1}, i-1, j+1} + \boldsymbol{W}_{h_1, \ldots, h_{n-1}, i+1}$

is satisfied for $1 \leq i \leq s_0 - 1$ and $0 \leq j \leq s_0 - 1 - i$.

PROOF. For $1 \leq i \leq s_0 - 1$ and $0 \leq j \leq s_0 - 1 - i$, we note that (24.10) of page 276, (26.45), (26.38), (24.11), and (26.46) yield

$$(26.48) \qquad A_i \equiv F_{i,j}\, r_{i,j}$$

and

$$(26.49) \qquad B_i \equiv -(m-1)\,(j+1)\, G_{i,j}\, r_{i-1,j+1}.$$

For the same i and j, we see that (24.22), (26.48), (26.49), and (26.39) give

$$(26.50) \quad J_{h_1,\ldots,h_{n-1},i+1} \equiv A_i\bigl(J^{(1)}_{h_1,\ldots,h_{n-1},i} + i\,\boldsymbol{b}_{m,1}\, J_{h_1,\ldots,h_{n-1},i}\bigr)$$
$$+ B_i\, \boldsymbol{b}_{m,2}\, J_{h_1,\ldots,h_{n-1},i-1} + W_{h_1,\ldots,h_{n-1},i+1}$$

$$\equiv F_{i,j}\bigl(r_{i,j}\, J^{(1)}_{h_1,\ldots,h_{n-1},i} + i\,\boldsymbol{b}_{m,1}\, r_{i,j}\, J_{h_1,\ldots,h_{n-1},i}\bigr)$$
$$- (m-1)\,(j+1)\, \boldsymbol{b}_{m,2}\, G_{i,j}\bigl(r_{i-1,j+1}\, J_{h_1,\ldots,h_{n-1},i-1}\bigr)$$
$$+ W_{h_1,\ldots,h_{n-1},i+1}$$

$$\equiv F_{i,j}\bigl(\mathfrak{S}^{(1)}_{h_1,\ldots,h_{n-1},i,j} + i\,\boldsymbol{b}_{m,1}\, \mathfrak{S}_{h_1,\ldots,h_{n-1},i,j}\bigr)$$
$$- (m-1)\,(j+1)\, \boldsymbol{b}_{m,2}\, G_{i,j}\, \mathfrak{S}_{h_1,\ldots,h_{n-1},i-1,j+1}$$
$$+ W_{h_1,\ldots,h_{n-1},i+1}.$$

Case 1. For $1 \leq i \leq s_0 - 2$ and $0 \leq j \leq s_0 - 1 - i$, we find that (26.39), (26.38), and (26.44) yield

$$(26.51) \quad \mathfrak{S}_{h_1,\ldots,h_{n-1},i+1,j} - \frac{2i+j+1}{m-1}\, E_{i,j}\, \mathfrak{S}_{h_1,\ldots,h_{n-1},i+1,j-1}$$

$$\equiv \left[r_{i+1,j} - \frac{2i+j+1}{m-1}\, E_{i,j}\, r_{i+1,j-1}\right] J_{h_1,\ldots,h_{n-1},i+1}$$

$$\equiv \left[\frac{(s_0 - i - 1 - j)(s_0 + i + j) + j\,(2i+j+1)}{(s_0 - i - 1)(s_0 + i)}\right] J_{h_1,\ldots,h_{n-1},i+1}$$

$$\equiv J_{h_1,\ldots,h_{n-1},i+1}.$$

By combining (26.51) and (26.50), we see that (26.47) is satisfied with respect to the preceding restriction.

Case 2. For the remaining situation, suppose that i and j satisfy $i = s_0 - 1$ and $j = 0$. Then, using $j = 0$ in (26.44) along with (26.41), we note that the left member of (26.47) reduces to $\mathfrak{S}_{h_1,\ldots,h_{n-1},s_0,0} \equiv J_{h_1,\ldots,h_{n-1},s_0}$. We combine this with a substitution of $i = s_0 - 1$ and $j = 0$ in (26.50) to see that (26.47) is valid for these values of i and j. This completes the proof. □

With reference to $\mathfrak{S}_{h_1,\ldots,h_{n-1},i,j}$ in (26.39)–(26.40) and $v_{i,j}$ in (26.12), we define $\mathfrak{R}_{h_1,\ldots,h_{n-1},i,j}$ in $\mathcal{S}_{m,n}$ through

$$(26.52) \qquad \mathfrak{R}_{h_1,\ldots,h_{n-1},i,j} \equiv v_{i,j}\, \mathfrak{S}_{h_1,\ldots,h_{n-1},i,j}, \quad \text{for } 0 \leq i \leq s_0, \text{ and } j \geq -i.$$

We employ (26.52), $v_{s_0,0} \equiv 1$ from (26.12), (26.41), and (26.42) to obtain

$$(26.53) \qquad \mathfrak{R}_{h_1,\ldots,h_{n-1},s_0,0} \equiv J_{h_1,\ldots,h_{n-1},s_0}$$

and

(26.54) $\mathfrak{R}_{h_1,\ldots,h_{n-1},s_0-s,s} \equiv 0, \quad \text{for } 1 \leq s \leq s_0.$

Also, (26.52) and (26.43) yield

(26.55) $\mathfrak{R}_{h_1,\ldots,h_{n-1},i,j} \equiv 0, \quad \text{for } 0 \leq i \leq 1, \text{ and } j \geq -i.$

We use (26.52) to define $\boldsymbol{Z}_{i,j}$ in $\mathcal{S}_{m,n}$ by means of

(26.56) $\boldsymbol{Z}_{i,j} \equiv \sum_{h_0=0}^{l_0} \cdots \sum_{h_{n-1}=0}^{l_{n-1}} \mathfrak{R}_{h_1,\ldots,h_{n-1},i,j}, \quad \text{for } 0 \leq i \leq s_0 \text{ and } j \geq -i.$

In particular, (26.56) and (26.53) give

(26.57) $\boldsymbol{Z}_{s_0,0} \equiv \sum_{h_0=0}^{l_0} \cdots \sum_{h_{n-1}=0}^{l_{n-1}} \boldsymbol{J}_{h_1,\ldots,h_{n-1},s_0}$

while (26.56) and (26.54) yield

(26.58) $\boldsymbol{Z}_{s_0-s,s} \equiv 0, \quad \text{for } 1 \leq s \leq s_0.$

We apply (26.56) and (26.55) to see that

(26.59) $\boldsymbol{Z}_{i,j} \equiv 0, \quad \text{for } 0 \leq i \leq 1 \text{ and } j \geq -i.$

PROPOSITION 26.9. *The identity*

(26.60) $\boldsymbol{Z}_{i+1,j} - \left[\dfrac{2i+j+1}{m-1}\right] \boldsymbol{Z}_{i+1,j-1}$

$\equiv \left[\begin{array}{l} \boldsymbol{Z}_{i,j}^{(1)} + i\,\boldsymbol{b}_{m,1}\,\boldsymbol{Z}_{i,j} - (m-1)(j+1)\,\boldsymbol{b}_{m,2}\,\boldsymbol{Z}_{i-1,j+1} \\ + v_{i+1,j} \displaystyle\sum_{h_0=0}^{l_0} \cdots \sum_{h_{n-1}=0}^{l_{n-1}} \boldsymbol{W}_{h_1,\ldots,h_{n-1},i+1} \end{array}\right].$

is valid for $1 \leq i \leq s_0 - 1$ *and* $0 \leq j \leq s_0 - 1 - i.$

PROOF. For $1 \leq i \leq s_0 - 1$ and $0 \leq j \leq s_0 - i - 1$, we use (26.12) and (26.44)–(26.46) to verify that $v_{i,j}$, $E_{i,j}$, $F_{i,j}$, and $G_{i,j}$ satisfy

(26.61) $v_{i+1,j}\, E_{i,j} \equiv v_{i+1,j-1},$

(26.62) $v_{i+1,j}\, F_{i,j} \equiv v_{i,j},$

and

(26.63) $v_{i+1,j}\, G_{i,j} \equiv v_{i-1,j+1}.$

For the same i and j, we multiply (26.47) by $v_{i+1,j}$ and apply (26.61)–(26.63) as well as (26.52) to obtain

$$0 \equiv v_{i+1,j}\, \mathfrak{S}_{h_1,\ldots,h_{n-1},i+1,j} - \frac{2i+j+1}{m-1}\, v_{i+1,j}\, E_{i,j}\, \mathfrak{S}_{h_1,\ldots,h_{n-1},i+1,j-1}$$
$$- v_{i+1,j}\, F_{i,j}\bigl(\mathfrak{S}^{(1)}_{h_1,\ldots,h_{n-1},i,j} + i\,\boldsymbol{b}_{m,1}\, \mathfrak{S}_{h_1,\ldots,h_{n-1},i,j}\bigr)$$
$$+ (m-1)(j+1)\,\boldsymbol{b}_{m,2}\, v_{i+1,j}\, G_{i,j}\, \mathfrak{S}_{h_1,\ldots,h_{n-1},i-1,j+1}$$
$$- v_{i+1,j}\, W_{h_1,\ldots,h_{n-1},i+1},$$

$$0 \equiv v_{i+1,j}\, \mathfrak{S}_{h_1,\ldots,h_{n-1},i+1,j} - \frac{2i+j+1}{m-1}\, v_{i+1,j-1}\, \mathfrak{S}_{h_1,\ldots,h_{n-1},i+1,j-1}$$
$$- v_{i,j}\bigl(\mathfrak{S}^{(1)}_{h_1,\ldots,h_{n-1},i,j} + i\,\boldsymbol{b}_{m,1}\, \mathfrak{S}_{h_1,\ldots,h_{n-1},i,j}\bigr)$$
$$+ (m-1)(j+1)\,\boldsymbol{b}_{m,2}\, v_{i-1,j+1}\, \mathfrak{S}_{h_1,\ldots,h_{n-1},i-1,j+1}$$
$$- v_{i+1,j}\, W_{h_1,\ldots,h_{n-1},i+1},$$

and

(26.64) $\quad 0 \equiv \mathfrak{R}_{h_1,\ldots,h_{n-1},i+1,j} - \dfrac{2i+j+1}{m-1}\, \mathfrak{R}_{h_1,\ldots,h_{n-1},i+1,j-1} - \mathfrak{R}^{(1)}_{h_1,\ldots,h_{n-1},i,j}$
$$- i\,\boldsymbol{b}_{m,1}\, \mathfrak{R}_{h_1,\ldots,h_{n-1},i,j} + (m-1)(j+1)\,\boldsymbol{b}_{m,2}\, \mathfrak{R}_{h_1,\ldots,h_{n-1},i-1,j+1}$$
$$- v_{i+1,j}\, W_{h_1,\ldots,h_{n-1},i+1}.$$

For $0 \le h_0 \le l_0,\ \ldots,\ 0 \le h_{n-1} \le l_{n-1}$, let $\mathfrak{W}_{h_1,\ldots,h_{n-1}}$ denote the right member of (26.64). We employ (26.64) and (26.56) to deduce that

(26.65) $\quad 0 \equiv \displaystyle\sum_{h_0=0}^{l_0}\cdots\sum_{h_{n-1}=0}^{l_{n-1}} \mathfrak{W}_{h_1,\ldots,h_{n-1}}$

$$\equiv Z_{i+1,j} - \frac{2i+j+1}{m-1}\, Z_{i+1,j-1} + (m-1)(j+1)\,\boldsymbol{b}_{m,2}\, Z_{i-1,j+1}$$
$$- Z^{(1)}_{i,j} - i\,\boldsymbol{b}_{m,1}\, Z_{i,j} - v_{i+1,j}\sum_{h_0=0}^{l_0}\cdots\sum_{h_{n-1}=0}^{l_{n-1}} W_{h_1,\ldots,h_{n-1},i+1},$$

for $1 \le i \le s_0 - 1$ and $0 \le j \le s_0 - 1 - i$. We rewrite (26.65) to obtain (26.60) and complete the proof of Proposition 26.9. $\quad\square$

THEOREM 26.10. *For (23.1) on Ω^{**} into which (21.1) on Ω is transformed by $y(z) = \rho(z)\,v(z)$ and $z = f(\zeta)$ according to the context of page 265, suppose that $d_{0,\ldots,0,1}(\zeta) \equiv d_{0,\ldots,0,2}(\zeta) \equiv 0$. Let $m \ge 2$, $n \ge 1$, $l_0 = 0$, and $l_1,\ldots,l_n$ be integers subject to $0 \le l_1 \le \cdots \le l_n \le m$ and either $l_{n-1} \ge 1$ or $l_n \ge 3$. Then, the coefficients of (23.1) satisfy*

(26.66) $\quad \displaystyle\sum_{j_1=0}^{l_1}\cdots\sum_{j_n=0}^{l_n} A^{l_1,\ldots,l_n}_{j_1,\ldots,j_n}\, d^{(j_1+\cdots+j_n)}_{l_1-j_1,\ldots,l_n-j_n}(\zeta) \equiv \bigl(f'(\zeta)\bigr)^{s_0} \mathcal{J}_{m,l_1,\ldots,l_n}\bigl(f(\zeta)\bigr),$

where $A^{l_1,\ldots,l_n}_{j_1,\ldots,j_n}$ is given by (23.34), $s_0 = l_1 + \cdots + l_n$, and $\mathcal{J}_{m,l_1,\ldots,l_n}\bigl(f(\zeta)\bigr)$ is obtained for $z = f(\zeta)$ by substituting $c^{(\kappa)}_{\mu_1,\ldots,\mu_n}(z)$ from (21.1) for $w^{(\kappa)}_{\mu_1,\ldots,\mu_n}$ in $\mathcal{J}_{m,l_1,\ldots,l_n}$ of (24.23) on page 277.

PROOF. We note that (26.12) yields $v_{i+1,-1} \equiv 0$, for $i \geq 0$. Due to (26.52), we have $\mathfrak{R}_{h_1,\ldots,h_{n-1},i+1,-1} \equiv 0$, for $0 \leq i \leq s_0 - 1$. We use (26.56) to obtain

$$\tag{26.67} \boldsymbol{Z}_{i+1,-1} \equiv 0, \quad \text{for } 0 \leq i \leq s_0 - 1.$$

We combine (26.29), (26.67), (26.30), and (26.60) to verify that

$$\tag{26.68} \boldsymbol{Z}_{i+1,j} \equiv \sum_{k=0}^{j} \Big[\epsilon_{i,j,k} \boldsymbol{Z}_{i+1,k} - \epsilon_{i,j,k-1} \boldsymbol{Z}_{i+1,k-1}\Big]$$

$$\equiv \sum_{k=0}^{j} \epsilon_{i,j,k}\Big[\boldsymbol{Z}_{i+1,k} - \frac{(2i+k+1)}{m-1}\boldsymbol{Z}_{i+1,k-1}\Big]$$

$$\equiv \sum_{k=0}^{j} \epsilon_{i,j,k} \left[\begin{array}{c} \boldsymbol{Z}_{i,k}^{(1)} + i\,\boldsymbol{b}_{m,1}\,\boldsymbol{Z}_{i,k} - (m-1)(k+1)\boldsymbol{b}_{m,2}\,\boldsymbol{Z}_{i-1,k+1} \\ + v_{i+1,k} \sum_{h_0=0}^{l_0} \cdots \sum_{h_{n-1}=0}^{l_{n-1}} \boldsymbol{W}_{h_1,\ldots,h_{n-1},i+1} \end{array}\right],$$

$$\text{for } 1 \leq i \leq s_0 - 1 \text{ and } 0 \leq j \leq s_0 - 1 - i.$$

In view of (26.32) on page 300 and (26.59), we see that the statement

$$\tag{26.69} \boldsymbol{Y}_{i,j} \equiv \boldsymbol{Z}_{i,j}, \quad \text{for } 0 \leq i \leq i_0 \text{ and } 0 \leq j \leq s_0 - i,$$

is true for $i_0 = 1$. Suppose that i_0 is an integer satisfying $1 \leq i_0 \leq s_0 - 1$ such that (26.69) is valid. Then, we use (26.33) on page 300 as well as (26.69) and (26.68) to deduce that $\boldsymbol{Y}_{i_0+1,j} \equiv \boldsymbol{Z}_{i_0+1,j}$, for $0 \leq j \leq s_0 - i_0 - 1$. Thus, we have

$$\tag{26.70} \boldsymbol{Y}_{i,j} \equiv \boldsymbol{Z}_{i,j}, \quad \text{for } 0 \leq i \leq s_0 \text{ and } 0 \leq j \leq s_0 - i.$$

We employ (26.31) on page 300 and (26.70) to see that

$$\tag{26.71} \boldsymbol{X}_{s_0,t} \equiv \boldsymbol{Y}_{s_0-t,t} \equiv \boldsymbol{Z}_{s_0-t,t}, \quad \text{for } 0 \leq t \leq s_0.$$

We observe that (26.71) and (26.58) yield

$$\tag{26.72} \boldsymbol{X}_{s_0,t} \equiv 0, \quad \text{for } 1 \leq t \leq s_0.$$

We substitute $c_{\mu_1,\ldots,\mu_n}^{(\kappa)}(z)$ from (21.1) for $w_{\mu_1,\ldots,\mu_n}^{(\kappa)}$ in (26.72), set $z = f(\zeta)$, and combine the resulting expressions with (26.9) on page 295 to deduce that

$$\tag{26.73} \sum_{j_1=0}^{l_1} \cdots \sum_{j_n=0}^{l_n} A_{j_1,\ldots,j_n}^{l_1,\ldots,l_n} d_{l_1-j_1,\ldots,l_n-j_n}^{(j_1+\cdots+j_n)}(\zeta) \equiv \big(f'(\zeta)\big)^{s_0} X_{s_0,0}\big(f(\zeta)\big).$$

Moreover, (26.71), (26.57), and (24.23) on page 277 give

$$\tag{26.74} \boldsymbol{X}_{s_0,0} \equiv \sum_{h_0=0}^{l_0} \cdots \sum_{h_{n-1}=0}^{l_{n-1}} \boldsymbol{J}_{h_1,\ldots,h_{n-1},s_0} \equiv \boldsymbol{\mathcal{J}}_{m,l_1,\ldots,l_n}.$$

After substituting $c_{\mu_1,\ldots,\mu_n}^{(\kappa)}(z)$ from (21.1) for $w_{\mu_1,\ldots,\mu_n}^{(\kappa)}$ in (26.74), we set $z = f(\zeta)$ and combine the resulting expression with (26.73) to obtain (26.66). This completes the proof of Theorem 26.10. □

CHAPTER 27

Basic Relative Invariants for $H_{m,n} = 0$ when $m \geq 2$

27.1. Preliminary results

PROPOSITION 27.1. *If P in $\mathcal{S}_{m,n}$ is a semi-invariant of the first kind for the equations (21.1), then P' is a semi-invariant of the first kind for such equations.*

PROOF. It suffices to make minor alterations in the arguments of pages 38–39 for Section 4.3. $\square$

PROPOSITION 27.2. *If P in $\mathcal{S}_{m,n}$ is a semi-invariant of the second kind of weight s for the equations (21.1) and $b_{m,1}$ is defined in $\mathcal{S}_{m,n}$ by (23.15) of page 267, then $P' + s\, b_{m,1} P$ is a semi-invariant of the second kind of weight $s+1$ for the same equations.*

PROOF. The proof on pages 45–46 for Proposition 4.17 can be repeated. $\square$

PROPOSITION 27.3. *Suppose that $i_1, \ldots, i_n$ satisfy $0 \leq i_1 \leq \cdots \leq i_n \leq m$. Then, the polynomial $L_{m,i_1,\ldots,i_n}$ defined in $\mathcal{S}_{m,n}$ by (24.6) is not a constant if and only if either $i_n \geq 3$ or $n \geq 2$ and $i_{n-1} \geq 1$. Moreover, when $L_{m,i_1,\ldots,i_n}$ is not a constant, it is a basic semi-invariant of the first kind having index $(i_1, \ldots, i_n)$.*

PROOF. For a given (21.1) on page 249 subject to (21.1.1) and a given (21.2) that transforms (21.1) on Ω into (21.3) on $\bar{\Omega}$ subject to (21.3.1), we substitute $c^{*(k)}_{j_1,\ldots,j_n}(z)$ from (21.3) for $w^{(k)}_{j_1,\ldots,j_n}$ in (24.6) to obtain

$$(27.1) \quad L^*_{m,i_1,\ldots,i_n}(z) \equiv \sum_{0 \leq j_1 \leq i_1,\ldots,0 \leq j_n \leq i_n} \left[\prod_{\nu=1}^{n} K^*_{m,i_\nu - j_\nu, j_\nu}(z)\right] c^*_{j_1,\ldots,j_n}(z).$$

We see that (21.2.1) and (21.2.2) give

$$(27.2) \quad c^*_{j_1,\ldots,j_n}(z) \equiv \sum_{0 \leq k_1 \leq j_1,\ldots,0 \leq k_n \leq j_n} \left[\prod_{\nu=1}^{n} \binom{m-k_\nu}{j_\nu - k_\nu} \frac{\rho^{(j_\nu - k_\nu)}(z)}{\rho(z)}\right] c_{k_1,\ldots,k_n}(z).$$

After introducing

$$(27.3) \quad \mathcal{H}^{j_1,\ldots,j_n}_{k_1,\ldots,k_n}(z) \equiv \prod_{\nu=1}^{n} \left[K^*_{m,i_\nu - j_\nu, j_\nu}(z) \binom{m - k_\nu}{j_\nu - k_\nu} \frac{\rho^{(j_\nu - k_\nu)}(z)}{\rho(z)}\right],$$

we combine (27.1)–(27.3) to deduce that

$(27.4) \quad L^*_{m,i_1,\ldots,i_n}(z)$

$$\equiv \sum_{0 \leq j_1 \leq i_1,\ldots,0 \leq j_n \leq i_n} \sum_{0 \leq k_1 \leq j_1,\ldots,0 \leq k_n \leq j_n} \mathcal{H}^{j_1,\ldots,j_n}_{k_1,\ldots,k_n}(z)\, c_{k_1,\ldots,k_n}(z)$$

$$\equiv \sum_{0 \leq k_1 \leq i_1,\ldots,0 \leq k_n \leq i_n} c_{k_1,\ldots,k_n}(z) \sum_{k_1 \leq j_1 \leq i_1,\ldots,k_n \leq j_n \leq i_n} \mathcal{H}^{j_1,\ldots,j_n}_{k_1,\ldots,k_n}(z).$$

We apply (27.3) to verify that

$$
(27.5) \quad \sum_{k_1 \leq j_1 \leq i_1, \ldots, k_n \leq j_n \leq i_n} \mathcal{H}_{k_1, \ldots, k_n}^{j_1, \ldots, j_n}(z) \equiv \sum_{0 \leq j_1 \leq i_1 - k_1, \ldots, 0 \leq j_n \leq i_n - k_n} \mathcal{H}_{k_1, \ldots, k_n}^{j_1 + k_1, \ldots, j_n + k_n}(z)
$$

$$
\equiv \prod_{\nu=1}^{n} \left[\sum_{j_\nu = 0}^{i_\nu - k_\nu} \binom{m - k_\nu}{j_\nu} \frac{\rho^{(j_\nu)}(z)}{\rho(z)} K_{m, i_\nu - j_\nu - k_\nu, j_\nu + k_\nu}^{*}(z) \right]
$$

$$
\equiv \prod_{\nu=1}^{n} R_{i_\nu - k_\nu, k_\nu}(z),
$$

where $R_{i,j}(z)$ is defined by

$$
(27.6) \quad R_{i,j}(z) \equiv \sum_{k=0}^{i} \binom{m-j}{k} \frac{\rho^{(k)}(z)}{\rho(z)} K_{m, i-k, j+k}^{*}(z), \quad \text{on } \Omega \text{ for all } i \text{ and } j.
$$

Proposition 23.10 on page 273 shows that $a_{m,2}$ in (23.14) is a semi-invariant of the first kind for the equations (21.1). Thus, we find that Lemma 8.2 on page 89 is valid for our present context and yields

$$
(27.7) \quad R_{i,j}(z) \equiv K_{m,i,j}(z), \quad \text{on } \Omega \text{ for } j \leq m \text{ and any } i,
$$

where $K_{m,i,j}(z)$ is the function obtained by substituting $c_{j_1, \ldots, j_n}^{(k)}(z)$ from (21.1) for $w_{j_1, \ldots, j_n}^{(k)}$ in (24.3)–(24.5). We combine (27.4)–(27.7) and then substitute $c_{j_1, \ldots, j_n}(z)$ from (21.1) for $w_{j_1, \ldots, j_n}$ in (24.6) to obtain

$$
(27.8) \quad L_{m, i_1, \ldots, i_n}^{*}(z) \equiv \sum_{0 \leq k_1 \leq i_1, \ldots, 0 \leq k_n \leq i_n} \left[\prod_{\nu=1}^{n} K_{m, i_\nu - k_\nu, k_\nu}(z) \right] c_{k_1, \ldots, k_n}(z).
$$

$$
\equiv L_{m, i_1, \ldots, i_n}(z).
$$

Hence, if $\boldsymbol{L}_{m, i_1, \ldots, i_n}$ is not a constant, then (27.8) shows that $\boldsymbol{L}_{m, i_1, \ldots, i_n}$ is a semi-invariant of the first kind.

Suppose that either $i_n \geq 3$ or $n \geq 2$ and $i_{n-1} \geq 1$. We apply (24.4) to rewrite (24.6) in the form

$$
(27.9) \quad \boldsymbol{L}_{m, i_1, \ldots, i_n} \equiv w_{i_1, \ldots, i_n} + \sum_{\substack{0 \leq j_1 \leq i_1, \ldots, 0 \leq j_n \leq i_n \\ (j_1, \ldots, j_n) \neq (i_1, \ldots, i_n)}} \left[\prod_{\nu=1}^{n} K_{m, i_\nu - j_\nu, j_\nu} \right] w_{j_1, \ldots, j_n}.
$$

We employ (24.3)–(24.5) of page 275 to see that each $\boldsymbol{K}_{m,i,j}$ is expressible as a differential-polynomial combination over $\mathbb{Q}$ of $\boldsymbol{w}_{0, \ldots, 0, 1}$ and $\boldsymbol{w}_{0, \ldots, 0, 2}$. In view of this and (27.9), we observe that the coefficient of $\boldsymbol{w}_{i_1, \ldots, i_n}$ in $\boldsymbol{L}_{m, i_1, \ldots, i_n}$ is 1. Thus, for $\boldsymbol{P} \equiv \boldsymbol{L}_{m, i_1, \ldots, i_n}$ and $l_k = i_k$, for $1 \leq k \leq n$, we note that condition (1) of Definition 21.5 on page 252 is satisfied. Also, we use (27.9) to see that (2) and (4) are satisfied. To check condition (3) of Definition 21.5, let $\boldsymbol{w}_{j_1, \ldots, j_n}$ be a symbol that appears in (27.9). Its subscripts satisfy $0 \leq j_k \leq i_k$, for $1 \leq k \leq n$; and there is a permutation π of $\{1, 2, \ldots, n\}$ such that $j_{\pi(1)} \leq j_{\pi(2)} \leq \cdots \leq j_{\pi(n)}$. For the variable $\boldsymbol{w}_{j_{\pi(1)}, \ldots, j_{\pi(n)}}$ of (21.6) on page 251, we have $\boldsymbol{w}_{j_1, \ldots, j_n} \equiv \boldsymbol{w}_{j_{\pi(1)}, \ldots, j_{\pi(n)}}$. Lemma 27.4 yields $j_{\pi(k)} \leq i_k$, for $1 \leq k \leq n$, and shows that (3) of Definition 21.5 is also satisfied. Thus, we conclude that $\boldsymbol{L}_{m, i_1, \ldots, i_n}$ is a basic polynomial of index $(i_1, \ldots, i_n)$. Moreover, it is a semi-invariant of the first kind.

27.1. PRELIMINARY RESULTS

For a fixed integer $n \geq 1$, we set

(27.10) $\quad \mathfrak{r}_0 \equiv \boldsymbol{w}_{\underbrace{0,\,\ldots,\,0,\,0}_{n-1}} \equiv 1, \quad \mathfrak{r}_1 \equiv \boldsymbol{w}_{\underbrace{0,\,\ldots,\,0,\,1}_{n-1}}, \quad \text{and} \quad \mathfrak{r}_2 \equiv \boldsymbol{w}_{\underbrace{0,\,\ldots,\,0,\,2}_{n-1}}.$

We use (24.3)–(24.5) on page 275, (23.14)–(23.15) on page 267, and computations similar to those for (1.43)–(1.44) on page 10 to verify that $\boldsymbol{K}_{m,0,j} \equiv 1$, for any j,

(27.11) $\quad \boldsymbol{K}_{m,1,j} \equiv -\dfrac{m-j}{m}\,\mathfrak{r}_1, \quad \text{for } m \geq 2 \text{ and } j \leq m,$

and

(27.12) $\quad \boldsymbol{K}_{m,2,j} \equiv \dfrac{\binom{m-j}{2}}{\binom{m+1}{2}} \left[-\dfrac{m+2j+1}{m-1}\,\mathfrak{r}_2 + j\,\mathfrak{r}_1^{(1)} + \dfrac{m+j+1}{m}\,(\mathfrak{r}_1)^2 \right],$

$$\text{for } m \geq 2 \text{ and } j \leq m.$$

When $n = 1$ and $i_1 = 1$, we apply (24.6) on page 275 as well as (27.11) to obtain

(27.13) $\quad \boldsymbol{L}_{m,1} \equiv \boldsymbol{K}_{m,0,1}\,\mathfrak{r}_1 + \boldsymbol{K}_{m,1,0}\,\mathfrak{r}_0 \equiv \mathfrak{r}_1 + \boldsymbol{K}_{m,1,0} \equiv 0.$

When $n = 1$ and $i_1 = 2$, we employ (24.6) and (27.11)–(27.12) to deduce that

(27.14) $\quad \boldsymbol{L}_{m,2} \equiv \boldsymbol{K}_{m,0,2}\,\mathfrak{r}_2 + \boldsymbol{K}_{m,1,1}\,\mathfrak{r}_1 + \boldsymbol{K}_{m,2,0}\,\mathfrak{r}_0 \equiv \mathfrak{r}_2 + \boldsymbol{K}_{m,1,1}\,\mathfrak{r}_1 + \boldsymbol{K}_{m,2,0} \equiv 0.$

If $n \geq 2$, $i_{n-1} = 0$, and $i_n = 1$, then a computation similar to that for (27.13) yields $\boldsymbol{L}_{m,0,\ldots,0,1} \equiv 0$. If $n \geq 2$, $i_{n-1} = 0$, and $i_n = 2$, then a computation similar to that for (27.14) gives $\boldsymbol{L}_{m,0,\ldots,0,2} \equiv 0$. Consequently, $\boldsymbol{L}_{m,i_1,\ldots,i_n}$ is a nonconstant in $\mathcal{S}_{m,n}$ if and only if either $i_n \geq 3$ or $n \geq 2$ and $i_{n-1} \geq 1$. This completes the proof. $\square$

LEMMA 27.4. *For integers $i_1, \ldots, i_n$ and $j_1, \ldots, j_n$ that satisfy*

(27.15) $\quad 0 \leq i_1 \leq i_2 \leq \cdots \leq i_n \quad \text{and} \quad 0 \leq j_k \leq i_k, \quad \text{for } 1 \leq k \leq n,$

let π be a permutation of $\{1, 2, \ldots, n\}$ such that

(27.16) $\quad j_{\pi(1)} \leq j_{\pi(2)} \leq \cdots \leq j_{\pi(n)}.$

Then, these integers satisfy $0 \leq j_{\pi(k)} \leq i_k$, for $1 \leq k \leq n$.

PROOF. We use (27.16) and (27.15) to obtain $j_{\pi(1)} \leq j_1 \leq i_1$ and

$$j_{\pi(k)} \leq \max(j_1, j_2, \ldots, j_k) \leq \max(i_1, i_2, \ldots, i_k) = i_k, \quad \text{for } 1 \leq k \leq n.$$

This completes the proof. $\square$

PROPOSITION 27.5. *Suppose that $i_1, \ldots, i_n$ satisfy $0 \leq i_1 \leq \cdots \leq i_n \leq m$. Then, the polynomial $\boldsymbol{V}_{m,i_1,\ldots,i_n}$ defined in $\mathcal{S}_{m,n}$ by (24.18) is not a constant if and only if either $i_n \geq 3$ or $n \geq 2$ and $i_{n-1} \geq 1$. Moreover, when $\boldsymbol{V}_{m,i_1,\ldots,i_n}$ is not a constant, it is a basic semi-invariant of the second kind having index $(i_1, \ldots, i_n)$.*

PROOF. For a given (21.1) on page 249 subject to (21.1.1) and a given (21.4) that transforms (21.1) on Ω into (21.5) on Ω^{**} subject to (21.5.1), we substitute $c^{**(k)}_{j_1,\ldots,j_n}(\zeta)$ from (21.5) for $\boldsymbol{w}^{(k)}_{j_1,\ldots,j_n}$ in (24.18) to obtain

(27.17) $\quad \boldsymbol{V}^{**}_{m,i_1,\ldots,i_n}(\zeta) \equiv \displaystyle\sum_{0 \leq j_1 \leq i_1,\,\ldots,\,0 \leq j_n \leq i_n} \left[\prod_{\nu=1}^{n} \boldsymbol{U}^{**}_{m,i_\nu - j_\nu, j_\nu}(\zeta) \right] c^{**}_{j_1,\ldots,j_n}(\zeta).$

In terms of $\alpha_{i,j}(\zeta)$ from (3.1)–(3.2) on page 29, we note that (21.4.1) and (21.4.2) give

(27.18) $\quad c^{**}_{j_1,\ldots,j_n}(\zeta)$
$$\equiv \sum_{0\le k_1\le j_1,\ldots,0\le k_n\le j_n} (f'(\zeta))^{k_1+\cdots+k_n} \left[\prod_{\nu=1}^{n} \alpha_{j_\nu-k_\nu,\,m-j_\nu}(\zeta)\right] c_{k_1,\ldots,k_n}(f(\zeta)).$$

After introducing

(27.19) $\quad \mathcal{K}^{j_1,\ldots,j_n}_{k_1,\ldots,k_n}(\zeta) \equiv \prod_{\nu=1}^{n} \left[U^{**}_{m,i_\nu-j_\nu,\,j_\nu}(\zeta)\, \alpha_{j_\nu-k_\nu,\,m-j_\nu}(\zeta)\right],$

we combine (27.17)–(27.19) to deduce that

(27.20) $\quad V^{**}_{m,\,i_1,\ldots,i_n}(\zeta)$
$$\equiv \sum_{0\le j_1\le i_1,\ldots,0\le j_n\le i_n}\;\sum_{0\le k_1\le j_1,\ldots,0\le k_n\le j_n} (f(\zeta))^{k_1+\cdots+k_n}\,\mathcal{K}^{j_1,\ldots,j_n}_{k_1,\ldots,k_n}(\zeta)\, c_{k_1,\ldots,k_n}(f(\zeta))$$
$$\equiv \sum_{0\le k_1\le i_1,\ldots,0\le k_n\le i_n} (f'(\zeta))^{k_1+\cdots+k_n}\, c_{k_1,\ldots,k_n}(f(\zeta)) \sum_{k_1\le j_1\le i_1,\ldots,k_n\le j_n\le i_n} \mathcal{K}^{j_1,\ldots,j_n}_{k_1,\ldots,k_n}(\zeta).$$

We apply (27.19) to verify that

(27.21) $\displaystyle\sum_{k_1\le j_1\le i_1,\ldots,k_n\le j_n\le i_n} \mathcal{K}^{j_1,\ldots,j_n}_{k_1,\ldots,k_n}(\zeta) \equiv \sum_{0\le j_1\le i_1-k_1,\ldots,0\le j_n\le i_n-k_n} \mathcal{K}^{j_1+k_1,\ldots,j_n+k_n}_{k_1,\ldots,k_n}(\zeta)$

$$\equiv \prod_{\nu=1}^{n}\left[\sum_{j_\nu=0}^{i_\nu-k_\nu} \alpha_{j_\nu,\,m-j_\nu-k_\nu}(\zeta)\, U^{**}_{m,i_\nu-j_\nu-k_\nu,\,j_\nu+k_\nu}(\zeta)\right] \equiv \prod_{\nu=1}^{n} S_{i_\nu-k_\nu,\,k_\nu}(\zeta),$$

where $S_{i,j}(\zeta)$ is defined by

(27.22) $\quad \displaystyle S_{i,j}(\zeta) \equiv \sum_{k=0}^{i} \alpha_{k,\,m-j-k}(\zeta)\, U^{**}_{m,i-k,\,j+k}(\zeta),\quad$ on Ω^{**} for all i and j.

Proposition 23.10 on page 273 shows that $\boldsymbol{b}_{m,2}$ in (23.16) is a semi-invariant of the second kind for the equations (21.1). Thus, we find that Lemma 10.2 on page 113 is valid for our present context and yields

(27.23) $\quad S_{i,j}(\zeta) \equiv (f'(\zeta))^{i}\, U_{m,i,j}(f(z)),\quad$ on Ω^{**} for $j\le m$ and any i,

where $U_{m,i,j}(z)$ is the function obtained by substituting $c^{(k)}_{j_1,\ldots,j_n}(z)$ from (21.1) for $\boldsymbol{w}^{(k)}_{j_1,\ldots,j_n}$ in (24.15)–(24.17). We combine (27.20)–(27.23) and use a substitution of $c_{j_1,\ldots,j_n}(z)$ from (21.1) for $\boldsymbol{w}_{j_1,\ldots,j_n}$ in (24.18) to obtain

(27.24) $\quad V^{**}_{m,\,i_1,\ldots,i_n}(\zeta)$
$$\equiv \sum_{0\le k_1\le i_1,\ldots,0\le k_n\le i_n} (f'(\zeta))^{k_1+\cdots+k_n} \left[\prod_{\nu=1}^{n} (f'(\zeta))^{i_\nu-k_\nu} U_{m,i_\nu-k_\nu,\,k_\nu}(f(\zeta))\right] c_{k_1,\ldots,k_n}(f(\zeta))$$
$$\equiv (f'(\zeta))^{i_1+\cdots+i_n}\, V_{m,\,i_1,\ldots,i_n}(f(\zeta)).$$

Hence, if $\boldsymbol{V}_{m,\,i_1,\ldots,i_n}$ is not a constant, then (27.24) shows that $\boldsymbol{V}_{m,\,i_1,\ldots,i_n}$ is a semi-invariant of the second kind of weight $i_1+\cdots+i_n$.

Suppose that either $i_n\ge 3$ or $n\ge 2$ and $i_{n-1}\ge 1$. We apply (24.16) to rewrite (24.18) in the form

$$\text{(27.25)} \quad \boldsymbol{V}_{m,i_1,\ldots,i_n} \equiv \boldsymbol{w}_{i_1,\ldots,i_n} + \sum_{\substack{0 \leq j_1 \leq i_1,\ldots,0 \leq j_n \leq i_n \\ (j_1,\ldots,j_n) \neq (i_1,\ldots,i_n)}} \left[\prod_{\nu=1}^{n} \boldsymbol{U}_{m,i_\nu-j_\nu,j_\nu}\right] \boldsymbol{w}_{j_1,\ldots,j_n}.$$

We employ (24.15)–(24.17) of page 276 to verify that each $\boldsymbol{U}_{m,i,j}$ can be expressed as a differential-polynomial combination over $\mathbb{Q}$ of $\boldsymbol{w}_{0,\ldots,0,1}$ and $\boldsymbol{w}_{0,\ldots,0,2}$. In view of this and (27.25), we note that the coefficient of $\boldsymbol{w}_{i_1,\ldots,i_n}$ in $\boldsymbol{V}_{m,i_1,\ldots,i_n}$ is 1. In the same way that (27.9) and Lemma 27.4 were used on page 308, we apply (27.25) and Lemma 27.4 to complete our argument here that $\boldsymbol{V}_{m,i_1,\ldots,i_n}$ is a basic polynomial of index $(i_1,\ldots,i_n)$. It is also a semi-invariant of the second kind.

For a fixed integer $n \geq 1$, we use (24.15)–(24.17), (23.15), (23.16), and (27.10) to deduce, for $m \geq 2$ and $j \leq m$, that $\boldsymbol{U}_{m,0,j} \equiv 1$,

$$\text{(27.26)} \quad \boldsymbol{U}_{m,1,j} \equiv -\frac{(m-j)(m-j-1)}{m(m-1)} \mathfrak{x}_1$$

and

$$\text{(27.27)} \quad \boldsymbol{U}_{m,2,j} \equiv \frac{(m-j)(m-j-1)j}{(m+1)m} \mathfrak{x}_1^{(1)} - \frac{(m-j)(m-j-1)(m+2j+1)}{(m+1)m(m-1)} \mathfrak{x}_2$$
$$+ \frac{(m-j)(m-j-1)(2m^3 - 4m^2 + j^2m - 3jm - 2m + j^2 + 5j + 4)}{2(m+1)m^2(m-1)^2} (\mathfrak{x}_1)^2.$$

When $n = 1$ and $i_1 = 1$, we apply (24.18) on page 276 as well as (27.26) to obtain

$$\text{(27.28)} \quad \boldsymbol{V}_{m,1} \equiv \boldsymbol{U}_{m,0,1} \mathfrak{x}_1 + \boldsymbol{U}_{m,1,0} \mathfrak{x}_0 \equiv \mathfrak{x}_1 + \boldsymbol{U}_{m,1,0} \equiv 0.$$

When $n = 1$ and $i_1 = 2$, we employ (24.18) and (27.26)–(27.27) to verify that

$$\text{(27.29)} \quad \boldsymbol{V}_{m,2} \equiv \boldsymbol{U}_{m,0,2} \mathfrak{x}_2 + \boldsymbol{U}_{m,1,1} \mathfrak{x}_1 + \boldsymbol{U}_{m,2,0} \mathfrak{x}_0 \equiv \mathfrak{x}_2 + \boldsymbol{U}_{m,1,1} \mathfrak{x}_1 + \boldsymbol{U}_{m,2,0} \equiv 0.$$

If $n \geq 2$, $i_{n-1} = 0$, and $i_n = 1$, then a computation similar to that for (27.28) yields $\boldsymbol{V}_{m,0,\ldots,0,1} \equiv 0$. If $n \geq 2$, $i_{n-1} = 0$, and $i_n = 2$, then a computation similar to that for (27.29) gives $\boldsymbol{V}_{m,0,\ldots,0,2} \equiv 0$. Thus, $\boldsymbol{V}_{m,i_1,\ldots,i_n}$ is a nonconstant in $\mathcal{S}_{m,n}$ if and only if either $i_n \geq 3$ or $n \geq 2$ and $i_{n-1} \geq 1$. This completes the proof. $\square$

PROPOSITION 27.6. *With $l_0 = 0$, let m, n, and $l_1,\ldots,l_n$ be integers that satisfy $m \geq 2$, $n \geq 1$, and (24.1) on page 275. Then, $\boldsymbol{\mathcal{I}}_{m,l_1,\ldots,l_n}$ in (24.14) is a basic semi-invariant of the first kind having index $(l_1,\ldots,l_n)$ and $\boldsymbol{\mathcal{J}}_{m,l_1,\ldots,l_n}$ in (24.23) is a basic semi-invariant of the second kind having index $(l_1,\ldots,l_n)$.*

PROOF. To establish that $\boldsymbol{\mathcal{I}}_{m,l_1,\ldots,l_n}$ satisfies (1) in Definition 21.5 on page 252, we first note that: if $n = 1$ and therefore $\boldsymbol{\mathcal{I}}_{m,l_1} \equiv \boldsymbol{I}_{m,l_1,l_1}$, for $3 \leq l_1 \leq m$, then formulas (24.3)–(24.13) show that the coefficient of $\boldsymbol{w}_{l_1}$ in $\boldsymbol{\mathcal{I}}_{m,l_1}$ is equal to the coefficient 1 of $\boldsymbol{w}_{l_1}$ in $\boldsymbol{L}_{m,l_1}$.

Suppose that $n \geq 2$ and $(h_1,\ldots,h_{n-1}) = (l_1,\ldots,l_{n-1})$. We use (24.13) to see that the coefficient of $\boldsymbol{w}_{l_1,\ldots,l_n}$ in $\boldsymbol{I}_{m,l_1,\ldots,l_n,l_1,\ldots,l_{n-1},l_1+\cdots+l_n}$ is equal to the coefficient of $\boldsymbol{w}_{l_1,\ldots,l_n}$ in $\boldsymbol{M}_{m,l_1,\ldots,l_n,l_1,\ldots,l_{n-1},l_1+\cdots+l_n}$. With $i = l_1 + \cdots + l_n$ and $h_n = l_n$, we find that (24.9) and (24.8) yield

$$\boldsymbol{M}_{m,l_1,\ldots,l_n,l_1,\ldots,l_{n-1},l_1+\cdots+l_n} \equiv \boldsymbol{H}^{m,l_1,\ldots,l_n}_{l_1,\ldots,l_n} \boldsymbol{L}_{m,l_1,\ldots,l_n} \equiv \boldsymbol{L}_{m,l_1,\ldots,l_n}.$$

Since Proposition 27.3 on page 307 shows that $\boldsymbol{L}_{m,l_1,\ldots,l_n}$ is a basic polynomial of index $(l_1,\ldots,l_n)$, the coefficient of $\boldsymbol{w}_{l_1,\ldots,l_n}$ in $\boldsymbol{L}_{m,l_1,\ldots,l_n}$ is 1. Thus, the coefficient of $\boldsymbol{w}_{l_1,\ldots,l_n}$ in $\boldsymbol{I}_{m,l_1,\ldots,l_n,l_1,\ldots,l_{n-1},l_1+\cdots+l_n}$ is 1.

Suppose that $n \geq 2$ and $(h_1, \ldots, h_{n-1}) \neq (l_1, \ldots, l_{n-1})$, where $h_1, \ldots, h_{n-1}$ are integers that satisfy $0 \leq h_\nu \leq l_\nu$, for $1 \leq \nu \leq n-1$. We use (24.13) to see that the coefficient of $\boldsymbol{w}_{l_1,\ldots,l_n}$ in $\boldsymbol{I}_{m,l_1,\ldots,l_n,h_1,\ldots,h_{n-1},l_1+\cdots+l_n}$ is equal to the coefficient of $\boldsymbol{w}_{l_1,\ldots,l_n}$ in $\boldsymbol{M}_{m,l_1,\ldots,l_n,h_1,\ldots,h_{n-1},l_1+\cdots+l_n}$. With $i = l_1 + \cdots + l_n$,

$$h_n = i - (h_1 + \cdots + h_{n-1}) = l_n + (l_1 - h_1) + \cdots + (l_{n-1} - h_{n-1}) > l_n,$$

and $H^{m,l_1,\ldots,l_n}_{h_1,\ldots,h_n} = 0$ from (24.8), we deduce $\boldsymbol{M}_{m,l_1,\ldots,l_n,h_1,\ldots,h_{n-1},l_1+\cdots+l_n} \equiv 0$ via (24.9). Thus, the coefficient of $\boldsymbol{w}_{l_1,\ldots,l_n}$ in $\boldsymbol{I}_{m,l_1,\ldots,l_n,h_1,\ldots,h_{n-1},l_1+\cdots+l_n}$ is 0.

We use the preceding results and (24.14) to conclude that: the coefficient of $\boldsymbol{w}_{l_1,\ldots,l_n}$ in $\boldsymbol{I}_{m,l_1,\ldots,l_n}$ is 1 and $\boldsymbol{I}_{m,l_1,\ldots,l_n}$ satisfies (1) in Definition 21.5.

To show that $\boldsymbol{I}_{m,l_1,\ldots,l_n}$ is an isobaric semi-invariant of the first kind whose weight is $l_1 + \cdots + l_n$, we begin by using Proposition 27.3 to see that: for any $\boldsymbol{L}_{m,i_1,\ldots,i_n}$ in (24.6) having $i_1 + \cdots + i_n \geq 1$, either it is identically zero or it is an isobaric semi-invariant of the first kind having weight $i_1 + \cdots + i_n$. Thus, for any $\boldsymbol{M}_{m,l_1,\ldots,l_n,h_1,\ldots,h_{n-1},i}$ in (24.9) having $i \geq 1$, either it is identically zero or it is an isobaric semi-invariant of the first kind having weight i. Since Proposition 23.10 shows that $\boldsymbol{a}_{m,2}$ is an isobaric semi-invariant of the first kind, we observe that: for any $\boldsymbol{I}_{m,l_1,\ldots,l_n,h_1,\ldots,h_{n-1},i}$ in (24.12) or (24.13), either it is identically zero or it is an isobaric semi-invariant of the first kind having weight i. In view of this and (24.14), we find that either $\boldsymbol{I}_{m,l_1,\ldots,l_n}$ is identically zero or it is an isobaric semi-invariant of the first kind having weight $l_1 + \cdots + l_n$. Since the coefficient of $\boldsymbol{w}_{l_1,\ldots,l_n}$ in $\boldsymbol{I}_{m,l_1,\ldots,l_n}$ is 1, we conclude that $\boldsymbol{I}_{m,l_1,\ldots,l_n}$ is an isobaric semi-invariant of the first kind having weight $l_1 + \cdots + l_n$. In particular, $\boldsymbol{I}_{m,l_1,\ldots,l_n}$ satisfies (2) in Definition 21.5.

To show that $\boldsymbol{I}_{m,l_1,\ldots,l_n}$ satisfies (3) in Definition 21.5 on page 252, suppose that $\boldsymbol{w}^{(k)}_{i_1,\ldots,i_n}$ is a variable from (21.6) of page 251 that is effectively involved in $\boldsymbol{I}_{m,l_1,\ldots,l_n}$. When $n = 1$, we find that

$$i_1 \leq i_1 + k = weight\big(\boldsymbol{w}^{(k)}_{i_1}\big) \leq weight\big(\boldsymbol{I}_{m,l_1}\big) = l_1.$$

Suppose that $n \geq 2$. It remains for us to establish that

(27.30) $\qquad\qquad i_\nu \leq l_\nu, \quad \text{for } 1 \leq \nu \leq n.$

Due to $l_n \geq 1$, (27.30) is satisfied when $\boldsymbol{w}^{(k)}_{i_1,\ldots,i_n} = \boldsymbol{w}^{(k)}_{0,\ldots,0,1}$. If $l_n = 1$, then the proof on page 318 for Proposition 27.14 shows that: for any $j \geq 0$, $\boldsymbol{w}^{(j)}_{0,\ldots,0,2}$ is not effectively involved in $\boldsymbol{I}_{m,l_1,\ldots,l_n}$. Thus, when $\boldsymbol{w}^{(k)}_{i_1,\ldots,i_n} = \boldsymbol{w}^{(k)}_{0,\ldots,0,2}$, we have $l_n \geq 2$ and see that (27.30) is satisfied. Next, we examine the remaining situation where $\boldsymbol{w}^{(k)}_{i_1,\ldots,i_n} \neq \boldsymbol{w}^{(k)}_{0,\ldots,0,1}$ and $\boldsymbol{w}^{(k)}_{i_1,\ldots,i_n} \neq \boldsymbol{w}^{(k)}_{0,\ldots,0,2}$. Then, there are integers $h_1, \ldots, h_{n-1}$ such that $0 \leq h_\nu \leq l_\nu$, for $1 \leq \nu \leq n-1$, and $\boldsymbol{I}_{m,l_1,\ldots,l_n,h_1,\ldots,h_{n-1},l_1+\cdots+l_n}$ effectively involves $\boldsymbol{w}^{(k)}_{i_1,\ldots,i_n}$. Since $\boldsymbol{a}_{m,2}$ and each $\boldsymbol{K}_{m,i,j}$ is a differential-polynomial combination over $\mathbb{Q}$ of $\boldsymbol{w}_{0,\ldots,0,1}$ and $\boldsymbol{w}_{0,\ldots,0,2}$, we use (24.6)–(24.13) to see that there is an integer i satisfying $2 \leq i \leq l_1 + \cdots + l_n$ such that $\boldsymbol{w}^{(0)}_{i_1,\ldots,i_n}$ is effectively involved in $\boldsymbol{M}_{m,l_1,\ldots,l_n,h_1,\ldots,h_{n-1},i}$. With respect to $h_n = i - (h_1 + \cdots + h_{n-1})$, we apply (24.9) and obtain

(27.31) $\qquad\qquad H^{m,l_1,\ldots,l_n}_{h_1,\ldots,h_n} \boldsymbol{L}_{m,h_1,\ldots,h_n} \equiv \boldsymbol{M}_{m,l_1,\ldots,l_n,h_1,\ldots,h_{n-1},i} \not\equiv 0.$

Due to $H^{m,l_1,\ldots,l_n}_{h_1,\ldots,h_n} \not\equiv 0$, (24.8) yields $l_n - h_n \geq 0$ and $0 \leq h_n \leq l_n$. Let π be a permutation of $\{1, 2, \ldots, n\}$ such that $0 \leq h_{\pi(1)} \leq h_{\pi(2)} \leq \cdots \leq h_{\pi(n)}$. In view of

27.1. PRELIMINARY RESULTS

this and $0 \leq h_\nu \leq l_\nu$, for $1 \leq \nu \leq n$, we employ Lemma 27.4 to deduce that

(27.32) $$0 \leq h_{\pi(\nu)} \leq l_\nu, \quad \text{for } 1 \leq \nu \leq n.$$

Because we have $\boldsymbol{L}_{m, h_1, \ldots, h_n} \equiv \boldsymbol{L}_{m, h_{\pi(1)}, \ldots, h_{\pi(n)}}$, Proposition 27.3 shows that $\boldsymbol{L}_{m, h_1, \ldots, h_n}$ is a basic polynomial of index $(h_{\pi(1)}, \ldots, h_{\pi(n)})$. Therefore, (27.31) and an application to $\boldsymbol{L}_{m, h_{\pi(1)}, \ldots, h_{\pi(n)}}$ of (3) in Definition 21.5 enable us to conclude that $\boldsymbol{w}_{i_1, \ldots, i_n}^{(0)}$ is effectively involved in $\boldsymbol{L}_{m, h_{\pi(1)}, \ldots, h_{\pi(n)}}$ and

(27.33) $$0 \leq i_\nu \leq h_{\pi(\nu)}, \quad \text{for } 1 \leq \nu \leq n.$$

By combining (27.33) and (27.32), we obtain $0 \leq i_\nu \leq l_\nu$, for $1 \leq \nu \leq n$, and conclude that $\boldsymbol{\mathcal{I}}_{m, l_1, \ldots, l_n}$ satisfies (3) in Definition 21.5.

To show that $\boldsymbol{\mathcal{I}}_{m, l_1, \ldots, l_n}$ satisfies (4) in Definition 21.5, it suffices to note that: (i) $\boldsymbol{a}_{m,2}$ and each of the $\boldsymbol{K}_{m,i,j}$ are elements of the ring $\mathbb{Q}\{\boldsymbol{w}_{0,\ldots,0,1}, \boldsymbol{w}_{0,\ldots,0,2}\}$ of differential polynomials over $\mathbb{Q}$ in $\boldsymbol{w}_{0,\ldots,0,1}$ and $\boldsymbol{w}_{0,\ldots,0,2}$; while, (ii) any nonzero term involved in (24.6)–(24.14) can have at most one factor that is a variable from (21.6) not in $\mathbb{Q}\{\boldsymbol{w}_{0,\ldots,0,1}, \boldsymbol{w}_{0,\ldots,0,2}\}$.

We combine the preceding results to see that $\boldsymbol{\mathcal{I}}_{m, l_1, \ldots, l_n}$ is a basic polynomial of index $(l_1, \ldots, l_n)$ as well as a semi-invariant of the first kind.

To verify that $\boldsymbol{\mathcal{J}}_{m, l_1, \ldots, l_n}$ satisfies (1) in Definition 21.5, we use arguments that are completely analogous to those given for $\boldsymbol{\mathcal{I}}_{m, l_1, \ldots, l_n}$. They are based on (24.15)-(24.23) in place of (24.3)–(24.9) and (24.12)–(24.14).

To show that $\boldsymbol{\mathcal{J}}_{m, l_1, \ldots, l_n}$ satisfies (2) in Definition 21.5 and is a semi-invariant of the second kind having weight $l_1 + \cdots + l_n$, we begin by using Proposition 27.5 to observe that: for any $\boldsymbol{V}_{m, i_1, \ldots, i_n}$ in (24.18) having $i_1 + \cdots + i_n \geq 1$, either it is identically zero or it is a semi-invariant of the second kind having weight $i_1 + \cdots + i_n$. Hence, for any $\boldsymbol{W}_{m, l_1, \ldots, l_n, h_1, \ldots, h_{n-1}, i}$ in (24.20) having $i \geq 1$, either it is identically zero or it is a semi-invariant of the second kind having weight i. Proposition 23.10 on page 273 shows that $\boldsymbol{b}_{m,2}$ is a semi-invariant of the second kind having weight 2. We combine the preceding observations with Proposition 27.2 on page 307 and use mathematical induction to deduce that: for any $\boldsymbol{J}_{m, l_1, \ldots, l_n, h_1, \ldots, h_{n-1}, i}$ in (24.21)–(24.22), either $\boldsymbol{J}_{m, l_1, \ldots, l_n, h_1, \ldots, h_{n-1}, i}$ is identically zero or it is a semi-invariant of the second kind having weight i. Thus, (24.23) shows that $\boldsymbol{\mathcal{J}}_{m, l_1, \ldots, l_n}$ is either identically zero or it is a semi-invariant of the second kind having weight $l_1 + \cdots + l_n$. However, the proof that $\boldsymbol{\mathcal{J}}_{m, l_1, \ldots, l_n}$ satisfies (1) in Definition 21.5 involved showing that the coefficient of $\boldsymbol{w}_{l_1, \ldots, l_n}$ in $\boldsymbol{\mathcal{J}}_{m, l_1, \ldots, l_n}$ is 1. Hence, $\boldsymbol{\mathcal{J}}_{m, l_1, \ldots, l_n}$ is a semi-invariant of the second kind having weight $l_1 + \cdots + l_n$. In particular, it satisfies (2) in Definition 21.5.

To verify that $\boldsymbol{\mathcal{J}}_{m, l_1, \ldots, l_n}$ satisfies (3) in Definition 21.5, suppose that $\boldsymbol{w}_{i_1, \ldots, i_n}^{(k)}$ is a variable from (21.6) of page 251 that is effectively involved in $\boldsymbol{\mathcal{J}}_{m, l_1, \ldots, l_n}$. After examining $n = 1$ separately, we assume that $n \geq 2$ and consider three cases according to whether $\boldsymbol{w}_{i_1, \ldots, i_n}^{(k)}$ is equal to one or the other or neither of $\boldsymbol{w}_{0, \ldots, 0, 1}^{(k)}$ and $\boldsymbol{w}_{0, \ldots, 0, 2}^{(k)}$. The arguments for this and for the proof that $\boldsymbol{\mathcal{J}}_{m, l_1, \ldots, l_n}$ satisfies (4) in Definition 21.5 are similar to the corresponding ones for $\boldsymbol{\mathcal{I}}_{m, l_1, \ldots, l_n}$. They are based on (24.15)-(24.23) in place of (24.3)-(24.9) and (24.12)-(24.14).

The preceding results show that $\boldsymbol{\mathcal{J}}_{m, l_1, \ldots, l_n}$ is a basic polynomial of index $(l_1, \ldots, l_n)$ as well as a semi-invariant of the second kind. This concludes the proof of Proposition 27.6. □

27.2. Principal results

THEOREM 27.7. *With $l_0 = 0$, let m, n, and $l_1, \ldots, l_n$ be integers that satisfy $m \geq 2$, $n \geq 1$, and (24.1) on page 275. Then, the polynomial $\mathcal{I}_{m, l_1, \ldots, l_n}$ in (24.14) is a basic relative invariant of index $(l_1, \ldots, l_n)$ for the equations (21.1). Moreover, $\mathcal{I}_{m, l_1, \ldots, l_n}$ and $\mathcal{J}_{m, l_1, \ldots, l_n}$ from (24.23) are related by*

$$(27.34) \qquad \mathcal{I}_{m, l_1, \ldots, l_m} \equiv \mathcal{J}_{m, l_1, \ldots, l_m}.$$

PROOF. For any given (21.1) on a region Ω, Proposition 23.4 on page 268 yields an analytic function $\rho_0(z) \not\equiv 0$ and a univalent analytic function $g_0(z)$ on a subregion $\mathcal{U}$ of Ω such that $\rho_0(z)$ and $g_0(z)$ are solutions on $\mathcal{U}$ of (23.19)–(23.20). Let $z = f_0(\zeta)$ on $\mathcal{V} = g_0(\mathcal{U})$ denote the inverse function for $\zeta = g_0(z)$ on $\mathcal{U}$. Theorem 23.1 on page 265 shows that the restriction of (21.1) to $\mathcal{U}$ is transformed by

$$y(z) = \rho_0(z)\, v(z), \quad z = f_0(\zeta), \quad \text{and} \quad w(\zeta) = (v \circ f_0)(\zeta)$$

into a unique differential equation

$$\left(w^{(m)}(\zeta)\right)^n + \sum_{\substack{0 \leq i_1, i_2, \ldots, i_n \leq m \\ (i_1, i_2, \ldots, i_n) \neq (0, 0, \ldots, 0)}} d_{i_1, i_2, \ldots, i_n}(\zeta) \prod_{\nu=1}^{n} w^{(m-i_\nu)}(\zeta) = 0, \quad \text{on } \mathcal{V},$$

whose meromorphic coefficients $d_{i_1, \ldots, i_n}(\zeta)$ satisfy

$$d_{i_{\pi(1)}, i_{\pi(2)}, \ldots, i_{\pi(n)}}(\zeta) \equiv d_{i_1, i_2, \ldots, i_n}(\zeta), \quad \text{for any permutation } \pi \text{ of } \{1, 2, \ldots, n\}.$$

Proposition 23.3 on page 267 yields $d_{0, \ldots, 0, 1}(\zeta) \equiv d_{0, \ldots, 0, 2}(\zeta) \equiv 0$. Thus, with respect to $A_{j_1, j_2, \ldots, j_n}^{i_1, i_2, \ldots, i_n}$ in (23.34), we apply Theorem 25.10 on page 290 to obtain

$$(27.35) \qquad \sum_{j_1=0}^{l_1} \cdots \sum_{j_n=0}^{l_n} A_{j_1, \ldots, j_n}^{l_1, \ldots, l_n}\, d_{l_1-j_1, \ldots, l_n-j_n}^{(j_1+\cdots+j_n)}(\zeta) \equiv \left(f_0'(\zeta)\right)^{s_0} \mathcal{I}_{m, l_1, \ldots, l_n}\bigl(f_0(\zeta)\bigr),$$

on $\mathcal{V}$, where $s_0 = l_1 + \cdots + l_n$ and where $\mathcal{I}_{m, l_1, \ldots, l_n}\bigl(f_0(\zeta)\bigr)$ is obtained for $z = f_0(\zeta)$ by substituting $c_{\mu_1, \ldots, \mu_n}^{(\kappa)}(z)$ from (21.1) for $w_{\mu_1, \ldots, \mu_n}^{(\kappa)}$ in $\mathcal{I}_{m, l_1, \ldots, l_n}$ of (24.14) on page 276. Moreover, Theorem 26.10 on page 304 yields

$$(27.36) \qquad \sum_{j_1=0}^{l_1} \cdots \sum_{j_n=0}^{l_n} A_{j_1, \ldots, j_n}^{l_1, \ldots, l_n}\, d_{l_1-j_1, \ldots, l_n-j_n}^{(j_1+\cdots+j_n)}(\zeta) \equiv \left(f_0'(\zeta)\right)^{s_0} \mathcal{J}_{m, l_1, \ldots, l_n}\bigl(f_0(\zeta)\bigr),$$

on $\mathcal{V}$, where $s_0 = l_1 + \cdots + l_n$ and where $\mathcal{J}_{m, l_1, \ldots, l_n}\bigl(f_0(\zeta)\bigr)$ is obtained for $z = f_0(\zeta)$ by substituting $c_{\mu_1, \ldots, \mu_n}^{(\kappa)}(z)$ from (21.1) for $w_{\mu_1, \ldots, \mu_n}^{(\kappa)}$ in $\mathcal{J}_{m, l_1, \ldots, l_n}$ of (24.23) on page 277. We equate the right members of (27.35) and (27.36), cancel the nonzero factor $\left(f_0'(\zeta)\right)^{s_0}$, substitute $\zeta = g_0(z)$, and use $f_0\bigl(g_0(z)\bigr) \equiv z$ to deduce that

$$(27.37) \qquad \mathcal{I}_{m, l_1, \ldots, l_n}(z) \equiv \mathcal{J}_{m, l_1, \ldots, l_n}(z),$$

on $\mathcal{U}$. Since the coefficients $c_{i_1, \ldots, i_n}(z)$ are meromorphic on Ω, (27.37) is valid on Ω. Thus, for the polynomial P defined in $\mathcal{S}_{m,n}$ by

$$P \equiv \mathcal{I}_{m, l_1, \ldots, l_n} - \mathcal{J}_{m, l_1, \ldots, l_n},$$

the substitution of $c_{\mu_1, \ldots, \mu_n}^{(\kappa)}(z)$ from (21.1) for $w_{\mu_1, \ldots, \mu_n}^{(\kappa)}$ in P yields $P(z) \equiv 0$, for any (21.1). We apply Corollary 21.10 on page 256 to conclude that $P \equiv 0$ and therefore (27.34) is valid.

27.2. PRINCIPAL RESULTS

Proposition 27.6 shows that $\mathcal{I}_{m,l_1,\ldots,l_n}$ is a basic semi-invariant of the first kind of index $(l_1, \ldots, l_n)$; and it shows that $\mathcal{J}_{m,l_1,\ldots,l_n}$ is a basic semi-invariant of the second kind of index $(l_1, \ldots, l_n)$. Since $\mathcal{I}_{m,l_1,\ldots,l_n}$ and $\mathcal{J}_{m,l_1,\ldots,l_n}$ are the same polynomial, we conclude that $\mathcal{I}_{m,l_1,\ldots,l_n}$ is a basic relative invariant of index $(l_1, \ldots, l_n)$ for the equations (21.1). This completes the proof. □

COROLLARY 27.8. *Theorem 23.8, as stated on page 271, is valid.*

PROOF. For $m \geq 2$, suppose that (21.1) satisfies

$$(27.38) \qquad c_{0,\ldots,0,1}(z) \equiv c_{0,\ldots,0,2}(z) \equiv 0, \quad \text{on } \Omega,$$

as well as (21.1.1); and suppose that the composite substitution $y(z) = \rho(z)\,v(z)$, $z = f(\zeta)$, and $w(\zeta) = (v \circ f)(\zeta)$ transforms this (21.1) into an equation (23.1) on Ω^{**} that satisfies

$$(27.39) \qquad d_{0,\ldots,0,1}(\zeta) \equiv d_{0,\ldots,0,2}(\zeta) \equiv 0, \quad \text{on } \Omega^{**},$$

in addition to (23.2). We shall establish that: for ζ in Ω^{**},

$$(27.40) \quad \sum_{j_1=0}^{i_1} \cdots \sum_{j_n=0}^{i_n} A^{i_1,\ldots,i_n}_{j_1,\ldots,j_n} d^{(j_1+\cdots+j_n)}_{i_1-j_1,\ldots,i_n-j_n}(\zeta)$$

$$\equiv \big(f'(\zeta)\big)^{i_1+\cdots+i_n} \sum_{j_1=0}^{i_1} \cdots \sum_{j_n=0}^{i_n} A^{i_1,\ldots,i_n}_{j_1,\ldots,j_n} c^{(j_1+\cdots+j_n)}_{i_1-j_1,\ldots,i_n-j_n}\big(f(\zeta)\big),$$

for $0 \leq i_1, \ldots, i_n \leq m$ and $2 \leq i_1 + \cdots + i_n$.

In view of (23.2), (23.34) on page 271, and (21.1.1), it suffices to establish (27.40) under the additional supposition that $0 \leq i_1 \leq \cdots \leq i_n \leq m$. For convenience, we introduce $i_0 = 0$. When $i_{n-1} = 0$ and $i_n = 2$, we see that (27.40) reduces to $0 \equiv 0$. For the remaining situation where $2 \leq i_1 + \cdots + i_n$ and either $i_{n-1} \neq 0$ or $i_n \neq 2$, we see that $i_{n-1} = 0$ implies $i_n \geq 3$. Since the identity substitution $y(z) = y(z)$, $z = z$ transforms (21.1) on Ω into (21.1) on Ω, Theorem 25.10 on page 290 yields

$$(27.41) \quad \sum_{j_1=0}^{i_1} \cdots \sum_{j_n=0}^{i_n} A^{i_1,\ldots,i_n}_{j_1,\ldots,j_n} c^{(j_1+\cdots+j_n)}_{i_1-j_1,\ldots,i_n-j_n}(z) \equiv \mathcal{I}_{m,i_1,\ldots,i_n}(z),$$

where $\mathcal{I}_{m,i_1,\ldots,i_n}(z)$ is the function on Ω obtained by substituting $c^{(\kappa)}_{\mu_1,\ldots,\mu_n}(z)$ from (21.1) for $w^{(\kappa)}_{\mu_1,\ldots,\mu_n}$ in the basic relative invariant $\mathcal{I}_{m,i_1,\ldots,i_n}$ of (24.14) having index $(i_1, \ldots, i_n)$. Since the identity substitution $w(\zeta) = w(\zeta)$, $\zeta = \zeta$ transforms (23.1) on Ω^{**} into (23.1) on Ω^{**}, we find that Theorem 25.10 on page 290 gives

$$(27.42) \quad \sum_{j_1=0}^{i_1} \cdots \sum_{j_n=0}^{i_n} A^{i_1,\ldots,i_n}_{j_1,\ldots,j_n} d^{(j_1+\cdots+j_n)}_{i_1-j_1,\ldots,i_n-j_n}(\zeta) \equiv \mathcal{I}^{\sharp}_{m,i_1,\ldots,i_n}(\zeta), \quad \text{on } \Omega^{**},$$

where $\mathcal{I}^{\sharp}_{m,i_1,\ldots,i_n}(\zeta)$ is the function on Ω^{**} obtained by substituting $d^{(\kappa)}_{\mu_1,\ldots,\mu_n}(\zeta)$ from (23.1) for $w^{(\kappa)}_{\mu_1,\ldots,\mu_n}$ in the basic relative invariant $\mathcal{I}_{m,i_1,\ldots,i_n}$ of (24.14) having index $(i_1, \ldots, i_n)$. For the composite transformation $y(z) = \rho(z)\,v(z)$, $z = f(\zeta)$, and $w(\zeta) = (v \circ f)(\zeta)$ of this (21.1) into this (23.1), the property that $\mathcal{I}_{m,i_1,\ldots,i_n}$ is a relative invariant of weight $i_1 + \cdots + i_n$ yields

$$(27.43) \qquad \mathcal{I}^{\sharp}_{m,i_1,\ldots,i_n}(\zeta) \equiv \big(f'(\zeta)\big)^{i_1+\cdots+i_n} \mathcal{I}_{m,i_1,\ldots,i_n}\big(f(\zeta)\big), \quad \text{on } \Omega^{**}.$$

We combine (27.42) (27.43), and (27.41) to obtain (27.40) and complete the proof. □

27.3. Some polynomials that are not relative invariants

For $m \geq 2$, let $\mathbb{Q}\{w_{0,\ldots,0,1}, w_{0,\ldots,0,2}\}$ denote the ring of polynomials over $\mathbb{Q}$ in the variables $w_{0,\ldots,0,1}^{(\kappa)}$, $w_{0,\ldots,0,2}^{(\lambda)}$ having $\kappa, \lambda \geq 0$.

PROPOSITION 27.9. *For any P in $\mathbb{Q}\{w_{0,\ldots,0,1}, w_{0,\ldots,0,2}\}$, P is not a relative invariant for the equations (21.1).*

PROOF. We modify the argument for Proposition 5.7 on page 52 by using Propositions 23.3–23.4 on pages 267–268 and Corollary 21.10 on page 256. □

COROLLARY 27.10. *Suppose that R in $\mathcal{S}_{m,n}$ is a relative invariant for the equations (21.1). Then, R must effectively involve at least one of the variables $w_{i_1,i_2,\ldots,i_{n-1},i_n}^{(k)}$ from (21.6) having either $i_n \geq 3$ or $n \geq 2$ and $i_{n-1} \geq 1$.*

PROOF. If R were to not involve any such variable, then R would be a polynomial in $\mathbb{Q}\{w_{0,\ldots,0,1}, w_{0,\ldots,0,2}\}$ and contradict Proposition 27.9. □

DEFINITION 27.11. For any two variables $w_{i_1,\ldots,i_n}^{(i_{n+1})}$ and $w_{j_1,\ldots,j_n}^{(j_{n+1})}$ in (21.6) on page 251, we define $w_{i_1,\ldots,i_n}^{(i_{n+1})}$ *precedes* $w_{j_1,\ldots,j_n}^{(j_{n+1})}$ to mean that there is an integer q satisfying $1 \leq q \leq n+1$ such that

$$i_k = j_k, \quad \text{when } 1 \leq k \leq q-1, \quad \text{and} \quad i_q < j_q.$$

THEOREM 27.12. *For integers $m \geq 2$, $n \geq 1$, $r \geq 0$, $p_0 = 0$, and $p_1, \ldots, p_n$ subject to $0 \leq p_1 \leq \cdots \leq p_n \leq m$ and either $p_{n-1} \geq 1$ or $p_n \geq 3$, suppose that an isobaric polynomial R in $\mathcal{S}_{m,n}$ of weight $s > p_1 + \cdots + p_n$ is expressible as*

(27.44) $$R \equiv S\, w_{p_1,\ldots,p_n}^{(r)} + T,$$

where S is a nonzero polynomial in $\mathbb{Q}\{w_{0,\ldots,0,1}, w_{0,\ldots,0,2}\}$ and T is a polynomial in $\mathcal{S}_{m,n}$ such that each variable $w_{i_1,\ldots,i_n}^{(k)}$ from (21.6) that is effectively involved in T precedes $w_{p_1,\ldots,p_n}^{(r)}$. Then, R is not a relative invariant for the equations (21.1).

PROOF. By making appropriate modifications at each step in the proof on pages 121–130 for Theorem 12.1, we obtain a proof of Theorem 27.12. In place of (2.5) on page 23, we use (21.2.1) on page 249; in place of (3.9) on page 30, we use (21.4.1) on page 250; and in place of Proposition 4.2 on page 36, we use Proposition 21.9 on page 256. □

27.4. Uniqueness of basic relative invariants

Each basic relative invariant for the equations (21.1) specifies a corresponding index $(l_1, \ldots, l_n)$ according to Definitions 21.5 on page 252. That each basic relative invariant for $m \geq 2$ is uniquely specified by its index is established next.

THEOREM 27.13. *For integers $m \geq 2$, $n \geq 1$, $l_0 = 0$, and $l_1, \ldots, l_n$ subject to*

$$0 \leq l_1 \leq l_2 \leq \cdots \leq l_n \leq m \quad \text{and either} \quad l_{n-1} \geq 1 \quad \text{or} \quad l_n \geq 3,$$

the basic relative invariant $\mathcal{I}_{m,l_1,\ldots,l_n}$ of Theorem 27.7 is the only basic relative invariant of index $(l_1, \ldots, l_n)$ for the equations (21.1).

PROOF. Let $\boldsymbol{P}_1$ and $\boldsymbol{P}_2$ in $\mathcal{S}_{m,n}$ be two basic relative invariants having the same index $(l_1, \ldots, l_n)$. To prove that $\boldsymbol{P}_1 \equiv \boldsymbol{P}_2$, suppose that $\boldsymbol{P}_1 \not\equiv \boldsymbol{P}_2$. Then the polynomial

(27.45) $$\boldsymbol{R} \equiv \boldsymbol{P}_1 - \boldsymbol{P}_2, \quad \text{in } \mathcal{S}_m,$$

is a relative invariant of weight $l_1 + \cdots + l_n$ for the equations (21.1). Since $\boldsymbol{P}_1$ and $\boldsymbol{P}_2$ are basic relative invariants of index $(l_1, \ldots, l_n)$, the coefficients of $\boldsymbol{w}_{l_1,\ldots,l_n}$ in both $\boldsymbol{P}_1$ and $\boldsymbol{P}_2$ are equal to 1. Hence, (27.45) shows that the coefficient of $\boldsymbol{w}_{l_1,\ldots,l_n}$ in $\boldsymbol{R}$ is 0. Also, (27.45) shows that any variable $\boldsymbol{w}_{i_1,\ldots,i_n}^{(k)}$ from (21.6) that is effectively involved in $\boldsymbol{R}$ must be effectively involved in at least one of $\boldsymbol{P}_1, \boldsymbol{P}_2$. Consequently, because $\boldsymbol{P}_1$ and $\boldsymbol{P}_2$ are basic relative invariants of index $(l_1, \ldots, l_n)$, any variable $\boldsymbol{w}_{i_1,\ldots,l_n}^{(k)}$ from (21.6) that is effectively involved in $\boldsymbol{R}$ must have

$$i_1 \leq l_1, \ldots, i_n \leq l_n, \quad i_1 + \cdots + i_n \leq l_1 + \cdots + l_n, \quad (i_1, \ldots, i_n) \neq (l_1, \ldots, l_n),$$

and $i_1 + \cdots + i_n < l_1 + \cdots + l_n$. Let p_1 be the greatest integer i_1 such that $\boldsymbol{w}_{i_1,\ldots,i_n}^{(k)}$ from (21.6) is effectively involved in $\boldsymbol{R}$ for some integers $i_2, \ldots, i_n$ and k; let p_2 be the greatest integer i_2 such that $\boldsymbol{w}_{p_1,i_2,i_3,\ldots,i_n}^{(k)}$ from (21.6) is effectively involved in $\boldsymbol{R}$ for some integers $i_3, \ldots, i_n$ and k; $\ldots$; let p_n be the greatest integer i_n such that $\boldsymbol{w}_{p_1,\ldots,p_{n-1},i_n}^{(k)}$ from (21.6) is effectively involved in $\boldsymbol{R}$ for some integer k; and let r be the greatest integer k such that $\boldsymbol{w}_{p_1,\ldots,p_n}^{(k)}$ is effectively involved in $\boldsymbol{R}$. Because $\boldsymbol{R}$ is a relative invariant, Corollary 27.10 shows that $\boldsymbol{w}_{p_1,\ldots,p_n}^{(r)}$ is not an element of $\mathbb{Q}\{\boldsymbol{w}_{0,\ldots,0,1}, \boldsymbol{w}_{0,\ldots,0,2}\}$. Since $\boldsymbol{P}_1$ and $\boldsymbol{P}_2$ are basic polynomials, we use (4) in Definition 21.5 on page 252 to see that the coefficient of $\boldsymbol{w}_{p_1,\ldots,p_n}^{(r)}$ in $\boldsymbol{P}_1$ and the coefficient of $\boldsymbol{w}_{p_1,\ldots,p_n}^{(r)}$ in $\boldsymbol{P}_2$ are polynomials in $\mathbb{Q}\{\boldsymbol{w}_{0,\ldots,0,1}, \boldsymbol{w}_{0,\ldots,0,2}\}$. Thus, there is a nonzero polynomial $\boldsymbol{S}$ in $\mathbb{Q}\{\boldsymbol{w}_{0,\ldots,0,1}, \boldsymbol{w}_{0,\ldots,0,2}\}$ and there is a polynomial $\boldsymbol{T}$ in $\mathcal{S}_{m,n}$ not effectively involving $\boldsymbol{w}_{p_1,\ldots,p_n}^{(r)}$ such that

(27.46) $$\boldsymbol{R} \equiv \boldsymbol{S}\, \boldsymbol{w}_{p_1,\ldots,p_n}^{(r)} + \boldsymbol{T}.$$

We have $p_1 + \cdots + p_n < s$, where $s = l_1 + \cdots + l_n$ is the weight of $\boldsymbol{R}$. The definition of $\boldsymbol{w}_{p_1,\ldots,p_n}^{(r)}$ shows that: if $\boldsymbol{w}_{i_1,\ldots,i_n}^{(k)}$ is any variable that is effectively involved in $\boldsymbol{T}$, then $\boldsymbol{w}_{i_1,\ldots,i_n}^{(k)}$ precedes $\boldsymbol{w}_{p_1,\ldots,p_n}^{(r)}$. Thus, $\boldsymbol{R}$ satisfies the hypotheses of Theorem 27.12. We use that result to conclude that $\boldsymbol{R}$ is not a relative invariant. This contradiction establishes that $\boldsymbol{P}_1 \equiv \boldsymbol{P}_2$ and completes the proof. □

27.5. The basic relative invariant of index $(l_1, \ldots, l_n)$ when $l_n = 1$

For there to be a basic relative invariant of index $(l_1, \ldots, l_n)$ having $l_n = 1$, we see that Definition 21.5 on page 252 requires $n \geq 2$ and $l_{n-1} = 1$. As in (21.15), we set $\mathfrak{w}_i \equiv \boldsymbol{w}_{0,\ldots,0,\underbrace{1,\ldots,1}_{i}}$ in $\mathcal{S}_{m,n}$, for $0 \leq i \leq n$.

PROPOSITION 27.14. *The basic relative invariants that correspond to the indices $(l_1, \ldots, l_n)$ having $l_n = 1$ are given by*

(27.47) $$\mathcal{I}_{m,\,\underbrace{0,\ldots,0}_{n-k},\underbrace{1,\ldots,1}_{k}} \equiv \sum_{\nu=0}^{k} (-1)^\nu \binom{k}{\nu} (\mathfrak{w}_1)^\nu \mathfrak{w}_{k-\nu}, \quad \text{when } 2 \leq k \leq n.$$

PROOF. For $m = 1$, we use Theorems 22.1 and 22.6 to obtain the assertion.

Suppose that $m \geq 2$. Now, Theorems 27.7 and 27.13 are applicable with $n \geq 2$. The corresponding basic relative invariants are represented by the left member of (27.47), for $2 \leq k \leq n$. For $0 \leq h_1, \ldots, h_n \leq 1$, we use (24.8) on page 275 to obtain

$$H^{m, \overbrace{0, \ldots, 0}^{n-k}, \overbrace{1, \ldots, 1}^{k}}_{h_1, \ldots, h_n} \equiv \begin{cases} 1, & \text{if } h_1 = \cdots = h_{n-k} = 0 \text{ and } h_{n-k+1} = \cdots = h_n = 1, \\ 0, & \text{otherwise.} \end{cases}$$

Thus, $\boldsymbol{M}_{m, \underbrace{0, \ldots, 0}_{n-k}, \underbrace{1, \ldots, 1}_{k}, h_1, \ldots, h_{n-1}, i}$ of (24.9) equals $\boldsymbol{L}_{m, \underbrace{0, \ldots, 0}_{n-k}, \underbrace{1, \ldots, 1}_{k}}$ when

$$(27.48) \qquad \Big(\underbrace{h_1, \ldots, h_{n-k}}_{n-k}, \underbrace{h_{n-k+1}, \ldots, h_{n-1}}_{k-1} \Big) = \Big(\underbrace{0, \ldots, 0}_{n-k}, \underbrace{1, \ldots, 1}_{k-1} \Big)$$

and $i = k$; however, it equals zero when either (27.48) is not satisfied or $i \neq k$. We combine this with (24.10)–(24.13) to see that: when (27.48) is not satisfied or when (27.48) is satisfied and $1 \leq i < k$, the expression $\boldsymbol{I}_{m, \underbrace{0, \ldots, 0}_{n-k}, \underbrace{1, \ldots, 1}_{k}, h_1, \ldots, h_{n-1}, i}$ is identically zero; and it equals $\boldsymbol{L}_{m, \underbrace{0, \ldots, 0}_{n-k}, \underbrace{1, \ldots, 1}_{k}}$ when (27.48) is satisfied and $i = k$. In regard to $2 \leq k \leq n$, $l_0 = 0$, $l_\mu = 0$, for $1 \leq \mu \leq n-k$, and $l_\nu = 1$, for $n - k + 1 \leq \nu \leq n$, we combine the preceding observations with (24.14), (24.6), (24.4), (21.7), and $\boldsymbol{K}_{m,1,0} \equiv -\boldsymbol{w}_1$ from (24.3)–(24.5) via (23.15) to verify that

$$\boldsymbol{\mathcal{I}}_{m, \underbrace{0, \ldots, 0}_{n-k}, \underbrace{1, \ldots, 1}_{k}} \equiv \sum_{h_0=0}^{l_0} \cdots \sum_{h_{n-1}=0}^{l_{n-1}} \boldsymbol{I}_{m, \underbrace{0, \ldots, 0}_{n-k}, \underbrace{1, \ldots, 1}_{k}, h_1, \ldots, h_{n-1}, k}$$

$$\equiv \boldsymbol{L}_{m, \underbrace{0, \ldots, 0}_{n-k}, \underbrace{1, \ldots, 1}_{k}}$$

$$\equiv \sum_{0 \leq p_1, \ldots, p_k \leq 1} \boldsymbol{K}_{m, 1-p_1, p_1} \cdots \boldsymbol{K}_{m, 1-p_k, p_k} \, \boldsymbol{w}_{\underbrace{0, \ldots, 0}_{n-k}, \underbrace{p_1, \ldots, p_k}_{k}}$$

$$\equiv \boldsymbol{w}_k + \binom{k}{1} \boldsymbol{K}_{m,1,0} \, \boldsymbol{w}_{k-1} + \binom{k}{2} (\boldsymbol{K}_{m,1,0})^2 \, \boldsymbol{w}_{k-2} + \cdots$$

$$\equiv \sum_{\nu=0}^{k} (-1)^\nu \binom{k}{\nu} (\boldsymbol{w}_1)^\nu \, \boldsymbol{w}_{k-\nu}.$$

This yields (27.47) and completes the proof. □

27.6. The number of basic relative invariants

For integers $m \geq 1$ and $n \geq 1$, let $\mathcal{N}_{m,n}$ denote the number of basic relative invariants for the differential equations (21.1). For $m = 1$ and $n \geq 1$, we use Theorems 22.1 and 22.6 to verify that $\mathcal{N}_{1,n} = n - 1$.

PROPOSITION 27.15. *For $m \geq 2$ and $n \geq 1$, the integer $\mathcal{N}_{m,n}$ is given by*

$$(27.49) \qquad \mathcal{N}_{m,n} \equiv \binom{m+n}{n} - 3.$$

27.6. THE NUMBER OF BASIC RELATIVE INVARIANTS

PROOF. In terms of integers $p \geq 1$ and $k \geq 0$, let $\mathcal{M}_{p,k}$ denote the number of p-tuples $(i_1, \ldots, i_p)$ that satisfy $0 \leq i_1 \leq \cdots \leq i_p \leq k$. Since we have $\mathcal{M}_{1,k} = k+1$, the formula

$$(27.50) \qquad \mathcal{M}_{p,k} \equiv \binom{p+k}{p}, \quad \text{for } k \geq 0,$$

is valid for $p = 1$. Suppose that (27.50) is valid for some positive integer p. Then, by counting the $(p+1)$-tuples corresponding to $i_{p+1} = 0, 1, \ldots, k$, we obtain

$$\mathcal{M}_{p+1,k} \equiv \sum_{\nu=0}^{k} \mathcal{M}_{p,\nu} \equiv \sum_{\nu=0}^{k} \binom{p+\nu}{p} \equiv \sum_{\nu=0}^{k} \left[\binom{p+\nu+1}{p+1} - \binom{p+\nu}{p+1} \right]$$
$$\equiv \binom{p+1+k}{p+1}, \quad \text{for } k \geq 0.$$

Thus, (27.50) is valid for $p \geq 1$.

We set $l_0 = 0$ and observe that the integer $\mathcal{N}_{m,n}$ is equal to the number of n-tuples $(l_1, \ldots, l_n)$ of integers that satisfy $0 \leq l_1 \leq \cdots \leq l_n \leq m$ and either $l_{n-1} \geq 1$ or $l_n \geq 3$. In particular, for $n = 1$, we note that

$$l_{n-1} = l_0 = 0, \quad 3 \leq l_n = l_1 \leq m, \quad \text{and} \quad \mathcal{N}_{m,1} = m - 2 \equiv \binom{m+1}{1} - 3.$$

Thus, (27.49) is valid for $m \geq 2$ and $n = 1$. For $m \geq 2$ and $n = 2$, we count the pairs (l_1, l_2) having $l_1 = \nu$, for $0 \leq \nu \leq m$, and obtain

$$\mathcal{N}_{m,2} \equiv (m-2) + m + (m-1) + \cdots + 1 \equiv m - 2 + \binom{m+1}{2} \equiv \binom{m+2}{2} - 3.$$

Thus, (27.49) is valid for $m \geq 2$ and $n = 2$. For $m \geq 2$ and $n \geq 3$, we use (27.50) and the identity

$$\sum_{\nu=0}^{a} \binom{a-\nu}{c} \binom{b+\nu}{d} \equiv \binom{a+b+1}{c+d+1}, \quad \text{for integers } a, b, c, d \geq 0 \text{ such that } d \geq b,$$

from [40, page 59, Formula (25)], to verify that

$$\mathcal{N}_{m,n} \equiv \overbrace{m-2}^{l_{n-1}=0} + \overbrace{\mathcal{M}_{n-2,1} \cdot m}^{l_{n-1}=1} + \overbrace{\mathcal{M}_{n-2,2} \cdot (m-1)}^{l_{n-1}=2} + \cdots + \overbrace{\mathcal{M}_{n-2,m} \cdot 1}^{l_{n-1}=m}$$
$$\equiv m - 2 + \sum_{\nu=1}^{m} \mathcal{M}_{n-2,\nu} \cdot (m-\nu+1)$$
$$\equiv (m-2) + \sum_{\nu=0}^{m+1} \binom{n-2+\nu}{n-2} \binom{m+1-\nu}{1} - (m+1)$$
$$= \sum_{\nu=0}^{m+1} \binom{m+1-\nu}{1} \binom{n-2+\nu}{n-2} - 3 \equiv \binom{m+n}{n} - 3.$$

Thus, (27.49) is valid as stated. This completes the proof. □

27.7. Relative invariants via basic ones for $m \geq 2$

Proposition 22.2 on page 259 shows that each relative invariant for the equations (21.1) having $m = 1$ is expressible as a differential-polynomial combination over $\mathbb{Q}$ of the basic relative invariants for such equations.

To present a corresponding result here for the equations (21.1) having $m \geq 2$, we first use $\boldsymbol{a}_{m,2}$ in (23.14) on page 267 to introduce

$$(27.51) \quad \boldsymbol{G}_{m,2} \equiv \binom{m+1}{3}\boldsymbol{a}_{m,2} \equiv \boldsymbol{w}_{\underbrace{0,\ldots,0}_{n-1},2} - \tfrac{m-1}{2}\boldsymbol{w}^{(1)}_{\underbrace{0,\ldots,0}_{n-1},1} - \tfrac{m-1}{2m}(\boldsymbol{w}_{\underbrace{0,\ldots,0}_{n-1},1})^2.$$

Now, the argument on pages 144–145 for Theorem 14.3 can be generalized.

THEOREM 27.16. *Each relative invariant for the equations* (21.1) *having order $m \geq 2$ is expressible as a differential-polynomial combination over $\mathbb{Q}$ of $\boldsymbol{G}_{m,2}$ and the basic relative invariants for such equations.*

PROOF. Let $\boldsymbol{P}$ be a relative invariant for the equations (21.1) of order $m \geq 2$. Then, $\boldsymbol{P}$ is expressible as a differential-polynomial combination over $\mathbb{Q}$ of those variables from (21.6) on page 251 given by

$$(27.52) \quad \boldsymbol{w}_{i_1,i_2,\ldots,i_n}, \quad \text{for } 1 \leq i_n \leq m \text{ and } 0 \leq i_1 \leq i_2 \leq \cdots \leq i_n.$$

We totally order these variables according to the scheme where $\boldsymbol{w}_{i_1,i_2,\ldots,i_n}$ precedes $\boldsymbol{w}_{j_1,j_2,\ldots,j_n}$ if and only if there is an integer q satisfying $1 \leq q \leq n$ such that $i_k = j_k$, for $q+1 \leq k \leq n$, and $i_q < j_q$. In the situation $n \geq 3$, this is indicated by

$$\boldsymbol{w}_{0,\ldots,0,1}, \; \boldsymbol{w}_{0,\ldots,1,1}, \; \ldots, \; \boldsymbol{w}_{1,\ldots,1,1}, \; \boldsymbol{w}_{0,\ldots,0,2}, \; \boldsymbol{w}_{0,\ldots,1,2}, \; \ldots, \; \boldsymbol{w}_{m,\ldots,m,m}.$$

We use (27.51) and the structure of basic relative invariants to obtain

$$(27.53) \quad \boldsymbol{w}_{\underbrace{0,\ldots,0}_{n-1},2} \equiv \boldsymbol{G}_{m,2} + \tfrac{m-1}{2}\boldsymbol{w}^{(1)}_{\underbrace{0,\ldots,0}_{n-1},1} + \tfrac{m-1}{2m}(\boldsymbol{w}_{\underbrace{0,\ldots,0}_{n-1},1})^2$$

$$\equiv \boldsymbol{G}_{m,2} + \boldsymbol{R}_{m,\underbrace{0,\ldots,0}_{n-1},2;0}$$

and

$$(27.54) \quad \boldsymbol{w}_{i_1,i_2,\ldots,i_n} \equiv \boldsymbol{I}_{m,i_1,i_2,\ldots,i_n} + \boldsymbol{R}_{m,i_1,i_2,\ldots,i_n;0},$$

$$\text{when } i_n \geq 3 \text{ or } n \geq 2 \text{ and } i_{n-1} \geq 1,$$

where $\boldsymbol{R}_{m,i_1,i_2,\ldots,i_n;0}$ in each of (27.54) and (27.53) is a differential-polynomial combination over $\mathbb{Q}$ of variables that precede $\boldsymbol{w}_{i_1,i_2,\ldots,i_n}$ in the ordering for (27.52). By repeatedly applying the derivation $'$ of $\mathcal{S}_{m,n}$ to (27.53) and (27.54), we find that: for $k \geq 0$,

$$(27.55) \quad \boldsymbol{w}^{(k)}_{\underbrace{0,\ldots,0}_{n-1},2} \equiv \boldsymbol{G}^{(k)}_{m,2} + \boldsymbol{R}_{m,\underbrace{0,\ldots,0}_{n-1},2;k}$$

and

$$(27.56) \quad \boldsymbol{w}^{(k)}_{i_1,i_2,\ldots,i_n} \equiv \boldsymbol{I}^{(k)}_{m,i_1,i_2,\ldots,i_n} + \boldsymbol{R}_{m,i_1,i_2,\ldots,i_n;k},$$

$$\text{when } i_n \geq 3 \text{ or } n \geq 2 \text{ and } i_{n-1} \geq 1,$$

where $\boldsymbol{R}_{m,i_1,i_2,\ldots,i_n;k}$ in each of (27.56) and (27.55) is a differential-polynomial combination over $\mathbb{Q}$ of the variables in (27.52) that precede $\boldsymbol{w}_{i_1,i_2,\ldots,i_n}$.

Let the ordering of the variables $\boldsymbol{w}_{i_1,i_2,\ldots,i_n}$ in (27.52) be represented by

$$(27.57) \quad \mathfrak{s}_1, \, \mathfrak{s}_2, \, \ldots, \, \mathfrak{s}_r.$$

In particular, we have $\mathfrak{s}_1 \equiv w_{0,\ldots,0,1}$ and $\mathfrak{s}_r \equiv w_{m,\ldots,m,m}$. Starting with $\boldsymbol{P}$ as a differential-polynomial combination of the variables in (27.57), we use (27.56) to eliminate from $\boldsymbol{P}$ the variables $\mathfrak{s}_r^{(k)}$, for each $k \geq 0$. Then, from the resulting expression, we use (27.56) or (27.55) to eliminate $\mathfrak{s}_{r-1}^{(k)}$, for each $k \geq 0$. We continue in this manner using either (27.56) or (27.55) until finally we eliminate the variables $\mathfrak{s}_2^{(k)}$, for each $k \geq 0$. The resulting expression is then a differential-polynomial combination over $\mathbb{Q}$ of $\mathfrak{s}_1$, $\boldsymbol{G}_{m,2}$, and the basic relative invariants for the equations (21.1). When $n \geq 3$, it is a differential-polynomial combination over $\mathbb{Q}$ of

$$w_{0,\ldots,0,1},\ \mathcal{I}_{0,\ldots,1,1},\ \ldots,\ \mathcal{I}_{1,\ldots,1,1},\ \boldsymbol{G}_{m,2},\ \mathcal{I}_{0,\ldots,1,2},\ \ldots,\ \mathcal{I}_{m,\ldots,m,m}.$$

We can rewrite it in the form

(27.58) $$\boldsymbol{P} \equiv \boldsymbol{P}_0 + \boldsymbol{Z}_0,$$

where: $\boldsymbol{P}_0$ is a differential-polynomial combination over $\mathbb{Q}$ of $\boldsymbol{G}_{m,2}$ and the basic relative invariants for the equations (21.1); and $\boldsymbol{Z}_0$ is a differential-polynomial combination over $\mathbb{Q}$ of $\mathfrak{s}_1$, $\boldsymbol{G}_{m,2}$, and the basic relative invariants for the equations (21.1) such that each term of $\boldsymbol{Z}_0$ possesses at least one factor of the form $\mathfrak{s}_1^{(k)}$, for some $k \geq 0$. Since $\boldsymbol{P}_0$ is a differential-polynomial combination over $\mathbb{Q}$ of semi-invariants of the first kind, Proposition 27.1 on page 307 shows that $\boldsymbol{P}_0$ is either 0 or a semi-invariant of the first kind.

For any given (21.1) on Ω, let $\rho_0(z) \not\equiv 0$ be a meromorphic function on a subregion $\mathcal{U}$ of Ω such that

(27.59) $$\rho_0^{(1)}(z) + \frac{c_{0,\ldots,0,1}(z)}{m}\rho_0(z) \equiv 0, \quad \text{on } \mathcal{U}.$$

Due to (27.59), an application of (21.2.1) on page 249 shows that the substitution $y(z) = \rho_0(z)\,v(z)$ transforms the restriction of (21.1) to $\mathcal{U}$ into a corresponding equation (21.3) on $\mathcal{U}$ having $c^*_{0,\ldots,0,1}(z) \equiv 0$. Thus, the substitution of $c^{*(k)}_{i_1,i_2,\ldots,i_n}(z)$ from this (21.3) for $w^{(k)}_{i_1,i_2,\ldots,i_n}$ in $\boldsymbol{Z}_0$ yields $\boldsymbol{Z}_0^*(z) \equiv 0$, on $\mathcal{U}$. We combine this with properties of $\boldsymbol{P}$ and $\boldsymbol{P}_0$ to obtain

$$P(z) \equiv P^*(z) \equiv P_0^*(z) + Z_0^*(z) \equiv P_0^*(z) \equiv P_0(z), \quad \text{on } \mathcal{U}.$$

Hence, we have $P(z) \equiv P_0(z)$, on Ω, for any (21.1). Now, we apply Corollary 21.10 on page 256 to the polynomial $\boldsymbol{P} - \boldsymbol{P}_0$ to conclude that $\boldsymbol{P} - \boldsymbol{P}_0 \equiv 0$. This yields $\boldsymbol{P} \equiv \boldsymbol{P}_0$ and completes the proof. $\square$

27.8. Rational semi-invariants for various classes of equations

For any fixed positive integer m, Chapter 20 developed the subject of rational semi-invariants and rational relative invariants for the class $\mathcal{C}_{m,2}$ specified by the equations of the form (1.1) having order m. In [24, Appendix D], analogous results were obtained for the class $\mathcal{C}_{m,1}$ of monic homogeneous linear differential equations having a fixed order $m \geq 3$. A theory about rational semi-invariants of both kinds and rational relative invariants can be developed for any class of equations that possesses polynomial semi-invariants of both kinds. In particular, such can be done for the class $\mathcal{C}_{m,n}$ specified by the equations (21.1) having a given m and n as well as for the classes specified by the equations considered in Chapters 28–30.

For the class $\mathcal{C}_{m,n}$, we would employ Proposition 21.9 on page 256 in place of Proposition 4.2 on page 36. The principal alterations would occur in Part (i) of the proof for Proposition 20.10 and in Part (i) of the proof for Propositioned 20.13.

As another example, we note that Theorem 28.7 on page 332 conveniently specifies the polynomial relative invariants for the class specified by the equations of the form (28.1). Thus, a simple characterization of the rational relative invariants for that class could be given.

Part 5

Additional Classes of Equations

CHAPTER 28

The Class of Equations Specified by $y''(z)\,y'(z)$

The equations $H_{m,n} = 0$ given by (21.1) on page 249 are the simplest ones that have $\left(y^{(m)}(z)\right)^n$ as a term and are transformed into ones of a similar type by any change of the dependent and independent variables. The representation (21.1) for them in terms of (21.1.1) was selected to obtain the reasonably formulas of Chapters 24 upon which Theorems 25.10 and 26.10 are based. Similarly, $y^{(m)}(z)$ specifies the equations $L_m = 0$ in (1.84) on page 17; $\left(y^{(m)}(z)\right)^2$ characterizes the equations $Q_m = 0$ in (1.1) on page 4; $\left(y''(z)\right)^2$ gives the equations $Q_2 = 0$ in (1.0–A) on page 3; and $\left(y'(z)\right)^2$ yields the equations (4.16) on page 39. Of the remaining monic monomials in $y''(z)$, $y'(z)$, $y(z)$ of order ≤ 2 and degree 2, only $y''(z)\,y'(z)$ specifies equations for which each nonconstant isobaric polynomial has positive weight. Equations specified by $y''(z)\,y'(z)$ were briefly considered by Paul Appell in [**3, 4**] of 1887 before his research that we described in Sections 1.1 and 1.6. Since we shall examine them without seeking generalizations, a special representation like (21.1.1) is not needed for (28.1). *The relative invariants for the equations (28.1) are given by Theorem 28.7 on page 332.*

28.1. Notation and terminology

The equations characterized by $y''(z)\,y'(z)$ are expressible as

(28.1) $$y''(z)\,y'(z) + a_{0,1}(z)\,y''(z)\,y(z) + a_{1,0}(z)\,\left(y'(z)\right)^2$$
$$+ a_{1,1}(z)\,y'(z)\,y(z) + a_{2,1}(z)\,\left(y(z)\right)^2 = 0$$

in terms of coefficients $a_{i,j}(z)$ that are meromorphic functions on a region Ω of the complex plane. For each meromorphic function $\rho(z) \not\equiv 0$ on Ω, there are unique meromorphic functions $a_{i,j}^*(z)$ on Ω such that the change $y(z) = \rho(z)\,v(z)$ of the dependent variable from y to v transforms (28.1) on Ω into

(28.2) $$v''(z)\,v'(z) + a_{0,1}^*(z)\,v''(z)\,v(z) + a_{1,0}^*(z)\,\left(v'(z)\right)^2$$
$$+ a_{1,1}^*(z)\,v'(z)\,v(z) + a_{2,1}^*(z)\,\left(v(z)\right)^2 = 0, \quad \text{on } \Omega.$$

For each univalent analytic function $\zeta = g(z)$ on Ω with inverse designated by $z = f(\zeta)$ on $\Omega^{**} = g(\Omega)$, there are unique meromorphic functions $a_{i,j}^{**}(\zeta)$ on Ω^{**} such that the change $z = f(\zeta)$ of the independent variable transforms (28.1) into

(28.3) $$u''(\zeta)\,u'(\zeta) + a_{0,1}^{**}(\zeta)\,u''(\zeta)\,u(\zeta) + a_{1,0}^{**}(\zeta)\,\left(u'(\zeta)\right)^2$$
$$+ a_{1,1}^{**}(\zeta)\,u'(\zeta)\,u(\zeta) + a_{2,1}^{**}(\zeta)\,\left(u(\zeta)\right)^2 = 0, \quad \text{on } \Omega^{**},$$

where $u(\zeta) = (y \circ f)(\zeta)$. Here, we assume that the symbols

(28.4) $$x_{i,j}^{(k)}, \quad \text{for } 0 \leq i \leq 2,\ (i-1)^2 \leq j \leq 1,\ \text{and } k \geq 0,$$

are algebraically independent variables over the field $\mathbb{Q}$ of rational numbers. We let $\mathfrak{R}_{2,1}$ denote the ring of polynomials in these variables over $\mathbb{Q}$ and we let $'$ denote the unique derivation for $\mathfrak{R}_{2,1}$ such that $\left(x_{i,j}^{(k)}\right)' = x_{i,j}^{(k+1)}$, for each $x_{i,j}^{(k)}$ in (28.4).

For any polynomial $\boldsymbol{P}$ in $\mathfrak{R}_{2,1}$, we let $P(z)$ on Ω, $P^*(z)$ on Ω, and $P^{**}(\zeta)$ on Ω^{**} denote the unique functions obtained by replacing $x_{i,j}^{(k)}$ in $\boldsymbol{P}$ with $a_{i,j}^{(k)}(z)$ from (28.1), $a_{i,j}^{*(k)}(z)$ from (28.2), and $a_{i,j}^{**(k)}(\zeta)$ from (28.3). We adapt Definitions 1.1–1.4 of page 6 by writing $\mathfrak{R}_{2,1}$, (28.1), (28.2), (28.3), $x_{i,j}^{(k)}$, $a_{i,j}^{(k)}(z)$, $a_{i,j}^{*(k)}(z)$, and $a_{i,j}^{**(k)}(\zeta)$ in place of $\mathcal{S}_m$, (1.1), (1.3), (1.5), $w_{i,j}^{(k)}$, $c_{i,j}^{(k)}(z)$, $c_{i,j}^{*(k)}(z)$, and $c_{i,j}^{**(k)}(\zeta)$, respectively. In particular, *the weight of* $x_{i,j}^{(k)}$ *in* (28.4) *is the positive integer* $i+j+k$; and we easily verify that *a semi-invariant of the second kind is necessarily isobaric.*

28.2. Principal formulas

The coefficients for (28.2) and (28.3) are given by

$$(28.5) \quad a_{0,1}^*(z) \equiv a_{0,1}(z) + \frac{\rho'(z)}{\rho(z)},$$

$$(28.6) \quad a_{1,0}^*(z) \equiv a_{1,0}(z) + 2\frac{\rho'(z)}{\rho(z)},$$

$$a_{1,1}^*(z) \equiv a_{1,1}(z) + 2\,a_{0,1}(z)\frac{\rho'(z)}{\rho(z)} + 2\,a_{1,0}(z)\frac{\rho'(z)}{\rho(z)} + 2\left[\frac{\rho'(z)}{\rho(z)}\right]^2 + \frac{\rho''(z)}{\rho(z)},$$

$$a_{2,1}^*(z) \equiv a_{2,1}(z) + a_{1,1}(z)\frac{\rho'(z)}{\rho(z)} + a_{1,0}(z)\left[\frac{\rho'(z)}{\rho(z)}\right]^2 + a_{0,1}(z)\frac{\rho''(z)}{\rho(z)} + \frac{\rho'(z)\,\rho''(z)}{(\rho(z))^2},$$

and

$$(28.7) \quad a_{0,1}^{**}(\zeta) \equiv f'(\zeta)\,a_{0,1}(f(\zeta)),$$

$$(28.8) \quad a_{1,0}^{**}(\zeta) \equiv f'(\zeta)\,a_{1,0}(f(\zeta)) - \frac{f''(\zeta)}{f'(\zeta)},$$

$$(28.9) \quad a_{1,1}^{**}(\zeta) \equiv (f'(\zeta))^2 a_{1,1}(f(\zeta)) - f''(\zeta)\,a_{0,1}(f(\zeta)),$$

$$(28.10) \quad a_{2,1}^{**}(\zeta) \equiv (f'(\zeta))^3 a_{2,1}(f(\zeta)).$$

We use the preceding formulas to verify that the polynomials defined in $\mathfrak{R}_{2,1}$ by

$$(28.11) \quad r_{1,0} \equiv x_{1,0} - 2\,x_{0,1},$$

$$(28.12) \quad R_{1,1} \equiv x_{1,1} - 2\,x_{0,1}\,x_{1,0} - x_{0,1}^{(1)} + (x_{0,1})^2,$$

and

$$(28.13) \quad R_{2,1} \equiv x_{2,1} - x_{0,1}\,x_{1,1} + (x_{0,1})^2\,x_{1,0},$$

satisfy

$(28.14) \quad r_{1,0}^*(z) \equiv r_{1,0}(z), \quad$ on Ω,

$(28.15) \quad R_{1,1}^*(z) \equiv R_{1,1}(z), \quad$ on Ω, and $\quad R_{1,1}^{**}(\zeta) \equiv (f'(\zeta))^2 R_{1,1}(f(\zeta)), \quad$ on Ω^{**},

as well as

$(28.16) \quad R_{2,1}^*(z) \equiv R_{2,1}(z), \quad$ on Ω, and $\quad R_{2,1}^{**}(\zeta) \equiv (f'(\zeta))^3 R_{2,1}(f(\zeta)), \quad$ on Ω^{**}.

THEOREM 28.1. *The polynomials $\boldsymbol{R}_{1,1}$ and $\boldsymbol{R}_{2,1}$ of (28.12) and (28.13) are relative invariants of respective weights 2 and 3 for the equations (28.1); while, $\boldsymbol{r}_{1,0}$ in (28.11) is an isobaric semi-invariant of the first kind having weight 1. Moreover, if $\boldsymbol{P}$ is a relative invariant of weight s for the equations (28.1), then the polynomial*

$$(28.17) \qquad \boldsymbol{I} \equiv \boldsymbol{P}^{(1)} + s\,\boldsymbol{r}_{1,0}\,\boldsymbol{P}$$

is a relative invariant of weight $s+1$ for such equations.

PROOF. The identities (28.15), (28.16), and (28.14) establish the assertions about $\boldsymbol{R}_{1,1}$, $\boldsymbol{R}_{2,1}$, and $\boldsymbol{r}_{1,0}$. Let $\boldsymbol{P}$ be a relative invariant of weight s. Since $\boldsymbol{P}$, $\boldsymbol{P}^{(1)}$, and $\boldsymbol{r}_{1,0}$ are isobaric semi-invariants of the first kind, we see that $\boldsymbol{I}$ in (28.17) is an isobaric semi-invariant of the first kind having weight $s+1$. The formulas for $a_{1,0}^{**}(\zeta)$ and $a_{0,1}^{**}(\zeta)$ yield

$$r_{1,0}^{**}(\zeta) \equiv a_{1,0}^{**}(\zeta) - 2a_{0,1}^{**}(\zeta) \equiv f'(\zeta)\left[r_{1,0}(f(\zeta))\right] - \frac{f''(\zeta)}{f'(\zeta)}.$$

We use this along with $P^{**}(\zeta) \equiv (f'(\zeta))^s P(f(\zeta))$ and

$$P^{**(1)}(\zeta) \equiv (f'(\zeta))^{s+1} P^{(1)}(f(\zeta)) + s(f'(\zeta))^s \frac{f''(\zeta)}{f'(\zeta)} P(f(\zeta))$$

to deduce on Ω^{**} that

$$I^{**}(\zeta) \equiv P^{**(1)}(\zeta) + s\,r_{1,0}^{**}(\zeta)\,P^{**}(\zeta)$$
$$\equiv (f'(\zeta))^{s+1}\left[P^{(1)}(f(\zeta)) + s\,r_{1,0}(f(\zeta))\,P(f(\zeta))\right]$$
$$\equiv (f'(\zeta))^{s+1} I(f(\zeta)).$$

Thus, $\boldsymbol{I}$ is a semi-invariant of the second kind. This shows that $\boldsymbol{I}$ is a relative invariant of weight $s+1$ for the equations (28.1) and completes the proof. □

OBSERVATION 28.2. Unlike $\boldsymbol{r}_{1,0}$ in Theorem 28.1, $\boldsymbol{b}_{m,1}$ for Proposition 4.17 on page 45 and $\boldsymbol{b}_{m,1}$ for Proposition 27.2 on page 307 are not semi-invariants of the first kind. For that reason, the expressions analogous to (28.17) in Propositions 4.17 and 27.2 are merely semi-invariants of the second kind for the contexts used there.

OBSERVATION 28.3. When the left member of (28.1) is regarded as a quadratic form in $y''(z)$, $y'(z)$, $y(z)$, the determinant of its matrix is given by

$$\begin{vmatrix} 0 & 1/2 & a_{0,1}(z)/2 \\ 1/2 & a_{1,0}(z) & a_{1,1}(z)/2 \\ a_{0,1}(z)/2 & a_{1,1}(z)/2 & a_{2,1}(z) \end{vmatrix} \equiv -\frac{1}{4} R_{2,1}(z),$$

where $R_{2,1}(z)$ is the function obtained by substituting $a_{i,j}(z)$ from (28.1) for $\boldsymbol{x}_{i,j}$ in $\boldsymbol{R}_{2,1}$ of (28.13). Thus, the left member of (28.1) has a nontrivial factorization with respect to some subregion of Ω if and only if $R_{2,1}(z) \equiv 0$.

PROPOSITION 28.4. *Each relative invariant for equations (28.1) is expressible as a differential-polynomial combination over $\mathbb{Q}$ of the relative invariants $\boldsymbol{R}_{1,1}$, $\boldsymbol{R}_{2,1}$ in (28.12)–(28.13) and the semi-invariant of the first kind $\boldsymbol{r}_{1,0}$, in (28.11).*

PROOF. Let $\boldsymbol{I}$ be a relative invariant for the equations (28.1). In particular, $\boldsymbol{I}$ is a differential-polynomial combination over $\mathbb{Q}$ of $\boldsymbol{x}_{2,1}$, $\boldsymbol{x}_{1,1}$, $\boldsymbol{x}_{1,0}$, and $\boldsymbol{x}_{0,1}$. We see that: for $k \geq 0$, (28.13) enables us to express $\boldsymbol{x}_{2,1}^{(k)}$ as a differential-polynomial combination over $\mathbb{Q}$ of $\boldsymbol{R}_{2,1}$, $\boldsymbol{x}_{1,1}$, $\boldsymbol{x}_{1,0}$, and $\boldsymbol{x}_{0,1}$. By using these relations to eliminate the variables $x_{2,1}^{(k)}$ appearing in I, we express I as a differential-polynomial combination over $\mathbb{Q}$ of $R_{2,1}$, $x_{1,1}$, $x_{1,0}$, and $x_{0,1}$. Next, we use (28.12) to eliminate the variables $x_{1,1}^{(k)}$ from I and express I as a differential-polynomial combination over $\mathbb{Q}$ of $R_{2,1}$, $R_{1,1}$, $x_{1,0}$, and $x_{0,1}$. Then, we employ (28.11) to eliminate the variables $x_{1,0}^{(k)}$ from I and express I as a differential-polynomial combination over $\mathbb{Q}$ of $R_{2,1}$, $R_{1,1}$, $r_{1,0}$, and $x_{0,1}$. Thus, we can write

(28.18) $$I \equiv I_1 + I_2,$$

where: I_1 is a differential-polynomial combination over $\mathbb{Q}$ of $R_{2,1}$, $R_{1,1}$, $r_{1,0}$; while I_2 is a differential-polynomial combination over $\mathbb{Q}$ of $R_{2,1}$, $R_{1,1}$, $r_{1,0}$, and $x_{0,1}$ such that each nonzero term of I_2 has at least one factor of the form $x_{0,1}^{(k)}$, for some $k \geq 0$. When (28.1) is given on Ω, there is a meromorphic function $\rho(z) \not\equiv 0$ on a subregion $\mathcal{U}$ of Ω such that

(28.19) $$\rho'(z) + a_{0,1}(z)\,\rho(z) = 0, \quad \text{on } \mathcal{U}.$$

Under the substitution $y(z) = \rho(z)\,v(z)$, the restriction of (28.1) to $\mathcal{U}$ is transformed into an equation (28.2) on $\mathcal{U}$. Due to (28.5) and (28.19), we see that this (28.2) satisfies $a_{0,1}^*(z) \equiv 0$, on $\mathcal{U}$. Thus, the substitution of $a_{i,j}^{*(k)}(z)$ from this (28.2) for $x_{i,j}^{(k)}$ in I_2 yields $I_2^*(z) \equiv 0$, on $\mathcal{U}$. Since I_1 is a polynomial combination over $\mathbb{Q}$ of semi-invariants of the first kind, I_1 is a semi-invariant of the first kind. Hence, we have $I_2(z) \equiv I(z) - I_1(z) \equiv I^*(z) - I_1^*(z) \equiv I_2^*(z) \equiv 0$, on $\mathcal{U}$. Thus, for any (28.1) on Ω, we obtain $I_2(z) \equiv 0$, on Ω. A result analogous to Proposition 4.2 on page 36 shows that $I_2 \equiv 0$. In view of (28.18), we obtain $I \equiv I_1$ and conclude that I is a differential-polynomial combination over $\mathbb{Q}$ of $R_{2,1}$, $R_{1,1}$, and $r_{1,0}$. This completes the proof. □

For Definition 1.5 on page 6, there is a natural modification that can be made in terms of which $R_{1,1}$ and $R_{2,1}$ are basic polynomials in $\mathfrak{R}_{2,1}$. It them follows that $R_{1,1}$ and $R_{2,1}$ are the only basic relative invariants for the equations (28.1).

OBSERVATION 28.5. An isobaric polynomial P in $\mathfrak{R}_{2,1}$ of weight 1 is expressible as $P \equiv A\,x_{1,0} + B\,x_{0,1}$, for some A, B in $\mathbb{Q}$ not both zero. We use (28.5)–(28.6) to see that P is a semi-invariant of the first kind if and only if it is related to $r_{1,0}$ in (28.11) by $P \equiv C\,r_{1,0}$, for some nonzero C in $\mathbb{Q}$. However, (28.7) and (28.8) show that $P \equiv C\,r_{1,0}$ is not a semi-invariant of the second kind. Consequently, *there are no relative invariants of weight 1 for the equations (28.1).*

28.3. The relative invariants of weight ≤ 9 for the equations (28.1)

For $s \geq 2$, there are linear independent relative invariants $G_{s,1}, \ldots, G_{s,n(s)}$ over $\mathbb{Q}$ of weight s for the equations (28.1) such that the relative invariants of weight s for the equations (28.1) are given by the linear combinations

(28.20) $$\sum_{i=1}^{n(s)} K_i\,G_{s,i}, \quad \text{with } K_1, \ldots, K_{n(s)} \text{ in } \mathbb{Q} \text{ and some } K_i \neq 0.$$

Here, for $2 \leq s \leq 9$, we present particular selections for suitable $G_{s,i}$ as differential-polynomial combinations over $\mathbb{Q}$ of $R_{1,1}$, $R_{2,1}$, and $r_{1,0}$. *That the corresponding linear combinations in (28.20) yield all the relative invariants of weight $s \leq 9$ for the equations (28.1)* is established by computations described in Section 28.4.

28.3. THE RELATIVE INVARIANTS OF WEIGHT ≤ 9 FOR THE EQUATIONS (28.1)

For $s = 2$, we have $n(2) = 1$ and select $\boldsymbol{G}_{2,1} \equiv \boldsymbol{R}_{1,1}$.

For $s = 3$, we have $n(3) = 2$ and select $\boldsymbol{G}_{3,1} \equiv \boldsymbol{G}_{2,1}^{(1)} + 2\boldsymbol{r}_{1,0}\boldsymbol{G}_{2,1}$, $\boldsymbol{G}_{3,2} \equiv \boldsymbol{R}_{2,1}$.

For $s = 4$, we have $n(4) = 3$ and select

$$\boldsymbol{G}_{4,1} \equiv (\boldsymbol{G}_{2,1})^2, \quad \boldsymbol{G}_{4,2} \equiv \boldsymbol{G}_{3,1}^{(1)} + 3\boldsymbol{r}_{1,0}\boldsymbol{G}_{3,1}, \quad \boldsymbol{G}_{4,3} \equiv \boldsymbol{G}_{3,2}^{(1)} + 3\boldsymbol{r}_{1,0}\boldsymbol{G}_{3,2}.$$

For $s = 5$, we have $n(5) = 4$ and select

$$\boldsymbol{G}_{5,1} \equiv \boldsymbol{G}_{2,1}\boldsymbol{G}_{3,1}, \qquad \boldsymbol{G}_{5,2} \equiv \boldsymbol{G}_{4,2}^{(1)} + 4\boldsymbol{r}_{1,0}\boldsymbol{G}_{4,2},$$

$$\boldsymbol{G}_{5,3} \equiv \boldsymbol{G}_{2,1}\boldsymbol{G}_{3,2}, \qquad \boldsymbol{G}_{5,4} \equiv \boldsymbol{G}_{4,3}^{(1)} + 4\boldsymbol{r}_{1,0}\boldsymbol{G}_{4,3}.$$

For $s = 6$, we have $n(6) = 8$ and select

$$\boldsymbol{G}_{6,1} \equiv (\boldsymbol{G}_{2,1})^3, \qquad \boldsymbol{G}_{6,2} \equiv (\boldsymbol{G}_{3,1})^2, \qquad \boldsymbol{G}_{6,3} \equiv \boldsymbol{G}_{2,1}\boldsymbol{G}_{4,2},$$

$$\boldsymbol{G}_{6,4} \equiv \boldsymbol{G}_{5,2}^{(1)} + 5\boldsymbol{r}_{1,0}\boldsymbol{G}_{5,2}, \quad \boldsymbol{G}_{6,5} \equiv \boldsymbol{G}_{3,1}\boldsymbol{G}_{3,2}, \quad \boldsymbol{G}_{6,6} \equiv (\boldsymbol{G}_{3,2})^2,$$

$$\boldsymbol{G}_{6,7} \equiv \boldsymbol{G}_{2,1}\boldsymbol{G}_{4,3}, \qquad \boldsymbol{G}_{6,8} \equiv \boldsymbol{G}_{5,4}^{(1)} + 5\boldsymbol{r}_{1,0}\boldsymbol{G}_{5,4}.$$

For $s = 7$, we have $n(7) = 10$ and select

$$\boldsymbol{G}_{7,1} \equiv (\boldsymbol{G}_{2,1})^2\boldsymbol{G}_{3,1}, \qquad \boldsymbol{G}_{7,2} \equiv \boldsymbol{G}_{3,1}\boldsymbol{G}_{4,2}, \qquad \boldsymbol{G}_{7,3} \equiv \boldsymbol{G}_{2,1}\boldsymbol{G}_{5,2},$$

$$\boldsymbol{G}_{7,4} \equiv \boldsymbol{G}_{6,4}^{(1)} + 6\boldsymbol{r}_{1,0}\boldsymbol{G}_{6,4}, \quad \boldsymbol{G}_{7,5} \equiv (\boldsymbol{G}_{2,1})^2\boldsymbol{G}_{3,2}, \quad \boldsymbol{G}_{7,6} \equiv \boldsymbol{G}_{3,2}\boldsymbol{G}_{4,2},$$

$$\boldsymbol{G}_{7,7} \equiv \boldsymbol{G}_{3,1}\boldsymbol{G}_{4,3}, \qquad \boldsymbol{G}_{7,8} \equiv \boldsymbol{G}_{3,2}\boldsymbol{G}_{4,3}, \qquad \boldsymbol{G}_{7,9} \equiv \boldsymbol{G}_{2,1}\boldsymbol{G}_{5,4},$$

$$\boldsymbol{G}_{7,10} \equiv \boldsymbol{G}_{6,8}^{(1)} + 6\boldsymbol{r}_{1,0}\boldsymbol{G}_{6,8}.$$

For $s = 8$, we have $n(8) = 17$ and select

$$\boldsymbol{G}_{8,1} \equiv (\boldsymbol{G}_{2,1})^4, \qquad \boldsymbol{G}_{8,2} \equiv \boldsymbol{G}_{2,1}(\boldsymbol{G}_{3,1})^2, \qquad \boldsymbol{G}_{8,3} \equiv (\boldsymbol{G}_{2,1})^2\boldsymbol{G}_{4,2},$$

$$\boldsymbol{G}_{8,4} \equiv (\boldsymbol{G}_{4,2})^2, \qquad \boldsymbol{G}_{8,5} \equiv \boldsymbol{G}_{3,1}\boldsymbol{G}_{5,2}, \qquad \boldsymbol{G}_{8,6} \equiv \boldsymbol{G}_{2,1}\boldsymbol{G}_{6,4},$$

$$\boldsymbol{G}_{8,7} \equiv \boldsymbol{G}_{7,4}^{(1)} + 7\boldsymbol{r}_{1,0}\boldsymbol{G}_{7,4}, \quad \boldsymbol{G}_{8,8} \equiv \boldsymbol{G}_{2,1}\boldsymbol{G}_{3,1}\boldsymbol{G}_{3,2}, \quad \boldsymbol{G}_{8,9} \equiv \boldsymbol{G}_{3,2}\boldsymbol{G}_{5,2},$$

$$\boldsymbol{G}_{8,10} \equiv \boldsymbol{G}_{2,1}(\boldsymbol{G}_{3,2})^2, \quad \boldsymbol{G}_{8,11} \equiv (\boldsymbol{G}_{2,1})^2\boldsymbol{G}_{4,3}, \quad \boldsymbol{G}_{8,12} \equiv \boldsymbol{G}_{4,2}\boldsymbol{G}_{4,3},$$

$$\boldsymbol{G}_{8,13} \equiv (\boldsymbol{G}_{4,3})^2, \qquad \boldsymbol{G}_{8,14} \equiv \boldsymbol{G}_{3,1}\boldsymbol{G}_{5,4}, \qquad \boldsymbol{G}_{8,15} \equiv \boldsymbol{G}_{3,2}\boldsymbol{G}_{5,4},$$

$$\boldsymbol{G}_{8,16} \equiv \boldsymbol{G}_{2,1}\boldsymbol{G}_{6,8}, \qquad \boldsymbol{G}_{8,17} \equiv \boldsymbol{G}_{7,10}^{(1)} + 7\boldsymbol{r}_{1,0}\boldsymbol{G}_{7,10}.$$

For $s = 9$, we have $n(9) = 24$ and select

$$\boldsymbol{G}_{9,1} \equiv \boldsymbol{G}_{2,1}^3\boldsymbol{G}_{3,1}, \qquad \boldsymbol{G}_{9,2} \equiv \boldsymbol{G}_{3,1}^3, \qquad \boldsymbol{G}_{9,3} \equiv \boldsymbol{G}_{2,1}\boldsymbol{G}_{3,1}\boldsymbol{G}_{4,2},$$

$$\boldsymbol{G}_{9,4} \equiv (\boldsymbol{G}_{2,1})^2\boldsymbol{G}_{5,2}, \qquad \boldsymbol{G}_{9,5} \equiv \boldsymbol{G}_{4,2}\boldsymbol{G}_{5,2}, \qquad \boldsymbol{G}_{9,6} \equiv \boldsymbol{G}_{3,1}\boldsymbol{G}_{6,4},$$

$$\boldsymbol{G}_{9,7} \equiv \boldsymbol{G}_{2,1}\boldsymbol{G}_{7,4}, \quad \boldsymbol{G}_{9,8} \equiv \boldsymbol{G}_{8,7}^{(1)} + 8\boldsymbol{r}_{1,0}\boldsymbol{G}_{8,7}, \quad \boldsymbol{G}_{9,9} \equiv (\boldsymbol{G}_{2,1})^3\boldsymbol{G}_{3,2},$$

$$\boldsymbol{G}_{9,10} \equiv (\boldsymbol{G}_{3,1})^2\boldsymbol{G}_{3,2}, \quad \boldsymbol{G}_{9,11} \equiv \boldsymbol{G}_{2,1}\boldsymbol{G}_{3,2}\boldsymbol{G}_{4,2}, \quad \boldsymbol{G}_{9,12} \equiv \boldsymbol{G}_{3,2}\boldsymbol{G}_{6,4},$$

$$\boldsymbol{G}_{9,13} \equiv \boldsymbol{G}_{3,1}(\boldsymbol{G}_{3,2})^2, \quad \boldsymbol{G}_{9,14} \equiv (\boldsymbol{G}_{3,2})^3, \quad \boldsymbol{G}_{9,15} \equiv \boldsymbol{G}_{2,1}\boldsymbol{G}_{3,1}\boldsymbol{G}_{4,3},$$

$$\boldsymbol{G}_{9,16} \equiv \boldsymbol{G}_{4,3}\boldsymbol{G}_{5,2}, \quad \boldsymbol{G}_{9,17} \equiv \boldsymbol{G}_{2,1}\boldsymbol{G}_{3,2}\boldsymbol{G}_{4,3}, \quad \boldsymbol{G}_{9,18} \equiv (\boldsymbol{G}_{2,1})^2\boldsymbol{G}_{5,4},$$

$$\boldsymbol{G}_{9,19} \equiv \boldsymbol{G}_{4,2}\boldsymbol{G}_{5,4}, \qquad \boldsymbol{G}_{9,20} \equiv \boldsymbol{G}_{4,3}\boldsymbol{G}_{5,4}, \qquad \boldsymbol{G}_{9,21} \equiv \boldsymbol{G}_{3,1}\boldsymbol{G}_{6,8},$$

$$\boldsymbol{G}_{9,22} \equiv \boldsymbol{G}_{3,2}\boldsymbol{G}_{6,8}, \quad \boldsymbol{G}_{9,23} \equiv \boldsymbol{G}_{2,1}\boldsymbol{G}_{7,10}, \quad \boldsymbol{G}_{9,24} \equiv \boldsymbol{G}_{8,17}^{(1)} + 8\boldsymbol{r}_{1,0}\boldsymbol{G}_{8,17}.$$

Theorem 28.1 shows that: for $2 \leq s \leq 9$ and $1 \leq i \leq n(s)$, $\boldsymbol{G}_{s,i}$ is a relative invariant of weight s for the equations (28.1).

The preceding evidence suggests that the relative invariants for the equations (28.1) are the polynomials of the set S_3 defined in terms of S_1 and S_2 as follows. Let S_1 denote the set consisting of $\boldsymbol{R}_{1,1}$ and $\boldsymbol{R}_{2,1}$ along with all the relative invariants obtainable from either $\boldsymbol{R}_{1,1}$ or $\boldsymbol{R}_{2,1}$ by repetitions of the construction (28.17); let S_2 denote the set consisting of the elements in S_1 along with the products of any two or more elements in S_1; and let S_3 consist of the nonconstant isobaric linear combinations over $\mathbb{Q}$ of the polynomials in S_2. Additional relative invariants are not obtained by applying (28.17) to elements in S_2. Namely, suppose that $\boldsymbol{I}$ is a relative invariant in S_2 of weight s. Then, there are relative invariants $\boldsymbol{P}_{p_1}, \boldsymbol{P}_{p_2}, \ldots, \boldsymbol{P}_{p_k}$ in S_1 of respective weights $p_1, p_2, \ldots, p_k$ such that $\boldsymbol{I} \equiv \boldsymbol{P}_{p_1} \boldsymbol{P}_{p_2} \cdots \boldsymbol{P}_{p_k}$, $s = p_1 + p_2 + \cdots + p_k$, and

$$\boldsymbol{I}^{(1)} + s\, \boldsymbol{r}_{1,0}\, \boldsymbol{I} \equiv \sum_{i=1}^{k} \left[\boldsymbol{P}_{p_i}^{(1)} + p_i\, \boldsymbol{r}_{1,0}\, \boldsymbol{P}_{p_i}\right] \left[\prod_{1 \leq j \leq k,\, j \neq i} \boldsymbol{P}_{p_j}\right].$$

This shows that $\boldsymbol{I}^{(1)} + s\, \boldsymbol{r}_{1,0}\, \boldsymbol{I}$ is a sum of elements in S_2.

Theorem 28.7 on page 332 was motivated by these observations.

28.4. Computational procedure for Section 28.3

Provided that we write $c_{i,j}(z)$ for $a_{i,j}(z)$ in (28.1) and $w_{i,j}^{(k)}$ for $x_{i,j}^{(k)}$ in (28.4), five alterations in the MATHEMATICA input of Section 16.4 on pages 172–177 yield a program that verifies the suitability for (28.20) of the $G_{s,i}$ on page 329 when $2 \leq s \leq 9$. First, in the formula for u[s_,p_,q_,r_] on page 172, replace {j,q+1,2} with {j,q+1,1} at its single location. Second, replace {j,Max[1,i],2} with {j,(i-1)^2,1} at each of its 14 occurrences. Third, delete the input line of page 173 containing c[j_,i_][z_] := c[i,j][z] /; j > i at its single location. Fourth, replace the command that defines cOneStar[i_,j_][z_] with the input

    cOneStar[0,1][z_] := c[0,1][z] + rho'[z]/rho[z]

    cOneStar[1,0][z_] := c[1,0][z] + 2*rho'[z]/rho[z]

    cOneStar[1,1][z_] := (c[1,1][z] + 2*c[0,1][z]*rho'[z]/rho[z]
                        + 2*c[1,0][z]*rho'[z]/rho[z]
                        + 2*(rho'[z]/rho[z])^2 + rho''[z]/rho[z])

    cOneStar[2,1][z_] := (c[2,1][z] + c[1,1][z]*rho'[z]/rho[z]
                        + c[1,0][z]*(rho'[z]/rho[z])^2 + c[0,1][z]*rho''[z]/rho[z]
                        + rho'[z]*rho''[z]/rho[z]^2)

that is based on the four equations beginning with (28.5). Fifth, replace the formula on page 173 for cTwoStar[i_,j_][zet_] with the four input statements

    cTwoStar[0,1][zet_] := f'[zet]*c[0,1][f[zet]]

    cTwoStar[1,0][zet_] := f'[zet]*c[1,0][f[zet]] -f''[zet]/f'[zet]

    cTwoStar[1,1][zet_] := (f'[zet]^2*c[1,1][f[zet]]
                        - f''[zet]*c[0,1][f[zet]])

    cTwoStar[2,1][zet_] := f'[zet]^3*c[2,1][f[zet]]

that are based on (28.7)–(28.10). After these alterations are made, the evaluation of the modified input followed by the evaluation of Do[invariants[s],{s,2,9}] yields integers $n(2), \ldots, n(9)$, as given on page 329, and output labeled Inv[s,i], for $2 \leq s \leq 9$ and $1 \leq i \leq n(s)$. For $2 \leq s \leq 9$, the expressions corresponding to Inv[s,1], ..., Inv[s,n(s)] are $n(s)$ linearly independent relative invariants over $\mathbb{Q}$ of weight s and their linear combinations give all of the relative invariants of weight s for the equations (28.1). Since machine computations show that each of them is a simple linear combination over $\mathbb{Q}$ of the polynomials $G_{s,1}, \ldots, G_{s,n(s)}$ from page 329, the polynomials of page 329 are suitable for (28.20) when $2 \leq s \leq 9$.

28.5. Laguerre-Forsyth reductions for the equations (28.1)

For any given (28.1) on Ω, there is a meromorphic solution $\rho(z) \not\equiv 0$ of

$$(28.21) \qquad \rho'(z) + a_{0,1}(z)\, \rho(z) = 0$$

on a subregion $\mathcal{U}_1$ of Ω; there is a meromorphic solution $h(z) \not\equiv 0$ of

$$(28.22) \qquad h'(z) + \bigl(a_{1,0}(z) - 2\,a_{0,1}(z)\bigr) h(z) = 0$$

on a subregion $\mathcal{U}_2$ of $\mathcal{U}_1$; and there is a univalent analytic function $g(z)$ on a subregion $\mathcal{U}$ of $\mathcal{U}_2$ such that $g'(z) = h(z)$, for each z in $\mathcal{U}$.

THEOREM 28.6. *For a given* (28.1) *on* Ω, *let* $\rho(z) \not\equiv 0$ *be a meromorphic solution of* (28.21) *on a subregion* $\mathcal{U}$ *of* Ω, *let* $g(z)$ *be a univalent analytic function on* $\mathcal{U}$ *for which* $h(z) \equiv g'(z)$ *is a solution of* (28.22) *on* $\mathcal{U}$, *and let* $z = f(\zeta)$ *on* $\mathcal{V} = g(\mathcal{U})$ *be the inverse function for* $\zeta = g(z)$ *on* $\mathcal{U}$. *Then, the restriction of* (28.1) *to* $\mathcal{U}$ *is transformed by the composite substitution*

$$(28.23) \qquad y(z) = \rho(z)\, v(z), \quad z = f(\zeta), \quad \text{and} \quad w(\zeta) = (v \circ f)(\zeta),$$

into the equation

$$(28.24) \qquad w''(\zeta)\, w'(\zeta) + d_{1,1}(\zeta)\, w'(\zeta)\, w(\zeta) + d_{2,1}(\zeta) \bigl(w(\zeta)\bigr)^2 = 0, \quad \text{on } \mathcal{V},$$

whose coefficients are given by

$$(28.25) \qquad d_{1,1}(\zeta) \equiv \bigl(f'(\zeta)\bigr)^2 R_{1,1}\bigl(f(\zeta)\bigr)$$

and

$$(28.26) \qquad d_{2,1}(\zeta) \equiv \bigl(f'(\zeta)\bigr)^3 R_{2,1}\bigl(f(\zeta)\bigr),$$

where $R_{1,1}(z)$ *and* $R_{2,1}(z)$ *are the meromorphic functions on* Ω *obtained when* $a_{i,j}^{(k)}(z)$ *from* (28.1) *is substituted for* $x_{i,j}^{(k)}$ *in* $R_{1,1}$ *and* $R_{2,1}$ *of* (28.12) *and* (28.13).

PROOF. The composite substitution (28.23) first transforms the restriction of (28.1) to $\mathcal{U}$ into an equation (28.2) on $\mathcal{U}$ and then transforms that equation into

$$(28.27) \quad w''\, w' + d_{0,1}(\zeta)\, w''\, w + d_{1,0}(\zeta)\, (w')^2 + d_{1,1}(\zeta)\, w'\, w + d_{2,1}(\zeta)\, w^2 = 0, \quad \text{on } \mathcal{V}.$$

Formula (28.21) yields

$$(28.28) \qquad \frac{\rho'(z)}{\rho(z)} \equiv -a_{0,1}(z) \quad \text{and} \quad \frac{\rho''(z)}{\rho(z)} \equiv -a_{0,1}^{(1)}(z) + \bigl(a_{0,1}(z)\bigr)^2, \quad \text{on } \mathcal{U}.$$

We use (28.28), (28.5)–(28.6), the formulas for $a_{1,1}^*(z)$ and $a_{2,1}^*(z)$ following (28.6), as well as (28.11)–(28.13) to obtain

(28.29) $\quad a_{0,1}^*(z) \equiv 0,$

(28.30) $\quad a_{1,0}^*(z) \equiv a_{1,0}(z) - 2\, a_{0,1}(z) \equiv r_{1,0}(z),$

(28.31) $\quad a_{1,1}^*(z) \equiv a_{1,1}(z) + 2\, a_{0,1}(z)\bigl(-a_{0,1}(z)\bigr) + 2\, a_{1,0}(z)\bigl(-a_{0,1}(z)\bigr)$
$$+ 2\bigl(-a_{0,1}(z)\bigr)^2 - a_{0,1}^{(1)}(z) + \bigl(a_{0,1}(z)\bigr)^2$$
$$\equiv a_{1,1}(z) - 2\, a_{0,1}(z)\, a_{1,0}(z) - a_{0,1}^{(1)}(z) + \bigl(a_{0,1}(z)\bigr)^2$$
$$\equiv R_{1,1}(z),$$

and

(28.32) $\quad a_{2,1}^*(z) \equiv a_{2,1}(z) + a_{1,1}(z)\bigl(-a_{0,1}(z)\bigr) + a_{1,0}(z)\bigl(-a_{0,1}(z)\bigr)^2$
$$+ \Bigl[-a_{0,1}^{(1)}(z) + \bigl(a_{0,1}(z)\bigr)^2\Bigr]\bigl[a_{0,1}(z) + \bigl(-a_{0,1}(z)\bigr)\bigr]$$
$$\equiv a_{2,1}(z) - a_{0,1}(z)\, a_{1,1}(z) + \bigl(a_{0,1}(z)\bigr)^2 a_{1,0}(z)$$
$$\equiv R_{2,1}(z).$$

Now, we apply (28.7)–(28.10), (28.29)–(28.32), and (28.22) to deduce that
$$d_{0,1}(\zeta) \equiv \bigl(f'(\zeta)\bigr)\, a_{0,1}^*\bigl(f(\zeta)\bigr) \equiv 0,$$
$$d_{1,0}(\zeta) \equiv f'(\zeta)\, a_{1,0}^*\bigl(f(\zeta)\bigr) - \frac{f''(\zeta)}{f'(\zeta)} \equiv f'(\zeta)\left[r_{1,0}\bigl(f(\zeta)\bigr) - \frac{f''(\zeta)}{\bigl(f'(\zeta)\bigr)^2}\right]$$
$$\equiv f'(\zeta)\left[r_{1,0}\bigl(f(\zeta)\bigr) + \frac{g''\bigl(f(\zeta)\bigr)}{g'\bigl(f(\zeta)\bigr)}\right] \equiv f'(\zeta)\left[r_{1,0}\bigl(f(\zeta)\bigr) + \frac{h'\bigl(f(\zeta)\bigr)}{h\bigl(f(\zeta)\bigr)}\right] \equiv 0,$$
$$d_{1,1}(\zeta) \equiv \bigl(f'(\zeta)\bigr)^2 a_{1,1}^*\bigl(f(\zeta)\bigr) - f''(\zeta)\, a_{0,1}^*\bigl(f(\zeta)\bigr) \equiv \bigl(f'(\zeta)\bigr)^2 R_{1,1}\bigl(f(\zeta)\bigr),$$

and
$$d_{2,1}(\zeta) \equiv \bigl(f'(\zeta)\bigr)^3 a_{2,1}^*\bigl(f(\zeta)\bigr) \equiv \bigl(f'(\zeta)\bigr)^3 R_{2,1}\bigl(f(\zeta)\bigr).$$

Thus, (28.27) is given by (28.24)–(28.26). This completes the proof. □

28.6. All of the relative invariants for the equations (28.1)

We use (28.11)–(28.13) to define $S_2, S_3, \ldots$ and $T_3, T_4, \ldots$ in $\mathfrak{R}_{2,1}$ through

(28.33) $\quad S_2 \equiv R_{1,1} \quad \text{and} \quad S_{k+1} \equiv S_k^{(1)} + k\, r_{1,0}\, S_k, \quad \text{for } k \geq 2$

as well as

(28.34) $\quad T_3 \equiv R_{2,1} \quad \text{and} \quad T_{k+1} \equiv T_k^{(1)} + k\, r_{1,0}\, T_k, \quad \text{for } k \geq 3.$

THEOREM 28.7. *A polynomial I in $\mathfrak{R}_{2,1}$ is a relative invariant for the equations (28.1) if and only if I is a nonconstant isobaric polynomial that is expressible as a polynomial combination over $\mathbb{Q}$ of $S_2, S_3, \ldots$ and $T_3, T_4, \ldots$.*

PROOF. Theorem 28.1 shows that: for $k \geq 2$, S_k is a relative invariant of weight k for the equations (28.1); and, for $k \geq 3$, T_k is a relative invariant of weight k for the equations (28.1). Thus, when I is a nonconstant isobaric polynomial in

$\mathfrak{R}_{2,1}$ that is expressible as a polynomial combination over $\mathbb{Q}$ of S_2, S_3, ... and T_3, T_4, ..., it is clear that I is a relative invariant for the equations (28.1).

Suppose that I in $\mathfrak{R}_{2,1}$ is a relative invariant of weight s for the equations (28.1). By altering the notation in the argument for Corollary 20.5 on page 222 or in the proof of [**24**, Theorem A.10, page 140], we deduce that I is a nonconstant isobaric polynomial in $\mathfrak{R}_{2,1}$. We use (28.33), (28.12), and (28.11) to obtain

(28.35) $$S_{k+2} \equiv x_{1,1}^{(k)} + \boldsymbol{\Phi}_k, \quad \text{for } k \geq 0,$$

where $\boldsymbol{\Phi}_k$ is a differential-polynomial combination over $\mathbb{Q}$ of $x_{1,1}$, $x_{1,0}$, $x_{0,1}$ that does not involve $x_{1,1}^{(j)}$, for $j \geq k$. We see that (28.34), (28.13), and (28.11) yield

(28.36) $$T_{k+3} \equiv x_{2,1}^{(k)} + \boldsymbol{\Psi}_k, \quad \text{for } k \geq 0,$$

where $\boldsymbol{\Psi}_k$ is a differential-polynomial combination over $\mathbb{Q}$ of $x_{2,1}$, $x_{1,1}$, $x_{1,0}$, $x_{0,1}$ that does not involve $x_{2,1}^{(j)}$, for $j \geq k$. If I effectively involves $x_{2,1}^{(\mu)}$, for some $\mu \geq 0$, and if μ_0 is the largest such μ, then we use the rewriting of (28.36) as

$$x_{2,1}^{(k)} \equiv T_{k+3} - \boldsymbol{\Psi}_k, \quad \text{for } k \geq 0,$$

to successively eliminate $x_{2,1}^{(\mu_0)}$, $x_{2,1}^{(\mu_0-1)}$, ..., $x_{2,1}^{(1)}$, $x_{2,1}^{(0)}$ from I. Therefore, we can write I as a polynomial combination of T_3, ..., T_{μ_0+3} for which the coefficients are differential-polynomial combinations over $\mathbb{Q}$ of $x_{1,1}$, $x_{1,0}$, $x_{0,1}$. If this expression for I effectively involves $x_{1,1}^{(\nu)}$, for some $\nu \geq 0$, and if ν_0 is the largest such ν, then we use the modification of (28.35) as

$$x_{1,1}^{(k)} \equiv S_{k+2} - \boldsymbol{\Phi}_k, \quad \text{for } k \geq 0,$$

to successively eliminate $x_{1,1}^{(\nu_0)}$, $x_{1,1}^{(\nu_0-1)}$, ..., $x_{1,1}^{(1)}$, $x_{1,1}^{(0)}$ from I. Therefore, we can express I as a polynomial combination of S_2, S_3, ... and T_3, T_4, ... for which the coefficients are differential-polynomial combinations over $\mathbb{Q}$ of $x_{1,0}$ and $x_{0,1}$. Consequently, we can write I in the form

(28.37) $$I \equiv I_1 + I_2,$$

where I_1 is a polynomial combination over $\mathbb{Q}$ of S_2, S_3, ... and T_3, T_4, ... while I_2 is a polynomial combination over $\mathbb{Q}$ of S_2, S_3, ..., T_3, T_4, ..., $x_{1,0}^{(0)}$, $x_{1,0}^{(1)}$, ..., and $x_{0,1}^{(0)}$, $x_{0,1}^{(1)}$, ... such that each term of I_2 has at least one factor of the form $x_{1,0}^{(j)}$, for some $j \geq 0$, or $x_{0,1}^{(j)}$, for some $j \geq 0$. For any meromorphic functions $a_{0,1}(z)$, $a_{1,0}(z)$, $a_{1,1}(z)$ and $a_{2,1}(z)$ on a region Ω and the corresponding (28.1) that they specify, let (28.23) be a composite substitution that transforms this (28.1), when restricted to some subregion $\mathcal{U}$ of Ω, into an equation (28.24) on a region $\mathcal{V}$. Let $I(z)$, $I_1(z)$, and $I_2(z)$ denote the functions on Ω obtained from I, I_1, and I_2 by replacing $x_{i,j}^{(k)}$ with $a_{i,j}^{(k)}(z)$ from this (28.1). Let $I^{\sharp}(\zeta)$, $I_1^{\sharp}(\zeta)$, and $I_2^{\sharp}(\zeta)$ denote the functions on $\mathcal{V}$ obtained by replacing $x_{i,j}^{(k)}$ in I, I_1, and I_2 with $d_{i,j}^{(k)}(\zeta)$ from the corresponding (28.24). Due to $d_{0,1}(\zeta) \equiv d_{1,0}(\zeta) \equiv 0$, we have $I_2(\zeta) \equiv 0$, on $\mathcal{V}$. If not identically zero, I_1 is an isobaric polynomial of weight s in $\mathfrak{R}_{2,1}$ that is a polynomial combination over $\mathbb{Q}$ of relative invariants. Consequently, we have $I_1^{\sharp}(\zeta) \equiv (f'(\zeta))^s I_1(f(\zeta))$, on $\mathcal{V}$. Since $\mathcal{I}$ is a relative invariant of weight s, we find

that
$$\begin{aligned}(f'(\zeta))^s I_2(f(\zeta)) &\equiv (f'(\zeta))^s I(f(\zeta)) - (f'(\zeta))^s I_1(f(\zeta))\\ &\equiv I^\sharp(\zeta) - I_1^\sharp(\zeta) \equiv I_2^\sharp(\zeta) \equiv 0, \quad \text{on } \mathcal{V}.\end{aligned}$$

Thus, we obtain $I_2(z) \equiv 0$ on $\mathcal{U}$ and therefore also on Ω. Due to the validity of $I_2(z) \equiv 0$ for any selection of the meromorphic functions $a_{i,j}(z)$, an argument similar to that for Corollary 4.3 on page 37 yields $\boldsymbol{I_2} \equiv 0$. Hence, (28.37) gives $\boldsymbol{I} \equiv \boldsymbol{I_1}$. This shows that $\boldsymbol{I}$ has the asserted structure and completes the proof. □

CHAPTER 29

Formulations of Greater Generality

The classes of equations throughout Chapters 1–28 correspond to the monic monomials $\left(y^{(2)}(z)\right)^2$, $\left(y^{(m)}(z)\right)^2$, $\left(y^{(m)}(z)\right)^n$, and $y^{(2)}(z)\,y^{(1)}(z)$. Other classes characterized by monic monomials are presented in Section 29.1. In each of these situations, the equations involve homogeneous polynomial combinations of $y(z)$ and its derivatives. We show in Section 29.2 that the study of relative invariants can be extended to various classes of nonhomogeneous equations.

29.1. Equations characterized by a single monic term

For any fixed integers $m_1, m_2, \ldots, m_n$ satisfying $m_1 \geq m_2 \geq \cdots \geq m_n \geq 1$, the monic monomial $y^{(m_1)}\,y^{(m_2)} \cdots y^{(m_n)}$ specifies equations having the form

$$(29.1) \quad y^{(m_1)}\,y^{(m_2)} \cdots y^{(m_n)} + \sum_{\substack{0 \leq j_1 \leq m_1,\, 0 \leq j_2 \leq m_2,\, \ldots,\, 0 \leq j_n \leq m_n \\ (j_1, j_2, \ldots, j_n) \neq (m_1, m_2, \ldots, m_n) \\ j_1 \geq j_2 \geq \cdots \geq j_n}} a_{m_1 - j_1,\, m_2 - j_2,\, \ldots,\, m_n - j_n}(z)\, y^{(j_1)}\, y^{(j_2)} \cdots y^{(j_n)} = 0$$

in terms of meromorphic functions $a_{i_1, i_2, \ldots, i_n}(z)$ on a region Ω of the complex plane. For any meromorphic function $\rho(z) \not\equiv 0$ on Ω, there are unique meromorphic functions $a^*_{i_1, i_2, \ldots, i_n}(z)$ on Ω such that the change $y(z) = \rho(z)\,v(z)$ of the dependent variable from y to v transforms (29.1) into

$$(29.2) \quad v^{(m_1)}\,v^{(m_2)} \cdots v^{(m_n)} + \sum_{\substack{0 \leq j_1 \leq m_1,\, 0 \leq j_2 \leq m_2,\, \ldots,\, 0 \leq j_n \leq m_n \\ (j_1, j_2, \ldots, j_n) \neq (m_1, m_2, \ldots, m_n) \\ j_1 \geq j_2 \geq \cdots \geq j_n}} a^*_{m_1 - j_1,\, m_2 - j_2,\, \ldots,\, m_n - j_n}(z)\, v^{(j_1)}\, v^{(j_2)} \cdots v^{(j_n)} = 0,$$

on Ω. Also, for any univalent analytic function $\zeta = g(z)$ on Ω whose inverse is designated by $z = f(\zeta)$ on $\Omega^{**} = g(\Omega)$, there are unique meromorphic functions $a^{**}_{i_1, i_2, \ldots, i_n}(\zeta)$ on Ω^{**} such that the change of the independent variable from z to ζ by means of $z = f(\zeta)$ and $u(\zeta) = (y \circ f)(\zeta)$ transform (29.1) on Ω into

$$(29.3) \quad u^{(m_1)}\,u^{(m_2)} \cdots u^{(m_n)} + \sum_{\substack{0 \leq j_1 \leq m_1,\, 0 \leq j_2 \leq m_2,\, \ldots,\, 0 \leq j_n \leq m_n \\ (j_1, j_2, \ldots, j_n) \neq (m_1, m_2, \ldots, m_n) \\ j_1 \geq j_2 \geq \cdots \geq j_n}} a^{**}_{m_1 - j_1,\, m_2 - j_2,\, \ldots,\, m_n - j_n}(\zeta)\, u^{(j_1)}\, u^{(j_2)} \cdots u^{(j_n)} = 0,$$

on Ω^{**}. We assume that the symbols

$$(29.4) \quad x^{(k)}_{m_1 - j_1,\, m_2 - j_2,\, \ldots,\, m_n - j_n}, \quad \text{for } 0 \leq j_1 \leq m_1,\, 0 \leq j_2 \leq m_2,\, \ldots,\, 0 \leq j_n \leq m_n,$$
$$(j_1, j_2, \ldots, j_n) \neq (m_1, m_2, \ldots, m_n),\; j_1 \geq j_2 \geq \cdots \geq j_n,\; \text{and } k \geq 0,$$

are algebraically independent variables over the field $\mathbb{Q}$ of rational numbers and we let $\mathfrak{R}_{m_1, \ldots, m_n}$ denote the ring of polynomials over $\mathbb{Q}$ in these variables. We introduce $'$ as the unique derivation for $\mathfrak{R}_{m_1, \ldots, m_n}$ such that

$$\left(x^{(k)}_{i_1, i_2, \ldots, i_n}\right)' = x^{(k+1)}_{i_1, i_2, \ldots, i_n}, \quad \text{for each } x^{(k)}_{i_1, i_2, \ldots, i_n} \text{ in (29.4)}.$$

The constants of $\mathfrak{R}_{m_1,\ldots,m_n}$ are the elements of $\mathbb{Q}$.

Given any polynomial $\boldsymbol{P}$ in $\mathfrak{R}_{m_1,\ldots,m_n}$, we let $P(z)$ denote the unique function on Ω obtained by replacing $\boldsymbol{x}^{(k)}_{i_1,i_2,\ldots,i_n}$ in $\boldsymbol{P}$ with $a^{(k)}_{i_1,i_2,\ldots,i_n}(z)$ from (29.1); we let $P^*(z)$ denote the unique function on Ω obtained by replacing $\boldsymbol{x}^{(k)}_{i_1,i_2,\ldots,i_n}$ in $\boldsymbol{P}$ with $a^{*(k)}_{i_1,i_2,\ldots,i_n}(z)$ from (29.2); and we let $P^{**}(\zeta)$ denote the unique function on Ω^{**} obtained replacing $\boldsymbol{x}^{(k)}_{i_1,i_2,\ldots,i_n}$ in $\boldsymbol{P}$ with $a^{**(k)}_{i_1,i_2,\ldots,i_n}(\zeta)$ from (29.3).

DEFINITION 29.1. *A polynomial $\boldsymbol{P}$ in $\mathfrak{R}_{m_1,\ldots,m_n}$ not in $\mathbb{Q}$ is a semi-invariant of the first kind for the equations (29.1) when*

(29.5) $$P^*(z) \equiv P(z),$$

for each (29.1) and each change $y(z) = \rho(z)\,v(z)$ of the dependent variable.

DEFINITION 29.2. *A polynomial $\boldsymbol{P}$ in $\mathfrak{R}_{m_1,\ldots,m_n}$ not in $\mathbb{Q}$ is a semi-invariant of the second kind for the equations (29.1) when there is an integer s such that*

(29.6) $$P^{**}(\zeta) \equiv \bigl(f'(\zeta)\bigr)^s P\bigl(f(\zeta)\bigr),$$

for each (29.1) and each change $z = f(\zeta)$ of the independent variable.

DEFINITION 29.3. *A polynomial $\boldsymbol{P}$ in $\mathfrak{R}_{m_1,\ldots,m_n}$ not in $\mathbb{Q}$ is a relative invariant for the equations (29.1) when it is both a semi-invariant of the first kind and a semi-invariant of the second kind for such equations.*

DEFINITION 29.4. *The weight of a nonzero element in $\mathbb{Q}$ is 0; the weight of $\boldsymbol{x}^{(k)}_{i_1,i_2,\ldots,i_n}$ in (29.4) is the positive integer $i_1 + i_2 + \cdots + i_n + k$; and the weight of a nonzero monomial in $\mathfrak{R}_{m_1,\ldots,m_n}$ is the sum of the weights of its factors. A nonzero $\boldsymbol{P}$ in $\mathfrak{R}_{m_1,\ldots,m_n}$ is isobaric of weight s when each of its nonzero terms has weight s.*

By modifying the argument of pages 220–222 for Corollary 20.5, we see that: *if $\boldsymbol{P}$ is a semi-invariant of the second kind for the equations (29.1), then there is a unique integer s with which $\boldsymbol{P}$ satisfies (29.6) and $\boldsymbol{P}$ is isobaric of weight s.*

EXAMPLE 29.5. For $n = 1$ and $m_1 = m$, the equations (29.1) are the monic homogeneous linear differential equations of order m.

EXAMPLE 29.6. For $n = 2$, $m_1 = 2$, and $m_2 = 1$, (29.1) reduces to (28.1) of page 325. In that situation, the relative invariants for (29.1) are the relative invariants for (28.1). They are given by Theorem 28.7 on page 332.

EXAMPLE 29.7. For $n = 2$ and $m_1 = m_2 = m$, (29.1) can be rewritten as (1.1) on page 4 with $c_{i,i}(z) \equiv a_{i,i}(z)$, for $1 \leq i \leq m$, and $c_{j,i}(z) \equiv c_{i,j}(z) \equiv a_{i,j}(z)/2$, for $0 \leq i < j \leq m$. The relative invariants for these particular equations (29.1) are obtained from the relative invariants for the equations (1.1) by replacing $\boldsymbol{w}^{(k)}_{i,i}$ with $\boldsymbol{x}^{(k)}_{i,i}$, for $1 \leq i \leq m$, and replacing $\boldsymbol{w}^{(k)}_{i,j}$ with $\boldsymbol{x}^{(k)}_{i,j}/2$, for $0 \leq i < j \leq m$.

EXAMPLE 29.8. For $n \geq 1$ and $m_\nu = m$, for $1 \leq \nu \leq n$, (29.1) can be rewritten in the form of (21.1) on page 249. Thus, for the variables $\boldsymbol{w}^{(k)}_{i_1,i_2,\ldots,i_n}$ of (21.6) on page 251, there are positive integers $\kappa_{i_1,i_2,\ldots,i_n}$ such that the relative invariants for these particular equations (29.1) are obtained from the relative invariants for the equations (21.1) by replacing $\boldsymbol{w}^{(k)}_{i_1,i_2,\ldots,i_n}$ with $\boldsymbol{x}^{(k)}_{i_1,i_2,\ldots,i_n}/\kappa_{i_1,i_2,\ldots,i_n}$.

Throughout Parts 1–4, our results depended on having simple formulas for the coefficients of the various transformed equations. For this reason, we developed the notation used with (1.1) on page 4 in place of that which Example 29.7 provides. Similarly, the notation used for (21.1) on page 249 yields simpler formulas than can be obtained by directly applying the context of Example 29.8.

29.2. Relative invariants for some nonhomogeneous equations

For $r \geq 1$, let $n, n_1, \ldots, n_r$ be distinct positive integers; let $m_1, m_2, \ldots, m_n$ be positive integers subject to $m_1 \geq m_2 \geq \cdots \geq m_n$; and, for $1 \leq \nu \leq r$, let $m_{\nu,1}, m_{\nu,2}, \ldots, m_{\nu,n_\nu}$ be fixed integers such that $m_{\nu,1} \geq m_{\nu,2} \geq \cdots \geq m_{\nu,n_\nu} \geq 0$. We consider differential equations having the form

$$(29.7) \qquad M_n + \sum_{\nu=1}^{r} N_{n_\nu} = 0,$$

where M_n is the left member of (29.1) on Ω given by

$$(29.8) \qquad M_n \equiv y^{(m_1)} \cdots y^{(m_n)} + \sum_{\substack{0 \leq j_1 \leq m_1, \ldots, 0 \leq j_n \leq m_n \\ (j_1, \ldots, j_n) \neq (m_1, \ldots, m_n) \\ j_1 \geq \cdots \geq j_n}} a_{m_1-j_1, \ldots, m_n-j_n}(z) \, y^{(j_1)} \cdots y^{(j_n)}$$

and $N_{n_1}, \ldots, N_{n_r}$ are defined by

$$(29.9) \qquad N_{n_\nu} \equiv \sum_{\substack{0 \leq j_1 \leq m_{\nu,1}, 0 \leq j_2 \leq m_{\nu,2}, \ldots, 0 \leq j_{n_\nu} \leq m_{\nu,n_\nu} \\ j_1 \geq j_2 \geq \cdots \geq j_{n_\nu}}} b_{n_\nu;\, m_{\nu,1}-j_1,\, m_{\nu,2}-j_2,\, \ldots,\, m_{\nu,n_\nu}-j_{n_\nu}}(z) \, y^{(j_1)} y^{(j_2)} \cdots y^{(j_{n_\nu})},$$

for $1 \leq \nu \leq r$,

in terms of given meromorphic functions $b_{n_\nu;\, i_1, i_2, \ldots, i_{n_\nu}}(z)$ on Ω. We note that the coefficient of $y^{(m_1)} y^{(m_2)} \cdots y^{(m_n)}$ in (29.7) and (29.8) is normalized to be 1.

PROPOSITION 29.9. *For any meromorphic function $\rho(z) \not\equiv 0$ on Ω, there are unique meromorphic functions $a^*_{i_1, i_2, \ldots, i_n}(z)$, $b^*_{n_\nu;\, i_1, i_2, \ldots, i_{n_\nu}}(z)$ on Ω such that the change $y(z) = \rho(z)\, v(z)$ of the dependent variable transforms (29.7) into*

$$(29.10) \qquad M_n^* + \sum_{\nu=1}^{r} N_{n_\nu}^* = 0, \quad \text{on } \Omega,$$

where M_n^* is the left member of (29.2) given as

$$(29.11) \qquad M_n^* \equiv v^{(m_1)} \cdots v^{(m_n)} + \sum_{\substack{0 \leq j_1 \leq m_1, \ldots, 0 \leq j_n \leq m_n \\ (j_1, \ldots, j_n) \neq (m_1, \ldots, m_n) \\ j_1 \geq \cdots \geq j_n}} a^*_{m_1-j_1, \ldots, m_n-j_n}(z) \, v^{(j_1)} \cdots v^{(j_n)}$$

and $N^*_{n_1}, \ldots, N^*_{n_r}$ are specified by

$$(29.12) \qquad N_{n_\nu}^* \equiv \sum_{\substack{0 \leq j_1 \leq m_{\nu,1}, 0 \leq j_2 \leq m_{\nu,2}, \ldots, 0 \leq j_{n_\nu} \leq m_{\nu,n_\nu} \\ j_1 \geq j_2 \geq \cdots \geq j_{n_\nu}}} b^*_{n_\nu;\, m_{\nu,1}-j_1,\, m_{\nu,2}-j_2,\, \ldots,\, m_{\nu,n_\nu}-j_{n_\nu}}(z) \, v^{(j_1)} v^{(j_2)} \cdots v^{(j_{n_\nu})},$$

for $1 \leq \nu \leq r$.

Moreover, if $\rho(z) \equiv 2$, then

$$(29.13) \qquad a^*_{i_1, \ldots, i_n}(z) \equiv a_{i_1, \ldots, i_n}(z) \quad \text{and} \quad b^*_{n_\nu;\, i_1, \ldots, i_{n_\nu}}(z) \equiv 2^{n_\nu - n}\, b_{n_\nu;\, i_1, \ldots, i_{n_\nu}}(z).$$

PROOF. We use (29.8) and the identity

$$(29.14) \qquad y^{(k)}(z) \equiv \rho(z) \sum_{\mu=0}^{k} \binom{k}{\mu} \frac{\rho^{(k-\mu)}(z)}{\rho(z)} v^{(\mu)}(z), \quad \text{for } k \geq 0,$$

to see that there are unique meromorphic functions $a^*_{i_1, i_2, \ldots, i_n}(z)$ on Ω for (29.11) such that the substitution of $y(z) = \rho(z) v(z)$ in M_n of (29.8) yields

$$(29.15) \qquad \left(M_n\right)_{y(z)=\rho(z)\,v(z)} \equiv \left(\rho(z)\right)^n M_n^*.$$

Thus, the differential equation $M_n^* = 0$ on Ω is the monic equation (29.2) into which the substitution $y(z) = \rho(z) v(z)$ transforms (29.1). By employing (29.9) and (29.14), we find that there are unique meromorphic functions $d_{n_\nu; i_1, i_2, \ldots, i_{n_\nu}}(z)$ on Ω such that the substitution $y(z) = \rho(z) v(z)$ in N_{n_ν} of (29.9) yields

$$(29.16) \qquad \left(N_{n_\nu}\right)_{y(z)=\rho(z)\,v(z)} \equiv D_{n_\nu}, \quad \text{for } 1 \leq \nu \leq r,$$

where

$$(29.17) \qquad D_{n_\nu} \equiv \sum_{\substack{0 \leq j_1 \leq m_{\nu,1}, \ldots, 0 \leq j_{n_\nu} \leq m_{\nu,n_\nu} \\ j_1 \geq \cdots \geq j_{n_\nu}}} d_{n_\nu; m_{\nu,1}-j_1, \ldots, m_{\nu,n_\nu}-j_{n_\nu}}(z)\, v^{(j_1)} \cdots v^{(j_{n_\nu})}.$$

For each of the coefficients $d_{n_\nu; i_1, i_2, \ldots, i_{n_\nu}}(z)$ in (29.17), we set

$$(29.18) \qquad b^*_{n_\nu; i_1, i_2, \ldots, i_{n_\nu}}(z) \equiv \left(\rho(z)\right)^{-n} d_{n_\nu; i_1, i_2, \ldots, i_{n_\nu}}(z).$$

Then, in terms of the differential polynomials $N^*_{n_1}, \ldots, N^*_{n_r}$ defined by (29.12) and (29.18), we see that (29.17), (29.18), and (29.12) yield

$$(29.19) \qquad D_{n_\nu} \equiv \left(\rho(z)\right)^n N^*_{n_\nu}, \quad \text{for } 1 \leq \nu \leq r.$$

We apply (29.15), (29.16), and (29.19) to deduce that

$$(29.20) \qquad \left[M_n + \sum_{\nu=1}^{r} N_{n_\nu}\right]_{y(z)=\rho(z)\,v(z)} \equiv \left(\rho(z)\right)^n \left[M_n^* + \sum_{\nu=1}^{r} N^*_{n_\nu}\right].$$

The coefficient of $v^{(m_1)} \cdots v^{(m_n)}$ is 1 in both M_n^* of (29.11) and in the left member of (29.10). Consequently, (29.20) shows that the substitution $y(z) = \rho(z) v(z)$ transforms (29.7) into (29.10). Since any two of $M_n^*, N^*_{n_1}, \ldots, N^*_{n_r}$ have no nonzero terms in common, the coefficients $a^*_{i_1, i_2, \ldots, i_n}(z)$ and $b^*_{n_\nu; i_1, i_2, \ldots, i_{n_\nu}}(z)$ for (29.10) are unique.

Suppose that $\rho(z) \equiv 2$. Then, we have $y^{(k)}(z) \equiv 2 v^{(k)}(z)$, for $k \geq 0$. In this situation, (29.15) is valid with

$$M_n^* \equiv v^{(m_1)} \cdots v^{(m_n)} + \sum_{\substack{0 \leq j_1 \leq m_1, \ldots, 0 \leq j_n \leq m_n \\ (j_1, \ldots, j_n) \neq (m_1, \ldots, m_n) \\ j_1 \geq \cdots \geq j_n}} a_{m_1-j_1, \ldots, m_n-j_n}(z)\, v^{(j_1)} \cdots v^{(j_n)}.$$

Thus, we have $a^*_{m_1-j_1, \ldots, m_n-j_n}(z) \equiv a_{m_1-j_1, \ldots, m_n-j_n}(z)$. Also, (29.9), (29.16), and (29.17) show that $d_{n_\nu; i_1, i_2, \ldots, i_{n_\nu}}(z)$ for (29.17) is given by

$$(29.21) \qquad d_{n_\nu; i_1, i_2, \ldots, i_{n_\nu}}(z) \equiv 2^{n_\nu} b_{n_\nu; i_1, i_2, \ldots, i_{n_\nu}}(z).$$

In view of (29.18) with $\rho(z) \equiv 2$, this yields the second formula in (29.13) and completes the proof. $\square$

29.2. RELATIVE INVARIANTS FOR SOME NONHOMOGENEOUS EQUATIONS   339

PROPOSITION 29.10. *For any univalent analytic function $\zeta = g(z)$ on Ω with inverse designated by $z = f(\zeta)$ on $\Omega^{**} = g(\Omega)$, there are unique meromorphic functions $a^{**}_{i_1, i_2, ..., i_n}(\zeta)$, $b^{**}_{n_\nu; i_1, i_2, ..., i_{n_\nu}}(\zeta)$ on Ω^{**} such that the change $z = f(\zeta)$ of the independent variable from z to ζ, with $u(\zeta) = (y \circ f)(\zeta)$, transforms (29.7) into*

$$(29.22) \qquad M_n^{**} + \sum_{\nu=1}^{r} N_{n_\nu}^{**} = 0, \quad \text{on } \Omega^{**},$$

where M_n^{**} is the left member of (29.3) given as

$$(29.23) \quad M_n^{**} \equiv u^{(m_1)} \cdots u^{(m_n)} + \sum_{\substack{0 \le j_1 \le m_1, ..., 0 \le j_n \le m_n \\ (j_1, ..., j_n) \ne (m_1, ..., m_n) \\ j_1 \ge \cdots \ge j_n}} a^{**}_{m_1-j_1, ..., m_n-j_n}(\zeta) \, u^{(j_1)} \cdots u^{(j_n)}$$

and $N_{n_1}^{**}, \ldots, N_{n_r}^{**}$ are specified by

$$(29.24) \quad N_{n_\nu}^{**} \equiv \sum_{\substack{0 \le j_1 \le m_{\nu,1}, 0 \le j_2 \le m_{\nu,2}, ..., 0 \le j_{n_\nu} \le m_{\nu,n_\nu} \\ j_1 \ge j_2 \ge \cdots \ge j_{n_\nu}}} b^{**}_{n_\nu; m_{\nu,1}-j_1, m_{\nu,2}-j_2, ..., m_{\nu,n_\nu}-j_{n_\nu}}(\zeta) \, u^{(j_1)} u^{(j_2)} \cdots u^{(j_{n_\nu})}, \quad \text{for } 1 \le \nu \le r.$$

PROOF. In terms of $u(\zeta) = (y \circ f)(\zeta)$ on Ω^{**}, Proposition 3.1 of page 29 shows that there are analytic functions $\alpha_{i,j}(\zeta)$ on Ω^{**} such that

$$(29.25) \quad y^{(i)}(f(\zeta)) \equiv \left(f'(\zeta)\right)^{-i} \left[u^{(i)}(\zeta) + \sum_{j=0}^{i-1} \alpha_{i-j,j}(\zeta) \, u^{(j)}(\zeta) \right], \quad \text{for } i \ge 0.$$

We use (29.25) to see that there are unique meromorphic functions $a^{**}_{i_1, i_2, ..., i_n}(\zeta)$ on Ω^{**} for (29.23) such that the substitution $z = f(\zeta)$ in M_n of (29.8) yields

$$(29.26) \qquad (M_n)_{z=f(\zeta)} \equiv \left(f'(\zeta)\right)^{-(m_1 + \cdots + m_n)} M_n^{**}.$$

Thus, the differential equation $M_n^{**} = 0$ on Ω^{**} is the monic equation (29.3) into which the substitution $z = f(\zeta)$ transforms (29.1). By employing (29.9) and (29.25), we see that there are unique meromorphic functions $e_{n_\nu; i_1, i_2, ..., i_{n_\nu}}(\zeta)$ on Ω^{**} such that the substitution $z = f(\zeta)$ in N_{n_ν} of (29.9) yields

$$(29.27) \qquad (N_{n_\nu})_{z=f(\zeta)} \equiv E_{n_\nu}, \quad \text{for } 1 \le \nu \le r,$$

where

$$(29.28) \quad E_{n_\nu} \equiv \sum_{\substack{0 \le j_1 \le m_{\nu,1}, ..., 0 \le j_{n_\nu} \le m_{\nu,n_\nu} \\ j_1 \ge \cdots \ge j_{n_\nu}}} e_{n_\nu; m_{\nu,1}-j_1, ..., m_{\nu,n_\nu}-j_{n_\nu}}(\zeta) \, u^{(j_1)} \cdots u^{(j_{n_\nu})}.$$

For each of the coefficients $e_{n_\nu; i_1, i_2, ..., i_{n_\nu}}(\zeta)$ in (29.28), we set

$$(29.29) \qquad b^{**}_{n_\nu; i_1, i_2, ..., i_{n_\nu}}(\zeta) \equiv \left(f'(\zeta)\right)^{(m_1 + \cdots + m_n)} e_{n_\nu; i_1, i_2, ..., i_{n_\nu}}(\zeta).$$

Then, in terms of the differential polynomials $N_{n_1}^{**}, \ldots, N_{n_r}^{**}$ defined by (29.24) and (29.29), we note that (29.28), (29.29), and (29.24) yield

$$(29.30) \qquad E_{n_\nu} \equiv \left(f'(\zeta)\right)^{-(m_1 + \cdots + m_r)} N_{n_\nu}^{**}, \quad \text{for } 1 \le \nu \le r.$$

We apply (29.26), (29.27), and (29.30) to deduce that

$$(29.31) \qquad \left[M_n + \sum_{\nu=1}^{r} N_{n_\nu}\right]_{z=f(\zeta)} \equiv \left(f'(\zeta)\right)^{-(m_1 + \cdots + m_r)} \left[M_n^{**} + \sum_{\nu=1}^{r} N_{n_\nu}^{**}\right].$$

The coefficient of $u^{(m_1)} \cdots u^{(m_n)}$ is 1 in both M_n^{**} of (29.23) and in the left member of (29.22). Consequently, (29.31) shows that the substitution $z = f(\zeta)$ transforms (29.7) into (29.22). Since any two of M_n^{**}, $N_{n_1}^{**}$, ..., $N_{n_r}^{**}$ have no nonzero terms in common, the coefficients $a_{i_1, i_2, ..., i_n}^{**}(\zeta)$ and $b_{n_\nu; i_1, i_2, ..., i_{n_\nu}}^{**}(\zeta)$ for (29.22) are unique. This completes the proof. □

Propositions 29.9 and 29.10 show that the equations (29.7) are transformed into ones of the same form by changes of the dependent and independent variables. To define semi-invariants and relative invariants for such equations, we assume that

(29.32) $\quad x_{m_1-j_1, ..., m_n-j_n}^{(k)}, \quad$ for $0 \leq j_1 \leq m_1, \ldots, 0 \leq j_n \leq m_n$,
$$j_1 \geq \cdots \geq j_n, \ (j_1, \ldots, j_n) \neq (m_1, \ldots, m_n), \ k \geq 0,$$

and

(29.33) $\quad x_{n_\nu; m_{\nu,1}-j_1, ..., m_{\nu,n_\nu}-j_{n_\nu}}^{(k)}, \quad$ for $1 \leq \nu \leq r, \ j_1 \geq \cdots \geq j_{n_\nu}$,
$$0 \leq j_1 \leq m_{\nu,1}, \ldots, 0 \leq j_{n_\nu} \leq m_{\nu,n_\nu}, \ k \geq 0,$$

are algebraically independent variables over $\mathbb{Q}$ and we let $\mathfrak{R}$ denote the ring of polynomials in these variables over $\mathbb{Q}$. We let $'$ be the derivation for $\mathfrak{R}$ such that

$$\left(x_{i_1, ..., i_n}^{(k)}\right)' = x_{i_1, ..., i_n}^{(k+1)} \quad \text{and} \quad \left(x_{n_\nu; i_1, ..., i_{n_\nu}}^{(k)}\right)' = x_{n_\nu; i_1, ..., i_{n_\nu}}^{(k+1)},$$

for the variables in (29.32) and (29.33). The ring $\mathfrak{R}_{m_1, ..., m_n}$ of polynomials over $\mathbb{Q}$ in the variables (29.32) that we considered for (29.4) can be viewed as a subring of $\mathfrak{R}$; and the restriction to $\mathfrak{R}_{m_1, ..., m_n}$ of the derivation $'$ for $\mathfrak{R}$ is the derivation for $\mathfrak{R}_{m_1, ..., m_n}$.

For any polynomial P in $\mathfrak{R}$, we let $P(z)$ denote the unique function on Ω obtained by replacing $x_{i_1, ..., i_n}^{(k)}$ and $x_{n_\nu; i_1, ..., i_{n_\nu}}^{(k)}$ in P respectively with $a_{i_1, ..., i_n}^{(k)}(z)$ and $b_{n_\nu; i_1, ..., i_{n_\nu}}^{(k)}(z)$ from (29.7); we let $P^*(z)$ denote the unique function on Ω obtained by replacing $x_{i_1, ..., i_n}^{(k)}$ and $x_{n_\nu; i_1, ..., i_{n_\nu}}^{(k)}$ in P respectively with $a_{i_1, ..., i_n}^{*(k)}(z)$ and $b_{n_\nu; i_1, ..., i_{n_\nu}}^{*(k)}(z)$ from (29.10); and we let $P^{**}(\zeta)$ denote the unique function on Ω^{**} obtained by replacing $x_{i_1, ..., i_n}^{(k)}$ and $x_{n_\nu; i_1, ..., i_{n_\nu}}^{(k)}$ in P respectively with $a_{i_1, ..., i_n}^{**(k)}(\zeta)$ and $b_{n_\nu; i_1, ..., i_{n_\nu}}^{**(k)}(\zeta)$ from (29.22). Definitions 29.1–29.3 on page 336 are applicable to our present context with (29.7) in place of (29.1) and $\mathfrak{R}$ in place of $\mathfrak{R}_{m_1, ..., m_n}$.

THEOREM 29.11. *Each semi-invariant of the first kind or the second kind for the equations* (29.1) *is a semi-invariant of the same kind for the equations* (29.7). *Moreover, the classes of equations* (29.1) *and* (29.7) *have the same semi-invariants of the first kind if and only if the condition*

(29.34) $\quad n > \max(n_1, n_2, \ldots, n_r) \quad \text{or} \quad n < \min(n_1, n_2, \ldots, n_r)$

is satisfied. Furthermore, the classes of equations (29.1) *and* (29.7) *have the same relative invariants if and only if* (29.34) *is satisfied.*

PROOF. Suppose that P is a semi-invariant for equations of the form (29.1). Then, P belongs to $\mathfrak{R}_{m_1, ..., m_n}$ and involves only the variables of (29.32). Thus, the functions $P(z)$, $P^*(z)$, and $P^{**}(\zeta)$ obtained by substituting coefficients from (29.7), (29.10), and (29.22) for the variables $x_{i_1, ..., i_n}^{(k)}$ and $x_{n_\nu; i_1, ..., i_{n_\nu}}^{(k)}$ in P are the same as the functions obtained by substituting coefficients from (29.1), (29.2), and

29.2. RELATIVE INVARIANTS FOR SOME NONHOMOGENEOUS EQUATIONS

(29.3) for the variables $x^{(k)}_{i_1,\ldots,i_n}$ in $\boldsymbol{P}$. Thus, each semi-invariant for the equations (29.1) is a semi-invariant of the same kind for the equations (29.7).

Suppose that (29.34) is satisfied and suppose that $\boldsymbol{P}_1$ is a polynomial in $\mathfrak{R}$ that does not belong to $\mathfrak{R}_{m_1,\ldots,m_n}$. Then, $\boldsymbol{P}_1$ effectively involves at least one of the variables in (29.33). Let $\boldsymbol{P}_2$ denote the polynomial in $\mathfrak{R}$ obtained from $\boldsymbol{P}_1$ by replacing each variable $x^{(k)}_{n_\nu;i_1,\ldots,i_{n_\nu}}$ from (29.33) in $\boldsymbol{P}_1$ with $2^{n_\nu - n} x^{(k)}_{n_\nu;i_1,\ldots,i_{n_\nu}}$. Due to (29.34), the integers $n - n_1, \ldots, n - n_r$ are either all positive or else all negative. Thus, each term $\boldsymbol{T}$ of $\boldsymbol{P}_1$ effectively involving at least one variable from (29.33) contributes a corresponding term $\neq \boldsymbol{T}$ to $\boldsymbol{P}_2$ that does not belong to $\mathfrak{R}_{m_1,\ldots,m_n}$; and linearly independent terms of this type in $\boldsymbol{P}_1$ contribute corresponding linearly independent terms to $\boldsymbol{P}_2$. Thus, we have $\boldsymbol{P}_2 \neq \boldsymbol{P}_1$. We set $\boldsymbol{D} \equiv \boldsymbol{P}_2 - \boldsymbol{P}_1$ and observe that $\boldsymbol{D} \not\equiv 0$. By applying a result that is completely analogous to Corollary 4.3 on page 37 or Corollary 21.10 on page 256, we obtain coefficients $a_{i_1,i_2,\ldots,i_n}(z)$, $b_{n_\nu;i_1,i_2,\ldots,i_{n_\nu}}(z)$ on a region Ω for a differential equation (29.7) such that the function $D(z)$ on Ω obtained by replacing $x^{(k)}_{i_1,i_2,\ldots,i_n}$ and $x^{(k)}_{n_\nu;i_1,i_2,\ldots,i_{n_\nu}}$ in $\boldsymbol{D}$ with $a^{(k)}_{i_1,i_2,\ldots,i_n}(z)$ and $b^{(k)}_{n_\nu;i_1,i_2,\ldots,i_{n_\nu}}(z)$ from this (29.7) satisfies $D(z) \not\equiv 0$. Consequently, the functions $P_1(z)$ and $P_2(z)$ on Ω obtained from $\boldsymbol{P}_1$ and $\boldsymbol{P}_2$ by substituting $a^{(k)}_{i_1,i_2,\ldots,i_n}(z)$ and $b^{(k)}_{n_\nu;i_1,i_2,\ldots,i_{n_\nu}}(z)$ from this (29.7) for $x^{(k)}_{i_1,i_2,\ldots,i_n}$ and $x^{(k)}_{n_\nu;i_1,i_2,\ldots,i_{n_\nu}}$ satisfy $P_1(z) \not\equiv P_2(z)$. For $\rho(z) \equiv 2$ and the equation (29.10) into which the substitution $y(z) = 2\,v(z)$ transforms this (29.7), Proposition 29.9 shows that (29.13) relates the coefficients of (29.10) to those of (29.7). Hence, the function $P_1^*(z)$ on Ω obtained by replacing $x^{(k)}_{i_1,i_2,\ldots,i_n}$ and $x^{(k)}_{n_\nu;i_1,i_2,\ldots,i_{n_\nu}}$ in $\boldsymbol{P}_1$ with $a^{*(k)}_{i_1,i_2,\ldots,i_n}(z)$ and $b^{*(k)}_{n_\nu;i_1,i_2,\ldots,i_{n_\nu}}(z)$ from this (29.10) is related to $P_2(z)$ by $P_1^*(z) \equiv P_2(z)$. Thus, we have $P_1^*(z) \not\equiv P_1(z)$, for this particular (29.7). This shows that $\boldsymbol{P}_1$ is not a semi-invariant of the first kind for equations of the general form (29.7). Consequently, when (29.34) is satisfied, each semi-invariant of the first kind for the equations (29.7) belongs to $\mathfrak{R}_{m_1,\ldots,m_n}$ and is therefore a semi-invariant of the first kind for the equations (29.1). In view of this, we see that: when (29.34) is satisfied, each relative invariant for the equations (29.7) belongs to $\mathfrak{R}_{m_1,\ldots,m_n}$ and is therefore a relative invariant for the equations (29.1). It remains for us to show that: when (29.34) is not satisfied, there is a relative invariant in $\mathfrak{R}$ for the equations (29.7) that does not belong to $\mathfrak{R}_{m_1,\ldots,m_n}$.

Supposes that (29.34) is not satisfied. Then, we have $r \geq 2$ and may assume the notation has been selected so that $1 \leq n_1 < n < n_2$. For this situation, the integers $p = n - n_1$ and $q = n_2 - n$ are positive. We observe that (29.9), (29.16), (29.14), and (29.17) yield

(29.35) $\qquad d_{n_k;0,\ldots,0}(z) \equiv \bigl(\rho(z)\bigr)^{n_k} b_{n_k;0,\ldots,0}(z), \quad \text{for } k = 1, 2.$

We employ (29.18) and (29.35) to deduce that

(29.36) $\qquad b^*_{n_1;0,\ldots,0}(z) \equiv \bigl(\rho(z)\bigr)^{n_1 - n} b_{n_1;0,\ldots,0}(z) \equiv \bigl(\rho(z)\bigr)^{-p} b_{n_1;0,\ldots,0}(z)$

and

(29.37) $\qquad b^*_{n_2;0,\ldots,0}(z) \equiv \bigl(\rho(z)\bigr)^{n_2 - n} b_{n_2;0,\ldots,0}(z) \equiv \bigl(\rho(z)\bigr)^{q} b_{n_2;0,\ldots,0}(z).$

Combining (29.36) and (29.37), we find that

(29.38) $\qquad \bigl(b^*_{n_1;0,\ldots,0}(z)\bigr)^q \bigl(b^*_{n_2;0,\ldots,0}(z)\bigr)^p \equiv \bigl(b_{n_1;0,\ldots,0}(z)\bigr)^q \bigl(b_{n_2;0,\ldots,0}(z)\bigr)^p.$

We observe that (29.9), (29.27), (29.25), and (29.28) give

(29.39) $\quad e_{n_k;0,...,0}(\zeta) \equiv (f'(\zeta))^{-(m_{k,1}+\cdots+m_{k,n_k})} b_{n_k;0,...,0}(f(\zeta)) \quad$ for $k=1,2$.

In view of (29.29) and (29.39), we have

(29.40) $\quad b^{**}_{n_1;0,...,0}(\zeta) \equiv (f'(\zeta))^{(m_1+\cdots+m_n)-(m_{1,1}+\cdots+m_{1,n_1})} b_{n_1;0,...,0}(f(\zeta))$

and

(29.41) $\quad b^{**}_{n_2;0,...,0}(\zeta) \equiv (f'(\zeta))^{(m_1+\cdots+m_n)-(m_{2,1}+\cdots+m_{2,n_2})} b_{n_2;0,...,0}(f(\zeta))$.

In terms of the integer

(29.42) $\quad \sigma = (p+q)(m_1+\cdots+m_n) - p(m_{2,1}+\cdots+m_{2,n_2}) - q(m_{1,1}+\cdots+m_{1,n_1})$,

we see that (29.40) and (29.41) give

(29.43) $\quad \left(b^{**}_{n_1;0,...,0}(\zeta)\right)^q \left(b^{**}_{n_2;0,...,0}(\zeta)\right)^p$
$$\equiv (f'(\zeta))^\sigma \left(b_{n_1;0,...,0}(f(\zeta))\right)^q \left(b_{n_2;0,...,0}(f(z))\right)^p.$$

For the polynomial I defined in $\mathfrak{R}$ by

(29.44) $\qquad\qquad I \equiv (x_{n_1;0,...,0})^q (x_{n_2;0,...,0})^p$,

(29.38) yields $I^*(z) \equiv I(z)$, for each (29.7) and each $y(z) = \rho(z) v(z)$; while (29.43) gives $I^{**}(\zeta) \equiv (f'(\zeta))^\sigma I(f(\zeta))$, for each (29.7) and each $z = f(\zeta)$. Thus, I is a relative invariant in $\mathfrak{R}$ for the equations (29.7); and I does not belong to $\mathfrak{R}_{m_1,...,m_n}$. This completes the proof. $\square$

The equations (29.7), (29.10), and (29.22) are normalized by their monic terms in the manner of (29.1), (29.2), and (29.3). For this reason, it is natural to define the weight of $x^{(k)}_{i_1,...,i_n}$ in (29.32) by Definition 29.4. However, a useful definition for the weight of $x^{(k)}_{n_\nu;i_1,...,i_{n_\nu}}$ in (29.33) is less obvious.

DEFINITION 29.12. For the ring $\mathfrak{R}$ of polynomials over $\mathbb{Q}$ in the variables of (29.32) and (29.33), the *weight* of a nonzero rational number is 0; the *weight* of $x^{(k)}_{i_1,...,i_n}$ in (29.32) is the positive integer

(29.45) $\qquad\qquad weight(x^{(k)}_{i_1,...,i_n}) = i_1 + \cdots + i_n + k$;

the *weight* of $x^{(k)}_{n_\nu;i_1,...,i_{n_\nu}}$ in (29.33) is the integer

(29.46) $\qquad weight(x^{(k)}_{n_\nu;i_1,...,i_{n_\nu}}) = m_1 + \cdots + m_n + k - \sum_{j=1}^{n_\nu}(m_{\nu,j} - i_j)$;

and the *weight* of a nonzero monomial in $\mathfrak{R}$ is equal to the sum of the weights of its factors. A nonzero polynomial in $\mathfrak{R}$ is *isobaric of weight s* when each of its nonzero monomials has weight s.

EXAMPLE 29.13. When $1 \leq n_1 < n < n_2$, the relative invariant I constructed in (29.44) is an isobaric polynomial in $\mathfrak{R}$ and its weight is the integer σ in (29.42). We note that σ may be positive, negative, or 0.

29.2. RELATIVE INVARIANTS FOR SOME NONHOMOGENEOUS EQUATIONS

THEOREM 29.14. *Suppose that $\boldsymbol{P}$ in $\mathfrak{R}$ is a semi-invariant of the second kind for the equations (29.7) and let s be an integer such that*

$$(29.47) \qquad P^{**}(\zeta) \equiv \bigl(f'(\zeta)\bigr)^s P\bigl(f(\zeta)\bigr),$$

for any (29.7) and any transformation $z = f(\zeta)$ of (29.7) into a corresponding (29.22). Then, $\boldsymbol{P}$ is an isobaric polynomial and s is equal to its weight.

PROOF. For any (29.7) on Ω and any nonzero rational number t, the function $g_t(z) \equiv (1/t)z$ on Ω has the univalent analytic inverse $f_t(\zeta) \equiv t\zeta$ on $\Omega^{**} = g(\Omega)$ that transforms (29.7) into a corresponding equation (29.22) repeated here as

$$(29.48) \qquad M_n^{**} + \sum_{\nu=1}^r N_{n_\nu}^{**} = 0, \quad \text{on } \Omega^{**}.$$

Due to $f_t'(\zeta) \equiv t$ and $f_t''(\zeta) \equiv 0$, the formulas (3.1) and (3.2) on page 29 for $\alpha_{i,j}(\zeta)$ in (29.25) yield $\alpha_{0,j}(\zeta) \equiv 1$ and $\alpha_{i,j}(\zeta) \equiv 0$, for $i \geq 1$ and any j. Thus, we see that (29.25) relates $y^{(k)}(z)$ in (29.7) and $u^{(k)}(\zeta) \equiv (y \circ f_t)^{(k)}(\zeta)$ in (29.48) through

$$(29.49) \qquad y^{(k)}\bigl(f_t(\zeta)\bigr) \equiv t^{-k} u^{(k)}(\zeta), \quad \text{on } \Omega^{**} \text{ for } k \geq 0.$$

Consequently, for the substitution $z = f_t(\zeta)$ in M_n of (29.8), we have

$$\bigl(M_n\bigr)_{z=f_t(\zeta)} \equiv t^{-m_1} u^{(m_1)}(\zeta) \cdots t^{-m_n} u^{(m_n)}(\zeta)$$
$$+ \sum_{\substack{0 \leq j_1 \leq m_1, \ldots, 0 \leq j_n \leq m_n \\ (j_1, \ldots, j_n) \neq (m_1, \ldots, m_n) \\ j_1 \geq \cdots \geq j_n}} a_{m_1-j_1, \ldots, m_n-j_n}\bigl(f_t(\zeta)\bigr) t^{-j_1} u^{(j_1)}(\zeta) \cdots t^{-j_n} u^{(j_n)}(\zeta)$$
$$\equiv t^{-(m_1+\cdots+m_n)} M_n^{**},$$

where M_n^{**} is given by (29.23) on page 339 with

$$a_{m_1-j_1, \ldots, m_n-j_n}^{**}(\zeta) \equiv t^{[(m_1-j_1)+\cdots+(m_n-j_n)]} a_{m_1-j_1, \ldots, m_n-j_n}\bigl(f_t(\zeta)\bigr).$$

Thus, for the coefficients $a_{i_1, \ldots, i_n}^{**}(\zeta)$ of M_n^{**} in (29.48), we find that

$$a_{i_1, \ldots, i_n}^{**}(\zeta) \equiv t^{\text{weight}(\boldsymbol{x}_{i_1, \ldots, i_n})} a_{i_1, \ldots, i_n}\bigl(f_t(\zeta)\bigr);$$

and they yield

$$(29.50) \qquad a_{i_1, \ldots, i_n}^{**(k)}(\zeta) \equiv t^{\text{weight}(\boldsymbol{x}_{i_1, \ldots, i_n}^{(k)})} a_{i_1, \ldots, i_n}^{(k)}\bigl(f_t(\zeta)\bigr), \quad \text{for } k \geq 0.$$

For the substitution $z = f_t(\zeta)$ in N_{n_ν} of (29.9), we use (29.49) to verify that

$$\bigl(N_{n_\nu}\bigr)_{z=f_t(\zeta)}$$
$$\equiv \sum_{\substack{0 \leq j_1 \leq m_{\nu,1}, \ldots, 0 \leq j_{n_\nu} \leq m_{\nu,n_\nu} \\ j_1 \geq \cdots \geq j_{n_\nu}}} b_{n_\nu; m_{\nu,1}-j_1, \ldots, m_{\nu,n_\nu}-j_{n_\nu}}\bigl(f_t(\zeta)\bigr) t^{-j_1} u^{(j_1)}(\zeta) \cdots t^{-j_n} u^{(j_{n_\nu})}(\zeta)$$
$$\equiv t^{-(m_1+\cdots+m_n)} N_{n_\nu}^{**},$$

where this $N_{n_\nu}^{**}$ for (29.48) is given by (29.24) on page 339 with respect to

$$(29.51) \quad b_{n_\nu; m_{\nu,1}-j_1, \ldots, m_{\nu,n_\nu}-j_{n_\nu}}^{**}(\zeta)$$
$$\equiv t^{(m_1+\cdots+m_n-j_1-\cdots-j_{n_\nu})} b_{n_\nu; m_{\nu,1}-j_1, \ldots, m_{\nu,n_\nu}-j_{n_\nu}}\bigl(f_t(\zeta)\bigr).$$

Due to (29.46), the exponent of t in (29.51) equals $weight(\boldsymbol{x}_{n_\nu; m_{\nu,1}-j_1, \ldots, m_{\nu,n_\nu}-j_{n_\nu}})$. Thus, the coefficients for $N_{n_\nu}^{**}$ in (29.48) are given by

$$b^{**}_{n_\nu; i_1, \ldots, i_{n_\nu}}(\zeta) \equiv t^{weight(\boldsymbol{x}_{n_\nu; i_1, \ldots, i_{n_\nu}})} \, b_{n_\nu; i_1, \ldots, i_{n_\nu}}\bigl(f_t(\zeta)\bigr)$$

and they yield

(29.52) $\quad b^{**(k)}_{n_\nu; i_1, \ldots, i_{n_\nu}}(\zeta) \equiv t^{weight(\boldsymbol{x}^{(k)}_{n_\nu; i_1, \ldots, i_{n_\nu}})} \, b^{(k)}_{n_\nu; i_1, \ldots, i_{n_\nu}}\bigl(f_t(\zeta)\bigr), \quad$ for $k \geq 0$.

Let $\boldsymbol{P}_h$ denote an isobaric polynomial of weight h in $\mathfrak{R}$. Then, there are monomials $\boldsymbol{T}_{h,1}, \ldots, \boldsymbol{T}_{h,q}$ in $\mathfrak{R}$ of weight h such that $\boldsymbol{P}_h \equiv \boldsymbol{T}_{h,1} + \cdots + \boldsymbol{T}_{h,q}$. For $1 \leq j \leq q$, $\boldsymbol{T}_{h,j}$ is a product of a nonzero rational number and various ones of the variables $\boldsymbol{x}^{(k)}_{i_1, \ldots, i_n}$, $\boldsymbol{x}^{(k)}_{n_\nu; i_1, \ldots, i_{n_\nu}}$ from (29.32), (29.33). To obtain $T^{**}_{h,j}(\zeta)$ on Ω^{**}, we replace each $\boldsymbol{x}^{(k)}_{i_1, \ldots, i_n}$ or $\boldsymbol{x}^{(k)}_{n_\nu; i_1, \ldots, i_{n_\nu}}$ in $\boldsymbol{T}_{h,j}$ with the corresponding $a^{**(k)}_{i_1, \ldots, i_n}(\zeta)$ or $b^{**(k)}_{n_\nu; i_1, \ldots, i_{n_\nu}}(\zeta)$ from (29.50) and (29.52). We obtain $T_{h,j}(z)$ on Ω by replacing each $\boldsymbol{x}^{(k)}_{i_1, \ldots, i_n}$ or $\boldsymbol{x}^{(k)}_{n_\nu; i_1, \ldots, i_{n_\nu}}$ in $\boldsymbol{T}_{h,j}$ with the corresponding $a^{(k)}_{i_1, \ldots, i_n}(z)$ or $b^{(k)}_{n_\nu; i_1, \ldots, i_{n_\nu}}(z)$ from (29.7). Since the weight h of $\boldsymbol{T}_{h,j}$ is equal to the sum of the weights of the factors of $\boldsymbol{T}_{h,j}$, we see that (29.50) and (29.52) yield

$$T^{**}_{h,j}(\zeta) \equiv t^h \, T_{h,j}\bigl(f_t(\zeta)\bigr), \quad \text{on } \Omega^{**} \text{ for } 1 \leq j \leq q.$$

Consequently, for any nonzero t in $\mathbb{Q}$, we have

(29.53) $\quad P^{**}_h(\zeta) \equiv \sum_{j=1}^{q} T^{**}_{h,j}(\zeta) \equiv t^h \sum_{j=1}^{q} T_{h,j}\bigl(f_t(\zeta)\bigr) \equiv t^h \, P_h\bigl(f_t(\zeta)\bigr), \quad$ on Ω^{**}.

With regard to s in (29.47) for the semi-invariant $\boldsymbol{P}$, there is a positive integer σ such that $-\sigma \leq s \leq \sigma$ and

(29.54) $$\boldsymbol{P} \equiv \sum_{h=-\sigma}^{\sigma} \boldsymbol{P}_h,$$

where, for each h satisfying $-\sigma \leq h \leq \sigma$, $\boldsymbol{P}_h$ is either 0 or an isobaric polynomial of weight h. With $f_t(\zeta) \equiv t\zeta$, we apply (29.54), (29.47), and (29.53) to verify that

$$t^s \sum_{h=-\sigma}^{\sigma} P_h(t\zeta) \equiv \bigl(f'_t(\zeta)\bigr)^s P\bigl(f_t(\zeta)\bigr) \equiv P^{**}(\zeta) \equiv \sum_{h=-\sigma}^{\sigma} P^{**}_h(\zeta) \equiv \sum_{h=-\sigma}^{\sigma} t^h P_h(t\zeta).$$

We replace $t\zeta$ with z in the preceding expression to obtain

(29.55) $$t^s \sum_{h=-\sigma}^{\sigma} P_h(z) \equiv \sum_{h=-\sigma}^{\sigma} t^h P_h(z), \quad \text{on } \Omega,$$

for each nonzero rational number t and each (29.7). We shall establish that

(29.56) $$\left[\sum_{h=-\sigma}^{\sigma} \boldsymbol{P}_h\right] t^s \equiv \sum_{h=-\sigma}^{\sigma} \boldsymbol{P}_h \, t^h, \quad \text{in } \mathfrak{R} \text{ for each nonzero } t \text{ in } \mathbb{Q}.$$

If there were a nonzero rational number t for which the formula in (29.56) is not valid, then by subtracting one member of this formula from the other, we would obtain a nonzero polynomial $\boldsymbol{D}$ in $\mathfrak{R}$ to which we could apply a result completely analogous to Corollary 4.3 on page 37. But, its conclusion that there is a particular (29.7) with coefficients that yield $D(z) \not\equiv 0$ when substituted into $\boldsymbol{D}$ would produce the contradiction that (29.55) is not valid.

29.2. RELATIVE INVARIANTS FOR SOME NONHOMOGENEOUS EQUATIONS

After multiplying (29.56) by t^σ, we regard the resulting left and right members as polynomial functions of t with coefficients from $\mathfrak{R}$. Since the coefficients of like powers of t are equal, we find that $\boldsymbol{P}_h \equiv 0$ when h satisfies $-\sigma \le h \le \sigma$ and $h \ne s$. We combine this with (29.54) to obtain $\boldsymbol{P} \equiv \boldsymbol{P}_s$. Thus, $\boldsymbol{P}$ is an isobaric polynomial and s is equal to its weight. This completes the proof of Theorem 29.14. $\square$

OBSERVATION 29.15. By permitting one of the integers $n_1, \ldots, n_r$ to be zero, we can include a term $N_0 \equiv b_{0;}(z)$ in (29.7). In that situation, (29.10) and (29.22) possess corresponding terms $N_0^* \equiv b_{0;}^*(z)$ and $N_0^{**} \equiv b_{0;}^{**}(\zeta)$. Propositions 29.9 and 29.10 remain valid. Moreover, by regarding the variables $\boldsymbol{x}_{0;}^{(k)}$, for $k \ge 0$, as the special case of $\boldsymbol{x}_{n_\nu; m_{\nu,1}-j_1, \ldots, m_{\nu,n_\nu}-j_{n_\nu}}^{(k)}$ in (29.33) when $n_\nu = 0$, we see that our previous notation permits replacements of $\boldsymbol{x}_{0;}^{(k)}$ with $b_{0;}^{(k)}(z)$ or $b_{0;}^{*(k)}(z)$ or $b_{0;}^{**(k)}(\zeta)$. Also, when $n_\nu = 0$, the sum $\big((m_{\nu,1} - i_1) + \cdots + (m_{\nu,n_\nu} - i_{n_\nu})\big)$ in (29.46) is vacuous and (29.46) therefore yields $weight\big(\boldsymbol{x}_{0;}^{(k)}\big) = m_1 + \cdots + m_n + k$. With these interpretations, Theorems 29.11 and 29.14 remain valid when one of $n_1, \ldots, n_r$ is permitted to be zero.

EXAMPLE 29.16. For the situation where $n = 2$, $m_1 = 2$, $m_2 = 1$, $r = 2$, $n_1 = 1$, $m_{1,1} = 1$, and $n_2 = 0$, we see that (29.7) has the form

$$(29.57) \quad y''y' + a_{0,1}(z)\,y''y + a_{1,0}(z)\,y'y' + a_{1,1}(z)\,y'y + a_{2,1}(z)\,y^2$$
$$+ b_{1;0}(z)\,y^{(1)} + b_{1;1}(z)\,y + b_{0;}(z) = 0.$$

Since (29.34) is satisfied, the equations (29.57) have the same semi-invariants of the first kind and the same relative invariants as the equations (28.1) of page 325. Their relative invariants are given by Theorem 28.7 on page 332.

The semi-invariants of the second kind for the equations (29.57) include all those for the equations (28.1) along with others. To indicate several of the latter, we note that the substitution $z = f(\zeta)$, $u(\zeta) = (y \circ f)(\zeta))$ transforms (29.57) into

$$u''u' + a_{0,1}^{**}(\zeta)\,u''u + a_{1,0}^{**}(z)\,u'u' + a_{1,1}^{**}(\zeta)\,u'u + a_{2,1}^{**}(\zeta)\,u^2$$
$$+ b_{1;0}^{**}(\zeta)\,u^{(1)} + b_{1;1}^{**}(\zeta)\,u + b_{0;}^{**}(\zeta) = 0,$$

where the coefficients are given by (28.7)–(28.10) in conjunction with

$$b_{1;0}^{**}(\zeta) \equiv \big(f'(\zeta)\big)^2 b_{1;0}\big(f(\zeta)\big),$$
$$b_{1;1}^{**}(\zeta) \equiv \big(f'(\zeta)\big)^3 b_{1;1}\big(f(\zeta)\big),$$

and

$$b_{0;}^{**}(\zeta) \equiv \big(f'(\zeta)\big)^3 b_{0;}\big(f(\zeta)\big).$$

Thus, for the equations (29.57), the polynomials $\boldsymbol{x}_{1;0}$, $\boldsymbol{x}_{1;1}$, and $\boldsymbol{x}_{0;}$ in $\mathfrak{R}$ are semi-invariants of the second kind of respective weights 2, 3, and 3.

CHAPTER 30

Invariants for Simple Equations unlike (29.1)

We have examined the classes of homogeneous quadratic differential equations of order $m \leq 2$ to which the context for (29.1) on page 335 can be applied. Namely, when the monic monomial is $y''(z)\,y''(z)$, the equations are those of (1.0–A) on page 3. When the monic monomial is $y''(z)\,y'(z)$, the equations are those in (28.1) on page 325. When the monic monomial is $y'(z)\,y'(z)$, the equations are those of (4.16) on page 39. As patterns for (29.1), the monic monomials $y''(z)\,y(z)$ or $y'(z)\,y(z)$ or $y(z)\,y(z)$ are excluded because the corresponding equations would have $y(z)$ as a factor. However, there are two remaining classes of homogeneous quadratic differential equations of order $m \leq 2$ that have not been previously considered. They are specified by (30.1) in Section 30.1 and by (30.28) in Section 30.2.

30.1. Equations without a dominant term

The form of the nonhomogeneous equations given by (29.7) on page 337 is not completely specified by their monic term $y^{(m_1)} \cdots y^{(m_n)}$. A simpler example is provided by the homogeneous differential equations of the type

$$(30.1) \qquad y''(z)\,y(z) + b_0(z)\bigl(y'(z)\bigr)^2 + b_1(z)\,y'(z)\,y(z) + b_2(z)\bigl(y(z)\bigr)^2 = 0,$$

where $b_0(z)$, $b_1(z)$, $b_2(z)$ are meromorphic functions on a region Ω.

For any meromorphic function $\rho(z) \not\equiv 0$ on Ω, the change $y(z) = \rho(z)\,v(z)$ of the dependent variable from y to v transforms (30.1) into

$$(30.2) \qquad v''(z)\,v(z) + b_0^*(z)\bigl(v'(z)\bigr)^2 + b_1^*(z)\,v'(z)\,v(z) + b_2^*(z)\bigl(v(z)\bigr)^2 = 0,$$

where the coefficients are given on Ω by

$$(30.3) \qquad b_0^*(z) \equiv b_0(z),$$

$$(30.4) \qquad b_1^*(z) \equiv b_1(z) + 2\,b_0(z)\,\frac{\rho'(z)}{\rho(z)} + 2\,\frac{\rho'(z)}{\rho(z)},$$

and

$$(30.5) \qquad b_2^*(z) \equiv b_2(z) + b_1(z)\,\frac{\rho'(z)}{\rho(z)} + b_0(z)\left(\frac{\rho'(z)}{\rho(z)}\right)^2 + \frac{\rho''(z)}{\rho(z)}.$$

For any univalent analytic function $\zeta = g(z)$ on Ω whose inverse is designated by $z = f(\zeta)$ on $\Omega^{**} = g(\Omega)$, the change $z = f(\zeta)$ of the independent variable from z to ζ and the notation $u(\zeta) = (y \circ f)(\zeta)$ transform (30.1) into

$$(30.6) \qquad u''(\zeta)\,u(\zeta) + b_0^{**}(\zeta)\bigl(u'(\zeta)\bigr)^2 + b_1^{**}(\zeta)\,u'(\zeta)\,u(\zeta) + b_2^{**}(\zeta)\bigl(y(\zeta)\bigr)^2 = 0,$$

347

where the coefficients are given on Ω^{**} by

(30.7) $$b_0^{**}(\zeta) \equiv b_0(f(\zeta)),$$

(30.8) $$b_1^{**}(\zeta) \equiv f'(\zeta)\, b_1(f(\zeta)) - \frac{f''(\zeta)}{f'(\zeta)},$$

and

(30.9) $$b_2^{**}(\zeta) \equiv (f'(\zeta))^2\, b_2(f(\zeta)).$$

30.1.1. Terminology. We assume that the symbols $x_i^{(j)}$, for $0 \leq i \leq 2$ and $j \geq 0$, are algebraically independent variables over $\mathbb{Q}$; we let $\mathcal{D}$ denote the ring of polynomials in these variables over $\mathbb{Q}$; and we let $'$ denote the derivation for $\mathcal{D}$ such that $(x_i^{(j)})' = x_i^{(j+1)}$, for $0 \leq i \leq 2$ and $j \geq 0$. In particular, the constants of $\mathcal{D}$ are the elements of $\mathbb{Q}$. Naturally, we write x_0, x_1, and x_2 for $x_0^{(0)}$, $x_1^{(0)}$, and $x_2^{(0)}$.

When any polynomial $\boldsymbol{P}$ in $\mathcal{D}$ is given, the functions $P(z)$ on Ω or $P^*(z)$ on Ω or $P^{**}(\zeta)$ on Ω^{**} are respectively obtained by replacing each $x_i^{(j)}$ in $\boldsymbol{P}$ with $b_i^{(j)}(z)$ from (30.1) or with $b_i^{*(j)}(z)$ from (30.2) or with $b_i^{**(j)}(\zeta)$ from (30.6).

A nonconstant polynomial $\boldsymbol{P}$ in $\mathcal{D}$ is a *semi-invariant of the first kind for the equations* (30.1) when the condition $P^*(z) \equiv P(z)$ is satisfied for each (30.1) and each substitution $y(z) = \rho(z)\, v(z)$.

A nonconstant polynomial $\boldsymbol{P}$ in $\mathcal{D}$ is a *semi-invariant of the second kind for the equations* (30.1) when there is an integer s such that the condition

(30.10) $$P^{**}(\zeta) \equiv (f'(\zeta))^s P(f(\zeta))$$

is satisfied for each (30.1) and each substitution $z = f(\zeta)$.

A nonconstant polynomial in $\mathcal{D}$ is a *relative invariant for the equations* (30.1) when it is both a semi-invariant of the first kind and a semi-invariant of the second kind for such equations.

The *weight* of a nonzero rational number is zero; the *weight* of $x_i^{(j)}$ is the nonnegative integer $i + j$; and the *weight* of a nonzero monomial in $\mathcal{D}$ is the sum of the weights of its factors. A nonzero polynomial in $\mathcal{D}$ is *isobaric of weight s* when each of its nonzero monomials has weight s. In particular, the weight of x_0 is 0.

30.1.2. Invariants for (30.1). Except for some situations that can occur in Section 29.2, we have not previously encountered a variable having weight 0.

PROPOSITION 30.1. *Suppose that $\boldsymbol{P}$ is a semi-invariant of the second kind for the equations* (30.1). *Then, $\boldsymbol{P}$ is an isobaric polynomial and the only integer s that satisfies* (30.10) *for $\boldsymbol{P}$ is the weight of $\boldsymbol{P}$.*

PROOF. There is an integer $p \geq 0$ such that

(30.11) $$\boldsymbol{P} \equiv \sum_{k=0}^{p} \boldsymbol{E}_k,$$

where $\boldsymbol{E}_p \not\equiv 0$ and, for $0 \leq k \leq p$, either $\boldsymbol{E}_k \equiv 0$ or $\boldsymbol{E}_k$ is an isobaric polynomial in $\mathcal{D}$ of weight k. For any (30.1), we restrict $z = f(\zeta)$ for (30.10) to the univalent analytic functions of the type $z = f_t(\zeta) = t\zeta$ in which t is a fixed nonzero rational number. With this, we see that (30.7)–(30.9) yield $b_i^{**}(\zeta) \equiv t^i b_i(t\zeta)$ and

(30.12) $$b_i^{**(j)}(\zeta) \equiv t^{i+j} b_i^{(j)}(t\zeta), \quad \text{for } 0 \leq i \leq 2 \text{ and } j \geq 0.$$

If $\boldsymbol{T}$ is a nonzero term of $\boldsymbol{E}_k$, then $\boldsymbol{T}$ is expressible as

(30.13) $$\boldsymbol{T} \equiv \beta \prod_{j=1}^{q} \boldsymbol{x}_{\mu_j}^{(\kappa_j)}, \quad \text{subject to} \quad \sum_{j=1}^{q} (\mu_j + \kappa_j) = k,$$

where: β is a nonzero rational number; q is a nonnegative integer; and μ_j, κ_j satisfy $0 \leq \mu_j \leq 2$ and $0 \leq \kappa_j$, for $1 \leq j \leq q$. We see that (30.13) and (30.12) give

(30.14) $$T^{**}(\zeta) \equiv \beta \prod_{j=1}^{q} b_{\mu_j}^{**(\kappa_j)}(\zeta) \equiv \beta \left[\prod_{j=1}^{q} t^{\mu_j+\kappa_j} \right] \left[\prod_{j=1}^{q} b_{\mu_j}^{(\kappa_j)}(t\zeta) \right] \equiv t^k T(t\zeta).$$

Let s be an integer for (30.10). Since each polynomial $\boldsymbol{E}_k$ in (30.11) is either zero or a sum of expression like (30.13) for which (30.14) is valid, we have

(30.15) $$t^s P(t\zeta) \equiv \bigl(f'_t(\zeta)\bigr)^s P\bigl(f_t(\zeta)\bigr) \equiv P^{**}(\zeta) \equiv \sum_{k=0}^{p} E_k^{**}(\zeta) \equiv \sum_{k=0}^{p} t^k E_k(t\zeta),$$

for any nonzero rational number t and any (30.1). We replace $t\zeta$ in (30.15) with z and use (30.11) to obtain

(30.16) $$t^s \sum_{k=0}^{p} E_k(z) \equiv \sum_{k=0}^{p} t^k E_k(z), \quad \text{for any rational number } t \text{ and any (30.1)}.$$

In order to establish that

(30.17) $$t^s \sum_{k=0}^{p} \boldsymbol{E}_k \equiv \sum_{k=0}^{p} t^k \boldsymbol{E}_k, \quad \text{for each rational number } t,$$

suppose that (30.17) is not true for $t = t_0$ in $\mathbb{Q}$. Then, the polynomial

$$\boldsymbol{D} \equiv (t_0)^s \sum_{k=0}^{p} \boldsymbol{E}_k - \sum_{k=0}^{p} (t_0)^k \boldsymbol{E}_k$$

in $\mathcal{D}$ satisfies $\boldsymbol{D} \not\equiv 0$. By applying a result completely analogous to Corollary 4.3 on page 37, we see that there is a differential equation (30.1) on some region Ω such that the substitution of $b_i^{(j)}(z)$ from that (30.1) for $\boldsymbol{x}_i^{(j)}$ in $\boldsymbol{D}$ yields $D(z) \not\equiv 0$. But, this contradicts (30.16). Hence, (30.17) is valid. For each $t \neq 0$ in $\mathbb{Q}$, the terms of weight p in the left and right members of (30.17) give $t^s \boldsymbol{E}_p \equiv t^p \boldsymbol{E}_p$. In view of $\boldsymbol{E}_p \not\equiv 0$, we obtain $s = p$. Regarding (30.17) as an equality of two polynomial functions in t over $\mathcal{D}$, we note that the coefficients of like powers of t must be equal. Thus, we deduce that $\boldsymbol{E}_k \equiv 0$, for $0 \leq k < p$, and $\boldsymbol{P} \equiv \boldsymbol{E}_p$. Consequently, $\boldsymbol{P}$ is an isobaric polynomial of weight p and $s = p$. This completes the proof. $\square$

PROPOSITION 30.2. *The nonconstant polynomial combinations over $\mathbb{Q}$ of $\boldsymbol{x}_0$ are the relative invariants of weight 0 for the equations (30.1). Moreover, $\boldsymbol{x}_0^{(1)}$ is a relative invariant of weight 1 for the equations (30.1).*

PROOF. We use (30.3) to see that x_0 is a semi-invariant of the first kind for the equations (30.1). Moreover, (30.7) shows that x_0 is a semi-invariant of the second kind of weight 0 for the equations (30.1). Consequently, a polynomial in $\mathcal{D}$ is a relative invariant of weight 0 for the equations (30.1) if and only if it is a nonconstant polynomial combination over $\mathbb{Q}$ of x_0.

We differentiate (30.3) and (30.7) to obtain
$$b_0^{*(1)}(z) \equiv b_0^{(1)}(z) \quad \text{and} \quad b_0^{**(1)}(\zeta) \equiv f'(\zeta)\, b_0^{(1)}(f(\zeta)).$$
This shows that $x_0^{(1)}$ is a relative invariant of weight 1 and completes the proof. $\square$

COROLLARY 30.3. *Let $\boldsymbol{P}$ denote a nonzero polynomial in x_0 over $\mathbb{Q}$ and let n denote a positive integer. Then, the polynomial $\left(x_0^{(1)}\right)^n \boldsymbol{P}$ is a relative invariant of weight n for the equations (30.1).*

PROOF. Proposition 30.2 shows that $\left(x_0^{(1)}\right)^n$ is a relative invariant of weight n for the equations (30.1). Since $\boldsymbol{P}$ is either a nonzero rational number or a relative invariant of weight 0, we obtain the assertion about $\left(x_0^{(1)}\right)^n \boldsymbol{P}$. $\square$

The equations (30.1) possess many relative invariants that are not given by Corollary 30.3. By modifying the MATHEMATICA instructions of pages 172–177 to our present context, we have found additional relative invariants. For instance, the polynomial

$$\begin{aligned}
(30.18) \quad \boldsymbol{\mathcal{I}}_4 \equiv\; & (x_1)^2 \left(x_0^{(1)}\right)^2 + x_0 (x_1)^2 \left(x_0^{(1)}\right)^2 - 4 x_2 \left(x_0^{(1)}\right)^2 \\
& - 8 x_0 x_2 \left(x_0^{(1)}\right)^2 - 4 (x_0)^2 x_2 \left(x_0^{(1)}\right)^2 - 2 x_1 \left(x_0^{(1)}\right)^3 \\
& + 2 \left(x_0^{(1)}\right)^2 x_1^{(1)} + 2 x_0 \left(x_0^{(1)}\right)^2 x_1^{(1)} - 2 \left(x_0^{(1)}\right)^2 x_0^{(2)} \\
& - 3 \left(x_0^{(2)}\right)^2 - 3 x_0 \left(x_0^{(2)}\right)^2 + 2 x_0^{(1)} x_0^{(3)} + 2 x_0 x_0^{(1)} x_0^{(3)}
\end{aligned}$$

in $\mathcal{D}$ is a relative invariant of weight 4 for the equations (30.1); the polynomial

$$\begin{aligned}
(30.19) \quad \boldsymbol{\mathcal{I}}_6 \equiv\; & + (x_1)^2 \left(x_0^{(1)}\right)^4 - 8 x_2 \left(x_0^{(1)}\right)^4 - 8 x_0 x_2 \left(x_0^{(1)}\right)^4 \\
& + 2 x_1 \left(x_0^{(1)}\right)^3 x_1^{(1)} + 2 x_0 x_1 \left(x_0^{(1)}\right)^3 x_1^{(1)} - 4 \left(x_0^{(1)}\right)^3 x_2^{(1)} \\
& - 8 x_0 \left(x_0^{(1)}\right)^3 x_2^{(1)} - 4 (x_0)^2 \left(x_0^{(1)}\right)^3 x_2^{(1)} - 2 (x_1)^2 \left(x_0^{(1)}\right)^2 x_0^{(2)} \\
& - 2 x_0 (x_1)^2 \left(x_0^{(1)}\right)^2 x_0^{(2)} + 8 x_2 \left(x_0^{(1)}\right)^2 x_0^{(2)} \\
& + 16 x_0 x_2 \left(x_0^{(1)}\right)^2 x_0^{(2)} + 8 (x_0)^2 x_2 \left(x_0^{(1)}\right)^2 x_0^{(2)} \\
& + 2 x_1 \left(x_0^{(1)}\right)^3 x_0^{(2)} - 4 \left(x_0^{(1)}\right)^2 x_1^{(1)} x_0^{(2)} - 4 x_0 \left(x_0^{(1)}\right)^2 x_1^{(1)} x_0^{(2)} \\
& + \left(x_0^{(1)}\right)^2 \left(x_0^{(2)}\right)^2 + 12 \left(x_0^{(2)}\right)^3 + 12 x_0 \left(x_0^{(2)}\right)^3 \\
& + 2 \left(x_0^{(1)}\right)^3 x_1^{(2)} + 2 x_0 \left(x_0^{(1)}\right)^3 x_1^{(2)} - 12 x_0^{(1)} x_0^{(2)} x_0^{(3)} \\
& - 12 x_0 x_0^{(1)} x_0^{(2)} x_0^{(3)} + 2 \left(x_0^{(1)}\right)^2 x_0^{(4)} + 2 x_0 \left(x_0^{(1)}\right)^2 x_0^{(4)}
\end{aligned}$$

as well as $\boldsymbol{\mathcal{J}}_6$ in (30.24) are relative invariants of weight 6. Also, for any nonzero polynomial $\boldsymbol{P}$ in x_0 over $\mathbb{Q}$, the polynomials $\boldsymbol{P}\boldsymbol{\mathcal{I}}_4$, $\boldsymbol{P}\boldsymbol{\mathcal{I}}_6$, and $\boldsymbol{P}\boldsymbol{\mathcal{J}}_6$ are relative invariants of respective weights 4, 6, and 6 for the equations (30.1).

PROPOSITION 30.4. *For the equations* (30.1), *the polynomial defined in* $\mathcal{D}$ *by*

$$(30.20) \quad u_2 \equiv (1+x_0)^2 x_2 + \frac{1}{2} x_0^{(1)} x_1 - \frac{1}{2}(1+x_0) x_1^{(1)} - \frac{1}{4}(1+x_0)(x_1)^2$$

is an isobaric semi-invariant of the first kind of weight 2. *Moreover, for any relative invariant* $\mathcal{I}$ *of these equations, there is an integer* $n \geq 0$ *such that* $(1+x_0)^n \mathcal{I}$ *is expressible as a differential-polynomial combination over* $\mathbb{Q}$ *of* x_0 *and* u_2.

PROOF. We use (30.3)–(30.5) to verify that (30.20) yields $u_2^*(z) \equiv u_2(z)$. Thus, we find that u_2 is an isobaric semi-invariant of the first kind of weight 2.

Let $\mathcal{I}$ in $\mathcal{D}$ be a relative invariant for the equations (30.1).

Case 1. Suppose that $\mathcal{I}$ effectively involves $x_2^{(j)}$, for some $j \geq 0$. Let k be the greatest integer j such that $x_2^{(j)}$ is effectively involved in $\mathcal{I}$. Let $\mathcal{E}$ be the quotient field for $\mathcal{D}$ and let $'$ be the unique derivation for $\mathcal{E}$ that is an extension of the derivation for $\mathcal{D}$. In $\mathcal{E}$, we rewrite (30.20) to obtain

$$(30.21) \quad x_2 \equiv \frac{u_2 - \frac{1}{2}x_0^{(1)} x_1 + \frac{1}{2}(1+x_0)x_1^{(1)} + \frac{1}{4}(1+x_0)(x_1)^2}{(1+x_0)^2}.$$

By repeatedly applying the derivation $'$ for $\mathcal{E}$ to (30.21), we obtain expressions for $x_2, x_2^{(1)}, \ldots, x_2^{(k)}$ as fractions in $\mathcal{E}$ with denominators that are integral powers of $(1+x_0)$. We successively substitute $x_2^{(k)}, x_2^{(k-1)}, \ldots, x_2$ in this order into $\mathcal{I}$ to eliminate each $x_2^{(j)}$, for $j \geq 0$; then, we rewrite the resulting expression in the form

$$(30.22) \quad (1+x_0)^n \mathcal{I} \equiv \mathcal{J} + \mathcal{K},$$

where $n \geq 0$, $\mathcal{J}$ is a differential-polynomial combination over $\mathbb{Q}$ of x_0 and u_2, while $\mathcal{K}$ is a differential-polynomial combination over $\mathbb{Q}$ of x_0, u_2, x_1 such that: if T is a nonzero term of $\mathcal{K}$, then T has a factor of the form $x_1^{(j)}$, for some $j \geq 0$.

Case 2. Suppose that $\mathcal{I}$ does not involve any $x_2^{(j)}$ having $j \geq 0$. Then, $\mathcal{I}$ clearly has a representation (30.22) with $n = 0$.

To complete the proof in terms of a given (30.22) for $\mathcal{I}$, it remains for us to show that $\mathcal{K} \equiv 0$. For an indirect argument, suppose that $\mathcal{K} \not\equiv 0$. Then, we see that $x_0, \mathcal{I}, \mathcal{J}$, and $\mathcal{K} \equiv (1+x_0)^n \mathcal{I} - \mathcal{J}$ are semi-invariants of the first kind. The reasoning for Corollary 4.3 on page 37 shows that there are meromorphic functions $b_0(z), b_1(z), b_2(z)$ on a region Ω such that the function $K(z)$ obtained by substituting $b_i^{(j)}(z)$ for $x_i^{(j)}$ in $\mathcal{K}$ satisfies $K(z) \not\equiv 0$. By replacing $b_0(z)$ with $b_0(z) + \epsilon z^p$ for some real number $\epsilon > 0$ and some positive integer p, we can assume that $b_0(z), b_1(z), b_2(z)$ on Ω satisfy $K(z) \not\equiv 0$ and $b_0(z) \not\equiv -1$. We consider the corresponding differential equation (30.1) that has them as coefficients. Let $\rho(z) \not\equiv 0$ be a meromorphic function on a subregion $\mathcal{U}$ of Ω that satisfies

$$(30.23) \quad (2b_0(z) + 2)\rho'(z) + b_1(z)\rho(z) \equiv 0, \quad \text{on } \mathcal{U}.$$

In view of (30.4) and (30.23), we note that the substitution $y(z) = \rho(z) v(z)$ transforms the restriction to $\mathcal{U}$ of our particular (30.1) into an equation (30.2) on $\mathcal{U}$ that has $b_1^*(z) \equiv 0$. Thus, the function $K^*(z)$ on $\mathcal{U}$ obtained by substituting $b_i^{*(j)}(z)$ from this particular (30.2) for $x_i^{(j)}$ in $\mathcal{K}$ satisfies $K^*(z) \equiv 0$. Since $\mathcal{K}$ is a semi-invariant of the first kind, we find that $K(z) \equiv K^*(z) \equiv 0$, on $\mathcal{U}$, and therefore have the contradiction $K(z) \equiv 0$, on Ω. This yields $\mathcal{K} \equiv 0$ and completes the proof. □

EXAMPLE 30.5. For $(1+x_0)^n \mathcal{I}$ in Proposition 30.4, some relative invariants require a positive integer n. The simplest example we have found is $\mathcal{I} \equiv \mathcal{J}_6$, where $\mathcal{J}_6$ is the relative invariant of weight 6 for the equations (30.1) given by

$$
\begin{aligned}
(30.24) \quad \mathcal{J}_6 \equiv & +4(x_1)^4(x_0^{(1)})^2 + 4x_0(x_1)^4(x_0^{(1)})^2 - 32(x_1)^2 x_2 (x_0^{(1)})^2 \\
& -64 x_0 (x_1)^2 x_2 (x_0^{(1)})^2 - 32(x_0)^2 (x_1)^2 x_2 (x_0^{(1)})^2 + 64(x_2)^2 (x_0^{(1)})^2 \\
& +192 x_0 (x_2)^2 (x_0^{(1)})^2 + 192 (x_0)^2 (x_2)^2 (x_0^{(1)})^2 + 64 (x_0)^3 (x_2)^2 (x_0^{(1)})^2 \\
& -16(x_1)^3 (x_0^{(1)})^3 + 64 x_1 x_2 (x_0^{(1)})^3 + 64 x_0 x_1 x_2 (x_0^{(1)})^3 + 124 x_2 (x_0^{(1)})^3 \\
& +16(x_1)^2 (x_0^{(1)})^2 x_1^{(1)} + 16 x_0 (x_1)^2 (x_0^{(1)})^2 x_1^{(1)} - 64 x_2 (x_0^{(1)})^2 x_1^{(1)} \\
& -128 x_0 x_2 (x_0^{(1)})^2 x_1^{(1)} - 64 (x_0)^2 x_2 (x_0^{(1)})^2 x_1^{(1)} - 62 x_1 (x_0^{(1)})^3 x_1^{(1)} \\
& +17 (x_0^{(1)})^2 (x_1^{(1)})^2 + 17 x_0 (x_0^{(1)})^2 (x_1^{(1)})^2 + 56 (x_0^{(1)})^3 x_2^{(1)} + 56 x_0 (x_0^{(1)})^3 x_2^{(1)} \\
& +14 (x_1)^2 (x_0^{(1)})^2 x_0^{(2)} - 48 x_2 (x_0^{(1)})^2 x_0^{(2)} - 48 x_0 x_2 (x_0^{(1)})^2 x_0^{(2)} \\
& -5 x_1 x_0^{(1)} x_1^{(1)} x_0^{(2)} - 5 x_0 x_1 x_0^{(1)} x_1^{(1)} x_0^{(2)} + 31 (x_0^{(1)})^2 x_1^{(1)} x_0^{(2)} + 10 x_0^{(1)} x_2^{(1)} x_0^{(2)} \\
& +20 x_0 x_0^{(1)} x_2^{(1)} x_0^{(2)} + 10 (x_0)^2 x_0^{(1)} x_2^{(1)} x_0^{(2)} - 20(x_1)^2 (x_0^{(2)})^2 \\
& -20 x_0 (x_1)^2 (x_0^{(2)})^2 + 80 x_2 (x_0^{(2)})^2 + 160 x_0 x_2 (x_0^{(2)})^2 \\
& +80 (x_0)^2 x_2 (x_0^{(2)})^2 + 45 x_1 x_0^{(1)} (x_0^{(2)})^2 - 40 x_1^{(1)} (x_0^{(2)})^2 - 40 x_0 x_1^{(1)} (x_0^{(2)})^2 \\
& -140 (x_0^{(2)})^3 + x_1 (x_0^{(1)})^2 x_1^{(2)} + x_0 x_1 (x_0^{(1)})^2 x_1^{(2)} - 31 (x_0^{(1)})^3 x_1^{(2)} \\
& -5 x_0^{(1)} x_0^{(2)} x_1^{(2)} - 5 x_0 x_0^{(1)} x_0^{(2)} x_1^{(2)} - 2(x_0^{(1)})^2 x_2^{(2)} - 4 x_0 (x_0^{(1)})^2 x_0^{(2)} \\
& -2(x_0)^2 (x_0^{(1)})^2 x_2^{(2)} + 15(x_1)^2 x_0^{(1)} x_0^{(3)} + 15 x_0 (x_1)^2 x_0^{(1)} x_0^{(3)} \\
& -60 x_2 x_0^{(1)} x_0^{(3)} - 120 x_0 x_2 x_0^{(1)} x_0^{(3)} - 60 (x_0)^2 x_2 x_0^{(1)} x_0^{(3)} \\
& -31 x_1 (x_0^{(1)})^2 x_0^{(3)} + 30 x_0^{(1)} x_1^{(1)} x_0^{(3)} + 30 x_0 x_0^{(1)} x_1^{(1)} x_0^{(3)} + 155 x_0^{(1)} x_0^{(2)} x_0^{(3)} \\
& +10 (x_0^{(3)})^2 + 10 x_0 (x_0^{(3)})^2 + (x_0^{(1)})^2 x_1^{(3)} + x_0 (x_0^{(1)})^2 x_1^{(3)} - 31 (x_0^{(1)})^2 x_0^{(4)} \\
& -10 x_0^{(2)} x_0^{(4)} - 10 x_0 x_0^{(2)} x_0^{(4)} + x_0^{(1)} x_0^{(5)} + x_0 x_0^{(1)} x_0^{(5)}.
\end{aligned}
$$

Machine computations based on the technique in the proof of Proposition 30.4 yield

$$
\begin{aligned}
(30.25) \quad (1+x_0) \mathcal{J}_6 \equiv & +64(u_2)^2 (x_0^{(1)})^2 + 64 u_2^{(1)} (x_0^{(1)})^3 - 2 (x_0^{(1)})^2 u_2^{(2)} \\
& -2 x_0 (x_0^{(1)})^2 u_2^{(2)} + 10 u_2^{(1)} x_0^{(1)} x_0^{(2)} + 10 x_0 u_2^{(1)} x_0^{(1)} x_0^{(2)} - 64 u_2 (x_0^{(1)})^2 x_0^{(2)} \\
& +80 u_2 (x_0^{(2)})^2 + 80 u_2 x_0 (x_0^{(2)})^2 - 140 (x_0^{(2)})^3 - 140 x_0 (x_0^{(2)})^3 - 60 u_2 x_0^{(1)} x_0^{(3)} \\
& -60 u_2 x_0 x_0^{(1)} x_0^{(3)} + 155 x_0^{(1)} x_0^{(2)} x_0^{(3)} + 155 x_0 x_0^{(1)} x_0^{(2)} x_0^{(3)} + 10 (x_0^{(3)})^2 \\
& +20 x_0 (x_0^{(3)})^2 + 10 (x_0)^2 (x_0^{(3)})^2 - 31 (x_0^{(1)})^2 x_0^{(4)} - 31 x_0 (x_0^{(1)})^2 x_0^{(4)} \\
& -10 x_0^{(2)} x_0^{(4)} - 20 x_0 x_0^{(2)} x_0^{(4)} - 10 (x_0)^2 x_0^{(2)} x_0^{(4)} + x_0^{(1)} x_0^{(5)} + 2 x_0 x_0^{(1)} x_0^{(5)} \\
& +(x_0)^2 x_0^{(1)} x_0^{(5)}.
\end{aligned}
$$

The right member of (30.25) is not divisible by $(1+x_0)$. Consequently, the positive integer $n = 1$ is needed by $\mathcal{J}_6$ for Proposition 30.4.

EXAMPLE 30.6. For both $\mathcal{I}_4$ in (30.18) and $\mathcal{I}_6$ in (30.19), the integer n in Proposition 30.4 may be selected as 0. Namely, we have the identities

$$\mathcal{I}_4 \equiv -4u_2(x_0^{(1)})^2 - 2(x_0^{(1)})^2 x_0^{(2)} - 3(1+x_0)(x_0^{(2)})^2 \tag{30.26}$$
$$+ 2(1+x_0) x_0^{(1)} x_0^{(3)}$$

and

$$\mathcal{I}_6 \equiv -4u_2^{(1)}(x_0^{(1)})^3 + 8u_2(x_0^{(1)})^2 x_0^{(2)} + (x_0^{(1)})^2 (x_0^{(2)})^2 \tag{30.27}$$
$$+ 12(x_0^{(2)})^3 + 12 x_0 (x_0^{(2)})^3 - 12 x_0^{(1)} x_0^{(2)} x_0^{(3)}$$
$$- 12 x_0 x_0^{(1)} x_0^{(2)} x_0^{(3)} + 2(x_0^{(1)})^2 x_0^{(4)} + 2 x_0 (x_0^{(1)})^2 x_0^{(4)}.$$

OBSERVATION 30.7. When the left member of (30.1) is regarded as a quadratic form in y'', y', y, the determinant of its matrix equals $-\frac{1}{4} b_0(z)$. Clearly, the left member of (30.1) has a nontrivial factorization if and only if $b_0(z) \equiv 0$.

Suppose that $b_0(z) \equiv -1$. Then, the local solutions of (30.1) are given by the local solutions of $y'(z) - w_0(z) y(z) = 0$ as $w_0(z)$ ranges over the local solutions of $w'(z) + b_1(z) w(z) + b_2(z) = 0$.

30.2. Another class of homogeneous quadratic equations

Here, we examine the class of differential equations having the form

$$(y'(z))^2 + b_0(z) y''(z) y(z) + b_1(z) y'(z) y(z) + b_2(z) (y(z))^2 = 0, \tag{30.28}$$

where $b_0(z)$, $b_1(z)$, $b_2(z)$ are meromorphic functions on a region Ω.

For any meromorphic $\rho(z) \not\equiv 0$ on Ω, the change $y(z) = \rho(z) v(z)$ of the dependent variable from y to v transforms (30.28) into

$$(v'(z))^2 + b_0^*(z) v''(z) v(z) + b_1^*(z) v'(z) v(z) + b_2^*(z) (v(z))^2 = 0, \tag{30.29}$$

where the coefficients are given on Ω by

$$b_0^*(z) \equiv b_0(z), \tag{30.30}$$

$$b_1^*(z) \equiv b_1(z) + 2 b_0(z) \frac{\rho'(z)}{\rho(z)} + 2 \frac{\rho'(z)}{\rho(z)}, \tag{30.31}$$

and

$$b_2^*(z) \equiv b_2(z) + b_1(z) \frac{\rho'(z)}{\rho(z)} + b_0(z) \frac{\rho''(z)}{\rho(z)} + \left(\frac{\rho'(z)}{\rho(z)}\right)^2. \tag{30.32}$$

For any univalent analytic function $\zeta = g(z)$ on Ω whose inverse is designated by $z = f(\zeta)$ on $\Omega^{**} = g(\Omega)$, the change $z = f(\zeta)$ of the independent variable from z to ζ and the notation $u(\zeta) = (y \circ f)(\zeta)$ transform (30.28) into

$$(u'(\zeta))^2 + b_0^{**}(\zeta) u''(\zeta) u(\zeta) + b_1^{**}(\zeta) u'(\zeta) u(\zeta) + b_2^{**}(\zeta) (u(\zeta))^2 = 0, \tag{30.33}$$

where the coefficients are given on Ω^{**} by

$$b_0^{**}(\zeta) \equiv b_0(f(\zeta)), \tag{30.34}$$

$$b_1^{**}(\zeta) \equiv f'(\zeta) b_1(f(\zeta)) - \frac{f''(\zeta)}{f'(\zeta)} b_0(f(\zeta)), \tag{30.35}$$

and

$$(30.36) \qquad b_2^{**}(\zeta) \equiv \bigl(f'(\zeta)\bigr)^2 b_2\bigl(f(\zeta)\bigr).$$

We have selected the notation for (30.28)–(30.36) so the context of Subsection 30.1.1 on page 348 can be used here without alteration.

PROPOSITION 30.8. *Suppose that $\boldsymbol{P}$ is a semi-invariant of the second kind for the equations (30.28). Then, $\boldsymbol{P}$ is an isobaric polynomial and the only integer s that satisfies (30.10) for $\boldsymbol{P}$ is the weight of $\boldsymbol{P}$.*

PROOF. It is sufficient to replace (30.1) and (30.7)–(30.9) with (30.28) and (30.34)–(30.36) throughout the statement and proof of Proposition 30.1. □

PROPOSITION 30.9. *The nonconstant polynomial combinations over $\mathbb{Q}$ of $\boldsymbol{x}_0$ are the relative invariants of weight 0 for the equations (30.28). Moreover, $\boldsymbol{x}_0^{(1)}$ is a relative invariant of weight 1 for the equations (30.28).*

PROOF. We use (30.30) to see that $\boldsymbol{x}_0$ is a semi-invariant of the first kind for the equations (30.28). Moreover, (30.34) shows that $\boldsymbol{x}_0$ is a semi-invariant of the second kind of weight 0 for the equations (30.28). Consequently, a polynomial in $\mathcal{D}$ is a relative of weight 0 for the equations (30.28) if and only if it is a nonconstant polynomial combination over $\mathbb{Q}$ of $\boldsymbol{x}_0$.

We differentiate (30.30) and (30.34) to obtain

$$b_0^{*(1)}(z) \equiv b_0^{(1)}(z) \quad \text{and} \quad b_0^{**(1)}(\zeta) \equiv f'(\zeta)\, b_0^{(1)}\bigl(f(\zeta)\bigr).$$

This shows that $\boldsymbol{x}_0^{(1)}$ is a relative invariant of weight 1 and completes the proof. □

COROLLARY 30.10. *Let $\boldsymbol{P}$ denote a nonzero polynomial in $\boldsymbol{x}_0$ over $\mathbb{Q}$ and let n denote a positive integer. Then, the polynomial $\bigl(\boldsymbol{x}_0^{(1)}\bigr)^n \boldsymbol{P}$ is a relative invariant of weight n for the equations (30.28).*

PROOF. Proposition 30.9 shows that $\bigl(\boldsymbol{x}_0^{(1)}\bigr)^n$ is a relative invariant of weight n for the equations (30.28). Since $\boldsymbol{P}$ is either a nonzero rational number or a relative invariant of weight 0, we obtain the assertion about $\bigl(\boldsymbol{x}_0^{(1)}\bigr)^n \boldsymbol{P}$. □

The equations (30.28) possess many relative invariants that are not given by Corollary 30.10. By modifying the MATHEMATICA instructions of pages 172–177 to our present context, we have found additional relative invariants.

For instance, the polynomial

$$(30.37) \qquad \begin{aligned}\mathcal{K}_4 \equiv\ & \bigl(\boldsymbol{x}_1\bigr)^2\bigl(\boldsymbol{x}_0^{(1)}\bigr)^2 + \boldsymbol{x}_0\bigl(\boldsymbol{x}_1\bigr)^2\bigl(\boldsymbol{x}_0^{(1)}\bigr)^2 - 4\boldsymbol{x}_2\bigl(\boldsymbol{x}_0^{(1)}\bigr)^2 \\ & - 8\boldsymbol{x}_0\boldsymbol{x}_2\bigl(\boldsymbol{x}_0^{(1)}\bigr)^2 - 4\bigl(\boldsymbol{x}_0\bigr)^2\boldsymbol{x}_2\bigl(\boldsymbol{x}_0^{(1)}\bigr)^2 - 2\boldsymbol{x}_0\boldsymbol{x}_1\bigl(\boldsymbol{x}_0^{(1)}\bigr)^3 \\ & + 2\boldsymbol{x}_0\bigl(\boldsymbol{x}_0^{(1)}\bigr)^2\boldsymbol{x}_1^{(1)} + 2\bigl(\boldsymbol{x}_0\bigr)^2\bigl(\boldsymbol{x}_0^{(1)}\bigr)^2\boldsymbol{x}_1^{(1)} + 2\boldsymbol{x}_0\bigl(\boldsymbol{x}_0^{(1)}\bigr)^2\boldsymbol{x}_0^{(2)} \\ & - 3\bigl(\boldsymbol{x}_0\bigr)^2\bigl(\boldsymbol{x}_0^{(2)}\bigr)^2 - 3\bigl(\boldsymbol{x}_0\bigr)^3\bigl(\boldsymbol{x}_0^{(2)}\bigr)^2 + 2\bigl(\boldsymbol{x}_0\bigr)^2\boldsymbol{x}_0^{(1)}\boldsymbol{x}_0^{(3)} \\ & + 2\bigl(\boldsymbol{x}_0\bigr)^3\boldsymbol{x}_0^{(1)}\boldsymbol{x}_0^{(3)}\end{aligned}$$

30.2. ANOTHER CLASS OF HOMOGENEOUS QUADRATIC EQUATIONS

in $\mathcal{D}$ is a relative invariant of weight 4 for the equations (30.28). Also, we have found that the polynomial

$$
\begin{aligned}
(30.38)\quad \mathcal{K}_6 \equiv\ & +(x_1)^2(x_0^{(1)})^4 - 8x_2(x_0^{(1)})^4 - 8x_0 x_2(x_0^{(1)})^4 - 2x_1(x_0^{(1)})^5 \\
& + 2x_1(x_0^{(1)})^3 x_1^{(1)} + 2x_0 x_1(x_0^{(1)})^3 x_1^{(1)} + 2(x_0^{(1)})^4 x_1^{(1)} + 2x_0(x_0^{(1)})^4 x_1^{(1)} \\
& - 4(x_0^{(1)})^3 x_2^{(1)} - 8x_0(x_0^{(1)})^3 x_2^{(1)} - 4(x_0)^2(x_0^{(1)})^3 x_2^{(1)} - 2(x_1)^2(x_0^{(1)})^2 x_0^{(2)} \\
& - 2x_0(x_1)^2(x_0^{(1)})^2 x_0^{(2)} + 8x_2(x_0^{(1)})^2 x_0^{(2)} + 16 x_0 x_2(x_0^{(1)})^2 x_0^{(2)} \\
& + 8(x_0)^2 x_2(x_0^{(1)})^2 x_0^{(2)} + 2x_0 x_1(x_0^{(1)})^3 x_0^{(2)} + 2(x_0^{(1)})^4 x_0^{(2)} \\
& - 4x_0(x_0^{(1)})^2 x_1^{(1)} x_0^{(2)} - 4(x_0)^2(x_0^{(1)})^2 x_1^{(1)} x_0^{(2)} - 10 x_0(x_0^{(1)})^2 (x_0^{(2)})^2 \\
& - 9(x_0)^2(x_0^{(1)})^2 (x_0^{(2)})^2 + 12(x_0)^2(x_0^{(2)})^3 + 12(x_0)^3(x_0^{(2)})^3 \\
& + 2x_0(x_0^{(1)})^3 x_1^{(2)} + 2(x_0)^2(x_0^{(1)})^3 x_1^{(2)} + 6x_0(x_0^{(1)})^3 x_0^{(3)} \\
& + 6(x_0)^2(x_0^{(1)})^3 x_0^{(3)} - 12(x_0)^2 x_0^{(1)} x_0^{(2)} x_0^{(3)} - 12(x_0)^3 x_0^{(1)} x_0^{(2)} x_0^{(3)} \\
& + 2(x_0)^2(x_0^{(1)})^2 x_0^{(4)} + 2(x_0)^3(x_0^{(1)})^2 x_0^{(4)}
\end{aligned}
$$

in $\mathcal{D}$ is a relative invariants of weight 6 for the equations (30.28). In addition, for any nonzero polynomial P in x_0 over $\mathbb{Q}$, the polynomials $P\mathcal{K}_4$ and $P\mathcal{K}_6$ are relative invariants of respective weights 4 and 6 for the equations (30.28).

PROPOSITION 30.11. *For the equations (30.28), the polynomial defined in $\mathcal{D}$ by*

$$(30.39)\quad v_2 \equiv (1+x_0)^2 x_2 + \frac{1}{2} x_0 x_0^{(1)} x_1 - \frac{1}{2}(1+x_0) x_0 x_1^{(1)} - \frac{1}{4}(1+x_0)(x_1)^2$$

is an isobaric semi-invariant of the first kind of weight 2. Moreover, for any relative invariant $\mathcal{I}$ of these equations, there is an integer $n \geq 0$ such that $(1+x_0)^n \mathcal{I}$ is expressible as a differential-polynomial combination over $\mathbb{Q}$ of x_0 and v_2.

PROOF. We use (30.30)–(30.32) to verify that (30.39) yields $v_2^*(z) \equiv v_2(z)$. Thus, we find that v_2 is an isobaric semi-invariant of the first kind of weight 2.

Let $\mathcal{I}$ in $\mathcal{D}$ be a relative invariant for the equations (30.28). In the quotient field $\mathcal{E}$ for $\mathcal{D}$ with the unique derivation $'$ that is an extension of the derivation for $\mathcal{D}$, we rewrite (30.39) to obtain

$$(30.40)\quad x_2 \equiv \frac{v_2 - \frac{1}{2} x_0 x_0^{(1)} x_1 + \frac{1}{2}(1+x_0) x_0 x_1^{(1)} + \frac{1}{4}(1+x_0)(x_1)^2}{(1+x_0)^2}.$$

By employing (30.40) in place of (30.21), we repeat the argument used for Case 1 and Case 2 in the proof of Proposition 30.4 to obtain

$$(30.41)\quad (1+x_0)^n \mathcal{I} \equiv \mathcal{J} + \mathcal{K},$$

where $n \geq 0$, $\mathcal{J}$ is a differential-polynomial combination over $\mathbb{Q}$ of x_0 and v_2, while $\mathcal{K}$ is a differential-polynomial combination over $\mathbb{Q}$ of x_0, v_2, x_1 such that: if T is a nonzero term of $\mathcal{K}$, then T has a factor of the form $x_1^{(j)}$, for some $j \geq 0$.

To complete the proof, it remains for us to show that (30.41) has $\mathcal{K} \equiv 0$. For an indirect argument, suppose that $\mathcal{K} \not\equiv 0$. Then, we see that x_0, $\mathcal{I}$, $\mathcal{J}$, and $\mathcal{K} \equiv (1+x_0)^n \mathcal{I} - \mathcal{J}$ are semi-invariants of the first kind. The form of (30.31) for (30.28) is the same as that of (30.4) for (30.1). Thus, we can replace (30.1) with (30.28) and (30.4) with (30.31) throughout the indirect argument on page 351 involving (30.23) to obtain a contradiction to the supposition $\mathcal{K} \not\equiv 0$. Thus, with $\mathcal{K} \equiv 0$ in (30.41), we see that the left member of (30.41) is expressible as a differential-polynomial combination over $\mathbb{Q}$ of x_0 and v_2. This completes the proof. $\square$

EXAMPLE 30.12. For $\mathcal{K}_4$ in (30.37) and $\mathcal{K}_6$ in (30.38), the technique indicated for the proof of Proposition 30.11 yields

$$\mathcal{K}_4 \equiv -4v_2(x_0^{(1)})^2 + 2x_0(x_0^{(1)})^2 x_0^{(2)} - 3(x_0)^2(x_0^{(2)})^2 - 3(x_0)^3(x_0^{(2)})^2 \\ + 2(x_0)^2 x_0^{(1)} x_0^{(3)} + 2(x_0)^3 x_0^{(1)} x_0^{(3)}$$

and

$$\mathcal{K}_6 \equiv -4v_2^{(1)}(x_0^{(1)})^3 + 8v_2(x_0^{(1)})^2 x_0^{(2)} + 2(x_0^{(1)})^4 x_0^{(2)} - 10x_0(x_0^{(1)})^2(x_0^{(2)})^2 \\ - 9(x_0)^2(x_0^{(1)})^2(x_0^{(2)})^2 + 12(x_0)^2(x_0^{(2)})^3 + 12(x_0)^3(x_0^{(2)})^3 \\ + 6x_0(x_0^{(1)})^3 x_0^{(3)} + 6(x_0)^2(x_0^{(1)})^3 x_0^{(3)} - 12(x_0)^2 x_0^{(1)} x_0^{(2)} x_0^{(3)} \\ - 12(x_0)^3 x_0^{(1)} x_0^{(2)} x_0^{(3)} + 2(x_0)^2(x_0^{(1)})^2 x_0^{(4)} + 2(x_0)^3(x_0^{(1)})^2 x_0^{(4)}.$$

Thus, for both $\mathcal{K}_4$ and $\mathcal{K}_6$, the procedure is applicable with $n = 0$. However, the equations (30.28) have a relative invariant of weight 6 that requires $n \geq 1$ and for which $n = 1$ is suitable. The details are similar to those for (30.24) and (30.25).

30.3. The character of x_0 as a polynomial absolute invariant

Rational relative invariants and absolute invariants for the equations (1.1) were formulated in Definition 20.3 on page 220 and in Definition 20.16 on page 243. These concepts can be similarly introduced for each of the other classes of equations that we have considered. In particular, their formulation for the equations (30.1) and the equations (30.28) requires the quotient field $\mathcal{E}$ of $\mathcal{D}$. In this regard, we see that the situation for Chapter 30 is quite unlike that for Chapters 1–28 where each polynomial relative invariant has weight ≥ 2 and each absolute invariant can not be a polynomial. Namely, the equations (30.1) and the equations (30.28) have x_0 as a polynomial relative invariant of weight 0. In other words, $\mathcal{D}$ contains x_0 as a polynomial absolute invariant.

Bibliography

1. P. Appell, *Sur la transformation des équations différentielles linéaires*, C. R. Acad. Sci. Paris **91** (1880), 211–214.
2. _____, *Mémoire sur les équations différentielles linéaires*, Ann. Sci. École Norm. Sup. (2) **10** (1881), 391–424.
3. _____, *Sur les équations différentielles algébriques et homogènes par rapport à la fonction inconnue et à ses dérivées*, C. R. Acad. Sci. Paris **104** (1887), 1776–1779.
4. _____, *Sur les invariants des équations différentielles*, C. R. Acad. Sci. Paris **105** (1887), 55–58.
5. _____, *Sur une classe d'équations réductibles aux équations linéaires*, C. R. Acad. Sci. Paris **107** (1888), 776–778.
6. _____, *Équations différentielles homogènes du second ordre à coefficients constants*, Ann. Fac. Sci. Toulouse Math. (1) **3** (1889), K1–K12.
7. _____, *Sur les invariants de quelques équations différentielles*, J. Math. Pures Appl. (4) **5** (1889), 361–423.
8. L. Bieberbach, *Theorie der gewöhnlichen Differentialgleichungen auf funktionentheoretischer Grundlage dargestellt*, Second Edition, Springer-Verlag, Berlin, 1965.
9. M. Bôcher, *Introduction to Higher Algebra*, Macmillan, New York, 1907 (reprinted in 1949).
10. N. Bourbaki, *Éléments de Mathématique, Livre II, Algèbre, Chapitres IV et V*, Hermann, Paris, 1950.
11. _____, *Elements of Mathematics, Algebra II, Chapters 4–7*, Springer-Verlag, Berlin, 1989.
12. C. L. Bouton, *Invariants of the general linear differential equation and their relation to the theory of continuous groups*, Amer. J. Math. **21** (1899), 25–84.
13. D. Caligo, *Sopra una classe di equazioni differenziali non lineari*, Mem. Accad. Sci. Torino Cl. Sci. Fis. Mat. Nat. (3) **1** (1952), 1–24.
14. _____, *Sulla integrazione delle equazioni differenziali del secondo ordine a riferimento razionale*, Rend. Mat. Appl. (5) **11** (1952), 299–314.
15. R. Campbell, *Les Intégrales Eulériennes et Leurs Applications: Étude approfondie de la fonction gamma*, Dunod, Paris, 1966.
16. R. Chalkley, *On the second-order homogeneous quadratic differential equation*, Math. Ann. **141** (1960), 87–98.
17. _____, *Explicit solutions of an algebraic differential equation*, J. Differential Equations **35** (1980), 275–290.
18. _____, *New contributions to the related work of Paul Appell, Lazarus Fuchs, Georg Hamel, and Paul Painlevé on nonlinear differential equations whose solutions are free of movable branch points*, J. Differential Equations **68** (1987), 72–117.
19. _____, *Relative invariants for homogeneous linear differential equations*, J. Differential Equations **80** (1989), 107–153.
20. _____, *The differential equation $Q = 0$ in which Q is a quadratic form in y'', y', y having meromorphic coefficients*, Proc. Amer. Math. Soc. **116** (1992), 427–435.
21. _____, *A formula giving the known relative invariants for homogeneous linear differential equations*, J. Differential Equations **100** (1992), 379–404.
22. _____, *Semi-invariants and relative invariants for homogeneous linear differential equations*, J. Math. Anal. Appl. **176** (1993), 49–75.
23. _____, *A persymmetric determinant*, J. Math. Anal. Appl. **176** (1994), 107–117.
24. _____, *Basic Global Relative Invariants for Homogeneous Linear Differential Equations*, Memoirs Amer. Math. Soc. **156** (2002), Number 744, 1–204.

25. J. Cockle, *Correlations of analysis*, The London, Edinburgh, and Dublin Philosophical Magazine and Journal of Science (4) **24** (1862), 531–534.
26. _____, *On a differential criticoid*, Philos. Mag. (4) **50** (1875), 440–446.
27. C. M. Cosgrove, *New family of exact stationary axisymmetric gravitational fields generalizing the Tomimatsu-Sato solutions*, J. Phys. A **10** (1977), 1481–1524.
28. _____, *A new formulation of the field equations for the stationary axisymmetric vacuum gravitational field I. General theory*, J. Phys. A **11** (1978), 365–382.
29. D. R. Curtiss, *On the invariants of a homogeneous quadratic differential equation of the second order*, Amer. J. Math. **25** (1903), 365–382.
30. G. Fano, *Ueber lineare homogene Differentialgleichlungen mit algebraischen Relationen zwischen den Fundamentallösungen*, Math. Ann. **53** (1900), 493–590.
31. A. R. Forsyth, *Invariants, covariants and quotient derivatives associated with linear differential equations*, Philosophical Transactions of the Royal Society of London **179** (1888), 377–489.
32. L. Fuchs, *Über Differentialgleichungen deren Integrale feste Verzweigungspunkte besitzen*, Sitzungsberichte Akad. Wiss. Berlin (1884), 699-710.
33. G.-H. Halphen, *Sur les invariants des équations différentielles linéaires du quatrième ordre*, Acta Math. **3** (1883), 325–380; C. Jordan, H. Poincaré, É. Picard, and E. Vessiot (eds.), *Oeuvres de G.-H. Halphen, Tome III*, Gauthier-Villars, Paris, 1921, pp. 463–514.
34. _____, *Mémoire sur la réduction des équations différentielles linéaires aux formes intégrables*, Mémoires présentés par divers savants à l'Académie des Sciences de l'Institut de France (2) **28** (1884), 1–301; C. Jordan, H. Poincaré, É. Picard, and E. Vessiot (eds.), *Oeuvres de G.-H. Halphen, Tome III*, Gauthier-Villars, Paris, 1921, pp. 1–260.
35. C. Hermit, H. Poincaré, and E. Rouché, *Oeuvres de Laguerre, Tome I*, Gauthier-Villars, Paris, 1898, pp. 420–423.
36. M. J. M. Hill and A. Berry, *On differential equations with fixed branch points*, Proc. London Math. Soc. (2) **9** (1910), 231–234.
37. O. Hölder, *Ueber die Eigenschaft der Gammafunction keiner algebraischen Differentialgleichung zu genügen*, Math. Ann. **28** (1887), 1-13.
38. E. L. Ince, *Ordinary Differential Equations*, Longmans, Green, and Co., London, 1927. (Reprinted by Dover, New York, 1956.)
39. C. Jordan, H. Poincaré, É. Picard, and E. Vessiot, *Oeuvres de G.-H. Halphen, Tome III*, Gauthier-Villars, Paris, 1921.
40. D. E. Knuth, *The Art of Computer Programming: Volume 1, Fundamental Algorithms*, Third Edition, Addison-Wesley, Reading, MA, 1997.
41. E. R. Kolchin, *Differential Algebra and Algebraic Groups*, Academic Press, New York, 1973.
42. E. Laguerre, *Sur les équations différentielles linéaires du troisième ordre*, C. R. Acad. Sci. Paris **88** (1879), 116–119; C. Hermit, H. Poincaré, and E. Rouché (eds.), *Oeuvres de Laguerre, Tome I*, Gauthier-Villars, Paris, 1898, pp. 420–423.
43. _____, *Sur quelques invariants des équations différentielles linéaires*, C. R. Acad. Sci. Paris **88** (1879), 224–227; C. Hermit, H. Poincaré, and E. Rouché (eds.), *Oeuvres de Laguerre, Tome I*, Gauthier-Villars, Paris, 1898, pp. 424–427.
44. S. Lang, *Algebra*, Revised Third Edition, Springer, New York, 2002.
45. J. Liouville, *Mémoire sur l'intégration d'une classe d'équations différentielles du second ordre en quantités explicites*, J. Math. Pures Appl. (1) **4** (1839), 423–456.
46. F. Neuman, *Global Properties of Linear Ordinary Differential Equations*, Kluwer Academic Publishers, Dordrecht/Boston/London, 1991.
47. J. R. Ritt, *Differential Algebra*, Amer. Math. Soc. Colloq. Publ., Vol. 33, Amer. Math. Soc., New York, 1950.
48. I. Schur, *Vorlesungen über Invariantentheorie*, Springer-Verlag, Berlin, 1968.
49. P. R. Vein and P. Dale, *Determinants, their derivatives and nonlinear differential equations*, J. Math. Anal. Appl. **74** (1980), 599–634.
50. G. Wallenberg, *Ueber nichtlinear homogene Differentialgleichungen zweiter Ordnung*, J. Reine Angew. Math. **119** (1898), 87–113.
51. Wolfram Research Incorporated, *Mathematica*, Version 3.0, Champaign, Illinois, 1996.

Index

$Q_m = 0$, as (1.1), 4
$c_{i,j}(z)$, coefficient of (1.1), 4
$c_{i,j}^*(z)$, coefficient of (1.3), 5
$c_{i,j}^{**}(\zeta)$, coefficient of (1.5), 5
$H_{m,n} = 0$, as (21.1), 249
$c_{i_1, i_2, \ldots, i_n}(z)$, coefficient of (21.1), 249
$c_{i_1, i_2, \ldots, i_n}^*(z)$, coefficient of (21.3), 250
$c_{i_1, i_2, \ldots, i_n}^{**}(\zeta)$, coefficient of (21.5), 250
m, order of (1.1) or (21.1), 4, 249
$y(z) = \rho(z)\,v(z)$, 5, 249, 325, 335, 337, 347, 353
$c_{i,j}^*(z)$, formula for, 23, 135
$c_{i_1, i_2 \ldots, i_n}^*(z)$, formula for, 249
$z = f(\zeta)$, 5, 250, 325, 335, 339, 348, 353
$u(\zeta) = (y \circ f)(\zeta)$, 5, 250, 325, 335, 339, 348, 353
$c_{i,j}^{**}(\zeta)$, formula for, 30, 33, 137
$c_{i_1, i_2, \ldots, i_n}^{**}(\zeta)$, formula for, 250
$f(\zeta)$, 5, 250, 325, 335, 339, 348, 353
$g(z)$, 5, 250, 325, 335, 339, 348, 353
$w(\zeta) = (v \circ f)(\zeta)$, 47, 265, 331
$d_{i,j}(\zeta)$, coefficient of (5.1), 47
  when $d_{0,1}(\zeta) \equiv d_{0,2}(\zeta) \equiv 0$, 49, 69, 93
$d_{i,j}^{(k)}(\zeta)$, when $d_{0,1}(\zeta) \equiv d_{0,2}(\zeta) \equiv 0$, 73, 97
$d_{i_1, i_2, \ldots, i_n}(\zeta)$, coefficient of (23.1), 265
  when $d_{0, \ldots, 0, 1}(\zeta) \equiv d_{0, \ldots, 0, 2}(\zeta) \equiv 0$, 279, 293
$d_{i_1, i_2, \ldots, i_n}^{(k)}(\zeta)$
  when $d_{0, \ldots, 0, 1}(\zeta) \equiv d_{0, \ldots, 0, 2}(\zeta) \equiv 0$, 280, 294
$c_j(z)$, 17
$c_j^*(z)$, 17
$c_j^{**}(\zeta)$, 17
$e_{i,j}$, 82, 287
$f_{i,j}$, 82, 287
$g_{i,j}$, 82, 287
$r_{i,j}$, 82, 106, 287, 301
$A(z)$ for $\boldsymbol{A}$ in $\mathcal{Q}_m$, 219
$A^*(z)$ for $\boldsymbol{A}$ in $\mathcal{Q}_m$, 219
$A^{**}(\zeta)$ for $\boldsymbol{A}$ in $\mathcal{Q}_m$, 219
$A_{i,j}(z)$, 186
A_i for $A_{l,n,i}$, 10, 82, 106

A_i for $A_{l_1, \ldots, l_n, i}$, 277, 288, 302
$A_{l,n,i}$, 8
$A_{l_1, \ldots, l_n, i}$, 276
$A_{p,q}^{i,j}$, 53, 54
$A_{j_1, j_2, \ldots, j_n}^{i_1, i_2, \ldots, i_n}$, 271
$B_{p,q}^{i,j}$, 54
$B_{j_1, j_2, \ldots, j_n}^{i_1, i_2, \ldots, i_n}$, 268
B_i for $B_{l,n,i}$, 10, 82, 106
B_i for $B_{l_1, \ldots, l_n, i}$, 277, 288, 302
$B_{l,n,i}$, 8
$B_{l_1, \ldots, l_n, i}$, 276
$B_{i,j}(z) \equiv c_{i,j}(z) - c_{0,i}(z)\,c_{0,j}(z)$, 186
$B_{i,m+1}(z) \equiv B_{m+1,j}(z) \equiv 0$, 186
$B_{m,i}(z)$, 211
$C_{j,q}$, 51, 53
$C^k(\mathcal{U})$, 135
$D_2(z)$, 3, 193
$D_k(z)$ from $\boldsymbol{D}_k$, 162, 164
$D_{p,q,r}^{i,j,k}$, 56
$D_{j_1, \ldots, j_n, r}^{i_1, \ldots, i_n, k}$, 269
$E_{i,j}$, 106, 301
$F_{i,j}$, 106, 301
$G_{i,j}$, 106, 301
$GL_2(\mathfrak{F}_0)$, 214
$GL_{n+1}(\mathfrak{F}_0)$, 214
$H_{p,q,s}^{i,j,\kappa,\lambda}$, 60
$M(z)$, 3, 207
M_A, 213
M_n, 337
N_n, 337
$P(z)$, for any $\boldsymbol{P}$ in $\mathcal{S}_m$, 5
$P^*(z)$, for any $\boldsymbol{P}$ in $\mathcal{S}_m$, 5
$P^{**}(\zeta)$, for any $\boldsymbol{P}$ in $\mathcal{S}_m$, 5
$P(z)$, for any $\boldsymbol{P}$ in $\mathcal{S}_{m,n}$, 251
$P^*(z)$, for any $\boldsymbol{P}$ in $\mathcal{S}_{m,n}$, 251
$P^{**}(\zeta)$, for any $\boldsymbol{P}$ in $\mathcal{S}_{m,n}$, 251
$P_{\kappa, \lambda, s}^{i,j}(\zeta)$, 62
$Q_{\kappa, \lambda, s}^{i,j}(\zeta)$, 60
Q_2, 3
$(Q_2)' + \lambda Q_2$, 4, 15
Q_m, 4, 157
$(Q_m)' + \lambda Q_m$, 185, 189

R_3, 205
$R_{i,j}(z)$, 88, 308
R_n, 208
S_2, 194
$S_{i,j}(\zeta)$, 112, 310
S_m, 157
T, matrix transposition, 161
$T_{i,j}(\zeta)$, 113
$T_{p,q,r}^{i,j,\mu,\nu}(\zeta)$, 59
$\mathbb{C}$, the field of complex numbers, 4
$\mathbb{Q}$, the field of rational numbers, 5
$\mathbb{Q}[\mathcal{A}]$, 143
$\mathbb{Q}[\mathcal{A}']$, 143
$\mathbb{Q}\{\mathcal{A}\}$, 143
$\mathbb{Q}\{\boldsymbol{w}_{0,1}, \boldsymbol{w}_{0,2}\}$, 52
$\mathbb{Q}\{\boldsymbol{w}_{0,\ldots,0,1}, \boldsymbol{w}_{0,\ldots,0,2}\}$, 316
$\mathbb{R}$, the field of real numbers, 140
$\mathcal{A}'$, for any subset $\mathcal{A}$ of $\mathcal{S}_m$, 143
$\mathcal{A}_m$, 131
$\mathcal{B}_{m,0}$, 8, 133
$\mathcal{B}_{m,k}$, 131
$\mathcal{D}$, 348, 354
$\mathcal{E}$, 351, 355
$\mathcal{H}_k$, 230
$\mathcal{I}_{3,3}(z)$, 209
$\mathcal{I}_{m,l,n}(z)$, read both 5, 8
$\mathcal{I}_{m,l,n}^*(z)$, read both 5, 8
$\mathcal{I}_{m,l,n}^{**}(\zeta)$, read both 5, 8
$\mathcal{I}_{m,l_1,\ldots,l_n}(z)$, read both 251, 276
$\mathcal{I}_{m,l_1,\ldots,l_n}^*(z)$, read both 251, 276
$\mathcal{I}_{m,l_1,\ldots,l_n}^{**}(\zeta)$, read both 251, 276
$\mathcal{J}_{m,l,n}(z)$, read both 5, 9
$\mathcal{J}_{m,l,n}^*(z)$, read both 5, 9
$\mathcal{J}_{m,l,n}^{**}(\zeta)$, read both 5, 9
$\mathcal{J}_{m,l_1,\ldots,l_n}(z)$, read both 251, 277
$\mathcal{J}_{m,l_1,\ldots,l_n}^*(z)$, read both 251, 277
$\mathcal{J}_{m,l_1,\ldots,l_n}^{**}(\zeta)$, read both 251, 277
$\mathcal{L}_3$, 15, 194
$\mathcal{M}_r$, 223
$\mathcal{N}$, 21
$\mathcal{N}_{m,n}$, 318
$\mathcal{P}(z)$, 70
$\mathcal{Q}(\zeta)$, 69, 279
$\mathcal{Q}_m$, 219
$\mathcal{R}(z)$, 94
$\mathcal{R}_m$, 17
$\mathcal{S}(\zeta)$, 93, 293
$\mathcal{S}_m$, 5, 219, 223
$\mathcal{S}_{m,n}$, 253
$\mathcal{S}_\mu$, 35
$\mathcal{W}_m$, 131
$\mathcal{X}_m$, 131
$\mathfrak{A}_t$, 138
$\mathfrak{B}_m(z)$, 161
$\mathfrak{D}$, 185, 194
$\mathfrak{E}$, 215
$\mathfrak{F}_\nu$, 159, 160

$\mathfrak{F}_\Omega$, xi, 3, 160, 185, 194
$\mathfrak{F}_{\Omega^{**}}$, xi, 165
$\mathfrak{L}_3$, 205
$\mathfrak{M}(\mathcal{W}_m)$, 132
$\mathfrak{M}(\mathcal{X}_m)$, 132
$\mathfrak{P}$, 223
$\mathfrak{Q}$, 223
$\mathfrak{R}_{2,1}$, 326
$\mathfrak{R}_{m_1,\ldots,m_n}$, 335
$\mathfrak{R}$, 340
$\mathfrak{S}$, 223
$\mathfrak{S}_\nu$, 36
$\mathfrak{T}$ and $\mathfrak{T}_k$, 223
$\alpha_{i,j}(\zeta)$, 29
$\beta_{i,j}(z)$, 33
$\gamma_{i,j,k}$, 76, 285
$\epsilon_{i,j,k}$, 100, 299
$\eta_{i,j}$, 93, 293
$\vartheta_{i,j}$, 80, 282
$\lambda(z)$, 15, 186, 194
$\xi_{i,j}$, 69, 279
$\sigma_{h,i,j}$, 76
$\tau_{h,i,j}$, 100
$\upsilon_{i,j}$, 104, 296
$\phi_{i,j}(z)$, 47
$\chi_{i,j}(z)$, 94
$\psi_{i,j}(z)$, 70
$\Gamma_{i,j}(z)$, 180
$\Delta(z)$, 202, 216
$\Theta_{i,j}^{r,s}(z)$, 186
$\Phi_{\kappa,\lambda}(\zeta)$, 60
Ω, 4, 249, 325, 335, 337, 347, 353
Ω^{**}, 5, 250, 325, 335, 339, 348, 353
$\boldsymbol{a}_{2,2}$, 13
$\boldsymbol{a}_{m,2}$, 7, 267
$\boldsymbol{b}_{m,1}$, 7, 267
$\boldsymbol{b}_{m,2}$, 7, 267
$\boldsymbol{r}_{1,0}$, 326
$\boldsymbol{t}_l$, 223
$(\boldsymbol{t}_l)' = \boldsymbol{t}_{l+1} - \boldsymbol{t}_1 \boldsymbol{t}_l$, 223
$\boldsymbol{w}_{i,j}^{(k)}$, 5
$\boldsymbol{w}_{i_1,i_2,\ldots,i_n}^{(k)}$, 251
$(\boldsymbol{w}_{i,j}^{(k)})' = \boldsymbol{w}_{i,j}^{(k+1)}$, 7
$(\boldsymbol{w}_{i_1,i_2,\ldots,i_n}^{(k)})' = \boldsymbol{w}_{i_1,i_2,\ldots,i_n}^{(k+1)}$, 253
$\boldsymbol{w}_{i,j}$, 5
$\boldsymbol{w}_{i_1,i_2,\ldots,i_n}$, 251
$\boldsymbol{w}_{0,0}^{(0)}$, 5
$\boldsymbol{w}_{0,0,\ldots,0}^{(0)}$, 251
$\boldsymbol{w}_i^{(j)}$, 17
$\boldsymbol{x}_{i,j}^{(k)}$, 325
$\boldsymbol{x}_{i_1,i_2,\ldots,i_n}^{(k)}$, 335, 340
$\boldsymbol{x}_{n_\nu;i_1,i_2,\ldots,i_{n_\nu}}^{(k)}$, 340
$\boldsymbol{x}_i^{(j)}$, 348, 354
$\boldsymbol{x}_j^{(k)}$, 215

INDEX

$B_{1,1}$, 14
$B_{1,2}$, 14
$B_{2,2}$, 14
$B_{i,j}$, 157
B_i, 255
D_1, 7
D_2, 14, 166
D_k, 162
E_6, 15
E_7, 15
$F_{i,j}$, 77, 281
$F_{s,i}$, 153
$G_{i,j}$, 101, 295
$G_{m,2}$, 143
H_7, 15
$I_{h,i}$ for $I_{m,l,n,h,i}$, 10, 82
$I_{h_1,\ldots,h_{n-1},i}$
  for $I_{m,l_1,\ldots,l_n,h_1,\ldots,h_{n-1},i}$, 277
$I_{m,l,n,h,i}$, 8
$I_{m,l_1,\ldots,l_n,h_1,\ldots,h_{n-1},i}$, 276
$J_{h,i}$ for $J_{m,l,n,h,i}$, 10, 106
$J_{h_1,\ldots,h_{n-1},i}$
  for $J_{m,l_1,\ldots,l_n,h_1,\ldots,h_{n-1},i}$, 277
$J_{m,l,n,h,i}$, 9
$J_{m,l_1,\ldots,l_n,h_1,\ldots,h_{n-1},i}$, 277
$K_{i,j}$ for $K_{m,i,j}$, 10, 277
$K_{m,i,j}$, 8, 275
$L_{i,j}$ for $L_{m,i,j}$, 10
$L_{i_1,\ldots,i_n}$ for $L_{m,i_1,\ldots,i_n}$, 277
$L_{m,i,j}$, 8
$L_{m,i_1,\ldots,i_n}$, 275
$M_{h,i}$ for $M_{m,l,n,h,i}$, 10, 81
$M_{h_1,\ldots,h_{n-1},i}$
  for $M_{m,l_1,\ldots,l_n,h_1,\ldots,h_{n-1},i}$, 277
$M_{m,l,n,h,i}$, 8
$M_{m,l_1,\ldots,l_n,h_1,\ldots,h_{n-1},i}$, 276
$N_{i,j}$, 75, 280
$P^{(1)} + s\,b_{m,1}P$, 45, 307
$P^{(r)}$, 7, 253
$P_{i,j}$, 76, 79, 286
$Q_{i,j}$, 84, 289
$R \equiv S\,w_{p,q}^{(r)} + T$, 121
$R \equiv S\,w_{p_1,\ldots,p_n}^{(r)} + T$, 316
$R_{h,i,j}$, 83
$R_{h_1,\ldots,h_{n-1},i,j}$, 288
$R_{i,1}$, 326
$S_{h,i,j}$, 82
$S_{h_1,\ldots,h_{n-1},i,j}$, 287
S_k, 332
$T[\rho;z]$, 224
$T[g';z]$, 225
T_k, 332
T_P, 225
$T[h;z]$, 224
$U_{i,j}$ for $U_{m,i,j}$, 10, 277
$U_{m,i,j}$, 9, 276
$V_{i,j}$ for $V_{m,i,j}$, 10

$V_{i_1,\ldots,i_n}$ for $V_{m,i_1,\ldots,i_n}$, 277
$V_{m,i,j}$, 9
$V_{m,i_1,\ldots,i_n}$, 276
$W_{h,i}$ for $W_{m,l,n,h,i}$, 10, 105
$W_{h_1,\ldots,h_{n-1},i}$
  for $W_{m,l_1,\ldots,l_n,h_1,\ldots,h_{n-1},i}$, 277
$W_{m,l,n,h,i}$, 9
$W_{m,l_1,\ldots,l_n,h_1,\ldots,h_{n-1},i}$, 276
$X_{i,j}$, 99, 294
X_P, 234
$Y_{i,j}$, 100, 103, 300
$Z_{i,j}$, 108, 303
$\widehat{I}_{m,n,n} \equiv \mathcal{I}_{m,0,n}$, 19
$\widehat{J}_{m,n,n} \equiv \mathcal{J}_{m,0,n}$, 20
$\mathcal{I}_{m,l,n} \equiv \mathcal{J}_{m,l,n}$, 119
$\mathcal{I}_{m,l_1,\ldots,l_m} \equiv \mathcal{J}_{m,l_1,\ldots,l_m}$, 314
$\mathcal{I}_{1,1,1} \equiv D_1$, 8
$\mathcal{I}_{2,1,1}$, 13
$\mathcal{I}_{2,1,2}$, 13
$\mathcal{I}_{2,2,2}$, 13
$\mathcal{I}_{m,1,1}$, 7, 11
$\mathcal{I}_{m,1,2}$, 11
$\mathcal{I}_{m,2,2}$, 13
$\mathcal{I}_{m,l,n}$, 8
$\mathcal{I}_{m,l_1,\ldots,l_n}$, 276
$\mathcal{J}_{m,l,n}$, 9
$\mathcal{J}_{m,l_1,\ldots,l_n}$, 277
$\mathcal{M}_{i,j,k,t}$, 73
$\mathcal{M}_{i_1,\ldots,i_n,k,t}$, 279
$\mathcal{W}_{i,j,k,t}$, 97
$\mathcal{W}_{i_1,\ldots,i_n,k,t}$, 293
$\mathfrak{s}_h^{(k)}$, 35, 256
$\mathfrak{w}_i$, 255, 317
$\mathfrak{r}_i$, 309
$\mathfrak{R}_{h,i,j}$, 108
$\mathfrak{R}_{h_1,\ldots,h_{n-1},i,j}$, 302
$\mathfrak{S}_{h,i,j}$, 106
$\mathfrak{S}_{h_1,\ldots,h_{n-1},i,j}$, 301
$\varGamma_n$, 215
$\varDelta_{i,j,k}$, 157
$\varDelta_{1,1,2} \equiv D_2$, 161

absolute invariants for $Q_m = 0$, 243, 244
Paul Émile Appell, xii, 3, 4, 10, 13, 15, 185,
    209, 325
Appell's condition for $Q_2 = 0$, 4, 22, 193
Appell's condition for $Q_m = 0$, 185
Appell-type solutions of $Q_2 = 0$, 4, 22, 196,
    197

basic relative invariants for $L_m = 0$
  as those for $H_{m,n} = 0$ when $n = 1$, 249,
    276, 314
  from $\mathcal{I}_{m,0,n}$, for $3 \leq n \leq m$, 19
  formulas equivalent to those in [24], 18
basic relative invariants for $Q_2 = 0$
  are defined for $Q_m = 0$ when $m = 2$, 6
  all three of them, 13

constructions using, 14, 16, 17, 153–156, 172–177
the one previously known, 13
basic relative invariants for $Q_m = 0$
  definitions about, 6
  Main Theorem about, 8
  are global in character, 7
  existence of
    when $m = 1$, 8, 42
    when $m \geq 2$, 8, 120
  uniqueness of
    when $m = 1$, 8, 42
    when $m \geq 2$, 8, 130
  number of, 21
  are algebraically independent over $\mathbb{Q}$, 8, 133
  constructions using
    when $m = 1$, 39, 41
    when $m \geq 2$, 143–146, 149, 166–169
  conditions involving, 179, 181
  for a real-valued context, 139
basic relative invariants for $H_{m,n} = 0$
  definitions about, 252
  existence of
    when $m = 1$, 257
    when $m \geq 2$, 314
  uniqueness of
    when $m = 1$, 264
    when $m \geq 2$, 316
  number of, 318
  constructions using
    when $m = 1$ and $n \geq 2$, 259
    when $m \geq 2$, 320
Ludwig Bieberbach, 206
binary forms of the second degree
  two invariants for, 199
binary forms of degree $n \geq 1$
  transformations of, 213
  an invariant Γ_n (verified for $1 \leq n \leq 5$), 217

coefficient field $\mathfrak{F}_\Omega$, remark about
  avoidance of abstract field extensions, 160
coefficients $c_{i,j}^*(z)$ of (1.3), 23
  in a real-valued context, 135
coefficients $c_{i,j}^{**}(\zeta)$ of (1.5), 30, 32
  in a real-valued context, 137
coefficients $d_{i,j}(\zeta)$ of (5.1)
  without restrictions, 47
  when $d_{0,1}(\zeta) \equiv d_{0,2}(\zeta) \equiv 0$, 69, 93
  identities for, 85, 109
coefficients $c_{i_1, i_2 \ldots, i_n}^*(z)$ of (21.3), 249
coefficients $c_{i_1, i_2, \ldots, i_n}^{**}(\zeta)$ of (21.5), 250
coefficients $d_{i_1, i_2, \ldots, i_n}(\zeta)$ of (23.1)
  without restriction, 265
  when $d_{0, \ldots, 0, 1}(\zeta) \equiv d_{0, \ldots, 0, 2}(\zeta) \equiv 0$, 279, 293
  identities for, 290, 304

condition that
  there is a factorization for
    Q_2, 3, 161
    $(Q_2)' + \lambda Q_2$, 4, 15, 16, 194
    Q_m, 159
    $(Q_m)' + \lambda Q_m$, 186
  $Q_2 = 0$ has Appell-type solutions, 4, 196
constants are the elements of $\mathbb{Q}$ for
  $\mathcal{S}_m$, 7
  $\mathcal{Q}_m$, 219
  $\mathfrak{P}, \mathfrak{Q}, \mathfrak{S},$ and $\mathfrak{T}$, 223
  $\mathcal{S}_{m,n}$, 253
  $\mathfrak{R}_{2,1}$, 326
  $\mathfrak{R}_{m_1, \ldots, m_n}$, 335
  $\mathfrak{R}$, 340
  $\mathcal{D}$, 348, 354
  $\mathcal{E}$, 351, 355

definition of
  a Laguerre-Forsyth canonical form for
    a linear equation of order $m \geq 2$, 51
    an equation (1.1) of order $m \geq 2$, 51
    an equation (21.1) of order $m \geq 2$, 268
  a differential-polynomial combination, 143
  Appell-type solutions for $Q_2 = 0$, 4
  a singular solution of $Q_m = 0$, 191
  a movable branch point for solutions, 206
  a movable pole for solutions, 206
derivation $'$ for
  $\mathcal{S}_m$, 7
  $\mathfrak{D}$, 185
  $\mathfrak{F}_\Omega$, 185
  $\mathfrak{E}$, 215
  $\mathcal{Q}_m$, 219
  $\mathfrak{Q}$ (and $\mathfrak{P}, \mathfrak{S}, \mathfrak{T}, \mathcal{S}_m, \mathcal{Q}_m$), 223
  $\mathcal{S}_{m,n}$, 253
  $\mathfrak{R}_{2,1}$, 326
  $\mathfrak{R}_{m_1, \ldots, m_n}$, 335
  $\mathfrak{R}$, 340
  $\mathcal{D}$, 348, 354
  $\mathcal{E}$, 351, 355
differential-polynomial combination
  definition of, 143
  usage of, 3, 8, 13, 143, 320

equations $Q_m = 0$, 4–246
  basic relative invariants
    definitions about, 6
    Main Theorem, 8
    existence when $m = 1$, 8, 41
    uniqueness when $m = 1$, 8, 41
    existence when $m \geq 2$, 8, 120
    uniqueness when $m \geq 2$, 8, 130
    number of, 21
  definition of
    $Q_m = 0$ as (1.1), 4
    the polynomial ring $\mathcal{S}_m$, 5
    the derivation $'$ for $\mathcal{S}_m$, 7
    the constants in $\mathcal{S}_m$, 7

an isobaric polynomial in $\mathcal{S}_m$, 6
the weight of an isobaric polynomial in $\mathcal{S}_m$, 6
a basic polynomial in $\mathcal{S}_m$, 6
the index of a basic polynomial in $\mathcal{S}_m$, 6
$P(z)$ for any $\boldsymbol{P}$ in $\mathcal{S}_m$, 5
$P^*(z)$ for any $\boldsymbol{P}$ in $\mathcal{S}_m$, 5
$P^{**}(\zeta)$ for any $\boldsymbol{P}$ in $\mathcal{S}_m$, 5
a semi-invariant of the first kind, 6
a semi-invariant of the second kind, 6
a relative invariant, 6
a basic relative invariant in $\mathcal{S}_m$, 6
a global relative invariant in $\mathcal{S}_m$, 7
a Laguerre-Forsyth canonical form when $m \geq 2$, 51
the quotient field $\mathcal{Q}_m$ for $\mathcal{S}_m$, 219
the derivation $'$ for $\mathcal{Q}_m$, 219
the constants in $\mathcal{Q}_m$, 219
an isobaric element in $\mathcal{Q}_m$, 222
the weight of an isobaric element in $\mathcal{Q}_m$, 222
a rational semi-invariant in $\mathcal{Q}_m$ of the first kind, 220
a rational semi-invariant in $\mathcal{Q}_m$ of the second kind, 220
a rational relative invariant in $\mathcal{Q}_m$, 220
an absolute invariant in $\mathcal{Q}_m$, 243
transformations of $Q_m = 0$, 5, 23–27, 29–34, 47–50
equations $H_{m,n} = 0$, 249–321
  basic relative invariants
    definitions about, 252
    existence when $m = 1$ and $n \geq 2$, 257
    uniqueness when $m = 1$ and $n \geq 2$, 264
    existence when $m \geq 2$, 314
    uniqueness when $m \geq 2$, 316
    number of, 318
  definition of
    $H_{m,n} = 0$ as (21.1), 249
    the polynomial ring $\mathcal{S}_{m,n}$, 251
    the derivation $'$ for $\mathcal{S}_{m,n}$, 253
    the constants in $\mathcal{S}_{m,n}$, 253
    an isobaric polynomial in $\mathcal{S}_{m,n}$, 252
    the weight of an isobaric polynomial in $\mathcal{S}_{m,n}$, 252
    a basic polynomial in $\mathcal{S}_{m,n}$, 252
    the index of a basic polynomial, 252
    $P(z)$ for any $\boldsymbol{P}$ in $\mathcal{S}_{m,n}$, 251
    $P^*(z)$ for any $\boldsymbol{P}$ in $\mathcal{S}_{m,n}$, 251
    $P^{**}(\zeta)$ for any $\boldsymbol{P}$ in $\mathcal{S}_{m,n}$, 251
    a semi-invariant of the first kind, 252
    a semi-invariant of the second kind, 252
    a relative invariant, 252
    a basic relative invariant, 252
    a Laguerre-Forsyth canonical form when $m \geq 2$, 268

transformations of $H_{m,n} = 0$, 249, 250, 265–268

Gino Fano, 179
Fano-type problems
  the simplest one for $L_m = 0$, 179
    its solution, 179, 180
  the simplest one for $Q_m = 0$, 180
    its solution in Theorem 17.2, 181
Andrew Russell Forsyth, ix, 50, 51, 146, 148, 149
Immanuel Lazarus Fuchs, 207

Georges-Henri Halphen, 146, 148, 149
Micaiah John Muller Hill, 207
homogeneous linear differential equations
  basic relative invariants for, 18
  inclusion of, 17–20

Edward Lindsay Ince, 207
isobaric polynomials in
  $\mathcal{S}_m$, 6
  $\mathcal{S}_{m,n}$, 252
  $\mathfrak{R}_{2,1}$, 326
  $\mathfrak{R}_{m_1,\ldots,m_n}$, 336
  $\mathfrak{R}$, 342
  $\mathcal{D}$, 348, 354
isobaric rational functions in $\mathcal{Q}_m$, 222

Edmond Laguerre, vii, ix, xii, 6
Laguerre-Forsyth canonical forms
  for $L_m = 0$ when $m \geq 2$, 51
  for $Q_m = 0$ when $m \geq 2$, 51
  for $H_{m,n} = 0$ when $m \geq 2$, 268
  for (28.1), 331
Laguerre-Forsyth-type identities
  for canonical forms of $L_m = 0$, 51
  for canonical forms of $Q_m = 0$, 58
  for canonical forms of $H_{m,n} = 0$, 271, 315
Joseph Liouville, 210

machine computations
  for relative invariants of $Q_2 = 0$, 172–177
  in terms of basic relative invariants for
    $\boldsymbol{D}_2$, 167, 168
    $\boldsymbol{E}_6$, 168, 169
    $\boldsymbol{E}_7$, 168, 169
  to check that $\boldsymbol{\mathcal{I}}_{m,l,n} \equiv \boldsymbol{\mathcal{J}}_{m,l,n}$, 169–171
  to represent $\boldsymbol{\mathcal{I}}_{m,l,n}$ or $\boldsymbol{\mathcal{J}}_{m,l,n}$, 169–171
Main Theorem, 8
  1. existence
     Theorem 11.2, 120
  2. uniqueness
     Corollary 12.3, 130
  3. algebraic independence of $\mathcal{B}_{m,0}$ over $\mathbb{Q}$
     Corollary 12.5, 133
  4. differential-polynomial combinations
     Corollary 14.3, 146
  5. for $m = 1$, Theorem 4.9, 41

MATHEMATICA, 167–177, 272, 273, 330, 350, 354
movable branch point of solutions, 206
movable pole of solutions, 206

František Neuman, ix
nonhomogeneous equations, 337–345
notational abbreviations, 10, 277

ordinary algebraic differential equations
  homogeneous with the monic monomial:
    $y^{(m)}(z)$, 17–20, 50, 51, 336
    $(y'(z))^2$, 39–42, 336
    $y''(z)\,y'(z)$, 325–334, 336
    $(y''(z))^2$, 3, 4, 13–17, 193–209, 336
    $(y^{(m)}(z))^2$, 4–246, 336
    $(y^{(m)}(z))^n$, when $m = 1$ and $n \geq 2$, 249–264, 336
    $(y^{(m)}(z))^n$, when $m \geq 2$, 249–256, 265–321, 336
    $y^{(m_1)}\,y^{(m_2)}\dots y^{(m_n)}$, 335–337
  homogeneous of other types, 347, 353
  nonhomogeneous, 337–345

problems about $Q_2 = 0$
  basic relative invariants, 3, 13
  conditions for $(Q_2)' + \lambda Q_2$ to have a nontrivial factorization, 4, 15, 194
  $\lambda(z)$ when $(Q_2)' + \lambda Q_2$ factors, 15, 194
  special form of solutions, 4, 196–205
problems about $Q_m = 0$
  basic relative invariants, 8
  conditions for $(Q_m)' + \lambda Q_m$ to have a nontrivial factorization, 186
  $\lambda(z)$ when $(Q_m)' + \lambda Q_m$ factors, 186
  special form of solutions, 189–192
problems about $H_{1,n} = 0$
  basic relative invariants
    existence, 257
    uniqueness, 264
problems about $H_{m,n} = 0$, for $m \geq 2$
  basic relative invariants
    existence, 314
    uniqueness, 316

quadratic binary forms
  two invariants for, 199

reference to work of
  Paul Émile Appell, ix, xii, 3, 4, 13, 185, 209, 325
  Maxime Bôcher, 192
  A. Berry, 207
  Ludwig Bieberbach, 206
  Nicolas Bourbaki, 7, 185, 219, 223, 253
  Charles Leonard Bouton, ix
  Domenico Caligo, 3
  Robert Campbell, 36, 42
  Roger Chalkley, ix, x, xii, 3, 4, 6, 9, 16, 18–20, 29, 43, 48–51, 85, 88, 109, 112, 149, 171, 179, 180, 219, 243
  James Cockle, 7
  Christopher M. Cosgrove, 3
  David Raymond Curtiss, ix, 3, 4
  P. Dale, 3
  Gino Fano, 179
  Andrew Russell Forsyth, ix, 50, 146, 149
  Immanuel Lazarus Fuchs, 207
  Georges-Henri Halphen, 6, 146, 149
  Micaiah John Muller Hill, 207
  Ludwig Otto Hölder, 36, 42
  Edward Lindsay Ince, 27, 207
  Donald Ervin Knuth, 64, 81, 319
  Ellis Robert Kolchin, 7, 52, 143, 191
  Edmond Laguerre, 6
  Serge Lang, 214, 216, 232
  Joseph Liouville, 210
  František Neuman, ix
  Joseph Fels Ritt, 191
  Issai Schur, 199
  P. R. Vein, 3
  Georg Jacob Wallenberg, 3
relative invariants as polynomials
  for the equations $Q_2 = 0$ of page 3
    are defined for $Q_m = 0$ when m is 2, 4, 6
    are necessarily isobaric, 6, 37
    their weight, 6
    the three basic relative invariants, 13
    $\mathcal{I}_{2,1,1}$, the basic one previously known, 13
    $\mathcal{I}_{2,1,2}$, the second basic one, 13
    $\mathcal{I}_{2,2,2}$, the third basic one, 13
    $\boldsymbol{D}_2$, 3, 14
    $\boldsymbol{E}_6$, 15
    $\boldsymbol{E}_7$, 15
    $\boldsymbol{D}_2$ in terms of basic ones, 14
    $\boldsymbol{E}_6$ in terms of basic ones, 16
    $\boldsymbol{E}_7$ in terms of basic ones, 17
    all those having weight ≤ 9, 153–155
relative invariants as polynomials
  for the equations $Q_m = 0$ in (1.1)
    definition of, 6
    are necessarily global, 7
    are necessarily isobaric, 6, 37
    their weight, 6
    basic relative invariants defined, 6
    index of a basic one defined, 6
    Main Theorem about basic ones, 8
    all relative invariants when $m = 1$, 41
    in terms of basic ones, 41, 143–146
    constructed from given ones, 146–153
    $\mathcal{I}_{m,1,1}$, a basic one for any $m \geq 1$, 11
    $\mathcal{I}_{m,1,2}$, a basic one for any $m \geq 2$, 11
    $\mathcal{I}_{m,2,2}$, a basic one for any $m \geq 2$, 13
    $\boldsymbol{D}_1$, for $m \geq 1$, 7, 34

D_2, for $m \geq 2$, 162
D_2 in terms of basic ones, 166
D_k, for $1 \leq k \leq m$, 162, 164
relative invariants as rational functions
  for the equations $Q_m = 0$ in (1.1)
    definition of, 220
    are elements of $\mathcal{Q}_m$, 220
    structure of, 243
relative invariants as polynomials
  for the equations $H_{m,n} = 0$ in (21.1)
    definition of, 252
    are necessarily isobaric, 252
    their weight, 252
    basic relative invariants defined, 252
    index of a basic one defined, 252
    in terms of basic ones, 259, 320
    existence of basic ones, 257, 314
    uniqueness of basic ones, 264, 316
    number of basic relative invariants, 318

semi-invariants of the first kind
  as polynomials for the equations $Q_m = 0$
    definition of, 6
    are elements of $\mathcal{S}_m$, 6
    all those for the case $m = 1$, 39
    isobaric ones of $weight = 2$, for $m \geq 2$, 43
    $\boldsymbol{a}_{m,2}$, for $m \geq 2$, 7, 43
    $\boldsymbol{L}_{m,i,j}$, for $m \geq 2$, 8, 87
semi-invariants of the first kind
  as rational functions for $Q_m = 0$
    definition of, 220
    are elements of $\mathcal{Q}_m$, 220
    structure of, 230
semi-invariants of the first kind
  as polynomials for $H_{m,n} = 0$
    definition of, 252
    are elements of $\mathcal{S}_{m,n}$, 252
    those for the case $m = 1$, 259
    isobaric ones of $weight = 2$, for $m \geq 2$, 273
    $\boldsymbol{a}_{m,2}$, for $m \geq 2$, 267, 273
    $\boldsymbol{L}_{m,i,j}$, for $m \geq 2$, 275, 307
semi-invariants of the second kind
  as polynomials for the equations $Q_m = 0$
    definition of, 6
    are elements of $\mathcal{S}_m$, 6
    are necessarily isobaric, 6, 37, 222
    their weight, 6
    ones of $weight = 2$, for $m \geq 2$, 43
    $\boldsymbol{b}_{m,2}$, for $m \geq 2$, 7, 43
    $\boldsymbol{V}_{m,i,j}$, for $m \geq 2$, 9, 111
semi-invariants of the second kind
  as rational functions for $Q_m = 0$
    definition of, 220
    are elements of $\mathcal{Q}_m$, 220
    are rationally isobaric, 220, 222
    their weight, 222

structure of, 240
semi-invariants of the second kind
  as polynomials for $H_{m,n} = 0$
    definition of, 252
    are elements of $\mathcal{S}_{m,n}$, 252
    are isobaric, 252
    their weight, 252
    ones of $weight = 2$, for $m \geq 2$, 273
    $\boldsymbol{b}_{m,2}$, for $m \geq 2$, 267, 273
    $\boldsymbol{V}_{m,i,j}$, for $m \geq 2$, 276, 309
singular solution, definition of, 191
solutions free of movable branch points, 206

the weight
  of an isobaric polynomial in $\mathcal{S}_m$, 6
  of an isobaric rational function in $\mathcal{Q}_m$, 222
  of an isobaric polynomial in $\mathcal{S}_{m,n}$, 252
  of an isobaric polynomial in $\mathfrak{R}_{2,1}$, 326
  of an isobaric polynomial in $\mathfrak{R}_{m_1,\ldots,m_n}$, 336
  of an isobaric polynomial in $\mathfrak{R}$, 342
  of an isobaric polynomial in $\mathcal{D}$, 348, 354
transformations of $Q_m = 0$ in (1.1)
  due to a new dependent variable, 5, 23
  due to a new independent variable, 5, 30, 33
  due to both variables altered, 47–49, 51
transformations of $H_{m,n} = 0$ in (21.1)
  due to a new dependent variable, 249, 250
  due to a new independent variable, 250, 251
  due to both variables altered, 265–271

Alexandre Théophile Vandermonde, 64, 81, 284, 298
Vandermonde's convolution, 64, 81
  a generalization of, 284, 298

Editorial Information

To be published in the *Memoirs*, a paper must be correct, new, nontrivial, and significant. Further, it must be well written and of interest to a substantial number of mathematicians. Piecemeal results, such as an inconclusive step toward an unproved major theorem or a minor variation on a known result, are in general not acceptable for publication.

Papers appearing in *Memoirs* are generally at least 80 and not more than 200 published pages in length. Papers less than 80 or more than 200 published pages require the approval of the Managing Editor of the Transactions/Memoirs Editorial Board.

As of July 31, 2007, the backlog for this journal was approximately 15 volumes. This estimate is the result of dividing the number of manuscripts for this journal in the Providence office that have not yet gone to the printer on the above date by the average number of monographs per volume over the previous twelve months, reduced by the number of volumes published in four months (the time necessary for preparing a volume for the printer). (There are 6 volumes per year, each usually containing at least 4 numbers.)

A Consent to Publish and Copyright Agreement is required before a paper will be published in the *Memoirs*. After a paper is accepted for publication, the Providence office will send a Consent to Publish and Copyright Agreement to all authors of the paper. By submitting a paper to the *Memoirs*, authors certify that the results have not been submitted to nor are they under consideration for publication by another journal, conference proceedings, or similar publication.

Information for Authors

Memoirs are printed from camera copy fully prepared by the author. This means that the finished book will look exactly like the copy submitted.

Initial submission. The AMS uses Centralized Manuscript Processing for initial submissions. Authors should submit a PDF file using the Initial Manuscript Submission form found at www.ams.org/cgi-bin/peertrack/submission.pl, or send one copy of the manuscript to the following address: Centralized Manuscript Processing, MEMOIRS OF THE AMS, 201 Charles Street, Providence, RI 02904-2294 USA. If a paper copy is being forwarded to the AMS, indicate that it is for it Memoirs and include the name of the corresponding author, contact information such as email address or mailing address, and the name of an appropriate Editor to review the paper (see the list of Editors below).

The paper must contain a *descriptive title* and an *abstract* that summarizes the article in language suitable for workers in the general field (algebra, analysis, etc.). The *descriptive title* should be short, but informative; useless or vague phrases such as "some remarks about" or "concerning" should be avoided. The *abstract* should be at least one complete sentence, and at most 300 words. Included with the footnotes to the paper should be the 2000 *Mathematics Subject Classification* representing the primary and secondary subjects of the article. The classifications are accessible from www.ams.org/msc/. The list of classifications is also available in print starting with the 1999 annual index of *Mathematical Reviews*. The Mathematics Subject Classification footnote may be followed by a list of *key words and phrases* describing the subject matter of the article and taken from it. Journal abbreviations used in bibliographies are listed in the latest *Mathematical Reviews* annual index. The series abbreviations are also accessible from www.ams.org/publications/. To help in preparing and verifying references, the AMS offers MR Lookup, a Reference Tool for Linking, at www.ams.org/mrlookup.

Electronically prepared manuscripts. The AMS encourages electronically prepared manuscripts, with a strong preference for $\mathcal{A}_{\mathcal{M}}\mathcal{S}$-LaTeX. To this end, the Society has prepared $\mathcal{A}_{\mathcal{M}}\mathcal{S}$-LaTeX author packages for each AMS publication. Author packages include instructions for preparing electronic manuscripts, samples, and a style file that generates the particular design specifications of that publication series. Though $\mathcal{AMS}$-LaTeX is the highly preferred format of TeX, author packages are also available in $\mathcal{AMS}$-TeX.

Authors may retrieve an author package from the AMS website starting from www.ams.org/tex/ or via FTP to ftp.ams.org (login as anonymous, enter username as password, and type cd pub/author-info). The *AMS Author Handbook* and the *Instruction Manual* are available in PDF format following the author packages link from www.ams.org/tex/. The author package can also be obtained free of charge by sending email to tech-support@ams.org (Internet) or from the Publication Division, American Mathematical Society, 201 Charles St., Providence, RI 02904-2294, USA. When requesting an author package, please specify $\mathcal{AMS}$-LaTeX or $\mathcal{AMS}$-TeX and the publication in which your paper will appear. Please be sure to include your complete mailing address.

After acceptance. The final version of the electronic file should be sent to the Providence office (this includes any TeX source file, any graphics files, and the DVI or PostScript file) immediately after the paper has been accepted for publication.

Before sending the source file, be sure you have proofread your paper carefully. The files you send must be the EXACT files used to generate the proof copy that was accepted for publication. For all publications, authors are required to send a printed copy of their paper, which exactly matches the copy approved for publication, along with any graphics that will appear in the paper.

Accepted electronically prepared files can be submitted via the web at www.ams.org/submit-book-journal/, sent via FTP, or sent on CD-Rom or diskette to the Electronic Prepress Department, American Mathematical Society, 201 Charles Street, Providence, RI 02904-2294 USA. TeX source files, DVI files, and PostScript files can be transferred over the Internet by FTP to the Internet node ftp.ams.org (130.44.1.100). When sending a manuscript electronically via CD-Rom or diskette, please be sure to include a message identifying the paper as a Memoir.

Electronically prepared manuscripts can also be sent via email to pub-submit@ams.org (Internet). In order to send files via email, they must be encoded properly. (DVI files are binary and PostScript files tend to be very large.)

Electronic graphics. Comprehensive instructions on preparing graphics are available at www.ams.org/jourhtml/. A few of the major requirements are given here.

Submit files for graphics as EPS (Encapsulated PostScript) files. This includes graphics originated via a graphics application as well as scanned photographs or other computer-generated images. If this is not possible, TIFF files are acceptable as long as they can be opened in Adobe Photoshop or Illustrator. No matter what method was used to produce the graphic, it is necessary to provide a paper copy to the AMS.

Authors using graphics packages for the creation of electronic art should also avoid the use of any lines thinner than 0.5 points in width. Many graphics packages allow the user to specify a "hairline" for a very thin line. Hairlines often look acceptable when proofed on a typical laser printer. However, when produced on a high-resolution laser imagesetter, hairlines become nearly invisible and will be lost entirely in the final printing process.

Screens should be set to values between 15% and 85%. Screens which fall outside of this range are too light or too dark to print correctly. Variations of screens within a graphic should be no less than 10%.

Inquiries. Any inquiries concerning a paper that has been accepted for publication should be sent to memo-query@ams.org or directly to the Electronic Prepress Department, American Mathematical Society, 201 Charles St., Providence, RI 02904-2294 USA.

Editors

This journal is designed particularly for long research papers, normally at least 80 pages in length, and groups of cognate papers in pure and applied mathematics. Papers intended for publication in the *Memoirs* should be addressed to one of the following editors. The AMS uses Centralized Manuscript Processing for initial submissions to AMS journals. Authors should follow instructions listed on the Initial Submission page found at www.ams.org/memo/memosubmit.html.

Algebra to ALEXANDER KLESHCHEV, Department of Mathematics, University of Oregon, Eugene, OR 97403-1222; email: ams@noether.uoregon.edu

Algebraic geometry and its application to MINA TEICHER, Emmy Noether Research Institute for Mathematics, Bar-Ilan University, Ramat-Gan 52900, Israel; email: teicher@macs.biu.ac.il

Algebraic geometry to DAN ABRAMOVICH, Department of Mathematics, Brown University, Box 1917, Providence, RI 02912; email: amsedit@math.brown.edu

Algebraic number theory to V. KUMAR MURTY, Department of Mathematics, University of Toronto, 100 St. George Street, Toronto, ON M5S 1A1, Canada; email: murty@math.toronto.edu

Algebraic topology to ALEJANDRO ADEM, Department of Mathematics, University of British Columbia, Room 121, 1984 Mathematics Road, Vancouver, British Columbia, Canada V6T 1Z2; email: adem@math.ubc.ca

Combinatorics to JOHN R. STEMBRIDGE, Department of Mathematics, University of Michigan, Ann Arbor, Michigan 48109-1109; email: FRS@umich.edu

Complex analysis and harmonic analysis to ALEXANDER NAGEL, Department of Mathematics, University of Wisconsin, 480 Lincoln Drive, Madison, WI 53706-1313; email: nagel@math.wisc.edu

Differential geometry and global analysis to LISA C. JEFFREY, Department of Mathematics, University of Toronto, 100 St. George St., Toronto, ON Canada M5S 3G3; email: jeffrey@math.toronto.edu

Dynamical systems and ergodic theory to AMIE WILKINSON, Department of Mathematics, Northwestern University, 2033 Sheridan Road, Evanston, IL 60208-2730; email: transactions@math.northwestern.edu

Functional analysis and operator algebras to DIMITRI SHLYAKHTENKO, Department of Mathematics, University of California, Los Angeles, CA 90095; email: shlyakht@math.ucla.edu

Geometric analysis to WILLIAM P. MINICOZZI II, Department of Mathematics, Johns Hopkins University, 3400 N. Charles St., Baltimore, MD 21218; email: trans@math.jhu.edu

Geometric analysis to MLADEN BESTVINA, Department of Mathematics, University of Utah, 155 South 1400 East, JWB 233, Salt Lake City, Utah 84112-0090; email: bestvina@math.utah.edu

Harmonic analysis, representation theory, and Lie theory to ROBERT J. STANTON, Department of Mathematics, The Ohio State University, 231 West 18th Avenue, Columbus, OH 43210-1174; email: stanton@math.ohio-state.edu

Logic to STEFFEN LEMPP, Department of Mathematics, University of Wisconsin, 480 Lincoln Drive, Madison, Wisconsin 53706-1388; email: lempp@math.wisc.edu

Partial differential equations to GUSTAVO PONCE, Department of Mathematics, South Hall, Room 6607, University of California, Santa Barbara, CA 93106; email: ponce@math.ucsb.edu

Partial differential equations and dynamical systems to PETER POLACIK, School of Mathematics, University of Minnesota, Minneapolis, MN 55455; email: polacik@math.umn.edu

Probability and statistics to KRZYSZTOF BURDZY, Department of Mathematics, University of Washington, Box 354350, Seattle, Washington 98195-4350; email: burdzy@math.washington.edu

Real analysis and partial differential equations to DANIEL TATARU, Department of Mathematics, University of California, Berkeley, Berkeley, CA 94720; email: tataru@math.berkeley.edu

All other communications to the editors should be addressed to the Managing Editor, ROBERT GURALNICK, Department of Mathematics, University of Southern California, Los Angeles, CA 90089-1113; email: guralnic@math.usc.edu.

Titles in This Series

890 **Steven Dale Cutkosky,** Toroidalization of dominant morphisms of 3-folds, 2007

889 **Michael Sever,** Distribution solutions of nonlinear systems of conservation laws, 2007

888 **Roger Chalkley,** Basic global relative invariants for nonlinear differential equations, 2007

887 **Charlotte Wahl,** Noncommutative Maslov index and eta-forms, 2007

886 **Robert M. Guralnick and John Shareshian,** Symmetric and alternating groups as monodromy groups of Riemann surfaces I: Generic covers and covers with many branch points, 2007

885 **Jae Choon Cha,** The structure of the rational concordance group of knots, 2007

884 **Dan Haran, Moshe Jarden, and Florian Pop,** Projective group structures as absolute Galois structures with block approximation, 2007

883 **Apostolos Beligiannis and Idun Reiten,** Homological and homotopical aspects of torsion theories, 2007

882 **Lars Inge Hedberg and Yuri Netrusov,** An axiomatic approach to function spaces, spec tral synthesis and Luzin approximation, 2007

881 **Tao Mei,** Operator valued Hardy spaces, 2007

880 **Bruce C. Berndt, Geumlan Choi, Youn-Seo Choi, Heekyoung Hahn, Boon Pin Yeap, Ae Ja Yee, Hamza Yesilyurt, and Jinhee Yi,** Ramanujan's forty identities for Rogers-Ramanujan functions, 2007

879 **O. García-Prada, P. B. Gothen, and V. Muñoz,** Betti numbers of the moduli space of rank 3 parabolic Higgs bundles, 2007

878 **Alessandra Celletti and Luigi Chierchia,** KAM stability and celestial mechanics, 2007

877 **María J. Carro, José A. Raposo, and Javier Soria,** Recent developments in the theory of Lorentz spaces and weighted inequalities, 2007

876 **Gabriel Debs and Jean Saint Raymond,** Borel liftings of Borel sets: Some decidable and undecidable statements, 2007

875 **C. Krattenthaler and T. Rivoal,** Hypergéométrie et fonction zêta de Riemann, 2007

874 **Sonia Natale,** Semisolvability of semisimple Hopf algebras of low dimension, 2007

873 **A. J. Duncan,** Exponential genus problems in one-relator products of groups, 2007

872 **Anthony V. Geramita, Tadahito Harima, Juan C. Migliore, and Yong Su Shin,** The Hilbert function of a level algebra, 2007

871 **Pascal Auscher,** On necessary and sufficient conditions for L^p-estimates of Riesz transforms associated to elliptic operators on $\mathbb{R}^n$ and related estimates, 2007

870 **Takuro Mochizuki,** Asymptotic behaviour of tame harmonic bundles and an application to pure twistor D-modules, Part 2, 2007

869 **Takuro Mochizuki,** Asymptotic behaviour of tame harmonic bundles and an application to pure twistor D-modules, Part 1, 2007

868 **Gelu Popescu,** Entropy and multivariable interpolation, 2006

867 **Vilmos Totik,** Metric properties of harmonic measures, 2006

866 **William Craig,** Semigroups underlying first-order logic, 2006

865 **Nathanial P. Brown,** Invariant means and finite representation theory of $C*$-algebras, 2006

864 **John M. Lee,** Fredholm operators and Einstein metrics on conformally compact manifolds, 2006

863 **M. Lübke and A. Teleman,** The Universal Kobayashi-Hitchin correspondence on Hermitian manifolds, 2006

862 **Alberto Canonaco,** The Beilinson complex and canonical rings of irregular surfaces, 2006

861 **Leon A. Takhtajan and Lee-Peng Teo,** Weil-Petersson metric on the universal Teichmüller space, 2006

TITLES IN THIS SERIES

860 **Thomas M. Fiore,** Pseudo limits, biadjoints and pseudo algebras: Categorical foundations of conformal field theory, 2006

859 **N. Arcozzi, R. Rochberg, and E. Sawyer,** Carleson measures and interpolating sequences for Besov spaces on complex balls, 2006

858 **Enrico Valdinoci, Berardino Sciunzi, and Vasile Ovidiu Savin,** Flat level set regularity of p-Laplace phase transitions, 2006

857 **Donatella Danielli, Nocola Garofalo, and Duy-Minh Nhieu,** Non-doubling Ahlfors measures, perimeter measures, and the characterization of the trace spaces of Sobolev functions in Carnot-Carathéodory spaces, 2006

856 **Vladimir Bolotnikov and Harry Dym,** On boundary interpolation for matrix valued Schur functions, 2006

855 **Yevgenia Kashina, Yorck Sommerhäuser, and Yongchang Zhu,** On higher Frobenius-Schur indicators, 2006

854 **Noam Greenberg,** The role of true finiteness in the admissible recursively enumerable degrees, 2006

853 **Joachim Krieger,** Stability of spherically symmetric wave maps, 2006

852 **Viorel Barbu, Irena Lasiecka, and Roberto Triggiani,** Tangential boundary stabilization of Navier-Stokes equations, 2006

851 **Jie Wu,** On maps from loop suspensions to loop spaces and the shuffle relations on the Cohen groups, 2006

850 **Siegfried Echterhoff, S. Kaliszewski, John Quigg, and Iain Raeburn,** A categorical approach to imprimitivity theorems for C^*-dynamical systems, 2006

849 **Katsuhiko Kuribayashi, Mamoru Mimura, and Tetsu Nishimoto,** Twisted tensor products related to the cohomology of the classifying spaces of loop groups, 2006

848 **Bob Oliver,** Equivalences of classifying spaces completed at the prime two, 2006

847 **Eric T. Sawyer and Richard L. Wheeden,** Hölder continuity of weak solutions to subelliptic equations with rough coefficients, 2006

846 **Victor Beresnevich, Detta Dickinson, and Sanju Velani,** Measure theoretic laws for lim–sup sets, 2006

845 **Ehud Friedgut, Vojtech Rödl, Andrzej Ruciński, and Prasad V. Tetali,** A Sharp threshold for random graphs with a monochromatic triangle in every edge coloring, 2006

844 **Amadeu Delshams, Rafael de la Llave, and Tere M. Seara,** A geometric mechanism for diffusion in Hamiltonian systems overcoming the large gap problem: Heuristics and rigorous verification on a model, 2006

843 **Denis V. Osin,** Relatively hyperbolic groups: Intrinsic geometry, algebraic properties, and algorithmic problems, 2006

842 **David P. Blecher and Vrej Zarikian,** The calculus of one-sided M-ideals and multipliers in operator spaces, 2006

841 **Enrique Artal Bartolo, Pierrette Cassou-Noguès, Ignacio Luengo, and Alejandro Melle Hernández,** Quasi-ordinary power series and their zeta functions, 2005

840 **Sławomir Kołodziej,** The complex Monge-Ampère equation and pluripotential theory, 2005

839 **Mihai Ciucu,** A random tiling model for two dimensional electrostatics, 2005

838 **V. Jurdjevic,** Integrable Hamiltonian systems on complex Lie groups, 2005

837 **Joseph A. Ball and Victor Vinnikov,** Lax-Phillips scattering and conservative linear systems: A Cuntz-algebra multidimensional setting, 2005

For a complete list of titles in this series, visit the AMS Bookstore at www.ams.org/bookstore/.